biology

biology

Grover C. Stephens
University of California,
Irvine

Barbara Best North
University of California,
Irvine

John Wiley & Sons, Inc.
New York
London
Sydney
Toronto

Cover: "Chimigramme 31/5/70 Cl" by Pierre Cordier. The chimigramme technique makes use of the immediate reaction of developer and fixing agent on areas of light-sensitive film, without the aid of a camera or light image.

This book was set in Trump Mediaevel by Progressive Typographers, Inc. It was printed by Murray Printing Co. and bound by Book Press, Inc. The designer was Jerome B. Wilke. The drawings were designed and executed by John Balbalis with the assistance of the Wiley Illustration department. Picture research was handled by Marjorie R. Graham. Thomas Hitchings was the depth editor and Joan E. Rosenberg supervised production.

Library of Congress Cataloging in Publication Data

Stephens, Grover C
    Biology.

  Includes bibliographies.
  1. Biology.   I. North, Barbara Best,
    joint author.
II. Title.   [DNLM: 1. Biology. QH308.2 S833g 1974]
QH308.2.S73   574   73-12161
ISBN 0-471-82206-X

Printed in the United States of America

10 9 8 7 6 5 4 3 2 1

preface

Living things are incredibly diverse, yet fundamental principles underlie and unify this diversity. Both the unity and the diversity are the subject matter of biology. Organisms are composed of atoms and molecules, and there is an unbroken chain that links the fundamental properties of atoms and chemical reactions to the structure and behavior of the individual organism, to complex assemblies of plants and animals, and, finally, to the whole layer of living things that blankets the Earth's surface.

The study of biology does more than simply provide the student with a body of information about organisms. It also leads to a world view, a comprehensive and internally consistent perspective on the nature of man and the world. In fact, biology, which is based on chemical and physical laws, is the only comprehensive and relevant perspective that is also consistent with the rational tradition of Western culture. Religious or mystical accounts of man are internally consistent and comprehensive, but these views depart from the tradition of rational thought: they have quite different criteria of evidence and canons of authority.

The possession of a world view is not a matter of choice: every person has one. A world view may be acquired piecemeal and by accident, in which case it will probably remain fragmentary and unexamined. However, it is also possible to exert some control over one's own perspective. A personal world view can be consciously considered and can be accepted or rejected on its merits. The advantage of having a consistent and examined perspective is that it helps us to cope with a complex environment.

Sometimes it helps by permitting us to control and modify the world to suit our taste. Thus, understanding disease may lead a person to seek out a vaccination, and vaccination does control smallpox. But more often our perspective does not let us modify the world directly. Understanding the biological basis of aging and death, or explaining the deterioration of the environment, does not make us immortal, nor does it stop the deterioration. When we are presented with such unpleasant realities, many of us feel personally responsible and guilty, or else we rage, suffering the delusion that we have been singled out for persecution. But these responses are paranoid and crippling. A consistent and examined world view can help us avoid such responses by providing a context and a perspective that bound our assumption of responsibility for unpleasant events and put limits on our expectations from the world. Biology provides such a perspective, one which is continuously expanding the universe of modifiable events, and which is also deepening our understanding of the inevitable.

In this book, we have organized biology around exchanges of energy and material between organisms and their environments. All of the major topics and concepts of biology are covered from this perspective. More basically, we have integrated the personal and social implications of the biological view of the world into our presentation of biology as a science. Our intent has not been to focus primarily on current population crises or environmental deterioration or public health. Such crises have existed throughout history, and concern about them waxes and wanes. Rather, our intention has been to provide a sound biological basis for understanding such problems. As a result, some standard topics in the field receive relatively brief treatment, while other matters that are normally discussed briefly or not at all receive extended attention. In essence, our aim has been to express a consistent view of biology as a science, and also to emphasize points of contact between biology and human concerns.

Grover C. Stephens
Barbara Best North

acknowledgments

Our sincere thanks to Tom Hitchings, for his thorough and often inspired in-depth editorial work; to John Balbalis and Marge Graham, for an expert illustration program; and to Bob Rogers, for his unfailing support and enthusiasm.

Our thanks also to Joe Arditti, Rez Darnell, Hans Gaffron, Gary Lynch, Jack Palmer, and Douglas Pratt for their careful and always constructive criticism; to Pat Wheeler and Sue Johnson for their excellent assistance; and most especially to Ann Long for her enthusiastic and valuable help in all phases of manuscript preparation.

Finally, our thanks to Wheeler North, for his constant objectivity and enduring patience.

G. C. S.
B. B. N.

contents

IX

4
the organism as an individual

5
micro-organisms

6
photo-synthesis

7 plant organization

8 plants and man

9 animal organization I: nutrition

10
animal organization II: the internal environment

11
animal organization III: behavior

12
reproduction

introduction

Throughout history man has attempted to define, describe, and analyze himself and his relation to the environment. Biology has made fundamental contributions to this effort. It has provided information about both the genetic and the environmental influences on man.

Genetic limitations of the individual are usually obvious. Any inherited trait is a genetic limitation. Sex, skin color, race and stature, intelligence, susceptibility to disease, and other important characteristics are all heritable to a greater or lesser extent. The currently popular theory that man has inherited the instincts of his prehistoric ancestors also emphasizes these genetic limitations, although the word "genetics" is not always used. Psychologists, biologists, and essayists have used this theory to explain aggression and violence in modern society, and to analyze human sexual behavior. Such thinkers tend to be politically conservative and pessimistic about the possibility of self-improvement. In fact, any genetic theory is equivalent to the doctrine of original sin, because it decrees that the characteristics of the individual are unalterable. Mankind is rigidly divided into the elect and the unregenerate.

Genetic factors may be obvious, but environmental influences on human development are equally important. Man is an animal and has physical requirements. He must take in a certain amount of food and water. He must clothe himself against the weather. He needs to be insulated from attack by disease agents. Obviously, such external factors influence the development of the individual. Many individuals and organizations advocate an equitable distribution of our food, goods, and services in order to satisfy these physical

1

requirements. They are convinced that improving the physical environment by means of equitable distribution alone will solve the ills of mankind. Political and social environments are also frequently cited as major factors in human development. Defenders of specific political systems often fall into this camp. They may support democracy, radical socialism, or dictatorship. But whatever the scheme, they believe that the problems of the individual can be resolved by creating an appropriate social structure. Such thinkers tend to be politically radical and optimistic about the effectiveness of social reform.

Another position that emphasizes the importance of the environment is based on studies of early development. The individual is known to be particularly susceptible to environmental influences during certain periods of his life. For a short time, the animal accepts key inputs from the environment, and these inputs then define its later behavior. This process is called imprinting. The imprinting process is quite spectacular in some animals, particularly birds. For example, a young bird can be induced to accept an adult of another species, first as a mother, and later as an appropriate sex object. (There is a sad case of a peacock in the Berlin zoo which spent its life courting the favors of an alligator.) The importance of early experience is also recognized in the case of humans. A psychoanalyst may observe that a flawed parent produces a warped child. But an alternative view is that the plasticity of early development gives humanity a golden opportunity for improvement. Some people feel that if the child can be placed in the kibbutz school or Project Headstart, we will have a brave new generation.

The common view that humans are inner-directed beings, capable of being modified by their own efforts, represents yet another statement about the importance of genetic and environmental influences. Man modifies himself by an exercise of his will and by self-discipline. A person can lift weights, study the piano, or practice yoga. The resulting change can be attributed either to new environmental influences (the piano, the weight, yoga instruction) or to genetic factors (tone and pitch discrimination, body type, suppleness). W. B. Yeats' idea that "man is plastic to his will" is an optimistic view of human potential. It should certainly be qualified a bit: man is plastic to his will within limits set by environmental and genetic factors. Self-actualization programs, whether from Norman Vincent Peale or from Maharishi Yogi, are attempts to reach and explore these limits.

Many world views emphasize either the genetic or the environmental influences on the individual. Each of these views can be based on sound information, but an overemphasis on any one can be dangerous. The result is distortion and occasionally bigotry. An organism functions as a unit in the context of an environment, and it is inseparable from that environment. Any rigid distinction made between organism and environment, or between instinct and learning, may be convenient but must also be quite artificial. This fact must be recognized when biological information is used to develop and support a world view. It is easy to misuse biological facts, in part because the facts themselves are often incomplete and approximate.

Biological information is always incomplete simply because there are many ways to study and describe organisms. The kind of description we can undertake, sometimes called the level of analysis, is closely related to the physical size of the system under investigation. Our description may involve whole populations of organisms or merely selected individuals. Alternatively, groups of cells may be investigated, or the activity of molecules at the chemical level. Yet, even when we confine ourselves to a single level of analysis, we are often forced to be satisfied with descriptions that are only approximate.

Sometimes this restriction is forced upon us by the very nature of the operation required to get the information. We can measure the length of a leaf or the weight of a branch, but we can do so only within a certain range of error. There is always a limit to the accuracy we can achieve. We can say the leaf is about half an inch long. We can look harder and say it is 0.513 inches long. But if we say it is 0.512768453 inches long, we are kidding ourselves and wasting ink on at least the last five digits. Our equipment is simply not that precise.

Even counting is typically an approximation. How many cells does our leaf have? Life is too short to count them individually. Anyway, a reasonable guess is usually adequate. And even if we do count them individually, there is no way to guarantee we have done so accurately. This fact surprises some people. But if you don't believe that counting is undependable and approximate, try this simple experiment: count the number of times the letter "f" appears on this page and persuade several other people to do the same. It will be quite astonishing if you all come up with the same number.

Since description is always incomplete and approximate, the real question is what kind of description is most useful for our purposes. A chemical analysis is not necessarily better or more useful than a population description. Different descriptions serve different functions.

The controversy over the potential health hazards of oral contraceptives in the United States is a good example of the way several different kinds of descriptions have been used for different purposes. The Food and Drug Administration, whose primary function is to protect the public, might use a simple population description. A survey may be conducted to determine how many women are stricken with cervical cancer or blood clots while using oral contraceptives. If the number of women is above the norm, the FDA may ban the "pill." But the drug companies could utilize a different approach, since their aim is to keep the product on the market. By collecting information about individuals, they might be able to demonstrate that only women over 40 are susceptible to harmful side effects. Thus, sales could be permitted to continue to women under forty. Both sides could join forces in experiments that utilize cellular and molecular levels of analysis. Animal studies would examine the influence of contraceptives on the formation of cancer cells and blood clots. If a cause–effect relation were found, chemists could attempt to modify the molecular structure of the drug to eliminate its harmful effects.

This example also illustrates that a natural consequence of the acquisition of information is its application. Man does not stop with descriptions and analyses of himself and his environment. The acquired information is applied, in the form of conscious manipulation of his surroundings and himself. In fact, from the viewpoint of society, biology and its applied fields such as medicine and agriculture are justifiable activities only because they provide information that is useful. According to Francis Bacon, the ultimate goals of science are:

The prolongation of life;
The restitution of youth in some degree;
The retardation of age;
The curing of diseases counted incurable . . .

A much longer and more sophisticated list might be prepared. But it can be summarized briefly by saying that the task of biology is to predict and make possible the control of the action of living things.

Description and prediction typically lie in the realm of biology (or other scientific disciplines). The control of the behavior of living systems may not. Very often, social and political considerations are most influential. This fact can be illustrated by another example. An abortion performed early in pregnancy under adequate medical care is less dangerous to the mother than child-bearing. This is a fact. It should be included in any discussion of legalized abortion, along with the fact that an abortion without adequate medical supervision is quite dangerous. But it would be very naïve to assume that these facts alone are important in determining public policy. Morals, folkways, and religious and political views are also involved. Nevertheless, the biological information should contribute to whatever decision is made.

The following chapters develop the theme that biology provides a framework for the description and analysis of the nature of man. We must devote sufficient attention to the background information to permit an adequate description at chemical and cellular levels. This information is particularly concentrated in Chapters 1 through 3. And because man is so intimately dependent on his biological environment, the organization of microorganisms and plants as well as animals is also discussed. But throughout the text, we have stressed biological information that is relevant to public policy and personal behavior.

The study of biology is an end in itself as well as a means for the manipulation of man and his environment. The organization of the book is designed to present and develop the major concepts in biology. Like any other field, biology has a structure and pattern built around principles that have proved valuable in organizing masses of factual information. Understanding this structure is crucial to an informed use of biology.

1 organisms and their environments

A closed system is a bounded and isolated region. Neither matter nor energy can enter or leave. Conversely, an open system is one that exchanges energy and material with the environment. All living things, or **organisms,** are open systems. Each organism is in intimate contact with its environment; every aspect of its existence, including its structure, functions, growth, reproduction, and evolution, reflects this open-system relationship.

The living organism accepts materials and energy from the environment, transforms them, and returns them to the environment. However, some organisms can survive for long periods without carrying on any obvious exchanges. Viruses can be crystallized and stored on the laboratory shelf. Seeds frozen for ten thousand years have been recovered and made to germinate. In some cases, exchanges with the environment continue only at very slow rates. And in other cases, the organism retains its structural integrity for long periods, but is hardly "alive" during that time: this state is called **cryptobiosis**— literally, "hidden life". When the organism reemerges from a cryptobiotic state, however, it resumes its exchanges with the environment (see Figure 1-1, next page).

energy from the environment

An organism is not a closed system; neither can it exist in a closed system. Almost any enclosed system will illustrate this point. A sealed house, for ex-

5

(a)

Figure 1-1 *A cryptobiotic
animal. (a) The microscopic
tardigrade, or "water bear,"
in a cryptobiotic state of
suspended animation. (b)
An active "water bear."*

(b)

ample, obviously cannot support life for long. Air, water, food, and heat must be exchanged between the house and its surroundings. The earth itself is an open system. Gases, heat, and energy from the sun are constantly exchanged between the earth and nearby space. Nonetheless, we can learn more about the open-system characteristics of organisms by trying to design a closed system containing organisms. We will begin with a balanced aquarium or terrarium containing both plants and animals. The plants provide oxygen and food for the animals through the process of photosynthesis; plant photosynthesis also consumes the carbon dioxide produced by the respiration of both plants and animals (Figure 1-2). Of course, such a system is not yet closed. Light must be provided for the plants, and we must keep heat from being transferred from the aquarium to the room.

We can include a battery-powered light source and close the system by surrounding it with a boundary that does not permit the passage of any matter or any form of energy. At first, the light provides energy for plant photosynthesis, and the closed system appears to operate smoothly (Figure 1-3). But consider what will happen as time passes. We have constructed a system in which various energy changes are going on. Chemical energy from the battery is being changed to electricity and then to light. The light energy in turn is converted to chemical form again by the plants, and so on. But if the light source is an ordinary incandescent bulb, only a small fraction of the electrical energy supplied to the filament actually appears as light; the rest is dissipated as heat. Then, when the light is absorbed by the aquarium and its contents, the plants change only a small amount of it to stored chemical energy in the form of plant products. The rest is again converted to heat. When the plants are consumed by animals, this stored chemical energy, too, is converted to heat, except for the small fraction that contributes to animal growth. Furthermore, the plants themselves give off heat as they utilize the organic compounds they produce in photosynthesis.

We can now conclude our experiment. If the power supply for the light source is not first exhausted, the temperature will increase beyond the tolerance limits of life. The system will cook. Heat production ultimately destroys the balance in the aquarium, and makes it impossible to maintain organisms in a closed system.

Light energy, electrical energy, and chemical energy are described as high-grade energy. These forms of energy are interconvertible, and all can be harnessed to do physical work. However, whenever energy conversion occurs, a large fraction of the energy always appears as heat. This process is called degradation of energy, because heat energy is actually a waste product. Heat, in the sense of a uniform increase in temperature, is low-grade energy: it cannot be used to do work.

This statement may seem odd, since many of our engines appear to work by adding heat to part of the system. For example, if we add heat to the cylinder of a gasoline engine, it will increase the gas pressure and move the piston—but only if a temperature difference is maintained between the inside and the outside of the cylinder; and high-grade energy is needed to maintain this temperature difference. The energy is supplied by fuels such as gasoline, diesel fuel, or coal. If the whole engine simply heats up uniformly, no work is done. (In fact, an overheated engine usually does less work, because the waste heat interferes with lubrication and other necessary processes.) High-grade energy, energy that can be harnessed to do work, must be available to every organism.

All organisms, whether plants, animals, or microorganisms, are energy converters of some sort. In general, organisms utilize and degrade either radiant energy or chemical energy. They survive because they can convert high-grade energy into forms that they can use. In fact, in the absence of life, these forms of energy would eventually be degraded to heat anyway. Organisms merely accelerate the natural chemical reactions and hasten energy conversions. (We can think of animals and plants as being like dams and water turbines lying between mountain streams and the ocean. The water will run down to the ocean in any case, but with the dam and turbine we can harness a portion of the energy lost by the water as it falls, converting it to electricity. Though this analogy is not perfect, it helps to make a fundamental point. Life interposes itself in an inevitable physical process, but it does not alter the final result.)

The most important form of high-grade energy is radiant energy from the sun. Green plants are able to convert this radiant energy into chemical compounds, which they use for their growth and activity. These compounds in turn supply most of the chemical energy degraded by other organisms. (The only exceptions are some microorganisms that can release the chemical energy stored in minerals and metals; for example, some bacteria obtain their energy from iron or sulfur.)

The energy conversion carried out by living things vastly exceeds any other type of energy con-

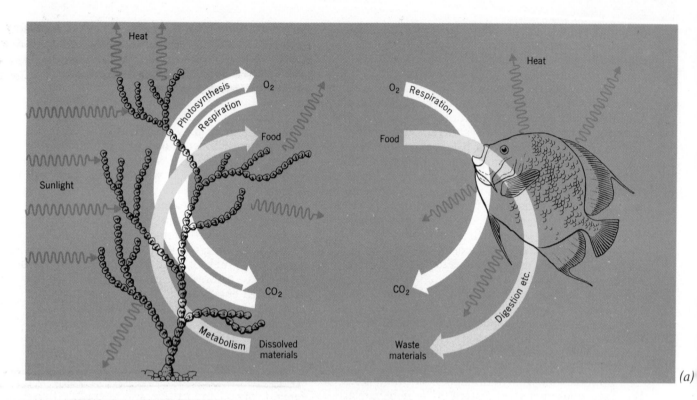

(a)

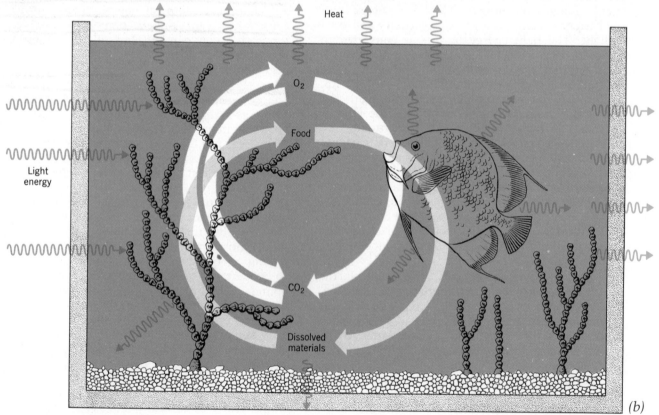

(b)

Figure 1-2 (facing page)   Exchanges between organisms and environment. Gas exchanges: most organisms consume oxygen and produce carbon dioxide during respiration; plants consume carbon dioxide and release oxygen during photosynthesis. Material exchanges: animals obtain necessary materials from plants and the water; plants use materials from the water, some of which are excreted by animals; both types of organisms return some materials to the environment as waste. Energy exchanges: all organisms obtain energy from the environment. Plants rely on light, and animals use chemical energy from food. Heat energy from both types of organisms moves into the environment as a waste product. (a) In an open system. (b) In a "balanced" aquarium.

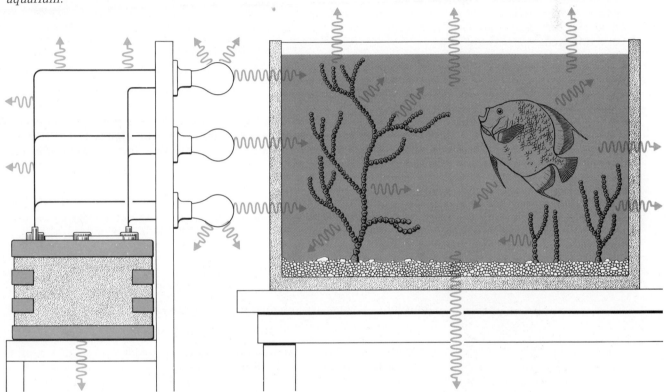

Figure 1-3 (above)   Energy flow in a closed system. Gas and material exchange between organisms and environment can be "balanced," but energy flow cannot. Energy moves from the battery (chemical energy) to light bulb (electrical energy) to light (light energy) to plants (chemical energy) to animals (chemical energy). At each conversion, some of the total energy is converted to heat energy. Heat cannot be converted back to any other form of energy. Thus, a closed system will eventually heat up.

version on earth. We can illustrate this fact by examining various kinds of energy conversions that take place each year (Table 1-1). Although the United States now accounts for about 40 percent of the world's consumption of energy for industrial purposes, this figure still is dwarfed by the energy turnover of green plants. In fact, the magnitude of the difference is even greater than our figures suggest, because the energy that is actually stored in organic material represents less than a tenth of all the radiant energy utilized by plants. The rest goes directly to support the activities of the plant (such as the movement of water and the burning of organic compounds for energy). The total energy turnover by all the living organisms in the world exceeds the conversion of energy sources for industrial and domestic purposes by at least a factor of one hundred.

Table 1-1

Annual World Energy Turnover

Total Radiation (available for photosynthesis)	2.5×10^{20} kcal[1]
Energy in Green Plants (organic material produced by photosynthesis)	4×10^{17} kcal
Fossil Fuel and Atomic Energy (World, 1970)	4×10^{16} kcal
Fossil Fuel and Atomic Energy (United States, 1970)	1.6×10^{16} kcal

[1] The kilocalorie (kcal) is a metric unit of heat. It is defined as 1,000 calories, or the amount of heat necessary to raise the temperature of 1,000 grams of water one degree. The "calories" commonly referred to in diet books and "calorie counters" are actually metric kilocalories.

Exponents

Exponents (also called logarithms) are a convenient and concise way to express very large or very small numbers. In the expression 10^2, the 2 is the exponent. It indicates the number of times that the 10's are to be multiplied together. In other words,

$$10^2 = 10 \times 10 = 100$$

Similarly,

$$10^3 = 10 \times 10 \times 10 = 1,000$$
$$10^4 = 10 \times 10 \times 10 \times 10 = 10,000$$

and so on. It is important to remember that a small change in an exponent,

from 10^4 to 10^5

represents a very large change in the whole expression,

from 10,000 to 100,000.

Positive exponents express numbers of 10 or larger. An exponent of zero means the expression is equal to 1, that is,

$$10^0 = 1,$$

and a negative exponent means the number is a fraction:

$$10^{-1} = 1/10 = 0.1$$
$$10^{-2} = 1/10 \times 1/10 = 0.01$$
$$10^{-3} = 1/10 \times 1/10 \times 1/10 = 0.001$$

and so forth. The utility of exponents is obvious in the case of extremely large or extremely small numbers. Not only is 10^{-16} less clumsy to write than 0.0000000000000001, but it is also much less likely to be miscopied or misread.

To express numbers that lie between the round powers of ten, the exponential number is multiplied by a second number. One can express 2,000 as 2×10^3 (that is, $2 \times 10 \times 10 \times 10$). Similarly,

$$3 \times_{10}{}^{-2} = 3 \times 1/10 \times 1/10 = 0.03$$
$$5.7 \times 10^3 = 5.7 \times 10 \times 10 \times 10 = 5{,}700.$$

A practical rule for converting exponential numbers into simple ones is to move the decimal point a number of places equal to the exponent. Move it to the right for positive exponents, to the left for negative ones:

$$2.\underline{75} \times 10^4 = 27{,}500$$
$$2.75 \times 10^{-4} = 0.000275.$$

Exponential notation is particularly useful when we want to compare numbers that cover a wide range of size, especially in graphs. A graph divided equally into logarithmic units (10^0, 10^1, 10^2, 10^3, and so on) is much more convenient than a graph based on the equivalent simple numbers (1, 10, 100, 1000, . . .). The latter graph quickly runs out of space.

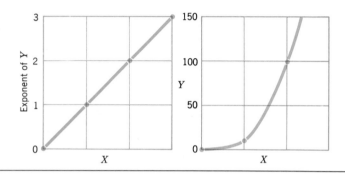

the composition of organisms

So far we have discussed only the energy exchanges between organisms and their environment. But organisms also accumulate and exchange materials. Plants and animals are solid and palpable, and they have characteristic compositions. Certain elements are present in high concentrations, and others less so. What is the usual elemental composition of the living substance, the **protoplasm** of a typical organism? Table 1-2 lists the major elements found in the human body and the most abundant elements in the environment.

Carbon, hydrogen, oxygen, and nitrogen are the most common elements found in organic compounds, the compounds characteristic of living things. Hydrogen and oxygen are especially abundant because a large percentage of living tissue is made up of water (H_2O). Some of the less common elements may still be extremely important. For instance, we need such minute quantities of cobalt

that it isn't even on the list. Yet cobalt is found in vitamin B_{12}, which plays a crucial role in many chemical reactions in the body.

The table makes another point. Though the elements found in the organism in large quantities are all readily available in the environment, there is no particular correspondence between the orders of abundance in the two cases. Silicon, one of the most plentiful elements in the environment, is not even on the list for organisms. It is important only in some single-celled plants, sponges, and a few other organisms. Conversely, elements such as phosphorous and carbon are concentrated in organisms and are necessary for their functions. Yet these elements are much less abundant in the environment. Finally, organisms can sometimes concentrate elements that are not really necessary for their structure or function. The antlers of deer may be relatively rich in gold simply because the an-

Table 1-2

The Composition of Man and His Environment

Figures are percentages, by weight. Trace elements have been omitted.[1]

Universe		Earth						Human Body	
		Crust		Oceans		Atmosphere			
H	91	O	47	H	66	N	76	H	63
He	9.1	Si	28	O	33	O	23	O	25.5
O	0.057	Al	7.9	Cl	0.33	Ar	1.0	C	9.5
N	0.042	Fe	4.5	Na	0.28			N	1.4
C	0.021	Ca	3.5	Mg	0.033			Ca	0.31
Si	0.003	Na	2.5	S	0.017			P	0.22
Ne	0.003	K	2.5	Ca	0.006			K	0.06
Mg	0.002	Mg	2.2	K	0.006			S	0.05
Fe	0.002	Ti	0.46	C	0.0014			Na	0.03
S	0.001	H	0.22	Br	0.0005			Cl	0.03
		C	0.19					Mg	0.01

[1] Explanation of Symbols:

Al	Aluminum	Ca	Calcium	K	Potassium	O	Oxygen
Ar	Argon	Cl	Chlorine	Mg	Magnesium	S	Sulfur
B	Boron	Fe	Iron	N	Nitrogen	Si	Silicon
Br	Bromine	H	Hydrogen	Na	Sodium	Ti	Titanium
C	Carbon	He	Helium	Ne	Neon		

imals have browsed on gold-accumulating plants.

Some of the differences between organisms and their immediate surroundings are spectacular. For instance, plants in an aquarium may have fully 100 to 1,000 times as much potassium in their cell water as there is in an equal volume of the water surrounding them. Yet, as far as we can tell, the potassium is free to move both into and out of the plant.

How can we account for large differences of this kind? We know that any element or compound dissolved in water will ultimately distribute itself equally throughout all the water available—when you put sugar in your coffee, you do not need to stir it. (It is true that without stirring you have to be very patient, because the movement of the molecules is slow; nevertheless, given enough time, the sugar will be equally distributed.) The tendency for net movement of a dissolved material from a region of high concentration into a region of low concentration is called **diffusion.** Unless it is prevented, diffusion will continue until the dissolved materials are equally distributed throughout the liquid.

There are two ways we could prevent equal distribution. We could place a barrier in the cup, dividing it permanently into two compartments. Alternatively, we could use some sort of pump that would move molecules back to the region of high concentration as rapidly as they diffuse away. Such a pump would require energy. Perhaps, then, the potassium is prevented from diffusing out of the cell by some sort of physical barrier. We can show directly that this is not true. If we add a radioactive form of potassium to the water, we can determine whether potassium moves into the plant simply by seeing whether the plant becomes radioactive. It does. Similarly, we can show that potassium also moves out of the plant into the surrounding water.

There is no barrier to the movement of materials between the plant and the surrounding water. Yet as long as the plant is alive, diffusion will not produce an equal distribution of potassium inside and outside. The plant therefore must be using some sort of energy-requiring pump. In fact, organisms are never in equilibrium with their environment. "Equilibrium" implies that soluble materials are equally distributed inside and out; that is certainly not the case with living organisms. An organism in equilibrium with its environment is one that is dead.

A live animal or plant maintains a relatively constant composition through a continuous process of dynamic interchange with its environment. This continuous interchange is called the **steady state.** It involves an active and energy-requiring accumulation of some materials and an active and energy-requiring extrusion of others. To give a picture of the steady state, an analogy is often drawn between organisms and a candle flame. The flame remains constant in appearance, but it is maintained by a continuous input of material (candle wax and oxygen) balanced by a constant output (carbon dioxide, water, soot, and heat).

structure and function

Organisms are composed of structures that keep them separate from the environment, but that still allow them to function as open systems. Structure and function are intimately related. Structures are not simply inert physical frameworks; they are constantly changing. These changes are usually rather slow, of course. The internal chemical reactions responsible for them are not obvious. But even seemingly static structures such as bone are continually being broken down and replaced (see Figure 1-4).

On the other hand, some chemical transformations occur very rapidly. For example, under a microscope the structure of a contracted muscle looks very different from the structure of a relaxed muscle. However, the changes in muscle structure during the process of contraction occur so rapidly and dramatically that we emphasize the "function" (that is, the contraction) of the muscle rather than the structural changes.

Distinctions between structure and function are useful but largely arbitrary. Structure in animals and plants is the product of functional requirements. In fact, most of our information about organisms is based on this assumption. A paleontologist is able to reconstruct the feeding habits of an extinct animal from a single fossil tooth because he knows that certain types of teeth are associated with certain kinds of feeding habits. Similarly, we assume that the liver in the wildebeest functions like the liver in a dog or a man simply because it is similar in structure and position. (This assumption usually works well, but it sometimes proves wrong. For example, we once believed that any plant that contains the green pigment chlorophyll carries on

photosynthesis. We now know that there are green plants living in the depths of the ocean where light never penetrates and, thus, photosynthesis cannot occur.)

We could even try to predict the structure of a hypothetical Venusian if we knew something about how it functions in the environment. For example, if we knew that it was photosynthetic and used radiant energy, we could immediately predict that the green man of Venus will need a large photosynthetic surface which can be exposed to light. This need can most easily be fulfilled by having many thin and flat or needle-like extensions from the body, which provide the maximum surface with the minimum weight. Still, despite this arrangement, the large number of surfaces required will be heavy and rather fragile. It would therefore be efficient to dispense with arm muscles altogether and attach the surfaces directly to bony projections from the skeleton—and in any case, muscles will no longer be required for food gathering. The bony extensions can be made hard and unpalatable as a protection against potential attackers.

Photosynthesis provides the Venusian with en-

Figure 1-4 A ring placed around the femur (thigh bone) of a young dog eventually finds its way into the marrow cavity of the bone. The bone is continually dissolved away at the border of the marrow cavity and deposited on the outside.

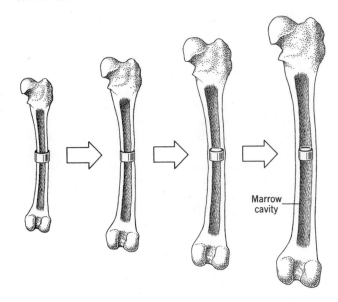

Marrow cavity

Figure 1-5 A sessile photosynthetic Venusian might very well resemble an earthly tree like the African baobab.

ergy and organic compounds, but he still needs a water supply, minerals, and a source of usable nitrogen. These things are all available in the soil. He can obtain them by producing spiky extensions from his feet and thrusting them into the ground. The spikes will have to penetrate deep into the earth and cover a large area to supply our green man. Thus he will probably choose a good spot and anchor to it permanently. Leg and body muscles are then unnecessary. Furthermore, our Venusian no longer needs his senses to detect potential food sources. With no muscles to control, he no longer needs a nervous system at all. He can substitute slower but simpler mechanisms of coordination, using chemicals that modify his growth and allow him to move slowly toward a source of light. Figure 1-5 shows the general structure of our green man from Venus. It happens to be a tree, but it would be very difficult to tell the difference between the two of them.

Our example has one obvious flaw. We began with information about structure and function in a tree. Next, we assumed that any organism which functions like a tree—that is, which uses radiant energy—must look like a tree. But the same type of argument could just as easily have produced a cactus or a sunflower. Actually, an enormous variety of widely different structures are all capable of exchanging energy and material with the environment. In fact, we know of more than a million different workable structures—that being the number of separate species of animals, plants, and microorganisms on earth.

The example of our green Venusian does make one very important point, nonetheless. Structure is closely related to function, and any structure that contributes nothing to energy exchange, the exchange of materials, or any other vital process, is as superfluous as muscles would be to a green Venusian or to a tree. The energy that maintains such a structure is wasted. Therefore, the structure will ultimately be weeded out.

growth and reproduction

Growth and reproduction, like other aspects of an organism's existence, depend on an open-system relationship between the organism and the environment. The general pattern of an organism's life cycle is inherited from its parents. The life cycle is predictable and is characteristic of that species alone. A specific set of instructions, or genes, is part of every organism's starting machinery at birth. These instructions, embodied in specific organic molecules, are passed from generation to generation during reproduction. This mechanism guarantees that the life cycle of each species will be stable and unique.

Despite these broad inherited controls, however, no stage of the life cycle will appear automatically or in isolation from the environment. Each organism interacts continuously with its environment, and the details of its growth and reproduction can be profoundly influenced by environmental factors. Stature and intelligence, for example, depend on nutrition as well as on heredity. There are many specific compounds that must be available in the diet to insure normal development. Vitamins are obvious examples. Vitamins are required for the body's normal operation, but the human body cannot itself synthesize most of them. They must be obtained from the environment.

Conversely, there are cases in which inputs from the environment must be avoided. Phenylketonuria (PKU) is a hereditary defect in man which can prevent the nervous system from developing normally. It arises because the enzyme that processes the amino acid phenylalanine is not synthesized by the body. An intermediate product accumulates in the blood. But if a person suffering from the defect is given a diet that contains no phenylalanine, he develops normally. The environment is at least as important as heredity in this case.

Inputs from the environment can take many other forms. As Figure 1-6 shows, an organism's development can be influenced by various phys-

Figure 1-6 *The environment influences plant development in various ways. (a) The position of the sun controls the direction of plant growth. (b) Gravity also influences the direction of growth. (c) Contact with hard surfaces directs growth. (d) Daylength determines whether or not a plant produces flowers. (e) Temperature influences seed germination.*

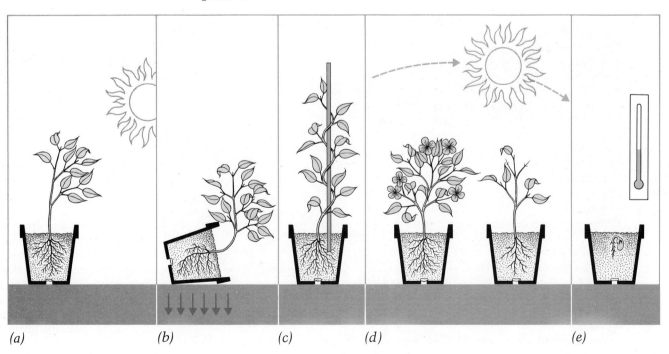

(a)   *(b)*   *(c)*   *(d)*   *(e)*

ical factors. The leafy portion of a plant grows toward the light and its roots grow downward; animals may breed at precise times of the year or enter resting phases during periods of cold or drought.

Many characteristics of organisms cannot be modified at all by the environment. But at the very least the environment must be permissive—it must allow these fixed characteristics to develop. Temperature, radiation, the availability of food, and other external factors all must be amenable to life. The example of eye color will make this clear. If the environment is permissive and your eyes develop at all, they will inevitably have the iris color that has been coded in your hereditary makeup.

evolution

Individual organisms interact with their environment during their lifetimes; similarly, whole populations of organisms interact with their environment over longer periods of time, over many generations. The environment slowly changes the population by eliminating some individuals before they can reproduce. Environmental influences such as the coming of an ice age, for example, may kill off individuals that cannot withstand cold temperatures. However, individuals in the population with thicker fur or more fat may survive and reproduce. Therefore, the next generation is more likely to inherit thick fur or fatty tissue. This kind of differential survival is called **natural selection;** the resulting changes in populations after many generations of natural selection is called **evolution.**

Evolutionary processes can take millions of years. But changes can also occur in a comparatively short time. For instance, moths collected recently near English industrial towns are generally darker and more heavily pigmented than specimens of the same species collected in the nineteenth century. They are also darker than specimens found in other localities (see Figure 1-7). This phenomenon is known as **industrial melanism,** since the pigment involved is a compound called melanin.

Industrial melanism is the result of the changing color of the environment. Chance variations in heredity continually produce some moths that are darker than most, and some that are lighter. As industry increasingly blackened the landscape with soot, darker moths had an increased chance to survive and reproduce. They were less conspicuous than their fellows, and therefore less likely to be eaten by birds and other predators. The dark moths

(a)

(b)

Figure 1-7 Industrial melanism. (a) The dark moth is nearly invisible on the soot-blackened tree trunk, and the white one stands out. (b) The situation is just the reverse on a normal tree trunk.

found recently are of the same species as the lighter forms collected earlier or found in other locations, but more of them are dark because of the darker environment. Evolution consists of long-term changes in populations of organisms as they encounter changes in their environment.

Evolutionary changes, caused by the environment, account for many features of organisms that are otherwise quite perplexing. For instance, large groups of plants and animals are very similar in their general organization. The hand of a man, the wing of a bat, and the flipper of a seal all have the same basic ground plan. This similarity is bewildering until we recognize that all three animals are historically related. All three are descended from some earlier animal with a five-fingered hand. From an engineering viewpoint, there are many ways to support a wing. However, no organism can generate a solution to a problem from scratch. Each animal represents a basic ground plan that has been modified through evolution to exploit a particular environment. (And at the same time, these very modifications limit the animal to that particular environment, as we acknowledge with such phrases as "a fish out of water.") Modification of the ground plan occurs through the continual operation of environmental forces, in the same way that produced the darker butterflies.

If we keep this historical and evolutionary perspective on questions of structure, we can also understand the persistence of "worthless" features in plants and animals. For instance, man's vermiform appendix, ear muscles, and tail bone seem more nuisances than anything else. There are many such structures. We now can interpret them as being the remnants of disappearing structures, which once had functions but no longer do. Otherwise, they would simply be puzzling nonsense.

We can also avoid philosophical questions about the origin and adaptations of a particular animal or plant. We see that each organism is closely adapted to its environment. Who adapted it? Did it adapt itself, after conscious analysis of the situation? If we try to pursue such questions, we are in difficulty. On the other hand, if we recognize that evolutionary changes are the inevitable result of both the open-system characteristics of organisms and the variety of environments, such questions become simply irrelevant.

We can understand how the tremendous variety of organisms arose if we appreciate how organisms are constantly interacting with their environments on a smaller scale. The colonization of a barren area, such as a recent lava flow or a sand dune, illustrates this process in microcosm. The first organisms that can take hold are those that can manage exchanges of energy and material on bare, exposed rock. But this situation is not static. As these first organisms grow and reproduce and die, they change the environment; soil is formed as their activities break down the rock and their bodies decay. The area can now support a greater variety of plants and animals. Other organisms which once could not have survived can now invade and exploit the area. Each new colonizer represents a potential source of food and material for further arrivals. This process continues until the bare rock or sand is transformed into a complicated community of plants and animals (see Figure 1-8).

On another time scale, one can visualize how the same kind of increasing variety and complexity arose through evolution. Every new organism, by the very fact of its being alive, inevitably altered its environment to some degree. This process, in turn, created opportunities that still newer organisms were able to exploit.

(a)

(b)

(c)

Figure 1-8 Lava flows in the vicinity of Craters of the Moon National Monument in Idaho. (a) New lava is devoid of life. But weathering and the activities of microorganisms capable of living in such an environment gradually produce soil suitable for larger and more conspicuous plants. Photographs (b) and (c) are later stages in the succession process. At each stage, the organisms modify the environment, permitting the growth of plants that would not have been able to survive at earlier stages of succession. The final result is rich and varied vegetation.

Dubos, Rene, *So Human an Animal.* New York: Scribner's (1968).

A synthesis of biology, anthropology, and sociology and a brilliantly written statement on the position of man in nature. It was awarded the 1969 Pulitzer Prize.

Science, published weekly by the American Association for the Advancement of Science, Washington, D.C.

Science publishes brief technical reports covering a wide variety of subjects. It also includes editorials, letters, and news of interest to the scientific community. The magazine is the principal channel of communication for the largest organization of professional scientists in the United States.

Scientific American, published monthly by Scientific American, Inc.

Scientific American presents a continuing review of recent advances in all scientific fields, including biology. Articles vary in sophistication but are intended for the general reader.

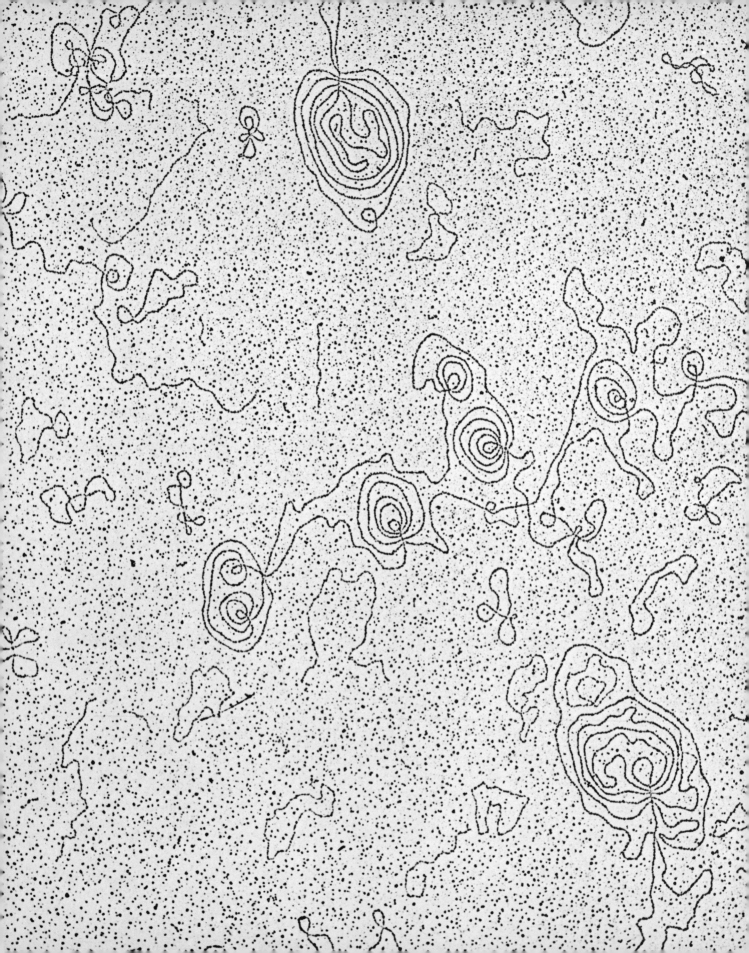

2 molecular machinery

College courses in biology often come to grief because of chemistry. Formulas and reactions are usually introduced early in the course, but the student sees little point in forcing his way through a discussion of chemistry; that is not what he anticipated. This is not a new problem. The eighteenth-century British physician John Hunter began his lectures to new medical students by listing the chemical constituents of organisms. And no doubt Hunter, too, had predecessors.

Chemistry is always an integral part of biology courses not merely because it is traditional, but because it supplies valuable information about organisms. In fact, the chemistry of organisms is surprisingly simple, because all life is composed chiefly of just a few major kinds of molecules. Some of them, such as proteins, are very large and occur in a tremendous variety. But even this variety results from a simple construction. Proteins are long chains of linked subunits called amino acids. There are only twenty different amino acids in all, but they can be joined together in an enormous number of ways. Each arrangement produces a unique protein. "Simple" does not mean easy, however. The chemistry of organisms, **biochemistry,** is challenging; it commands the best efforts of brilliant men.

As we discussed in the introduction, biological phenomena are not always analyzed at the chemical level. In fact, chemical information may add very little to a description of what a whole organism is doing at a particular time. For some purposes it might be more useful to describe an organism in physical rather than in chemical terms, by referring to heat flow, for example. However, in practice, chemical descriptions are so useful and so fundamental that they cannot be ignored or long postponed.

DNA from the chromosomes of a pea plant

21

Water is the most abundant constituent of living things, but the actual water content of an organism can vary considerably. For example, the water content of man depends on age. It ranges from 74 percent in a newborn child to about 63 percent at maturity. Animals such as jellyfish may be 95 percent water. Large plants may have a relatively low water content because much of their bulk is dead tissue, but in their active portions water is abundant.

The properties of water are so intimately related to the functions of organisms that no life could exist without it. Water has several crucial physical characteristics. First, it is a relatively good heat conductor; it limits local heating and prevents damage to cells and tissues by conducting heat away from high-temperature regions. Water also has a high heat capacity: more energy is required to raise the temperature of a given weight of water than to heat almost any other liquid comparably. This property protects organisms from sudden temperature changes, because water absorbs heat with a minimal increase in temperature.

Even the behavior of water when it freezes is beneficial. As it cools, water becomes more dense, reaching its highest density at 4° C. But with further cooling, its density begins to decrease again, and water freezes at 0° C. Thus, ice at 0° or below is lighter than the warmer, 4° C water. The ice floats. As a result, lakes and oceans freeze from the top down rather than from the bottom up. The surface layer of ice insulates the water below from cold air, enabling aquatic animals to survive the winter.

Water Solutions

There are further reasons for water's importance to life. Water is as close to a universal solvent as anything we know. More substances dissolve in it than in any other liquid. Water's versatility as a solvent is due in part to another of its properties: its polarity. In a polar molecule such as water, the protons (each with a positive electric charge) and the electrons (each with a negative charge) are distributed unequally. The reason is that oxygen atoms are strongly electronegative; that is, they have a strong attraction for electrons. When hydrogen is joined to oxygen in a water molecule, the shared electron pair is pulled toward the electronegative oxygen atom. This attraction gives the oxygen a slight negative charge and the hydrogens a slight positive charge. The molecule is thus polar, even though the total number of electrons and protons is equal and, overall, the molecule is electrically neutral.

Since like charges repel each other and unlike ones attract, water molecules will be attracted to other molecules that have unequal charge distributions. They may be other water molecules, or molecules of various other compounds. For example, the negative end of an alcohol molecule (the oxygen atom) will be attracted to the positive region in a water molecule (the hydrogen atoms). This kind of weak chemical bond is called a **hydrogen bond.** Many molecules with unequal charge distributions dissolve easily in water in the same way, by interacting with the polar water molecules (see Figure 2-1).

Water also promotes the ionization of many compounds. In water solution, negative and positive ions interact with the polar water molecules. Ions acquire water shells. Positive ions attract the negative regions of water molecules, and negative ions attract the positive regions (Figure 2-2). The dissociation of many compounds into ions, as we shall see later, is essential to such diverse processes as the transmission of nerve impulses, muscle contraction, and photosynthesis.

Acidity

Acidity is another crucial property of a water solution. It depends on the fact that water itself dissociates to form ions. The water molecule separates into a hydrogen ion, which is positively charged, and a hydroxyl ion, which is negatively charged. The symbols for these two ions are H^+ and OH^-, respectively.

$$H_2O \rightleftharpoons H^+ + : OH^-$$

Not very much water is dissociated—only about one molecule in 550 million in pure water. But hydrogen ions and hydroxyl ions can be provided by other compounds. The acidity of a water solution is an expression of the concentration of hydrogen ions in the solution. An **acid** is any compound that increases the concentration of hydrogen ions; a **base** is any compound that decreases their concentration.

In the equation above, the water molecule and its ions are linked by a double arrow, indicating that some ions can reassociate to form water, while other water molecules ionize. These two opposite reactions happen with equal frequency, so a dynamic

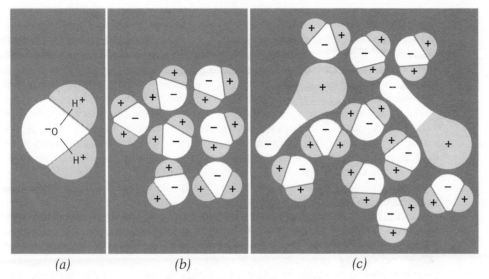

(a)   (b)   (c)

Figure 2-1   Polarity in water molecules. (a) The oxygen atom (O) attracts electrons more strongly than do the two hydrogen atoms (H). Thus, the molecule is polar, slightly negative at the oxygen end and slightly positive at the other end. (b) Water molecules form hydrogen bonds, becoming oriented so that positive and negative regions are adjacent. (c) Other polar molecules are easily dispersed in water because of hydrogen bonds.

balance is maintained between the ions and the molecules, and their relative concentrations remain constant. If we add equal amounts of a compound that provides hydrogen ions and a compound that provides hydroxyl ions, we will increase the concentration of both ions on the right-hand side of the equation. However, many of the newly-provided ions will simply associate to form water molecules. There will be no change in the overall concentration of hydrogen ions, and therefore no change in acidity.

If we add unequal amounts of hydrogen and hydroxyl ions, however, the result is different. Suppose we add a compound that provides only hydrogen ions. Some of these ions will associate with hydroxyl ions in the solution to form water molecules. But there are not enough hydroxyl ions for all the newly-added hydrogen ions. The concentration of hydrogen ions therefore increases, while the concentration of hydroxyl ions decreases. Because the hydrogen ion concentration has increased, the acidity of the solution has increased. The compound we added has behaved as an acid.

It is important to realize that an acid need not be

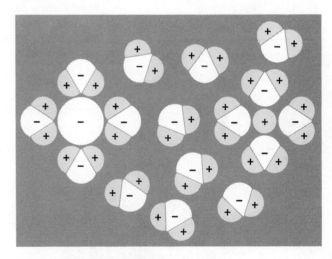

Figure 2-2   Water promotes the separation of molecules into positive and negative ions. Water molecules form a shell around the ions so that positive and negative regions are adjacent.

The Structure of Matter

All matter, whether solid, liquid, or gas, is composed of atoms. Atoms are made up of three basic kinds of subatomic particles. The central nucleus is composed of protons and neutrons, and it is surrounded by one or more electrons. A proton has a positive electrical charge, and an electron carries an equal but negative charge. Neutrons have no charge.

There are over a hundred different kinds of atoms, corresponding to the various chemical elements. The atoms of each element have a specific number of protons and electrons and various numbers of neutrons. No matter how many neutrons there are in an atom, the number of protons and electrons is equal, so the atom is electrically neutral.

Atoms are usually linked together to form molecules, and the linkages that join the atoms are **chemical bonds.** These bonds are broken, rearranged, and reestablished between atoms during chemical reactions. All chemical bonds involve the sharing of electrons by two atoms. Electrons normally occur in pairs, and a chemical bond is constructed of one or more pairs, never of a single electron. Since electrons are negatively charged, they can be attracted to the positive protons of adjacent atoms; in other words, the electrons are shared by the atoms. A chemical bond results.

The strongest kind of chemical bond, a **covalent bond,** occurs when an electron pair is equally shared by the protons in two adjacent atoms. Covalent bonds are very stable; that is, they are difficult to break. A second kind of chemical bond is the **ionic bond.** In this case, the electron pair is attracted more strongly to one atom in the pair (that atom is said to be more **electronegative**). For example, common table salt is composed of sodium chloride (NaCl) molecules joined by ionic bonds. The sodium and chloride atoms in salt crystals form a regular arrangement. But when salt dissolves, the ionic bonds break and the atoms in the crystals separate. For reasons of stability the electron pair remains intact and stays with the most electronegative

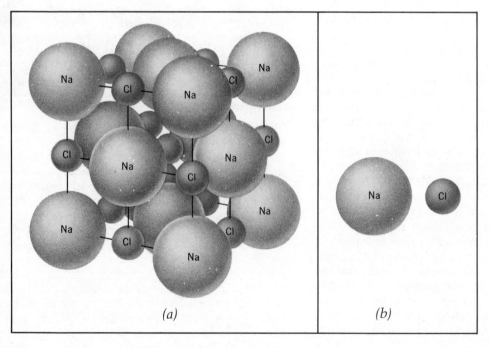

(a)   *(b)*

Structure of sodium chloride (NaCl). (a) A salt crystal of sodium chloride. (b) A single sodium chloride molecule.

atom. This process is called **ionization,** and the electrically-charged atoms are **ions.** An ion is an atom or molecule that has either acquired one or more extra electrons, giving it a net negative charge, or lost one or more electrons, giving it a net positive charge. In this case, chlorine retains the electron pair. The chlorine atom now has an extra electron and is a negative ion. The sodium atom now lacks an electron and is a positive ion.

$$Na:Cl \underset{\ce{H_2O}}{\rightleftharpoons} Na^+ + :Cl^-$$

There are several additional kinds of chemical bonds that consist of weak attractive forces between atoms. In these cases, an electron pair stays tightly bound to one atom, but it exerts an attraction for another one. Such weak bonds help to stabilize the structure of large biological molecules.

---

a direct source of hydrogen ions. A compound that binds hydroxyls increases the acidity of the solution equally well, since it produces a corresponding increase in hydrogen ions. In the same way, a base can either add hydroxyl ions directly to the solution or trap and bind hydrogen ions. In either case, the result is a decrease in hydrogen ion concentration, and therefore a decrease in acidity. The acidity of a particular solution is controlled by all the compounds in it that add or remove hydrogen and hydroxyl ions.

Acidity is closely regulated in both plants and animals. The shapes of protein molecules, and thus all the biological activities that depend on protein molecules, are very sensitive to changes in acidity. The properties of the mineral constituents of organisms also are affected by acidity. Calcium carbonate (found in bone) dissolves in acid solution but becomes insoluble as the solution becomes more basic. Water is more than a solvent in which various reactions occur. It is part of the structure and functioning machinery of life.

Colloids

The size of the molecules or particles that are dissolved or suspended in water also has a significant effect on the system's properties. When the molecules are small, the water and the dissolved material are indistinguishable; they form a true solution because the system consists of a single phase. (A phase is any homogeneous and physically distinct part of a system which is separated from other parts of the system by definite surfaces. A system can contain any number of phases, and they need not consist of different substances. Ice in water, for instance, represents a separate phase. Individual phases can be either **continuous,** such as a layer of oil on top of water, or **dispersed,** such as oil droplets in the water after shaking.)

Larger molecules do not form true solutions. When the molecules are large, the system contains two separate phases: water, and dispersed molecules or particles. The phases remain distinct in their physical and chemical properties. If the particles are so large that they will eventually settle out, the system is called a **suspension.** However, if the particles are large enough to behave as a separate phase but small enough so they do not settle out, the system is a **colloidal solution.**

Molecules such as proteins show colloidal properties. Gelatin is an example. When gelatin, a protein, is dissolved in water, it forms a colloidal solution, or **sol.** Individual protein molecules are dispersed in the water. (The protein molecules form a dispersed phase, and the water is a continuous phase.) When more gelatin is added, or water is removed, the protein molecules link up and become a continuous phase. This process confers the familiar Jell-O-like consistency on the solution, which is now in the **gel** state. A gel is a haystack arrangement of protein molecules linked together; the water would be analogous to the air in the haystack (Figure 2-3).

Colloidal solutions are extremely important in biology. The living substance of cells is colloidal, and the best evidence suggests that it is constantly undergoing localized transformations between the sol and gel states. The junctions of the phases of a colloidal solution provide surfaces that aid chemical reactions in the cell. Some small molecules tend to accumulate at these surfaces, while others may be repelled by them. For example, molecules that are relatively insoluble in water will accumulate at the surfaces of the solution. But sub-

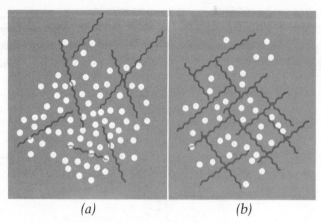

Figure 2-3 A colloidal solution of protein and water molecules. (a) In the sol state, the water molecules form a continuous phase and the protein molecules are dispersed. (b) In the gel state, the protein molecules link up to form a continuous phase.

stances such as sodium and chloride ions, which bind to water molecules, are scarce on the same surfaces.

A colloidal solution provides a tremendous surface area where such accumulation or scarcity can occur. This effect results from the small size of the dispersed particles. The smaller the particles, the greater will be their surface area in proportion to their volume. If this relationship does not seem obvious, think of a cube one inch on a side. It has a total surface area of six square inches. But if the same cube is divided into 100 cubes, each one a tenth of an inch on a side, the combined surface area is increased to 60 square inches. If we keep dividing our original cube until the new cubes are one millionth of an inch on a side (in the size range of a colloidal particle), the total surface area becomes over 40,000 square feet. Minute particles have a vastly greater combined surface area than do large masses of the same volume.

The effect of this increased surface area on chemical reactions can be very dramatic. To take just one example, the decomposition of hydrogen peroxide into water and oxygen can be hastened by the presence of platinum, which absorbs the peroxide molecules on its surface. Platinum is a **catalyst** for the reaction; that is, it accelerates the chemical reaction without itself being consumed. If a smooth plate of platinum is put into a solution of hydrogen peroxide, nothing happens. But if the surface area of

the platinum is increased by scratching the surface, bubbles of oxygen slowly form. If platinum black (finely divided platinum) is deposited on the plate, the bubbling is much more vigorous. And if a solution of platinum in colloidal dispersion is added to hydrogen peroxide, the reaction is so fast that the mixture explodes.

Many biological reactions also involve catalysts. Proteins called **enzymes** catalyze a great variety of chemical reactions between small molecules. The enzymes are large enough to behave as a separate phase. Small molecules are bound to the surface of an enzyme molecule, where they undergo a chemical reaction. The process is similar to the interaction between platinum and hydrogen peroxide.

Colloidal solutions may contain clusters of molecules as well as large individual ones. The most common example in organisms is a colloidal solution of small fat droplets. Since fat droplets are non-polar, they are not easily dissolved or dispersed in a water solution. Instead, the droplets tend to leave the water and accumulate on the surface of the solution. This so-called phase separation can be prevented by **emulsifiers,** compounds that have polar and non-polar regions on the same molecule. The charged polar regions become linked to water molecules, while the non-polar regions tend to associate with the non-polar fat droplets. Thus, the emulsifier stabilizes the colloidal solution by binding with both water and fat. Soaps and detergents are the most familiar emulsifying agents. Milk is actually a colloid of butterfat droplets in water, with milk proteins serving as the emulsifiers. In the cells of organisms, proteins and other large molecules act as emulsifiers.

Living cells actually contain several phases (for example, various proteins, very small fatty droplets, and water). Small molecules are in solution in one or more of these phases and may be distributed unequally among the different phases. Larger particles or droplets are suspended in this complex solution. These larger particles in turn grade into the yet larger functional units of the cell, which will be described in the next chapter.

inorganic compounds

The chemical compounds found in organisms and in the environment are usually divided into two broad categories, inorganic and organic. (Water is either classified as inorganic or put in a separate

Table 2-1

Concentration of Major Ions in Blood Plasma and Sea Water

Units give the relative numbers of atoms per liter of water.

Ion	Symbol	Concentration in Plasma	Concentration in Sea Water
Sodium	Na^+	139	460
Chloride	Cl^-	103	535
Potassium	K^+	2	10
Calcium	Ca^{++}	3	10
Magnesium	Mg^{++}	1	53

category altogether.) This division does not make much sense, since all these compounds are found in organisms, and all are necessary for life. However, the distinction is that inorganic compounds are found naturally in both living and nonliving systems, while organic ones are always associated with living systems. Organic compounds in the environment, such as petroleum products, are not exceptions; they too were produced by organisms.

The inorganic compounds in organisms often occur as ions in solution. Table 2-1 lists the ions found in sea water and in human blood serum. The similarities in the two lists are evident. This correspondence has impressed some biologists so much that they have described man as an ambulatory skin full of sea water. However, the total concentration of ions in human blood is only about one third that in sea water.

Inorganic compounds are intimately involved in both the functions and the structures of biological systems. For instance, the shapes, charge distributions, and solubilities of protein molecules all depend on the surrounding inorganic ions. Specific ions are required for enzyme activity. An unequal distribution of ions inside and outside the cells is the basis of the electrical activities of organisms, and these electrical activities in turn are responsible for sensation, coordination, and the control of movements.

Some inorganic compounds found in organisms are part of insoluble structures. Calcium phosphate and calcium carbonate form the inorganic matrix of bone. Silicon is used in some plants to provide rigidity and support. There are many other ex-

amples. Inorganic compounds are vital parts of living tissue, and the inorganic composition of an organism is usually regulated within narrow limits.

organic compounds

Organic chemistry is, basically, the chemistry of carbon compounds. All organic compounds contain the element carbon, and the properties of that element determine the nature of biochemical reactions. The carbon atom has six electrons; two of them are tightly bound to the nucleus, and the remaining four less so. As a result, the four outer electrons are available for sharing with other atoms in the formation of chemical bonds; in chemical formulas, the available bonding sites are represented by four lines radiating out from "C," the symbol for carbon.

Carbon is capable of bonding with up to four separate atoms, but it can also share two or even three of its bonds with a single atom (forming **double** or **triple bonds**). Carbon atoms can bond with each

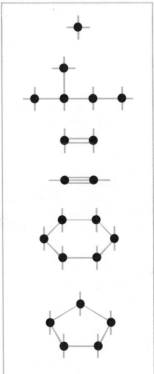

Figure 2-4 Chemical bonds between carbon atoms. Each line is a covalent bond, formed when one electron pair is shared by two carbon atoms (represented by dots). The vacant bonding sites are usually filled by hydrogen, oxygen, nitrogen, sulfur, or phosphorus.

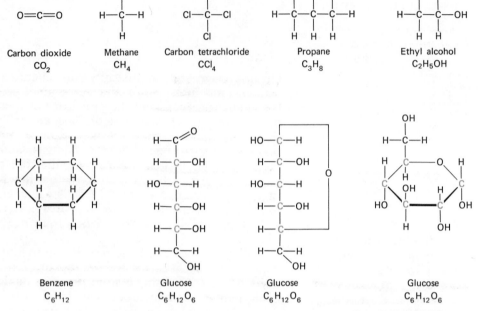

Benzene
C_6H_{12}

Glucose
$C_6H_{12}O_6$

Glucose
$C_6H_{12}O_6$

Glucose
$C_6H_{12}O_6$

Carbon dioxide
CO_2

Methane
CH_4

Carbon tetrachloride
CCl_4

Propane
C_3H_8

Ethyl alcohol
C_2H_5OH

Figure 2-5 (above) Types of carbon compounds. Carbon atoms can form a wide variety of molecules, some of which occur in different configurations. Glucose may take on a chain or a ring shape, and the ring form can be drawn in two ways, as shown. (The carbon atoms shown in color are "asymmetric"; that is, they can be joined to the other atoms in two different mirror-image arrangements.)

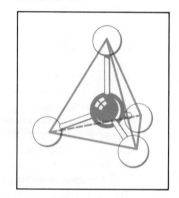

Figure 2-6 (below) The four bonding sites of a carbon atom are at the corners of a tetrahedron.

other to form chains or rings (Figure 2-4). Unlike ionic bonds, the bonds between carbon atoms are very strong, and organic compounds are not pulled apart by polar forces in water solution.

Some examples of organic compounds are diagrammed in Figure 2-5. Notice that each carbon atom is always involved with four bonds. Similarly, oxygen (O) always shares two bonds; nitrogen (N), three; and hydrogen (H), only a single bond. The simple sugar glucose occurs in two forms, as a chain or as a ring. Linking the oxygen atom to the five-carbon chain produces a six-part ring structure. (But notice that both forms have the same number of C, H, and O atoms.)

In chemical diagrams, a chemist attempts to provide as much information as possible in a shorthand form. However, in one respect the diagrams are quite misleading. The diagram ignores the fact that the bonds do not lie in a single plane: they

have a three-dimensional arrangement. If three of the hydrogens attached to the carbon in methane were to lie touching this page, then the fourth hydrogen atom would either stick up in the air or lie below the surface of the page. The bonds are actually at the four corners of a tetrahedron, a four-sided pyramid (Figure 2-6).

Any time four different atoms or groups of atoms are attached to the four bonding sites of a carbon atom, two different arrangements are possible. Consider the following analogy. A glove has distinct fingers and thumb. But if the back and the palm are identical, the glove can be worn on either hand. However, if the back and palm are different, the glove must be either a left or a right glove. The same is true of a carbon atom. If any two or more of the groups or atoms attached to the carbon are the same (in other words, if the "back" is identical to the "palm"), there will be only one form of the compound. But if the carbon atom is attached to four different groups, two distinct forms can exist. The carbon is said to be **asymmetric.** (The glucose molecules diagrammed in Figure 2-5 have several asymmetric carbon atoms.)

The two possible arrangements of an asymmetric carbon are identified by the letters "L" and "D" (from the Latin *levo* and *dextro*, left and right) (see Figure 2-7). This phenomenon is of interest because biological systems are often very fussy about which form is acceptable. For example, natural proteins are composed entirely of L-amino acids. The compounds of the other configuration, the D-amino acids, are rarely incorporated into protein as it is synthesized by living creatures.

The versatile carbon atom produces a very large number of different organic compounds. Usually, however, only specific groups of atoms participate in biochemical reactions. Since the carbon-carbon bonds are strong covalent bonds, the carbon skeleton or framework itself is relatively stable. But

Figure 2-7   Whenever a carbon is bonded to four different atoms or groups of atoms (a, b, c, and d), there are two possible mirror-image arrangements, labelled L- *and* D-.

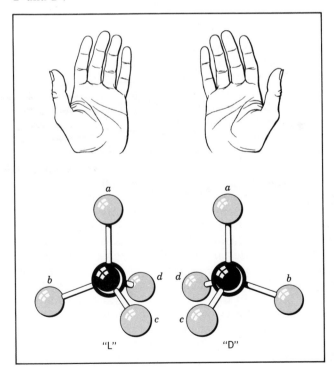

"L"   "D"

Table 2-2

Composition of Human Tissue
Figures are percentages, by weight.

Water	65%
Protein	16
Fat and Lipid	13
Carbohydrates	1
Minerals	5

some of the atoms or groups of atoms bonded to carbon are much more reactive than the carbon itself; these reactive groups are called **functional groups.** Some of them are diagrammed in Figure 2-8.

The first three functional groups pictured can undergo many chemical reactions. For example, they can all lose a hydrogen ion to a basic solution, and they can all gain one from an acidic solution. (The terms "proton" and "hydrogen ion" are interchangeable in this context. The hydrogen atom consists of one proton and one electron. When the electron is stripped away, the resulting H^+ ion thus consists of the proton alone.) These functional groups can also easily combine with other atoms and molecules in chemical reactions. The other two functional groups, hydroxyl and sulfhydryl, do not give up hydrogen ions in water solution. However, they can give them up during other chemical reactions.

The three most abundant classes of organic compounds are the carbohydrates, the lipids, and the proteins. Table 2-2 gives the approximate percentages of these types of compounds in various organs and tissues of human beings. A fourth class, the nucleic acids, is less abundant but equally important, because it ultimately controls all biochemical reactions.

Carbohydrates

Carbohydrates consist of sugars and sugar derivatives. They are probably most familiar to us as energy sources—the various sugars and starches—but the most common carbohydrate of all is a structural component of green plants, cellulose. Carbohydrate molecules all contain C, H, and O. Their basic structure is a carbon chain (or ring) with H's, OH's and O's attached to it.

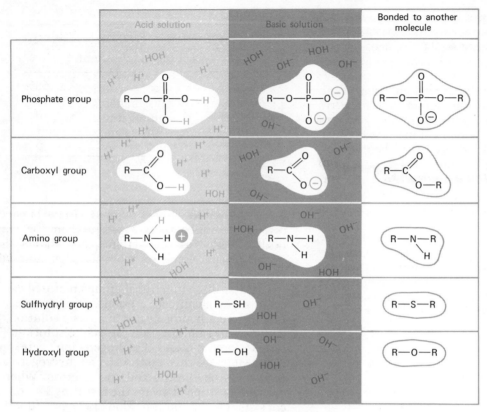

	Acid solution	Basic solution	Bonded to another molecule
Phosphate group			
Carboxyl group			
Amino group			
Sulfhydryl group			
Hydroxyl group			

Figure 2-8 Functional groups in carbon compounds. The phosphate, carboxyl, and amino groups react in acid water solution by binding with protons; in basic solution they lose protons. Sulfhydryl and hydroxyl groups do not lose or gain protons in water, but all these groups react with other molecules under the proper conditions. (The "R's" represent the skeletons of the molecules. "R" is a conventional symbol for one or more atoms bound together.)

Sugars are the simplest carbohydrates. The sugar glucose ($C_6H_{12}O_6$) is a common energy source in both plants and animals. There are many other sugars with the same overall atomic formula as glucose. But they are different molecules because each has a unique arrangement of H, OH, and O along the carbon chain, and because the carbon chain itself is arranged in various ways. And, since there are several asymmetric carbon atoms in these sugars, some of the arrangements are mirror images. In addition to this great variety of six-carbon sugars, there exist other sugars that have anywhere from three to seven carbon atoms in their skeletons. Two five-carbon sugars, ribose and deoxyribose, are an important part of the structure of nucleic acids.

All of these carbohydrate molecules are termed simple sugars. The simple sugars can be used to produce more complex carbohydrates in two ways. In the first, functional groups can be substituted for an H or an OH attached to the carbon skeleton of a sugar; many such sugar-like compounds exist. The second method for producing larger carbohydrates is to link many simple sugars together. Compounds ten thousand times the size of glucose are common. Fortunately for the sanity of the biochemist, these large molecules are all constructed as **polymers**: in other words, they are built up by repetition of rather simple units (called **monomers**). Carbohydrate polymers are called **polysaccharides** ("many sugars"), while the monomers of which they are made are called **monosaccharides** ("one sugar").

The three most important polysaccharides are starch, glycogen, and cellulose. All three of these polymers are formed from monomers of the simple sugar glucose. They differ only in the type of linkage between the glucose units (Figure 2-9).

In cellulose, the bonds between adjacent glucose molecules are formed between the oxygen of an OH group in one molecule and a carbon in the next. This type of bonding produces a long, unbranched chain. In the process an H and an OH are removed from the two carbons involved in the linkage (the equivalent of a molecule of water).

The nature of the bond in cellulose is extremely important. This linkage joins the first carbon in one glucose to the fourth carbon in the second glucose; for this reason it is called a 1,4 linkage. Also, the glucose units joined are in only one of the two possible mirror-image arrangements, called α and β. The cellulose linkage is a β-1,4 linkage. This linkage cannot be split by most animals. Hence, cellulose is indigestible to them.

Cellulose is the most abundant organic compound in organisms. It contains more than half of the organic carbon in plants and, with other polymers, forms plant cell walls and wood. Cotton is about 90 percent cellulose. Small amounts of it also appear in animals, including man. The presence of the β-1,4 linkage means that most of the organic carbon in plants is protected from direct digestion by animals.

Starch and glycogen are composed of glucose units in the α form. This linkage usually joins carbons one and four (α-1,4), but occasionally an α link is made between carbons one and six (α-1,6). This possibility allows branches in the chain (see Figure 2-10). Through a similar linkage, two or three different sugar units can be joined together to form specific di- and trisaccharides. The most common example is sucrose, table sugar. Sucrose is a disaccharide of two simple sugars, glucose and fructose. The α linkages are easily split, allowing the sugar units to be used for energy. Consequently,

Figure 2-9 *Important long-chain carbohydrates. Cellulose is the chief structural material in plants. Starch and glycogen serve for food storage in plants and animals, respectively.*

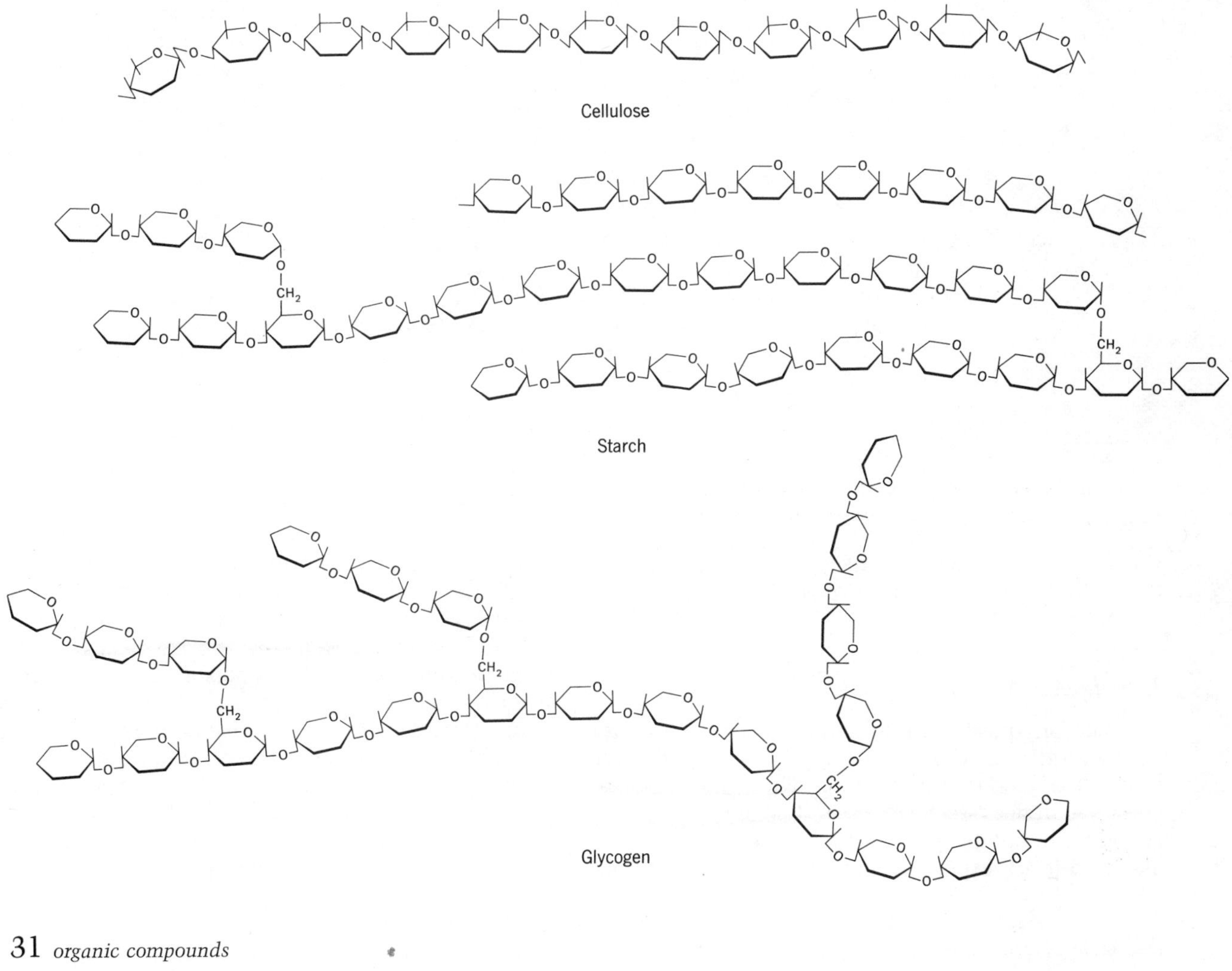

Cellulose

Starch

Glycogen

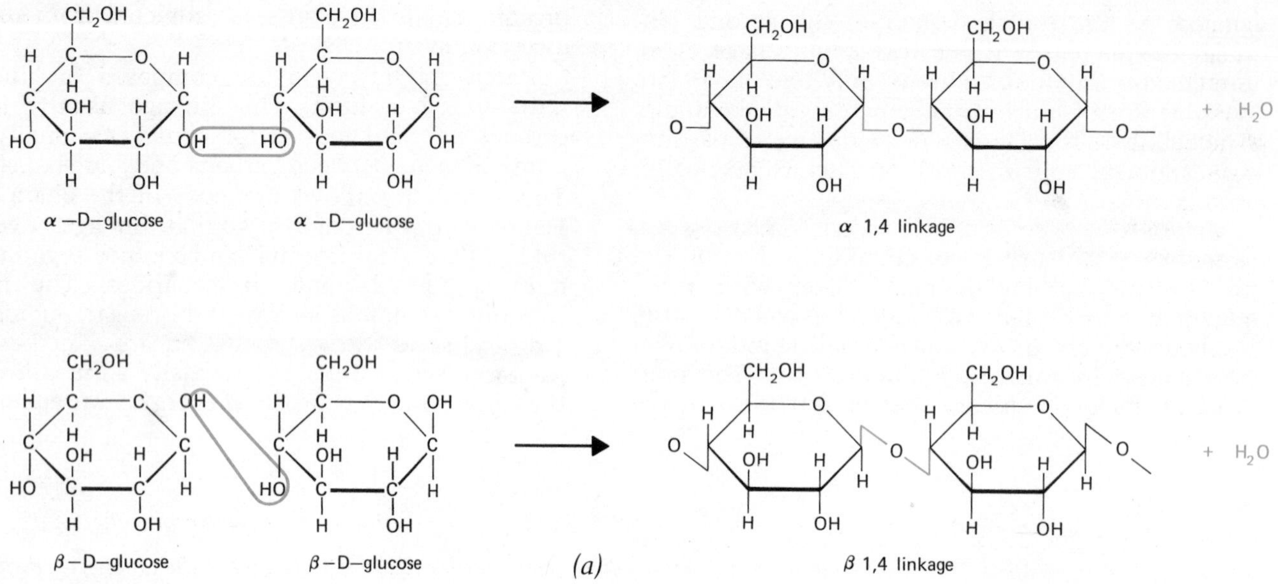

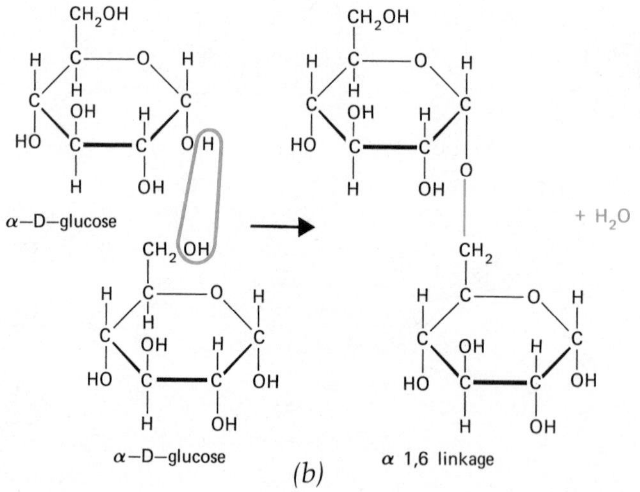

Figure 2-10 *Linkages between glucose molecules. The molecules are joined in a condensation reaction, the equivalent of removing a molecule of water. (a) The linkage between carbons 1 and 4 can be in an α or a β configuration. (b) The linkage between carbons 1 and 6 is an α link.*

starch and glycogen are used for storing energy rather than for structural purposes.

Lipids

The carbohydrates can be defined by the composition and structure of their molecules. All of them have clearly related structures. The **lipids,** by contrast, are defined by their solubility characteristics. If you treat a plant or an animal with a solvent such as chloroform, ether, or benzene, some compounds in the tissues will dissolve. They are the lipids. Lipids are all soluble in such non-polar solvents but relatively insoluble in water.

One class of lipids is the **fatty acids.** A typical fatty acid molecule consists of a long chain of carbon atoms. (Since most or all of these carbons are linked to hydrogen atoms, it is called a hydrocarbon chain.) A carboxyl group (–COOH) is attached to one end of the chain. This group gives the molecule its acid properties.

The acid group is polar and tends to be water-soluble; the hydrocarbon portion can be various lengths. As the hydrocarbon chains get longer, the molecules become less and less soluble, depending on the lengths of the chains. Acetic acid is the smallest fatty acid. It has only a single carbon in the hydrocarbon portion of its molecule, and hence it is quite soluble in water. (Vinegar is about five percent acetic acid.) By contrast, fatty acids with twelve or fourteen carbons will not dissolve at all in water, even though the carboxyl group on the end still attracts water molecules. Instead, they will form a layer on the surface of the water, with the carboxyl groups in the water and the hydrocarbon chains out of it (Figure 2-11).

Many kinds of fatty acids are produced by orga-

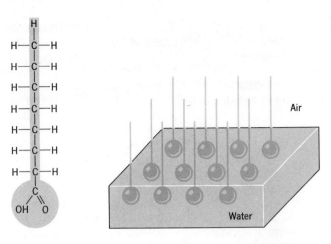

Figure 2-11 (above) Long-chain fatty acids oriented on a water surface. The carboxyl group, represented by the circle, is water-soluble; the hydrocarbon chain, represented by the straight line, is not.

Figure 2-12 Common fatty acids, their chemical formulas, and their sources in nature. Unsaturated double bonds are shown in color.

Name	Formula	Source
Acetic acid CH_3COOH	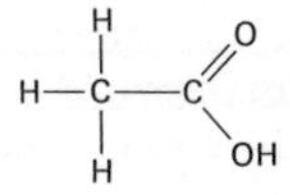	vinegar
Butyric acid $CH_3(CH_2)_2COOH$	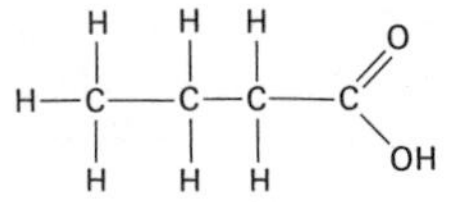	butter, milk
Capric acid $CH_3(CH_2)_8COOH$	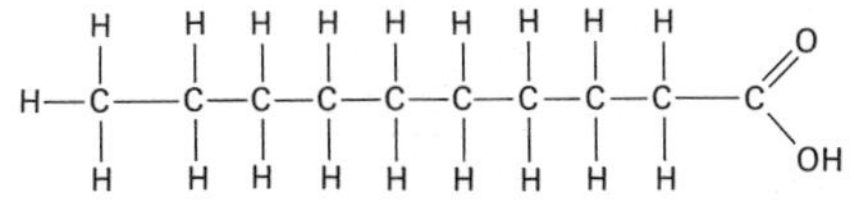	coconut oil
Palmitic acid $CH_3(CH_2)_{14}COOH$	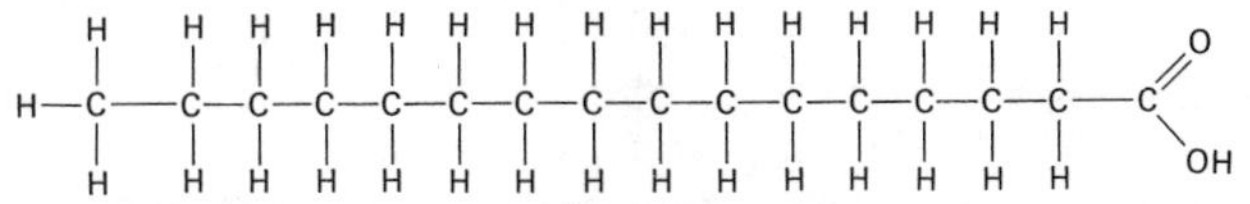	animal, plant, and bacterial fat
Behenic acid $CH_3(CH_2)_{20}COOH$	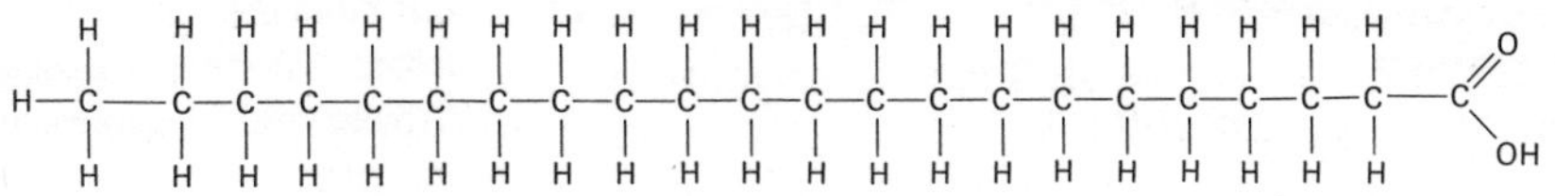	peanut oil
Oleic acid $CH_3(CH_2)_7CH{=}CH(CH_2)_7COOH$	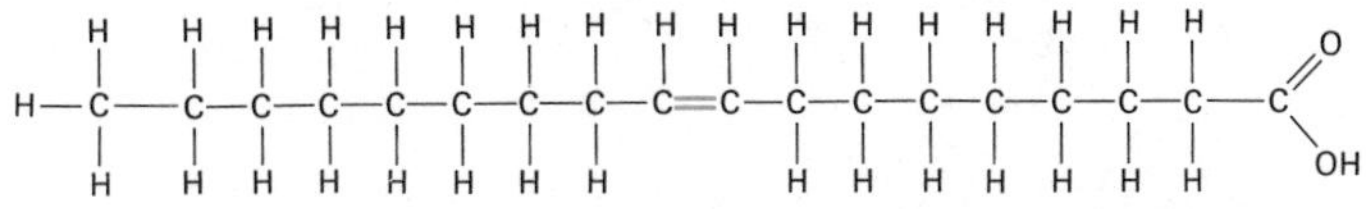	animal, plant, and bacterial fat
Linolenic acid $CH_3(CH_2CH{=}CH)_3(CH_2)_7COOH$	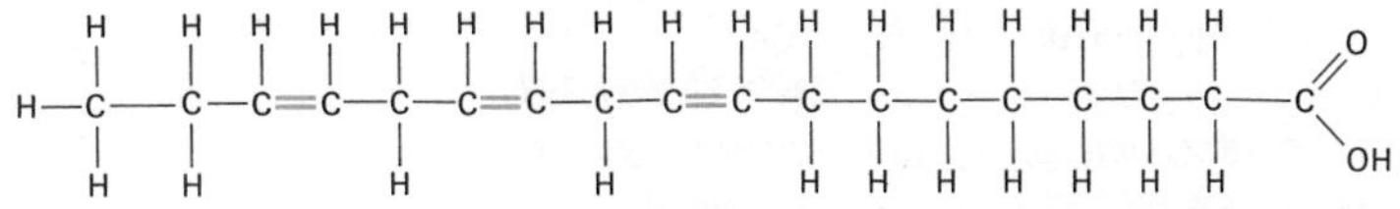	linseed oil

nisms (Figure 2-12). Some are saturated; that is, each carbon atom in the chain is linked to two hydrogen atoms. Others are unsaturated. In other words, a few of the carbons in the chain are joined together by double bonds rather than single ones (and hence are not "saturated" by hydrogen atoms). The amount of saturation in a fatty substance is reflected in its melting point. For instance, vegetable oils that are liquid at room temperature are high in unsaturated fatty acids. Hard vegetable fats, margarines, and animal fats have more saturated fatty acids than oils. Hydrogenated vegetable oils have been deliberately saturated with hydrogen to make them solid at room temperature. Butter, unlike most animal fat, is soft at room temperature; this fact means that it contains some unsaturated fatty acids.

A **fat** molecule consists of three fatty acids joined to a glycerol molecule. The joint is made at the carboxyl ends of the fatty acids, inactivating the carboxyls as acid groups. Hence, these lipids are called "neutral" fats (Figure 2-13). The removal of the carboxyls also makes fats essentially insoluble in water. Therefore, they occur in organisms as droplets or globules. Fats are important as energy reserve compounds. They store energy more efficiently than glucose or polysaccharides because they contain more available energy per unit weight. A gram of glucose provides only about half the energy that is made available by a gram of fat. For this reason it is fat, not sugar, that is accumulated by birds prior to migration and by many animals just before winter dormancy. Though fat is useful for energy storage, it has to be broken down and converted to other compounds before it can enter reactions that provide energy. Glucose is less efficient for energy storage, but it can enter these reactions directly.

Various other lipids are worth noting. Some fat-like compounds serve as insulating sheaths for nerve fibers. Others, called **sterols** and **steroids,** are ring-shaped lipids that function as important animal hormones. Steroids are produced from another lipid molecule, cholesterol (Figure 2-14). Cholesterol is found throughout the body.

Proteins

Proteins represent half to two-thirds of the dry weight of most living tissues. They perform two basic functions in an organism. First, they function as biological catalysts, or **enzymes.** An enzyme participates in the intermediate steps of a chemical

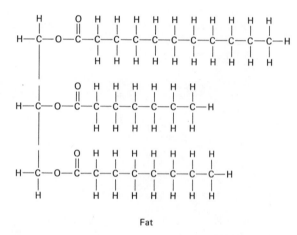

Glycerol   Fatty acids

Fat

Figure 2-13 Neutral fats are composed of three fatty acids joined to a glycerol molecule. Three water molecules are removed in the process.

reaction but emerges intact, that is, chemically unchanged. Enzymes control the chemical reactions that take place in cells, and therefore control the entire chemistry of the organism. All enzymes are proteins, or at least have protein portions which are necessary for their specific action. Second, proteins function as structural material. Hair, fingernails, and cartilage all contain the fibrous protein collagen. Structural proteins also coat the surfaces of cells, or are formed into muscle fibers. These two functions of proteins are really quite different. The enzymatic function is basic to all living organisms. The structural function is more important in animals than in plants. (Plants tend to use carbohy-

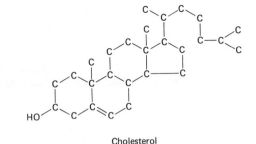

Cholesterol

Figure 2-14   Some ring-shaped lipids are steroids, such as the male hormone testosterone and the female hormone progesterone. Both compounds are synthesized from cholesterol.

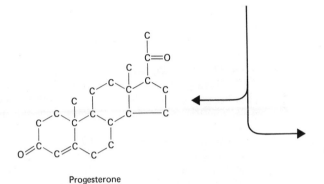

Progesterone

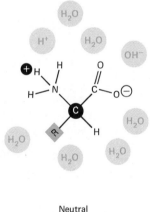

Testosterone

Figure 2-15   An amino acid gains or loses protons, depending on the acidity of the environment.

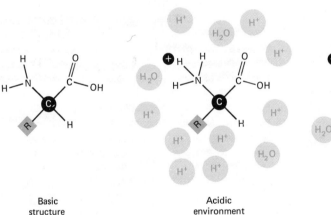

Basic
structure

Acidic
environment

Neutral
environment

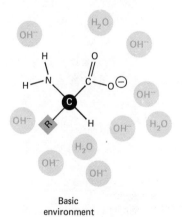

Basic
environment

drates such as cellulose as their structural material.)

Protein molecules are extremely large, and the clarification of their structure has been one of the major triumphs of modern science. Proteins are polymers, made up of a class of subunits called **amino acids.** As the name implies, each amino acid has both a basic, amino group (–NH₂) and an acidic, carboxyl group (–COOH). In a neutral water solution, the carboxyl group can ionize and lose a proton, leaving the group with a negative charge. In

the same solution the amino group can accept a proton and acquire a positive charge. Thus, each amino acid has both acidic and basic properties. In a neutral water solution, the hydrogen atom of the carboxyl group is transferred to the basic amino group (Figure 2-15).

Each amino acid always includes one carbon to which are bonded the carboxyl group, the amino group, and a hydrogen atom. The amino acids differ only in what is attached at that carbon's fourth bond, symbolized by "R" in Figure 2-16. "R" can be

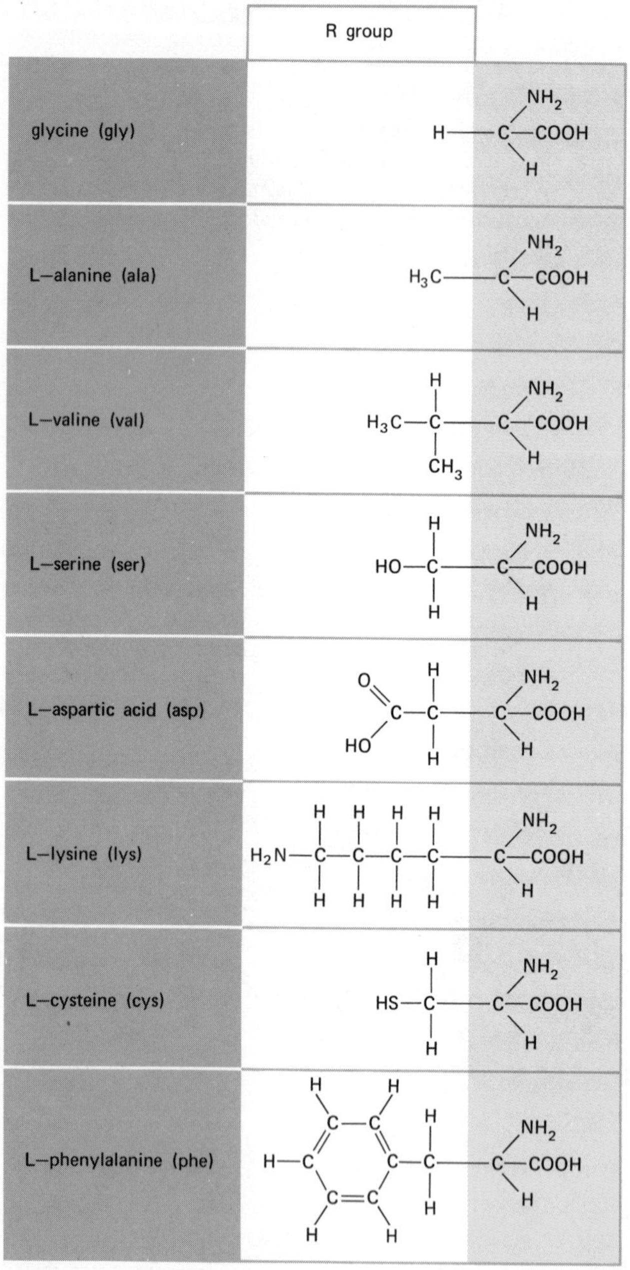

	R group	
glycine (gly)		H—C(NH₂)(H)—COOH

Figure 2-16. All amino acids have one carbon that is bonded to an –H, an –NH₂, and a –COOH. The various amino acids differ only in the structure of the "R" group attached to that carbon's fourth bonding site.

simply a hydrogen atom, in which case we have the simplest amino acid, glycine. But in more complex amino acids "R" represents a carbon chain or ring, which may include amino, carboxyl, or sulfhydryl functional groups. There are about twenty common amino acids found in nature (in other words, about twenty different "R" groups). In all of the amino acids but glycine the central carbon is asymmetric, because it is bonded to four different groups, and each of these amino acids therefore exists in both L- and D- forms. However, only L-amino acids are normally found in proteins, and their biological properties are quite different from those of their mirror-image forms.

In a protein molecule, each amino acid is linked to the next one by a chemical bond between its carboxyl group and the amino group of the adjacent amino acid. This linkage is called a **peptide bond** (Figure 2-17). A single protein molecule may contain thousands of amino acids linked together in this manner. Each position in the chain can be occupied by any one of the twenty amino acids, and changing a single amino acid can alter the properties of the entire protein. Thus, a huge number of different proteins can be constructed from various combinations of the twenty simple building blocks.

Actually, the structure of a protein involves more than the order of its amino acid units. That order is termed the primary structure of the protein (or more properly, of the **peptide chain**—a single protein may contain more than one peptide chain). But there are secondary and tertiary structural characteristics as well. The secondary characteristics depend on the coiling of each peptide chain into a helix (a spiral of fixed diameter: that is, a corkscrew shape), and also on the association of several chains to form sheets. The secondary characteristics are produced by hydrogen bonding between oxygen and hydrogen atoms near the peptide linkage. The tertiary structure is the folding of the helix or sheet at characteristic places.

The complex, three-dimensional structure of a protein molecule is produced by various types of chemical bonds between "R" groups on widely separated amino acids. These bonds keep the protein molecules folded or bent into stable structures. Sulfhydryl groups form particularly important cross-linkages (see Figure 2-18).

The final structure of a protein in solution is brutally complicated. The folded structure of the peptide chains in myoglobin (the pigmented protein that carries oxygen in muscle) is depicted in Figure 2-19. Very few proteins have yet been analyzed in such detail.

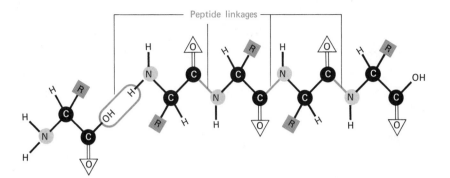

Figure 2-17 The peptide bond is formed between the amino group on one amino acid and the carboxyl group on another amino acid. Peptide bonds link the amino acids in a peptide chain.

The coiling and folding in proteins is not static. Proteins are not like bricks or building blocks; they are capable of changing shape. Only a fraction of a peptide chain may be coiled at any one time. The tertiary folding is stabilized by bonds between the side chains or "R" groups along the peptide chain, but these bonds are usually weak and easy to break. The association between peptide chains in a larger molecule is also easily broken.

Protein structure is therefore extremely sensitive to changes in the environment. If the temperature increases, the energy of molecular vibration or random motion may disrupt some of the weak bonds that maintain protein structure. This is what happens when an egg is boiled or meat is cooked; heat disrupts the protein structure. If the hydrogen ion concentration changes, so does the charge on the exposed "R" groups, and the whole pattern may shift. Inorganic salts may have a similar effect. Not only do the different sequences of amino acids make many different kinds of proteins, but a particular protein itself may exist in many different forms because of changes in the forces that underlie its coiling and folding.

Enzymes are large protein molecules. In solution they behave as colloids, representing a physically different phase from the surrounding solution and having definite surfaces. Enzymes act as catalysts for a vast number of biological reactions. Consider a small molecule of the substance which is to undergo a change or a reaction (usually called the **substrate**). The substrate molecule moves about randomly in the solution until it happens to collide with the enzyme at a specific point, to which it attaches. The attached molecule may now react with another small molecule similarly attached to the enzyme surface. Attachment may rearrange or weaken the chemical bonds in the substrate molecule, or it may change its charge distribution—whatever is appropriate to the reaction. In any case, these changes facilitate the reaction and accelerate the rate at which it occurs. When the reaction is complete, the small molecules are detached. The enzyme returns to its original condition, ready to begin again (see Figure 2-20).

The attachment between the enzyme and small molecules depends on the specific structure of the enzyme surface. For this reason, enzyme proteins are particularly sensitive to the environment. For example, enzyme action is strongly influenced by temperature. As the temperature starts to rise, so does the rate of the reaction. (The random motion of the molecules increases with temperature. Therefore, so does the number of collisions between molecules.) However, as temperature continues to increase, the rate of the enzyme-catalyzed reactions falls off again, because the secondary and tertiary structure of the enzyme is disrupted. Most enzyme configurations are also sensitive to acidity and often to concentrations of specific inorganic ions. This sensitivity is by no means a defect or a disadvantage. Rather, it provides a very delicate system of control which can be used to modify protein shape, and therefore protein function, to suit the needs of the organism.

Though many different proteins could be constructed from random arrangements of amino acids, the enzyme proteins in different organisms are surprisingly similar. This similarity results partly from evolutionary relations. For example, man is more closely related to monkeys than to horses. Not surprisingly, these relationships are reflected in their respective protein structures: cytochrome c (a respiratory protein) in man is identical to cytochrome c in monkeys except for one amino acid. However, horse cytochrome c differs from the human protein by twelve amino acids. The more distant the relationship, the larger are the dif-

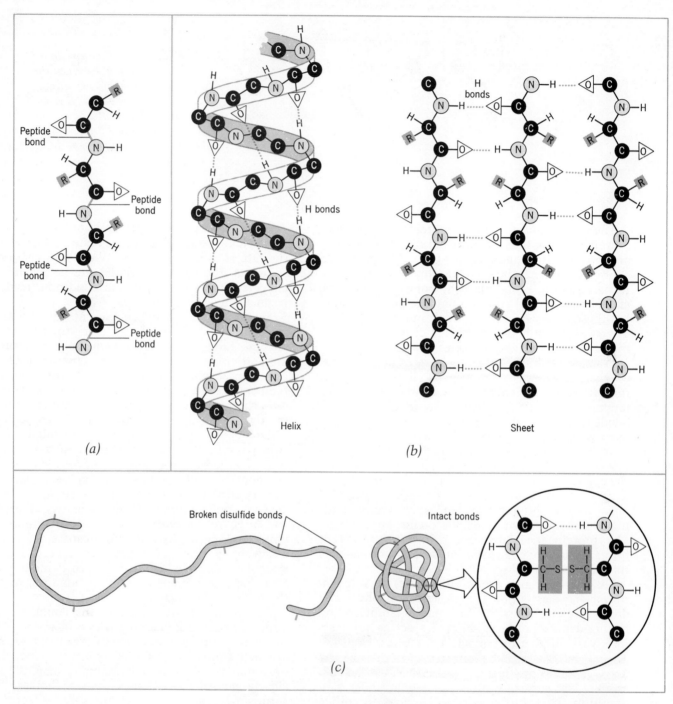

Figure 2-18 Protein structure. (a) Primary structure is a linear sequence of amino acids, joined by peptide bonds and forming a peptide chain. (b) Secondary structure is a helical or parallel arrangement of the peptide chains, joined by hydrogen bonds. (c) Tertiary structure is the additional folding of the molecule into a complex, irregular, three-dimensional form. The folding is caused by disulfide bonds between sulfhydryl groups that link distant portions of the molecule.

ferences between analogous proteins (see Table 2-3). Nevertheless, these differences would be much larger still if the amino acid sequence in proteins were truly random.

Similarities in protein structure are also dictated by chemical necessity. A protein must have a particular shape in order to catalyze a specific reaction. Many reactions in diverse organisms are identical. Therefore, enzymes that perform the same function in different plants and animals must have analogous structures. The fact that variations in structure do occur demonstrates only that other portions of the molecules are not needed for catalytic activity and are thus free to change. The existence of non-specific regions as well as specific surface sites in the enzyme can be demonstrated by removing the amino acids from an enzyme molecule a few at a time. The enzyme may still remain functional until a large portion of it has been destroyed, that

Table 2-3

Cytochrome C in Various Organisms

Species Compared	Number of Differences	Time Elapsed Since Divergence (in years)
Man–Monkey	1	5×10^7
Man–Horse	12	7×10^7
Mammal–Chicken	15	28×10^7
Vertebrate–Yeast[1]	45	100×10^7

[1] Note that though the relationship between yeast and vertebrates is extremely distant, they still share 58 common amino acids in their sequences for cytochrome c, a much larger number than can be due to chance.

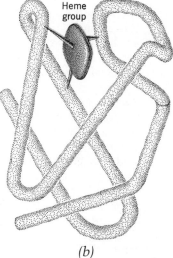

(a)

(b)

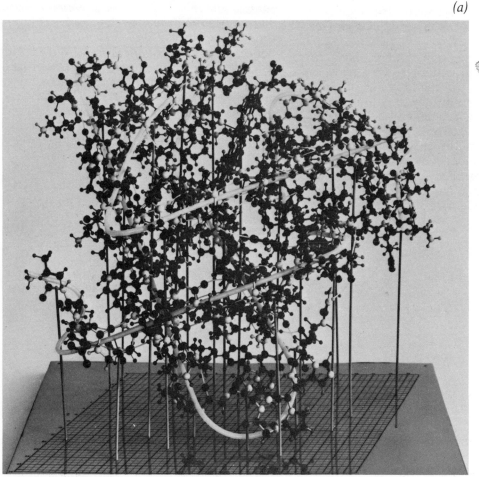

Figure 2-19 Tertiary structure of the protein myoglobin. The large central portion—the heme group—contains an iron atom that combines with oxygen and later releases it in muscle cells for use in respiration. (a) A detailed three-dimensional model showing the location of all the atoms. (b) A simplified drawing of the same molecule.

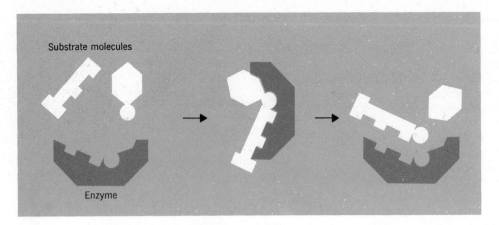

Figure 2-20 *Enzyme action. Two substrate molecules collide with and bind to an enzyme molecule. The binding forces and the proximity of the two molecules facilitate a chemical reaction. In this case, a small group of atoms on one molecule is transferred to the second. The enzyme molecule itself is unchanged.*

Substrate molecules

Enzyme

is, until the surface site is affected.

Proteins are sometimes compared to words spelled with amino acid "letters." In cytochrome c, for example, the spelling (the amino acid sequence) at the active site must be accurate. However, the other letters do not carry a message and can be changed at random.

Proteins serve many functions in organisms, as enzymes and as structural materials. Yet a particular organism contains only a small number of kinds compared to the almost infinite variety that would be produced by random combination of amino acids. What directs and controls the synthesis of proteins in the organism? Because proteins are so essential for such a tremendous variety of functions and activities, this is equivalent to asking how the entire organism is synthesized and defined. The structure and individuality of an organism is established by its genetic material, the nucleic acids.

the nucleic acids

The search for the genetic material in organisms was a major preoccupation of biology for many years. Such material must perform two main functions. First, it must carry information peculiar to one organism alone—the information that makes that organism an individual. The genetic material indirectly controls the synthesis and activity of all the compounds in a particular organism. And it is the nature and distribution of these organic molecules that produce and define the organism—whether a bacterium, a pine tree, or an elephant. The genetic material achieves complete control of the organism's chemistry simply by directing and controlling the synthesis of proteins. Protein enzymes direct cell growth and maintenance, processing chemicals from the environment. They also direct the chemical reactions that oxidize compounds for energy, build cellular substances, and carry on many other essential activities. Other proteins serve structural functions. Still other proteins function as hormones (such as insulin) or respiratory proteins (hemoglobin). But the primary role of proteins is to control cellular activity.

The second requirement for the genetic material is that it must be able to replicate itself and be passed on to daughter cells, or progeny, during reproduction. In this way, the genetic information in the progeny will be identical to that contained in the parent.

The nucleic acids, DNA and RNA, have been generally recognized as the genetic material for about twenty years. Biology can now give a good account of how they function as genetic material, both in replication and in the control of protein synthesis.

Nucleic Acid Structure

The nucleic acids, like the proteins, are large, polymeric molecules. Their name stems from their high concentration in the nuclei of cells and from the presence in their structure of numerous phosphoric acid groups (H_3PO_4). The repeating monomer in a nucleic acid is called a **nucleotide;** each nucleotide consists of one sugar molecule, one phosphoric acid, and one **base,** a ring-shaped molecule that contains nitrogen and has basic properties (Figure 2-21). The nucleotide monomers are linked by phosphate bonds between the sugar molecules.

There are two fundamental types of nucleic acids in organisms: **DNA** (deoxyribonucleic acid), and **RNA** (ribonucleic acid). These compounds have two basic differences in structure. First, the sugar in RNA is ribose, while that in DNA is deoxyribose. The two sugars are identical except that an OH group of the ribose is replaced by an H in deoxyribose (hence the "deoxy," which means "without oxygen"). Second, the bases differ. The four types of nitrogenous bases in DNA are adenine, thymine,

cytosine, and guanine. But in RNA, thymine is replaced by the closely related uracil; the bases in RNA are adenine, uracil, cytosine, and guanine.

The differences in nucleic acids depend on the order of these few bases. The precise linear structure of each molecule is defined by the arrangement of just four compounds. Certain pairs of bases readily form hydrogen bonds with each other. Because of slight differences in the molecular structures of the bases, guanine (G) bonds only with cytosine (C), and adenine (A) only with thymine (T) in DNA or uracil (U) in RNA. These bonded "base pairs"—C–G, A–T, and A–U—are specific and stable. They are responsible for a kind of secondary structure in nucleic acid polymers.

DNA usually occurs as a helical molecule of two

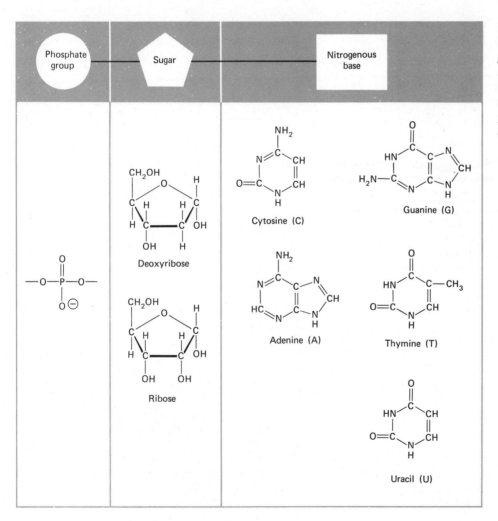

Figure 2-21 *Each nucleotide contains a phosphate group, a sugar molecule, and a ring-shaped nitrogenous base. The nucleotides are the monomers from which DNA and RNA are built.*

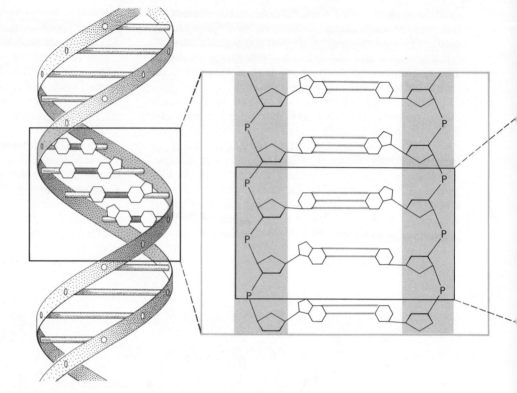

Figure 2-22 Structure of a DNA molecule. DNA usually occurs as a double helix. The two nucleic acid strands are joined by hydrogen bonds between bases.

interwoven strands. The phosphate-sugar molecules form a backbone on the outside, and the bases face inward. The two strands are connected by hydrogen-bonded base pairs stretching across like the rungs of a ladder (Figure 2-22).

The specificity of the base pairs means that the two DNA strands of each double helix will be different but complementary. The numbers of the individual bases will vary considerably from molecule to molecule. But when a double-stranded molecule is analyzed, the number of adenine residues will always be found to equal the number of thymines, and the guanines will equal the cytosines. Hydrogen bonding can also occur between bases in a single strand. This mechanism produces complicated and irregular secondary structure in some RNA molecules.

Replication

The structure of the DNA molecule was first deduced by James D. Watson and F. H. C. Crick in 1953 (Figure 2-23). Their discovery convinced them

that DNA is the primary genetic material, because the specificity of the base pairs and the double-helical configuration of the molecule suggested a simple and accurate method of replication. Their theory of DNA replication has since been confirmed and elaborated upon.

As the first step in replication, the relatively weak hydrogen bonds between the base pairs of the two DNA strands are broken. Previously synthesized nucleotide units then move in and form hydrogen bonds with the exposed bases on both of the old strands. Once the nucleotides are in position, an enzyme called DNA polymerase directs the formation of phosphate bonds between them. As a result, there are ultimately two identical DNA molecules where only one existed before, each molecule containing one old and one new strand. In short, each half of the double chain serves as a template for the organization of a new complementary strand (see Figure 2-24).

Occasionally DNA occurs as a single strand (in some viruses), or as a double-stranded loop (in viruses and bacteria), but the general scheme for replication is similar. There are a few microor-

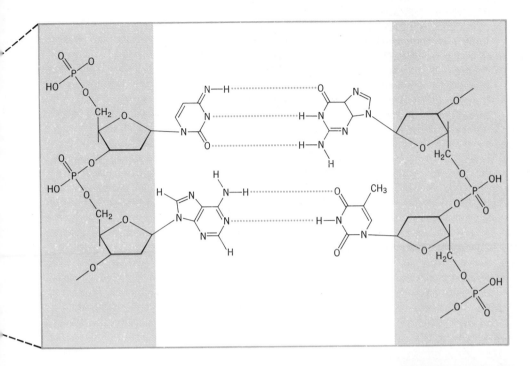

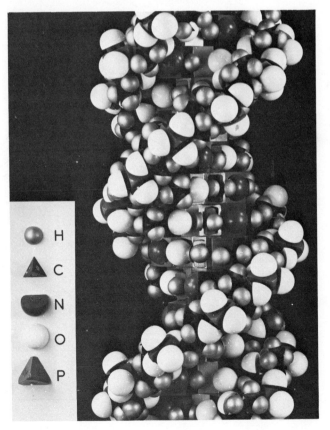

Figure 2-23 (below) A detailed model of a portion of a DNA molecule.

ganisms (plant viruses) in which no DNA is found. Instead, RNA supplies their primary genetic information. The most important and universal function of RNA, however, is to control protein synthesis.

Protein Synthesis

Protein synthesis is a complicated process, but it is usually summarized as:

$$\text{DNA} \xrightarrow{\text{transcription}} \text{RNA} \xrightarrow{\text{translation}} \text{PROTEIN}$$

In other words, the information contained in a DNA molecule is first coded, or **transcribed,** onto a new RNA molecule. Later, it is **translated** from the RNA molecule into a specific amino acid sequence, and therefore a particular protein. The genetic information that is contained in DNA depends on the order of the bases in the molecule. In other words, the "letters" of the genetic code are A, T, G, and C in DNA. They are transcribed into U, A, C, and G, respectively, in RNA by means of base pairing.

43 *the nucleic acids*

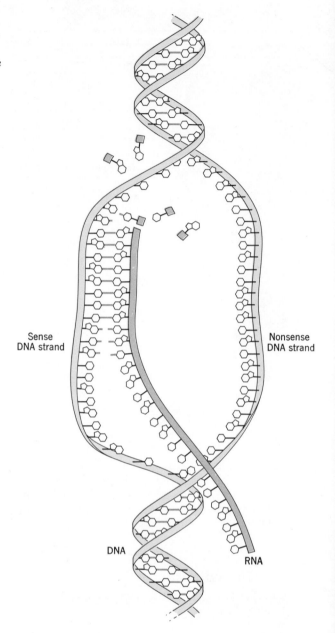

Figure 2-25 Transcription of DNA onto an RNA molecule. The DNA helix unfolds, and nucleotides move in and bond with the "sense" DNA strand. The nucleotides are bonded together by RNA polymerase to form RNA.

Sense
DNA strand

Nonsense
DNA strand

DNA

RNA

Figure 2-24 DNA replication begins when the strands separate. Nucleotide units then move into position; they are joined by DNA polymerase.

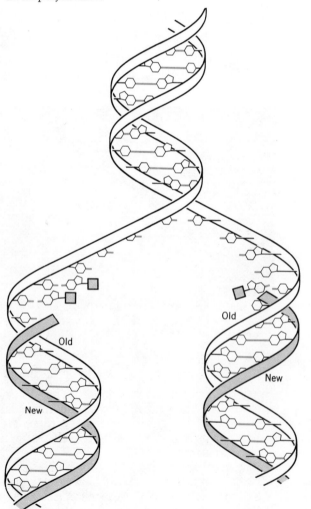

Old

Old

New

New

It is now known that each amino acid is coded for by at least one three-letter "word." A three-base code on a nucleic acid ultimately becomes an instruction to insert a specific amino acid into a protein molecule during protein synthesis. There are 64 possible unique "code words" or **codons** in all (four possible bases, in each of three positions, or 4×4×4). But there are only twenty important amino acids. In some cases, several codons may

Table 2-4

The Genetic Code[1]

1st ↓ 2nd →	U	C	A	G	↓3rd
	PHE	SER	TYR	CYS	U
	PHE	SER	TYR	CYS	C
U	LEU	SER	Terminator[2]	Terminator[2]	A
	LEU	SER	Terminator[2]	TRP	G
	LEU	PRO	HIS	ARG	U
C	LEU	PRO	HIS	ARG	C
	LEU	PRO	GLUN	ARG	A
	LEU	PRO	GLUN	ARG	G
	ILEU	THR	ASPN	SER	U
A	ILEU	THR	ASPN	SER	C
	ILEU	THR	LYS	ARG	A
	MET, Initiator[3]	THR	LYS	ARG	G
	VAL	ALA	ASP	GLY	U
G	VAL	ALA	ASP	GLY	C
	VAL	ALA	GLU	GLY	A
	Initiator[3]	ALA	GLU	GLY	G

[1] Modified from F. H. C. Crick. Note that many amino acids are coded for by more than one codon. The third nucleotide is usually the one that is changed. The amino acids are abbreviated as follows:

ALA	Alanine	GLUN	Glutamine	PRO	Proline
ARG	Arginine	GLY	Glycine	SER	Serine
ASP	Aspartic Acid	HIS	Histidine	THR	Threonine
ASPN	Asparagine	ILEU	Isoleucine	TRP	Tryptophan
CYS	Cystine	LEU	Leucine	TYR	Tyrosine
GLU	Glutamic Acid	LYS	Lysine	VAL	Valine
		MET	Methionine		
		PHE	Phenylalanine		

[2] These "nonsense" codons do not code for any amino acid; they terminate the chain.
[3] In bacteria, the chain is initiated by a derivative of methionine.

correspond to a single amino acid; for example, both AAA and AAG code for lysine. Other extra sequences are "nonsense" codons, since they do not correspond to any amino acid. Still others may serve specialized functions by directing where the transcription of a particular DNA segment should begin and end. Table 2-4 describes the genetic code as it was worked out in 1965.

The first step in protein synthesis is the transcription of the information carried by DNA onto RNA molecules. The process is similar to the self-replication of DNA. The DNA strands separate, and RNA nucleotides move in. They are base-paired in the usual way, except that the adenine on the DNA now bonds with uracil rather than with thymine. In this case, the formation of the backbone is catalyzed by the enzyme RNA polymerase. However, transcription, unlike DNA replication, involves the copying of only one strand of the original DNA double helix. This strand is called the "sense" strand, since these nucleotide sequences are eventually translated into protein. (The other DNA strand, the "nonsense" strand, is never used in RNA synthesis; it functions only to produce another sense strand during the self-replication of DNA.) (See Figure 2-25.)

The complementary RNA molecule, synthesized off the DNA sense strand, is called **messenger RNA,** or mRNA. A DNA template sequence of TTT will be transcribed as the complementary

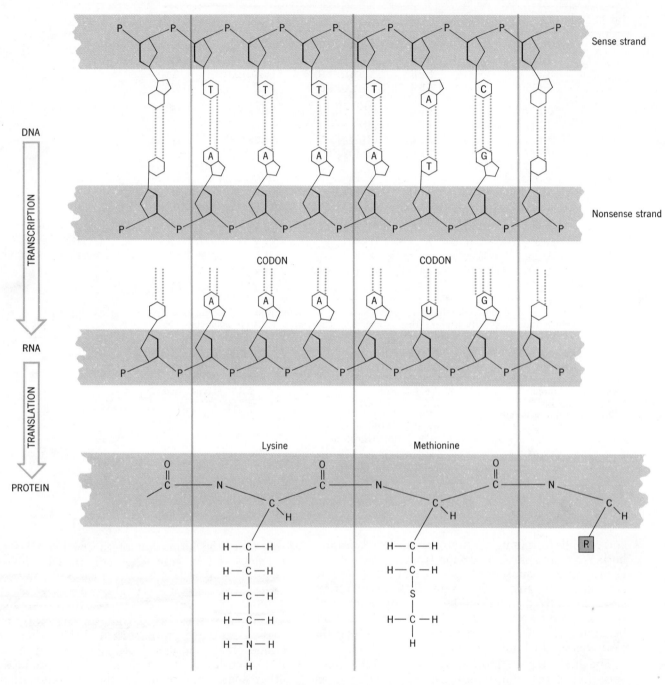

Figure 2-26 *Reading the genetic code, from DNA to protein. DNA is transcribed onto RNA, and the RNA is translated into protein. A three-base sequence on DNA corresponds to a complementary three-base sequence on RNA (the codon). The codon, in turn, corresponds to a single, specific amino acid in a protein. Only one of the two DNA strands is used in protein synthesis.*

46 *molecular machinery*

mRNA sequence of AAA. AAA is the three-base mRNA codon for the amino acid lysine. Similarly, a sequence of TAC on the DNA sense strand produces a mRNA sequence of AUG, which is the codon for methionine. Therefore a sequence of TTT-TAC on the original DNA would ultimately direct the insertion of a lysine-methionine sequence into the final protein (see Figure 2-26).

The codons that act as start and stop signals are extremely important elements of the code. Their existence means that only a portion of the entire DNA molecule needs to be transcribed at any one time in order for protein synthesis to occur. A DNA segment bounded by start and stop signals is called a **gene.** Each gene codes for one protein, or a small group of related proteins. Individual genes can be transcribed independently of the rest of the DNA molecule. When the cell requires large quantities of a particular protein, transcription occurs on the appropriate gene. Several mRNA molecules are transcribed simultaneously within a gene, one slightly behind the other (see Figure 2-27).

With the transcription of the code onto mRNA, the second phase of protein synthesis can begin: the translation of the code into protein. First, small, spherical particles attach to the mRNA strand. These particles, the **ribosomes,** provide a surface for protein synthesis. Each of the ribosomal units actually consists of two spheres, which bind to the end of the mRNA molecule (Figure 2-28). Ribosomes are composed of protein closely bound to another kind of RNA, **ribosomal RNA** (rRNA).

Each ribosome contains two attachment sites, arranged so that two consecutive codons of the mRNA molecule will be associated with the ribosome at any time. As protein synthesis begins, the appropriate amino acid moves to its codon, located at the first attachment site. A second amino acid moves in to its codon at site 2. An enzyme called peptidyl transferase then joins the amino acids by a peptide bond.

When the two molecules have been joined, the ribosome releases the first one from site 1 and moves one step along the mRNA strand, so that the attachment sites now encompass the second and third codons of the mRNA. The second amino acid is now in site 1, and a third amino acid is moved to its codon at site 2. By the same process, the new

Figure 2-27   Transcription of RNA on a single gene. This electron micrograph shows fibrils of RNA attached to a central DNA strand. The shortest RNA fibrils are just starting to form; the longest ones are almost complete. The small spheres on the DNA strand at the base of each RNA fibril are molecules of RNA polymerase. Magnification 55,000×.

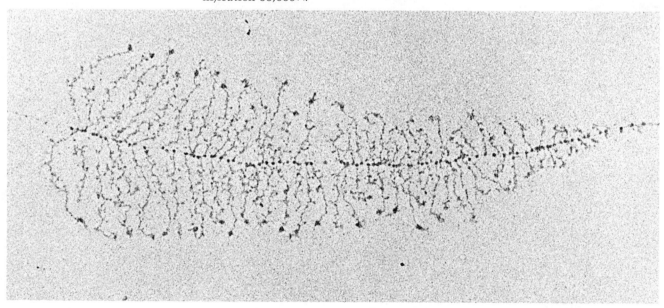

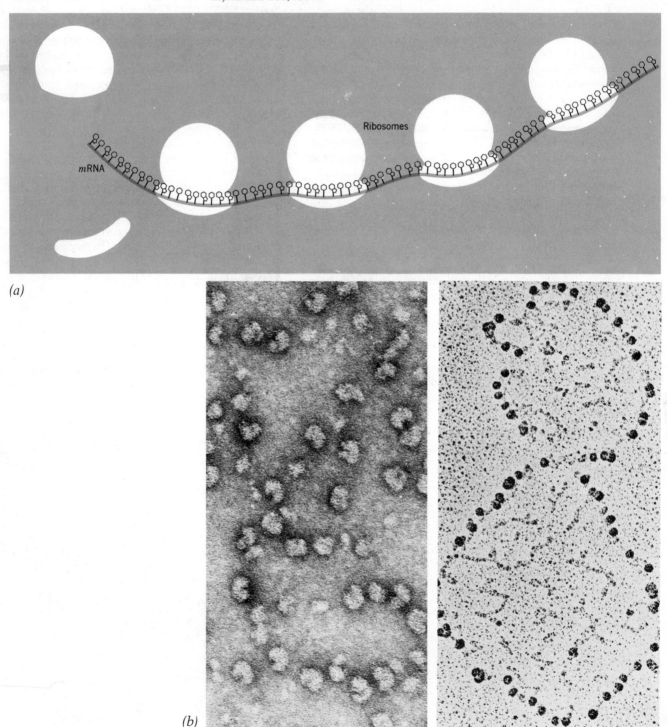

Figure 2-28 Ribosomes. (a) Several ribosomes become attached to a mRNA strand during protein synthesis. Each ribosome has two subunits, which become dissociated when detached from RNA. (b) Ribosomes from yeast cells. Magnification 200,000×. (c) Ribosomes attached to a strand of mRNA from a bacterial cell. Magnification 100,000×.

(a)

(b)

amino acid is bonded to the growing peptide chain. The ribosome has been visualized as moving down the mRNA molecule "reading" and translating each successive codon into the appropriate amino acid and adding it to the chain. A new protein is synthesized by the addition of one amino acid at a time. The ribosome's function is to coordinate and position the mRNA and the participating amino acids (Figure 2-29). Protein synthesis cannot occur in the absence of ribosomes.

How does an amino acid "recognize" its particular codon? In fact, amino acids cannot bind directly to the three-base sequences of the mRNA, because many amino acids are structurally too similar to one another. A glycine codon, for instance, would not be able to distinguish between the amino acids glycine and alanine on the basis of hydrogen bonds alone, because their structures have similar hydrogen-bonding characteristics. Therefore, an adaptor molecule with two specific sites is required for accurate translation. One portion of the adaptor must attach to a specific amino acid while a second

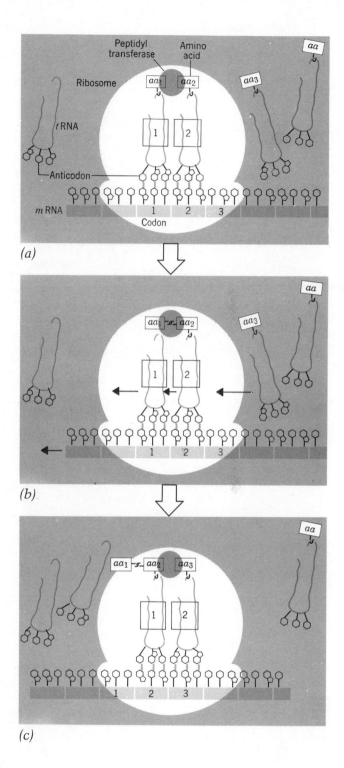

Figure 2-29   The information on the mRNA molecule is translated into protein with the aid of ribosomes. (a) mRNA attaches to the smaller ribosomal subunit at codons 1 and 2. Amino acids 1 and 2 attach to the larger ribosomal subunit at sites 1 and 2, by means of the appropriate tRNA's. (b) Amino acid 1 is released from its tRNA and linked to amino acid 2 with a peptide bond; the responsible enzyme is peptidyl transferase. (c) tRNA$_1$ is released from site 1; tRNA$_2$ moves from site 2 to site 1; tRNA$_3$, with amino acid 3, now moves into position in site 2. Simultaneously, the mRNA moves along so that codons 2 and 3 are now in place.

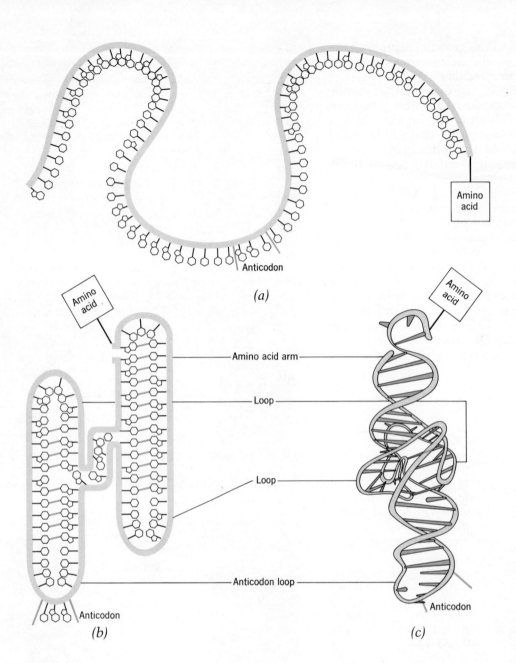

Figure 2-30 Structure of transfer RNA. One end of each tRNA molecule contains an anticodon, which becomes linked to the mRNA codon for an amino acid. The other end is linked to the appropriate amino acid. The amino acid–tRNA complex becomes linked to the attachment site on the ribosome by yet another part of its structure. (a) The linear structure of a tRNA molecule. (b) A simplified representation of folded tRNA. (c) A drawing of the actual three-dimensional structure of a tRNA.

portion accurately selects the corresponding mRNA codon. (This second site on the adaptor molecule is called an **anticodon;** it consists of a base sequence that is complementary to the codon.) The adaptor molecule is yet another species of RNA, **transfer RNA** (tRNA).

Transfer RNA's are rather complex molecules with many unusual bases and elaborate three-dimensional structures resembling cloverleaves (Figure 2-30). There are at least twenty different tRNA's, corresponding to the twenty amino acids. Each tRNA contains the appropriate anticodon at one end and a site at the other end to which the appropriate amino acid is attached. This last task is accomplished by a set of enzymes called amino acyl synthetases, which can distinguish between similar amino acid structures. Through this elaborate mechanism, accurate translation of the code is assured.

In spite of its complexity, protein synthesis is quite fast. A peptide may grow at the rate of 25 amino acids a second. It assumes its final three-dimensional shape as it is formed. Protein synthesis is summarized in Figure 2-31.

Figure 2-31 Protein synthesis. DNA directs the synthesis of at least three kinds of RNA: rRNA, mRNA, and tRNA. The rRNA combines with protein to form ribosomes. Ribosomes attach to specific codons on mRNA. Two tRNA's with attached amino acids position themselves on the ribosome, bonded to the appropriate codon. The amino acid on one tRNA is linked to the amino acid on the second tRNA. The sequence is then repeated until a protein is synthesized. When the protein is complete, it is released from the ribosome and assumes its secondary and tertiary structure.

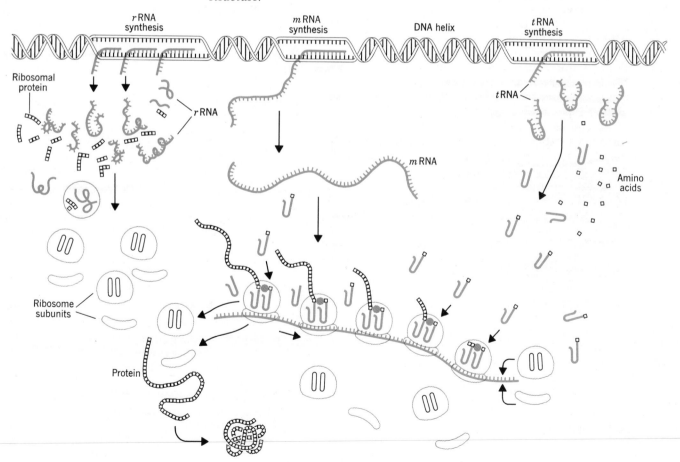

respiration

Chapter 1 outlined some of the major interactions between organisms and their environments. The most fundamental of these interactions is the exchange and conversion of energy. So far, we have described the major categories of organic compounds without making any reference to energy requirements. However, the synthesis of polymers such as carbohydrates, proteins, and nucleic acids will not proceed at all if the monomers are simply mixed—even if we include enzymes. These reactions all require the input of energy, which must be supplied by other chemical reactions that take place in organisms. Energy may also be needed for movement, nerve impulses, and a number of other activities. But in all cases the molecular machinery that provides energy is much the same. The energy is derived from the burning, or oxidation, of organic compounds, and this process is known as **respiration.**[1]

In air, sugars (or other molecules) burn and liberate heat. Carbon from the sugar combines with oxygen in air to form carbon dioxide. In this process water is also formed. Burning in the presence of oxygen is one form of oxidation, but there are others that do not actually require oxygen. In fact, any reaction that removes electrons from an element or a compound is called an oxidation reaction, and the element or compound is said to be **oxidized.** In some elements, such as iron, electrons can be removed directly from the atom. In compounds composed of many atoms, breaking and rearranging the bonds that join the atoms may also furnish electrons for oxidation. The electrons have to go somewhere. They are transferred to an acceptor element or compound, which is now said to be **reduced.** Reduction and oxidation always go together as complementary processes. The two terms simply refer to the transfer of electrons from a source compound to an acceptor compound.

It would be useless for an organism simply to burn sugar, since all the liberated energy would then appear immediately as heat. As we have already noted, heat has no work content. Energy liberated as heat cannot be used by the organism—either for chemical reactions or for the other processes that depend on energy. Furthermore, the high temperature produced by uncon-

trolled burning is incompatible with life. Thus, oxidation in organisms must be carefully controlled so that a portion of the energy liberated through oxidation can be stored rather than dissipated as heat.

We can write an equation for a reaction between sugar (glucose) and oxygen that yields carbon dioxide, water, and heat:

$$C_6H_{12}O_6 + 6\ O_2 \longrightarrow 6\ CO_2 + 6\ H_2O + \text{heat}$$

Although this equation is an adequate account of the uncontrolled burning of glucose in air, it represents merely a summary of the much more complicated set of controlled reactions that takes place in organisms. In fact, at least twenty-five intermediate steps are involved in the oxidation of glucose. Only six of these are actually oxidation steps. The rest are rearrangements of the intermediate molecules, reactions with water, or reactions in which carbon dioxide is lost (see Figure 2-32). These steps allow the controlled oxidation steps to occur.

In each oxidation step a pair of hydrogen atoms is removed from the organic molecule being oxidized. Since the hydrogen atom is composed of a proton and an electron, a pair of electrons is removed in the process; this reaction fits the definition of oxidation. The electrons and the protons are first transferred to an acceptor molecule called NAD (nicotinamide adenine dinucleotide). This molecule contains two nucleotides linked together; one contains adenine as a base, the other contains a base called nicotinamide. When NAD accepts a proton and two electrons, it becomes reduced. This state is symbolized by the expression $NADH_2$ (Figure 2-33).

Next, the two electrons and the proton are transferred from $NADH_2$ through a series of compounds, collectively called the **respiratory chain** (see Figure 2-34). The electrons and the protons are processed separately; ultimately they are reunited in combination with oxygen to form water. Many molecules in the respiratory chain contain iron, which is important in biological systems for the same reason that it is an excellent electrical conductor: electrons are easily added to or removed from the iron atom. As electrons are passed along the respiratory chain, a portion of the energy that would otherwise be lost as heat participates in the formation of so-called **high-energy phosphate** compounds. Biologically, this is the payoff step which justifies all the rest.

The most important high-energy phosphate in biological systems is formed from adenosine diphosphate (ADP). ADP consists of an organic molecule, adenosine, to which two phosphate groups are linked in a linear arrangement. In respiration, a por-

[1] Respiration at the molecular level is only distantly related to respiration in the sense of breathing. The word is commonly used in both senses, however.

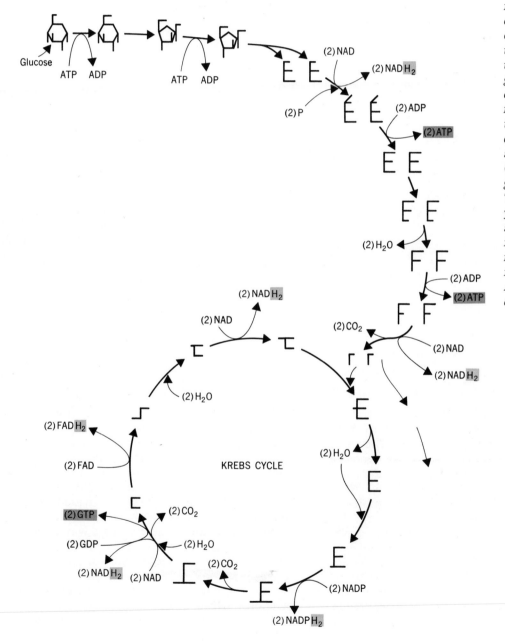

Glucose

ATP ADP

ATP ADP

(2) NAD

(2) NAD H₂

(2) P

(2) ADP

(2) ATP

(2) H₂O

(2) ADP

(2) ATP

(2) CO₂

(2) NAD

(2) NAD H₂

(2) NAD H₂

(2) NAD

(2) H₂O

(2) H₂O

(2) FAD H₂

(2) FAD

KREBS CYCLE

(2) H₂O

(2) GTP

(2) CO₂

(2) GDP

(2) H₂O

(2) NAD H₂   (2) NAD

(2) CO₂

(2) NADP

(2) NADP H₂

Figure 2-32 Organisms release the energy in a glucose molecule in a controlled series of chemical reactions. The glucose ring is rearranged and then split into two identical small molecules. These molecules then undergo more reactions before entering the Krebs Cycle. In this cycle of reactions, each small molecule combines with a larger one. The larger molecule undergoes more reactions until the small molecule is given off, piece by piece, as carbon dioxide. The larger molecule is regenerated in the process, and it can now combine with another small molecule. The Krebs Cycle turns twice for each glucose molecule oxidized. The six oxidation steps that yield electrons and associated protons are shown in color. High-energy ATP molecules are used up in initial reactions, but more ATP (and GTP) are produced a few steps later.

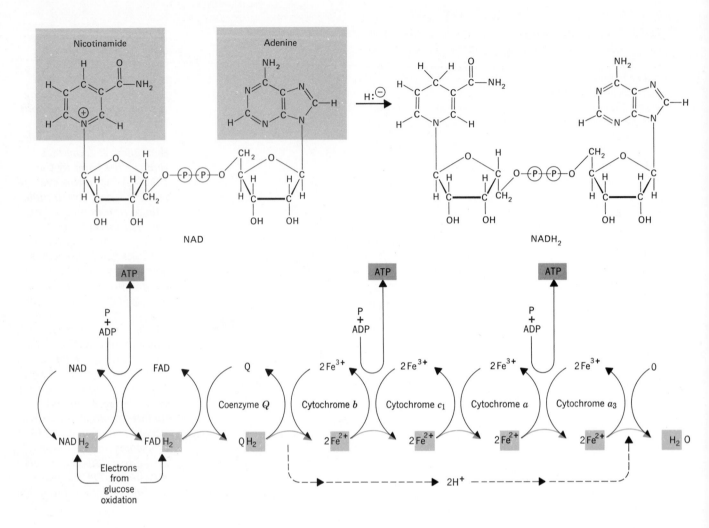

tion of the energy released in the electron transfer is used to bond a third phosphate group to ADP, forming adenosine triphosphate (ATP). The extra phosphate group of ATP is particularly reactive, hence the name "high-energy." (ATP and related compounds are diagrammed in Figure 2-35.) When this terminal phosphate is split off the molecule again, the considerable amounts of energy that went into attaching it are released. In effect, a portion of the energy in the original sugar molecule can be stored in the form of ATP instead of appearing directly as heat in the process of oxidation. About half the energy originally present in the glucose is bound into ATP during the biological oxidation of the sugar.

ATP can also be formed directly from intermediates of the sugar. A phosphate group can be added directly to one of the intermediate compounds. Then, after internal rearrangements of the molecule, the phosphate is transferred directly from the sugar to ADP. The amount of ATP formed in this way is small compared to that provided by the reactions of the respiratory chain. However, it is still important, because it represents a way in which the energy of organic compounds can be tapped in the absence of oxygen. Some organisms are able to live in **anaerobic** (oxygen-free) environments using this part of the oxidation reaction. But they waste a lot of energy because they cannot use the respiratory chain to produce ATP. The chain works only if ox-

Figure 2-33 (facing page, top) The chief electron transfer compound, NAD (nicotinamide adenine dinucleotide). NAD is composed of two nucleotides, one consisting of adenine, ribose, and phosphate, and the other of nicotinamide, ribose, and phosphate. NAD becomes reduced to $NADH_2$ by accepting a proton with a pair of electrons ($H:^-$). (Even though it accepts only one proton, the reduced form is designated $NADH_2$ to emphasize the fact that two electrons are transferred.) ℗ represents the phosphate group, $PO_3^=$.

Figure 2-34 (facing page, center) The respiratory chain. A considerable amount of energy is produced as electrons from $NADH_2$ (or $FADH_2$) move through the compounds of the respiratory chain. Each reduced molecule from glucose oxidation supplies one pair of electrons (and associated protons) to the respiratory chain. The electrons and protons are passed along as a unit until they reach the cytochromes. From there, electrons are passed along the chain by binding to iron (Fe^{+++} + electron $\leftrightarrows$ Fe^{++}). The protons are released to the environment, but they eventually rejoin the electrons, combining with oxygen to form water.

Figure 2-35 (below) High-energy phosphate compounds. These compounds consist of bases (adenine, guanine, or uracil) linked to ribose and two or three phosphates. The diphosphates (ADP, GDP, and UDP) become triphosphates (ATP, GTP, and UTP) by the addition of a phosphate group. High-energy bonds are shown in color.

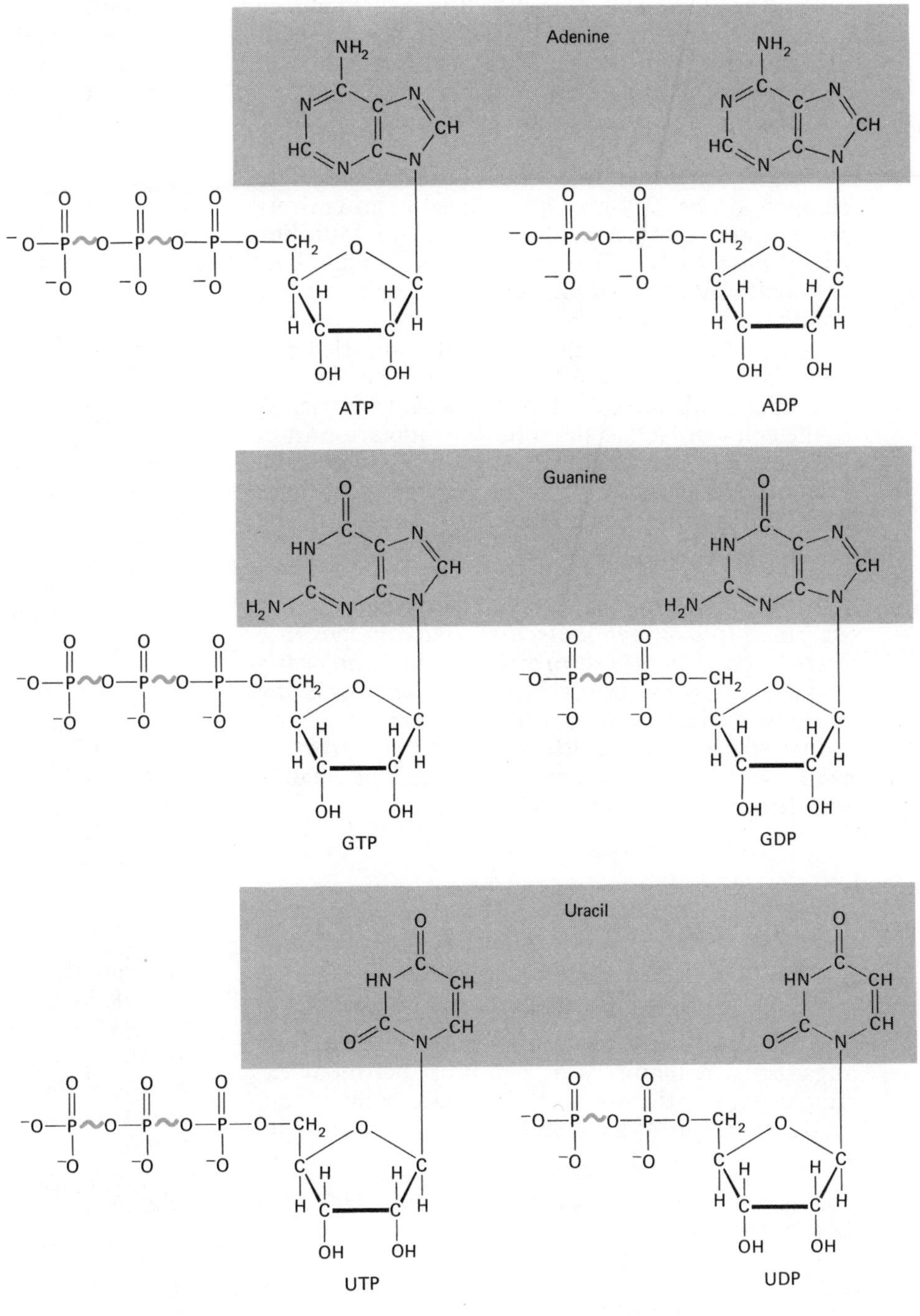

ygen is present to combine with protons and electrons, forming water.

We can summarize these events as follows. The oxidation of sugar takes place in a number of steps, each involving the transfer of a pair of electrons to NAD, producing $NADH_2$. The electron pairs in $NADH_2$ are then passed along the respiratory chain in a series of oxidation and reduction reactions and are finally coupled with oxygen. These transfers generate energy-rich phosphate compounds by converting ADP to ATP. Some ATP is also produced directly by early oxidation steps. In this way, about half the energy in the sugar molecule is stored in numerous ATP molecules.

Other kinds of organic molecules can be oxidized in a similar way. They may join the glucose cycle by being converted into one of its intermediate compounds, or they may be oxidized in an entirely separate set of reactions. In addition, UTP (uridine triphosphate) and GTP (guanosine triphosphate) occasionally function as high-energy phosphate compounds.

The value of the complicated but controlled respiration that generates ATP should now be obvious. The energy liberated by burning would be useless to the cell—in fact, as the following quotation from Rabinowitch and Govindjee suggests, it would be extremely destructive. It is the energy trapped in ATP and similar compounds that drives all the energy-requiring processes in living things.

All organic matter on earth is surrounded by a swarm of free oxygen molecules, like maidens in a castle wooed by a host of suitors. Only the castle walls . . . assure the precarious existence of living matter in contact with air. The respiration enzymes stealthily open little doors in the activation walls, and lead the organic molecules, one by one, into the embrace of oxygen.[2]

chemical energy

Molecules of ATP, derived from the oxidation of the original glucose molecule, can now be used by organisms in many ways. The most pertinent example is the utilization of phosphate bond energy in the synthesis of large molecules. Complex mole-

cules for use in growth and repair must be constructed from simpler compounds. The energy content of the more complicated molecule is usually greater than that of the starting material. Thus, in order for synthesis to occur, energy must be added to the starting material; no set of reactions can lead spontaneously from low-energy to high-energy compounds. In fact, *excess* chemical energy must be supplied, in order to compensate for the irreversible heat loss that always occurs during reactions. In biological systems the surplus chemical energy is often supplied by phosphate bond energy in ATP.

The energy requirements for the synthesis of carbohydrate polymers are a good illustration. Glucose is stored in organisms in the form of long chains linked together to form starch (in plants) or glycogen (in animals). The energy content of these chains is higher than that of the constituent glucose molecules. We can symbolize a glycogen chain by the expression "$glucose_n$"—which represents some number (n) of glucose molecules linked together. The reaction we want to consider is the addition of a new glucose molecule to a glycogen chain, or

$$glucose + glucose_n \longrightarrow glucose_{n+1} + heat \quad (2\text{-}1)$$

The glycogen chain is formed by repeated additions of single glucose units. However, this reaction will not occur spontaneously, since the chemical energy of the product is greater than that of the starting materials. The following reaction *can* occur:

$$glucose{\sim}UDP + glucose_n \rightarrow$$
$$glucose_{n+1} + UDP + heat \quad (2\text{-}2)$$

Here, excess energy is supplied by the phosphate bond between glucose and UDP. Now our problem is to form glucose $\sim$ UDP. We said earlier that the terminal phosphate on compounds like ATP has high energy and is reactive. In fact, the following reaction will proceed very well:

$$glucose + UTP \rightarrow$$
$$glucose{\sim}UDP + inorganic\ phosphate$$
$$+ heat \quad (2\text{-}3)$$

Our problem is solved. We can make glucose $\sim$ UDP by transferring the phosphate bond energy from UTP to glucose. It is then possible to add another unit of glucose to our glycogen molecule by means of the reaction in Equation 2-2. We have not violated any rules about energy exchange. All three reactions can be combined and written as one:

$$glucose + glucose_n + UTP \longrightarrow$$
$$glucose_{n+1} + UDP + heat \quad (2\text{-}4)$$

[2] Rabinovitch, E. B., and Govindjee, *Photosynthesis.* New York: John Wiley & Sons, 1969.

The UTP will be regenerated by the oxidation of more glucose. The overall scheme is shown in Figure 2-36.

This outline of reactions is a general model for the synthesis of complex molecules from simple precursors. The synthesis of proteins from amino acids is entirely analogous. In that case, energy is added to the system by activating the amino acid with ATP energy. The amino acid is activated even before it is attached to the tRNA.

Phosphate bond energy is the common currency which can be spent by both plants and animals to synthesize complex molecules. It is also required for muscle contraction, for the maintenance of electrical charge differences, and for other energy-requiring events in organisms.

Figure 2-36 Synthesis of glycogen from glucose. Glucose first reacts with ATP to form a glucose—phosphate molecule. The phosphate is then transferred to another part of the glucose molecule. A second high-energy phosphate molecule, UTP, now becomes involved, and glucose-phosphate is converted to UDP~glucose. Finally, UDP~glucose is linked to the growing glycogen molecule.

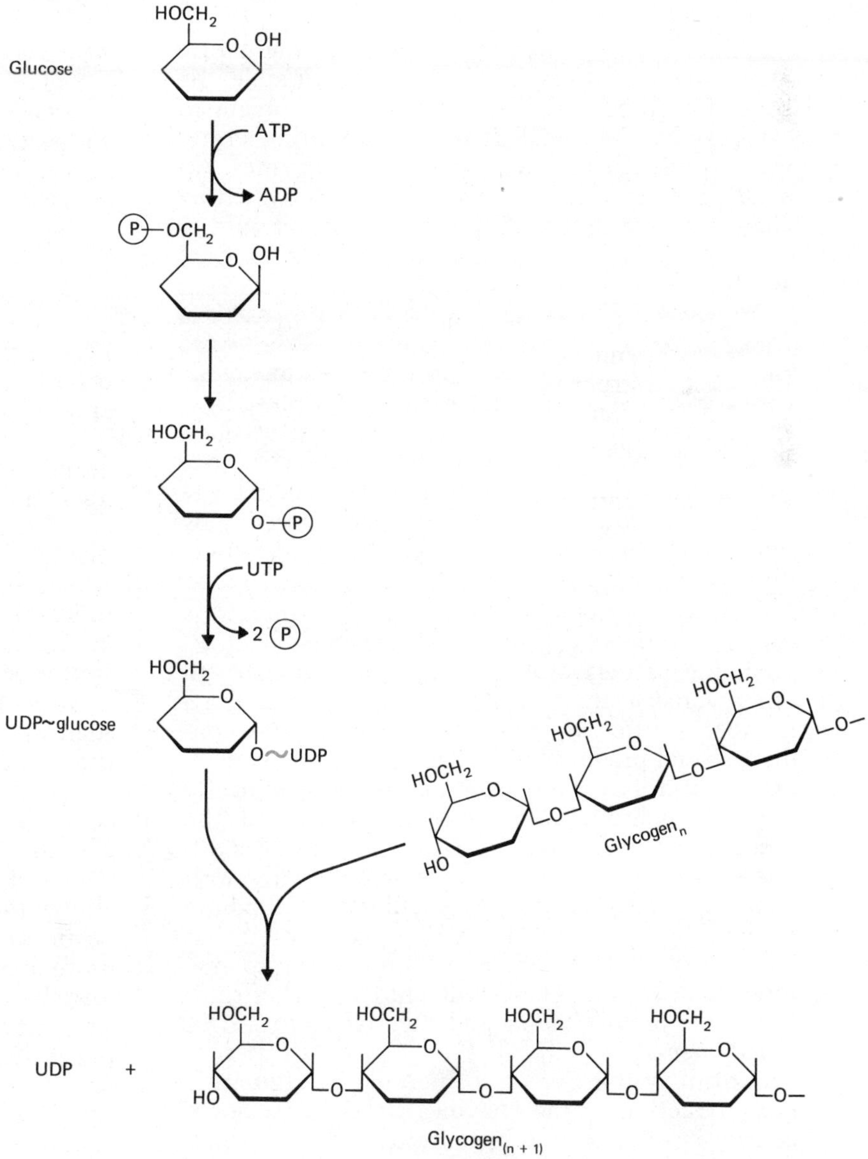

metabolism and control systems

Metabolism is a general term which refers to the sum of all the reactions that take place in an organism. An organism's metabolism does not run free; it is very carefully governed. If the enzymes for all the steps in metabolic reactions were active at all times, the system would be out of control. Metabolic processes would be at the mercy of whatever starting materials might be available. Furthermore, the organism might waste phosphate bond energy in constructing enzymes for which no substrate was present.

We are learning more and more about the nature of metabolic controls in organisms. In principle, the control methods are rather simple. The organism can regulate its metabolism by controlling either the activities or the quantities of key enzymes. An existing enzyme can be activated or inactivated by changing its structure. Alternatively, the amount of enzyme present may be changed by controlling its synthesis.

The production of many amino acids in bacteria is regulated by inactivating some of the enzymes that are involved in their synthesis. This process is called **end-product inhibition.** In the simplest case, the presence of the amino acid (the end product) would inactivate the enzyme that catalyzes its formation from the substrate (Figure 2-37). Thus, the end product inhibits its own synthesis—possibly by binding to the enzyme and changing its three-dimensional structure.

In actual control systems, this simple situation is usually complicated by two additional factors. First, the pathways of many reactions branch. In other words, a single substrate may be turned into a number of different end products. Under these conditions, end-product inhibition can provide considerable control over the system by acting specifically at the branch points in the sequence. Second, the product of one chemical reaction may inhibit a different, separate reaction. Or it may equally well facilitate such a reaction by blocking branch points in the other compound's reaction pathway, so that all the available substrate is funnelled into one reaction sequence. These different patterns of inhibition can provide varied and effective control over metabolism.

Control systems can also function by regulating the production of the enzymes themselves. Some control systems in the bacterium *Escherichia coli*

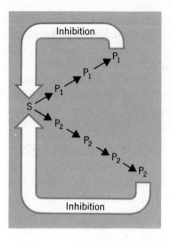

Figure 2-37   Feedback inhibition. A substrate molecule, S, can go through two different reaction sequences, forming products P_1 and P_2. These products inactivate the enzymes that catalyze early reactions in the sequence and thereby inhibit the further synthesis of the products.

behave in this way. *E. coli* can use many different organic compounds as energy sources. In the presence of the sugar lactose, *E. coli* produces large quantities of a pair of enzymes that enable it first to accumulate lactose, and then to split it into products that can be oxidized for energy. A normal *E. coli* cell grown in the absence of lactose might contain just a couple of these molecules. But when lactose is supplied, thousands of the required enzyme molecules soon appear in each cell (Figure 2-38). Somehow the presence of lactose has "induced" the synthesis of these enzymes.

A theoretical model accounts for the behavior of such systems. The particular genes that code for the two enzymes are situated next to three control sites on the DNA molecule. Each of these sites is essential to the operation of the overall system. The first site, called the **promoter site,** provides a point of attachment for RNA polymerase and allows the transcription of the necessary genes to begin. The second, the **regulator site,** synthesizes a protein called the **repressor.** If this molecule is allowed to reach the third site, called the **operator site,** its binds tightly. With the repressor molecule in place, the RNA polymerase at the promoter site cannot progress beyond this complex, and transcription of the genes is prevented—the enzymes cannot be synthesized. Therefore, in order for enzyme synthesis to occur, the repressor must somehow be kept from reaching the operator site.

This is exactly what happens when lactose is present. One particular product of lactose metabolism is able to bind strongly with the repressor protein. In this way, the repressor is prevented from attaching to the operator site. The operator region is not blocked, so RNA polymerase can proceed normally with the synthesis of the appropriate mRNA. The enzymes are rapidly synthesized (see Figure 2-39).

Most of our information about control systems comes from the study of such microorganisms. Genes from *E. coli* have been isolated from the rest of the DNA and are used for elegant and sophisticated experimentation. Ultimately, it may be possible to do the same with human genes—even to manipulate them at the molecular level. But at present much less is known about control systems in more complex organisms. For one thing,

mammals may have a thousand times as much DNA per cell as *E. coli*. In addition, the DNA in higher organisms is not present in simple strands but is located in larger structures called chromosomes, where it is tightly interwoven with proteins and yet another species of RNA (nuclear RNA).

Still another problem adds difficulty—and interest—to the study of control systems in higher plants and animals: the question of differentiation and development. How does a single cell or fertilized egg eventually develop into a complex organism with highly specialized tissues and organs, containing many different kinds of cells? After all, the genetic material is presumably the same in all the cells. This is the subject matter of developmental biology, which we cover in Chapter 13. The mechanisms involved in bacterial metabolism are simple by contrast, yet they demonstrate how the environment can interact with and control the organism.

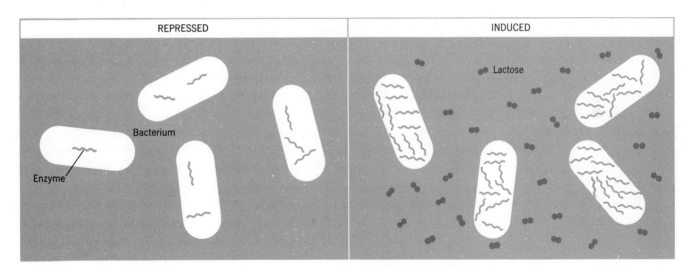

| REPRESSED | INDUCED |

Figure 2-38 *Enzyme induction in the bacterium* Escherichia coli. *When lactose is absent from the surrounding medium, the bacterium contains very few enzymes for lactose utilization. When lactose is present, enzyme molecules are quickly synthesized.*

Figure 2-39 Control of protein synthesis in bacteria. The lac operon consists of regulator, promoter, and operator sites and genes for the synthesis of enzymes. (a) The regulator synthesizes repressor proteins. When a repressor combines with the operator site, synthesis of the enzymes is repressed; RNA polymerase cannot move beyond the operator to genes Z and Y. (b) When inducer molecules are present they combine with the repressor, preventing the repressor from binding with the operator. RNA polymerase can then move to genes Z and Y, and protein sysnthesis is induced. (c) Electron photomicrograph shows the repressor bound to the operator region of the lac operon.

REPRESSED

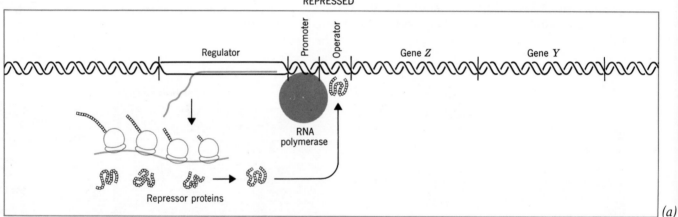

(a)

INDUCED

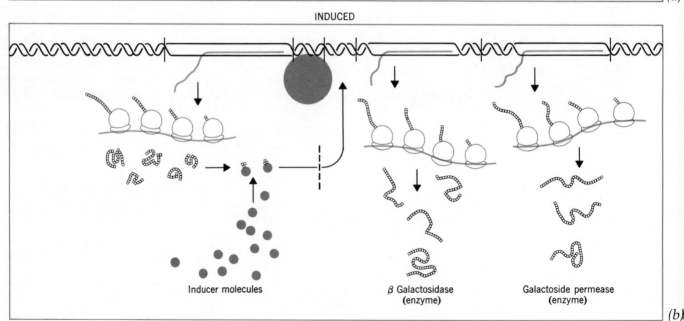

(b)

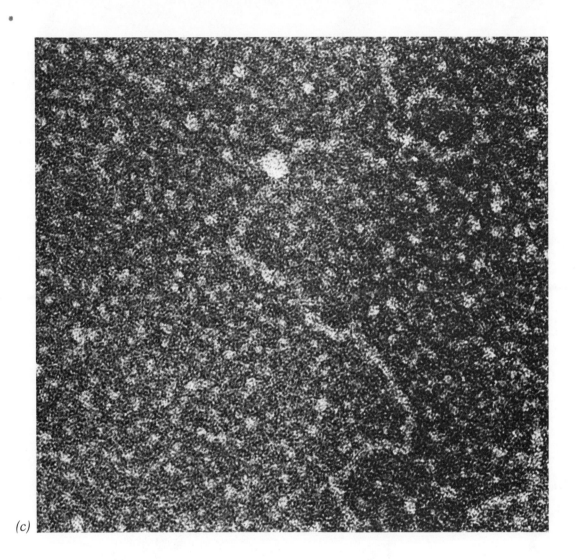

(c)

Recommended Reading

Klotz, I. M., *Energy Changes in Biochemical Reactions.* New York: Academic Press (1967).

A small paperback monograph designed to give students a reading knowledge of the language of energy relations and an introduction to simple calculations of biochemical energetics.

Watson, James D., *The Double Helix.* New York: Atheneum (1968).

A personal account, direct and candid, of the discovery of the structure of DNA.

Watson, James D., *The Molecular Biology of the Gene* (2nd ed.). New York: W. A. Benjamin (1970).

This is a general book on molecular biology. It is beautifully organized, lucid, and readable.

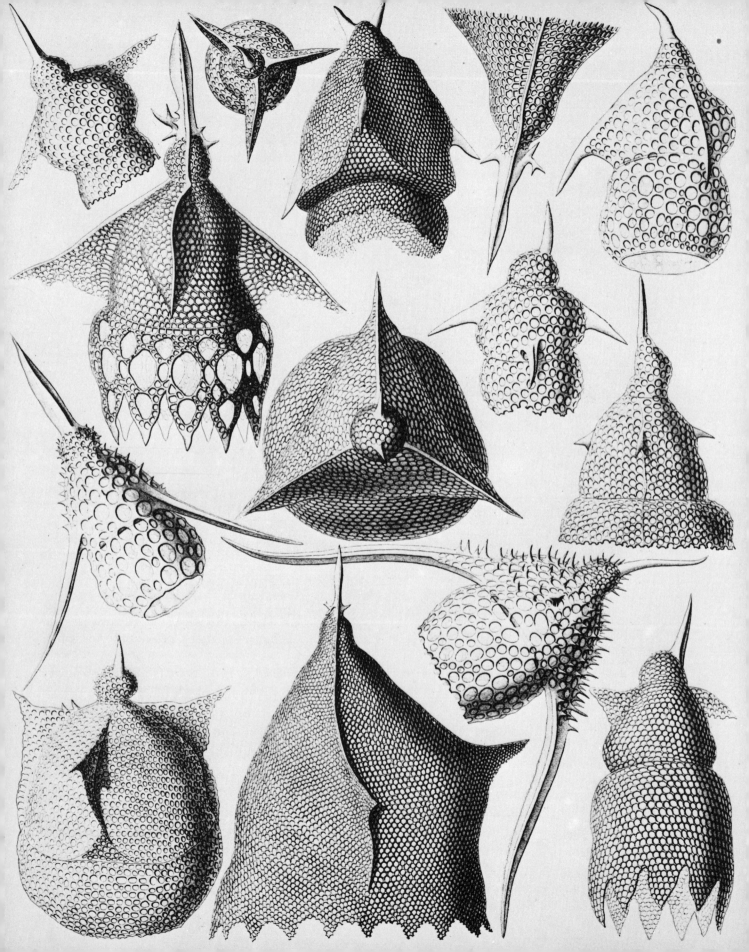

3 cellular machinery

All organisms are composed of cells and cell products. This generalization has now been accepted for about 160 years. It is probably familiar to most people, but its implications are far-reaching. **Cells** are the basic units of life, in both structure and function. Structurally, a cell is composed of a colloidal suspension of substances, called the **cytoplasm;** various subcellular particles and structures within the cytoplasm; and a system of membranes that surrounds the whole cell and extends into the cytoplasm.

Functionally, a cell has the same problems as the organism has as a whole. It must preserve its structural integrity, maintain a steady-state relation with the environment, exchange materials with the environment, and utilize some form of energy. The structural and functional independence of a cell as a basic unit of life can be demonstrated in many ways—but perhaps the best way is to point out that many organisms are composed of simply a single cell; that is, they are **unicellular.**

Larger plants and animals have to be **multicellular;** they must be composed of many cells in order to maintain adequate exchanges of energy and material with the environment. Consider a spherical organism which is oxidizing glucose to obtain energy. It will require oxygen. If oxygen enters the organism by diffusion across the surface, when the organism reaches a diameter of one millimeter (approximately 1/25 of an inch), no oxygen will reach the center of it. Oxygen is consumed by respiration more rapidly than it can diffuse in from the surface. This example provides us with a rough limit for the size of a cell. In fact, most cells are considerably smaller than one millimeter in diameter. Larger organisms are composed of more cells, not of larger cells (see Figure 3-1).

Drawings of radiolaria, single-celled-marine organisms, by the 19th-century naturalist Ernst Haeckel.

63

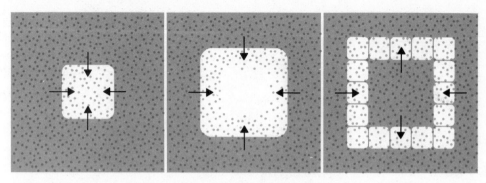

Figure 3-1 *Oxygen molecules and other substances to be used in metabolism diffuse into cells. If the cell is small, oxygen moves throughout the cell. If the cell were larger, oxygen would be consumed before it reached the center. But if the surface/volume ratio were increased by dividing the large cell into many small cells, oxygen could again diffuse to all parts of each cell.*

It is interesting to look at exceptions that prove or test the rough limit we have established. There are a few cells very much larger than a millimeter in diameter. The yolk of a hen's egg is a single cell. However, it is composed almost entirely of stored food. The actively metabolizing portion of the cell is confined to a minute plate of protoplasm on its surface. In this case, the entire cell can be very large because most of it is simply a storage vat; no material exchange with the environment is required. A nerve cell also appears to violate the size rule. A single nerve may extend from your spinal cord to your calf. However, the cell is only about 2/100 of a millimeter thick throughout most of this length. Similarly, some primitive plants are quite large though they are not multicellular. However, their form is always that of a flat sheet, a starlike network, or a thin branched structure; no portion is very far from a surface. Perhaps a better way to state the size limit, therefore, is to say that no active portion of an organism can be more than half a millimeter from an exchange surface. Whatever the form, all the active portions of the organism must be close to a surface that can carry on exchanges with the environment.

Many aspects of cellular organization are related to simple facts about the linear dimension, surface area, and volume of a solid figure. (The linear dimension is anything that can be measured with a tape measure, such as the diameter of a sphere or the length of a cube.) Any solid figure obeys two basic rules: (1) the surface is porportional to the square of the linear dimension, and (2) the volume is porportional to the cube of the linear dimension.

Figure 3-2 lists the length of a side, the surface area, and the volume, for cubes of various sizes. Notice that as the length increases, the surface increases rapidly, but the volume increases even more rapidly. In other words, as size *increases*, the surface/volume ratio *decreases*. In the context of gas exchanges, this relationship means that as the cube gets larger, less surface would be available to carry out exchanges for a given mass of tissue.

The body of a large organism is composed of incredible numbers of cells. A small drop of blood contains tens of millions of red blood cells. A drop a quarter of an inch in diameter will have more cells in it than there are people in the United States. There are roughly seven billion cells in the outer layer of your brain, almost twice as many cells as there are people on earth. The total number of cells in your body is too large to count directly but it is estimated at a few hundred trillion. At the cellular level there is little difference between mice and elephants—elephants just have more cells.

The classical study of cells gives the impression that they are static structural units. The usual procedure is to kill an organism and quickly preserve it to minimize changes in the appearance of its cells. A piece of dead tissue is embedded in paraffin or some other suitable material and thinly sliced. Tissue slices are then exposed to stains and dyes that react with some cell components and not others. Finally, cells in the tissue slice are viewed through a microscope. This technique is a very powerful one. For example, by using specific dyes, a particular structure or organelle or compound —DNA, protein, starch, lipid—can be localized

	Length	Surface	Volume	Surface/volume
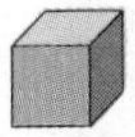	1	1	1	1
	2	4	8	$\frac{1}{2}$
	3	9	27	$\frac{1}{3}$
	4	16	64	$\frac{1}{4}$

Figure 3-2 As a cube's dimensions are increased, its surface area (length of a side squared) increases faster, and its volume (length of a side cubed) increases faster still. As a result, as objects get larger, their surface/volume ratios get smaller.

in the cell and its concentration estimated. However, it must always be remembered that dead cells are being examined, not living ones. Prepared and stained slides are inexpensive and easily stored, so they are staple items for the study of elementary biology. But it requires an imaginative effort to realize that living cells are not static units like bricks in a wall.

Living cells are very active. Some single-celled organisms move around in a very striking and obvious way. The tiny unicellular organisms found in pond water move so quickly that it is difficult to keep them in focus and in the field of the microscope. Cells of many types show internal churning movements.

Even cells that appear to be static and uninteresting are in fact very active. Two tools are particularly valuable for observing the less obvious cell activities. The first is the **phase-contrast microscope.** It is a light microscope with an optical arrangement that emphasizes small differences in the way light passes through different structures.

As a result, structures that normally blend in with the background become visible (Figure 3-3). The second tool is **time-lapse photography.** A movie camera is set to expose the frames of film at rather long intervals, say one every ten or fifteen seconds. Then, after the film has been developed, it is projected at the normal speed of 24 frames per second. As a result, movements appear to be greatly speeded up.

When time-lapse photography is applied to living cells in conjunction with the phase-contrast microscope, a variety of cellular activities can be observed. Some cells show creeping movements, others produce long extensions. Particles inside the cell are in constant motion and may change shape continuously. The impression of constant dynamic change is overwhelming. Of course, not all kinds of cells are equally active, nor are they all active in the same way; some cells are quite stable and sedate. But still, the cell certainly is not analogous to a simple building block. No living cell is completely static.

Metric Units

The metric system of weights and measures has been officially adopted by (almost) every country in the world except the United States. Scientists from all countries have long used metric units to express and communicate their results. The metric system is superior to English measures for several reasons. First, metric units are internally consistent and easily interconvertible. For example, all units of length are related by factors of 10. English inches, yards, miles, and furlongs have no such consistent relation. Also, the units of length, weight, and volume are rationally related to one another:

1 cubic centimeter = 1 milliliter = 1 gram (of water at 4° C).

Another important advantage of the metric system is that it provides units for measuring very small objects. This feature is particularly important for discussions of cells. Size is an important aspect of the overall structure and function of any organism. If you become familiar with the units necessary to discuss size, your understanding of biology will be enhanced. The metric system has no value unless you establish a relation between metric units and your own experience. The following table relates metric units to each other and to more familiar objects or measures. However, only use will make these new units familiar to you.

	Metric Unit	Symbol	Metric Equivalent	Physical Equivalent
Units of Length				
	Angstrom	Å	10^{-10} m	the distance between atoms in a molecule
	Micron	μ	10^{-6} m	a small bacterium
	Millimeter	mm	10^{-3} m	a period (.)
	Centimeter	cm	10^{-2} m	about ⅓ inch (1 cm = 0.39 in)
	Meter	m	10^{0} m	about 1 yard (1 m = 1.1 yd)
	Kilometer	km	10^{3} m	about ⅝ mile (1 km = 0.62 mile)
Units of Volume				
	Microliter	μl	10^{-6} l	about 1/50 of a drop
	Milliliter	ml	10^{-3} l	about 20 drops
	Liter	l	10^{0} l	about 1 quart (1l = 0.91 qt)
Units of Weight				
	Microgram	μg	10^{-6} g	the weight of a bacterium
	Milligram	mg	10^{-3} g	the weight of a few grains of sand
	Gram	g	10^{0} g	about 1/30 ounce (1 g = 0.035 oz)
	Kilogram	kg	10^{3} g	about 2 pounds (1 kg = 2.2 lb)

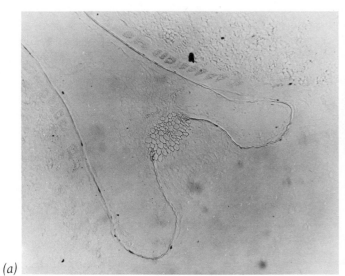
(a)

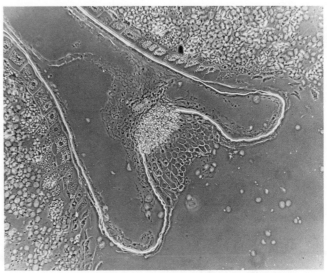
(b)

Figure 3-3   The phase-contrast microscope. (a) A normal light microscope produces a light, low-contrast image. (This subject is a cross section of a wheat grain.) (b) Viewed through a phase-contrast microscope, the same specimen has bright contours outlined against a dark background.

Cell Structure

Classical studies also gave the impression that a cell has a simple structure. Under a light microscope, a cell looks like a membrane-bounded sac which contains liquid, some stray particles, and a distinct central area. The outside membrane was called the **plasma membrane,** the internal liquid was named the **cytoplasm,** and the central region was the **nucleus.** However, the electron microscope quickly proved that the interior of the cell is very complex and highly structured. This discovery should not have been too surprising, since the cell does carry out many functions, and it is difficult to understand how this could be accomplished in a single solution inside a membrane. In fact, the cell actually has many compartments where different functions are carried out. Such compartmentalization is produced by a complex system of membranes within the cell (see Figure 3-4).

The plasma membrane not only surrounds the cell, but at some places it may fold into the cytoplasm to join with internal membranes called the **endoplasmic reticulum** (meaning, roughly, "network within the cytoplasm"). The endoplasmic reticulum may be so abundant that the whole cell

seems to be packed with membranes (see Figure 3-15). These membranes provide a large surface area on which metabolic reactions can take place.

The cell nucleus is also bounded by a membrane, which separates it from the cytoplasm. The nuclear membrane is actually a double membrane. Both the inner and the outer layers may be perforated with pores (Figure 3-5). In some cells, the outer membrane of the nucleus may fold out into the cytoplasm to join with the endoplasmic reticulum or the plasma membrane (Figure 3-6). The entire membrane system is visualized as a complicated set of folds extending from the surface of the cell to the nucleus.

Cell compartmentalization and the separation of cell functions also depends on small structures in the cytoplasm called **organelles** (literally, "little organs"). Various organelles are responsible for DNA and RNA synthesis, chemical oxidation and energy production, protein synthesis, cell movements, food intake, the excretion of wastes, and other cell functions. The structures of many of these organelles also depend on membranes. Some organelles are surrounded by membranes, and they

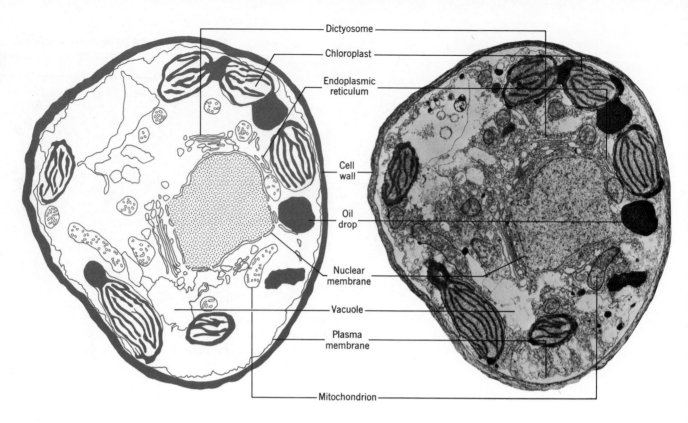

Dictyosome

Chloroplast

Endoplasmic
reticulum

Cell
wall

Oil
drop

Nuclear
membrane

Vacuole

Plasma
membrane

Mitochondrion

may have internal membranes as well. Finally, the
cytoplasm contains still other structures, including
various granules, droplets, and crystals. Cells show
a great variety of internal structure, but some
organelles and structures are found in many cells
and should be described in detail.

plasma membrane

The plasma membrane, or cell membrane, is found
in all living cells. It is the boundary between a cell
and its environment. Cells may lay down structures outside the cell membrane—the cell walls of
plants and the hard matrix of bone are examples—but the plasma membrane is the functional cell boundary because it restricts and controls exchanges of materials. This membrane also
accounts for the cell's surface properties, such as its
adherence to adjacent cells and its ability to recognize foreign material.

Before the advent of the electron microscope, scientists studied cell membranes primarily by ob-

*Figure 3-4 (above) Most
cells, like this plant cell,
contain many kinds of
membranes. All the struc-
tures shown in the diagram
(except the cell wall and
the oil drops) are composed
of membranes. Magnifica-
tion 9,000 ×.*

*Figure 3-5 (facing page, top)
The nucleus of this onion
root cell is surrounded by a
double membrane per-
forated by pores. Magnification 25,000 ×.*

*Figure 3-6 (facing page,
bottom) The outer nuclear
membrane in this primitive
plant cell is continuous
with the plasma mem-
brane. Magnification
15,000×.*

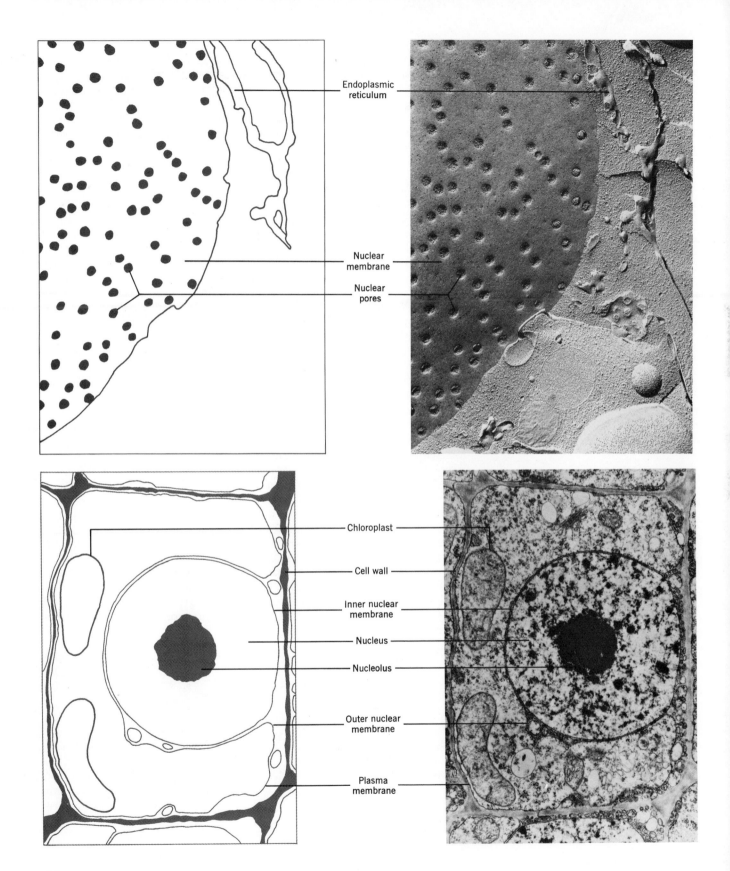

Endoplasmic reticulum

Nuclear membrane

Nuclear pores

Chloroplast

Cell wall

Inner nuclear membrane

Nucleus

Nucleolus

Outer nuclear membrane

Plasma membrane

serving the rates at which various materials were able to penetrate into cells. They found that molecules soluble in lipids such as fats or oils penetrate more rapidly than water-soluble molecules. There were many anomalies, however; some molecules penetrate much more rapidly than we would predict on the basis of their ability to dissolve in lipids. Nevertheless, a general picture of the structure of the cell membrane began to emerge. The cell membrane was visualized as a layer of lipid. Lipid would form a good barrier around the cell because relatively few molecules are soluble enough in lipid to move through it. Most cells exist in watery environments, and neither water itself nor most of the substances dissolved in it can penetrate a lipid barrier.

The Electron Microscope

An ordinary light microscope uses lenses to magnify an image which is produced by light passing through an object. Since the image is usually examined visually, the illumination must be provided by light that can be seen with the naked eye. However, the kind of light used limits the resolution that can be obtained. (**Resolution** is the ability to distinguish between separate objects of small size.) In the ordinary light microscope, if two dots lie closer than two or three tenths of a micron apart, they will be seen as one dot no matter how great the magnification may be.

The electron microscope (Figure A, facing page, top) uses a beam of electrons rather than a beam of visible light for illumination. Beams of electrons can provide resolution of about five Angstrom units—about a thousand times better than that achieved by a light microscope (see Figure B, below).

The image in the electron microscope is magnified by magnetic lenses which are analogous to the glass lenses of the light microscope (see Figure C). Since an electron beam is not visible to the naked eye, the image has to be focused on a fluorescent viewing screen comparable to that of a television set. Some experiments have been undertaken to develop a proton microscope that should in principle provide even greater resolution. The results, however, have not been promising.

Electron microscopy does have disadvantages. The electrons must travel in a vacuum so they will not be deflected by air molecules. Similarly, the sections of tissues to be examined must be sliced very thin so the electron beam can penetrate them and form an image. It is sometimes difficult to decide what features of the image are real and what features have been added or distorted by the process of preparing the material for examination. But in spite of these difficulties, the electron microscope has become a routine instrument for the investigation of cell structure, and it has generated an explosive burst of new information.

Figure B   A cell from an onion root is photographed with (a) a light microscope, and (b) an electron microscope. The difference in resolution is obvious.

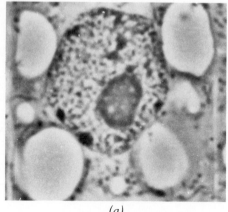

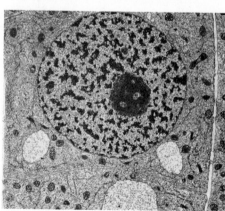

(a)

(b)

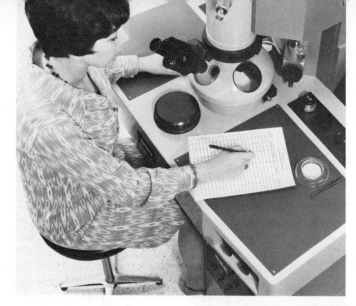

Figure A   An electron microscope.

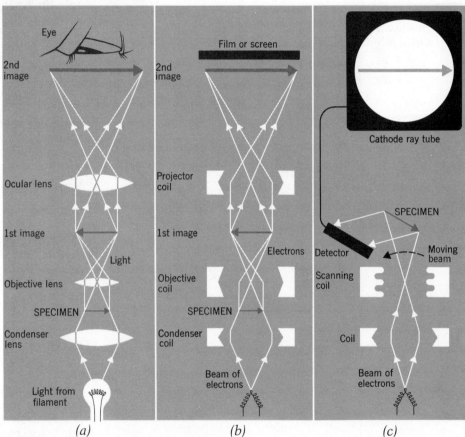

Figure C   Three kinds of microscopes. (a) The light, or optical, microscope (LM) focuses light rays on the specimen with glass lenses. Rays that pass through the specimen are collected and focused into an image that can be seen with the eye.

(b) The electron microscope (EM) focuses a beam of electrons on the specimen with magnetic coils. Again, rays that pass through the specimen are collected and focused into an image. This image cannot be seen directly, because our eyes are not sensitive to electron beams. However, the electron beam can be converted to a visual image by projecting it onto film or a fluorescent screen. (c) The scanning electron microscope (SEM) moves a beam of electrons back and forth over the specimen and scans its surface. Electrons are deflected from the surface and collected by a detector. This detector sends a signal to a cathode ray tube resembling a television screen. The image produced has a three-dimensional appearance.

Figure 3-7 *Measurement of a lipid monolayer with a Langmuir trough. (a) Lipids are floated on the water between a compression bar and a float. As the bar is moved, the lipid layer becomes compressed and exerts pressure against the float. As the float and attached yoke move, they twist the torsion wire and the indicator needle. The balance wheel moves in the opposite direction, until the indicator returns to a vertical position. The reading on the balance wheel is a measure of the force exerted by the monolayer. (b) The graph shows the pressure exerted by the lipid monolayer as it is compressed. (c) The compression of the monolayer is shown schematically. The arrows indicate the point at which a continuous monolayer is formed.*

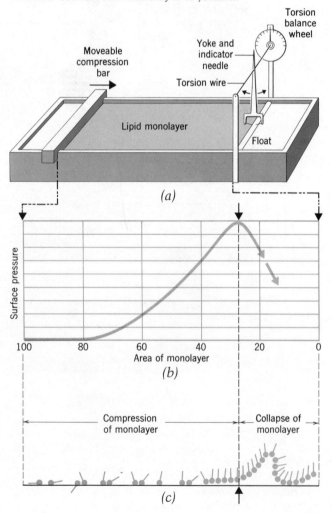

More information about the lipid layer in plasma membranes comes from studies of their behavior in water. Such experiments are usually performed on lipids extracted from red blood cells. (A red blood cell is a simple membranous sac containing the protein hemoglobin.) First, the area of the red blood cell is calculated. Next, all the lipids in the cell membranes are extracted with a lipid solvent. An apparatus called a Langmuir trough is then used to measure the area occupied by a single layer of the extracted lipid (Figure 3-7). The measured surface area of a blood cell turns out to be just half the area of the monolayer. This evidence suggests that the membranes in cells are a double layer of lipid. Other experiments indicate that the carboxyl groups of the molecules face outward, while the hydrocarbon chains are apposed. This picture is astonishingly close to the image produced by the electron microscope.

Figure 3-8 shows an electron micrograph of a cell membrane, and a diagram of the postulated structure of the membrane. Notice that the membrane in the photograph consists of two dark lines separated by a light area. The light line is assumed to be the double layer of fatty material. The dimensions of these lines support this interpretation; a double layer of lipid would be just thick enough to account for the distance between the two dark lines. The dark lines themselves were thought to represent two protein coats. Thus, the membrane was now visualized as a protein sandwich around a lipid filling.

More recent evidence suggests that the proteins are scattered along the lipid, are globular in shape, and may penetrate through the whole membrane. Certain kinds of electron micrographs support this conception of the membrane. If tissues are frozen and then shattered, the membrane may split in half through the middle of the double lipid layer. This procedure reveals globular particles embedded in the lipid. The particles could be proteins (Figure 3-9).

The presence of proteins in the cell membrane would help to explain some of the anomalies about the way molecules penetrate into cells. Water molecules, for example, go through cell membranes much more rapidly than would be predicted on the basis of their solubility in lipids. The proteins apparently provide watery patches or channels in the membrane. Proteins also explain a second anomaly—the fact that some organic molecules move across the membrane much more easily than other molecules which are chemically closely related to them. Thus, glucose molecules may enter cells quite rapidly, while other, similar sugar molecules are either excluded altogether or enter very slowly. The membrane protein apparently acts as a selective agent. It somehow "recognizes" glucose and facilitates its entry. At the same time it rejects molecules of similar size but slightly different structure.

Let us examine the movement of glucose into a cell as we change its concentration in the surrounding liquid. As the concentration of glucose outside the cell is increased, the rate at which it enters the cell also increases. However, at still higher concentrations, there is no corresponding increase in the rate of entry. Apparently, the glucose molecules can enter the cell only at specific sites in the proteins on the membrane. Because these sites have a particular shape and chemical structure, they bind with one kind of molecule—such as glucose—and not another. When the molecule comes into contact with such a site, it binds with it, is transferred into the cell, and is released. When we increase the glucose concentration on the outside slightly, more sites become occupied and the entry rate increases. But when all the sites on the cell membrane are occupied, the rate at which molecules can be transported into the cell has reached a maximum. The sites are now said to be **saturated**, and a further increase in outside concentration has no effect (see Figure 3-10). The existence of specific protein receptor sites would provide a very satisfactory explanation of the

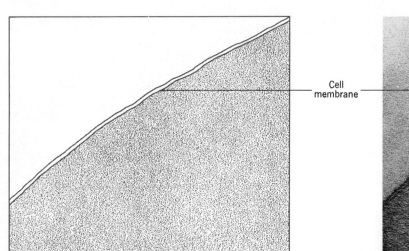

(a)

Cell membrane

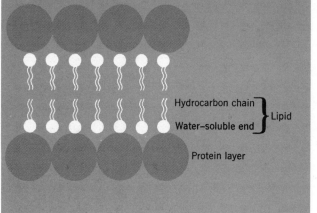

(b)

Figure 3-8 Membrane structure. (a) Electron micrograph shows a cross section of the plasma membrane of a red blood cell. Magnification 96,000 ×. (b) In the membrane, outer protein layers surround an inner lipid layer.

Hydrocarbon chain
Water-soluble end
} Lipid

Protein layer

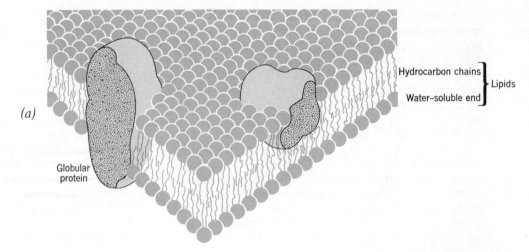

(a)

Hydrocarbon chains } Lipids
Water–soluble end }

Globular
protein

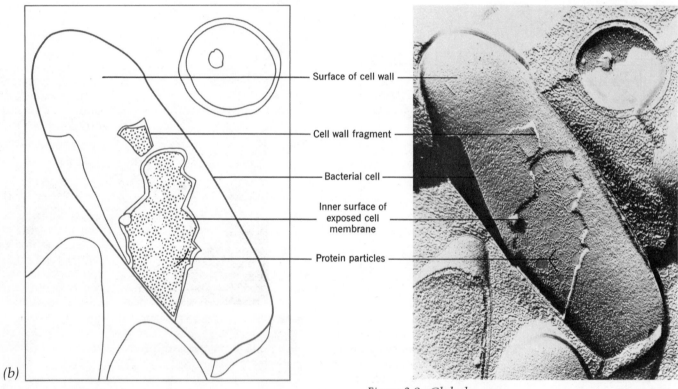

(b)

Surface of cell wall

Cell wall fragment

Bacterial cell

Inner surface of
exposed cell
membrane

Protein particles

Figure 3-9 Globular pro-
tein molecules are thought
to be scattered throughout
the cell membrane. (a)
Diagram of membrane
structure. (b) The inner
membrane surface under
the cell wall of a bacterium
appears to be studded with
globular proteins. Mag-
nification 8,750 ×.

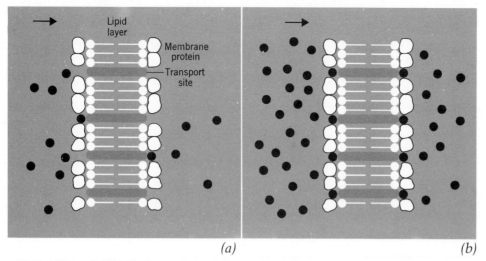

Figure 3-10 *Transport of molecules across a cell membrane. (a) When the concentration of molecules is low, transport sites are not all filled. (b) When the concentration is high, transport sites are saturated, and molecules are moved through the membrane at the fastest rate possible.*

(a)

(b)

way a membrane can exclude some materials, and yet permit others to enter, even when they are similar in size and in their solubility in lipids.

Another anomaly in the movement of molecules across membranes is that some cells can accumulate or extrude some solutes against a concentration gradient. In other words, they can move molecules in a direction opposite the one that would be predicted on the basis of diffusion. The movement of a solute against a concentration gradient requires metabolic energy, since work must be done to counteract the forces of diffusion. Such transport of solutes against a concentration gradient is called **active transport.** As an example, most animal cells actively extrude sodium into the surrounding medium against a considerable concentration gradient. But if we interfere with the energy metabolism of the cell by interrupting the production of ATP, sodium extrusion stops. The energy required for active transport is presumably used to regulate the shape of the protein molecule. The protein binds with the ion on one side of the membrane, and then releases it again on the opposite side.

Active transport is also selective; the membrane can discriminate between sodium—which is extruded—and a very similar ion, potassium. Potassium is retained by the cell, and may actually be transported into it. The movement of many small solutes across the plasma membrane is accomplished by such active transport. In each case, the binding site is probably a specific protein in the protein coat of the membrane.

organelles

The plasma membrane separates the cell from its environment and regulates the movement of various substances. Within the boundaries of the plasma membrane, other cellular activities are localized in organelles.

The Nucleus

The nucleus is usually the most conspicuous organelle in a cell, and most cells contain at least one. The shape of the nucleus is often characteristic of the particular kind of cell. In single-celled organisms it may be a long helix, a chain of beads, or a kidney-shaped mass. In human blood cells, the nucleus takes the form of several lobes joined by narrow bars. But most often, the nucleus is spherical in shape (see Figure 3-11).

The nucleus carries the instructions for the synthesis of proteins: 95 percent or more of the DNA in the cell is found in the nucleus. This nuclear DNA directs the synthesis of more DNA, and of the RNA that directs protein synthesis. When the nucleus is removed, the cell will continue to function only as long as it retains enough mRNA for protein synthesis in the cytoplasm. New mRNA cannot be produced without DNA, and sooner or later the existing supply will be exhausted. An amoeba whose nucleus has been removed with a fine glass needle will continue to move around at first. The organism can still engulf and digest food

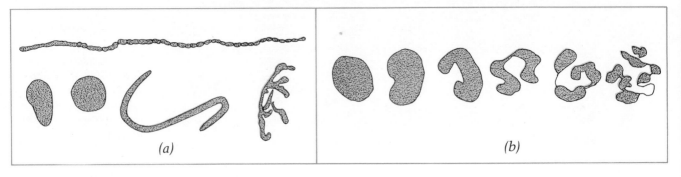

Figure 3-11 Nuclei from various cells. (a) The cell nucleus is often spherical, but it may also be elongated, branched, beaded, or multilobed. (b) The shape of a nucleus may change as the cell ages. This sort of progression is characteristic of white blood cells.

(a)

(b)

for a time, but it does not grow. Within a couple of days the amoeba disintegrates. Other cells last longer without nuclei. Red blood cells in man naturally lose their nuclei in the late stage of their development. Their rate of metabolism is extremely low and their mRNA is long-lived, so they survive for several months. However, the nucleus always is required for sustained protein synthesis and for cell reproduction.

The DNA in the nucleus is bound up with protein molecules and organized into strands. These strands, in turn, are folded and packaged into long tubular structures called **chromosomes.** ("Chromosome" means "colored body." They were given this name because they are easily stained with certain kinds of dyes.) The chromosomes in cells that are about to divide are very short and thick, presumably because the strands of DNA and protein are tightly coiled and folded. At other times, the chromosomes unwind into longer, thinner strands that look like spots and patches. In this state they are called **chromatin** (Figure 3-12).

The internal structure of chromosomes is being investigated by trying to unwind them. The most successful experiments show that chromosomes contain double fibers that are highly twisted around each other. The fibers appear to be two-stranded DNA molecules coated with proteins called **histones** (Figure 3-13). This association between DNA and histones may be related to the control of protein synthesis in the cell. The nucleus of each cell contains a complete set of DNA instructions for a rather complicated organism, but not every cell needs to manufacture every protein in the organism. In fact, it would be disastrous if it did. The

histone coating may serve to inactivate the instructions for proteins that are not needed by a particular cell.

A spherical structure, the **nucleolus,** is often found associated with a chromosome in the nucleus (Figure 3-14). It is composed of DNA, RNA, and protein. The current theory is that the DNA in the nucleolus consists of many identical segments end to end, and that every segment directs the synthesis of one of the rRNA molecules used to construct ribosomes. Other DNA segments in the nucleolus may direct the synthesis of ribosomal protein. Thus, the nucleolus appears to be responsible for the synthesis of the ribosomes.

This description of the nucleus, chromosomes, and nucleolus does not apply to all cells. Unicellular organisms such as bacteria lack membrane-bounded nuclei. Instead, their DNA is found in central "nuclear regions." All the DNA is located in a single molecule, which is packaged into a single circular chromosome along with relatively few protein molecules.

Ribosomes

A **ribosome** is a small particle composed of two spherical subunits (see Figure 2-28). Each subunit contains molecules of protein and ribosomal RNA (rRNA). Ribosomal protein and rRNA appear to be synthesized in the nucleolus.

Ribosomes are found in the cytoplasm, where they direct protein synthesis. In some cells they appear to be scattered throughout the cytoplasm. However, in other cells, the ribosomes are attached

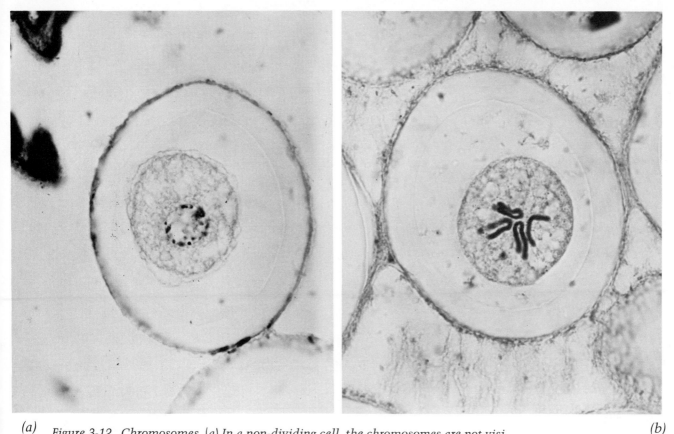

(a)   (b)

Figure 3-12   Chromosomes. (a) In a non-dividing cell, the chromosomes are not visible with a light microscope. They occur as thin strands called chromatin. (b) In a cell
that is about to divide, chromosomes wind and coil into short, thick segments.

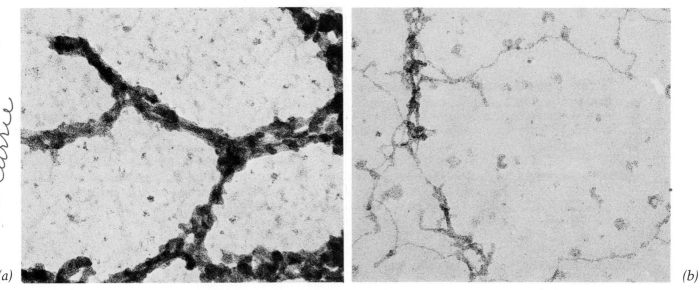

(a)   (b)

Figure 3-13   (a) This partially unwound chromosome, from a cell of a salamander,
is seen to consist of two interwoven strands. Each strand is composed of a central
DNA molecule coated with histone proteins. (b) The DNA molecules are revealed by
treatment with protein-digesting enzymes, which remove the histones. Magnifications 160,000 ×.

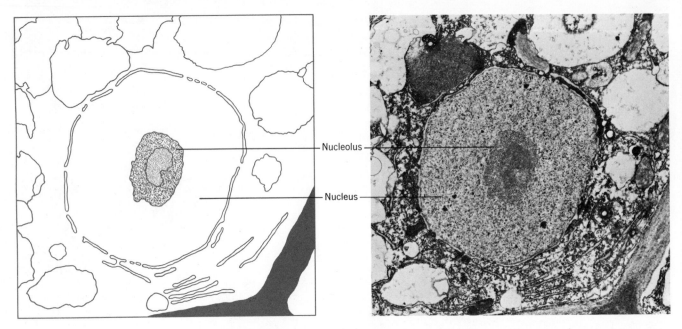

Figure 3-14 *The nucleolus in this plant cell has a central, fibrous core, surrounded by a granular layer rich in RNA. Magnification 9,000 ×.*

to the endoplasmic reticulum. ER covered with ribosomes has a rough, granular appearance and hence is called **rough ER** (Figure 3-15).

Dictyosomes

In some cells, synthesized proteins are transported to the cell surface and secreted to the exterior. This is how digestive enzymes are produced. The **dictyosomes** are organelles that function in the secretion of cell products.

A dictyosome is a stack of membranous sacs and nearby **vesicles,** small membrane-bounded spheres (Figure 3-16). Secretion occurs in several steps. First, protein is synthesized by ribosomes on the ER and packaged into pouches of the ER membrane. These pouches then pinch off and migrate to the region of the dictyosomes. There, the pouches either transfer their contents to the dictyosomes or fuse with them. In the dictyosomes, the protein product may undergo chemical changes before it is collected into the vesicles which surround the dictyosome region. The vesicles migrate to the cell membrane, merge with it, and extrude their contents (Figure 3-17). Dictyosomes perform a similar

secretory function in the formation of plant cell walls. Dictyosome vesicles transport cellulose molecules to the region where a cell wall is forming.

Mitochondria and Chloroplasts

In most organisms the conversion of energy to usable form takes place in two different types of organelle. The first is the **mitochondrion.** It is the site of chemical oxidation and the production of high-energy phosphate compounds. The **chloroplast** is the site of photosynthesis, the process that converts light energy to chemical energy. The two organelles have many features in common. Both are membranous in their internal organization, both have enzymes bound to these membranes, both are separated from the remainder of the cell by membranes, and both contain DNA.

Mitochondria are rather small—only a few tenths of a micron in diameter. In living cells, they twist and bend and move around in the cytoplasm of the cell. The mitochondrion is covered by a single, outer membrane. Its internal structure is produced by an inner membrane that is folded into flat sheets or tubules, which penetrate toward the interior of

Figure 3-15   Endoplasmic reticulum and ribosomes. (a) Rough ER is usually found in cells that are actively engaged in protein synthesis, such as this cell from the pancreas of a rat. (b) Drawing shows the three-dimensional structure of rough ER.

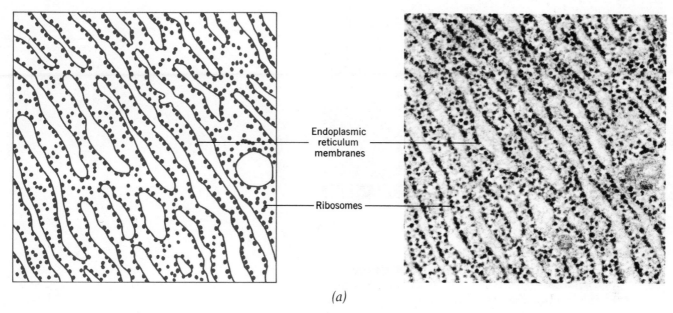

Endoplasmic reticulum membranes

Ribosomes

(a)

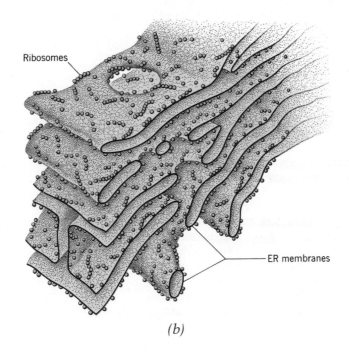

Ribosomes

ER membranes

(b)

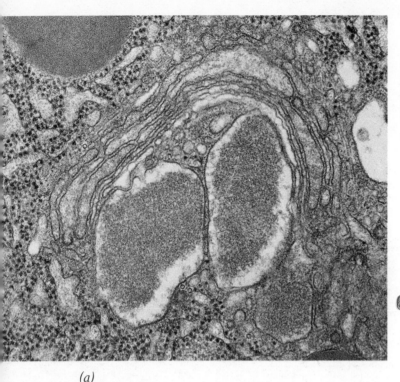

(a)

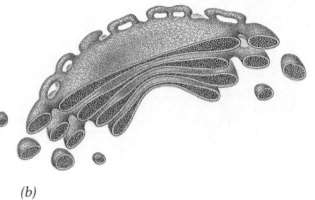

(b)

Figure 3-16   Dictyosomes. (a) A dictyosome from a mammalian cell; magnification 37,000 ×. (b) Three-dimensional structure of a dictyosome.

the mitochondrion. These folds or tubules are called **cristae** (Figure 3-18). The enzymes that catalyze the intermediate steps in the oxidation of glucose are present in the fluid that bathes the cristae. In addition, lollypop-shaped granules are attached to the membrane surfaces of the cristae. These granules are the sites of the final steps of chemical oxidation. They contain the electron transport system that consumes oxygen and produces ATP.

Chloroplasts are the sites of photosynthesis. An individual chloroplast is bounded by double membranes. Lying within the double membrane is a set of membranous sacs that are flattened and apposed to each other like a stack of pancakes. An electron micrograph of a corn plant chloroplast shows numerous regions in which ten to twenty membrane sacs are closely packed together. These regions are called **grana** (Figure 3-19).

The surfaces of the internal sacs have a cobblestone-like structure. The units that form the cobblestone pattern have been termed **quantasomes** (Figure 3-20). A single quantasome may be equivalent to a complete photosynthetic unit. That is, it may contain all the molecules necessary for the first chemical reactions of photosynthesis. Enzymes associated with other aspects of photosynthesis are

Figure 3-17 (facing page) Secretion by dictyosomes. (a) "Goblet cells" in the intestine secrete large amounts of mucus. Amino acids and sugars enter the cell. The amino acids are converted to protein on ribosomes at the base of the cell, and the protein and sugars enter the dictyosome, where they combine to form mucus. Dictyosome vesicles, filled with mucus, migrate to the end of the cell and empty into the intestine. (b) Electron micrograph shows a mucus vesicle forming. Magnification 56,000 ×.

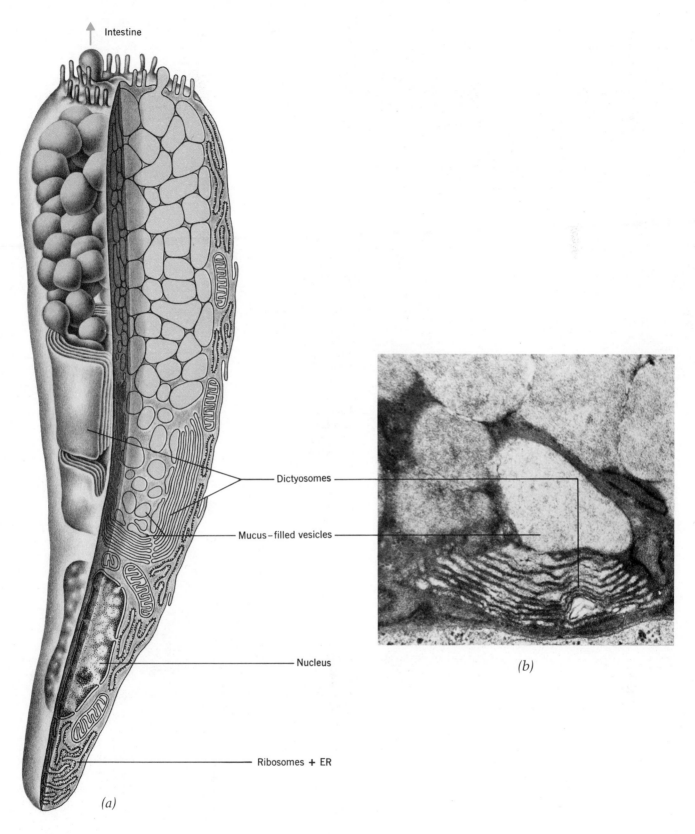

Intestine

Dictyosomes

Mucus-filled vesicles

Nucleus

Ribosomes + ER

(a)

(b)

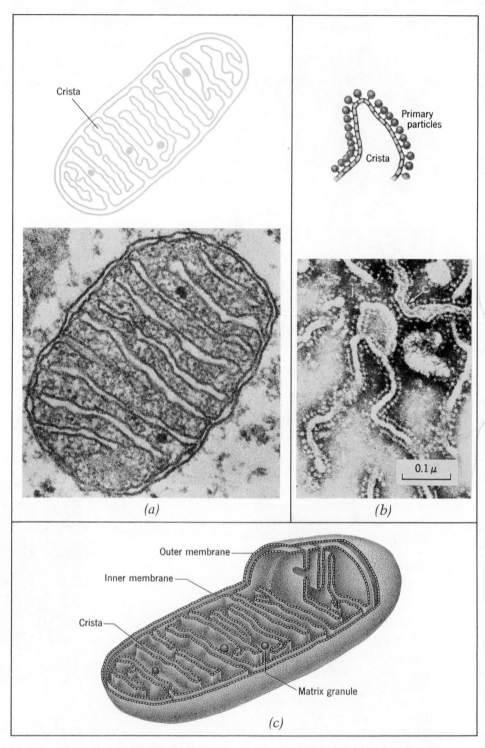

Figure 3-18 Mitochondria. (a) Cross section of a mitochondrion. (b) Cross section of cristae with attached primary particles. (c) Drawing of a whole mitochondrion, cut away to show internal features. The several large matrix granules contain calcium and lipids, but their function is unknown.

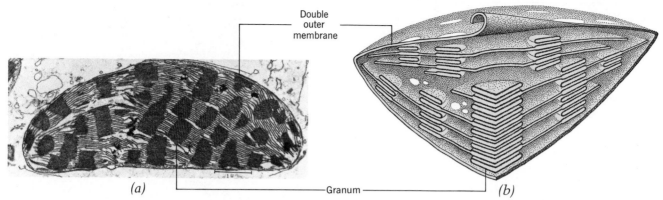

Double
outer
membrane

Granum

(a)

(b)

*Figure 3-19 (above) Chloroplasts. (a) Cross section of a
corn chloroplast, showing stacks of grana and the outer
double membrane. Magnification 10,000×. (b) Drawing of
a whole chloroplast, cut away to show internal structure.*

*Figure 3-20 (below) Surfaces of grana in chloroplasts
are covered with closely-packed particles, the quantasomes. (b) Electron micrograph shows quantasomes
on the surface of a granum membrane. Magnification
85,000×.*

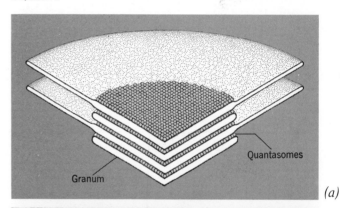

Quantasomes

Granum

(a)

Granum
surface

Quantasomes

1000 Å

(b)

in solution within the chloroplast. Chloroplasts
also contain free ribosomes, which presumably are
associated with protein synthesis. Some plants
possess chloroplasts with different internal structures, but the details cannot be too crucial since
these plants, too, are capable of photosynthesis.

The chemical products of oxidation and photosynthesis are transported through the bounding
membranes for use by the rest of the cell. The outer
membrane also controls the passage of raw materials and waste products in and out of the organelles.
However, both organelles can function in isolation
from other cell components. Isolated mitochondria
can carry out oxidation and create high-energy
phosphates in a test tube. Similarly, isolated chloroplasts can photosynthesize—that is, transform
light energy into reduced chemical compounds. Chloroplasts and mitochondria both contain their own
DNA and can duplicate themselves within a cell.

Lysosomes

Some animal cells contain another kind of organelle called a **lysosome.** A lysosome is simply a
droplet bounded by a membrane (Figure 3-21). The
lysosome droplet contains a number of enzymes
which can lyse, or split, the major components of a
cell, including the nucleic acids, proteins, phosphate compounds, and polysaccharides. Lysosomes
are very important in animals such as the amoeba,
which engulfs its prey. The cells of higher animals
may contain lysosomes that break down parts of
the cell that need periodic replacement.

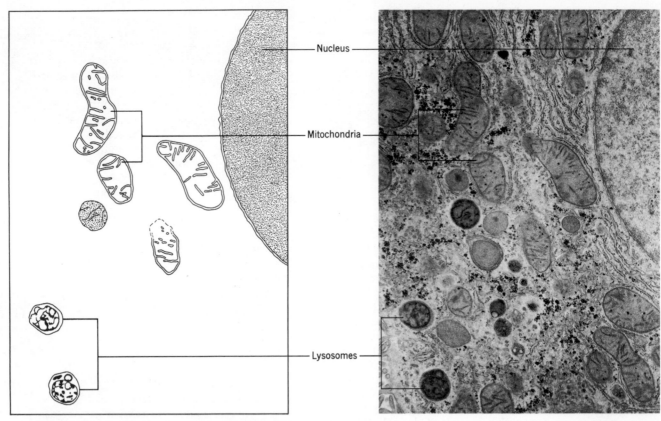

Nucleus

Mitochondria

Lysosomes

Figure 3-21   Lysosomes in a liver cell; magnification 10,500 ×.

Figure 3-22   Role of lysosomes. Cells that digest other cells (phagocytes) contain numerous lysosomes. In the development of the chick embryo, phagocytes are thought to digest cells in the region between the digits on the embryo's limb. This process frees the digits from one another. Magnification 5,000×.

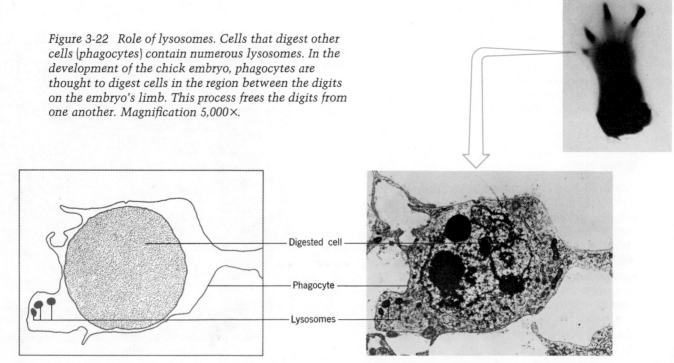

Digested cell

Phagocyte

Lysosomes

In some cases, lysosomes may aid in the removal of whole cells. This function may be important in the formation of structures from dead cells, as in the growth of feathers, or for the destruction of tissues during normal development. For example, the resorption of a tadpole's tail or the removal of tissue between the digits of vertebrate embryos may be caused by lysosomes (see Figure 3-22).

Vacuoles and Inclusions

A **vacuole** is a membrane-bounded region within the cytoplasm that contains various substances or solutions. This vague definition covers a wide range of organelles of diverse functions and structural complexity. The term was originally applied to a number of different organelles before the days of electron microscopy, when their structures and functions were still largely unknown. Among the organelles referred to as vacuoles are some complicated functional units such as the contractile vacuoles of the protozoa (unicellular fresh-water animals). This type of vacuole pumps excess water out of the cell. A second kind of vacuole is the food vacuole, found in the unicellular amoeba. When an amoeba detects food, it engulfs it and encloses it. Digestive enzymes from a lysosome are then secreted into the vacuole, and the food is broken down into molecules that can be used for energy. (An organelle such as a lysosome is technically a vacuole, too, since it is a membrane-bounded unit, but it is not usually called by that name.)

Other types of vacuoles simply provide places for storing enzymes or other substances. In plant cells,

Figure 3-23 *Most plant cells contain one or more large vacuoles, bounded by single membrane. The vacuoles contain a complex solution of inorganic and organic molecules. Magnification 3,500×.*

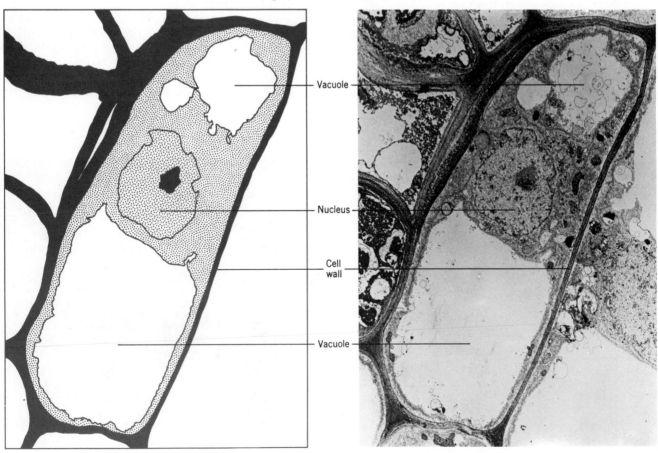

Vacuole

Nucleus

Cell wall

Vacuole

there are often so many large vacuoles that they form another major division of the cell, along with the cytoplasm and the nucleus (Figure 3-23).

The cytoplasm may also include various crystals, granules, or droplets that are not surrounded by membranes. Plant cells contain starch granules, produced by photosynthesis. In animal cells, fat droplets and glycogen granules are common. Other granules or crystals are composed of waste products.

Cilia and Flagella

Cilia and **flagella** are threadlike organelles that extend out from a cell's surface and undulate to and fro. Flagellar and ciliary movement is so rapid that the eye cannot follow it; photography must be used (Figure 3-24). Cilia and flagella propel many small organisms through the water. They are also found on the exposed surfaces of sheets of cells in plants and animals; there they serve to move fluid over the surface of the sheet, rather than to move the organism through the water. The principle is the same in both cases, however.

Cilia and flagella differ only in that flagella are larger and often occur singly or in small numbers on a single cell, while cilia occur in groups. Both structures are a few tenths of a micron in diameter, and they range in length from about five microns for small cilia to about 50 microns for large flagella.

Structurally, each cilium or flagellum is made up of nine double protein fibrils surrounding two central ones. This arrangement of protein fibrils is characteristic of almost all cilia and flagella (Figure 3-25). Electron microscopists have found it duplicated in rather unexpected places. For instance, the light receptor cells in our eyes (the rods and cones) contain ciliary segments. Likewise, an organelle that functions during cell division, the centriole, also shows the ciliary structural pattern.

cell types

The cell is the smallest fully-functioning unit of an organism. All cells must deal with the basic functional problems of maintaining life, and all cells are potentially autonomous units. However, cells within an organism are not necessarily interchangeable, nor are they necessarily similar in different kinds of organisms. The cells within a complex organism, such as a mammal or a tree, differ depending on their location and functional roles. These differences are reflected in the sizes and shapes of the cells and in the membranes, particles, and organelles they contain. A cell specialized for secretion has an extensive endoplasmic reticulum and prominent dictyosomes, but it lacks locomotor organelles such as cilia and flagella. Muscle cells are filled with contractile protein fi-

Figure 3-24 The motion of the flagellum of a sea urchin sperm is captured on film through multiflash photography. Magnification 1,400 ×.

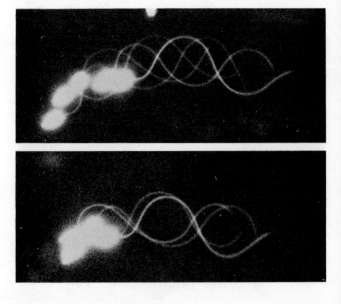

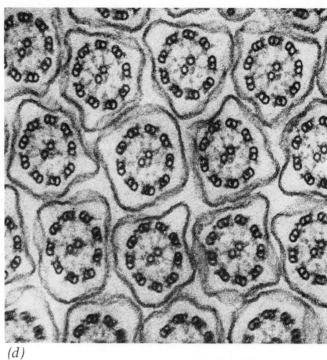

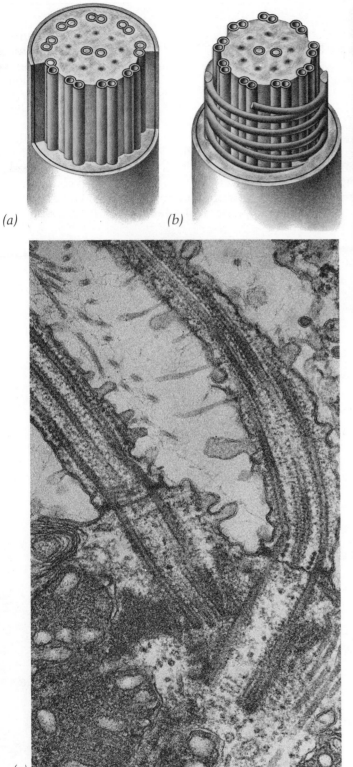

(a) *(b)*

(d)

Figure 3-25   Cilia and flagella. (a) A cilium has nine pairs of peripheral filaments and two central filaments embedded in a matrix. (b) Flagella may also have peripheral strands wrapped around the central core. (c) Longitudinal section of two flagella; magnification 63,000×. (d) Cross section of many cilia; magnification 81,500×.

(c)

Protein Fibrils

Many of the structures that are responsible for cell movements contain protein fibrils. These fibrils are often composed of subunits called **microtubules** (Figure A). A microtubule is a hollow tube formed from a single layer of stacked protein molecules. The double protein fibrils in cilia and flagella are composed of cylinders of such microtubules. Microtubules also help to maintain the shape of some cells, and they direct the creeping and gliding movements of others. They seem to function as a "skeleton" inside the cell.

Microtubules are often associated with another kind of protein fibril, the **microfilaments** (Figure B). Microfilaments are probably contractile and may serve as an internal cell "muscle" in conjunction with the microtubule "skeleton." This combination may allow cells to move about and to change shape. Muscle cells in animals are packed with yet another kind of contractile protein fibril, which occurs in a regular, parallel arrangement (Figure C).

Figure A Microtubules. These tiny tubules, composed of protein subunits, are important structural elements in many organelles. They make up the axial filaments of flagella and cilia, as well as intracellular support rods in various kinds of cells.

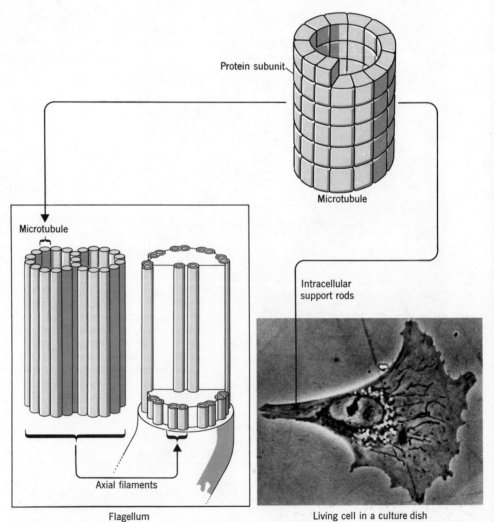

Protein subunit

Microtubule

Microtubule

Intracellular support rods

Axial filaments

Flagellum

Living cell in a culture dish

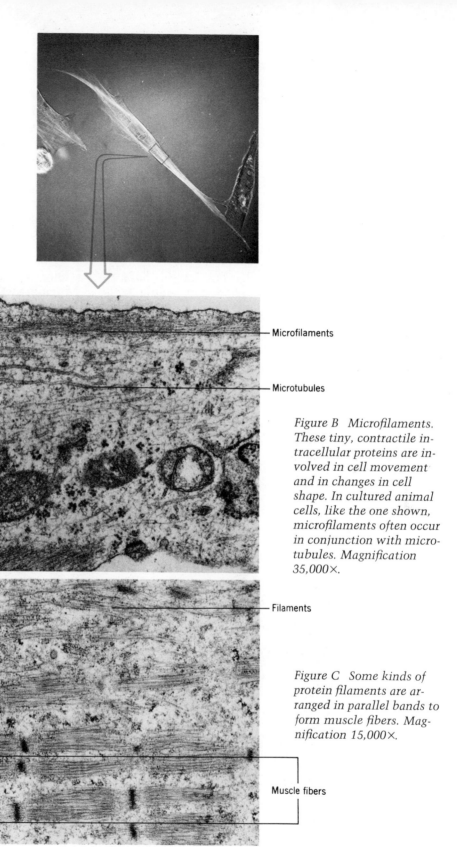

Microfilaments

Microtubules

Figure B Microfilaments. These tiny, contractile intracellular proteins are involved in cell movement and in changes in cell shape. In cultured animal cells, like the one shown, microfilaments often occur in conjunction with microtubules. Magnification 35,000×.

Filaments

Figure C Some kinds of protein filaments are arranged in parallel bands to form muscle fibers. Magnification 15,000×.

Muscle fibers

brils and mitochondria for energy production, but they lack many other organelles. Chloroplasts are found in leaf or stem cells but not in roots.

Even though a multicellular organism contains many kinds of cells, we can still talk about a "typical" or "generalized" plant or animal cell to emphasize a few characteristic differences (Figure 3-26). Plant cells usually have cell walls and possess chloroplasts, for instance. Animal cells do not. Most plant cells also possess large central vacuoles. These features reflect the functional differences between plants and animals. Plants are photosynthetic, while animals use chemical energy from the environment. Plants receive structural support from their cell walls, but animals have to have other kinds of structural support; animal cells may secrete cartilage or bone. Still, in spite of all these differences, the general organization of the cells of plants and animals is surprisingly similar. Many of the organelles are identical, or nearly so.

A much more fundamental difference is found in cells of procaryotic organisms. The **procaryotes** (the name means "before nucleus") are the bacteria and some related organisms. As their name implies, they lack membrane-bounded nuclei. Procaryotes also lack mitochondria, chloroplasts, and complicated internal membrane systems such as dictyosomes and ER. Instead, the procaryotes have simple internal membranes and a central area that contains DNA (Figure 3-27). The characteristics of procaryotic cells are considered primitive; that is, biologists think that the first cells that arose resembled the procaryotes. All the other organisms (the protozoa, molds, animals, and most plants) are termed **eucaryotes** ("true nucleus"). Their cells contain distinct nuclei, and they possess numerous membrane-bounded organelles and complex internal membrane systems.

This difference is really a fundamental one. Eucaryotic cells are much more compartmentalized, they contain more sophisticated machinery, and they can therefore carry out more complex activi-

Figure 3-26 Cells of plants and animals. An animal cell is bounded by a flexible membrane and contains numerous organelles. A plant cell usually contains the same organelles except for centrioles. In addition, the plant cell contains chloroplasts, starch grains, large vacuoles. It is bounded by a rigid cell wall.

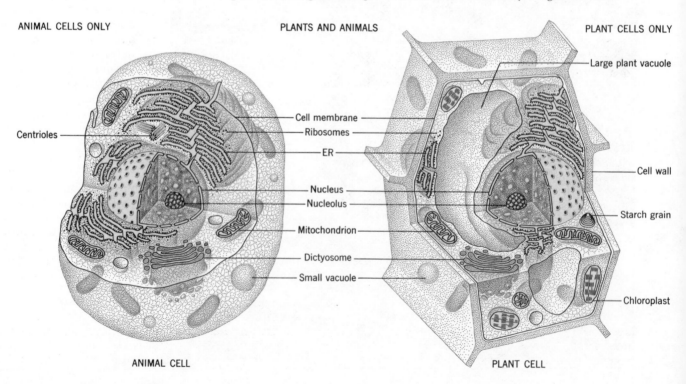

ANIMAL CELLS ONLY

PLANTS AND ANIMALS

PLANT CELLS ONLY

Large plant vacuole

Centrioles

Cell membrane
Ribosomes
ER

Nucleus
Nucleolus

Mitochondrion

Dictyosome
Small vacuole

Cell wall

Starch grain

Chloroplast

ANIMAL CELL

PLANT CELL

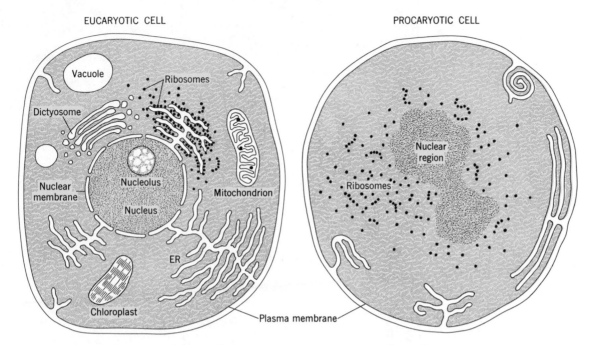

Figure 3-27 *Eucaryotic cells have complex internal membranes and membrane-bounded organelles. Procaryotic cells have comparatively simple internal structures.*

ties. Eucaryotes presumably developed from procaryotes during evolution, but we can only speculate about the actual steps by which a simple cell suddenly became complex. Evolutionary theories about the origin of eucaryotic cells are discussed in Chapter 17.

All cells living today are either distinctly procaryotic or distinctly eucaryotic—they either have no mitochondria at all, or they have complex and highly structured ones. Cells have never been found with a simple mitochondrion that might represent a transition between procaryotic and eucaryotic. The distinction between these two cell types is so fundamental that it forms the basis for modern classification of organisms.

matrix

So far we have looked exclusively at cells and their machinery. But as we said at the outset, organisms are composed of cells and cell products. Cell products, in the form of matrix material, are often more conspicuous than the cells themselves. **Matrix** is a general term for material that is secreted external to a cell and in which the cell is embedded. Matrix may be a fluid, a colloidal gel, or a solid. Some examples will illustrate the importance of these secretions. The wall of a plant cell is a matrix; in fact, wood matrix comprises the overwhelming bulk of a tree. Blood consists of cells suspended in a fluid matrix. Bone and cartilage, hooves and hair, teeth and scales are all matrix products secreted by special cells.

Matrix is not simply tossed out of the cell. It often has a complicated and characteristic form. A plant cell wall, for example, is built up in layers that reflect the history and activities of the cell.

A cell wall begins forming at the end of cell division. Dictyosome vesicles containing cellulose and other polysaccharides begin to congregate at the dividing line between the newly formed daughter cells. The contents of the vesicles are secreted and built up into a cell wall, which cuts the original cell into two daughter cells. The new cell wall is thickened by the addition of more cellulose and other secreted materials. New cell wall material is aligned in precise patterns and layers (Figure 3-28), probably because new cellulose molecules have

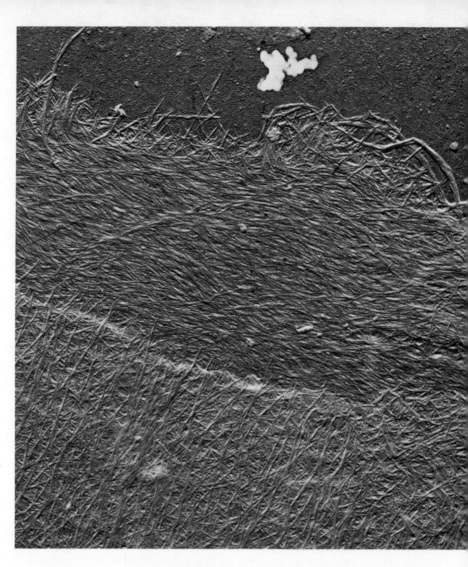

Figure 3-28 The cellulose fibrils in a plant cell wall are arranged in complex layers. Magnification 35,000 ×.

"self-assembly" properties. That is, certain bonding arrangements between new and preexisting molecules are favored.

Matrix production is not fully understood in all cases. However, it is clear that such products are a critical part of the cellular organization of living things.

cell water

Cell membranes exercise control over the entry and exit of many kinds of molecules. Water, however, behaves differently and requires a separate discussion, because all membranes are permeable to water. Furthermore, with very few exceptions, water cannot be actively transported from a region of low concentration to a region of high concentra-

tion. Nonetheless, cells often can maintain internal concentrations that are significantly different from those of their environments.

The term **concentration** usually refers to the quantity of salt, sugar, or other solutes in a known volume of water. However, the water itself also varies in concentration. Consider a solution of table salt. In addition to water, it contains ions of sodium and chlorine. Because of the volume occupied by the ions, there will be less water in a liter of this solution than in a liter of pure water. A higher concentration of salt results in a lower concentration of water.

The water concentration inside cells is often lower than it is outside. This means that water tends to diffuse into the cell, and solute molecules tend to diffuse out. Though most of the solute molecules cannot move through a membrane, water can. The tendency of water to move into or out of a cell—that is, from one solution to another across a

92 cellular machinery

membrane—is called **osmosis.** Osmosis depends largely upon the concentration of solute particles present. It doesn't make much difference what kind of particles are in solution—they may be large or small, charged or uncharged. It is the number of particles, not their characteristics, that is important.

The concentration of particles in a solution is called the **osmotic concentration.** Cells maintain an osmotic concentration higher than that of their environment (and a lower water concentration) by keeping solutes in and keeping water out. The cell maintains a high concentration either by actively transporting solutes into the interior of the cell, or by having a membrane that is impermeable to them.

Cells keep water out by one of two methods. The first works only in cells that are surrounded by tough and semirigid walls. A plant cell usually possesses such a wall. If we place a plant cell in fresh water, water will tend to enter across the membrane. As it does so, the cell begins to swell and press against the surrounding cell wall. However, the expansion of the cell is countered by the elastic properties of the wall. When these two forces are balanced, net movement of water stops. The pressure exerted by the cell against the wall is called **turgor pressure** (Figure 3-29). The effect of turgor pressure is demonstrated whenever lettuce is placed in cold water to make it crisp. Water enters the cells and presses against the cell walls. The maintenance of turgor is important for the normal appearance and function of many plants.

Some animal cells use a second method to keep water out, because they lack rigid walls. If we place a unicellular animal in water, it will begin to swell. Water will enter the cell because of the difference in osmotic concentration between inside and outside. No counter force is exerted, however, because the plasma membrane is very weak and there is no cell wall. The plasma membrane is therefore kept as impermeable to water as possible. This condition slows down the rate of water's entry, but it does not stop it altogether. Periodically, excess water has to be expelled by a special contractile vacuole system.

Figure 3-30 is a drawing of the contractile vacuole of the single-celled organism *Paramecium*. It operates in several steps. First, some of the cell fluid is filtered into the vacuole. Salts and other materials are then removed from the fluid in the vacuole and returned to other parts of the cell. The relatively pure water remaining in the vacuole is then emptied to the exterior, and the cycle begins again. Energy is required to recover the solutes tochondria associated with the vacuole system. When the organism is exposed to a poison such as cyanide, which interferes with the energy supply from the mitochondria, its vacuole ceases to con- from the vacuole. This energy is provided by mi-

Figure 3-29 *Water movement into plant cell. (a) When the concentration of solutes inside the cell is higher than in the surrounding medium, water tends to move into the cell because of osmotic forces. (b) The cell increases in volume and begins to press against the cell wall. (c) When the force exerted against the wall by the cell is equal to the pressure of the wall against the cell, the cell no longer increases in size.*

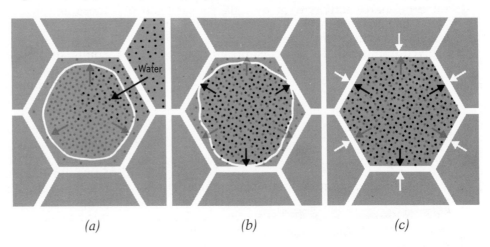

(a) (b) (c)

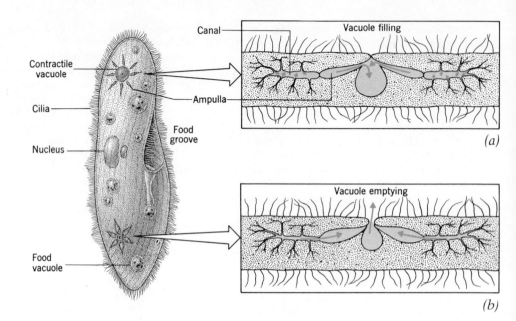

Figure 3-30 The contractile vacuole of a protozoan, Paramecium. (a) The vacuole fills with water from channels called ampullae. At the same time, water enters the end canals from tiny tubules spread throughout the organism.

(b) The vacuole contracts, forcing water out through the pore. Simultaneously, the canals contract, moving more water into the ampullae. The cycle repeats.

Labels in figure: Canal, Vacuole filling, Contractile vacuole, Ampulla, Cilia, Food groove, *(a)*, Nucleus, Vacuole emptying, Food vacuole, *(b)*

tract. The organism swells and bursts.

Animal cells, then, in order to maintain a constant size, must expel fluid at the same rate that water flows in. Salts may or may not be withdrawn from the fluid before it is lost. But if they are not, active transport of salts into the cell from the surrounding medium has to replace the salts that are lost when the fluid is expelled.

Many multicellular animals avoid problems of regulating cell water by controlling the immediate environment of the individual cell. The osmotic concentration of the fluid around each cell is carefully regulated to match the concentration inside the cell. In fact, many cells in multicellular animals have lost the capacity to survive changes in the concentration of the extracellular medium.

Multicellularity

Multicellularity is a form of organization based on the coordinated activity of many cells, yet each cell performs many functions independently. It is widely assumed that multicellular organisms evolved from small, single-celled organisms.

origin of multicellular organisms

Two major theories attempt to explain the origin of multicellular organisms. The first account, the **colonial theory,** proposes that multicellular organisms arose through the formation of **colonies** of single-celled organisms. The word "colony" is

vaguely defined. It can refer to associations that are loose and accidental, such as a group of bacteria growing on a culture plate. At the other extreme, it can refer to organisms in which the individuals comprising the colony have become so specialized that they are no longer self-sufficient.

Many simple algae show the rudiments of colonial organization. **Algae** are relatively simple plants. They may be unicellular or multicellular, but even the multicellular forms have few specialized cells or tissues. They all lack the roots, stems, and leaves characteristic of more complex plants. Some unicellular algae form colonies when they undergo cell division and the two daughter cells simply fail to separate. The process can continue until a chain of cells is produced (Figure 3-31). This organization is a very simple colony. Another species always forms colonies of sixteen cells. If a

Figure 3-31 a simple colony. The freshwater alga Spirogyra *forms a chain of cells.*

single cell is teased out of such a mass, it will divide and produce the characteristic colony again (Figure 3-32). A much more complicated colonial organization is found in the alga *Volvox* (Figure 3-33). Single cells in the *Volvox* colony are specialized. Sexual reproduction in *Volvox* is limited to certain cells.

All these colonial plants exist today, but we have arranged them in an evolutionary series to illustrate how a relatively complex organism could have evolved from a colony of single cells. The colonial theory simply extrapolates forward from this kind of series. It suggests that by continuing the same process, a multicellular organism with many specialized cells might have arisen from a complex colonial ancestor.

Another line of evidence for the colonial theory is provided by a group of organisms called **cellular slime molds.** They spend most of their lifetime as single-celled individuals that creep about and feed. The unicells may increase in numbers through simple cell division. However, cellular slime molds undergo a second type of reproductive process, and this is where their chief interest lies. First, certain cells called initiators secrete a substance that attracts neighboring cells. Cells within range of this stimulus move toward the initiator. The cells aggregate, and eventually they form a slug-like mass which begins to migrate. While the mass migrates, the cells begin to change and develop the specialized characteristics needed for the reproductive process. After migrating some distance, the slug-like mass stops and the specialized cells produce a reproductive structure. Some cells contribute to the base, others to the stalk, and still others to the capsule at the top. This capsule produces and dis-

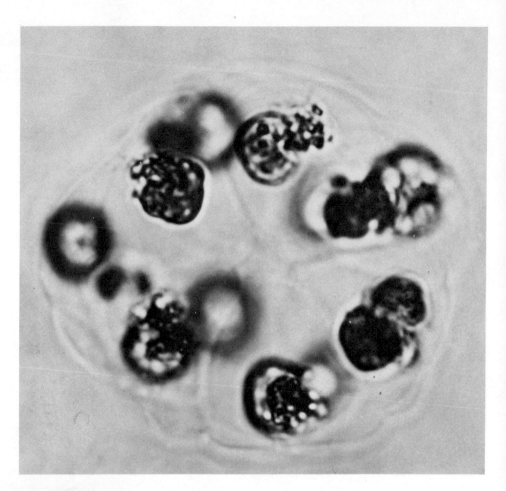

Figure 3-32 A simple colony. The freshwater alga Pandorina forms regular colonies of sixteen cells. Magnification 1,500×.

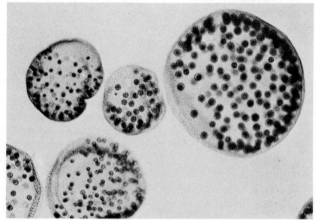

Figure 3-33 A complex colony. The freshwater alga Volvox forms large spherical colonies. The small dark spheres inside each colony are reproductive structures.

charges single-celled individuals, and the cycle begins again (see Figure 3-34). The slime mold colony resembles many multicellular plants in the complexity of its organization and in the specialization of its cells.

A second proposal for the origin of multicellularity involves protozoa. Protozoa are unicellular, but they can be very complicated in structure. They have well-defined "head" and "tail" regions, cilia or flagella, and special organelles for feeding, excretion, light reception, and other functions. Some protozoa have many nuclei, which are scattered about in the body of the cell. If each nucleus and its surrounding protoplasm were to be enclosed with a cell membrane, the result would be a multicellular animal as complex as some existing multicellular forms (Figure 3-35). This process, called **cellularization,** would not merely produce many cells from one, but the cells themselves would already be specialized. The animal would

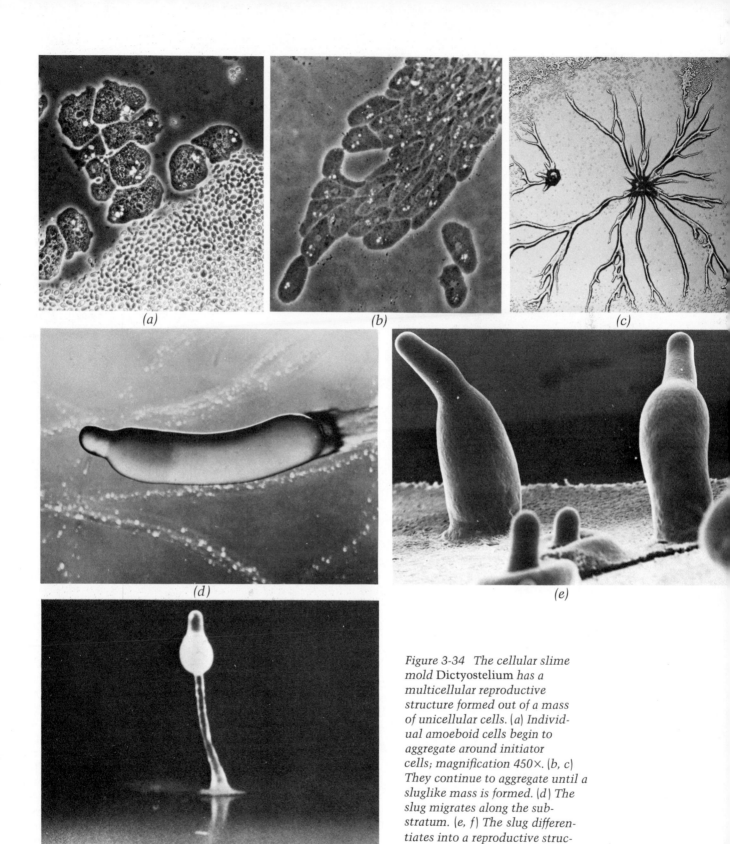

Figure 3-34 The cellular slime mold Dictyostelium *has a multicellular reproductive structure formed out of a mass of unicellular cells. (a) Individual amoeboid cells begin to aggregate around initiator cells; magnification 450×. (b, c) They continue to aggregate until a sluglike mass is formed. (d) The slug migrates along the substratum. (e, f) The slug differentiates into a reproductive structure with base, stalk, and cup.*

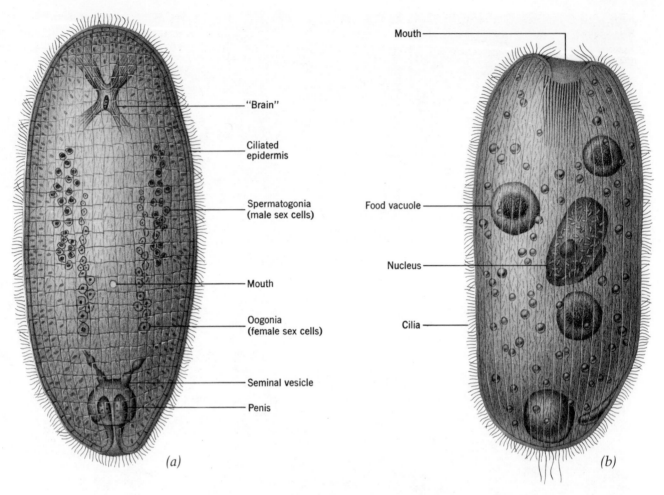

Labels in figure (a): "Brain" / Ciliated epidermis / Spermatogonia (male sex cells) / Mouth / Oogonia (female sex cells) / Seminal vesicle / Penis

Labels in figure (b): Mouth / Food vacuole / Nucleus / Cilia

(a)   *(b)*

Figure 3-35 The first multicellular animals may have resembled primitive flat-worms like Afronta *(a). They may have been descended from a ciliated protozoan like* Prorodon *(b). Multicellular plants, however, (and possibly multicellular animals as well) were almost certainly derived from colonial aggregates of individual flagellated cells.*

contain "head" cells, "tail" cells, cells with attached cilia, excretory cells, and others. The cellularization process could produce a multicellular organism in one step. By contrast, the colonial theory of multicellularity invokes a two-step process: first the aggregation of identical cells, and then their differentiation.

The colonial and cellularization hypotheses are both speculative, but both are reasonable. In fact, each mechanism may have produced different multicellular organisms many times in evolutionary history.

size

According to the fossil record, multicellular organisms have become larger in the course of evolution. This observation suggests that increased size must confer some advantage on an organism. Sponges, which have a very simple multicellular organization, provide an example of one advantage of increased size.

A simple sponge is a vase-shaped organism (Figure 3-36). The wall of the vase is pierced with many fine pores. Cells lining the inside of the vase have flagella that beat back and forth, moving water in through the pores in the wall. The effect is analogous to having fish trapped in a net with their tails all pointing the same way. As the fish attempt to swim, they produce a current of water through the net. The water current drawn in through pores of the sponge leaves through the large opening at the top of the vase, the **osculum.** The water jet that shoots out of the osculum is quickly slowed down again by the friction of the surrounding motionless water. This interaction sets up a spherical circulation pattern around the sponge. The size of the sphere of circulating water is important to the sponge, because water provides food and oxygen and removes carbon dioxide and other waste products. The greater the volume of circulated water, the better chance the sponge has of obtaining new food and oxygen.

More complex sponges are organized as though they were colonies of simple sponges sharing a common osculum (Figure 3-37). Let us suppose that we have a colony of 100 simple sponges, and that all are pumping water at the same rate they would as individuals, but that they now are all emptying through one osculum. Since the water jets exit together, they present less of a total surface to the surrounding still water than they would separately. The combined jet therefore encounters less resistance, and it shoots about 10 times farther into the surrounding water. This increase creates a sphere of water circulation that is 1,000-fold greater. That is, the 100 sponges circulate 1,000 times more water than a single sponge. Simply by forming an aggregation of this kind, each sponge now has ten times as much water available to it for food and oxygen as it did as an isolated individual. The rate at which water moves through the sponge does not change, but the colony is able to bring a greater volume of water into circulation, instead of recirculating the same small volume over and over.

Increased size confers another kind of advantage on organisms like birds and mammals because it reduces their metabolic rate. **Metabolic rate** is a measure of the overall rate of energy utilization by an organism. It can be determined by measuring the organism's production of carbon dioxide or heat or its consumption of oxygen over a measured time interval. The metabolic rate is often expressed as heat produced per gram of animal per hour. The

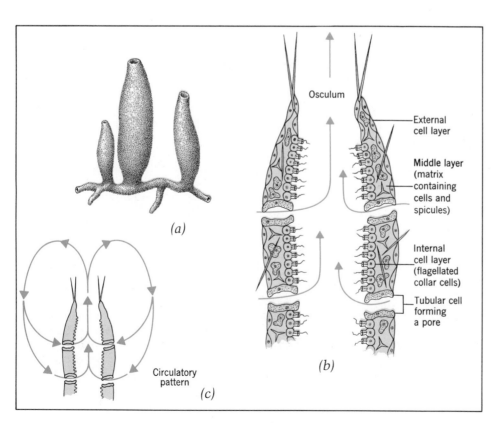

Figure 3-36 Simple sponges. (a) A group of simple sponges. (b) Internal structure of a simple sponge. Flagellated collar cells create a water current through the pores. (c) A spherical volume of water is circulated by the sponge.

Osculum

External cell layer

Middle layer (matrix containing cells and spicules)

Internal cell layer (flagellated collar cells)

Tubular cell forming a pore

(a)

(b)

Circulatory pattern

(c)

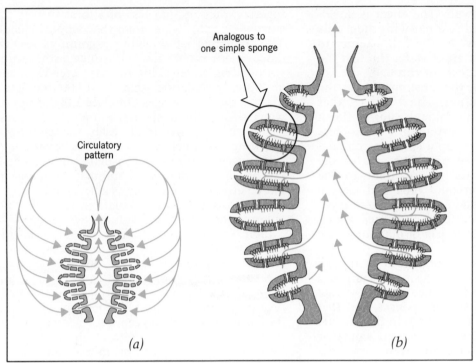

Figure 3-37 *Complex sponges are constructed like colonies of simple sponges. (a) A simple sponge circulates a spherical volume of water. (b) A complex sponge also circulates a spherical volume of water. But more water is circulated per individual sponge.*

organism's metabolism is affected by two major factors: exchanges of heat or respiratory gases across all its surfaces, and its total volume (or weight) of tissue. Therefore, the metabolic rate is dependent on a surface/volume ratio.

As an animal gets larger, its surface/volume ratio decreases. So does its metabolic rate. Figure 3-38 lists the rates of metabolism of several mammals. Very small mammals, such as shrews, have such high metabolic rates that they have to be voracious eaters. Shrews normally consume a third of their weight in food each day. A humingbird must lower its body temperature at night, to decrease its metabolic rate while not feeding.

Larger animals tend to be stronger and faster than small ones (Figure 3-39). Increased size is obviously beneficial to animals that feed on smaller prey. And large size is beneficial to land plants as well. A tree, by virtue of its size and organization, lives in two distinct environments at the same time. Its root system absorbs water and minerals from the soil, but its leaves are exposed to air and light for photosynthesis. Obviously there is some lower size limit below which the plant could not span the two necessary environments.

Other aspects of increased size, however, may not be advantageous. The following example is

quoted from Julian Huxley's essay, "The Size of Living Things." He is discussing the relation between size and the rate of falling.

The greater the amount of surface exposed relative to weight, the greater the resistance of the air. So that it comes about that the spores of bacteria or ferns or mushrooms, or the pollen-grains of higher plants, are kept up by the feeblest of air currents; and even in still air they cannot fall fast. They float down, like Alice down the well, rather than fall. If a mouse is dropped down the shaft of a coal mine, the acceleration due to gravity soon comes up against the retardation due to air resistance, and after a hundred feet or so a steady rate is reached, which permits it to reach the bottom dazed but unhurt, however deep the shaft. A cat, on the other hand, is killed; a man is not only killed, but horribly mangled; and if a pit pony happens to fall over, the speed at the bottom is so appalling that the body makes a hole in the ground, and is so thoroughly smashed that nothing remains save a few fragments of the bones and a splash on the walls.[1]

[1] Julian Huxley, "The Size of Living Things," in *Man Stands Alone.* New York: Harper & Row (1941).

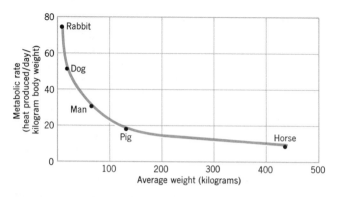

Figure 3-38 *Metabolic rates of animals of various weights. The rates are lower for heavier animals, because their surface/volume ratios are lower.*

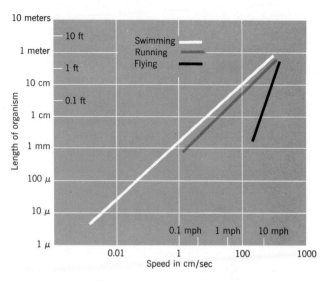

Figure 3-39 *As the size of an organism increases, so does its speed. This relationship holds true for all kinds of locomotion: swimming, running, or flying.*

Gravity becomes a threatening force for large organisms.

Larger organisms face yet another difficulty because of gravitational force. Consider the problem of building a mouse the size of an elephant. If the mouse were increased in all dimensions by a factor of 100, the resulting monster would be the size of an elephant. However, it wouldn't be able to move a step, because its skeleton and muscles would be too weak. The mouse's weight would be increased by a factor of a million (100×100×100). But the strength of its supporting skeleton, which depends on its cross-sectional area, would increase by a factor of only 10,000 (100×100). The enlarged skeleton would thus be 100 times weaker in relation to the animal's weight. The bones of our monster mouse would crumple under the tremendous load. An elephant has a much stronger skeleton, and its legs are particularly strong because they are so thick. Elephants are so heavy that they may knock down trees, but they never jump fences.

Finally, size has an effect on the time scale on which an organism lives. Figure 3-40 is a plot of the weights of various organisms against the times required for them to mature and develop the capacity to reproduce. With a few exceptions, all plants and animals fall along the same straight line. The range is from the virus, which weighs a few millionths of a gram and can reproduce every twenty minutes or so, to sequoia trees, which weigh hundreds of tons and produce seeds only after hundreds of years. The general correspondence between size and the time required to mature is not surprising. All organisms start as single cells, and larger organisms should require longer development

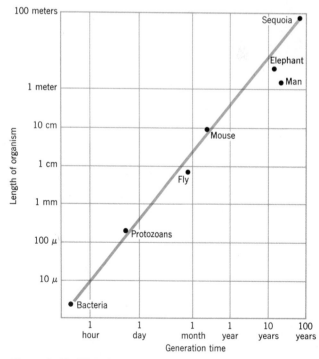

Figure 3-40 *The time required for an organism to reach reproductive maturity is related to its size.*

periods. It is surprising, however, that all living things fall so neatly along the same line.

Other relations to size might be a subject for speculation but are difficult to demonstrate. Larger size requires more cells, and might therefore be thought to imply greater intelligence in comparable organisms. There is some evidence to support this contention for mice and rats. But in man, there is no correlation between size and intelligence except when small size is the result of childhood malnutrition.

specialization

Multicellularity permits cell specialization as well as an increase in size. When large numbers of cells are organized into a single functioning unit, it is no longer necessary for each cell to retain all the capacities necessary for the survival of the organism as a whole. Collectively, the organism encounters the same problems of reproduction, energy conversion, and maintenance of the steady state that confront a single cell. But these problems may no longer be faced by each individual cell. Therefore, the cells in a multicellular organism can become specialized to carry out particular functions.

A single type of cell may be specialized to perform some process individually. For instance, a red blood cell is specialized to carry oxygen. More often, however, specialized cells of one type are bound together to form a **tissue,** and then several different tissues are organized into large functional units. Such units are called **organs** or **organ systems.** Examples are the kidney, which is responsible for excretion and water balance, and the leaf, which conducts photosynthesis and gas exchange. These organs carry out specific functions—just like the organelles of a single-celled organism.

The cell specialization of multicellular organisms and their increased size permits an even greater range of organization types. A multicellular plant or animal can live in environments in which a single-celled organism could not survive. The reverse statement, however, is equally valid. Small unicellular organisms can thrive where no large organisms exist.

Recommended Reading

Jensen, W. A., and R. B. Parke, *Cell Ultrastructure.* Belmont, Cal. Wadsworth (1967).

A collection of superior electron micrographs of cells, selected to illustrate the structure of cell and organelle.

Novikoff, A. B., and Eric Holtzman, *Cells and Organelles.* New York: Holt, Rinehart, and Winston (1970).

An excellent treatment of the cell, including both structural and functional aspects of cells and organelles.

Swanson, Carl P., *The Cell.* Englewood Cliffs, N.J.: Prentice-Hall (1969).

The structure, function, and normal behavior of cells and cell organelles in plants and animals.

Drawing by Chas. Addams; © 1946.
The New Yorker Magazine, Inc.

4 the organism as an individual

We have said that organisms are constructed of cells and cell products, and that the cell membrane forms a boundary between the organism and its environment. Sometimes, however, the relationship between organism and environment is so intimate that it is difficult to tell where the organism ends and the environment begins. Many organisms modify portions of their environment to such an extent that the environment really becomes part of them.

Many animals have behavior patterns that make it quite difficult to discern the limits of the individual. For instance, aquatic worms and insect larvae may modify their immediate environment by collecting sand grains and cementing them together to build tubes. The tubes may provide protection, or they may be used to direct a flow of water which supplies oxygen and food. Since the tube is necessary for the organism's normal behavior or for its survival, it seems more reasonable to consider the tube as part of the animal rather than part of the environment. Similar examples can be found in behavior of plants. Some plants secrete materials that prevent the growth of other plants in their immediate vicinity. Almost all organisms modify their environment in some way. When these modifications are striking, they are considered part of the organism.

Man is an extreme example of an organism that modifies its local environment. Some of man's modifications are critical for survival; clothing and shelter are obvious examples. Even when clothing is not required for survival, it is often a psychological necessity. The individual surrounds himself with an incredible range of external possessions. Most of the items acquired by humans in affluent societies are only distantly related to survival, if at all.

105

Nevertheless, they are part of the way an individual interacts with and alters his immediate environment. This tendency is found in other animals and plants; it is not a uniquely human aberration.

Other organisms are an important part of the individual's environment. Associations between different species are extremely common. Some associations are beneficial to all the organisms involved; others benefit only one species without seriously inconveniencing the other. However, there are some associations against which an organism must defend itself. Any plant or animal is a rich source of food. It attracts large predators, parasites, and a multitude of microorganisms, all seeking a source of energy and materials. The individual must be able to defend itself against attack or invasion by all these organisms.

defenses against invasion of the individual

The entire external surface of an organism cannot be inpenetrable, because it must conduct exchanges with the environment. Respiratory gases and both organic and inorganic compounds must penetrate the surface to support the organism's metabolism. But at the same time the surface must resist penetration by microorganisms. A single-celled organism may simply secrete a protective layer of material at its surface. In larger organisms, the defenses may include more complicated structures, such as the bark of a tree or the skin of an animal. Nonetheless, even the most complex surface defenses are often breached. The external covering may be punctured or scarred. Microorganisms may enter along with food ingested from the environment. Invasion may occur across respiratory gas-exchange surfaces. Some bacteria, fungi, and other small organisms inevitably will enter the organism, and many of them will present problems.

In many plants, the defense against invasion consists of an array of organic molecules which the plant can tolerate but which interfere with the metabolism of the invading organisms. These compounds are produced as long as the plant is alive. After its death, however, the plant will succumb to various molds and will rot as the protective compounds wear off. Woods such as redwood are resistant to rot for a very long time; this property makes redwood an excellent building material. (We use an analogous defense mechanism when we treat wood with creosote to retard rotting.) The defense fails, of course, if an invading organism is able to tolerate the toxic compound.

Other plants produce toxic compounds only when an invader is actually present. Such compounds are called **phytoalexins.** Beans, carrots, and a number of other plants respond to attack in this way. Some metabolic compound from the invader triggers the synthesis of the phytoalexin, which is toxic to the invader. Many lower animals have similar chemical defense systems. That is, they may produce defense compounds in response to invasion.

Animals have another characteristic form of defense called **phagocytosis.** Animals often have cells in their body fluids and tissues that are capable of ingesting foreign material. The cells are **phagocytes,** or, literally, "eating-cells." Phagocytes were first observed a century ago by the Russian zoologist Élie Metchnikoff. He was studying water fleas, or *Daphnia*, small pond animals whose transparency makes them very suitable for observation (Figure 4-1). Metchnikoff noticed that there are amoeba-like cells in *Daphnia*. When he introduced small numbers of yeast cells into the body of *Daphnia*, these amoeboid cells acted as phagocytes and engulfed and digested the invaders. But when larger numbers of yeast cells were injected, the yeasts overwhelmed the phagocytes and multiplied, until the water flea eventually died. Phagocytosis is a common phenomenon in animals. Some of the white blood cells in man are phagocytes, which engulf bacteria and other microorganisms (Figure 4-2).

Animals may combine phagocytosis with chemical defenses to resist invasion by microorganisms. This combination can be illustrated by simple experiments with human blood cells. If phagocytes are separated from blood **serum** (the liquid portion of blood) and placed in a salt solution, they will survive and crawl around in a normal fashion. But when presented with bacteria, the phagocytes will ingest only a few of them. By contrast, when phagocytes are presented with the same bacteria in the presence of serum, they will ingest 20 to 50 times as many. The phagocytes will become jammed with the foreign cells. If the bacteria are first exposed to serum in the absence of phagocytes and then washed in salt solution, they will still be engulfed as eagerly as they were in whole blood. Clearly, some chemical compound in the serum has attached to the surfaces of the bacteria or modified them in some way that makes them more susceptible to phagocytosis (Fig-

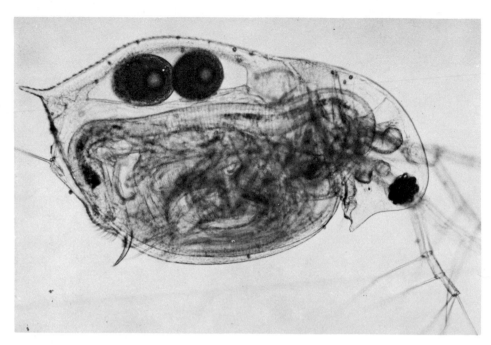

Figure 4-1 *The water flea,* Daphnia; *magnification 40×.*

ure 4-3).

Serum from an animal that has already been attacked once by a particular type of bacterium is much more effective in promoting phagocytosis than is serum from a naive animal (that is, one that has not previously been attacked). This fact indicates that the body has specific means of recognizing particular microorganisms and specific defenses against them. Specificity is what distinguishes an immune response from other kinds of defenses. Phagocytosis and the chemical defense agents in the blood of a naïve animal are general responses. The response is the same regardless of the identity of the invader. Immune responses, however, are specific.

Immunity

Many diseases attack an individual only once. If you survive an initial case of measles or smallpox, you are protected against a recurrence. In a word, you are **immune.**

The typical immune response depends on interactions between an antigen and antibodies. The **antigen** is a chemical compound associated with the invading organism that triggers the response. It may be a protein, a large carbohydrate molecule, or a smaller organic molecule in combination with one of these. It may be free, or it may be bound to the surface of the invading organism. The **antibodies** are specific protein molecules produced by the attacked organism in response to the presence of the antigen. The plural form should be used, since more than one kind of antibody can be produced in response to an antigen. Antibodies remove the antigen from the body, or kill or inactivate the invading organism that carries the antigen.

Antibodies are named on the basis of their action. **Precipitins** form a precipitate with a soluble antigen and thus remove it or permit it to be phagocytized. **Opsonins** specifically promote the phagocytosis of an invader carrying the antigen. (The word comes from a Greek root meaning "dainty food.") **Lysins** promote the lysing, or breaking open, of the invader. **Agglutinins** produce agglutination, or clumping together. There are other categories as well. The organism may synthesize several kinds of antibodies, with different actions, in response to a single antigen. In man and other vertebrates, antibodies circulate in the blood. But they also appear in other places: tissue fluids, milk, saliva, and sweat.

Precipitins are relatively easy to study, because a precipitate can readily be recognized. If the cells are removed from whole blood, the serum still contains the protein antibodies. If a new antigen, one to

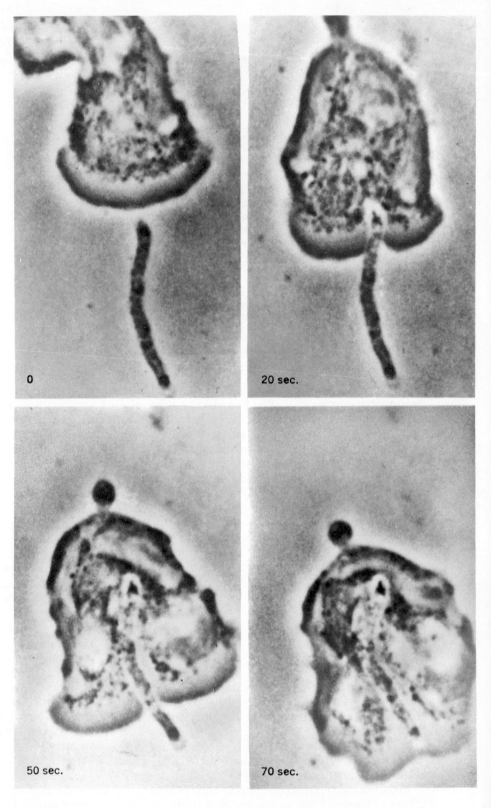

Figure 4-2 A sequence showing phagocytosis of a bacterium by a human white blood cell. Magnification 1,100×.

which the animal has not formed antibodies, is added to the serum, no precipitate will form. However, if the antigen is one that previously had been injected into the animal's bloodstream, the serum already contains antibodies against it. The antibodies react with the antigen to form a cloudy precipitate. The amount of precipitate depends on the concentration of the appropriate antibodies.

Using the amount of precipitate as a guide, we see that the antibody concentration increases with time after the first injection, as antibodies are synthesized. A second injection of the antigen produces an increase in antibody concentration, with a shorter delay, a more rapid rate of synthesis, and higher final levels (Figure 4-4). The increased response to later injections is called **immunological memory.** This term implies that the organism retains some of its original defense and can therefore synthesize antibodies more rapidly when the antigen returns a second time.

There are many differences between individuals in their capacity to respond to antigens. A low level of circulating antibodies may be maintained in the blood, or it may not. It may take several injections of the antigen to produce a response, or just one. The essentials of the response, however, are

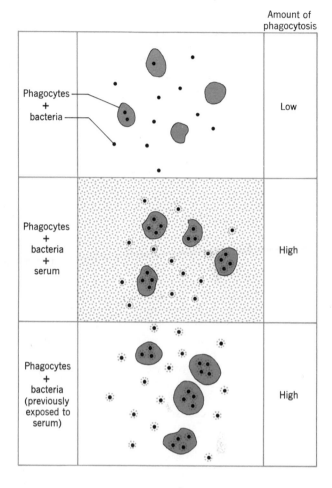

Figure 4-3 (above) Effect of blood serum on phagocytosis. (a) Suspended in salt solution, white blood cells ingest few bacteria. (b) Suspended in blood serum, white blood cells ingest many bacteria. (c) Pretreatment of bacteria with serum makes them susceptible to phagocytosis in a salt solution.

Figure 4-4 Immunological memory. The injection of an antigen into an organism results in the production of antibodies. After a second injection, the change in serum antibody concentration is more rapid, and the level reached is higher.

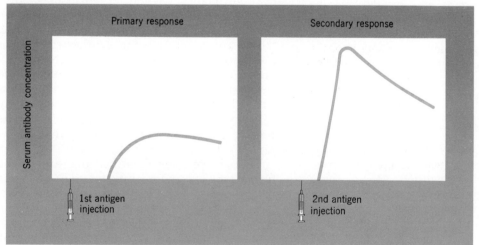

always the same. First, the foreign material is identified by the body. Then the body synthesizes a specific set of protein antibodies that will react with it.

Immunity is never absolute. The protection conferred by a previous attack of chicken pox or mumps usually prevents reinfection, but almost any immunity can be overcome by massive doses of the infecting agent. Diseases such as influenza or gonorrhea confer little immunity against another attack. On the other hand, immunity against one disease agent may provide immunity against attack by another if their antigens are similar.

The specificity of the antibody-antigen response has been used by biologists to gather information about evolutionary relationships between different kinds of organisms. When rabbits are injected with the blood serum of oxen, they form a precipitin against it. If this precipitin is then added to the sera of other animals, the reactions vary widely. Table 4-1 shows the extent of the reaction between the precipitin and the sera of several kinds of animals. Predictably, the strongest reaction is produced by the original antigen—the serum of oxen. The serum from sheep, which are distantly related to oxen, reacts much more weakly. Presumably the extent of the reaction is a measure of the similarity of the serum proteins. The similarity of the proteins in turn is a measure of the degree to which the animals are related. Elaborate and detailed "family trees" have been prepared in this way for both plants and animals.

Antibodies are extremely valuable medically. A wide variety of organisms produce poisons, or **toxins,** and antibodies can be prepared commercially to counteract their effects. First, an animal is injected with the toxin. The animal then produces an antibody (or **antitoxin**) that inactivates the toxic material. If these antitoxins are isolated and injected into an afflicted individual they afford relief from the toxic agent. Antitoxins have been prepared against toxins from microorganisms such as the diphtheria bacillus and from larger forms such as poisonous snakes.

Antibody Formation

Antibodies are formed many places in the body: spleen and lymph nodes, intestine, bone marrow, lung, blood cells, and elsewhere. All of these regions contain lymphoid tissue, which is the site of antibody formation. When an antigen challenges the system, antibody formation can be observed in almost all of the lymphoid tissues in the body. The

Table 4-1

Immune Response

Units reflect the amount of precipitate formed.

Serum From	Response to Rabbit Antibodies Against Ox Serum
Ox	100
Sheep	66
Goat	50
Pig	16
Horse	16
Dog	16
Human	8
Wild Rat	less than 1

specific cells from the lymphoid tissue that form antibodies are termed **immunologically competent.**

A particularly elegant technique is used to determine just which lymphoid cells are competent to form antibodies. First, an antigen is administered to an animal so that the animal produces antibodies to it. These antibodies (call them antibodies$_1$) can be isolated from samples of blood serum. The antibodies$_1$ are then administered to a second animal. This animal in turn produces its own antibodies, using antibody$_1$ as the antigen. Call this second group antibodies$_2$. At this point a fluorescent organic compound is attached to antibodies$_2$. If we then inject the first animal with more of the original antigen, we will stimulate the synthesis of more antibodies$_1$. Tissues of this animal are now cut into thin sections and flooded with the fluorescent antibodies$_2$. Wherever antibodies$_1$ are being formed in the tissues of the first animal, the antibodies$_2$ will react with them. We can see precisely where antibody$_1$ is being synthesized by viewing the tissue slices under ultraviolet light, which makes the antibody$_1$–antibody$_2$ complexes fluoresce (see Figure 4-5). Such experiments have shown that antibody production occurs in special lymphoid cells called **plasma cells.** Plasma cells contain extensive ER and many ribosomes, to facilitate the rapid synthesis of antibody protein.

Though we have identified the site of antibody formation, a more fundamental question remains—how does an organism synthesize antibodies that are specific for a given antigen? Two decades ago, before much was known about protein synthesis, the most popular idea concerning an-

Figure 4-5 Indentification of immunologically competent cells. A rabbit is injected with an antigen, stimulating it to produce antibody$_1$. These antibodies are then isolated and used as antigens in another rabbit to form antibody$_2$. Antibody$_2$ is then isolated and labelled with a fluorescent dye. Lymphoid tissues of a rabbit injected with the original antigen can then be cut into thin sections and treated with fluorescent antibody$_2$. Since fluorescent antibody$_2$ will bind specifically with antibody$_1$ in the tissues, any cells that are synthesizing antibody$_1$ will fluoresce under ultraviolet light.

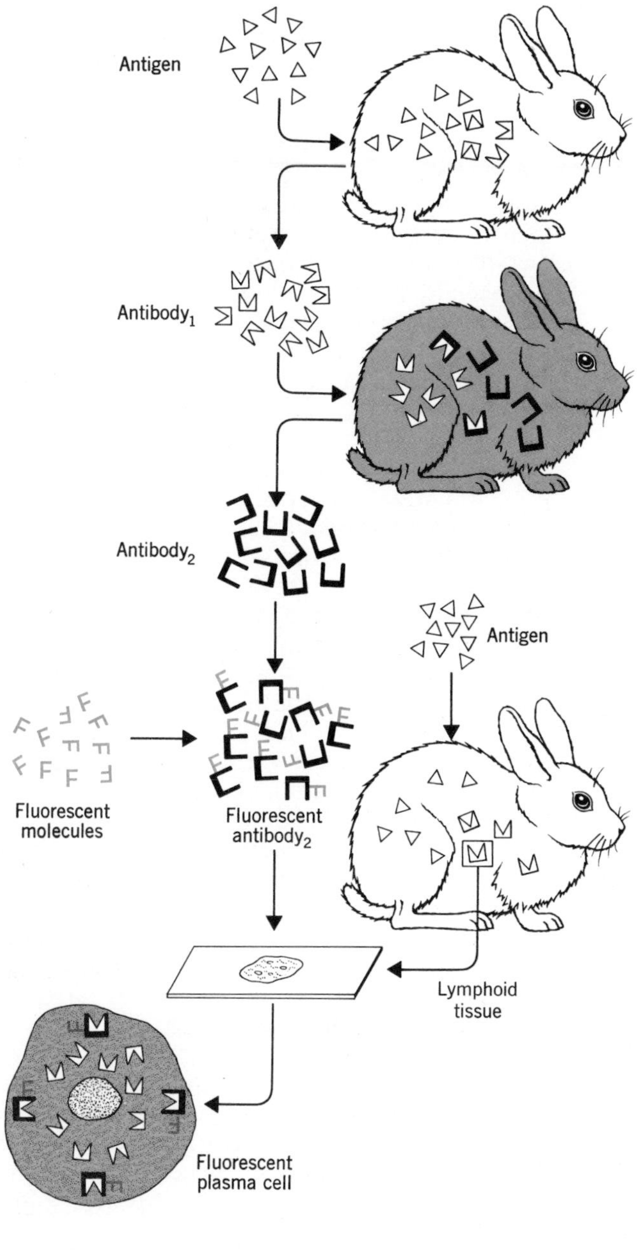

Antigen

Antibody$_1$

Antibody$_2$

Fluorescent
molecules

Fluorescent
antibody$_2$

Antigen

Lymphoid
tissue

Fluorescent
plasma cell

tibody formation was the **instructional theory.** It was argued that antibodies are effective because they acquire precise shapes, fitting tightly around the antigen. In other words, it was assumed that the antigen acts as a set of instructions for the formation of the antibodies. When the antigen is introduced into the organism, it comes in contact with a preexisting protein and the protein molds itself around the antigen molecule. The protein, now a specific antibody, is then able to direct the synthesis of more antibodies with an identical configuration (Figure 4-6).

This theory does not fit with our present knowledge about protein synthesis. A protein's primary shape is controlled by its specific amino acid sequence. This amino acid sequence in turn is directed by RNA templates, not by other proteins. Our ideas about antibody formation have therefore had to be revised to agree with new information. The specificity of an antibody depends upon a specific sequence of amino acids. But how does the organism synthesize a specific antibody when it has had no previous contact with the antigen?

The situation is not entirely clear at present. There is considerable evidence for the theory that the proper antibodies are selected from a large group of preexisting possibilities. This **selection theory** holds that there is a variety of im-

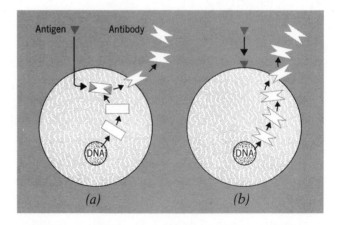

Figure 4-6 Two theories of antibody formation. (a) In the instructional theory, the antigen directs, or provides a set of instructions for, the synthesis of antibodies. (b) In the selection theory, the antigen selectively activates the specific immunologically competent cells that are capable of synthesizing the appropriate antibodies.

munologically competent cells in the organism, and that each type is capable of making a different kind of antibody. It is believed that an antigen activates only those cells that will produce the proper antibodies; the other immunologically competent cells in this large population do not actually produce antibodies until an appropriate antigen stimulates them (see Figure 4-6). Immunological memory is explained in the following manner: the cells that are competent to produce the required antibodies have already been activated once, and thus are ready to synthesize protein when the antigen appears a second time. They are brought into action more rapidly to increase the concentration of antibodies.

The selection theory of immunity, then, assumes the existence of a very large population of immunologically competent cells. Each type of cell has to carry a different code for the production of antibodies. The assumption is that the cells of the lymphoid tissue represent a pool of different DNA sequences, each of them ready to be activated by a particular antigen. But since the range of possible antigens is very large, the pool must be comparably large, so that differences in the composition of the antigens can be matched by differences in amino acid sequences of the antibodies. Yet our information about DNA replication says that all the cells in an organism must contain the same DNA. Furthermore, present evidence suggests that all the embryonic cells that are destined to produce lymphoid cells begin with the same DNA. However, certain parts of the DNA sequences that direct synthesis of antibody proteins may be quite unstable. As the lymphoid cells grow and divide, these DNA segments may change because of errors in DNA replication. In any event, the growing animal acquires a large group of cells with slightly different DNA sequences.

At this point, we can test the selection hypothesis by seeing whether different blood antibodies do in fact have different amino acid sequences. Blood antibodies are part of a group of blood proteins called the globulins. For this reason, blood antibodies are referred to as **immunoglobulins.** There are at least four different kinds of immunoglobulins. But they are all similar in their basic construction. Each immunoglobulin contains two light and two heavy polypeptide chains.

Analysis of the specific amino acid sequences of these chains, in one class of immunoglobulins, shows that one end of both the light and the heavy chains is variable—the other parts of the chains are quite similar. To consider only the light chain, there are about 30 places where the substitution of different amino acids is possible. This diversity, in combination with similar variations in the heavy chains, gives a tremendous number of possible variants, enough to account for the entire range of antigen-antibody responses. Light and heavy chains bond together to make a rod-shaped antibody. Each end of the rod has a site that combines with an antigen (Figure 4-7).

recognition of self

We have just described the means by which an organism recognizes an invader. But this ability is not always a necessary part of defense processes. Some organisms simply cannot distinguish between their own tissues and foreign objects or tissues. Many plants, for example, cannot make this distinction. Some insects will accept transplants of tissues from other, unrelated insects. Many other organisms, however, *can* recognize their own tissues. The ability of an organism to tolerate a tissue graft from another organism is a key index to its power of self-recognition.

Grafts are a matter of great economic and medical interest. In horticulture and agriculture, grafting is a routine and valuable procedure. Various techniques are used to establish a plant graft, some of which are diagrammed in Figure 4-8. In general the shoot, or leafy portion, of one plant is attached to the rootstock of another, so that the conducting tissues just below the bark can establish continuity. They must be joined in order to allow the transport of materials between the stock and the graft.

There are a number of practical reasons for grafting. The plant that furnishes the graft may have a poor root structure for the available soil. Grafting can remedy this defect. Grafting may produce a dwarf plant that flowers earlier or bears fruit more rapidly. The flavor and texture of a fruit may be improved by grafting. Pears produced by grafting pear branches to quince rootstock have a better flavor and a higher sugar content than ordinary pears.

In general, grafts and rootstocks must come from closely related species in order to survive. But there are many exceptions. Some grafts between varieties of the same species are unsuccessful, while in other cases (such as the quince and the pear) the graft and

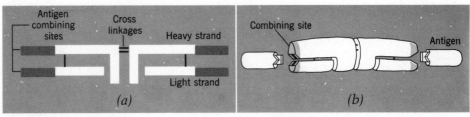

Antigen combining sites | Cross linkages | Heavy strand

Light strand

(a)

Combining site | Antigen

(b)

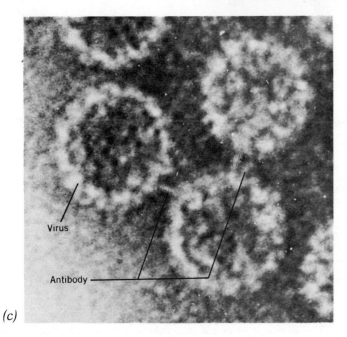

Virus

Antibody

(c)

Figure 4-7 *Antibody structure. (a) An antibody is composed of two light and two heavy strands. (b) The light and heavy chains at each end of the antibody form a combining site that can bind with an antigen. (c) Electron micrograph shows an antibody–antigen complex. The antigens are virus particles. Magnification 750,000×.*

the rootstock may be rather distantly related. The failure of grafts is not well understood. In some cases it occurs because tissues do not line up or do not adhere at the junction. Such difficulties can sometimes be bypassed by interposing a section of a third plant that is compatible with both partners. In other cases some chemical factor that is transported along the conducting tissues is responsible for graft rejection.

Our knowledge of tissue recognition in man and many other warm-blooded animals is greater. Tissue transfer between humans is actually an old procedure; blood is a tissue, and blood transfusions were common in the nineteenth century. The first recorded transfusion was done in the 1660's and was successful. But the physician who performed it soon undertook several others, one of which proved fatal. As a result, the operation was declared illegal in Europe. When interest in transfusions reawakened, the procedure again proved to be dangerous and was used only as a last resort.

In 1900, Karl Landsteiner identified and described the ABO blood groups in man. His discovery at last provided a basis for choosing compatible blood for transfusion. We now know at least nine independent factors in blood that are used to determine compatibility, but the ABO system is still by far the most important.

The blood-group interactions demonstrated by Landsteiner are very similar to an immune response to an antigen. The major difference is that the antibodies in this case are already present in the serum of an individual prior to exposure to the antigen. The antibodies are agglutinins; they promote the clumping of any red blood cells that carry the corresponding antigen (Figure 4-9).

There are two different antigens in the Landsteiner system, A and B. The blood groups are named for the antigens. Thus, a person with type A blood has the A antigen on his red blood cells; one with AB has both; and one with type O, neither. Corresponding to the two antigens, there are two sets of antibodies in blood serum, anti-A and anti-B. An individual's blood contains antibodies that react with any antigen not present on his own red cells. A person with type O blood would have both anti-A and anti-B. These relations are indicated in Table 4-2.

The ABO groups are the most important compatibility factors in blood. Most of the others rarely cause major problems. But an exception is the **Rh factor,** also discovered by Landsteiner, in 1940 — 40 years after his original description of the ABO system. The Rh antigen was originally discovered in the red blood cells of rhesus monkeys, hence its name. Most people are Rh positive. In other words, their blood cells carry the Rh antigen. An Rh nega-

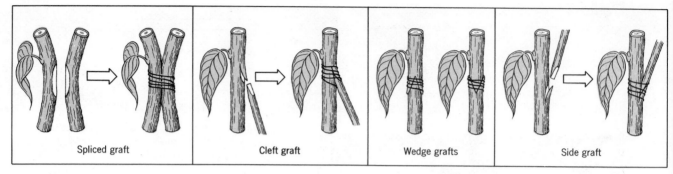

Figure 4-8 *Various types of plant grafts.*

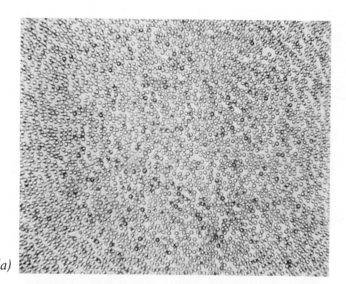

(a)

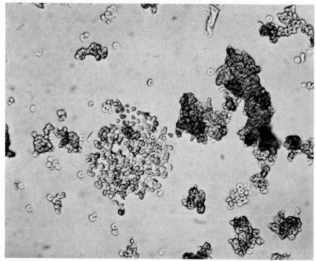

(b)

Figure 4-9 *(a) Normal and (b) agglutinated human red blood cells. Magnification 80×.*

tive individual does not have the Rh antigen and will produce antibodies against it after being exposed to Rh positive blood. An embryo with an Rh+ father and an Rh− mother may be Rh+; that is, its blood cells may carry the antigen. If the blood cells carrying the antigen accidently pass into the mother's circulation during the pregnancy, she will produce anti-Rh antibodies, which then circulate back into the embryo. These antibodies may agglutinate or lyse the embryo's red blood cells, so that it is stillborn or very anemic at birth. The anemia was usually fatal to the infant until the 1940's, when a technique for complete blood transfusion was developed. Firstborn children usually escape the anemia, because not much antigen has leaked into the mother's circulation. Anemia becomes more and more likely in succeeding pregnancies.

A direct transfusion of Rh+ blood to an Rh− individual is always to be avoided, since it stimulates the production of anti-Rh antibodies. A second, later transfusion of Rh+ blood would then result in agglutination of the Rh+ red blood cells by the antibodies in the recipient's blood.

Table 4-2

The ABO Blood Groups

Blood Group	Antigen(s)	Antibodies	Percentage of U.S. Population
A	A	Anti-B	41
B	B	Anti-A	10
AB	A and B	None	4
O	None	Anti-A and Anti-B	45

Organ Transplants

Blood transfusions are now routine and quite safe if done under adequate supervision. But blood is perhaps the only tissue that can be "transplanted" between unrelated individuals with routine success. Other tissue and organ transplants behave more selectively. Table 4-3 lists the record of success for kidney transplants, according to the relationship between the donor and the host, the person receiving the transplant. Clearly, the more distant the genetic relation, the less successful the transplant.

Skin grafts have been studied extensively in experimental animals, particularly in mice. Strains of mice have been inbred to the point that they can be considered genetically identical, like identical twins. Skin grafts between members of an inbred strain are very successful. A normal blood supply is established to the new skin, and it appears to be entirely normal. Such grafts persist indefinitely. If a skin graft is attempted between individuals of two different strains, however, the results are different. Initially the graft appears to be accepted by the recipient. But after a period of time the graft is rejected—the blood supply is shut off and the grafted skin is sloughed. If a second graft is later attempted from the same donor, it is rejected even more rapidly. The sequence is entirely comparable to our earlier account of immune response: immunological competence and memory are involved in both cases. Antigens present in the foreign tissue initiate rejection.

Table 4-3 can now be interpreted in terms of immune response. Identical twins share the same tissue antigens, so transplants between them should be successful because there is no foreign tissue antigen to trigger antibody production and rejection. But when the genetic relation between recipient and donor is more distant, the difference in tissue antigens is greater. Such grafts are therefore less successful. No one with a transplanted kidney, except one from an identical twin, is able to maintain normal kidney function without continuing treatment with drugs that suppress the immune response.

Despite much publicity about heart transplants, the results have not been very happy, because immune responses cannot be controlled. By January 1973, 200 heart transplants had been performed. There was 90 percent mortality in the first six months, and 95 percent after the first year.

The problem of suppressing the host reaction to the grafted organ is formidable. Several methods are

Table 4-3

The Survival of Patients with Kidney Transplants[1]

Donor Group	Total	Percent Living After One Year
Twins		
Identical	6	100
Fraternal	2	100
Other Relatives		
Parents	274	77
Siblings	279	79
Other Blood Relatives	25	88
Non-related Donors	23	48

[1] Transplants performed in 1968–1969.

currently being used. All of them work by preventing cell division in immunologically competent cells. Normally, when these cells respond to an antigen, they begin to synthesize antibodies. They also start to divide, generating a large number of identical cells, all of which are capable of synthesizing the same antibodies. The rejection of grafts can therefore be prevented by suppressing the division of these immunologically competent cells. Whole-body radiation will have this effect, but the required doses are rather high. A dose of 400 roentgens prolongs the survival of a kidney transplant, but it does not permanently inhibit rejection. (A lethal dose is about 500 roentgens.)

More success is obtained with drugs. The common agents are 6-mercaptopurine and azothioprine (Imuran). These drugs are mitotic inhibitors; that is, they prevent cell division. They are administered in very nearly lethal doses at the time the graft is made, and high doses must be maintained for some time. Under their influence the immunologically competent cells in the lymphoid tissue, which would normally respond to the antigen by dividing and synthesizing antibodies, are unable to do so. Of course all the other cells in the body are also prevented from dividing. Unfortunately, the treatment must be continued indefinitely—as soon as it is stopped, the graft will be rejected. The suppression of cell division and antibody production leaves the body quite defenseless to invaders. Even microorganisms that are normally harmless may cause a serious infection. Thus, a patient who manages to retain a transplant because his immune response has been suppressed is very likely to succumb to something like pneumonia.

Of course, additional research into immune defenses may give us greater control over the responses. But as yet, we cannot restore an individual to normal health through organ transplants.

The Development of Individuality

Organisms have tissue antigens that are capable of producing an immune response in any number of other organisms. So what prevents these antigens from provoking an immune response in the organism itself? It is certainly not just a happy accident that individuals with type A blood do not happen to have anti-A antibodies in their serum.

Many observations suggest that an individual organism adapts to its own tissue in the course of development. Cattle embryos, for instance, sometimes develop in such a way that twins share the same circulation. They are not identical twins, and their blood types may be different, but no problems arise from the shared circulation. The same thing can happen, though more rarely, in humans. As adults, such individuals have blood with a combination of characteristics. For example, assume that one twin originally had blood of type O, and the other, type A. The blood that is circulated by the embryos, then, is a mixture of type A (with A antigens) and type O (with no antigens). We would expect the O-type individual to produce anti-A antibodies, but he does not. Clearly the adult has been influenced by the common blood circulation during embryonic development.

This occurrence is itself suggestive, but a definitive experiment done in 1968 makes the situation completely clear. Cells produced by the first few divisions of a fertilized mouse egg can be separated, recombined, and implanted in the uterus of another mouse. They develop normally. The same thing can be done with a mixture of cells from the eggs of two different strains of mice, even though the recombined egg cells have two different sets of tissue antigens. An adult mouse produced from the combination of cells also still has both sets of tissue antigens. Yet both sets of antigens will be accepted as "self." Clearly, the presence of foreign tissue early in development produces a tolerance to that tissue's antigens. Similarly, if cells from a foreign strain are injected into the developing mouse embryo, the adult can be made to accept grafts of

something as far removed as rat tissue by this procedure. Such "induced tolerance" is an example of the way a developing organism establishes its self-identity. This process of self-recognition is essential: it insures that antigens present in early development will not activate an immune response.

We can outline the process of self-recognition in terms of the selection theory. When an antigen is introduced during the early development of an organism, it combines with or is recognized by specific immunologically competent cells. At this early stage, however, the immunologically competent cells are not stimulated to make antibodies. Rather, they are destroyed, or their protein synthesis is inhibited so they will not make antibodies.

Direct evidence for this hypothesis comes from an experiment in which a mouse of strain A is made to tolerate a graft from strain B. Cells from B are injected into a developing embryonic A mouse. Skin from strain B can later be successfully grafted onto the A mouse. However, after the graft has taken, lymphoid cells from another A mouse—one that has not been made tolerant of B tissue—are injected into the graft recipient. This injection will include immunologically competent cells. The new lymphoid cells react to the graft as foreign tissue, and the once-tolerant mouse now rejects the graft.

The Thymus Gland

Recognition of self early in development is accomplished by the thymus gland. The thymus lies in the throat and extends down into the chest (Figure 4-10). It is a prominent organ in childhood but becomes smaller with increasing age.

The importance of the thymus is shown by removing it from newborn mice. The operation is called thymectomy. Early removal of the thymus interferes with the production of immunologically competent cells. As a result, thymectomized mice can tolerate skin grafts from other strains of mice (see Figure 4-11). These thymectomized mice cannot resist invasion by microorganisms; they often thrive for a time and then develop a "wasting" or "runting" disease and die. But mice raised and maintained under germ-free conditions (Figure 4-12) do not show these effects. The conclusion is that just after birth the thymus activates the immune system.

The thymus gland continues to function throughout life, but at a lower level. If it is removed later in life, immune responses continue for a while without it, presumably because the organism has a

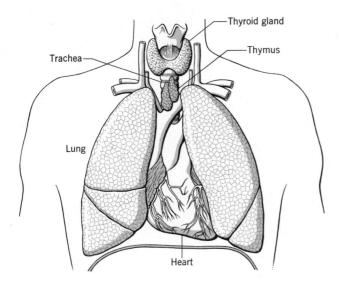

Figure 4-10 The human thymus gland.

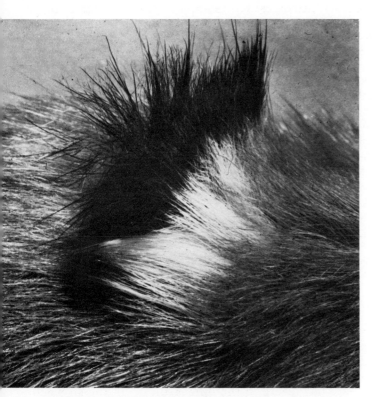

Figure 4-11 As an adult, a mouse that was thymectomized just after birth tolerates skin grafts that it
would normally reject. The black hair is skin from an
unrelated mouse, and the white hair is skin from a rat.

supply of mature immunologically competent cells
stored in various tissues. But eventually the supply
runs out and the ability to synthesize antibodies
becomes impaired.

The thymus gland is composed of lymphoid
tissue, but it does not itself seem to be a major
source of immunologically competent cells. Instead, the thymus behaves as a processing station
for immunologically competent cells that migrate
to it from other parts of the body—mainly immature lymphoid cells from lymphoid tissue in the
bone marrow. The cells mature in the thymus and
then are distributed to the other lymphoid tissues
in the body. When a foreign antigen appears, the
lymphoid cells divide and produce plasma cells;
these plasma cells then divide and produce antibodies.

At least part of the processing and maturation of
lymphoid cells is accomplished by a soluble substance produced by the thymus. The action of the
substance can be shown by placing whole thymus
tissue in a membrane-bounded sac, so that contact
between cells is prevented but soluble materials
can move in and out. If the sac is implanted into an
animal that has been thymectomized, the animal's
immune system functions almost as well as it
would with an intact thymus.

The action of the thymus is selective, however. It
appears to release only the lymphoid cells that
make antibodies against foreign antigens. Cells that
would produce antibodies against self tissues are
presumably lysed and destroyed in the thymus itself. The thymus' decision as to which antigens are
foreign and which are not depends, as we have seen,
on which antigens are present during early development. Any antigen molecules that are in contact
with immature cells in the thymus at this time are
recognized as self. Recognition cannot occur later,
because the thymus apparently develops some kind
of physical barrier that prevents contact with circulating antigens.

If this hypothesis is correct, then self tissues that
stay out of the general circulation early in development should be treated as foreign tissues by the
adult immune system. The lens of the eye is one
such structure. And, in fact, if we remove one eye
lens from an experimental animal, extract the protein in it, and inject this protein back into the
animal, the animal produces antibodies against it.
An extract of the second lens will then produce pre-
cipitation and other immune responses. Clearly, if
tissues are isolated from the rest of the organism
during early development, they are not recognized
as self.

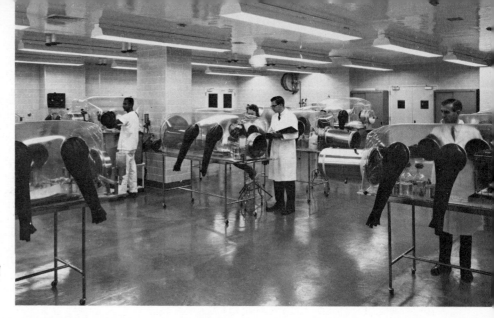

Figure 4-12 A germ-free environment. This laboratory at Ohio State University is elaborately equipped for maintaining animals in a sterile condition.

We can summarize as follows. Early in its development, an animal does not respond to antigens. It tolerates any tissue, whether foreign or self. At some time from shortly before to shortly after birth, the thymus plays a critical role in activating the immune system. The animal now responds to new antigens by synthesizing antibodies. But any antigens that were present in the animal when the immune system was activated are treated as native. They do not stimulate the production of antibodies.

Immunological competence and immunological memory are properties of sharks, fishes, frogs, reptiles, and birds, as well as of mammals. An organ that functions like the thymus, called the bursa of Fabricius, appears in birds but not in lower animals. This organ can still be found in mammals and may complicate the interpretation of thymectomy experiments. Its presence would suggest that, in mammals, several immunity systems may be operating simultaneously. Such a situation is actually not unusual. Very often animals develop a functional system that supersedes an earlier one, without fully discarding the more primitive apparatus.

Immunological Defects

The complicated multiple defense systems in humans can go astray in many ways. The so-called **autoimmune** diseases are one result. In these conditions, antibodies are produced against antigens that are normally present in the organism. The immunological tolerance to self has failed. Rheumatoid arthritis is one such disease.

A more common kind of immunological difficulty is **hypersensitivity,** or extreme sensitivity to

certain antigens. The most violent form of hypersensitivity is called **anaphylactic shock.** It is an antigen-antibody reaction, so it will occur only in an animal that has already had a chance to synthesize antibodies. For example, if a guinea pig is injected with a minute amount of egg albumin, no immediate reaction is observed. But a second injection a few weeks later produces anaphylactic shock, which may result in death.

In anaphylaxis, the antibody-antigen reaction causes cells in the connective tissue to release histamine. Histamine is a small molecule that is synthesized from one of the common amino acids. It has several powerful effects when released into the blood. It dilates blood vessels and increases their permeability, so liquid moves from the blood into the tissues and causes them to swell. Histamine also stimulates the contraction of certain muscles. Dogs in anaphylactic shock, for example, die when muscle in the blood vessel leading away from the liver contracts; as much as 60 percent of the dog's blood may be found in the liver after death. In the guinea pig, death results from the contraction of muscle in the air passages; the animal cannot get any air, and it asphyxiates.

Allergy is a milder form of hypersensitivity that involves a localized response between antibodies and an allergy agent, called an **allergen.** The antibodies to the allergen may be localized in the mucous membranes of the nose and throat (as in hay fever), in the lung (asthma), in the stomach or intestine (food allergies), or on the skin (rashes and contact allergies). In each case the allergen combines with the antibody on the surface of cells in the area involved. Histamine or other agents are then released, producing the allergy's characteristic symptoms. The best course for an individual who is seriously hypersensitive to an allergen is simply to avoid the agent entirely, if at all possible.

Some allergens act as weak antigens directly, but other allergens apparently combine with and modify the body's own proteins, turning them into antigens. Penicillin is an example. It combines with proteins in such a way as to produce an antigen. Likewise, the oily secretions of poison ivy and poison sumac plants modify skin proteins. A similar mechanism gives rise to the sensitivity of some biologists to formalin.

Some kinds of cancer can be interpreted as immunological defects. A virus called *Polyoma* produces many tumors in mice, but only when administered to newborn mice. In older animals, which have developed immunological competence, tumors can be produced only occasionally. This experiment suggests that some kinds of tumors produce antigens and can therefore be arrested by the immune process. Cancers induced by coal tar derivatives also produce antigens that stimulate an immune response. Many cancerous growths are characterized by such foreign antigens. In these cases the actual growth of cancer may represent the failure or the overwhelming of the body's immune defense mechanisms. In fact, an examination of the tissues of a normal person reveals much more evidence of arrested cancer than would be expected on the basis of clinical records. That is, most cancers are arrested early in their development, presumably because the body treats them as foreign tissue.

The search for a specific antigen carried by tumor cells is one of the major themes of current cancer research. If such an antigen is found, then antibodies could be produced, and individuals could be immunized and protected against the tumor. The involvement of the immune system in cancer is evident, but as yet no clear success has attended the search for specific antigens. In some cases, of course, the uncontrolled growth of cancerous cells may not produce any abnormal proteins or other antigens. In those cases, nothing would stimulate the immune defenses.

One more factor in the operation of the immune system needs to be considered. The emotional state of the individual seems to be important in the whole range of immune responses. Susceptibility to infectious disease, the development of allergies, auto-immune diseases, and the occurrence of cancer have all been correlated with psychosocial factors. Such observations are actually not new. The Greek physician Galen (A.D. 130–200), for instance, thought that melancholy women are especially susceptible to cancer. Self-alienation and problems of self-identification in psychological terms run parallel to difficulties in self-identification at the cellular level. Scholarly papers are now appearing with such titles as "Cancer as an Alternative to Psychosis." This connection may seem fanciful. Nevertheless, there actually is considerable evidence to support the view that emotional depression has a profound effect on the whole immune system. The existence of emotional factors does not, of course, lessen the reality and the importance of exposure to antigens or allergens. But in immune reactions, especially, the susceptibility of the host cannot be ignored.

Recommended Reading

Fuller, Watson (ed.), *The Social Impact of Modern Biology*. London: Routledge and Kegan Paul (1971).

A series of articles resulting from a meeting held by the British Society for Social Responsibility in Science. It contains good papers on immunology and cancer.

Sondheimer, E., and J. B. Simeone (eds.), *Chemical Ecology*. New York: Academic Press (1970).

A collection of papers dealing with chemical defenses and interactions among organisms. The chapters on plants and the interrelations between plants and insects are of particular interest.

Titmuss, Richard, *The Gift Relationship*. New York: Random House (1971).

A discussion of the implications of blood transfusions and blood banking for the sociology of medicine.

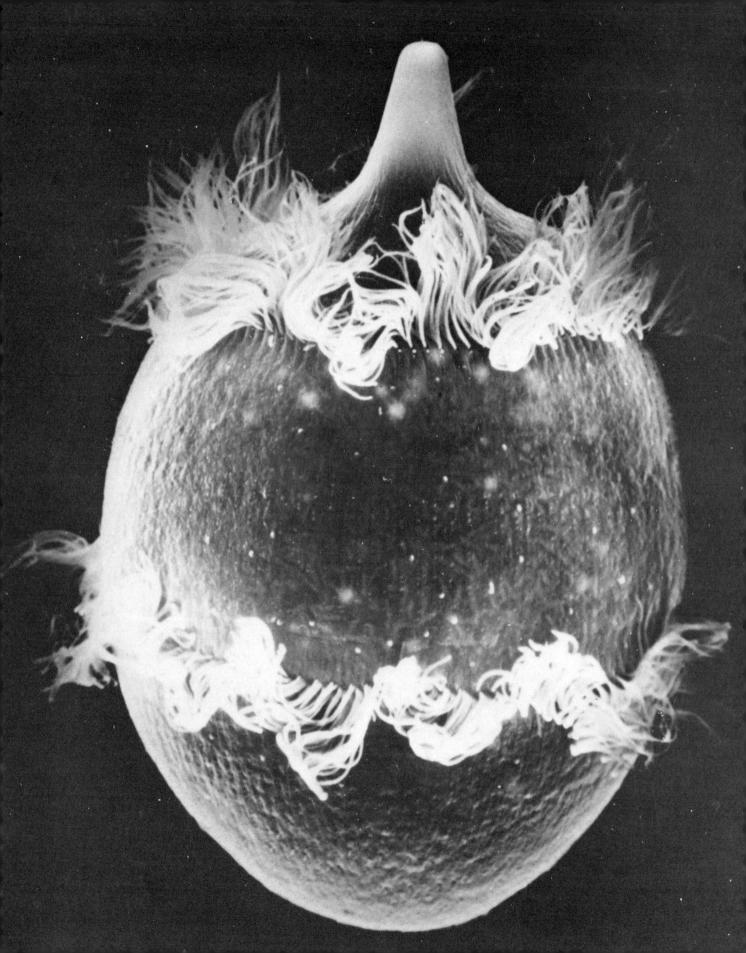

5 microorganisms

The elaborate defense mechanisms described in Chapter 4 are necessary because plants and animals are always vulnerable to invasion by microorganisms. Microorganisms are everywhere, and in large numbers. If one could magically dissolve everything in the world except the soil microorganisms, the shapes of all the major land features of the world would still be visible, outlined in a ghostlike film a few inches to a few feet thick. If other kinds of microorganisms were left behind as well, then all the large organisms, plant and animal, would have their places in this spectral world. There would be very few empty spaces.

Microorganisms can survive in all kinds of environments. As a prelude to extraterrestrial exploration, the National Aeronautics and Space Administration supported studies of the microorganisms in desert regions of the earth. Researchers collected samples from 26 of the most desolate and forbidding areas known. Bacterial life was found in all but two of the samples.

By strict definition, any organism too small to be seen with the unaided eye is a microorganism. Many kinds of organisms fit this definition. A handful of soil and a few drops of pond water will contain a staggering variety of microorganisms. The smallest identifiable forms are the bacteria, some of which are in motion and some of which are sessile. Assorted fungi will also be apparent. Some of them, such as yeasts, are simple spheres. Others are more elaborate, consisting of long strands and various reproductive structures that produce spores. There will be a bewildering number of unicellular animals, the protozoa. Among these may be found the familiar, creeping amoeba and a variety of fast-swimming cells equipped with flagella or cilia. In addition,

Didinium nasutum,
a ciliated protozoan

121

there will be a multitude of flagellated algae, which can be distinguished from the protozoa by their green, gold, or brown coloration. Other algae present in pond water or soil may be immotile or creeping unicells, colonies, and filaments or sheets of cells. Microscopic multicellular organisms such as roundworms, and tiny insects such as mites and springtails, will also be common. A community of microorganisms contains an immense variety of life.

Classes of Microorganisms

Before the advent of techniques for studying microorganisms, all organisms were traditionally placed in two major groups, the plant kingdom and the animal kingdom. But after the discovery of the microscope, it became painfully obvious that many microorganisms did not conform to this classification scheme. *Euglena* (see Figure 5-1) is a good example. This flagellated unicell occurs in several forms that are identical except for their modes of nutrition. One variety would be considered a "plant," since it contains the photosynthetic pigment chlorophyll and gets its energy through photosynthesis. But other forms contain no chlorophyll. They absorb nutrients across their surfaces or ingest food particles through simple gullets, and thus would be classified as "animals." A plant-like *Euglena* can be converted to an animal-like *Euglena* by treating it with ultraviolet light or streptomycin or high temperatures. All of these treatments cause the irreversible destruction of the cell's chlorophyll. In such cases the "plant" becomes an "animal" in the laboratory.

To overcome such difficulties, scientists have proposed classification schemes containing three, four, or five kingdoms. The system followed in this text divides organisms into five kingdoms:

1. **Monera:** bacteria and blue-green algae. These organisms are unicellular and procaryotic; that is, they lack membrane-bounded nuclei and other organelles.
2. **Protista:** unicellular algae and the protozoa. These microorganisms are also unicellular, but they are all eucaryotic cells, so their internal structure is more highly organized.
3. **Plantae:** all the familiar green plants and the multicellular algae, such as seaweeds.
4. **Fungi:** the common molds, mushrooms, and related organisms. They are non-photosynthetic and dwell mainly in the soil.
5. **Animalia:** All multicellular animals.

In addition, the **viruses** form a separate category of living material. They are usually discussed separately because they are not true cells. They are still

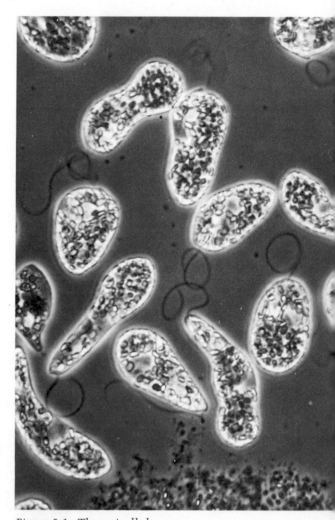

Figure 5-1 The unicellular Euglena *occurs in both photosynthetic and non-photosynthetic forms. Magnification 1,200×.*

considered part of the biological world, however, because they contain genetic material and can reproduce themselves.

Two of the five biological kingdoms—the Monera and the Protista—as well as the viruses, contain microorganisms exclusively. But each of the other kingdoms also has microscopic members. Thus, the term microorganism says nothing except that the organism is small relative to man. Consequently, microbiology—the study of microorganisms—is a very general field in terms of the kinds of organisms studied. A microbiologist may specialize in bacteriology, virology, mycology (the study of fungi), algology, or protozoology. Yet in spite of the diversity of the organisms involved, the concept of a "microorganism" and a "microbiologist" is convenient and practical. When an organism is small, special techniques must be used to study it. Many experimental techniques require elaborate and expensive equipment. These methods and procedures can be applied with equal success to a culture of protozoa or algae. In experimental terms, therefore, the size of the organism is more important than its identity. Even today, comparatively little is known about the structure of most small organisms, simply because conventional microscopes cannot reveal fine details. The more powerful electron microscope, however, is rapidly becoming available to most researchers. Hundreds of microorganisms are described each year. Electron microscopy is producing a renaissance in microbiology.

Microorganisms are much more than just a diverse group of interesting but inconspicuous organisms. The total number of microorganisms in various environments is often so large that their activities have wide-reaching effects. It has been estimated that the total dry weight of microorganisms in the soil community is about 1,200 pounds per acre of soil. This organic matter is almost as much as would be contained in a harvest of wheat from the same area. Since the surface–volume ratio in small organisms is very large, their metabolic rates are extremely high. The combined metabolic activities of such huge numbers of small organisms have a profound impact on the environment in general and on man in particular.

The activities of microorganisms will be discussed in more detail in the second half of this chapter. First we will examine the major classes of microbes.

viruses

Viruses are small particles of nucleic acid and protein. They are significant disease agents in both plants and animals. Outside of cells they can survive, but they cannot grow. In fact, they can grow and reproduce *only* inside the living cells of other organisms, because they lack the cellular machinery necessary for an independent existence. They have none of the organelles and enzyme systems required for respiration, protein synthesis, or DNA replication. For this reason, viruses are not considered true cells and are usually excluded from classification schemes.

Viruses are small particles, 100–300 A in diameter. Each virus consists of a single strand of DNA or RNA surrounded and protected by a protein coat. Most of our information about viral activity and

Figure 5-2 A bacteriophage virus. The virus is composed of protein units around a central strand of DNA. Magnification 350,000×.

structure comes from two sources, electron micros-copy and biochemical studies of a series of viruses that attack the bacterium *Escherichia coli.* These viruses are classified as **bacteriophages** (literally, "bacteria-eaters") or simply **phages.**

The most complex bacteriophage viruses are composed of protein "head" and "tail" regions (Figure 5-2). The DNA strand is coiled up inside the head. When a phage comes into contact with a bacterium, fibers on its tail attach to specific sites on the cell wall of the host. Enzymes weaken the wall enough for the hollow phage tail to be inserted. Then the viral DNA strand is literally injected into the host (see Figure 5-3). At this point, the protein coat becomes expendable, and the bacterium's machinery for protein synthesis and energy production comes under the control of the phage DNA. The genetic instructions from the viral DNA direct the host's machinery to synthesize more viral DNA and proteins. The protein molecules assemble into protein coats around the new viral DNA strands. One hundred to two hundred new viruses are produced. As a final step, the new phages direct the lysis of the host bacterial cell and are released to infect additional bacteria (see Figure 5-4). The entire process takes only about 25 minutes.

The host cell is not always lysed after infection by a phage. Sometimes the injected viral DNA becomes attached to the host's chromosome and remains inactive. But each time the host's DNA is replicated, the viral DNA is copied as well. In this way it is passed on to the host's daughter cells. Sooner or later the passive cycle ends–the phage DNA becomes active and lyses the host cell.

The viral parasite probably lives in equilibrium with the host cell most of the time. This fact makes viruses especially difficult to study, because it is virtually impossible to detect them with present techniques when they are being carried as small DNA segments in individual bacterial cells. The only way we can study viruses is to grow them in culture until we have enough to work with. Simple liquid culture media are inadequate, since viruses will not grow in the absence of host cells. Therefore, the host cells themselves must be cultured. This is easily done for viruses that have bacterial hosts, but it is much more difficult for viruses that live in the complex tissues of higher plants and animals. In spite of the difficulties, however, viral parasites of animals and plants are well worth studying because of the great practical importance of viral diseases.

the kingdom monera

Bacteria and blue-green algae are set off from other organisms and placed in the kingdom Monera. They are procaryotic organisms and, therefore, are relatively undifferentiated. They lack complex internal membrane systems, membrane-bounded nuclei, chloroplasts, and mitochondria. Monera do have a few folds of the cell membrane extending into the cell. DNA is located in a central nuclear region. Protein synthesis occurs on ribosomes scattered throughout the cytoplasm. Photosynthetic pigments, when present, occur in small tubular or flat membranes, rather than in chloroplasts.

Bacteria

Bacteria are about the size of the mitochondria found in eucaryotic cells. The only feature of bacteria that is easily observed with the light microscope is their shape, so shape is the basis of traditional classification schemes. A spherical bacterium is called a **coccus;** a rod-shaped one is a **bacillus;** and a helical one, a **spirillum** (see Figure 5-5). Some bacteria are equipped with flagella, others are immotile. Their relatively simple internal structure, as revealed by the electron microscope, is shown in Figure 5-6.

In addition to shape, the bacteriologist relies on chemical and nutritional properties to characterize and identify bacteria. For instance, the staining properties of the bacterial cell wall depend on its chemical structure. Staining, therefore, is a useful classification tool. A procedure called **Gram staining** distinguishes between two fundamental types of bacteria. The technique involves smearing the cells on a glass slide and treating them with iodine and crystal violet, a purple dye. The slide is then immersed in a solvent, such as alcohol, that dissolves lipid. If a bacterium has lipids in its walls, the dye will be leached out when the lipids are dissolved. These bacteria are termed **Gram-negative.** Bacteria that hold the purple stain because they have no lipid in their walls are called **Gram-positive.**

The nutritional characteristics of bacteria are also variable and form another basis for classifica-

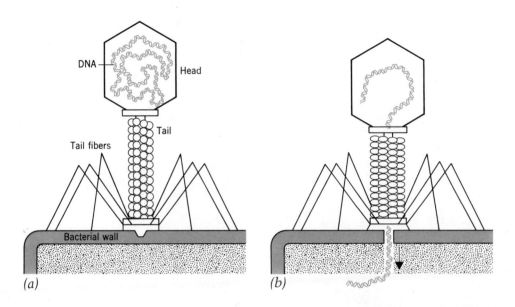

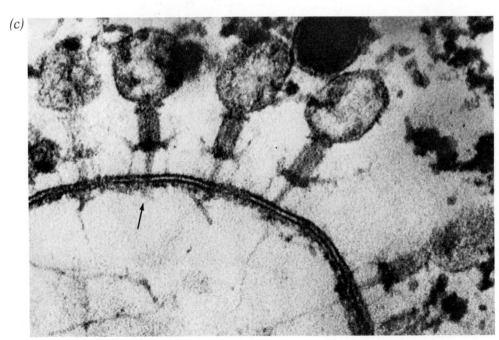

Figure 5-3 Injection of phage DNA into a bacterium. (a) The virus attaches to the bacterial cell wall by means of its tail. The base of the tail contains enzymes that weaken the bacterial wall. (b) The tail contracts, injecting the viral DNA into the cell. (c) In this electron micrograph, several phage viruses are shown attached to a bacterial cell. The tip of the tail of one virus (indicated by the arrow) can be seen protruding through the cell wall. The thin fibrils extending into the bacterium are probably phage DNA.

125 *the kingdom monera*

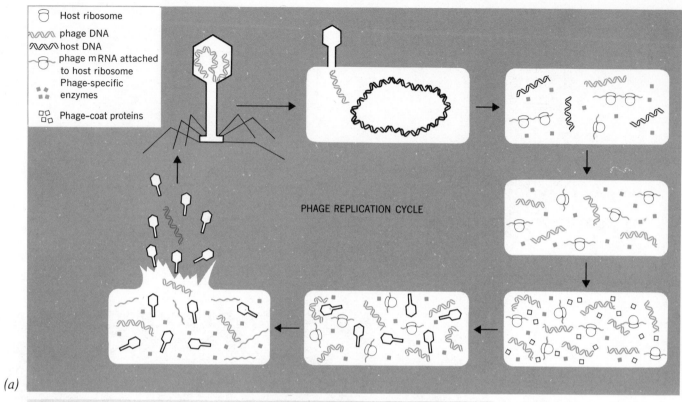

Host ribosome

phage DNA

host DNA

phage mRNA attached to host ribosome

Phage-specific enzymes

Phage–coat proteins

PHAGE REPLICATION CYCLE

(a)

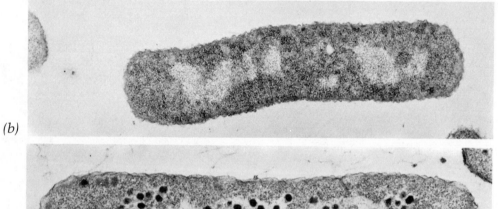

(b)

(c)

Figure 5-4 Replication of bacteriophage viruses. (a) The replication cycle. (b, c) Early and late stages of phage development in the bacterial cell. Magnification 32,500×. (d, facing page) New phage particles dissolve the bacterial wall and spill out to continue the cycle. Magnification 55,000×.

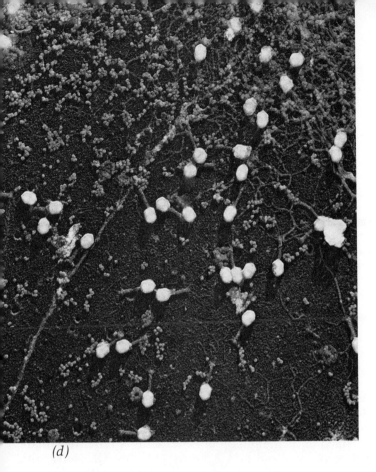

(d)

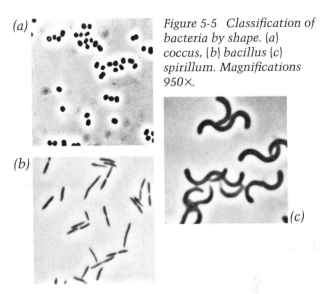

(a)

(b)

(c)

Figure 5-5 Classification of bacteria by shape. (a) coccus, (b) bacillus (c) spirillum. Magnifications 950×.

tion. Some bacteria are photosynthetic. Some obtain energy in the usual way—by oxidizing sugar (or other organic compounds) in the presence of oxygen. They are called **aerobes** because they use oxygen. Others, called **anaerobes,** obtain energy by oxidizing organic compounds in the absence of oxygen. The ability of a particular bacterium to oxidize glucose in the absence of oxygen may be the key to determining whether it is a harmless form or a disease agent.

Size, shape, staining properties, and nutrition are all important aids to description and identification of bacteria. But chemical staining and nutrition tests cannot be performed on a single cell. Like the virologist, the bacteriologist must be able to grow the organism in large quantities in order to perform his tests.

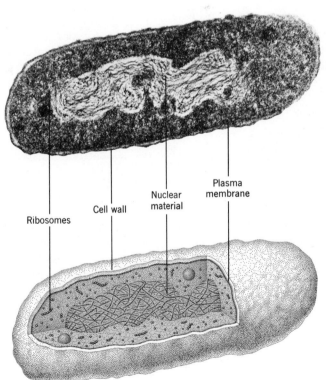

Ribosomes

Cell wall

Nuclear material

Plasma membrane

Figure 5-6 Structure of a bacterium, Escherichia coli, common in the human intestine. Electron micrograph, magnification 42,000×.

Cultures of Microorganisms

Bacteria and other microorganisms are grown in bottles or flasks containing liquid medium, in petri dishes containing nutrient medium solidified with agar, or in culture vessels to which liquid medium is added continuously.

If a sample of soil is simply added to a rich culture medium, a mixed culture of organisms will appear. Such cultures are interesting in their own right, since they undergo complicated changes with time as the original organisms exhaust some nutrients and produce others. But the conditions of growth are constantly changing, and the dominant microorganisms wax and wane in an unpredictable way. The aim of the microbiologist is to produce pure cultures of single types of organisms. Such cultures are termed **axenic** — literally, "with no strangers."

In practice, a variety of techniques can be used to isolate a single organism. A crude mixed culture may be inoculated into a medium that lacks critical nutrients for most of the organisms. Thus, only the few kinds of organisms that can synthesize the missing nutrients will grow. Repetitions of this process—removing samples from these actively growing cultures and transferring them to still other media—may weed out all but one species. The result is an axenic culture. Sometimes various antibiotics can be used to kill the more sensitive contaminating organisms while leaving the desired form unharmed. In still other cases it is possible to remove a single organism from a drop of water with a fine glass capillary tube. It can then be transferred to a sterile medium. There are many such techniques, but the choice of the proper one and its successful use require skill and patience.

Once an axenic culture has been created, it still must be carefully manipulated to prevent accidental contamination. The term **sterile technique** denotes the practical rules for handling laboratory cultures. Of course, the basic nutrient media and glassware must be sterile to begin with. They are sterilized by being heated to a temperature high enough to kill any unwanted microorganisms and their spores—that is, a temperature above the boiling point of water. The device used for sterilization is an **autoclave,** which is simply a glorified pressure cooker. (Unfortunately, microbiologists often must use media that decompose at high temperatures; in such cases the sterilization process becomes vastly more difficult.)

Airborne contaminants must be kept out of the sterile culture. Bacteria, algae, protozoa, and a variety of dormant spores are always present in the air and on every surface. All manipulations of the culture, therefore, have to be carried out in a closed space, free of air currents. Cultures are opened to the air for as short a time as possible. Glassware is sterilized in advance and kept in closed containers. The principles are simple enough, but the actual manipulations can become very complicated. The additional protective steps that are required for dealing with an airborne human pathogen (disease agent) or with an organism that cannot survive exposure to oxygen can be left to the imagination of the reader.

The fact that microorganisms must be cultured before they can be studied gives us a very biased view of microorganisms in nature. Some microorganisms grow very poorly in the laboratory, and many do not grow at all. No doubt, many microorganisms remain unknown simply because they cannot be cultured in the laboratory. Another difficulty is that a culture medium may be so different from the organism's natural environment that its growth and metabolism may be very abnormal. Medical microbiologists have been very successful at culturing many bacterial disease agents in solutions of blood and proteins. As a result, many microbes are routinely grown in such rich culture media, even though their normal habitat may be soil, or sediment, or lake water, rather than blood. Hence the behavior of the organism in laboratory culture probably bears little relation to that of the microbe in nature. For all these reasons, the study of bacteria, and of microbes in general, can be very difficult.

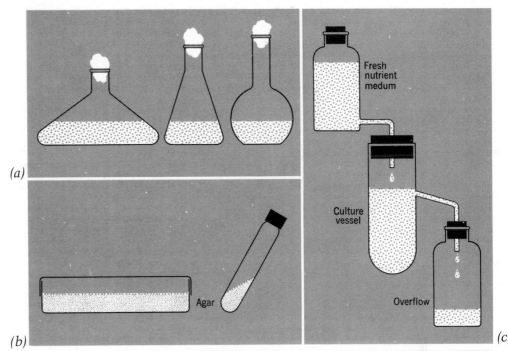

Culture of microorganisms. Microorganisms are grown in: (a) liquid medium in flasks, (b) medium solidified with agar and placed in petri dishes or test tubes, (c) continuous culture flasks to which fresh medium is constantly added as the organisms multiply.

Blue-Green Algae

The blue-green algae have a procaryotic cell structure which contains an extensive series of photosynthetic membranes (see Figure 5-7). Some are unicellular, while others form long filaments or colonies (see Figure 5-8). Blue-green algae get their color from phycocyanin, a blue photosynthetic pigment.

Blue-greens are truly ubiquitous. They occur in soil, water, deserts, and hot springs. They are often found in areas devoid of other forms of life, and were probably the first organisms to colonize the land in the course of evolution. Their ability to survive in bleak environments reflects their very simple nutritional requirements and their tolerance of extreme physical conditions. Carbon dioxide and a few simple minerals are all that they require for growth. Many blue-green algae can use molecular nitrogen (N_2) from the atmosphere as a nitrogen source. Since the air is 80 percent molecular ni-

trogen, these algae are never troubled by nitrogen scarcity. Although all blue-greens are capable of photosynthesis, some have been found growing deep in mud, where no light could possibly penetrate. It thus seems probable that these algae also have the ability to utilize chemical substrates as an energy source.

the kingdom protista

The kingdom Protista includes various unicellular algae and the protozoa. Like the monera, they are primarily unicellular. However, their structure is eucaryotic; they contain membrane-bounded organelles such as nuclei and mitochondria.

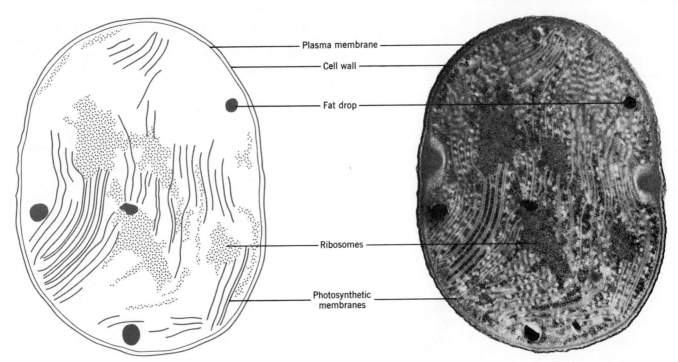

Figure 5-7 *Structure of a blue-green alga,* Gleocapsa. *Magnification 40,000×.*

The labels on the figure read, from top to bottom: Plasma membrane, Cell wall, Fat drop, Ribosomes, Photosynthetic membranes.

The protozoa were once thought of as single-celled animals. But their cells are much more complex than the average animal cell, so they are now placed in a separate kingdom altogether. In fact, protozoa are often described as **acellular** organisms (organisms "without cells"), to emphasize the difference between their structure and that of the cells of multicellular organisms.

A general description of the protozoa is difficult. They are variable in size, shape, and internal structure (see Figure 5-9). This diversity allows them to live in widely different habitats. For decades the protozoa were classified according to their mode of locomotion—amoeboid, flagellated, ciliated, immobile. However, the electron microscope is forcing a revision of this scheme. The structures of flagella and cilia seem to be so similar that all protozoa using these means of propulsion are probably closely related. On the other hand, the immotile protozoans are so various that they may have evolved from half a dozen separate ancestors.

Three groups of unicellular algae belong to the Protista; green flagellates such as *Euglena* (see Figure 5-1), brown flagellates, and golden algae. The sizes and shapes of their cells are diverse; the brown flagellated algae and the golden algae have very elaborate and sculptured cell walls (Figure 5-10). These micro-algae are widely distributed in moist and aquatic environments. They are particularly abundant in the floating microscopic community of lakes and seas, the **plankton.** Though they are abundant, they are usually inconspicuous. However, when conditions in oceans or lakes become favorable, their growth is stimulated, and huge "blooms" of algae appear. The great red tides that occasionally appear in coastal waters are caused by algal blooms.

the kingdom fungi

The fungi are non-photosynthetic organisms, usually found in the soil. Some, such as yeasts, are unicellular. Other fungi are typically composed of a network of branched filaments called **hyphae.** The cytoplasm in the hyphae is usually continuous,

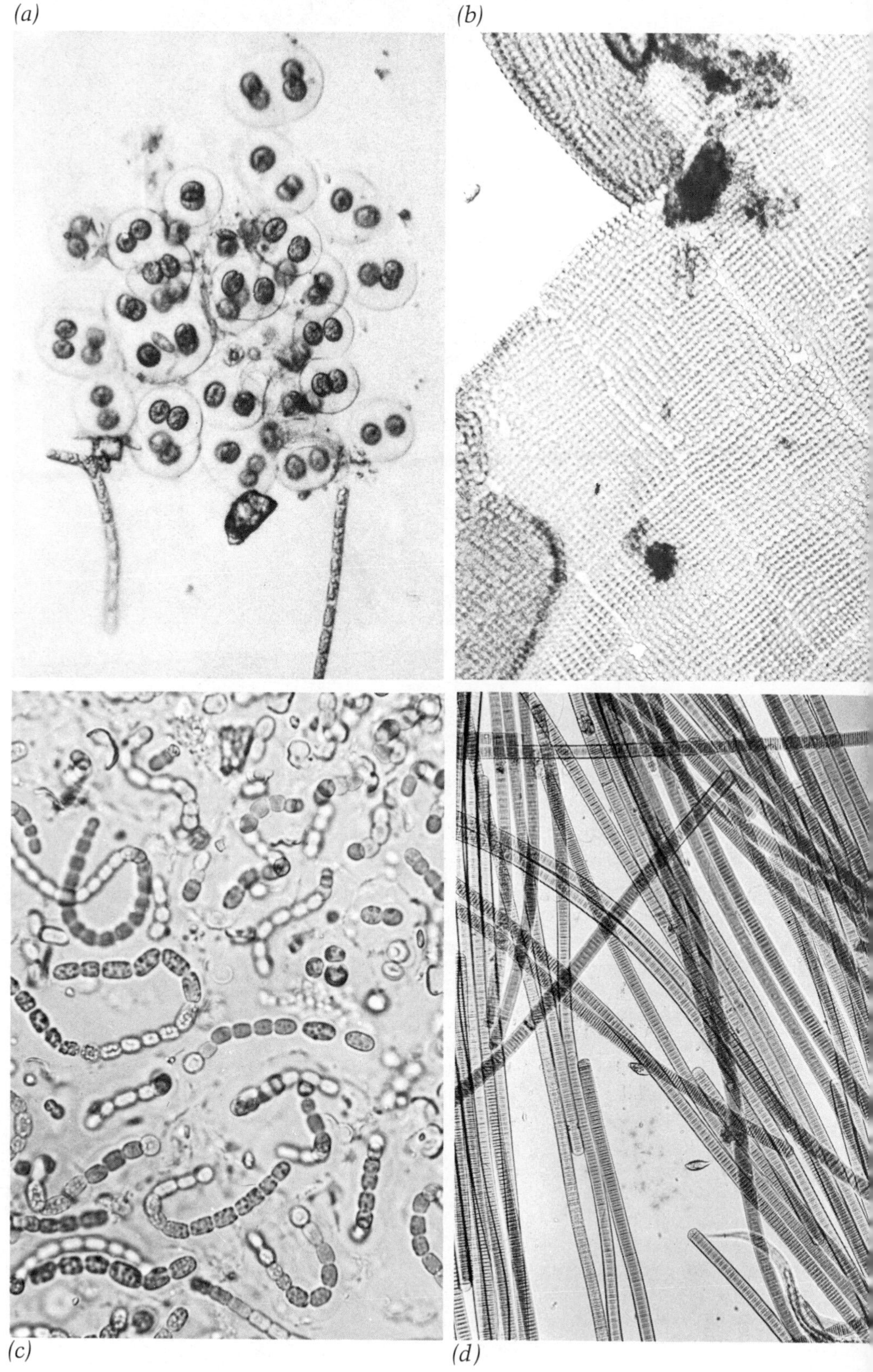

Figure 5-8 Various blue-green algae. (a) Gleocapsa. *Magnification 450×. (b)* Merismopedia. *Magnification 225×. (c)* Nostoc. *(d)* Oscillatoria. *Magnification 170×.*

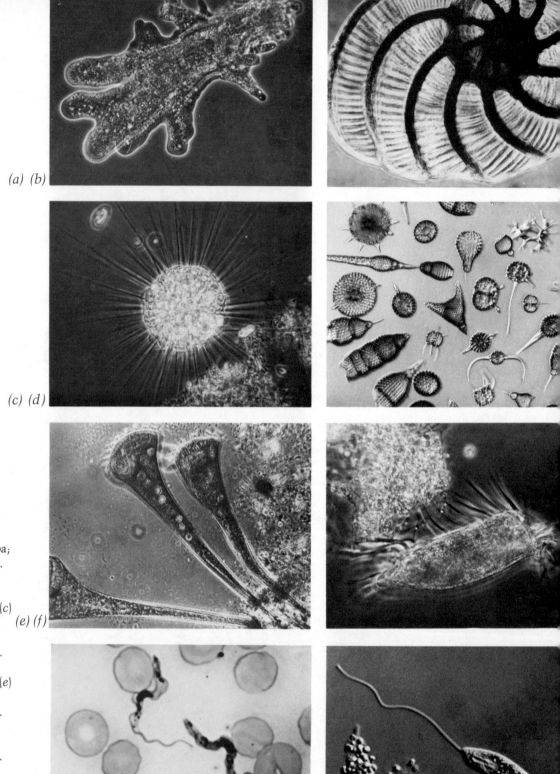

Figure 5-9 *Various protozoa.* (*a*) Amoeba; *magnification 130×.* (*b*) *A foraminiferan,* Peneroplis; *magnification 10×.* (*c*) *A heliozoan,* Acinospharerium; *magnification 130×.* (*d*) *Radiolarians; magnification 70×.* (*e*) *The ciliate* Stentor; *magnification 145×.* (*f*) *The ciliate* Euplotes; *magnification 250×.* (*g*) *The flagellate* Trypanosoma; *magnification 1,600×.* (*h*) *The flagellate* Perinema; *magnification 720×.*

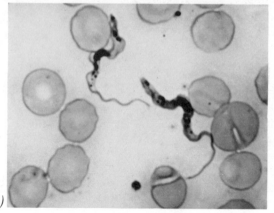

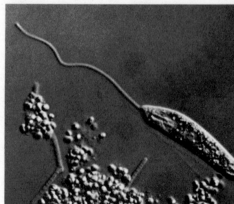

(a)

(b)

(c)

(d)

Figure 5-10 Scanning electron micrographs of the cell walls of various unicellular algae. (a) The dinoflagellate Ornithocercus; magnification 545×. (b) The dinoflagellate Ceratium; magnification 720×. (c) The diatom Stephanodiscus; magnification 4,650×. (d) The diatom Cymbella; magnification 1,250×.

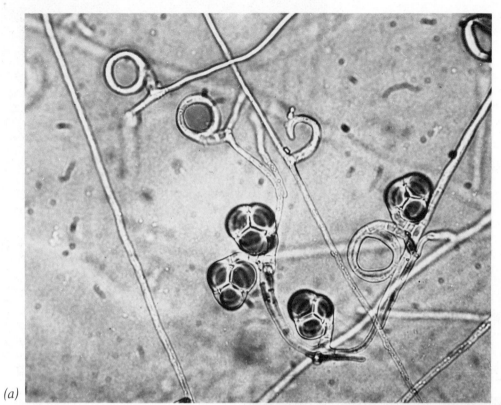

(a)

Figure 5-11 Noose fungus. (a) This soil fungus has loops made of three hyphal cells each. (b) When a nematode worm enters a loop, the cells swell, and the worm is trapped and digested. Magnification 500×.

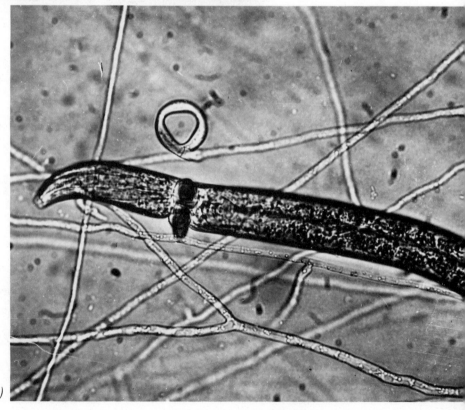

(b)

with few or no crosswalls, and it usually contains many nuclei. Growth occurs at the tips of the hyphal filaments, much as it does at the root tips of higher plants.

Many fungi live in water or in soil. They obtain energy by degrading the organic material around them. First, the hyphae synthesize and secrete digestive enzymes that split long-chain biological molecules such as starch or cellulose into smaller units. These smaller molecules are then transported into the fungal hyphae, where they are oxidized.

Some fungi obtain energy by preying on other microorganisms in soil. The noose fungus, for instance, captures small roundworms in special hyphal loops (see Figure 5-11). The tip of a hypha then penetrates the body of the worm and digests its contents. Specialized fungi can also capture and digest live protozoa. Predation by fungi may actually control the number of some protozoa and roundworms in the soil.

Food molds, mushrooms, and yeasts are probably the most familiar fungi. The mushroom is actually a reproductive structure produced by underground hyphae. The dark "gills" on the underside of the mushroom cap release millions of reproductive spores. Another fungus of special interest is *Penicillium*, which synthesizes and secretes the antibiotic penicillin (see Figure 5-12).

There are thousands of species of fungi, of which a small number cause disease in man. Fungal diseases cause fewer fatalities than bacterial pathogens, but they are nonetheless significant. Over half of all males in the United States at some time contract athlete's foot, on fungal disease of humans. Most plant diseases are of fungal origin.

the kingdom plantae

This kingdom includes red, brown, and green algae in addition to the larger and more familiar land plants. These algae are classified with the plants rather than the protists because most of them are multicellular, and some are large and relatively complex. The microscopic green algae may be either unicellular or grouped into small colonies, filaments, or sheets of cells. Sometimes they are so abundant that they discolor or form a scum on the waters of lakes and ponds. They also occur abundantly in damp soil. A few filamentous brown and red algae are microscopic, but most are much larger.

Figure 5-12 Types of fungi. (a) Yeast cells. (b) Penicillium. Cells form a network of hyphae; spores are generated in specialized regions. (c) A mushroom. The cap is a reproductive structure made of packed hyphae; spores are borne on the underside on "gills".

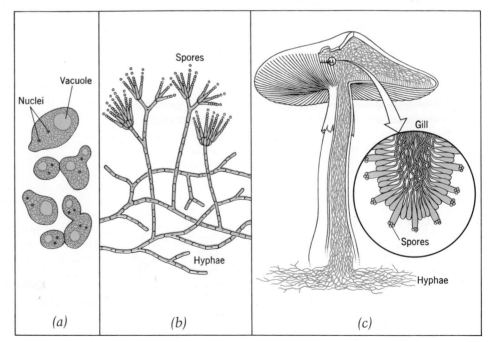

the kingdom animalia

Microscopic animals occur both on land and in aquatic environments. The most numerous small animals in terrestrial environments are the insects, mites and spiders, and roundworms that dwell in the soil. Animal microorganisms in marine habitats are much more diverse. Many microscopic animals are found in the plankton of the sea. Animal plankton feed upon the microscopic bacteria and algae that also dwell in the plankton community. Many plankton animals are merely immature forms of larger animals, and they are only transient members of the plankton. After a period of feeding they settle to the bottom and metamorphose into adult animals such as starfish, barnacles, clams, or worms. Other small animals are permanent microscopic members of the drifting planktonic community; microscopic relatives of shrimps and crabs are particularly prominent (see Figure 5-13). Microscopic animals are also abundant on the ocean bottom.

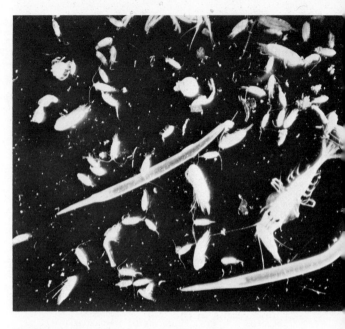

Figure 5-13 Microscopic animals of the plankton community. Crustaceans (shrimps and their relatives) and chaetognaths (arrow worms) are shown in this photograph. Magnification 6.5×.

Activities of Microorganisms

Microorganisms are numerous and diverse, and they interact with man in many ways. Perhaps the most spectacular interaction is the role of microorganisms in human disease.

disease

The term **disease** is applied to most structural or functional impairments in an organism. It includes various forms of mental and psychological dysfunction as well as metabolic impairments, but it arbitrarily excludes maiming or crippling accidents.

The Germ Theory of Disease

We now know that many diseases result from invasion by microorganisms, or "germs." But the relation between microorganisms and disease was still unknown by the mid nineteenth century, even though long experience had shown a correlation between some diseases and poor hygiene. The awareness that disease and hygiene were somehow

related manifested itself at various times in history and in various ways. The long-standing practice of isolating lepers and the traditional fear of plague victims both are evidence of some practical understanding of contagion. By 1850 the Hungarian physician Ignaz Semmelweis had demonstrated that cleanliness in maternity wards decreased the occurrence of childbed fever. But the role of microorganisms in disease was still unclear. Rules of hygiene seemed to work, but no one had a clear idea of why they did.

The chief problem was that the reproduction and growth of microorganisms were simply not understood. The continuity of one generation with the next was clear in the larger animals and plants. But microorganisms are small and difficult to observe. They seemed to appear mysteriously and spontaneously. If a broth was sterilized by boiling and then allowed to stand, it soon became a teeming culture of microorganisms. As a result, it was widely believed that microorganisms appeared through a process of **spontaneous generation.** In other words, their organization was considered to be so simple that they could be created out of inert matter.

The theory of spontaneous generation was finally

disproved, largely by Louis Pasteur (1822–1895). In a brilliant series of experiments, Pasteur proved that microorganisms arise only from other microorganisms. First, he showed that microorganisms would not appear in sterilized broth if the air reaching it was first filtered through guncotton. But organisms identical to those found in contaminated broth could be found in the cotton. And when Pasteur placed the cotton plug in the broth, a culture of microorganisms developed rapidly. Pasteur concluded that organisms in contaminated broth came from the air.

This experiment was convincing, but some skeptics felt that guncotton or other plugs kept air from reaching the broth, and that air was necessary for spontaneous generation to take place. Pasteur next showed that a broth could be kept completely open to the air and still remain sterile, but only if microbes were prevented from reaching it. First he drew out the necks of a series of flasks into long S-shapes. Then he poured in the broth and boiled it. Air was completely free to pass in and out, but any microorganisms in the air settled out along the tortuous walls of the necks before they could reach the broth. The broth remained sterile (see Figure 5-14). But when Pasteur later tipped the flasks, sloshing broth into the necks, microorganisms again appeared. In 1861, Pasteur published a monograph on spontaneous generation that effectively settled the issue.

Pasteur's demonstration set the stage for the germ theory of disease. Once it was clear that microbes were not the *product* of decaying medium but rather *invaders* of it, it became reasonable to ask whether disease was caused by microorganisms that invaded the body. Pasteur soon became involved in a series of studies on disease. He first worked with diseases of silkworms, because natural silk production was an important industry in France at the time. Later, he turned to investigations of the roles of microorganisms in chicken cholera and in anthrax, a disease of cattle and man. However, the definitive demonstration of the relation between microorganisms and disease was actually performed by Robert Koch, another bacteriologist. Koch developed techniques for growing disease agents on artificial culture media. He then stated that all of the following criteria must be satisfied in order to demonstrate that a particular microorganism is the cause of a disease:

1. The agent must always be present in every case of the disease.

Figure 5-14 Pasteur's experiments demonstrate that microorganisms come from other microorganisms, not from spontaneous generation. (a) Filtering air through gun cotton removes microorganisms and prevents the contamination of the sterile nutrient medium. The presence of microbes on the gun cotton can be demonstrated by placing the cotton in a sterile medium. Microbes quickly multiply. (b) A nutrient medium remains sterile if airborne microorganisms are prevented from reaching it by sealing the flask, or by using a flask with an S-shaped neck, so that microorganisms settle out before reaching the medium.

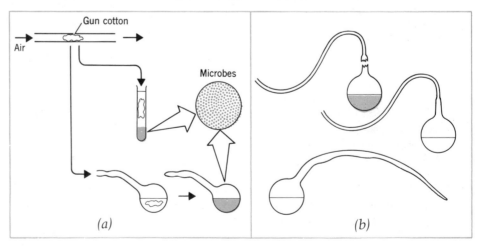

2. The microorganism should be isolated from a diseased individual and raised in pure culture.
3. Microorganisms from the pure culture should produce the disease when administered to a susceptible animal.

Koch successfully applied these criteria for the first time in 1876 when he proved that anthrax is caused by a specific rod-shaped bacterium. Koch's criteria are not always fulfilled, however. Even today, many diseases are attributed to specific organisms in spite of the fact that Koch's requirements have never been met. Many viral agents, for instance, have not yet been cultured. And disease organisms can sometimes be "carried" by individuals who never show clinical symptoms themselves. Nevertheless, Koch's view of disease was extremely persuasive and is still valid in many cases.

An infectious disease such as tuberculosis or cholera clearly results from an attack by an infectious agent. The diseased individual represents a complicated culture solution which supports the growth of the invading microorganism. In infectious diseases, the invader lives at the expense of the host. The host may be damaged in numerous ways. Vital tissues, such as those of the brain and liver, may serve as a direct food source for the disease agent. The agent may produce toxins that destroy the host's tissue or impair his metabolism. Damage can also occur from the way the diseased individual responds to the presence of the agent: by allergic responses, or by walling the microorganism off in cysts.

The same microorganism may attack different individuals in different ways. The damage caused by the invading microorganism may be so drastic that the host dies. On the other hand, the host may muster his immunological defenses and kill the invader. In diseases such as malaria, the result may often be a stand off in which the host and the agent coexist. The host continues to live, though his functions are impaired. Table 5-1 lists some common infectious diseases and their agents.

Koch and Pasteur were pioneers in a golden age of medical microbiology. Their discovery that disease agents could be isolated and studied had far-reaching implications. One area of particular importance was the prevention of disease. Prior to Pasteur, the relation between public health and sanitation was evident, but the extent of the relation was not clear. The realization that a specific disease bacterium will be present in the feces of diseased individuals gave great force and relevance to sanitary measures. The prevention of the contamination of food and water by feces was clearly essential. As a result, standards of personal cleanliness and the collection and disposal of sewage were identified as crucial measures for disease control.

A second consequence of the study of disease agents was the realization that a disease could be cured or controlled by killing or inactivating the agent. This realization led to a search for specific chemicals that would attack disease agents without causing undue damage to the patient. Compounds containing mercury and arsenic were effective against some agents, but they were dangerous to the patient as well. Eventually, effective antimicrobial drugs such as quinine and modern antibiotics were discovered.

Attempts to prevent and cure disease also led to the development of vaccines. A **vaccine** is a preparation of disease-causing microorganisms that have been inactivated or weakened in some way. The weakened form confers immunity on an individual by stimulating antibody production, but it does not produce the most serious symptoms of the disease. In many cases, vaccines are prepared from disease agents that have been killed. The antigens in the dead microbes are sufficiently similar to those carried by the living ones to stimulate the production of antibodies. The vaccinated individual is then protected against infection by the living disease agent.

"Live" vaccines are also produced. They are made by culturing a disease agent until it changes some of its characteristics. After several generations in culture a microorganism may produce only mild symptoms when injected into a susceptible animal. Yet its antigenic properties may still confer immunity against attack by the original agent. The original, **virulent** disease agent is said to be **attenuated** by the process of repeated culturing.

The term "vaccination" comes from the Latin word *vacca*, which means "cow." The British physician Edward Jenner coined the word to describe his work with smallpox and a related disease of cattle, cowpox. Cowpox can be transmitted from cows to humans, in whom it is a relatively minor disease. Jenner recognized that people who contracted cowpox did not get smallpox. They were immune. Jenner succeeded in producing the same immunity in a boy by introducing liquid taken from a cowpox sore into his bloodstream.

Jenner's first vaccination was performed in 1796. Eighty years later, Pasteur recognized that Jenner's procedure might work for other diseases as well. It was necessary merely to find a weak form of the disease microorganism, which would produce immunity but not kill the animal. Pasteur was able to produce an attenuated form of the chicken cholera bacillus by exposing it to air. Later, he attenuated the anthrax bacillus by heating it. In 1881, Pasteur performed a public experiment to prove that his attenuated anthrax bacillus would protect livestock from anthrax. He first innoculated 31 animals with his vaccine. Three weeks later, he innoculated the same animals with a highly virulent anthrax culture. He also injected anthrax into control animals that had not been vaccinated. Two days later, spectators gathered to watch. All the control animals soon died, but the vaccinated animals remained healthy. Pasteur's success was so spectacular that 85,000 animals were vaccinated within a year.

Pasteur went on to develop a vaccine that would cure rabies. The rabies virus has an incubation period of several weeks. Pasteur showed that if a person who has been bitten by a rabid animal is given daily injections of vaccine, he develops immunity before the virus can produce the disease. By 1905 the treatment had been used on more than 100,000 people, and it is still used today.

As a result of Pasteur's studies, culture techniques and vaccines for most of the well-known bacterial diseases have been available for half a century. However, vaccines against most viral agents are much more recent. Viruses can be cultured only in living cells, and techniques of cell and tissue culture are comparatively recent advances.

Resistance and Virulence

We have already seen that a disease organism may become attenuated in culture, providing a source of live vaccine. Disease agents can also change their virulence when they move from host to host. Influenza is the most familiar example. Some strains of flu virus may be comparatively harmless, while others are responsible for millions of deaths in major epidemics. These are instances in which changes in the virulence of the organism have caused corresponding changes in the disease.

Historical records suggest that some diseases have changed with the passage of time. Leprosy, a bacterial infection, was the scourge of western Europe during the Middle Ages—there were thousands of leper hospitals all over Europe. But by the fourteenth and fifteenth centuries they stood empty, and the disease practically disappeared from Europe. There are also examples of trivial diseases suddenly becoming significant causes of death. In nineteenth-century England, for instance, measles

was a comparatively mild disease. But by the middle of the century it was the most important infectious disease of children and accounted for about five percent of all the deaths in the population.

A change in the virulence of the disease agent is no doubt an important factor in such cases. The resistance of the population is also a critical factor. A disease introduced into an unexposed population often proves devastating. The population of Hawaii at the time of Captain Cook's first visit was probably about 300,000. Within 80 years, the population fell to 37,000, primarily as a result of measles, venereal disease, scarlet fever, and tuberculosis—all introduced by the Europeans. In a similar situation, smallpox ravaged the Indian population of North America and played a significant role in the Europeans' conquest of the continent. In at least one case, Indians were deliberately given blankets that had been used by smallpox victims—an early case of biological warfare. European diseases wreaked devastation upon the Aztecs and contributed to the success of Cortez and the Spanish. All of these populations were very susceptible to these diseases because they had never been exposed to them before. By contrast, when contagious diseases are **endemic**—constantly present—the resistance of a population increases with time.

The virulence of an infectious disease may also depend on the susceptibility of the individual to attack. Many disease agents are widely spread in human populations, yet the incidence of their symptoms may be quite low and sporadic. So why does a particular individual succumb to a disease at a particular time? Poor nutrition may increase the susceptibility of the affected individual and increase the seriousness of otherwise trivial infections. Constant emotional or physical stress may lead to hormonal changes that favor disease. Psychological factors play a large role in an individual's susceptibility to infectious disease. For these reasons, successful treatment may depend as much on the general condition of the host as it does on the identification of the disease agent. The diseased individual must be evaluated as a whole person and not just as a "case" of a particular disease—be it athlete's foot or tuberculosis.

Many biologists have suggested that serious symptoms actually indicate that the disease organism is poorly adapted to its host. It would be more profitable for the microorganism to maintain its host in good condition. In the ideal adaptation, the microorganism maintains a small population so free of side effects that the host is not harmed. Virulence and resistance come into balance.

Table 5-1

Major Human Diseases

Virus Diseases	Transmission Route	Control Measures
Smallpox	Exhalation droplets; direct contact; contaminated articles	Vaccination with living cowpox virus
Yellow fever	Man-to-man transmission by the mosquito *Aedes aegypti*	Destruction of mosquitoes; vaccination with attenuated live vaccine
Measles	Exhalation droplets	Short-term protection afforded by injection of gamma globulin from pooled serum of normal adult humans
Poliomyelitis	Portal of entry is believed to be the mouth. Virus reaches central nervous system via the bloodstream and destroys certain types of nerve cells, causing, paralysis	Vaccines, consisting of inactivated virulent virus (Salk vaccine) or an active, attenuated virus (Sabin vaccine)
Rabies	Transmitted by the bite of an infected animal, usually a dog	Quarantine and vaccination of dogs, destruction of stray dogs, vaccination of bitten persons
Mumps	Exhalation droplets, direct contact, contaminated articles	Vaccines available, but disease is rarely serious enough to warrant use
Influenza	Exhalation droplets	None
Common cold	Exhalation droplets	None
Infectious hepatitis	Fecal contamination of water or objects	Sanitation measures, surveillance of food handlers. Gamma globulin injections can give temporary protection

Bacterial Diseases	Organism	Transmission Route	Remarks
Typhoid fever	*Salmonella typhi (typhosa)*	Fecal contamination of water or food	Most severe of the Salmonella infections
Bacillary dysentery	*Shigella dysenteriae* *Shigella flexneri (paradysenteriae)* *Shigella sonnei,* and others	Fecal contamination of water or food	Most severe of the dysentery infections
Streptococcal infections	*Streptococcus pyrogenes*	Respiratory	Common throat inhabitant but invades many other tissues
Pneumonia	*Streptococcus pneumoniae*	Respiratory	Localizes in lungs
Gonorrhoea	*Neisseria gonorrhoeae*	Direct contact	Venereal disease. Newborn infants can be infected during birth

Table 5-1 (*Continued*)

Bacterial Diseases	Organism	Transmission Route	Remarks
Epidemic meningitis	*Neisseria meningitidis* (*intracellularis*)	Respiratory	Throat inhabitants that invade brain covering via blood stream
Bubonic plague	*Pasteurella pestis*	From rat to man by flea bite	Localizes in lymph glands, causing swelling (buboes)
Whooping cough	*Hemophilus pertussis*	Respiratory	Disease of very young children; vaccine provides effective control
Asiatic cholera	*Vibrio cholerae* (*comma*)	Fecal contamination of water or food	Infection of lower bowel; high mortality due to loss of body fluids
Tuberculosis	*Mycobacterium tuberculosis*	Respiratory	Most commonly a lung infection, but other tissues may be invaded
Syphilis	*Treponema pallidum*	Direct contact	Venereal disease

Disease and Culture

Disease and man have interacted throughout two million years of human cultural development. During this time the character of human diseases has changed along with the pattern of human existence.

During most of his existence as a species, man lived by hunting and gathering. The world population was very small (a few million individuals) and the prevalent social unit was the small, nomadic group. Waste disposal was not a problem, so there were no pollution-related diseases such as cholera. Furthermore, the groups were too small to be reservoirs of contagious community diseases such as measles and smallpox. In a small population, all the susceptible individuals are likely to contract the disease in a short period of time. But as soon as they recover—or die—the disease agent disappears because it has run out of hosts. The same agent will persist in a large population because there are more susceptible individuals to begin with. It takes longer for all of them to become exposed. Furthermore, individuals are always being added to the population, so the disease agent has a continuing supply of new hosts.

Parasites were probably the most troublesome disease agents in small human groups: lice, pinworms, and similar pests. Man would also contract diseases from other animals through being bitten by insects, handling game, or eating contaminated meat. Tularemia, a bacterial disease of rabbits and squirrels, may have been transmitted to man this way. Tularemia is now rare simply because we now have relatively little contact with wild animals.

With the development of agriculture, beginning about 10,000 years ago, man increased in number and grouped into larger, more sedentary units. Rising concentrations of people in restricted areas created new disease problems. The accidental contamination of food supplies with human waste became more frequent and was a path for the transmission of various new diseases such as cholera and dysentery. The beginnings of agricultural and herding activities created further opportunities for disease agents. The domestication of animals exposed man to anthrax and tuberculosis, diseases of sheep and cattle. The cultivation of land caused the spread of malaria by driving wild apes back into the forest. The apes were the natural hosts for the mosquitoes that carry malaria. As the apes disappeared, the mosquitoes transferred their attention to the invading men. Malaria also spread because cultivation required ponds and water-filled ditches, which served as breeding sites for the mosquitoes. Since the agricultural population was sedentary and concentrated, malaria became an endemic disease.

The rise of crowded preindustrial and industrial cities increased problems of waste disposal and food supply. Waste is usually transported away from towns in the water supply. This practice provides ideal conditions for the spread of pollution diseases. Furthermore, the crowding, squalor, and appalling conditions of hygiene in city slums promoted the spread of such diseases as typhus and plague. Populations were now so large that community diseases such as measles and smallpox were constantly present. Increasing personal contact favored the transmission of infectious diseases.

Despite modern public health technology, these problems are still with us. The integrity of our water supply and public hygiene must be monitored continually. The outbreaks of pollution diseases and contagious diseases that occur during wartime conditions are a reminder that disease is held in check by easily-disrupted social arrangements. Modern rapid transportation also creates new problems. Disease carriers can move halfway around the world in a day, spreading infection with an efficiency that was impossible just a few years ago.

Habit and tradition in human society can also foster or prevent disease. Malaria again provides an example, this time in Vietnam. Malarial mosquitoes rarely occur in the lowland deltas but do occur in the hills. The natives who farm the terraced hills build their houses on stilts and keep their livestock under the house. The malarial mosquitoes come out at dawn and dusk, but they stay close to the ground. Human inhabitants are usually up in their houses at this time, so the population is not afflicted with malaria. The Vietnam war, however, displaced people from the delta to the hills. The delta farmers traditionally built their houses on the ground, and they continued to do so in the hills. They promptly came down with malaria. The displaced delta people came to believe that there are evil forces in the hills warning them not to settle there.

Venereal Disease

Syphilis and gonorrhea are the most serious and widespread venereal diseases in the United States. Both are currently reaching epidemic stature. Control of these diseases is essential because they are extremely serious—even fatal—if untreated. But detection and control are difficult for two reasons. First, their initial symptoms may be mild and easily ignored. Second, mores and customs concerning sexual activity often discourage the individual from seeking treatment. Effective control of these diseases requires a clear public understanding of their causes, means of transmission, symptoms, and course.

The **venereal diseases** are so called because they are often transmitted through sexual contact. (The word "venereal" is derived from Venus, the goddess of love.) Syphilis can be transmitted by normal intercourse, by extravaginal intercourse (homosexual or heterosexual), or by kissing. Toilet seats and doorknobs are extremely unlikely vehicles of transmission, however, since the infecting organism is sensitive to temperature change and drying. It can survive outside the body for only a few moments. In advanced and hygienic societies syphilis is almost always transmitted by contact between moist mucous surfaces. Gonorrhea, too, is usually spread by sexual activity, but it can also be passed to a child at birth. As the infant moves through the birth canal, its eyes may be infected by the bacteria present in the mother's vagina. The resulting infection in the cornea of the eye once was a common cause of blindness. This phenomenon is now rare because of the almost universal practice of putting a few drops of dilute silver nitrate or an antibiotic in the eyes of newborn babies.

Syphilis

Syphilis is caused by a spirochaete, a corkscrew-shaped bacterium, called *Treponema* (see Figure 5-15). The same bacterium is also responsible for two other diseases, **yaws** and **pinta**. Both are skin diseases that occur in humid tropical areas. The three diseases—syphilis, yaws, and pinta—can be distinguished only by their symptoms. All are caused by identical forms of *Treponema*. Changes in climate change the symptoms of these diseases so that they mimic one another. It is the environment, not the microorganism, that determines which symptoms will appear.

Since spirochaetes are often found in decaying organic matter, man probably first acquired *Treponema* by handling or eating decaying flesh. As long as humans lived in small groups, *Treponema* was probably responsible for a contagious skin disease among children, spread by intimate contact during play. Low levels of personal hygiene and humid tropical climates would facilitate its transmission. Under these conditions, the disease would appear as yaws. As man migrated into temperate regions, the disease would still be a contagious

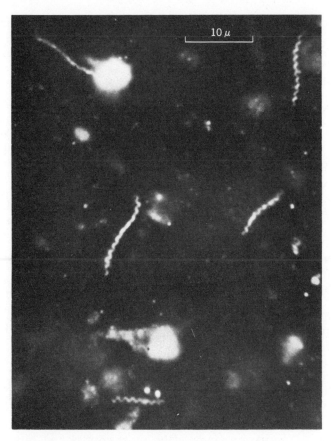

Figure 5-15 Treponema, *the spirochaete that causes syphilis. Magnification 1,800×.*

Treponema now depended on social customs. Prostitution and the leisure and inclination for extramarital sexual activity are much more characteristic of a town and city culture than of a village culture. The result is that syphilis became a venereal disease simply because sexual activity was now the primary or exclusive form of intimate physical contact. It became a disease of adults and pre-adults because the level of hygiene was sufficient to prevent its spread among children.

This account suggests that venereal syphilis arises from a reservoir of contagious diseases (yaws, pinta, and endemic syphilis) whenever social conditions are proper and a source of the *Treponema* is introduced. The usual historical account of the origin of syphilis in Europe is somewhat different, however. It credits Columbus and his crew with introducing the disease to Europe upon his return from America. It is true that syphilis raged through Europe beginning in the late fifteenth century. However, there is evidence that syphilis was present in Europe before Columbus. Syphilitic bone damage has been found in human skeletons from both the Old and New Worlds, dating from long before Columbus. Some combination of the two theories, therefore, is most likely. It may well be that endemic syphilis had been present all along, but that new and more virulent strains may have been introduced during the age of exploration.

Syphilis usually goes through three stages. First, the spirochaete multiplies, producing a lesion at the site of entry after a few weeks. The lesion is usually painless and can easily escape attention, particularly in women. This initial lesion disappears in a few weeks. The second stage occurs when the spirochaetes are carried in the blood to new locations, where they produce secondary sores or lesions. These sores are usually found on the skin and mucus surfaces. But in three fourths of the infected individuals, no secondary stage appears. In either case, the disease then enters a latent period which can last for a few months or for decades.

The third, or tertiary, stage produces a variety of symptoms. Any tissue may be involved. The spirochaetes may attack the nervous system, producing blindness, gross interference with intentional movements, or insanity. Syphilitic infection of major arteries is another serious tertiary symptom of the disease. It produces about ten percent of all circulatory disease. Another late manifestation is the appearance of tumor-like growths, **gummata,** which destroy surrounding tissue. They may affect skin, mucus surfaces, or bone. About one fourth of the individuals contracting syphilis develop lethal or

children's disease, but its symptoms would change to resemble those of syphilis because of the difference in climate. The skin lesions that are characteristic of yaws would retreat to the crotch or other moist regions where the organisms would not dry out. The disease would now be called **endemic syphilis,** because it is still spread by contact between children, not just by sexual activity among adults.

Further developments awaited the evolution of the village and improvements in hygiene. Even very primitive hygiene—occasional bathing and the use of soap—could wipe out endemic syphilis. As village society evolved and hygiene improved, endemic syphilis would virtually disappear. But reservoirs would remain in backward villages and undeveloped regions. This is still the case: some experts argue that endemic syphilis afflicts more people in the modern world than venereal syphilis.

As villages were transformed into cities and towns, hygiene improved still more. The fate of

incapacitating tertiary symptoms if they receive no medical treatment. Because the symptoms are so variable, diagnosis is based on finding *Treponema* in the blood.

Syphilis is one of the few diseases that can be transmitted from the mother to the developing embryo. Congenital syphilis is the result. For this reason, most states require blood tests prior to marriage, to detect syphilitic women before they become mothers.

Many pivotal historical figures including such rulers as Ivan the Terrible, Henry VIII, and Bloody Mary (Mary I of England), may have been syphilitic. The list of probable sufferers also includes the philosopher Friedrich Nietzsche and the sculptor Benvenuto Cellini. Unfortunately, tertiary syphilis is not just a historical curiosity. In 1950 more than twenty million cases of syphilis were reported in the world; the actual number then, and the present number, must greatly exceed this figure.

In the United States, syphilis is a major killer. In 1969 it was responsible for the deaths of about 100,000 people. This number represents about five percent of all the deaths in the population. But most of these deaths were officially attributed to heart disease rather than to syphilis. In fact it would actually be quite inhumane to inform patients that they are suffering from tertiary syphilis, because this stage of the disease usually shows up in old age, and it cannot be cured. Widespread knowledge of the early symptoms would greatly reduce the human suffering caused by syphilis.

Gonorrhea
Gonorrhea is caused by a bacterium, *Neisseria gonorrhoeae*. The symptoms of the disease are different in men and women. In the male, the disease begins with a painful swelling of the urethra a few days after exposure. Urination is painful. The onset of the disease is accompanied by a discharge of pus from the tip of the penis. These symptoms may subside without treatment after two or three weeks. If the infection reaches the prostate gland or the testicles, it is extremely painful, and the individual will probably seek medical help. If such an infection is not treated, sterility may result.

In women, the early symptoms of gonorrhea may be so mild that the disease goes unnoticed. A few days after infection there may be a mild vaginal burning, with or without discharge. The infection then passes up the uterus and reaches the oviducts or the ovaries. Discharge of pus into the body cavity might produce violent illness, but it might also be mild enough to be passed off as an upset stom-

ach. After these initial symptoms, a chronic infection continues. Many women are unsuspecting carriers in this stage. The most common serious consequence is sterility.

V.D. Control
In the United States the taboo against discussing venereal diseases was finally broken in the 1930's, with the publication of a book by Dr. Thomas Parran, *Shadow on the Land*. Dr. Parran served as Surgeon General of the United States from 1936 to 1948. During World War II, there was a major campaign to eradicate venereal disease in the United States. An intensive program of education and treatment was begun in the armed forces, and there was widespread public support for treatment with new antibiotics. The pearly teeth of movie stars beamed from billboards that urged one and all to see the doctor and stamp out V.D. As a result, venereal disease declined from two million cases in 1941 to a minimum of about 400,000 in 1956. But in the last few years syphilis and gonorrhea have climbed back to the two-million mark and continue to rise. For example, the California Public Health Department estimates that one of every ten people in California under the age of 25 had venereal disease in 1970.

In the United States, venereal disease control units are operated by most county and municipal public health departments. Yet the diseases continue to spread, and they have reached epidemic proportions. Ignorance of the symptoms and a lack of publicity about agencies for control and treatment are the major factors. Communities that have sex education programs in the schools have a markedly lower incidence of V.D. than those that do not. The experience of communities in southern California is illustrative. In 1963 a program of sex education was introduced in the high schools of the Los Angeles area. In 1964–1965, while the V.D. rate was rising steadily in the United States as a whole, it dropped by 58 percent among the teenagers in Los Angeles County. However, two communities in the county refused to participate in the program. Their teenage V.D. rates rose 500 percent and 700 percent respectively. Table 5-2 summarizes the incubation periods, symptoms, and treatments of the common venereal diseases.

food

A second major interaction between man and microorganisms involves food—its production and preservation. Man has always unwillingly shared his food with a host of other organisms. As civilized man increased his food production through agriculture, predators and pests continued to take their tithe. Loss of livestock to large predators such as wolves and cougars is spectacular, but most food losses are caused by microorganisms. About a tenth of the total food production in the United States is taken by microorganisms, despite efforts at prevention.

Crop Diseases

Agriculture creates conditions that are often ideal for the uncontrolled growth of microorganisms on crops. When man converts natural grassland or woodland into a tilled field, he destroys a varied collection of organisms. A natural area is populated by many kinds of plants and animals, including plant disease agents such as worms, insect larvae, bacteria, fungi, and viruses. Most of these destructive agents can attack only a single type of plant. When one type of plant is hit hard by a disease agent, it may be destroyed. But the disease agent then declines because it runs out of food. Other kinds of plants in the plant community will replace the affected individuals. The natural community is diverse and therefore relatively stable.

A farm field, on the other hand, is usually planted with a single crop. Other plants are treated as weeds and eliminated. A single-plant crop, therefore, represents a concentrated food resource that can be devastated by a single kind of microbe. For this reason, plant diseases can be disastrous in regions where the human population is dependent on a single crop. A classic example is the Irish potato famine. By 1840, the potato had become the staple food of the majority of the Irish people. But at about that time a fungus disease known as the potato blight was introduced into Europe. It was caused by a fungus, *Phytophthora infestans,* which attacks the leafy portion of the plant and the potato tuber itself. In 1846 the blight spread throughout Ireland. At that time the Irish population was eight and a half million; but in just six years it was cut to six and a half million, through starvation, disease, and emigration.

Crop losses through disease are a continuing problem. Smuts and rusts (fungus diseases) parasitize our cereal grains and potatoes. Insect pests, such as the citrus scale and the apple coddling moth, destroy fruits. The damage is varied and extensive.

Spoilage

Once the grain is harvested (or the animals slaughtered) further depredations occur. Insect and fungus pests take a tremendous toll of stored grain. In some undeveloped countries, spoilage after harvest decreases food supplies by 40 percent. Attacks on meat by microorganisms are also costly. In modern countries, refrigeration, cooking, and canning, together with inspection and control procedures, protect the consumer from rotten meat. Even so, losses are not eliminated. In undeveloped countries, losses are much greater because meat preservation is quite expensive.

One inexpensive way to avoid food losses through spoilage is to consume rotten food. This is merely a matter of taste and custom. Eskimos are fond of *titmuch,* a food consisting of fish that has been buried and allowed to rot until it is semiliquid. The unpleasant flavor of spoiled meat can be masked by the pungent tastes and strong odors of spices and condiments. For this reason, spices were in extremely great demand in Europe before the Industrial Revolution, and they continue to be heavily used in tropical countries.

Food Poisoning

Food poisoning, like food spoilage, is the result of the action of bacteria or other microorganisms on man's food. However, a distinction must be carefully drawn between the two conditions. Spoilage changes the odor, taste, or consistency of the food, making it unpalatable but not poisonous. **Food poisoning,** on the other hand, renders food harmful and even lethal, but it does not usually alter its taste or appearance.

The most dangerous food poisoning is caused by the bacterium *Clostridium botulinum.* This organism synthesizes a powerful toxin and then releases it to the environment. *C. botulinum* is common in soil and in the ocean. When it gets into a favorable environment, such as food, it grows rapidly and liberates the toxin. Even very small amounts are dangerous—especially so because they are odorless and tasteless, and hence undetectable in food. Fortunately, contaminated food often contains other bacteria that produce foul odors; the odor acts as a danger signal.

Poisoning by the toxin is called **botulism.** It most often results from eating undercooked canned food, or food that has been prepared and set aside

Table 5-2

Major Venereal Diseases

Disease	Pathogen	Incubation Period	Clinical Characteristics
Syphilis	*Treponema pallidum*	7–90 days, usually 3 weeks	Primary: chancre present Secondary: rashes or mucous patches Latent: no symptoms, but positive tests Relapse: recurrence of infectious lesions after disappearance of secondary lesions Tertiary: cardiovascular and central nervous system damage, gummata
Gonorrhea	*Neisseria gonorrhoeae*	2–14 days, usually 3 days	Discharge Burning pain Swelling of genitalia and glands
Chancroid	*Hemophilus ducreyi*	2–6 days	Pustular ulcer on genitalia or inguinal region Suppurative inguinal lesions
Lymphogranuloma venereum	LGV virus	5–30 days	Vesicle on genitalia Enlarged inguinal lymph glands Fistula with pus
Granuloma inguinale	*Donovania granulomatis*	Probably 8–90 days	Eroded papule on genitalia Inguinal or perineal ulcers

without refrigeration. A few hours at room temperature are enough for the organism to grow and produce the toxin. Vomiting and diarrhea begin within 24 hours. The toxin soon affects the oculomotor nerve (the nerve that controls the muscles positioning the eye), causing double vision. About 70 percent of all cases of botulism are fatal, with death usually resulting from damage to the nerves that control breathing. Though an antitoxin exists, it must be administered quickly. The toxin is destroyed by heating, but the bacterium itself may survive as a spore. Suspect food should be boiled for fifteen to twenty minutes.

The bacterium *Staphylococcus aureus* is responsible for a second kind of food poisoning. *S. aureus* produces a toxin that poisons the intestine. It causes vomiting, cramps, and diarrhea. These symptoms are very unpleasant but almost never fatal. *S. aureus* normally is found on the skin, so it often gets into food during handling and preparation. *S. aureus*, like *C. botulinum*, grows well at room temperature and in a few hours can contaminate food that is left out. Its toxin cannot be destroyed by boiling, however.

A third kind of bacterium, *Salmonella*, is sometimes mistakenly associated with food poisoning.

the intestine. Other food-borne diseases are cholera, dysentery, tapeworm, and trichinosis.

Diagnostic Procedures	Treatment

Microscopic examination
Serologic test
Spinal fluid test
Clinical and contact
  histories
	Penicillin
	Broad-spectrum
	  antibiotics
	Tetracyclines

Culture
Smear
Clinical and contact
  histories
	Penicillin
	Sulfonamides
	Broad-spectrum anti-
	  biotics, including
	  tetracyclines

Microscopic examination
  for ducrey bacillus
Clinical and contact
  histories
	Sulfonamides
	Broad-spectrum anti-
	  biotics, including
	  streptomycin

Frei Skin Test
Clinical and contact
  histories
	Broad-spectrum anti-
	  biotics
	Sulfonamides

Tissue scraping or biopsy
Clinical and contact
  histories
	Broad-spectrum
	  antibiotics

Salmonella is a disease agent that grows in the intestine of man and other animals. Some kinds of *Salmonella* may cause local infections of the intestinal tract. Other species of *Salmonella* may move from the intestine into other parts of the body and cause a general infection and fever called typhoid or paratyphoid. Since *Salmonella* grows in the intestine, it is present in fecal matter. The *Salmonella* diseases are usually spread in food or water that has been contaminated by feces. *Salmonella* does not cause food poisoning, because it does not act by secreting a poison. Rather it is one of a number of food-borne diseases that invade the body through

Food Preservation

There are several ways to make foodstuffs unavailable to microorganisms and yet still fit for human consumption. Food can be preserved by killing any microorganisms that may be present and then preventing new ones from getting at it: this is the principle behind canning. Canned goods are cooked at temperatures and pressures high enough to kill the spores of dangerous microorganisms such as *Clostridium botulinum*. After heating, the cans are then sealed against further contamination. *C. botulinum* spores are unusually hardy: they can survive boiling for as much as five hours. Spores of harmless bacteria that survive at temperatures higher than the *C. botulinum* spores may therefore be included in test batches. If the cooking kills the harmless forms, it is certain that the botulism organism has been eliminated. Since the entire content of the can has been treated in this way, it remains unspoiled as long as the can is intact. But if a can is dented or punctured or discolored in any way it should always be discarded—there is a good chance that microorganisms are present.

It is not always necessary to kill all the organisms in food, just the dangerous ones. The **pasteurization** process, for example, consists of heating a liquid to 145–150° F for 30 minutes, or to 160° for one minute, and then cooling it rapidly. Pasteurization is used to control the quality of milk, because the relatively low temperatures involved do not disrupt the protein structure of the milk. "Raw" and pasteurized milk are nutritionally identical, but the heating is sufficient to kill all active bacteria that are dangerous to man. However, since the temperatures are not high enough to kill everything, pasteurized milk must still be refrigerated so that the bacterial spores and other microorganisms will not begin to grow again.

Refrigeration will not kill microorganisms, but it does slow their growth. If foods are relatively free of microorganisms to start with, refrigeration will keep them safe for a while. But if foods have already become contaminated, refrigeration will not make them any better.

A second basic approach to food preservation is to reduce the water content of the food. Microorganisms can grow only if water is available. The easiest way to deprive them of it is simply to dry the food. Parching corn, sun-drying wheat, and smoking meat are familiar procedures. The sun-drying of grain is not always successful, however. The grain must be dried uniformly to avoid wet spots: a wet spot can allow a fungus to get started, and once the fungus begins to grow it can generate its own water. The fungus oxidizes carbon compounds with the use of oxygen, yielding carbon dioxide and also water. This so-called **water of oxidation** is sufficient to allow the spread of a mold once it starts. Water of oxidation is used even more effectively by insect pests. Insects are beautifully adapted to dry conditions. Many of them can get along on water of oxidation alone and require no liquid water. Drying is therefore no protection at all against grain beetles.

Freezing is another effective way to remove usable water from food. Freezing thus differs in principle from refrigeration, which merely retards spoilage by slowing down the metabolism of microorganisms. Another way to reduce a food's water content is to raise its osmotic concentration. This is what happens when fruit is made into jelly of jam. The sugar concentration of the jelly is very high, and its water content is thus correspondingly lower than the concentration of water in microbial cells. Water will be osmotically removed from any microorganisms that may land on the jelly. Mold can survive only on the surface of a jelly. It does not contaminate the bulk of the jar because it cannot survive submerged in the sugar. Bees use the same principle in making honey. They collect nectar from flowers and spread it out in compartments in the hive. Then they fan the compartments with their wings, promoting the evaporation of water and increasing the sugar concentration. The process of pickling vegetables or meat in strong salt solutions is another way of protecting foods by raising their osmotic concentration.

Fermentation is yet another method of food preservation, one that uses microorganisms as allies. **Fermentation** is defined as the oxidation of sugar without the use of oxygen. In fermentation, the sugar molecules are converted to alcohol and other organic compounds. Just as aerobic microorganisms obtain energy from the oxidation of sugar to carbon dioxide and water in the presence of oxygen, anaerobic organisms obtain energy by means of fermentation. Some organisms can carry on both processes.

The production of cheese, wine, beer, and a

Brewing

Beer is usually made from barley, but similar beverages have been prepared from all the other cereal grains, and from honey, artichokes, turnips, and other foods.

Modern beer-making consists of several steps. Barley is first converted into **malt** by allowing the barley grains to germinate. The germinating seed produces enzymes which begin to break down the starches and proteins stored in the seed. These enzyme-rich grains represent the malt. Next, the barley malt is mixed with water and extra starch to form a **mash.** The mash is then cooked at progressively higher temperatures, each one being the optimum temperature for the activity of one particular enzyme. The enzymes break down the barley proteins into amino acids, and the starch into various sugars, including maltose. The activity of each specific enzyme, and thus the quality and composition of the product, can be determined by careful regulation of the cooking temperatures.

The mash is finally heated to a temperature that inactivates all the barley enzymes. At this time, hops or other flavoring herbs are added. The mash is then cooled and filtered, and brewer's yeast is added. The growing yeast uses the amino acids as a nitrogen source and converts the maltose to alcohol and carbon dioxide. The latter substance provides the carbonation in the final product. The yeast is then filtered out. Of course, the steps of this involved process were first worked out by trial and error, long before the roles of enzymes or yeasts were understood.

The conversion of grain into beer is actually a very efficient way to preserve the energy value of the grain. Two thirds of the original caloric content of the grain winds up in the beer.

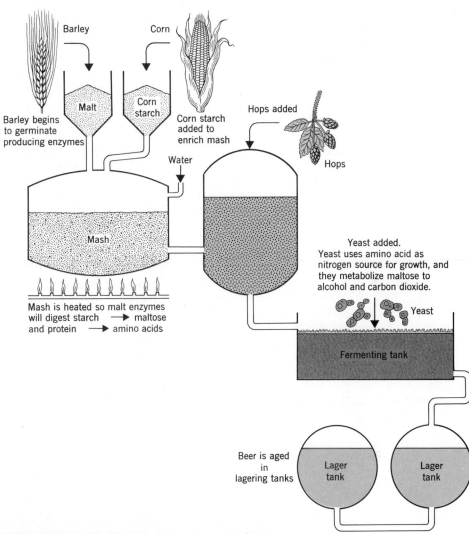

Figure A Steps in beer production.

Barley

Corn

Malt

Corn starch

Barley begins to germinate producing enzymes

Corn starch added to enrich mash

Hops added

Hops

Water

Mash

Mash is heated so malt enzymes will digest starch ⟶ maltose and protein ⟶ amino acids

Yeast added.
Yeast uses amino acid as nitrogen source for growth, and they metabolize maltose to alcohol and carbon dioxide.

Yeast

Fermenting tank

Beer is aged in lagering tanks

Lager tank

Lager tank

Figure B Open fermenting tanks for brewing.

number of other products depends on fermentation by certain kinds of yeasts, molds, and bacteria. Such fermented products actually keep better than the original food source. In wine and cheese-making, microbes produce alcohol and other products during their growth. These products increase in concentration until eventually they halt the growth of the microbes that synthesize them. At these levels they also prevent the growth of other kinds of microorganisms. The products of microbial fermentation also add variety and zest to food that would otherwise be bland and uniform.

The use of fermenting microorganisms in food production is a traditional craft which the biochemist is just beginning to understand. Beer-making, for instance, occurs in all civilizations that have grain-producing agricultures. Records of beer-making date back to 6000 B.C. In Babylonia, sixteen different kinds of beer were known by 4000 B.C. Originally, beer was made by crumbling underbaked bread into water. The viable yeast cells in the bread crumbs fermented the carbohydrates in the bread to produce beer. In ancient Egypt this mixture was served to workmen, crumbs and all. (It was filtered for the upper classes.) Such beer was a significant source of food: the Egyptian laborers were paid four loaves of bread and two jugs of beer a day.

sewage disposal

Microorganisms are scavengers and decomposers. They can oxidize almost any organic material for energy and return it to the environment in the form of carbon dioxide, water, and simple inorganic compounds. The process of sewage disposal is largely a matter of making use of these scavenging abilities.

Any organic compound that can be oxidized by microorganisms is said to be **biodegradable.** The amount of biodegradable material in sewage, before and after treatment, is estimated by measuring its **biochemical oxygen demand** (BOD). This figure represents the amount of oxygen consumed when the organic matter in solution is completely oxidized by bacteria and other microorganisms. Since the quantity of oxygen required depends on the concentration of the organic material and on its chemical structure, the BOD is a good measure of the amount of organic matter present. In practice, the BOD is found by taking a sample of water and measuring the oxygen content, letting the water sit in a tightly stoppered jar for several days, and then measuring the oxygen content again. The difference between the initial and final oxygen contents of the water is the BOD. Such determinations are used routinely by public health workers to estimate the organic content of polluted waters.

Because of its high BOD, raw sewage that is dumped into a stream or lake can be very harmful to organisms that need high concentrations of oxygen. Upstream of a sewage outfall, the oxygen content of the water is near maximum and will support various fishes as well as smaller organisms such as snails or worms. But at the outfall, the BOD suddenly increases. Bacteria and molds begin to oxidize the organic material in the sewage, consuming oxygen. The oxygen content drops. Downstream, the heaviest particles of sewage settle out and form a sludge. Under such circumstances, a stream may become entirely anaerobic. When the oxygen is completely exhausted, the microorganisms cannot complete the oxidation of the organic material to carbon dioxide and water. Instead, the material is partially oxidized by means of fermentation, so that various intermediate products are formed: organic acids, alcohols, methane, and hydrogen sulfide. Few animals can survive under these conditions, but those that do—some worms and sewage mosquitoes—flourish in the rich and nutritious sludge. Some of these animals can tolerate waters low in oxygen. Others are able to obtain their oxygen from the air. (A mosquito larva, for instance, breathes by means of an air tube.)

Further downstream there is no more sludge, but the oxidation of dissolved organic compounds continues. The oxygen content of the water remains low. The normal oxygen content of the water and the community of organisms associated with clean water may not be restored for hundreds of miles. If a second community discharges its sewage into the stream somewhere along the line, the restoration process is further delayed.

Sewage disposal plants process sewage in several ways. In the first step, called **primary treatment,** the raw sewage is filtered to remove non-degradables such as plastic and rubber goods, and large particles are allowed to settle out as sludge. The next step is **secondary treatment.** The liquid is sprinkled over crushed rock containing a large population of microorganisms. The microorganisms oxidize the remaining organic matter. Water that has undergone secondary treatment in this manner is clean enough for reuse. Some communities sprinkle treated waste water on fields, where it filters through the soil and replenishes underground water supplies. Sludge may also be dried and used as fertilizer (see Figure 5-16).

The volume of sewage in the United States is so large that such reuse, or **reclamation,** of sewage water for use in agriculture and industry is already common. The United States alone produces more than a billion gallons of liquid waste a day. Massive recycling will be forced upon us as we simply run out of water to carry away our sewage. We may hope that the increased demand for reclaimed water will bring about the routine secondary treatment of sewage. All too often at present, untreated sewage is simply discharged after only the largest solid material has been filtered out.

The water that is produced by secondary treatment still has particles and sediment that cannot be oxidized. A further treatment, or **tertiary treatment,** clarifies the liquid before its final discharge, by removing non-degradable matter in various ways. **Chlorination** (the addition of chlorine) helps to kill any pathogens that survive the earlier stages of sewage treatment.

Public health agencies must continually monitor bodies of water to determine whether they contain dangerous disease agents. The usual method is to take water samples and determine the number of *Escherichia coli* bacteria that are present. This figure is the sample's **coliform count.** Though found in feces, *E. coli* itself is not a disease agent. However, it is easy to identify and it is hardier than disease-causing organisms. Thus, the presence of *E. coli* in a water sample indicates that the water has been contaminated by feces. If *E. coli* is not present, fecal contamination has disappeared, and we can be certain that disease agents that are carried in fecal material are no longer present.

Even when the BOD has been lowered and fecal contamination has been eliminated, there are further waste problems. The oxidation of organic material may leave the resulting waste water high in inorganic compounds such as nitrates (NO_3^-) and phosphates ($PO_4^{\equiv}$), compounds that stimulate plant growth. This condition may be an advantage if the water is to be used in agriculture. However, if it is

Figure 5-16 Sewage treatment. Primary treatment consists of filtering the raw sewage and allowing particles to settle out as sludge. Secondary treatment consists of sprinkling the sewage over rock beds, where organic matter is oxidized by bacteria. Finally, the effluent is treated with chlorine to kill remaining microorganisms and discharged. The sludge is treated, dried, and used for fertilizer. Chlorination is not necessarily part of secondary treatment; it can occur at any time.

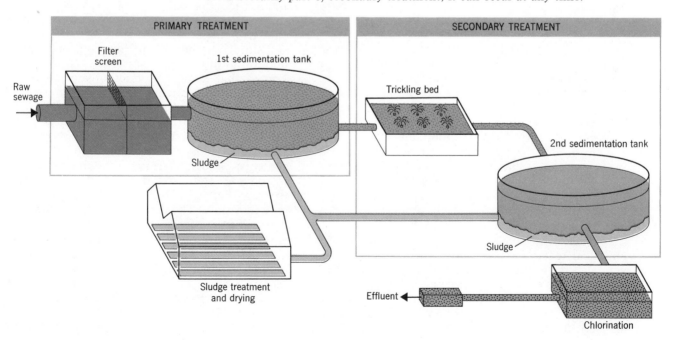

simply discharged into a stream, the nitrates and phosphates may stimulate the growth of algae to the point that the water becomes anaerobic. The effects of these compounds are much the same as those of high-BOD sewage.

Microorganisms oxidize most organic material, but there are some wastes that they cannot easily attack. More than half a million different chemicals are discharged into lakes and streams and ultimately the sea. Many of them are not biodegradable. At best, they are simply diluted. But in some cases, mere dilution is not enough to deprive them of their toxic effects. The insecticide DDT, for example, is now found world wide, in oceans, lakes, and soils. Even though the DDT is "dilute" in the environment, it can become concentrated in the tissues of many plants and animals. Dilute DDT in the ocean is absorbed by microscopic marine algae. The algae are consumed in large quantities by small marine animals, which therefore accumulate DDT in their tissues. These animals in turn are consumed by fish. The DDT becomes more concentrated at each step. Finally, the brown pelican, which feeds on the fish, accumulates so much DDT in its tissues that the formation of hard egg shells is disrupted. During incubation, the weight of the adult crushes the weakened egg and destroys the young. Many other animals, including humans, are continually absorbing DDT. In most cases, the effects (if any) have not yet been detected.

soil

We have discussed microorganisms as disease agents, as agricultural pests, and as the agents of fermentation and waste disposal. These topics are of particular interest to man, but they give a very narrow view of the importance of microorganisms. The most important function of microorganisms is to cycle nutrients in the environment.

Communities of soil microorganisms decompose organic material, breaking complex biological molecules down into smaller units that can be absorbed by plants. In this way they cycle important nutrients such as nitrogen and phosphorus from the soil to plants, to animals, and back to the soil. In many cases, the microorganisms merely accelerate chemical reactions that would occur anyway, but very slowly. Without these small decomposers, the surface of the earth would become littered with the bodies of plants and animals. Since nutrients would remain tied up in corpses, none would be available for plant production. New growth would cease.

Soil Microorganisms

Soil is the material in which land plants grow. This definition emphasizes the biological events that take place in soil, rather than its physical nature. Soil contains a good deal of space that can be occupied by air, water, and small organisms. In fact, pores or spaces represent about half its total volume. Some of the physical characteristics of soil depend on the underlying rock and available minerals, but soil derives many of its properties from the large plants that grow in it. As the plants die or drop their leaves, a large quantity of plant litter is deposited in the soil. This material is a source of energy and nutrients for a complicated community of soil microorganisms.

The number of organisms in the soil community is difficult to convey. One gram of rich soil may contain a billion bacteria. There is an equal weight of fungi, although the number of individual organisms is smaller. In each square foot there may be a million or more minute roundworms and tens of thousands of microscopic insects, mites, and spiders. Algae and protozoa also exist in immense numbers. If we include the giants of the soil community, the snails and slugs and earthworms, there may be 10,000 pounds of soil organisms per acre of rich land.

The study of individual soil organisms is extremely difficult, because they are hard to culture. One problem is that many soil organisms cannot survive even small changes in their environment. Since habitats in soil are very diverse, it is difficult to select the particular culture conditions that will support the growth of a specific microorganism. The study of **nitrification,** or nitrate production by microorganisms, is typical. It was long known that some soil microorganisms converted ammonia to nitrate in a two-step process. First, ammonia (NH_3) is oxidized to nitrite (NO_2^-). Then the nitrite is further oxidized to nitrate (NO_3^-). Decades passed, however, before Sergius Winogradsky, in 1890, isolated the responsible microorganisms. He demonstrated that the bacterium *Nitrosomonas* could oxidize ammonia to nitrite, and that another bacterium, *Nitrobacter*, oxidized nitrite to nitrate. The discovery of these bacteria was very difficult, because they had to be grown in the dark in anaerobic cultures that contained no organic compounds. These conditions were the only ones under which *Nitrosomonas* could work. Many previous attempts to find these organisms had failed because the culture conditions were wrong. It is really a hopeless task to produce an inventory of soil microorganisms and their characteristics by trying to reproduce the conditions that occur in soil. There is a great range of conditions in even a small sample.

There is an even more fundamental difficulty in the culture of soil organisms. Many simply cannot grow alone. For example, there are two types of soil fungi that must grow together, because both require vitamin B_1 for growth but neither one can synthesize it by itself. A vitamin B_1 molecule is composed of two major subunits, pyrimidine and thiazole. One fungus can synthesize pyrimidine but not thiazole; it can make vitamin B_1 only if thiazole is present in the environment. The other fungus synthesizes thiazole but not pyrimidine; it makes vitamin B_1 only in the presence of pyrimidine. Both fungi secrete the portion of the vitamin that they do synthesize into the medium. When they are grown together, therefore, their vitamin requirement is satisfied. But to grow either of them separately, the vitamin must be added to the culture medium. Interactions of this kind can become very much more complicated. Unfortunately, we are a long way from understanding the activities of most of the organisms found in the soil.

It is easiest to think of soil microorganisms as a whole assembly capable of carrying out many chemical conversions. Many of the reactions are oxidative, supplying energy to the microbes. Others convert one type of molecule into another that is easier to transport into the cell or to oxidize. Still others are reductive, acting as electron acceptors for the oxidized compounds.

Individual species of soil microorganisms are, in effect, "ecological enzymes," because each one accelerates a specific oxidative chemical reaction. A particular species carries out only a small portion of the total scheme. The decomposition of organic matter occurs in many stages. At each stage, the decomposing material furnishes energy and raw material for certain microorganisms. They, in turn, secrete compounds to the environment or are themselves consumed, dead or alive, by other organisms. Step by step, the organic matter entering the soil is oxidized and converted to simpler compounds, including nitrates and other important plant nutrients. The nutrients are absorbed by plants, incorporated into plant tissues, and later returned to the soil as plant litter.

In a stable community the sum of all these processes is a steady state. Organic material is decomposed as rapidly as it is produced. For the individual organisms, however, life in the soil is feast or famine. Almost all soil microorganisms have resting stages, such as spores, that can survive unfavorable conditions for long periods of time. When conditions are favorable, they become active again, repro-

Earthworms

Earthworms can hardly be considered microorganisms; the largest may be six or seven feet long and an inch in diameter. Nonetheless, they are best discussed as part of the soil community. Earthworms literally eat their way through the ground by feeding upon the organic material and microorganisms in the soil. The number of earthworms varies, but may be as high as several million per acre. Worms are found surprisingly deep in the soil. The ordinary earthworm has been found eight feet below the surface, and the deep-diving title holder is a Russian species that is said to go down as far as 26 feet.

Earthworms play a vital role in soil development. Their burrows provide for water drainage and improve the aeration of the soil. They also accelerate the decomposition of surface plant litter. The worms drag dead leaves into their burrows and break them into fragments during feeding. These small fragments are more accessible to attack by microscopic forms. By bringing new material to the surface, they deepen and renew the soil. (There is little to be gained from adding earthworms to one's garden, however. If conditions are suitable, they will already be there; if not, any imported worms will die.)

Charles Darwin was particularly taken with the importance of earthworms in soil formation. He drew attention to their role in burying rocks and archeological ruins. He calculated that earthworms brought ten or fifteen tons of soil per acre to the surface of the earth each year. This amounts to over an inch of soil each decade.

duce rapidly to take advantage of the food source, and then lie back and enter the resting stage until the next windfall.

Chemical Cycles

The decomposing activities of microorganisms in soil, sediments, and the oceans ensure that important components of living tissues will be cycled—in other words, reconverted to usable form. Carbon is, of course, a key substance in biological molecules. The overall pattern of movement of carbon between organisms and the environment is known as the **carbon cycle.** The earth has large reservoirs of carbon, in the form of carbon dioxide in the atmosphere and carbonate ($CO_3^=$) in both the ocean and rocks and soil. Carbon dioxide is reduced to organic form by photosynthesis, which uses light energy. Some of this organic carbon is returned directly to the atmosphere in the course of animal and plant respiration. Most of it, however, is returned through the microbial decomposition of tissues (see Figure 5-17).

The **nitrogen cycle** is somewhat more complicated. Nitrogen, in the form of nitrogen gas (N_2), makes up nearly 80 percent of the earth's atmosphere. But, though most organisms are surrounded by it, few are able to utilize it. Most plants use nitrogen in the form of either ammonia (NH_3) or nitrate ions (NO_3^-). Many plants can also use organic nitrogen, in the form of amino acids or other small molecules. But the ability to use nitrogen gas is limited to a few groups of microorganisms, called **nitrogen-fixing** microorganisms.

Nitrogen, as NH_3, NO_3^-, or N_2, is absorbed and incorporated into the growing tissues of plants. Animals acquire nitrogen by feeding on plant tissues and on other animals. When plants and animals die, the nitrogen is returned to the environment by the activity of microorganisms. As the nitrogen compounds in plant and animal remains are oxidized by microorganisms, some of the nitrogen is bound up in the microbes and some is excreted as ammonia. If conditions are anaerobic, nitritication may then occur. In this process, ammonia is converted to nitrite, and nitrite to nitrate.

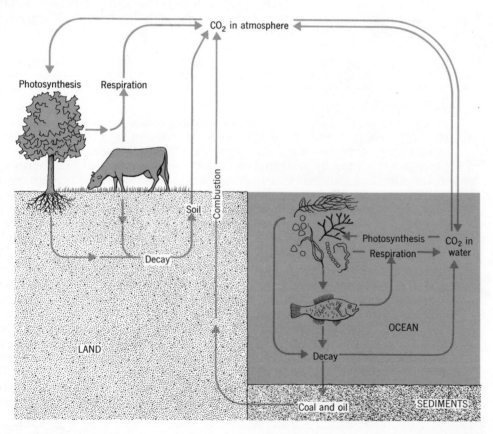

Figure 5-17 The carbon cycle.

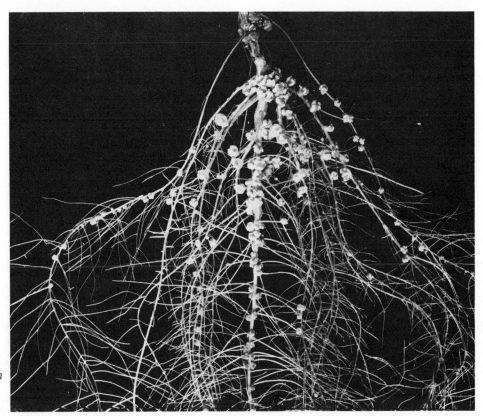

Figure 5-18 Root nodules of a soybean plant. The nodules contain the bacterium Rhizobium, *which converts nitrogen gas to other compounds of nitrogen that can be utilized by the bean plant.*

This part of the nitrogen cycle is not 100-percent efficient, however. In other words, not all of the nitrogen in biological compounds is taken up by plants again. There are always some losses. Animals excrete some gaseous nitrogen. Some nitrate is reduced to nitrogen gas by bacteria, some is lost in water runoff, and some is locked up in minerals. In order for the system to stay in balance, therefore, usable nitrogen must continually be added. To some extent, rainwater adds nitrogen by bringing down small amounts of ammonia and nitrate. Volcanic gases also contain ammonia. But the most important inputs by far are those made by the nitrogen-fixing microorganisms. These organisms convert nitrogen gas to ammonia or organic nitrogen.

The best-known nitrogen-fixing organisms are bacteria associated with plants such as clover and alfalfa. The bacterium *Rhizobium* invades the root of the plant, proliferates, and forms a nodule within the root (see Figure 5-18). This unit is an efficient device for the conversion of nitrogen gas to organic nitrogen, which is then incorporated into the plant.

Some free-living microorganisms in the ocean and in the soil also fix nitrogen. So can various other bacteria and blue-green algae, and possibly some fungi.

The importance of nitrogen fixation can be illustrated in several ways. When the corn is harvested from an acre of cornfield, 50 pounds of nitrogen are removed. This nitrogen must be replenished for the next crop, either by adding fertilizer or by encouraging nitrogen fixation by microorganisms. With proper cultivation, the soil microbes alone can make up the entire deficiency. In rice paddies, and probably in the open ocean as well, nitrogen is continually provided by blue-green algae. Nitrogen fixation also proceeds rapidly in forest and grassland. One report estimates that microbes fix 640 pounds of nitrogen per acre per year in African bushland. The nitrogen cycle is outlined in Figure 5-19.

Microorganisms play a similar role in other natural cycles. Phosphorus and sulfur, for instance, are also required for plant growth. These elements, too, cycle through microbes, soil, water, and higher organisms.

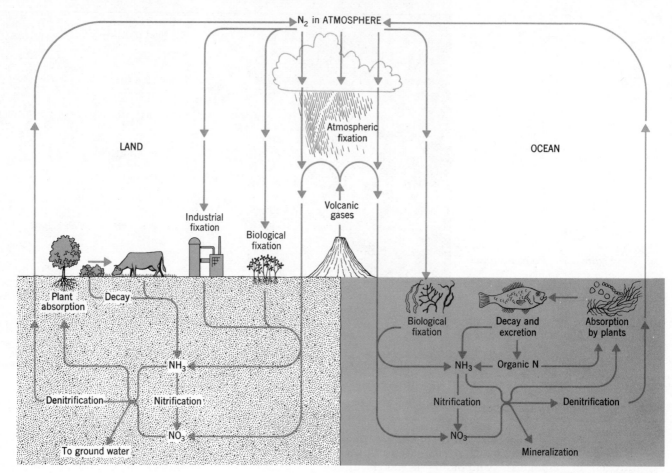

Figure 5-19 *The nitrogen cycle.*

The chemical activities of microorganisms are essential to the continued existence of larger and more conspicuous organisms, including man. The adaptability, voraciousness, and ubiquity of microorganisms are curses when they are functioning as pests or disease agents. But it is precisely these characteristics that permit them to carry out their roles in the soil and oceans. They fill in the gaps in the world's basic chemical cycles and keep the world running.

Recommended Reading

Burnet, Sir Macfarlane, and D. O. White, *The Natural History of Infectious Disease.* New York and London: Cambridge University Press (1972).

A discussion of the relationship between the nature of infectious disease agents and the phenomenon of infectious disease in human societies.

Dubos, René, *Man, Medicine, and Environment*. New York: Praeger (1969).

A modern discussion of health and disease, and a vision of the potential role of medicine in society.

Jackson, R. M., and Frank Raw, *Life in the Soil*. London: Edward Arnold (1966).

A brief discussion of the variety and activities of the microorganisms in soil.

Rosebury, Theodor, *Microbes and Morals*. New York: Viking (1971).

A discussion of the history, sociology, and biology of venereal disease.

Woodham-Smith, Cecil, *The Great Hunger*. New York: Harper & Row (1962).

A biological and historical account of the Irish famine of 1846–1849.

6 photosynthesis

In Chapter 1 we introduced the idea that organisms are open systems, which interact continually with their environments. The energy flow between the organism and its environment is a crucial part of this interaction. Organisms need high-grade energy from the environment to maintain their structural integrity, and the only significant source of this energy on earth is sunlight.

Photosynthetic organisms trap sunlight and convert it to chemical energy, which is stored in the chemical bonds of organic compounds. In plants, some of this stored chemical energy is oxidized to support their own metabolism, but the excess is used to make the tissues of the plant. It is this excess, of course, that feeds animals, supports animal metabolism, and makes the very existence of animals possible. The photosynthetic process also produces free oxygen. In fact, it is the major source of oxygen on earth. Without photosynthesis, the oxygen consumed by animals would not be replenished, and the oxygen content of the atmosphere would slowly decrease. Because photosynthesis is so fundamental to man's survival, it is a vital area of research. Studies of photosynthesis are popular also because the process itself is so intriguing: in many ways, plants appear to produce something for nothing. A seed needs only a little soil, water, and light (and photosynthesis) to grow into a large plant.

As a result of the combined efforts of chemists, physicists, and biologists, the photosynthetic process can now be described in very sophisticated terms. We will go into some detail partly because photosynthesis is so important to all life on earth, but also because it is a key factor in the overall structure and organization of plants. Most characteristics of plants are in one way or another related to their photosynthetic function.

159

In this chapter, the term "plants" will refer to all photosynthetic organisms, regardless of their classification. This procedure does not imply that our five-kingdom classification system is unsound; it is simply convenient to treat all of these organisms together. Plants, in this broad sense, are a very diverse group of organisms. They include photosynthetic bacteria, algae, and all the familiar green plants. The giants, sequoia trees, are 10^{21} times larger than the smallest photosynthetic bacteria. Plant life spans vary from a few hours to thousands of years. Nonetheless, we can treat all photosynthesizing organisms together because they all use light energy for the synthesis of organic compounds in a similar way. In fact, most of our information comes from the photosynthetic bacteria and the algae rather than from higher plants. Their small size, diversity, and rapid growth in culture have made them ideal experimental subjects.

Photosynthesis in most photosynthetic organisms is often described by a simple equation:

$$CO_2 + H_2O \xrightarrow{plant} CH_2O + O_2$$

where CH_2O is a general formula for carbohydrates. Written in this condensed form, photosynthesis is simply reverse respiration: carbon dioxide and water react to form carbohydrate and oxygen gas. More specifically, the carbon dioxide and the hydrogens of the water molecule react to form carbohydrates, which are stored in plant tissues and used to support the plant's metabolism. The oxygen from the water molecule is released as oxygen gas. The whole reaction is an "uphill" one. In other words, if the materials were simply combined, nothing would happen; complex cellular apparatus and large inputs of energy are required to make the reaction proceed. In photosynthesis this energy is supplied in the form of light, and some of the energy is eventually deposited in the chemical bonds of the carbohydrates. In respiration—the reverse, "downhill" reaction—this energy is released. There are many other complementary pairs of reactions in biological systems and in the physical world as well. The rusting of metal and the refining of metal ore are a good example. Rusting is a spontaneous downhill oxidation reaction. Ore refining, the reverse process, produces pure metals from natural oxides. The latter reaction is uphill, and it requires a complex technology and tremendous inputs of energy.

The overall photosynthetic reaction can be divided into three basic phases:

1. Light energy is absorbed and trapped by plant pigments such as chlorophyll.
2. This trapped energy is converted to a stable chemical form, ATP and $NADPH_2$.[1]
3. ATP and $NADPH_2$ then drive the reactions that incorporate carbon dioxide into carbohydrate and other compounds.

Before we turn to the individual steps of photosynthesis, we will first consider sunlight as an energy source.

radiant energy

The sun produces energy by nuclear reactions that emit immense quantities of electromagnetic radiation. The most obvious form of this radiation is light. Electromagnetic radiation can be considered to be a stream of particles traveling at 186,000 miles per second (the speed of light). The particles themselves, called **photons** or **quanta,** are analogous to packets of energy. All streams of radiant energy travel at the same speed, but the size of the individual photons varies. Some streams are composed of large packets of energy, and others of very small ones.

In other contexts, the streams of radiation behave as moving waves of energy, with each form having a characteristic wavelength. (Wavelength is the distance from peak to peak in the wave.) In a particular stream of radiation, the wavelength and the size of the quanta are inversely related. In other words, radiation with long wavelengths has small quanta (small packets of energy), and radiation with short wavelengths has large quanta. Thus, each form of radiation has its own particular energy content and wavelength.

The range of electromagnetic radiation is very broad. At one end are very long, low-energy radio waves, with wavelengths measured in miles. At the other end are high-energy gamma rays, with wavelengths of only a few Angstroms. Somewhere in between is visible light, with wavelengths between 3,500 and 7,600 Å (see Figure 6-1). Overall, the entire electromagnetic spectrum covers more than 60 octaves, each octave representing a doubling of wavelength. But what we know as visible light repre-

[1] NADP is identical to NAD except that it has an extra phosphate in its structure. Like NAD, NADP can accept an electron pair into its structure and become the reduced form, $NADPH_2$.

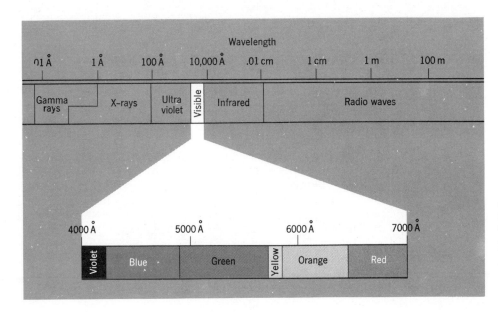

Figure 6-1 *The spectrum of radiant energy from the sun. The visible spectrum (wavelengths 4,000 to 7,000 Å) has been expanded to show the wavelengths of the different colors.*

sents just one of the octaves of the spectrum.

Most photosynthetic organisms use visible light as their source of energy, though some photosynthetic bacteria can absorb infrared light at wavelengths up to 9,500 Å as well. Only a small fraction of the sunlight that reaches the earth finally ends up as chemical energy stored in plant tissues. In fact, only about one part per million of the total energy output of the sun is trapped in vegetable matter. Yet this trapped energy supports almost all life on earth.

light absorption

The first step in photosynthesis is the absorption of the appropriate radiant energy by plant tissues. This energy is trapped by various pigments in the cells. A **pigment** is a chemical compound that absorbs certain wavelengths of light. Red paint, for example, looks red because the paint pigments have absorbed most *non*-red wavelengths of visible light. The red light is free to be transmitted and reflected.

The way a colored body absorbs light is often described by an **absorption spectrum,** a simple graph showing the percentage of light absorbed at each wavelength (color). An absorption spectrum for red paint would show high absorbance of blue and green light and very low absorbance (that is, very high transmittance or reflection) of red light. Similarly, green objects such as plants have very low absorption in the green wavelengths and high absorption in the blue and red (see Figure 6-2).

The absorption spectra of different photosynthetic organisms vary widely. This discovery is not surprising, since the organisms contain various pigments and actually appear to be different colors. But when pigments are extracted from various plants and separated, chlorophyll *a* is always found to be present. It is the chief photosynthetic pigment, and it appears to be a necessary part of all photosynthetic mechanisms.[1] A variety of other pigments, collectively called **accessory pigments,** are also present. Some of them are similar to chlorophyll *a* and are called chlorophylls *b, c,* and so forth. They are green; that is, they absorb blue and red light. Other accessory pigments absorb light in the middle regions of the visible spectrum (see Figure 6-3). These pigments account for the wide variety of colors of photosynthetic organisms. Obviously, light must be absorbed before it can be used in a photochemical process such as photosynthesis, and the absorption spectra of plant pigments provide information about the kind of light absorbed by plants.

[1] Bacteria are exceptions. They contain slightly different forms of chlorophyll, called **bacteriochlorophylls.**

Photochemistry

Photochemistry is the study of chemical changes caused by light. Obviously, a biologist who studies photosynthesis must be familiar with some principles of photochemistry.

Photochemists have found that when photons strike pigment molecules, their energy can be transferred to the molecules in several ways. In a single hydrogen atom, light energy is transferred from the photon to the atom's one electron, and the electron is said to be **excited.** The electron can be visualized as oscillating back and forth in the region of the nucleus. Like a guitar string, the electron can vibrate only at certain frequencies, and it can absorb only the specific levels of energy that will cause it to oscillate in a permitted frequency. In the jargon of photochemists, electrons can jump only to certain discrete energy states. Thus, not all wavelengths can be absorbed by a hydrogen atom. The absorption spectrum for hydrogen gas is therefore simply a series of lines, which correspond to the various wavelengths that have been absorbed to produce electronic transitions in the individual atoms of the gas.

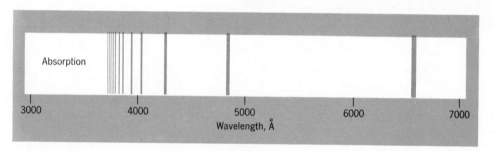

The absorption spectrum of hydrogen. The wavelengths of the absorption lines correspond to the energies of the various possible electronic transitions.

In molecules, however, other kinds of energy transitions occur. Molecules contain two or more atomic nuclei, and the relations between the electrons and the nuclei are vastly more complex. Energy transitions in molecules involve changes in the vibration and rotation of the various atoms in addition to changes in the individual electron energies. The more complex a molecule is, the more allowable energy states it will have. Consequently, the absorption spectra for large pigment molecules are complex curves, not simple lines.

The absorption spectra of pigments can be measured using a white light, a series of colored filters that pass different wavelengths of light, and a light-measuring device like a photographic light meter. A filter is placed in front of the light source, and the amount of colored light that passes through the filter is measured. This filtered light is then passed through the pigment and the amount is measured again. The difference between the first and second measurements is the amount of light absorbed by the pigment. The process is repeated using other filters, until enough values are obtained to plot the pigment's absorption over a broad spectrum of wavelengths. This plot is the pigment's **absorption spectrum.** This technique has been elaborated and refined to the point where it is now all done automatically by a machine called a spectrophotometer. The machine scans across the spectrum automatically and records the variations in absorption that occur at different wavelengths.

The spectrophotometer is an elegant and sensitive tool for the biologist. It can also be used to study and detect changes in molecules caused by factors other than light. Some biological compounds undergo shifts in absorption spectra related to their functional roles. For example, molecules in the respiratory chain undergo changes in their absorption spectra when they accept electrons and become reduced. The hemoglobin in red blood cells undergoes a detectable change in its absorption spectrum when it associates with an oxygen molecule.

Several things can happen to an excited molecule that has absorbed a quantum of light energy.

1. The excited molecule may lose the absorbed energy immediately by reemitting a photon through **fluorescence.** The emitted light exhibits a characteristic wavelength, usually slightly longer than that of the absorbed light, because the emitted light particle has slightly less energy. Each emission spectrum is just as characteristic of a specific molecule as its absorption spectrum. This fact has been important in tracing the path of energy transfer from molecule to molecule in pigment systems.
2. The excited molecule can also emit photons more gradually, through **phosphorescence.** The excited molecules are first converted to a slightly lower energy condition that is more stable. Then they slowly return to their original condition by emitting photons.
3. The absorbed energy may excite an electron enough to strip it from the molecule. Ionization is the result.
4. The energy can be transferred to a nearby molecule.
5. Most often, all the absorbed energy is lost as heat.

Energy transfer and ionization are probably the most important processes in photosynthesis.

Absorbed light does not necessarily cause a photochemical reaction, however—the light may simply be converted to heat. Thus, by itself, an absorption spectrum tells us nothing about the wavelengths that actually produce photosynthesis. To identify the important wavelengths we need some way to detect a chemical change, an "action" caused by light absorption. One well-known photochemical "action" is the formation of an image on photographic film. Black-and-white film is a suspension of silver salt crystals. When struck by light, the salt crystals are changed chemically so they turn black when exposed to developing chemicals. Not all light produces such an action on photographic film. Most film, for example, is insensitive to red light; that is to say, red light will not produce the critical photochemical action. (This is why a red "safelight" is often used in a darkroom.) We can plot the action spectrum of the film by graphing the relationship between the wavelength of light and its effectiveness in blackening film. In doing so we are plotting some action of radiant energy—in this case the sensitization of silver salts—against the wavelength of the radiant energy.

There are several ways to detect action in photosynthesis. Perhaps the simplest technique is to look for evolved oxygen. If a plant produces oxygen when it is illuminated, we know that photosynthesis has occurred. In 1950, this technique was used to compare action and absorption spectra in several marine algae. Action spectra parallel absorption spectra fairly closely in most regions of the visible spectrum. This evidence suggests that most of the absorbed light is active in photosynthesis. However, all the algae tested have less action than absorption in the red region of the spectrum, at wavelengths greater than 6,800 Å. This discrepancy between absorption and action is called the **red drop.**

Attempts to explain the red drop have produced some of the most interesting results in the field of photosynthesis. For instance, the red drop can be eliminated if the plant is illuminated with two light beams simultaneously: long-wavelength red light supplemented by a light with a shorter wavelength (Figure 6-4). This experiment led to the hypothesis that *two* photochemical absorption steps are necessary for normal photosynthesis, and that the steps occur in two different groups of pigment molecules, called **photosystems I** (PS I) and **II** (PS II). Additional experiments showed that PS I contains mostly chlorophyll *a,* and probably some chlorophyll *b.* The chlorophyll *a* molecules occur in slightly different forms. One form absorbs long wavelength red light more efficiently than the others. PS II also contains chlorophyll *a,* but it lacks the long-wavelength form. However, it does

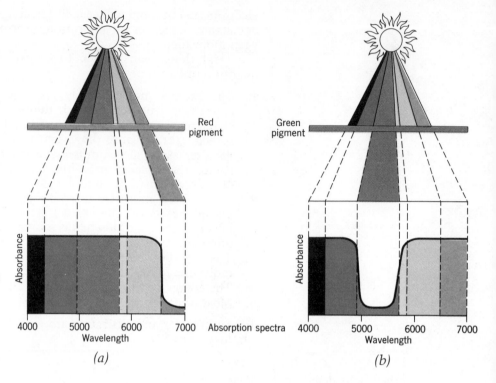

Figure 6-2 Light absorption by pigments. (a) A red pigment absorbs all wavelengths of visible light (that is, all colors) except red. (b) A green pigment absorbs all non-green wavelengths.

Absorption spectra

contain chlorophyll *b* and probably other accessory pigments as well. Therefore, at wavelengths greater than 6,800 Å, PS I absorbs much more light than PS II, causing photosynthesis to slow down. This slow down is the red drop. Apparently, both systems need to absorb equivalent amounts of light energy in order for the overall photosynthetic reaction to proceed efficiently. Illumination with the second light beam supplies this energy to PS II, and the red drop is eliminated.

The hypothesis goes on to suggest that the photosynthetic pigment molecules in a photosystem do not act independently. Instead, several hundred of them cooperate to trap quanta of light energy. They then transfer the trapped energy to a **reaction center,** which contains the necessary enzymes for the subsequent steps of photosynthesis. The absorbing molecules of the photosystems and the reaction center combined are called a **photosynthetic unit.**

Whenever energy is trapped by any one of the pigment molecules in a photosynthetic unit, it is immediately passed along the array of closely-packed pigment molecules to the reaction center. Photochemists believe that this kind of transfer could occur when an electron transition in one molecule induces a similar transition in a close neighbor. Such transfers could occur even though the absorption bands of the two molecules were quite different. An energy quantum could first be absorbed by an accessory pigment. Then, when the excited molecule had dissipated some of the absorbed energy, it could transfer a smaller energy quantum to chlorophyll *a*.

One implication of this hypothesis is that the energy transfer should be unidirectional. That is, the energy can move only to a pigment that has an absorption band at lower energy (longer wavelengths). One would expect energy transfers to occur in the following direction:

Accessory Pigment 1 ⟶
(absorbs blue light)

        Accessory Pigment 2 ⟶ Chlorophyll *a*
        (absorbs green light)   (absorbs red light)

An elegant proof of this hypothesis involves measurements of the fluorescence of chlorophyll *a*. The light emitted by chlorophyll *a* during fluorescence is distinctive—it has a characteristic spectrum of wavelengths. Whenever this spectrum is detected, therefore, it means that chlorophyll *a* is fluor-

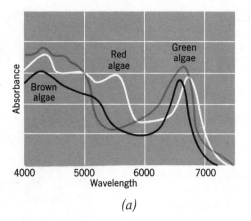

Figure 6-3 Light absorption by photosynthetic organisms. (a) Different plants have different absorption spectra, because they have different kinds of pigments. (b) Each plant pigment has its own characteristic absorption spectrum.

(a)

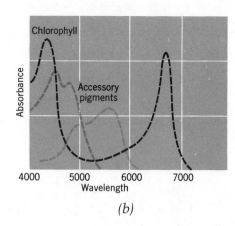

(b)

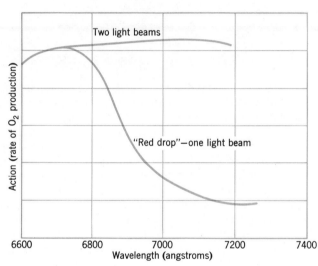

Figure 6-4 The red drop in photosynthesis. A plant's rate of photosynthesis falls off abruptly under red light of wavelength longer than 6,800 Å. If the red light is supplemented with another light of shorter wavelength, the photosynthetic rate is greatly increased.

escing because it has absorbed energy. When plants are illuminated with blue or green light—wavelengths that are absorbed mainly by the accessory pigments and not by chlorophyll *a*—they produce almost as much chlorophyll *a* fluorescence as plants that are illuminated with red light, which is absorbed directly by chlorophyll *a*. This discovery suggests that the energy absorbed by accessory pigments is efficiently transferred to chlorophyll *a*. Presumably this kind of transfer goes on in both photosystems. In PS I, energy is transferred from chlorophyll *b* through the various forms of chlorophyll *a*, and finally to the long-wavelength form that absorbs red light so efficiently. In PS II, energy is passed along from various accessory pigments to chlorophyll *a* in similar fashion, but the long-wavelength form is absent.

production of ATP and NADPH₂

Once the absorbed energy has reached chlorophyll *a* it is transferred to a reaction center, where it is used to excite an electron. The electron is passed along a series of compounds that eventually produce ATP and NADPH$_2$. This process is the second basic step in photosynthesis. As it has been worked out, the reaction center associated with PS II supplies the energy required to remove an electron from a compound called simply "compound Z," and to transfer it uphill to another compound, "compound Y." Compounds Z and Y are still hypothetical. Their existence has been inferred from many observations and experiments, but their structures are still unknown.

Compound Y then passes the electron through a downhill sequence of reactions, which is analogous to electron transport in the respiratory chain. This downhill sequence drives the formation of ATP from ADP. At this point, however, compound Z is missing an electron, so it removes one from a water molecule by splitting it:

$$2H_2O \longrightarrow 4H^+ + 4e^- + O_2.$$

This reaction is the one that releases oxygen gas from a water molecule.

At this point, the reaction center in PS I takes over. It supplies energy to move the electron uphill again to another hypothetical compound called X. The electron is passed downhill again, and this time it is accepted by NADP, which then becomes reduced to NADPH₂.

These photochemical events make up the so-called **Z scheme** (Figure 6-5). In summary, the net movement of an electron from water to NADPH₂ is uphill, driven by energy from the reaction centers associated with PS II and PS I. ATP and oxygen are also produced in the process.

We are now in a position to understand the red drop more fully. When long-wavelength light is used, net photosynthesis slows down because PS I receives more energy than PS II; therefore PS II cannot keep PS I supplied with electrons. However, when PS II receives additional energy from a second light beam with a shorter wavelength, it transfers more electrons. Photosynthesis speeds up again.

PS I can operate alone by cycling electrons from compound X back through the electron transport chain. ATP is formed with each cycle, and many ATP molecules can be formed by recycling the same electrons (see Figure 6-5). Photosynthetic bacteria use this method to produce ATP, because they contain PS I exclusively. All the algae and the more complex green plants possess both photosystems.

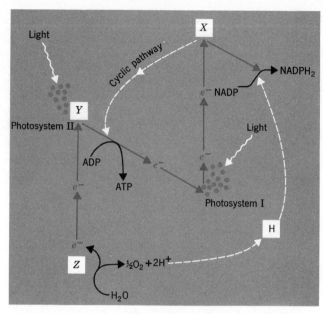

Figure 6-5 The Z scheme of photochemical reactions in photosynthesis. The path of the electron is represented by the green arrows. The vertical position of each arrow represents the energy content at each intermediate step in the sequence. Electrons are transferred from water to compound Z and then to compound Y, with the aid of photochemical energy from photosystem II. Oxygen is produced in this step. The electron then passes "downhill" through a series of electron transport compounds and arrives at photosystem I. ATP is generated in this process. With the aid of photochemical energy provided by photosystem I, the electron is transferred "uphill" again to compound X. Two electrons and a proton are then combined with NADP to give NADPH₂.

carbon dioxide fixation

Since, as we have seen, ATP and NADPH₂ are the common energy currency for many metabolic reactions, it is not surprising that they also supply the energy for the final set of reactions in the photosynthetic process, the conversion of carbon dioxide into carbohydrate. The detailed steps of this process were elucidated by Melvin Calvin and his coworkers at the University of California, Berkeley. Their work was one of the first major biological discoveries to take advantage of the radioactive tracer technique.

In the early 1940's, a long-lived radioactive form of carbon (C¹⁴) was discovered, and a way was found to replace the normal C¹² in carbon dioxide with the radioactive form. In this way the carbon dioxide was "labelled." A plant could be supplied with the radioactive carbon dioxide and allowed to photosynthesize. The labelled carbon dioxide would be incorporated into organic molecules in the plant cells, and these organic molecules themselves would become radioactive. The carbon that came originally from the carbon dioxide could thus be distinguished from all the other carbon atoms in the plant.

Calvin and his coworkers exposed cells of a green alga, *Chlorella*, to labelled carbon dioxide for very short periods. The carbon compounds were then separated by paper chromatography. Calvin's group found that the reactions were so rapid that dozens of different compounds were labelled after just a few minutes. It was impossible to tell the sequence in which the compounds became labelled, and nothing could be deduced about the individual metabolic steps. They had to design equipment that allowed them to stop all reactions after only a few seconds. At last they were able to follow C^{14} as it appeared in different compounds, and the path of carbon in photosynthesis could be traced. This path has become known as the Calvin-Benson cycle (Figure 6-6).

The cycle begins with acceptor molecules that

Figure 6-6 Calvin and Benson's study of the path of carbon in photosynthesis by Chlorella. *(a) The algae are pumped rapidly down a narrow tube into boiling alcohol, which kills them and quickly stops all photosynthetic reactions. At some point before they die, radioactive CO_2 is injected into the tube to allow them to photosynthesize briefly with it. By varying the location of the injection, the reactions can be stopped after as little as one second. (b) The photosynthetic products after five seconds of photosynthesis. The compounds were separated by paper chromatography. Abbreviations are as follows: PGA, phosphoglyceric acid; TP, triose phosphate; RP, ribulose phosphate; RDP ribulose diphosphate; FP, fructose phosphate; SP sedoheptulose phosphate; GP, glucose phosphate; PEP, phosphoenol pyruvate.*

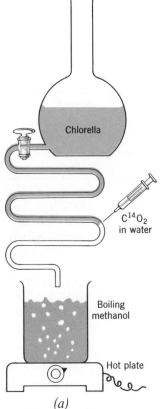

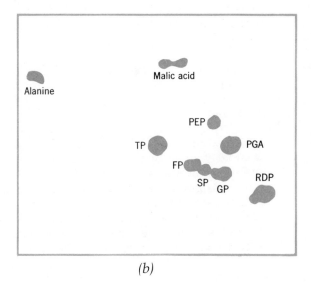

(a)

(b)

Chromatography

The compounds in any mixture must be separated before their individual properties can be studied. **Chromatography** is one method that is frequently used for separating small quantities of substances. It has been particularly valuable in studies of photosynthesis.

The first step is to remove the substances from the cells, usually by immersing the cells in a solvent such as alcohol or acetone or water. Next, the solution containing the dissolved substances is concentrated by evaporating most of the solvent. Finally, a small quantity of the concentrated solutions is passed through a system consisting of two phases, where the individual components are separated. A common chromatographic system consists of a sheet of paper as one phase and organic solvents as the other. The sample to be analyzed is placed on the paper, and the solvent is added. As the solvent moves along the paper it carries the sample with it.

The rate of movement of each substance in the sample depends on its solubility in the liquid and on how tightly it binds to the solid phase. One substance may bind strongly to the liquid molecules (in other words, it may be highly soluble in it), while interacting only weakly with the molecules of the paper. In that case the substance will migrate rapidly, probably as fast as the liquid itself. Another substance may bind tightly to the paper and consequently move slowly or not at all. Because of their different properties the various compounds of the sample become separated on the paper (see Figure 6-6).

By choosing different liquid and solid phases, the rate of movement can be changed. In this way the separation of particular types of molecules can be maximized. But for each specific two-phase system, the rate at which each substance moves is constant. This rate can be expressed as a ratio, called the **relative front** (Rf):

$$Rf = \frac{\text{distance moved by substance}}{\text{distance moved by solvent}}.$$

The compound is not only separated from the rest of the mixture, but it can also be identified by the value of its ratio.

Paper chromatography is only one system. Column chromatography uses a glass column packed with powder as one phase; a liquid is run through the column for the second phase (Figure A). Other systems are used as well, but the principle is the same in all of them.

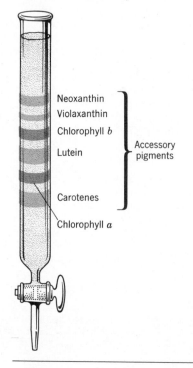

Figure A Column chromatography. Pigments from a flowering plant are dissolved in methyl alcohol and separated on a column of sucrose.

Neoxanthin
Violaxanthin
Chlorophyll *b*
Lutein
Carotenes
Chlorophyll *a*

Accessory pigments

combine with carbon dioxide. These molecules then undergo a complex series of reactions that does two things: it regenerates the acceptor molecules, and at the same time it converts three carbon dioxide molecules to a single molecule with three linked carbon atoms. This molecule can then enter other metabolic pathways. Glucose, synthesized from two of the three-carbon molecules, is the major carbohydrate product of the overall photosynthetic reaction; it is the basic unit of energy storage. Amino acids and fats can also be quickly synthesized from the three-carbon compound.

Both ATP and $NADPH_2$ are required to drive the Calvin cycle. The photochemical reactions must supply these molecules: three ATP's and two $NADPH_2$'s for each carbon dioxide molecule assimilated into the Calvin-Benson cycle. A diagram of the essential reactions of the Calvin-Benson cycle in *Chorella* is shown in Figure 6-7.

This picture of photosynthesis depends heavily on exhaustive analyses of a few photosynthetic organisms. When other species are examined, we find many variations in the way light energy is utilized:

1. The first photochemical steps can be carried out by many combinations of photoactive pigments.
2. The Z scheme for the production of ATP and $NADPH_2$ is by no means universal. As we

mentioned earlier, bacteria may use only one photosystem. Electron transport in different plants may be handled by a variety of different compounds. In other plants, molecules such as H_2S and H_2 may provide the electrons that are usually supplied by water, so oxygen will not be produced.

3. The Calvin-Benson cycle is not the only possible route of carbon dioxide fixation. In fact, carbon fixation is not even limited to carbon dioxide. Organisms living in aquatic environments may fix certain small organic molecules dissolved in the surrounding water.

Because of these variations, the overall photosynthetic equation can be rewritten in many ways by substituting other electron donors for water, and other carbon sources for carbon dioxide. Each equation represents a different solution to the problem of utilizing the primary energy source on earth, solar radiation.

the chloroplast

As we have seen, a highly organized physical arrangement of pigments, electron transport molecules, and enzyme systems is necessary to carry on photosynthesis. In photosynthetic bacteria, these pigments are located in small membrane-bounded sacs, or flattened or tubular membranes. In primitive blue-green algae, the photoactive pigments are associated with a system of flattened membranes called **lamellae.** In protistan algae and the higher plants these lamellae are more highly organized. They are arranged in stacks, and each stack is surrounded by a double membrane. This overall structure constitutes the chloroplast, the site of all photosynthetic reactions from light absorption to the carbohydrate product. The variety of photosynthetic apparatus is shown in Figure 6-8. Many unicellular algae have just one chloroplast per cell;

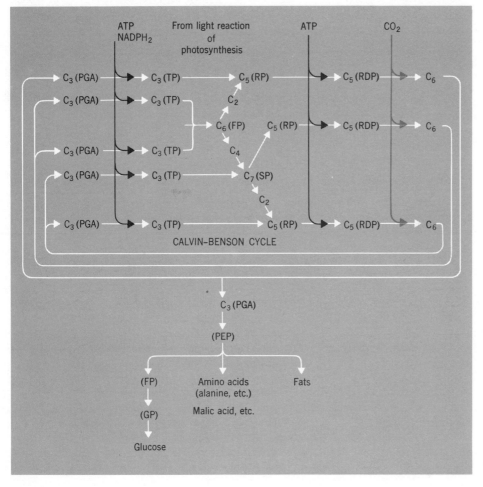

Figure 6-7 The Calvin-Benson cycle is the sequence of reactions in which CO_2 is assimilated into reduced organic compounds. Five C_3 (three-carbon) molecules are rearranged to produce three C_5's. Each C_5 combines with a CO_2 and is split into two C_3's. The result is that $3C_5 + 3CO_2 = 6C_3$. Five of the C_3's are used again in the cycle, but the sixth can enter various metabolic pathways to make sugar, amino acids, fats, and other compounds. Energy is supplied by the ATP and $NADPH_2$ generated by photosystems II and I. (The abbreviations of the carbon compounds are explained in Figure 6-6.)

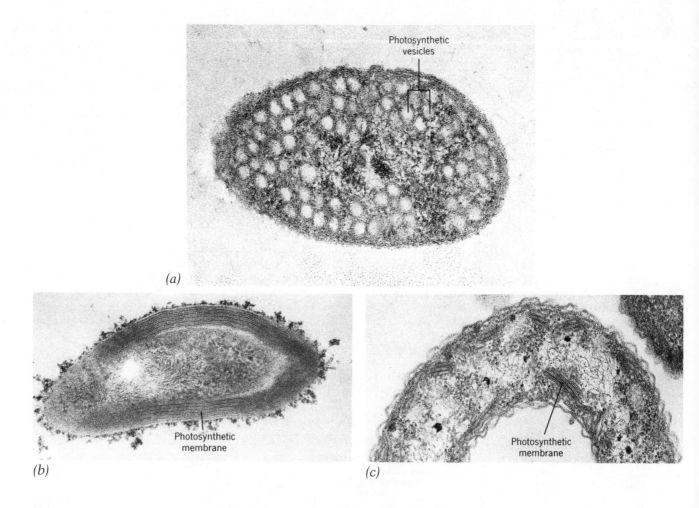

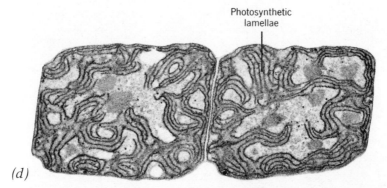

Figure 6-8 Photosynthetic apparatus in various organisms. (a, b, c) Vesicles and internal membranes in photosynthetic bacteria. (d) Photosynthetic lamellae in a blue-green alga, Nostoc. (e) Chloroplast of a green alga, Chlamydomonas. (f) Several chloroplasts in a higher plant cell. (g) Phycobilisomes in red alga.

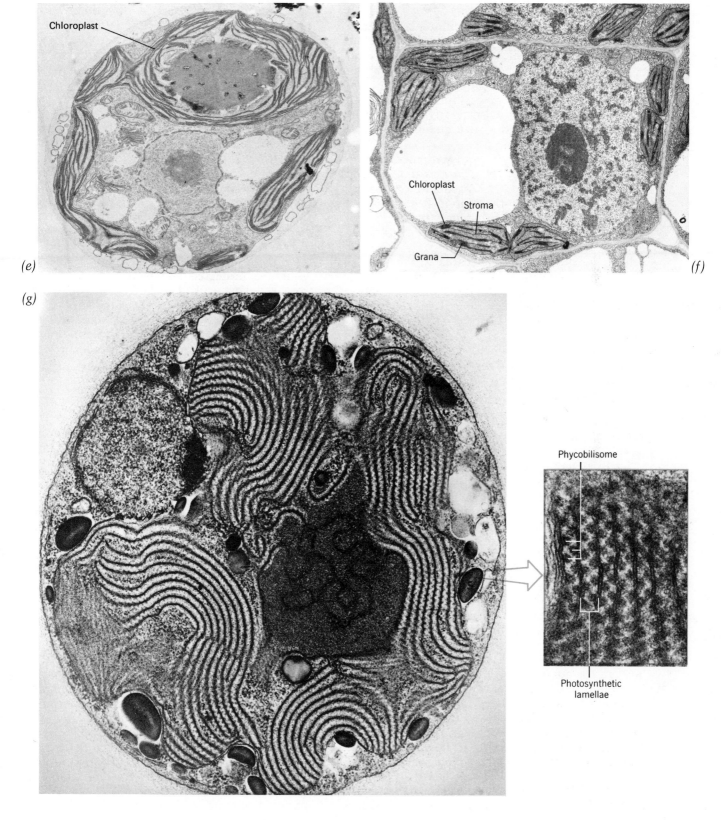

(e)

(f)

Chloroplast

Chloroplast

Stroma

Grana

(g)

Phycobilisome

Photosynthetic
lamellae

higher plants typically have many in each cell.

An electron micrograph of a chloroplast from a higher plant shows that some of the lamellae are tightly packed in bundles, called **grana.** The clear regions between the lamellae are filled with a protein matrix substance called **stroma.** Some electron micrographs show lamellar membrane surfaces in the grana covered with tiny, regularly arranged particles. These particles are called **quantasomes** (see Figure 3-20).

The three parts of the photosynthetic reaction (that is, light absorption by PS I and II, ATP and NADPH$_2$ production by the Z scheme, and carbohydrate production by the Calvin-Benson cycle) seem to be localized in different structures. The electron carriers of the Z scheme are probably bound to the lamellar surfaces; the Calvin-Benson cycle enzymes are probably located in the stroma. The pigment molecules of PS I and PS II may be associated with lamellar surfaces, but the situation is still very speculative. Chlorophyll molecules are particularly concentrated in the lamellar membranes that form the grana. The chlorophyll molecule itself has two portions: a large "head," composed of a complex ring structure that dissolves in water or bonds to proteins, and a "tail," a carbon chain that dissolves in lipids. The chlorophyll molecules may be sandwiched between the protein and lipid layers in the lamellae, with their heads bound to the protein and their tails embedded in the lipid layer.

The particles on the membrane surfaces of the grana, the quantasomes, are just about large enough to hold all the pigment molecules that make up the photosystem of one photosynthetic unit. Some researchers, therefore, have speculated that the quantosomes and the photosynthetic unit may be one and the same. There are many difficulties with this suggestion, however. When quantosomes are broken up, photosynthesis still proceeds. A second problem involves the location of the accessory pigments that are supposed to be part of PS I. In the red and blue-green algae they appear to be localized in small independent particles called phycobilisomes, attached to the lamellae (see Figure 6-8). How can the quantosomes on the lamellae be complete photosynthetic units when some of their pigment molecules are localized in separate particles?

Chloroplasts contain the entire photosynthetic apparatus, it is clear, even though the precise location of the reactions is uncertain. The chloroplast is a self-contained organelle in other ways. It possesses considerable quantities of DNA, RNA, and ribosomes. These components of the chloroplast appear to be fully functional, and they control the synthesis of chloroplast's own proteins. In some plants, chloroplast DNA replicates at a different time from nuclear DNA; and recently, isolated chloroplasts have been observed to divide in culture.

In many respects, in fact, the chloroplast of a eucaryotic plant (the protists and plantae) resembles the cell of a primitive blue-green alga. For this reason, many biologists believe that chloroplasts result from a relationship that may have evolved a billion years ago. They suggest that primitive procaryotic cells, similar to present-day forms of blue-green algae, first developed photosynthetic capabilities. Non-photosynthetic organisms fed on the procaryotes, or came in contact with them in some other way. In some cases, the ingested cells were able to survive inside the host cells (Figure 6-9). Such an arrangement would be advantageous to both forms. The photosynthetic cell would be protected and would receive nutrients from the host cell. The host, in turn, would be supplied with energy-rich products of photosynthesis, such as glucose. According to this theory, the chloroplasts of higher plants may have originated in a primitive interdependence between photosynthetic procaryotes and non-photosynthetic cells.

photosynthesis in nature

Most of the plants used in photosynthetic studies are studied in controlled environments. Temperature, light, and other conditions are held constant. Such controlled conditions are useful because they minimize the variability of laboratory results, but they do not give an accurate account of photosynthesis in nature. The intensity and quality of light, the length of the day, temperature, the carbon dioxide concentration of the air, and the availability of water are only some of the normal environmental variables. Variation in such factors produces corresponding changes in an individual plant's photosynthesis.

On a longer time scale, the photosynthetic apparatus of a plant becomes adapted to a particular environment, through evolution. For example, the variety of photosynthetic pigments in a given plant is related to the quality of sunlight available to it. This relationship is quite obvious in the seaweeds. Sea water itself absorbs sunlight very effectively, but the absorption is selective. Red light is strongly absorbed, while the shorter (green and blue) wavelengths penetrate to greater depths. It is not surprising, therefore, that green algae, which depend

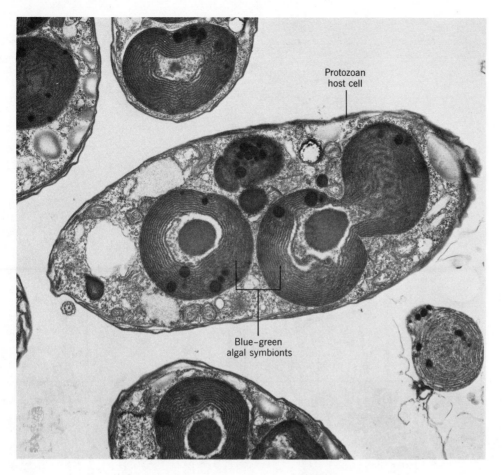

Protozoan
host cell

Blue-green
algal symbionts

*Figure 6-9   Blue-green
algae living symbiotically
inside a protozoan host cell.
Magnification 13,700×.*

on the absorption of red light by chlorophylls, tend to be found near the surface. Brown and red algae, by contrast, often occur in the deeper waters. Their chlorophyll is ineffective in absorbing blue and green light, but their accessory pigments are well adapted to make use of the blue-green light of the submarine world. Differences in pigment concentration, as well as in composition, are also common. Many land plants can be divided into two categories, depending on their concentration of chlorophyll. "Shade plants" have high chlorophyll concentrations and can grow in very weak light. "Sun plants" require higher light intensities for efficient growth and have less chlorophyll.

Further discussion of photosynthesis as it occurs in nature draws us into considering the organization of the whole plant. Many plant structures are directly concerned with supplying carbon dioxide and water for photosynthesis, exposing the chloroplasts to light, and transporting photosynthetic products to other parts of the plant. All of these systems are part of the photosynthetic process. However, these same plant structures must serve other needs of the organism as well. Some of the features of a plant are only distantly related to photosynthesis. But all of them, collectively, give plants their unity and integrity as organisms.

Recommended Reading

Rabinowitch, E., and Govindjee, *Photosynthesis.* New York: John Wiley & Sons (1969).

A modern account of photosynthesis, clearly written. It progresses from a simple discussion to a sophisticated analysis of recent research.

7 plant organization

Many aspects of a plant's organization are related to its use of light as an energy source. Plants must be structured so that their photosynthetic apparatus is orientated toward sunlight. They must be organized so that carbon dioxide and water can reach the photosynthetic cells. Plant metabolism must maintain the photosynthetic apparatus and provide for storing the energy-rich compounds produced in photosynthesis. The waste materials from these cellular activities need to be removed.

These photosynthetic requirements are met in diverse ways in different kinds of plants. The structures that support photosynthesis in a small aquatic plant are very different from those in a tree. A small marine plant floats in a water solution containing ample supplies of carbon dioxide, so it needs no special apparatus for obtaining water and carbon dioxide for photosynthesis. But it does need some way to regulate the movement of salts into its cell. Also, it must float or be motile in order to stay in surface waters that are bathed in sunlight. A forest tree has very different problems. Its leaves must be raised above the ground to intercept the sunlight. Water must be extracted from the ground by an elaborate root system and transported to the leaves for photosynthesis. Carbon dioxide must be obtained from the air, and water must be conserved. The whole plant is large and heavy and requires massive mechanical support.

A plant has other functional problems to solve, in addition to maintaining photosynthesis. It must protect itself from destructive elements (predators, parasites, and harsh weather) while still maintaining an exchange of materials with its environment. Plants also require sensory mechanisms that allow

Albrecht Dürer, "The Great Piece of Turf"

175

them to respond to changes in the environment. Reproductive processes must provide for new individuals. Structural solutions to all these requirements must be coordinated and integrated into a functioning organism. Each plant species has a specific set of solutions to the functional requirements for its existence.

There are fundamental differences between plants adapted to aquatic environments and those that survive on land. Almost all the plants that live in water are algae, and almost all the more familiar and complex green plants are terrestrial. The algae are either simple unicells or relatively unspecialized multicellular organisms.

Aquatic Plants

Algae are quite diverse in many ways. They are usually divided into seven or more broad groups, in three different kingdoms, defined according to their pigment composition. But nearly all the algae are found in aquatic or moist environments, and they all share features that allow them to survive in a watery medium.

single-celled algae

Most unicellular algae are part of the **phytoplankton,** the plant members of the floating plankton community. As such, they face very special problems. Phytoplankton are not mere curiosities; rather, they are the mainstay of life in the ocean. Though seaweeds along rocky shores may be more striking, most of the photosynthesis in the sea is carried out by the myriad tiny plants of the open ocean.

The **dinoflagellates** are the most conspicuous algae in the phytoplankton. They belong to the Kingdom Protista. They are brown in color, and each is equipped with a pair of flagella. Dinoflagellates sometimes occur in such huge numbers that they cause so-called "red tides" in coastal waters (see Figure 7-1). The red tides often appear in the spring, when water conditions become particularly favorable for algal growth. The days become longer, temperature increases, and nutrients are added to surface waters either by runoff from the land or by wind and currents that mix surface waters with deeper, nutrient-rich layers. Red tides often reappear sporadically throughout the summer and autumn, whenever nutrients become abundant. The dinoflagellates that create the red tides secrete a powerful toxin into the water. When it becomes concentrated it may kill fish, crustaceans, clams, and other animals. The toxin affects the nervous system.

Humans can get paralytic shellfish poisoning by eating seafood contaminated by dinoflagellate toxin, and the condition is sometimes fatal. As a result, red tides can be very damaging economically. In 1972, for example, New England suffered a major red tide. Dozens of people were poisoned by contaminated shellfish (fortunately, none fatally), and many seafood restaurants and seafood dealers were financially ruined.

Diatoms are golden algae, also part of the Protista. They are the most abundant organisms of the phytoplankton. In addition, diatoms can be found along the shore and on rocks, sand, seaweeds, and wood pilings. Some even thrive on whales. (In fact, the blue whale is also known as the sulfur-bottom whale because of the yellow diatoms on its belly.) Individual diatom cells or colonies may be so abundant that they cover everything with a fine yellow or brown fuzz. Any bucket of sea water will contain a variety of them (Figure 7-2).

The cell membrane of a diatom is not directly exposed to the surrounding sea water. Instead, it lies beneath an elaborate cell wall called the **frustule.** The frustule is constructed like a petri dish with two halves, one of which fits inside the other. This external skeleton is made of amorphous silica—in other words, glass. The frustule is pierced by elaborate pores and pits, which are just barely visible in a good light microscope. (In fact, microscope slides of diatoms used to be a standard test object to check the resolution of a light microscope: if pores could be seen clearly, the resolution was considered excellent.) The electron microscope has revealed the existence of several kinds of pores and also complex sculpturing of the frustule (Figure 5-10). The pores keep the cell membrane in communication with

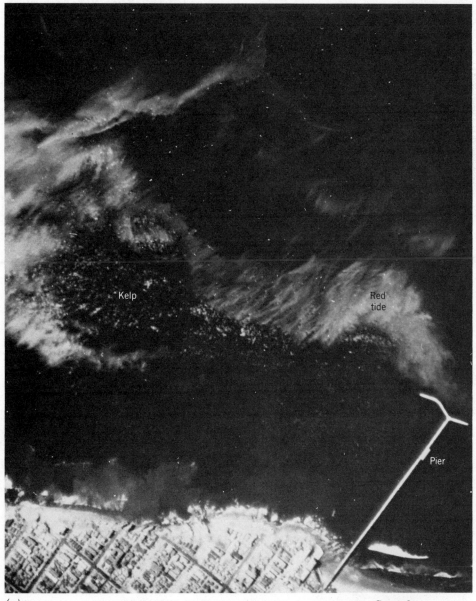

(a)

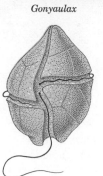

Gonyaulax

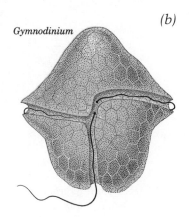

Gymnodinium

(b)

Figure 7-1 *Red tides. (a) A red tide off the coast of California shows up as light streaks in this aerial infrared photograph. (A kelp bed is also visible). (b) Red tides are caused by tremendous numbers of dinoflagellates, such as Gonyaulax or Gymnodinium.*

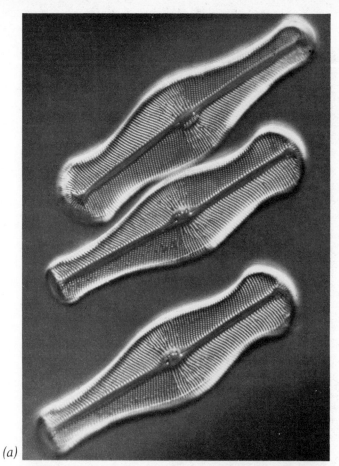

(a)

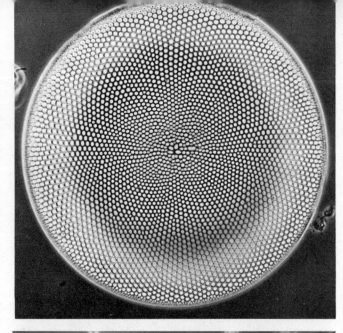

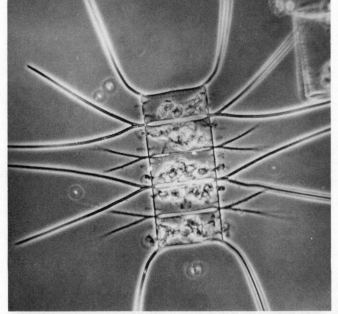

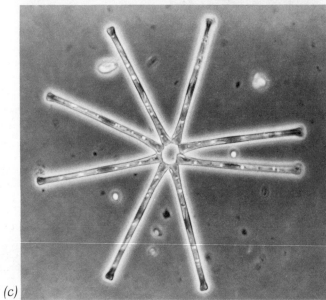

(c)

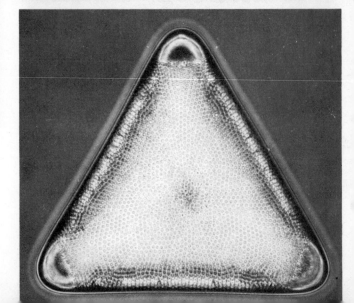

Figure 7-2 *Various diatoms.* (*a*) Gomphonema; *magnification 350×.* (*b*) Odontodiscus; *magnification 820×.* (*c*) Asterionella; *magnification 1,040×.* (*d*) Chaetoceros; *magnification 1,340×.* (*e*) Triceratium; *magnification 335×.*

the environment, and the frustule protects the cell.

The cell wall offers protection, but it is rigid and limits growth. When the diatom divides, each half of the frustule becomes the external part of a new frustule. One of the new cells is therefore smaller than the parent cell, since the small half becomes the outside half. The other remains the same size (see Figure 7-3). One line of cells gets smaller and smaller after repeated division. This sequence finally culminates in the formation of a different kind of cell, an **auxospore,** which escapes the rigid glass frustule and enlarges to restore the original size. On the death of the diatom, the frustule falls to the sediments on the sea floor. This material eventually forms a so-called **diatomaceous ooze,** which covers large areas of the sea bottom.

The frustule is protective, but it is also heavy, and planktonic diatoms must remain near the surface of the sea in order to photosynthesize: light does not penetrate very far into the ocean, even in the clearest water. Though traces of visible light have been reported at a depth of about 2,000 feet, the plant must stay in shallower waters if it is to obtain enough light energy for photosynthesis: within 50 feet of the surface in coastal waters and about 200 feet in the clearest water of the open ocean.

Diatoms use several devices to remain in the lighted surface waters. The cell wall may have long bristles or spines that increase its surface area. Since its rate of fall depends on surface friction, these projections retard its downward drift. At the same time they increase the chance that the cell will be returned to the surface in an upwelling current. Several types of diatoms extrude strands of protoplasm through the pores of the frustule, also increasing the surface area. Another flotation device in diatoms involves the products of photosynthesis. Diatoms store energy in oil droplets that are lighter than sea water. As the droplets get larger, the cell becomes lighter.

Diatom cells must obtain all their necessary materials from sea water. The carbon dioxide required for photosynthesis is present primarily in the form of bicarbonate (HCO_3^-) ions. Dissolved oxygen, as well as oxygen from photosynthesis, supports their respiration. Other necessary inorganic substances include magnesium for chlorophyll; nitrogen for protein, chlorophyll, and nucleic acids; and phosphorus for high-energy phosphates and nucleic acids. The silica that is used for the cell walls also comes from the sea water.

Most of these materials are present in abundance, but some are scarce. Though nitrogen gas from the

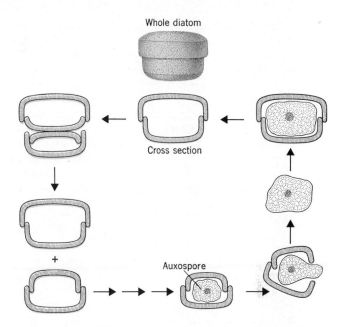

Figure 7-3 *Cell division in diatoms. The two unequal halves of the cell wall separate and become the outer halves of the two daughter cells. At each division, one of the two daughters will be smaller than the parent cell. Eventually an auxospore is formed, which escapes from the small cell wall and forms a new large cell.*

air is dissolved in the water, diatoms cannot use it. They require nitrogen in the form of dissolved nitrate ions (NO_3^-), ammonia (NH_3), or organic substances such as urea (NH_2CONH_2) or amino acids ($RCHNH_2COOH$). These nitrogenous compounds occur only in very low concentrations. Plants use phosphorous in the form of phosphate ($PO_4^\equiv$), which can also be scarce in the ocean. Iron, manganese, copper, or other elements, or organic compounds such as vitamins, may also be required in small amounts. In areas of the ocean where these critical nutrients are brought up by deep-water currents or carried out to sea by rivers or sewage outfalls, the plants become abundant. But if any of the required materials is scarce, the plant population will be correspondingly small. The factors that limit growth in any particular case are sometimes difficult to discover. Most biologists believe that nitrogen and phosphate are often in short supply.

The flotation of diatom cells may be related to their problems of nutrition, as well as to their light requirements. A sinking cell is moving; it therefore comes into contact with more dissolved nutrient molecules than does a stationary cell. The ability to regulate sinking is thus a particularly valuable trait in the nutrient-deficient regions of the open ocean.

Some aspects of diatom cell shape and composition actually increase their rate of sinking, presumably in order to increase their absorption of nutrients. For instance, many planktonic diatoms fail to separate after cell division. Instead, they form chains, decreasing their surface–volume ratios and reducing flotation. Other species become heavier when starved for nutrients. Such starved cells apparently accumulate compounds that increase their weight. Diatoms in rich culture media sink slowly or not at all. Cells from an old, nutrient-poor culture or from a natural oceanic population sink faster. Such rapid sinking, however, may carry a cell below the lighted region so rapidly that vertical currents do not return it to the surface. But in general, weight and shape appear to be carefully regulated to strike a fine balance between flotation for photosynthesis and sinking for nutrient absorption.

multicellular algae

Algae with a multicellular organization—the green, red, and brown algae—are included in the kingdom Plantae. A few multicellular algae may lead planktonic lives in certain environments. Floating masses of tangled green algae filaments often are found in quiet ponds, lakes, and estuaries. Sheets of sea lettuce (*Ulva*) six feet in diameter may drift about in sheltered bays and mudflats. However, most multicellular algae are **benthic;** that is, they are normally attached to some fixed object, called their **substratum.** Rock, shells, sand, and even mud are suitable substrata. Many multicellular algae attach to animals and other plants. The multicellular algae in the ocean are usually called **seaweeds.**

Benthic plants face an environmental situation quite different from that of planktonic ones. Many characteristics of the phytoplankton appear to be related to the movement of the cell through the water and the acquisition of dissolved nutrients. By contrast, benthic plants are fixed, and water flows past them. Their cellular surfaces act like molecular sieves, straining nutrient molecules from the moving water mass. Many benthic algae are thick plants with very low surface–volume ratios. As a result, they probably cannot absorb nutrients ef-

ficiently enough to survive in still waters or as plankton.

In the ocean, phytoplankton can occur anywhere in lighted surface waters. But the seaweeds are restricted to reefs and inshore regions, where sufficient light penetrates to support photosynthesis. As a result, multicellular algae account for only about one thousandth of the total algal biomass in the oceans of the world. The rest are unicellular algae.

Green Algae

The green algae, or **Chlorophyta,** have a wide range of structures. The Chlorophyta include single cells, filaments, and sheets of cells (see Figure 7-4). Some, particularly tropical species, have an elaborate multinucleate structure—that is, the plant body contains nuclei that are not separated by cell walls. In spite of their diversity, however, no green alga

Figure 7-4 The Chlorophyta, or green algae, include unicells and a variety of colonial forms.

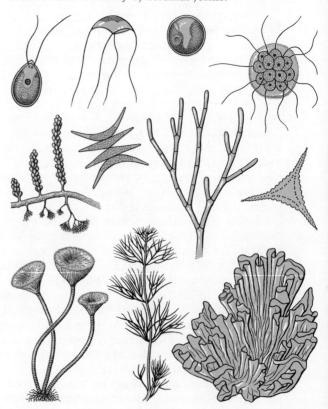

Figure 7-5 Green algae form a conspicuous scum on the surface of a lake.

Figure 7-6 Tropical green algae. A submerged colony of Caulerpa in Jamaica.

achieves the multicellular complexity of the red
and brown algae and higher plants. Often all the
cells are similar in appearance and function, even
in the larger forms.

Over 90 percent of the different Chlorophyta are
found in fresh water. Though they are mostly
single-celled or simple filamentous colonies, they
may still be the major form of plant life in fresh
water (see Figure 7-5). The marine Chlorophyta
tend to be larger, but they are rarely as abundant as
the red and brown algae, except in tropical regions
(see Figure 7-6).

Red Algae

Over 4,000 species of red algae, or **Rhodophyta,**
have been described to date. The number is increasing rapidly, in large part because of continuing sub-
tidal collections by divers. Very few red algae are
found in fresh water. Delicate, filamentous, and
highly branched species of red algae are characteristic of the tropical oceans. More massive, leafy
plants are common in cooler temperate waters (see
Figure 7-7, page 184). Another group of Rhodophyta,
the **crustose corallines,** is distributed throughout

Eutrophication

The accumulation of plant nutrients in natural waters is called **eutrophication.** This term comes from a Greek word meaning "well-nourished." Eutrophication is a natural process that occurs in many bodies of water. Lakes, for example, gradually pass through a lifespan leading from oligotrophy (little nourishment) to eutrophy (rich nourishment). An oligotrophic lake has clear water and a sandy or stony bottom. The water is nutrient-deficient and does not support extensive algal growth; such lakes are attractive for recreation and often contain desirable game fish. However, they slowly fill in. Silt and nutrients gradually convert the lake into a richer environment. Aquatic plants flourish, the bottom becomes muddy, and the fish population changes character. Eventually the lake fills with silt and disappears altogether. This process may be completed in a few decades or it may take thousands of years, depending on the particular geological features of the lake.

Human intervention usually hastens eutrophication. Artificial lakes and reservoirs created by dams have extremely short life histories. They may pass from oligotrophy through eutrophy to marshland in a few decades. Similarly, the addition of phosphates, nitrates, and other materials to lakes and rivers accelerates the progression. The practical effects of accelerated eutrophication can be extremely unpleasant. Some of the chemicals dumped into our lakes and rivers can stimulate algal growth so much that the environment deteriorates. The large mass of plant material produces a rich debris for bacterial decomposers. Oxygen may be depleted by the microbes, and noxious substances such as ammonia are released.

Phosphate pollution is a problem of eutrophication. Our use of phosphate $(PO_4^=)$ has increased much more rapidly than our population, partly because detergents have replaced soap for household cleaning: many detergents contain phosphates. Though the detergent molecule itself may be biodegradable and can be removed almost completely by secondary treatment in sewage disposal, the phosphates survive treatment. When discharged into a body of water, they become nutrients for algae.

In theory, it should be a simple matter to determine whether the added phosphates do indeed increase algal growth in a particular situation, or if some other combination of physical factors is actually responsible. For example, if the input of phosphate into a body such as Lake Erie is reduced, will the rampant growth of vegetation be reversed? In fact, this question is very difficult to resolve. Other nutrients besides phosphates have also increased in Lake Erie and in similar bodies of water. Even a 90-percent reduction in the input of phosphate might not halt the growth of undesirable organisms. Yet such a reduction would represent a major social and economic undertaking.

An experimental investigation of the problem may be very difficult. Half a dozen important nutrients could be selected and cultures set up. Each culture would contain a specific amount of each nutrient, and the amount of algal growth could be measured to estimate the interactions between nutrient concentration and growth. However, as the number of variables is increased, the number of cultures quickly becomes unmanageably large. Furthermore, each organism behaves differently. If we confine our attention to one organism that happens to be annoying, we may find ourselves faced with excessive growth of another organism and another problem, after seemingly solving the first. The situation is not hopeless, but it is certainly complicated enough to require considerable research before large reclamation projects are initiated.

Some bodies of water can recover from artificial enrichment. A reduction of nutrient input has produced a dramatic improvement in water quality in some cases. But even when reduced nutrient input causes improvement we still cannot be sure which particular compound was responsible for the algal growth in the first place. In any case, it is safe to say that accelerated eutrophication almost always decreases environmental quality and should be avoided whenever possible.

(a)

(b)

Figure A   The eutrophication of a lake. (a) Initially, the lake is poor in nutrients and supports little algal growth. (b) As nutrient levels increase through runoff from the surrounding land, vegetation becomes more dense. (c) The lake is finally filled in with sediment and detritus and becomes a bog or marsh.

(c)

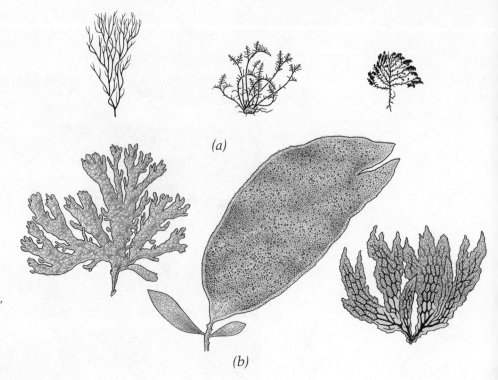

(a)

Figure 7-7 *The Rhodo-phyta, or red algae. (a) Species from tropical seas tend to be small and delicate (b) Species from cooler, temperate areas are larger and more massive in organization.*

(b)

the world. These plants deposit calcium carbonate ($CaCO_3$) in their cell walls. The resulting plant often resembles a lump of pale pink coral, hence its name. Crustose corallines are an essential part of the structure of many atolls and coral reefs. Reef margins are held together by tough cementing layers of coralline algae. These algae flourish in the heavy surf of reef margins, where brittle corals would be fragmented (see Figure 7-8).

The larger, leafy red algae of the temperate regions often dominate rocky coastlines, along with brown algae. Certain species have been used for centuries as food and medicine by human populations on the shores of the Pacific, particularly the Japanese, Chinese, Hawaiians, and American Indians. Europeans collect red algae for use as goat and sheep fodder. The cultivation and harvesting of a red alga called *nori* is a major industry in Japan. *Nori* is a thin, purple-colored rhodophyte that is used as a food supplement. Japanese ocean farmers grow *nori* on nets attached to bamboo poles in quiet bays. The plants are plucked from the nets at low tide (see Figure 7-9). Another important product derived from Rhodophyta is agar, a substance extracted from the cell walls of several species. Agar is a polysaccharide with many of the proper-

ties of gelatin. It has many uses, but it is particularly valuable in microbiology. The agar is mixed with water and nutrients, heated, and then allowed to cool and gel. The gel provides a solid but moist surface for the growth of microorganisms.

Brown Algae

The brown algae, or **Phaeophyta,** comprise fewer species than the preceeding groups, but they are by far the largest and most conspicuous seaweeds in cooler waters. Nearly all of them are marine. None is unicellular, and only a few of the brown algae are simple filaments.

The most spectacular brown algae are the huge kelps. In shallow rocky areas, kelp plants may become so abundant that their beds are analogous to forests on land. A single kelp plant may provide food, shelter, and a colonizing surface for hundreds of fishes, crabs, and snails, and for tens of thousands of smaller marine animals.

Perhaps the most complex kelp, and certainly the largest, is *Macrocystis,* the giant kelp (see Figure 7-10). It grows up to the surface from depths as

Figure 7-8 (*left*) Coralline algae thrive in heavy surf and are important components of coral reefs.

Figure 7-9 (*right*) Nori. These red algae are grown on nets that are exposed at low tide.

Figure 7-10 (*left*) Young plants of the giant kelp, Macrocystis.

great as 130 feet, and growth rates of two feet per day have been measured. Its base is a massive rootlike holdfast, with small tendrils that grow around rocks and into crevices to anchor the plant. Above the holdfast the "stem," or **stipe**, branches repeatedly to give hundreds of secondary stipes, each one of which is clothed with dozens of leaf-like blades. At the junction of each blade and the stipe is an inflated gas bladder. These gas bladders serve to float the stipes toward lighted surface waters.

The floating parts of the plant form a canopy which effectively shades everything in the water below, including its own holdfast and younger, shorter stipes. Since these tissues cannot photosynthesize, organic material from the actively photosynthesizing blades must be transported downward to support the growth and metabolism of the holdfast cells and the shorter stipes. For this purpose, long cells that have many of the characteristics of the transport cells of higher plants run down the stipe. In many of the large brown algae, the stipe is a food reserve area as well as a transport system.

Some of the brown algae are sweet in taste and are used as human food. However, they are good only as dietary supplements and not as staples. Their carbohydrates are not readily digestible; they sweeten and thicken food, but have little nutritional value for man. *Macrocystis* is harvested for fertilizer and is also a source of algin, a gelatinous compound like agar, widely employed as a food thickener.

Another kelp, *Sargassum*, is as important in the tropics as *Macrocystis* is in temperate waters, in terms of abundance. *Sargassum* is shorter and bushier than giant kelp. It has smaller gas bladders and blades shaped like small oak leaves. There are dozens of benthic species of *Sargassum* in coastal areas (see Figure 7-11).

Sargassum gave its name to the Sargasso Sea, a region of the Atlantic Ocean east of Florida that is covered by a mass of floating kelp. This aggregation probably originated from fragments of kelp that broke away from the coast, were carried out to sea, and became trapped in a giant eddy. At the present time this floating *Sargassum* comprises the largest stretch of uniform vegetation in the world. The plants have accumulated over a geological time span and now grow continuously in open ocean. Specially adapted fishes, crabs, shrimps, and other animals live with the weed.

Figure 7-11   Sargassum, *a brown alga.*

the intertidal environment

With few exceptions (such as *Sargassum*) large multicellular algae grow only in shallow waters, because they require both light and a substratum for attachment. The closer inshore the plant moves, the more favorable the environment becomes, at least in some respects. The average light intensity increases, and nutrients are supplied by runoff from the land. Yet the environment also becomes more complex and more hazardous. The water temperature changes more rapidly along the shore, particularly in intertidal regions (the areas between the high and low-tide marks). Rain may dilute the salts in shallow water, or evaporation may concentrate them. But a plant that can tolerate the violent changes from submergence to exposure escapes competition from all those forms that cannot adjust. *Fucus*, a common brown alga, is a good example. This alga clings to rocks with a disc-shaped holdfast capable of withstanding violent wave action on unprotected coasts. It is often exposed to air at low tide, and may lose three fourths of its water through evaporation, yet it still recovers when the tide comes in again. During low tide, the film of water trapped between its blades loses its dissolved oxygen very quickly as the plant respires. During the day, more oxygen is supplied by photosynthesis. But when a low tide occurs at night, almost all of the plant's tissue must survive without any oxygen until the tide returns.

Each seaweed tolerates a particular range of temperature, drying, lack of oxygen, and wave action. This tolerance limits its spread toward the high-tide mark. But as it spreads in the other direction, toward the low-tide mark, it encounters more and more competition for space and light from other plants. These contrary pressures produce **zonation;** in other words, the intertidal portion of the shore shows clear lines of demarcation, or zones, where first one and then another plant dominates and monopolizes the available space. Zonation may sometimes be quite spectacular and obvious (Figure 7-12). It is not restricted to the intertidal region but extends on down beyond the low-tide mark. Many factors, such as the nature of the substratum, currents and waves, and the presence of animals, help determine which plants occur in which zones. As we saw in the last chapter, differences between the photosynthetic pigments of different plants may also influence the relative occurrence of green, red, and brown algae at various depths. Pigment is another factor in zonation.

the origin of land plants

The brown algae have developed some of the functional attributes of land plants. Their growth patterns are comparable to those of land plants, and they have specialized cells and tissues for photosynthesis, transport, attachment, and reproduction. Also, they can tolerate exposure to air and drying to some extent. Yet the brown algae are almost certainly not the plants that gave rise to our present land plants. Instead, land plants probably arose from types that had adapted first to fresh water and later to moist soil; it is unlikely that plant life reached land directly from the ocean shore.

The pigment composition, storage products, and cell wall structure of green algae suggest that their ancestors, rather than the browns, were the progenitors of land plants. In addition, the green algae are the most successful group of algae in both fresh water and soil. Many greens penetrate shallow bays or river mouths and seem equally happy in both fresh and marine environments; by contrast, few red and brown algae can survive dilution of sea water.

Figure 7-12 Intertidal zonation. The brown alga called rockweed forms a distinctive zone on intertidal rocks; a zone of red algae called Irish moss occurs lower down, at the water's edge. The zone of human feet is transient.

Barnacles   Rockweed   Mussels   Irish moss

Terrestrial Plants

Unicellular and multicellular algae are the dominant plants in water. Some algae occur in moist soils, but they lack the adaptations necessary to exist in relatively dry environments. Plants on land are organized very differently. They have made various kinds of adaptations that help them exist in environments where the water supply is limited. These adaptations are the basis of land-plant classification.

mosses and liverworts

The simplest land plants are the liverworts and mosses, called the Bryophyta. They resemble multicellular algae in that they have a relatively simple internal organization. Like algae, they have no real roots or other specialized cells or tissues for water transport. They also lack tissues for transporting photosynthetic products any distance from the site of photosynthesis. However, the Bryophyta have developed specialized hair-like cells that function as primitive roots; they anchor the plant and absorb water and nutrients (see Figure 7-13).

Photosynthesis occurs in cells on the upper sur-faces of the plant. Since they have no conducting tissue, the Bryophyta are necessarily rather small; all parts of the plant body must be close to the sites of water absorption and energy production. The Bryophyta also need liquid water for sexual reproduction, to enable the sperm to swim to the egg. Consequently, Bryophyta are restricted to damp habitats. They are probably very similar to the first plants that invaded the land.

The remainder of terrestrial plants have all developed **vascular tissues,** tissues for conducting water and nutrients throughout the plant body. The three most important groups are the ferns, the conifers, and the flowering plants.

(a)

Figure 7-13 *Bryophyta.* (a) Riccia, *a common liverwort.* (b) *Its internal structure. The rhizoids serve as primitive roots.*

Chlorophyll–bearing cells

Air chambers

Upper surface

Storage tissue

Lower surface

Rhizoid

(b)

ferns

The ferns, or Filicinae, have small roots, underground stems, and upright fronds divided into small leaflets (Figure 7-14). Ferns can be quite large, because they are able to transport water from the ground to their aerial parts, and photosynthetic compounds from leaves to roots. In past ages, ferns were numerous, and many attained large size. Some present-day ferns are the size of small trees (see Figure 7-15). However, like the Bryophyta, the ferns require liquid water in their environment for sexual reproduction.

In the past, ferns and their allies were the dominant form of land vegetation. Forests composed of tree ferns formed the major coal beds three hundred million years ago. The ferns were slowly displaced by plants that were better adapted to terrestrial conditions—the seed plants. At present ferns persist in moist environments: along streams, in bogs, or in damp tropical forests.

Stem

Roots

Figure 7-14 (left) Structure of a fern.

Figure 7-15 (below) A tree fern.

seed plants

The seed plants possess a means of reproduction that represents the final step in the adaptation of plants to land existence. They do not require liquid water for reproduction. The sperm does not swim to the egg; instead, it is encased in a pollen grain that can be transferred by wind or by insects to an egg in another plant. The pollen fertilizes the egg, which develops into an embryo; and the embryo becomes encased in a seed. Water is supplied by the parent seed plant, not by the environment. Seed plants have various mechanisms to conserve the water present in their tissues so it does not all evaporate into the air.

There are two main groups of seed plants: conifers (and related plants) and the flowering plants. The conifers and their relatives are collectively called **gymnosperms** (literally, "naked seeds"). The name indicates that their seeds are not protected by external coverings supplied by the parent plant. The flowering plants are called **angiosperms** ("vessel of seeds"), because their seeds are enclosed in vessel-shaped containers that are derived from the flower (see Figure 7-16). In this chapter we will use the structure of flowering plants to illustrate the basic problems of life on land. Much of the discussion applies with equal validity to the gymnosperms.

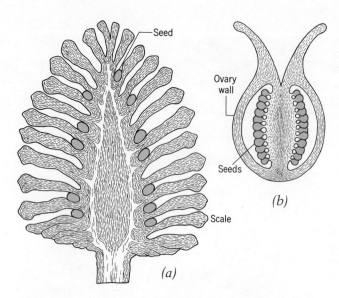

Figure 7-16 Reproductive structures of seed plants. (a) The cone of a gymnosperm has "naked" seeds. (b) The flower of an angiosperm has tissues that enclose the seeds in a "vessel."

flowering plants

There are perhaps 285,000 species of flowering land plants. Presumably each one of them has evolved as an adaptation to a particular environment. Most often, a flowering plant can be described simply as a cluster of photosynthetic organs (usually leaves) connected by a vascular system (the stem) to absorptive organs in the soil (usually roots). The leaves and the roots are separated by large distances in terms of the size of the individual cells. The leaves must be supplied with inorganic ions and water from the roots, and the roots must be supplied with sugar and other nutrients from the leaves. Since diffusion cannot serve as an efficient transport mechanism beyond a few millimeters, the vascular system has to be rather sophisticated in order to conduct water and dissolved substances over such great distances (Figure 7-17).

The Leaf

The typical leaf is thin and flattened, so that a large area can be exposed to sunlight. Each leaf is at-

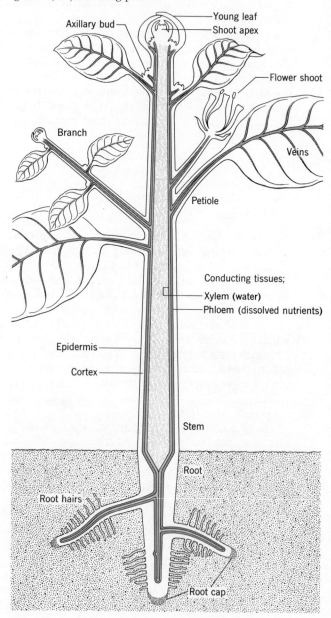

Figure 7-17 Schematic diagram of a flowering plant.

tached to the stem at a specialized region at its base called a **petiole.** The petiole is the connecting point between the vascular conducting tissues of the stem and the leaf. Conducting tissue in the leaf itself is often visible as a network of veins.

Leaves are so conspicuous and so variable that their form is used as an aid in identification of plant species. An extensive list of terms has been generated to describe leaves. The important features are size, overall shape (elliptical, heart-shaped, lanceolate, and so on), the nature of the leaf edge (smooth, saw-toothed, deeply lobed), the conspicuousness of the petiole, the arrangement of leaves on the stem, and so forth. There are also simple and compound leaves. A **simple leaf** may be complicated in shape, but it always consists of a single blade borne by the petiole. A **compound leaf,** by contrast, is a series of leaflets on a single petiole. Superficially these leaflets look like individual leaves, but one can always tell them apart by remembering that a leaf is attached directly to a stem by the petiole (see Figure 7-18). The angle formed by the stem and the petiole is called the **axil** (from the Latin word for "armpit") and it always bears a bud, called the **axillary bud.**

The cell structure of leaves reflects their requirements for light absorption, gas exchange with the environment, and water conservation. The leaf's surface is covered with a layer of flattened, transparent cells called the **epidermis** (Greek for "outer skin"). These cells lack chloroplasts. They secrete a waxy layer, or **cuticle,** which is impermeable to water. Light passes through the epidermis to the underlying layers, called the **mesophyll** (literally, "middle of the leaf").

The upper region of the mesophyll is an orderly array of elongated cells. They look a bit like a stake fence and hence are called the **palisade layer.** Beneath the palisade layer is a more diffuse arrangement of mesophyll cells called the **spongy mesophyll.** The cells of the mesophyll are packed with chloroplasts. The cells of both the palisade layer and the spongy mesophyll are all in contact with an internal air space. The leaf is basically just a transparent box, covered by waterproof wax and containing air spaces and a large number of chloroplast-bearing cells (Figure 7-19).

The air space within the leaf is saturated with water vapor, since it is in intimate contact with the moist mesophyll cell walls. The leaf's waxy surface prevents excessive water loss across the surface. However, the internal air space must be open to the surrounding environment in some way, in order to

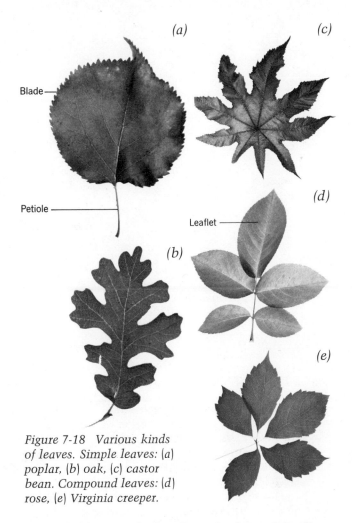

Figure 7-18 *Various kinds of leaves. Simple leaves:* (a) *poplar,* (b) *oak,* (c) *castor bean. Compound leaves:* (d) *rose,* (e) *Virginia creeper.*

provide the mesophyll cells with the carbon dioxide required for photosynthesis and the oxygen for respiration. But any mechanism that permits gas exchange also permits the loss of water vapor. To survive in air, therefore, a leaf must have an organization capable of conserving water while at the same time providing for gas exchange.

The major device for water conservation in plants is the system of **stomata** (Greek for "mouths"). Stomata are small pores that pierce the surface of the leaf at intervals. They allow for efficient gas movement between the air spaces near the mesophyll cells and the surrounding air (see Figure 7-20). But even though the stomata are small relative to total leaf area, considerable quantities of water are lost when they are open. The reason is that each small hole represents a point source of saturated air, which can diffuse out in all directions. By contrast, if there were just one large hole the size of all the small ones combined, diffusion would occur in only

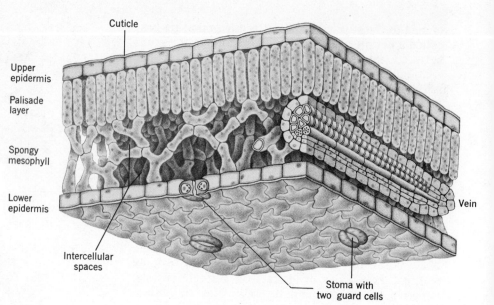

 labels: Cuticle, Upper epidermis, Palisade layer, Spongy mesophyll, Lower epidermis, Intercellular spaces, Vein, Stoma with two guard cells

Figure 7-19 *The internal structure of a foliage leaf (approximately 20× life size).*

one direction—away from the surface—except around the hole's edges. Because water can diffuse in all directions from each small hole, much more water is lost through a series of small holes than through a single large one (Figure 7-21). However, for the same reason, the efficiency of exchanges of carbon dioxide and oxygen is high, and therefore beneficial.

The stomata may open to permit free diffusion or close to isolate the internal air space and prevent excessive water loss. Specialized epidermal cells, called **guard cells,** border each stoma and control its opening and closing. The mechanism involves the turgor of the guard cells. Contrary to what one might expect, the stomata open when the guard cells swell. The reason is that the guard-cell wall is much thicker on the surface facing the pore. Thus, as the cell takes up water and swells, its thin wall curves outward, stretching the thick walls and opening the air channel (see Figure 7-22).

The behavior of the guard cells depends on several factors in the environment. The presence of light tends to cause the stomata to open. A low carbon dioxide concentration in the air space of the leaf also produces opening. A low water supply causes them to close. Sometimes these environmental influences act in concert and sometimes they oppose each other, so stomatal behavior is not always easy to predict.

Several theories attempt to explain the control of stomatal opening at the molecular level. Perhaps the most cogent hypothesis suggests that the controlling factor is the carbon dioxide concentration in the guard cells. When it is reduced (during photosynthesis, for example) the quantity of organic acids in the cell also goes down, as they are broken down to produce more carbon dioxide by the reaction

$$R-\underset{\substack{\| \\ \text{O}}}{C}-OH \rightleftharpoons R-H + \underset{\substack{\| \\ \text{O}}}{C}=O$$

organic acid   organic acid   carbon
                  precursor   dioxide

The breakdown of organic acids lowers the acid content of the cell. The acidity, in turn, controls the conversion of starch polymers to glucose monomers. When acidity is low, the enzyme that breaks down starch to glucose becomes very active. The resulting conversion of a few large polymers into many smaller units increases the osmotic concentration of the guard cell. The increased concentration in turn causes the cell to absorb water and swell. The stoma opens. Thus, whenever the carbon dioxide concentration is lowered, the stomata open, providing that the plant has an ample supply of water. If the plant is suffering from water stress, the guard cells may not be able to obtain enough water to open, in spite of the regulatory mechanism.

This theory of regulation is consistent with many experimental facts. Nevertheless, there are some plants that behave quite differently. Cacti and other plants adapted to dry conditions are a good ex-

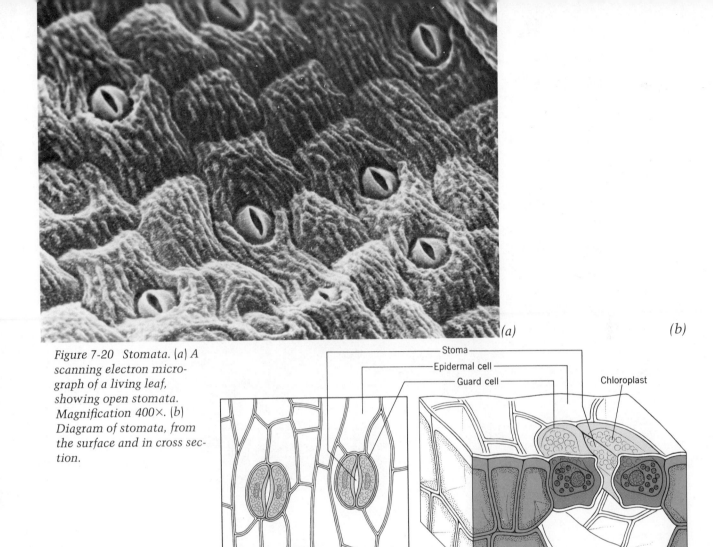

Figure 7-20   Stomata. (a) A scanning electron micrograph of a living leaf, showing open stomata. Magnification 400×. (b) Diagram of stomata, from the surface and in cross section.

Stoma
Epidermal cell
Guard cell
Chloroplast
Air space

Figure 7-21   Overall, more water evaporates from an open water surface (a) than through pores (b). However, less water evaporates per unit of exposed water surface.

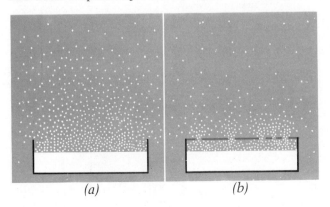

(a)   (b)

ample. Such plants open their stomata at night and keep them closed during photosynthesis in the daytime. They use a metabolic pathway that fixes carbon dioxide at night rather than during daylight photosynthesis. The carbon dioxide is trapped in the form of organic acids. Then, during the day, these acids are converted to carbohydrate with the aid of the $NADPH_2$ and ATP produced in photosynthesis. At night, when organic acid concentration is high, the stomata are open. This system appears to be completely the opposite of the behavior described above. However, the concentration of organic acids is so high at night that the acid molecules themselves increase the guard cells' osmotic concentration (and thus their water uptake). The

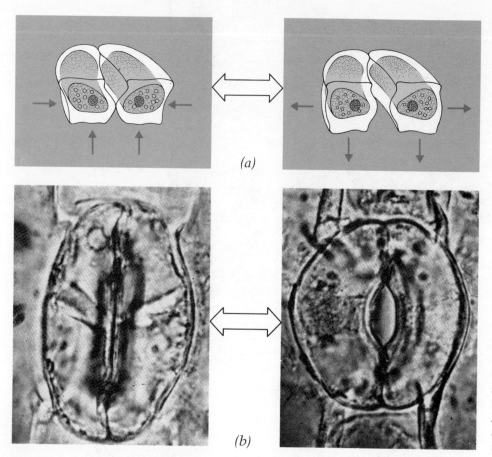

Figure 7-22 Action of the stomata. (a) As the guard cells take up water, their inner surfaces curve outward to open the stoma. (b) A stoma of an onion leaf in a closed and open condition.

conversion of the organic acids to carbohydrate during photosynthesis allows the stomata to close during daylight hours. This mechanism is excellent for conserving water; the internal tissues are exposed to the environment only at night, when temperatures are low and humidity is high.

The more typical plants open their leaf stomata during daylight hours and are very prodigal with water as a result. Table 7-1 lists the number of gallons of water required for the growth of a pound of each of several crop plants. More graphically, Figure 7-23 shows the total water consumption of a typical plant.

Water continuously passes through the plant and is lost to the air in the process called **transpiration.** The quantity of water passed through the leaves in this way is a major climatic factor. Rainfall that might otherwise run off to the ocean is trapped by plants and returned to the atmosphere. In fact, the

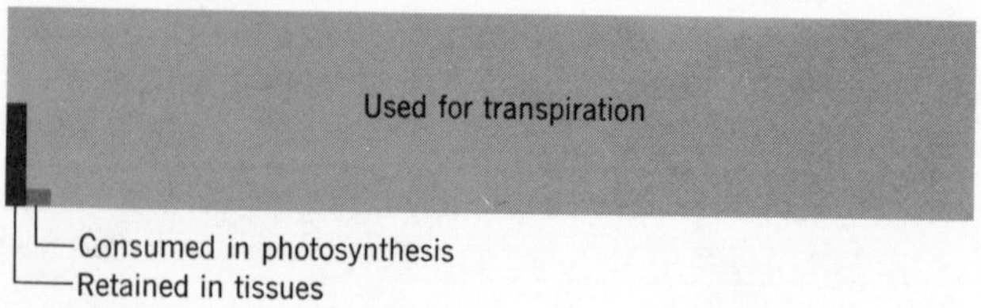

Used for transpiration

Consumed in photosynthesis
Retained in tissues

Figure 7-23 The use of water by a corn plant. Most of the water is transpired from the soil to the atmosphere. A small fraction (about 2%) is retained in the plant, and only a fraction of that is used in photosynthesis.

Table 7-1

Water Requirements for Production of Various Foods

	Pounds of Water Needed to Produce One Pound of Dry Food
Millet	225
Wheat	500
Potato	800
Rice	2,000
Milk	10,000
Meat	20,000–50,000

atmosphere receives more water from plants in warm, dry areas of the land than it receives by evaporation from the ocean surface.

The Root

The source of water for higher plants is the soil. But not all of the water in the soil is available to the plant. Consider what happens when dry soil is thoroughly wetted by rain or irrigation. Some water drains away under the force of gravity. Some is bound weakly by capillary attraction. The remainder is held in the soil by strong binding forces between charged soil particles and polar water molecules. Plants can obtain water from the soil only when it is present above a critical concentration. A plant may wilt (that is, fail to obtain enough water) even though there may still be a great deal of water in the soil. This is the case particularly if the soil particles are very fine. When water is below a certain concentration, the forces binding the water to the soil particles are so strong that the water is no longer available to the plant.

The mechanism that absorbs water and minerals is, of course, the root system. At the center of the root lies the **stele**, which contains the conducting tissues. The stele is surrounded by a tube of cells called the **endodermis.** Around this layer is the thick **cortex** (From a Latin word meaning "bark"). The cortex, in turn, is covered by an **epidermis,** or outer skin of cells. Some cells of the epidermis form extensions near the tip of the root. These so-called **root hairs** increase the surface available for absorption and are the main sites of water absorption (see Figure 7-34).

Water and minerals from the environment penetrate freely into the space between the cells of the epidermis and the cortex, as far as the endodermis. But the cell walls of the endodermis are tightly joined, forming a waterproof barrier. Water and solutes must therefore enter the stele by another route. Nutrient ions in the soil water between the cells outside the endodermis are transported across the cell membranes and become concentrated in the cytoplasm of the endodermal cells. (Since the ions are moved against a concentration gradient, their transport into the cell cytoplasm requires energy.) The cytoplasm of adjacent cells is connected through pores in the cell walls by minute extensions of the cytoplasm called **plasmodesmata.** The accumulated ions can diffuse freely through plasmodesmata from cell to cell. In this way, nutrients that cannot pass between the endodermal cells can move through them into the cytoplasm of cells in the stele.

Some of the nutrient ions pass out of the stele cells into the surrounding fluid, inside the endoderm barrier. The osmotic concentration of the fluid in the stele is therefore higher than that in the soil water. As a result, water moves across cell membranes into the stele (see Figure 7-24). The inward water movement tends to pump water upward in the plant. This so-called **root pressure** is occasionally quite high, but it is probably not very important in moving water through the plant even when high. Probably a more important function of the endoderm layer is to control the passage of ions and other molecules in and out of the stele. For example, the endodermis may facilitate the entry of potassium while rejecting sodium. The endoderm monitors the flow of inorganic nutrients into the stele and the rest of the plant body.

Conducting Tissue

The stele contains two major types of conducting tissues, xylem and phloem (Figure 7-25). These tissues make up the circulatory system of the plant, also called the plant's vascular skeleton.

Water Transport

Xylem is responsible for water movement throughout the plant. Water moves through two types of xylem conducting elements: **vessels** and **tracheids.** Xylem vessels are tubular cells joined end to end; their end walls have dissolved away to produce long, continuous tubes. The tracheids are

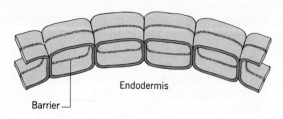

Endodermis

Barrier

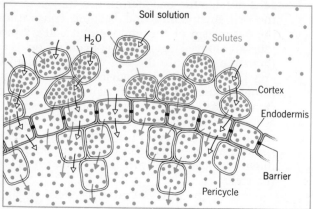

Soil solution

H₂O

Solutes

Cortex

Endodermis

Barrier

Pericycle

Figure 7-24 Movement of water into the root. Water and dissolved materials penetrate into the intercellular spaces of the root cortex. A waterproof barrier, formed by the cell walls of the endodermis, prevents them from moving further. Solutes must therefore pass through the endodermal cells rather than between them. The osmotic concentration of solutes becomes higher in the stele than outside, and water enters osmotically.

elongated elements with closed ends, but water can move through their walls into adjacent tracheids. Most of the xylem in trees such as pines is composed of tracheids. Vessels are more common in flowering plants (see Figure 7-26).

The living cells that secrete the tubular wall of a vessel or a tracheid die before the wall is used for water transport. Thus, the tubular elements of the vessels and tracheids are analogous to the inert pipes in plumbing. Xylem is continuous from root to leaf. Each tiny vein in a leaf contains xylem elements. These tubules join with major leaf veins, pass through the petiole to the stem, and thence down to the tip of the root. The continuous chain of xylem elements from leaf system to roots contains an unbroken column of water.

Since the xylem elements are simply dead cell walls that form conduits to the leaf, how do we account for the tremendous water movement that occurs in transpiration? Most evidence suggests that evaporation of water from the leaf draws water

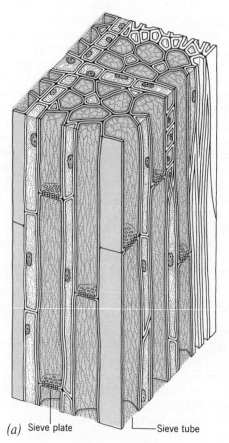

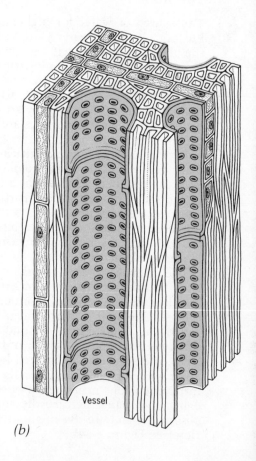

Figure 7-25 Conducting tissues in plants. (a) Phloem tissue transports dissolved compounds; it is composed of sieve tubes and accompanying companion cells. (b) Xylem vessels and tracheids conduct water through the plant from soil to leaf.

(a) Sieve plate — Sieve tube

(b) Vessel

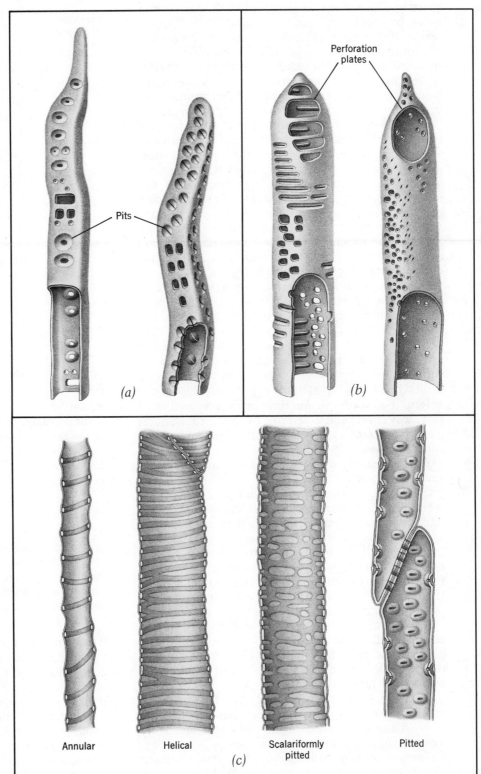

Figure 7-26 Xylem elements. (a) Tracheids are long cells with pointed ends. The cell walls may have pits communicating with adjacent tracheids. (b) Vessel elements are shorter cells that lie end to end and communicate by holes or perforated plates, forming continuous tubular vessels. (c) Both tracheids and vessels develop various kinds of secondary wall thickening and pitting.

Pits

Perforation plates

Annular   Helical   Scalariformly pitted   Pitted

(a)   (b)   (c)

up through the plant. As we mentioned, the air space within the leaf is saturated with water vapor. In other words, the relative humidity within the leaf is 100 percent. Outside, the surrounding atmosphere usually has a lower relative humidity. Since molecules move from regions of high concentration to regions of low concentration, water vapor diffuses out of the leaf whenever the stomata are open. The water that diffuses out of air spaces in the leaf is replaced by water evaporating from the walls of the mesophyll cells bordering the air spaces. These cells, in turn, absorb water from neighboring cells that have a higher water content. Water is ultimately pulled from the xylem elements in the leaf veins. Since the xylem itself contains unbroken columns of water from leaf to root, water moves all the way upward from the soil through the roots, stem, and leaves, all because of suction generated by the diffusion and evaporation of water from the leaf (Figure 7-27).

The great suction forces in xylem have been measured. First, the stem of a plant is cut. Suction pulls the water in the xylem away from the cut surface of the stem. The cut stem with attached leaves is then placed in a closed container, and pressure is applied until the water in the xylem just returns to the cut surface. At this point, the pressure in the container should be exactly equal to the suction pressure in the xylem (Figure 7-28). Suction pressures measured this way may equal ten to twenty atmospheres (that is, ten to twenty times the standard atmospheric pressure).

This account of water movement in plants sounds very mechanical. With a little effort, one could design a plastic plant that would work as well. Although xylem cells function as inert pipes, one must remember that the plant is an organism, and the flow of water and inorganic material is under control. In most cases, the leaf has overall control of the flow rate. But the entry of solutes into the root is controlled by the root endoderm layer and other structures. Furthermore, once inorganic ions have entered the stele and become part of the xylem stream, they are not always carried passively throughout the plant. Potassium and phosphorus are extremely mobile in the plant; other solutes, such as calcium, move much more slowly and tend to stay put.

Many biologists have argued that transpiration, though interesting, is not in itself beneficial to the plant. They suggest that excessive water loss is no more than an unhappy consequence of gas exchange. However, others feel that transpiration may benefit the plant in several ways. Water evapo-

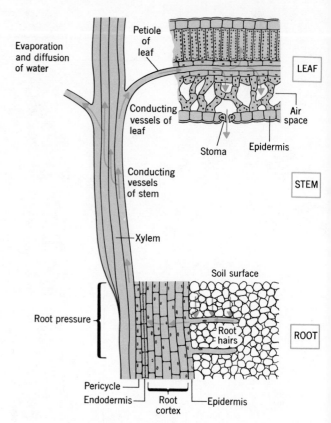

Figure 7-27 The major force drawing water through the plant is evaporation at the leaf. Since the water column in the xylem is continuous from leaf to root, evaporation at the leaf draws water upward. Root pressure is less important in water movement and cannot be detected at all in some plants.

ration from leaf surfaces has an important cooling effect. Without transpiration, the leaf might be damaged by high temperatures. The transpiration stream also speeds the transport of inorganic solutes from the soil.

Transport of Organic Compounds
Organic materials move through the other major conducting tissue, the **phloem.** The phloem system is even more complex than the xylem. Some compounds move only downward, from the stem to the roots. Others move in different directions, depending on circumstances. For example, sugar may move from a photosynthesizing leaf both toward the roots and upward, toward a developing bud at the tip of the stem.

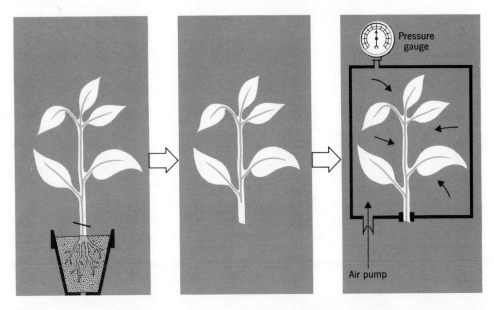

Figure 7-28 *When the plant is cut, water in the xylem of the shoot portion pulls away from the cut surface. The plant is placed in a pressurized container, and the pressure is increased until the fluid returns to the cut surface; at this point, the positive pressure in the container is just equal to the negative pressure that had been exerted by evaporation of water at the leaves.*

The basic functional element in phloem, called a **sieve tube,** begins its development as an elongated cell with a nucleus and the usual organelles. But as the cell matures, it becomes a pipe-like element. Each sieve tube is connected to the adjacent sieve tube by an end plate that is pierced by pores or slits, which give it its name. The protoplasm of mature phloem elements no longer contains a nucleus, mitochondria, or vacuoles. Next to the sieve tube elements are cells called **companion cells.** They are more normal, retaining their nuclei and other organelles. Companion cells presumably communicate with the sieve tube elements through protoplasmic connections. The companion cells may carry on metabolic functions for the sieve tubes as well as for themselves (Figure 7-29).

Three centuries ago, Marcello Malpighi demonstrated that materials move in the phloem, by simply removing a ring of bark from a tree. This procedure cut a segment out of all the phloem that moved through the stem at that level, but it left the xylem intact. Below the ring the bark withered away. The tissue above, however, swelled and remained healthy. Removing a ring of bark—called **girdling**—has many practical uses. Girdling is used on the stems of grapevines to ripen the grapes; it prevents the transfer of sugar through phloem to the roots; excess sugar then accumulates in the fruit. The sap of the sugar maple, used for maple syrup and sugar, is also phloem exudate. But despite such practical information, the details of phloem transport have only recently begun to emerge.

Phloem cells are very difficult to study. They are thin-walled, fragile, and quite sensitive to changes in their surrounding conditions. Two important innovations in technique have facilitated studies of phloem. The first is the use of radioactive tracers. As soon as these compounds became available, about 25 years ago, plant physiologists used them to trace the movement of material in plants. Much of tracer techniques. The second innovation is the use of aphids, or plant lice. These insects feed on the phloem stream by inserting a needle-like proboscis into a single phloem cell. A biologist places an aphid on the stem and allows it to puncture a phloem cell and begin feeding. Then the insect is anaesthetized so that it can be decapitated. The

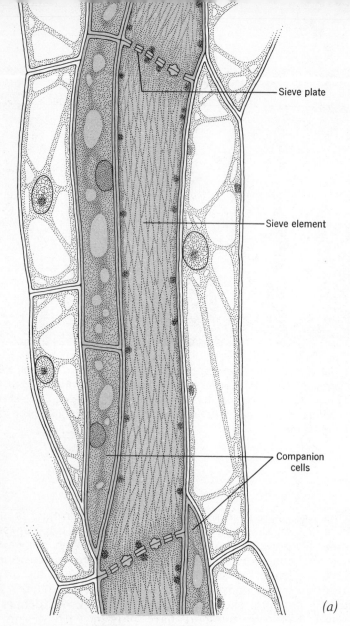

Sieve plate

Sieve element

Companion cells

(a)

proboscis remains in the stem and serves as a tap into a single phloem cell. Phloem exudate drips out of the cut tip to be collected and analyzed (see Figure 7-30).

Using these techniques we have learned that ten to 25 percent of phloem sap is carbohydrate, while a smaller amount consists of amino acids. The carbohydrate is mostly sucrose. The aphid digests the nitrogenous amino acids but ignores the abundant sucrose. It passes most of the sucrose through its digestive tract, producing what are known as **frass drops.** Dry frass drops may be the "manna" of the Old Testament; they are still collected and eaten in some parts of the world.

Aphids have also been used to measure the movement of material in the phloem. First, radioactive compounds are supplied to a leaf. As the material moves down the phloem, radioactivity begins to drip from aphid probosci placed at various points down the stem. The time it takes for radioactivity to appear along the stem gives a measure of how fast the phloem is moving. This rate is actually slow compared to that of xylem transport, which may be several hundred feet per hour. Nonetheless, phloem sap circulation in plants is too fast to be caused by diffusion alone.

Phloem movement is best explained by the pressure-flow scheme first proposed in the 1930's. Consider two sacs of semipermeable material connected by a tube. One is filled with a relatively concentrated solution of dissolved substances, the other with a more dilute solution. If the two sacs are placed in an external solution that is also more dilute than the first sac, osmotic pressure will

Figure 7-29 Phloem. (a) A sieve tube element and companion cells. (b) Electron micrograph of pores in a sieve plate. Protein filaments and endoplasmic reticulum are sometimes found near the pores; their function is unknown. Magnification 38,000×.

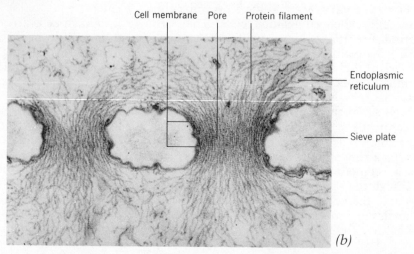

Cell membrane   Pore   Protein filament

Endoplasmic reticulum

Sieve plate

(b)

(a)

(b)

Figure 7-30 Use of aphids to study phloem transport. (a) An aphid feeds on the phloem stream by means of a needlelike proboscis inserted through the bark. The droplet of fluid at the tip of the abdomen (a frass drop) is rich in sugar. (b) After the aphid is decapitated, the detached proboscis exudes phloem sap for hours. (c) Micrograph shows an aphid proboscis piercing a single phloem cell. Magnification 600×.

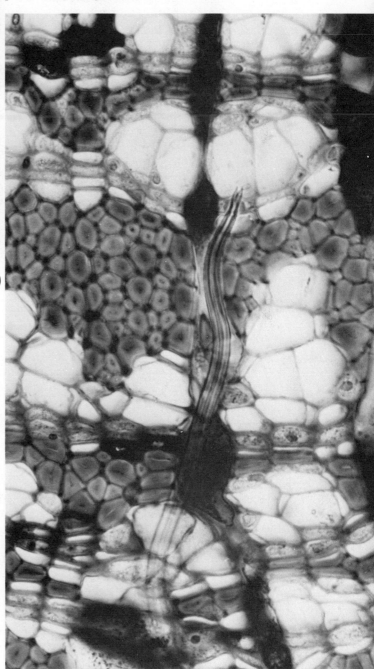

(c)

move material from the more concentrated sac to the less concentrated one. In plants, a photosynthesizing leaf cell is equivalent to the sac with the concentrated solution. The cells in the root, stem, and fruit that do not photosynthesize are analogous to the dilute sac. Phloem elements represent the connecting system, and the whole is bathed in the dilute water solution found between the cells and in the xylem (see Figure 7-31).

The leaf cell will maintain a high osmotic concentration as long as sugars and other compounds are produced in photosynthesis. The non-photosynthetic root cells represent a "sink" region. That is, they are the area where sugar is metabolized or removed from solution for storage. Thus, the non-photosynthesizing cells maintain an osmotic concentration lower than that of the leaf. In fact, the leaf cell is about ten times more concentrated than a root cell. And so long as the concentration difference is maintained, a force is generated which pumps fluid from leaf cell to root cell via the phloem.

The model provides a very satisfactory explanation for the rapid rates of material transport in

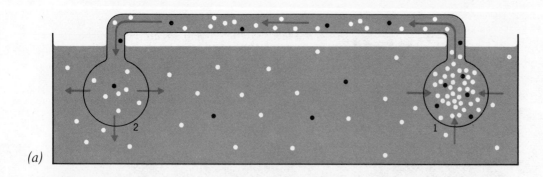

(a)

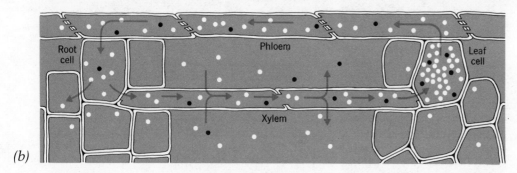

(b)

Figure 7-31 The pressure-flow scheme for the movement of substances in the phloem. (a) In this hypothetical model, 1 and 2 are semipermeable sacs connected
by a tube and suspended in a common medium. The
osmotic concentration in sac 1 is higher than that of
the medium and higher than that of sac 2. Therefore,
water enters sac 1 osmotically and is driven out at sac
2 producing a bulk flow between the two sacs. This
flow continues as long as the concentration differences
are maintained. (b) The elements of the model may
correspond to the plant tissues as shown.

phloem. This scheme also accounts for the flow of
material in different directions. Whenever one
plant cell is less concentrated than another cell,
material is moved from one to the other via the
phloem. A leaf may supply photosynthetic products
to roots or to fruit or to developing buds.

Let us summarize the distribution of water and
solutes in plants. A flow of water, containing dilute
inorganic salts, passes from the root to the leaves
along the continuous tubes of the xylem. The
driving force is the evaporation of water at the leaf.
Such evaporation produces sufficient suction to
draw water from the soil and up through the xylem.

The total flow is controlled by the opening and
closing of the leaf stomata. The nature of the solutes dissolved in the water is controlled by the endoderm layer in the root.

Organic compounds, produced by the photosynthetic tissues, flow through the phloem
system. The tissues of the root, fruit, and buds consume organic compounds. All these tissues are
linked by the phloem system. Osmotic differences
between the different tissues provide the force for a
bulk flow of material through the phloem. The rate
and direction of phloem transport vary according to
the needs of the plant.

Plant Skeleton

The plant is supported by various kinds of cells and tissues. A major part of the plant's support is supplied by xylem tissue. The cellulose walls of xylem vessels and tracheids must be thick to withstand the suction forces in the xylem stream. These thick walls also help to support the weight of the plant. The skeleton of xylem elements is also reinforced by elongated fibers and by thick-walled **stone cells,** which occur as strengthening elements throughout the plant.

Wood is made up of such xylem tissue. The mechanical properties of wood are quite remarkable. It is high in tensile strength, compliance under continued stress, and resistance to compression. These properties enable it to support the great weight of a large land plant, to bend under wind forces but return to its original shape, and to keep the plant upright against the force of gravity. The tensile strength of wood, in terms of the breaking load per square inch, is comparable to that of softer metals such as aluminum or sterling silver. And in terms of breaking load per pound, wood compares very favorably with steel. "Soft" and "hard" woods differ in their resistance to shearing forces, in surface hardness, and in other properties, but they all have similar tensile strength.

The low density of wood results from a system of air spaces running through the stem and root system of the plant. In the stem, these air spaces communicate with the outside air by means of surface pores, which are comparable to leaf stomata in their function. Gas exchange through these pores assures a continuous oxygen supply to the living cells of the stem. It also allows the carbon dioxide produced in respiration to be released to the surroundings. Most of the stem is protected by a waxy covering to reduce water loss.

Air spaces may also be found in the roots, but their function is much less clear. Plants such as cypress trees, which live in swampy areas, may use air spaces in the root for respiratory gas exchanges. Cypress trees whose roots are subject to flooding grow **knees**—root extensions that protrude above the water line (see Figure 7-32). Similar plants living in dry areas have no such extensions. Since the knees have external pores that communicate with air spaces down in the root, they seem to be an adaptation to increase the oxygen supply during flooding. But physiological measurements do not support this idea. The carbon dioxide level in the air spaces is high, and gas does not seem to circu-

Figure 7-32 Cypress knees protrude above the water.

late. Though it is certainly tempting to interpret the air spaces in plant roots as a system to facilitate gas exchange, the matter has not been settled. Whatever their physiological function, the air spaces are certainly important in reducing the total weight of the plant.

Growth

Many aspects of plant growth represent responses and adjustments to the environment. Looked at in another way, many of the environmental responses of a plant depend on its capacity to change its growth pattern. In fact, the plant itself can be interpreted as a structural record of the growth patterns that produced it.

A typical higher plant is analogous to an elongated rod or column capable of growth at either end—at both the root tip and the apex of the shoot. Sometimes the rod can increase in diameter as well. And this simple rod-like model can be complicated by the development of branches.

Shoot

The elongation of plant shoots is usually produced by cell division in a small region at the tip. This dividing region is called the **apical meristem:** "apical" because it is at the apex, or the tip, of the shoot, and "meristem" from a Greek word meaning "dividing." **Meristematic** is an adjective applied to any tissue in the plant which retains its capacity to divide.

We can follow the development of the cells produced by the apical meristem simply by looking at successive slices of the shoot farther and farther from the tip. The meristematic cells are thin-walled and approximately equal in all dimensions. But below the tip the cells have changed gradually to form the various specialized tissues of the plant. In other words, the cells produced by the meristem have **differentiated** into specialized cells.

Surface cells become flattened and develop into a **cuticle.** The cuticle surrounds another layer of cells, the **cortex.** A central mass of cells forms the **pith,** or center of the stem. The cells of the **procambium** appear as bundles between the pith and the cortex. Procambium tissue itself differentiates into the chief conducting tissues, primary xylem and primary phloem. At various points along the shoot, many intermediate stages of differentiation can be distinguished.

In the stem, some of the cells produced by the apical meristem may not differentiate: instead they remain meristematic. In grasses, masses of these meristematic cells, known as **intercalary meristems,** occur at intervals along the stems. The intercalary meristems allow grasses to elongate very rapidly and continuously, even after the apical meristem has been removed—accidentally or by mowing the lawn. The stems of plants that have apical meristems alone can still elongate for a time even though the meristem has been removed. The cells produced by the tip continue to grow and change in shape. But the stems of such plants no longer can grow through cell division.

Growth in length by means of apical or intercalary meristems is called **primary growth.** The diameter of plants that are capable only of primary growth is limited by the size of the apical meristem itself, and by the potential size increase of the individual cells. But in some cases the result is impressive. A palm has only one broad, dome-like apical meristem. Yet palms may become 100 feet tall and several feet thick from primary growth alone. In general, grasses and herbaceous plants such as alfalfa are more typical examples of forms produced by primary growth. They tend to be tall and thin and relatively short-lived. In life-span, too, palms are exceptions. They may live to be 200 years old and apparently survive very well despite having to transport materials in 200-year-old phloem cells.

Secondary growth is the usual solution to the limitations on diameter and longevity inherent in primary growth alone. In secondary growth, some of the cells produced by the apical meristem become secondary meristems. These meristems produce an increase in the girth of the stem. The most important such meristematic tissues are the vascular cambium and the cork cambium.

When the bundles of procambium tissue begin to differentiate, not all the cells in the bundle form primary vascular tissue. Instead, a sheet of thin-walled cells remains, separating the primary phloem from the primary xylem. The cells in the sheet retain their capacity to divide. Pith cells between the bundles also remain meristematic. A complete tube of meristematic tissue, the **vascular cambium,** is the result.

The cells in the vascular cambium divide lengthwise and differentiate to form additional xylem or phloem elements. New, or secondary, xylem is added on the inside of the cambium. Secondary phloem cells are added by the outer surface of the vascular cambium, but inside the ring of older phloem. Older phloem cells are pushed out from the cambium, die after two or three years, and are crushed as they are pushed farther from the center.

A second cambium layer, the **cork cambium,** lies just beneath the epidermis. The original cells of the epidermis cannot continue to cover the stem as it increases in size. The cork cambium therefore continues to divide to cover the enlarging stem. It produces tubes of cells at the periphery of the stem, and these cells take over the protective function of the cuticle. The cork cells are thin-walled and closely packed. The walls are impregnated with a fatty material that prevents water loss. This material also isolates the protoplasm of the cork cells from nutrients and gas exchange, so the cells die shortly after they are formed. New layers of cork are added by the cork cambium as the earlier cork is worn away. A particular cork cambium is relatively short-lived. Each year, a new one is formed on the inside of the old one and embedded in the layer of secondary phloem. The structure of a shoot is illustrated in Figure 7-33.

The layers of an old woody stem can now be

identified. At the exterior is a layer that contains cork cells and cork cambium alternating with secondary phloem. (In older stems, the cortex and the primary phloem growth are often impossible to identify, since they have been pushed out from the center by the vascular cambium and crushed.) Inside of the secondary phloem is the vascular cambium. All the layers from the outside down to the vascular cambium make up the **bark.** Inside the vascular cambium is the secondary xylem. Primary xylem and pith are at the very center of the stem, though their contribution to the structure is dwarfed by the great mass of secondary xylem. Nearly all of an old woody stem is composed of layers of secondary xylem produced by the vascular cambium.

Seasonal differences in the activity of the cambium produce layers of **spring wood** and **summer wood.** Growth in the spring is vigorous, and it produces a wide layer of xylem. Summer growth is slower, so less xylem is produced and the xylem elements tend to have thicker, darker walls. Spring wood is a wide, light-colored layer, while summer wood is thinner and darker. The alternating layers form the well-known **growth rings.** Trees that live in tropical regions where there is no seasonal change in climate lack growth rings.

Root

Roots are similar to shoots in their pattern of primary growth, though some details differ. An apical meristem at the root tip produces cells on both of its surfaces. The cells formed by the lower surface produce a **root cap,** which covers and protects the root tip as it is forced through the ground by the elongation of the root. As might be expected, the root cap is absent in many aquatic plants. Cells from the upper surface form the body of the root. They elongate and differentiate into epidermis, cortex, and strands of vascular tissue.

The arrangement of tissue layers in roots is somewhat different from that in the shoot. There is no pith, so the vascular tissue is in the center. Primary phloem is again outside the primary xylem, and the two are separated by strands of vascular cambium. A layer of cells called the **pericycle** is formed around the xylem and phloem. This layer functions in branching. The endodermis, just outside the pericycle, serves as a barrier to passive diffusion. The cortex of a root is thicker than the cortex of a stem; it functions as a storage tissue. Finally, the epidermis covers the surface of the root.

The root epidermis has no waterproof cuticle. Its

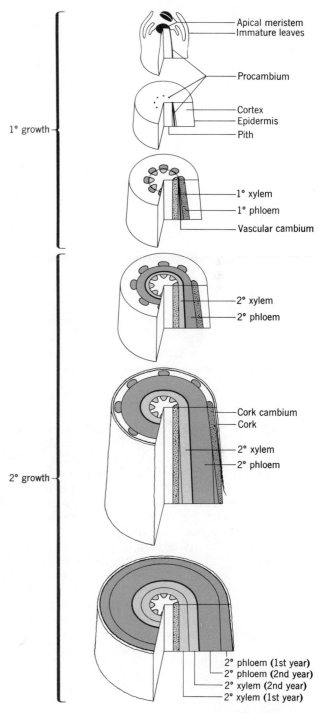

Figure 7-33 *The stem of a woody plant. Primary (1°) growth—elongation—results from the activity of the apical meristem, which also gives rise to primary xylem and phloem. During secondary (2°) growth—increase in stem diameter—the vascular cambium produces secondary xylem and phloem. The cork cambium also appears and produces cork.*

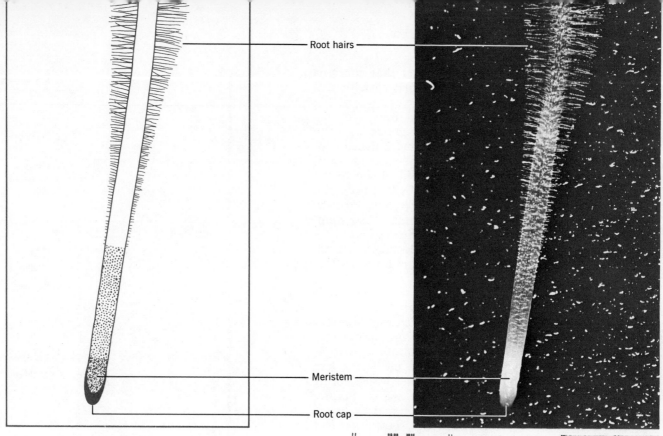

Root hairs

Meristem

Root cap

cells are thin-walled and specialized for absorption. The epidermal cells near the apical meristem produce root hairs to increase the surface available for absorption. Millions of such root hairs are produced each day in an actively growing plant. New root hairs constantly appear at the growing tip of the root, persist for a few days or weeks, and then disappear farther back from the tip.

The secondary growth of roots is entirely comparable to that of shoots. In both cases a ring of vascular cambium is formed between the primary xylem and the primary phloem. New xylem is then laid down internal to the cambium, and new phloem external to it. Cork cambium produces a protective layer over the surface. As the root enlarges, its role in absorption diminishes. Absorption is confined to the finer branches and root hairs (see Figure 7-34).

Branching

Primary and secondary growth of the shoot and root alone would give rise to a tapered structure rather like a telephone pole. But plants clearly have characteristic arrangements of leaves on the stem, and various branching patterns. These growth patterns

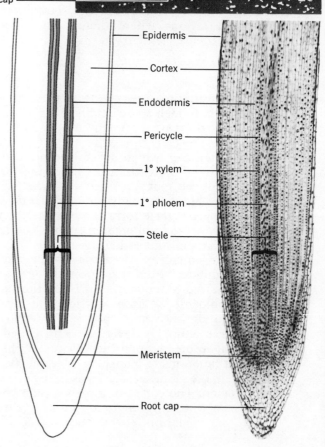

Epidermis

Cortex

Endodermis

Pericycle

1° xylem

1° phloem

Stele

Meristem

Root cap

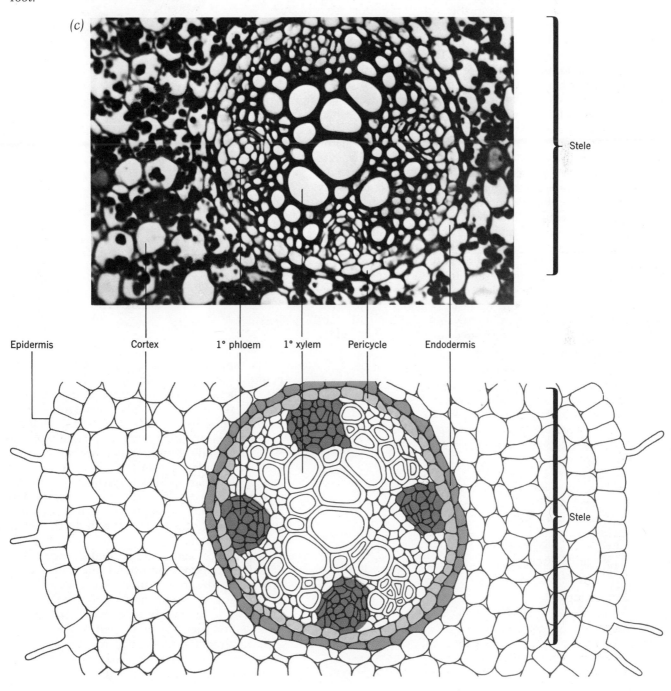

Figure 7-34 Root structure. (a) The whole root. (b) Longitudinal section of a root. (c) Cross section of a root.

(c)

Stele

Epidermis   Cortex   1° phloem   1° xylem   Pericycle   Endodermis

Stele

are controlled by buds. A **bud** is a complicated terminal structure which encloses the apical meristem. Buds of woody plants are protected by **leaf scales** during the dormant or winter phase of their annual growth cycle. Underneath this covering the apical meristem is surrounded by tiny embryonic leaves. More properly, these are **leaf primordia,** which will later develop into leaves (Figure 7-35).

When growth begins in the spring, the protective scales break off, exposing the bud. The tissues formed by the apical meristem begin to elongate into a stem and differentiate. The elongation of the stem separates the leaf primordia while they are developing into leaves. The original pattern of leaf primordia in the bud is now reflected in the leaf arrangement on the newly-formed stem. The leaves may be borne first on one side and then the other, or they may occur as clusters of two or more at the same point on the stem. In any case, the bud is equivalent to a condensed stem with very short spaces between the leaves. Spring growth merely begins the elongation of the stem and expands the pattern already established.

The activity of buds also determines the pattern of branching. The apical meristem produces cells that organize into new buds at the base of each leaf. An axillary bud is present in the axil of each leaf (Figure 7-36). These buds may produce either flowers or a branch. Buds do not always develop, however. In some cases, the growth of a bud into a stem or a flower is predetermined. In others, it is under environmental control.

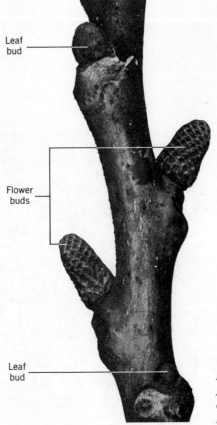

Figure 7-36
*Axillary buds
of a walnut
branch.*

Leaf bud

Flower buds

Leaf bud

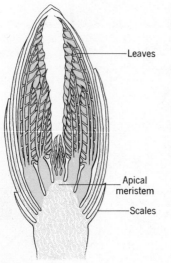

Figure 7-35 Longitudinal section of a bud.

Leaves

Apical meristem

Scales

The apical meristem itself is permanently juvenile—that is, capable of cell division. But once the cells it produces have undergone primary growth, the form of the plant is largely fixed. It can be thickened by secondary growth or deformed by environmental forces, but its plasticity is limited. An axillary bud, however, can become a new apical meristem under appropriate circumstances. In this way, a plant can modify its growth to respond to its environment. Such buds can either replace the apical meristem or supplement it, providing additional regions of primary growth. In the latter case, branching results.

The leaves and axillary buds produced by the apical meristem contain all the types of tissues previously described, including a vascular system continuous with the one in the main stem. The primary and secondary growth of branches proceeds in a fashion identical with that of the stem.

Branching in roots originates somewhat differently. The pericycle initiates branch roots. A group of cells in the pericycle layer becomes meristematic, pushes laterally through the cortex, forms a root cap, and develops as a secondary root. Unlike branching in the shoot, branching in the roots is not confined to regions adjacent to the apical meristem. It can occur in regions far removed from the area of primary growth.

Control of Growth

Growth and branching provide the plant with a greater surface for exposure to sunlight and for absorption of water and minerals. These mechanisms also give each plant its characteristic form. Some aspects of growth and branching depend upon hereditary instructions that are relatively independent of environmental influences; the arrangement of leaves and the pattern of bud formation are examples. However, the hereditary instructions also provide for physiological responses that can modify growth. Such responses determine when lateral buds will become active, or how the plant will respond to changes in temperature, light, and other environmental factors.

Interactions between growth and the environment can be illustrated by the response of a seedling to light. A seed contains stored food and an embryo. The embryo is simply a small plant, consisting of a root and a shoot. As the seed germinates, the embryo uses the stored food for its initial growth. By the time this food is exhausted, the plant should have raised the shoot into the light and anchored its roots near a supply of water and minerals. Otherwise it will die.

If the growth of the embryo root and shoot were to proceed at random, the chance of success would be very small. In fact, the embryo's growth is not random. The shoot tip grows upward and the root down, regardless of the initial orientation of the seed. When the shoot tip emerges from the soil, it grows toward a source of light.

Charles Darwin carried out the first systematic studies of the responses of plants to light, and his initial experiments are still worth describing. He worked with seedlings of oats grown in the dark. When the seedling was exposed to a single light source, its tip bent toward the light. When Darwin covered the tip of the seedling with an opaque cap, he found that it no longer bent. But when he placed an opaque sheath over the region where the bending actually occurred, leaving the tip itself exposed to the light, bending still occurred. Darwin concluded that the photosensitive tip was transmit-

ting information to the region below it to produce bending. This conclusion has been borne out by subsequent work.

The linkage between the photosensitive tip of the plant and the bending region is chemical, as was demonstrated by F. W. Went in a simple experiment. Went cut off the tip of an oat seedling and placed it on a block of agar. After some time, he removed the tip and placed the agar on top of another decapitated seedling. He found that seedlings treated in this way grew better than similar plants capped with blocks of pure agar. Clearly, some material that diffuses from the tip into the agar promotes growth in the rest of the seedling. Since no cell division occurs below the growing tip, the material in the agar must cause cell elongation. This material was given the name **auxin,** from the Greek word *auxien,* meaning "to increase."

Darwin's observations can be explained in terms of auxin distribution. If auxin is produced by the shoot tip and becomes distributed unequally in the stem, bending will result: some cells will be stimulated to elongate more than others. This hypothesis can be checked by exposing an agar block to the cut tip of a seedling and then placing it on one side of the cut end. Bending occurs as predicted. The auxin in the agar causes greater elongation on the side of the stem under the block (see Figure 7-37).

"Auxin" actually turns out to be a group of compounds, of which the most common is indole-3-acetic acid (IAA) (see Figure 7-38). IAA can now be synthesized and used by plant physiologists for investigation of the effects of auxin at the cellular level. The role of IAA in cell enlargement has been studied extensively. Plant cells are surrounded by walls of cellulose fibers. IAA promotes the breakage of the bonds between these fibers, loosening the structure and permitting the cell to enlarge.

If a section of an oat seedling is placed in a medium that is of precisely the same osmotic concentration as the cells, and to which auxin has been added, there will be no net movement of water into the cells. Under these circumstances, auxin fails to cause the cells to enlarge. But when the same seedlings are later transferred to pure water, their cells do enlarge. Clearly, the auxin causes the loosening of the structure of the cell wall, permitting the cells to elongate. Turgor pressure is necessary, however, to produce the actual extension of the cell wall.

Auxin has no effect on the mechanical properties of dead cells. It is believed, therefore, that auxin does not weaken the cellulose linkage itself but rather promotes the formation of enzymes that do.

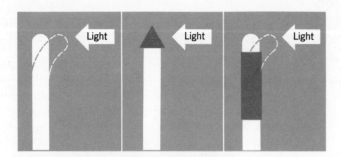

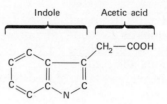

CH_2—COOH

Figure 7-38 *The most common auxin, indole-3-acetic acid.*

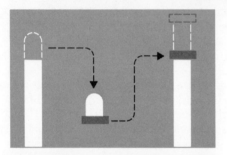

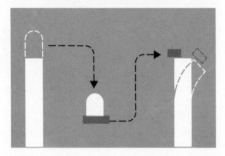

Figure 7-37 *Role of auxins in the response of oat seedlings to light. (a) The plant grows toward a single light source. Darwin found that this response is prevented when the tip is shielded from the light, but that it still occurs when the shield covers the lower region — where the bending actually takes place. (b) F. W. Went demonstrated that the tip of the oat seedling produces a substance that promotes the elongation. Auxin diffuses from the cut tip into a block of agar, which then promotes the elongation of a stem whose tip has been removed. (c) In further experiments, Went demonstrated that if a block of agar containing auxin is placed on one side of a cut seedling, the cells beneath it elongate. The resulting curvature mimics the bending response to light.*

It has been found that auxin does stimulate RNA and protein synthesis. Also, auxin's effects can be blocked by compounds that inhibit RNA or protein synthesis. These facts support the idea that auxin is involved in enzyme action; however, the nature of the enzymes involved and the precise point at which auxin controls the process are still under investigation.

Auxin has a number of other effects on plant growth and development that are not related to cell enlargement. For example, auxin promotes flowering and the formation of roots. Also, it accelerates cell division in the cambium. Furthermore, at concentrations above some optimum level, auxin can inhibit both cell enlargement and cell division (see Figure 7-39). Typically there is a level of auxin below which a plant tissue fails to respond, then a region of increasing concentration through which plant growth is stimulated, and finally a high concentration at which growth is inhibited.

Figure 7-39 *The effect of different concentrations of auxin on the growth of various plant parts.*

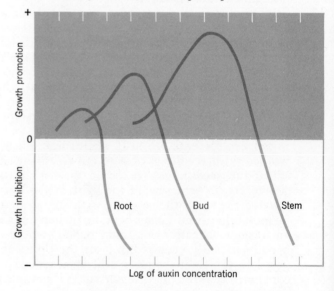

Herbicides

An important byproduct of auxin research was the discovery of very powerful and specific herbicides, or weed-killers. In an attempt to discover which parts of the IAA molecule have biological activity, scientists have synthesized compounds that are chemically similar to it. One series is made by adding fatty acid chains to the organic compound 2,4-dichlorophenol. All the

2,4 dichlorophenol

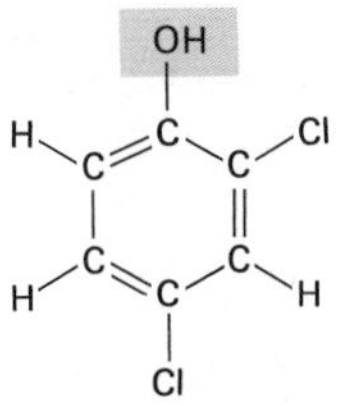

Acetic (2C) derivative

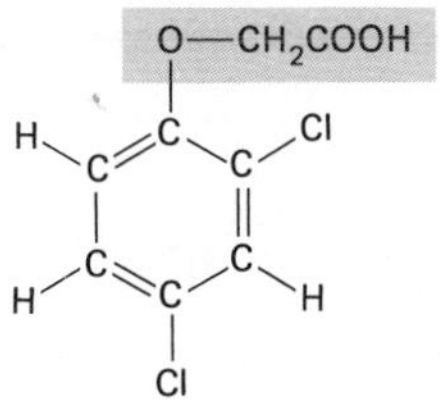

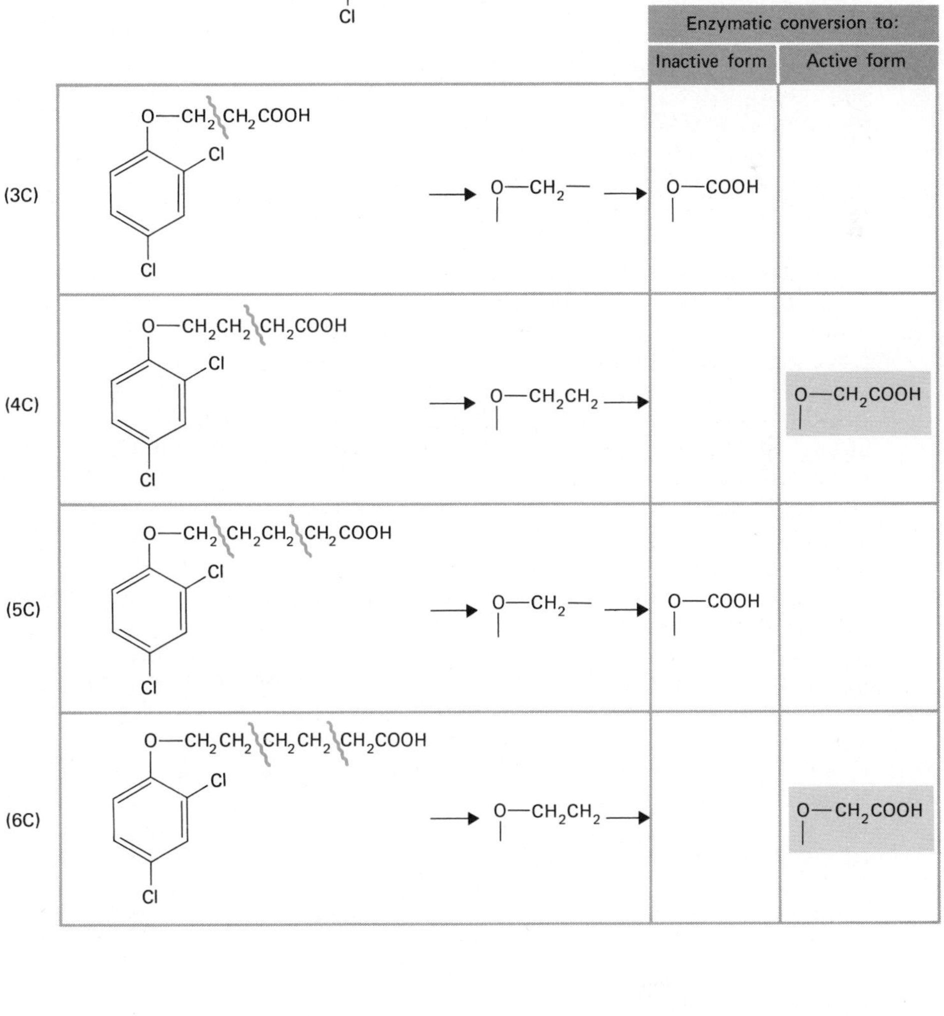

2,4-dichlorophenol and its acetic derivative are active as auxins. Derivatives with longer chains are inert, but some plants have enzymes that split off two carbons at a time from the chains. If the original compound has an even number of carbons in its chain, the final product will be an active auxin. A 4C or 6C derivative will function as an auxin (and therefore as a herbicide) for plants that have the enzymes. Other plants are not affected.

		Enzymatic conversion to:	
		Inactive form	Active form
(3C)	O—CH₂CH₂COOH ... Cl Cl → O—CH₂— →	O—COOH	
(4C)	O—CH₂CH₂CH₂COOH ... Cl Cl → O—CH₂CH₂— →		O—CH₂COOH
(5C)	O—CH₂CH₂CH₂CH₂COOH ... Cl Cl → O—CH₂— →	O—COOH	
(6C)	O—CH₂CH₂CH₂CH₂CH₂COOH ... Cl Cl → O—CH₂CH₂— →		O—CH₂COOH

derivative compounds with either one carbon or more than two carbons in the chain are inactive as auxins. However, the acetic (two-carbon) derivative is active. When present in sufficient concentration it will inhibit growth.

Some plants have enzymes that will split long-chain fatty acid derivatives, lopping off two carbons at a time. When presented with butyric (four-carbon) or caproic (six-carbon) derivatives, they remove an even number of carbon atoms and convert the longer-chain derivatives to the active acetic form. When the acetic form reaches toxic levels the plants die. But when presented with derivatives with an odd number of carbons, such as proprionic acid (three carbons) or valeric acid (five carbons), the lopping off of two carbons at a time produces the inactive form, and no response occurs. It turns out that not all plants have the enzymes for this reaction. Most broad-leafed weeds have the enzyme, while clover and similar plants do not. Therefore, if a field of clover is sprayed with the butyric derivative of 2,4-dichlorophenol, the weeds oxidize it to acetic form and are killed, while the clover is unaffected.

We have discussed one aspect of environmental control of plant growth: namely, the relation between light and growth. The plant's response to light is mediated by the activity of auxin. Auxin is also responsible for growth responses to other environmental factors. Under the influence of gravity, auxin moves to the lower side of the growing stem. The stem thus bends upward and maintains the plant in a vertical position. In some plants, auxin also accumulates in response to mechanical stimuli. For example, if the shoot tip contacts an obstacle it may curl around it, as pea tendrils do. All such directional responses to environmental stimuli are called **tropisms;** the word comes from a Greek root meaning "to turn." Shoot tips are usually negatively geotropic (that is, they turn away from the force of gravity) and positively phototropic (they turn toward light).

Other chemical control agents, in addition to auxin, have been isolated and described: gibberellins and cytokinins, abscissic acid, and ethylene are some other compounds that have profound effects on plant growth. All of these agents may interact with each other. The timing of the major events in the life cycle of the plant—germination, growth, and flowering—is under the control of still other chemical agents.

Our treatment of the environmental control of plant growth is oversimplified, but it is still substantially correct. The plant responds primarily by modifying its growth pattern. That modification is triggered by changes in the production or distribution of chemical agents. Those changes in turn are triggered by environmental stimuli.

Chemical control agents such as auxin are called plant **hormones.** The word is Greek and means "an exciter." It was coined by animal physiologists to refer to chemical agents, produced in one tissue and carried in the blood stream, that affect some distant tissue. "Exciter" is misleading, since hormones can both enhance and inhibit; perhaps a better definition would be "chemical messenger." By extension, "hormone" now refers to analogous chemical agents in plants and lower animals.

Movements

Plants' growth responses to environmental stimuli can be surprisingly rapid. Responses to gravity occur in a few hours or less. Tendrils can wrap around a thread or a wire in a minute. And some plants are capable of even more rapid movements. We have already mentioned the beating of flagella and cilia in unicellular plants. Some higher plants, too, show extremely rapid movements, reversible responses, or even almost explosive changes of form. These rapid responses are all based on changes in turgor.

The opening and closing of leaf stomata, caused by changes in the internal pressure of the guard cells, is one such phenomenon. In specialized cases a similar mechanism permits the movement of leaves or spines. The walls of the cells at the base of the leaf are unequally thickened. Thus, as turgor increases or decreases, the thinner cell wall expands or folds in. These changes are translated into folding or expanding leaf movements, or movements that position the leaf more favorably to intercept light.

Turgor responses can be quite dramatic. The leaf-

Figure 7-40 The mimosa (a) responds rapidly to touch by folding its leaves (b).

lets of *Mimosa pudica* fold up when touched, and the entire leaf drops down toward the stem (see Figure 7-40). In the mimosa, specialized cells extrude fluid in response to mechanical and heat stimuli. The total movement depends partly on the folding of the thin cell walls and partly on the continuing turgor in the remaining tissue that supports the leaf.

The same mechanism is responsible for the leaf movements of a Venus's-flytrap. This unusual plant traps insects that alight on the leaf surface. The leaf's response is triggered by six sensory bristles, three on each half of the leaf. When the bristles are touched repeatedly, turgor changes occur in a strand of tissue that runs down the central vein of the leaf. The marginal spines bend in, and the leaf folds up, forming a cage around the insect. Digestive glands on the surface of the leaf then break down the insect tissues and absorb the juices (see Figure 7-41).

The explosive discharge of seeds or spores in some plants is accomplished through a related mechanism. As the seed pod dries, water is lost from its cells. Unequal tension develops along the thickened portion of the cell walls. This tension may be opposed by the cohesive force of water or by shrinkage of the thinner portion of the wall. But as drying proceeds, there is a sudden release of tension along the inner side of the structure, producing an explosive opening movement. Seeds or spores may be thrown several feet.

Protection and Defenses

In addition to responding to climate and the physical environment, plants must protect themselves from predation by grazing animals and destruction by parasites, and they must compete with other plants for growing room and light and nutrients. We dealt with chemical protection against invasion by microorganisms in Chapter 4. Here we will look only at the devices plants use to counter an attack by animals.

Some protection is afforded by the same structures that prevent dessication. The waxy cuticular layer that covers leaves and young stems, and the cork layers of older stems, both serve to discourage small grazing animals. Cellulose itself, which constitutes much of the material of the plant, is extremely resistant to digestion by animals. Plants may also produce structures that make browsing painful or difficult. Spines and thorns are obvious examples. Some thorns are derived from leaves and still have axillary buds at their bases. All cactus

(a)

(b)

Figure 7-41 *When an insect touches sensory bristles on the inner surface of a leaf of a Venus's-flytrap (a), the leaf folds and traps the insect (b).*

spines are such modified leaves. (Photosynthesis is carried out by the stem tissues.) Other thorns may be stems (as in the honey locust) or outgrowths of the stem (as in the rose).

Most higher plants produce protective hairs, or **trichomes** (Greek for "hair"). These structures are derived from epidermal cells. They are extremely varied in shape and in function. Trichomes may cover the leaf surface, reducing evaporation or the intensity of the light that reaches the mesophyll. They may also produce knobs or bristles that are impregnated with calcium or silica. The fishhook-shaped trichomes of the prickly pear cactus are of this design, and they are extremely effective protection, as anyone knows who has tried to pick a prickly pear barehanded. Trichomes of the stinging nettle are hollow and filled with a poison or irritant. The rounded tip of the hair covers a needle-sharp point that resembles a hypodermic needle (see Figure 7-42). Brushing against the surface of a nettle breaks the tip off of the trichomes, so that a myriad of tiny hypodermics penetrate the skin and inject the irritant. The chemistry of the trichome poisons is varied; some injected material can be very painful or even lethal. Trichomes of other plants form glands that produce sticky mucilage for trapping insects; insects provide the plant with nitrogen and minerals (see Figure 7-43). Protection has become predation, and the biter, bit.

The protection strategy is sometimes reversed,

Stinging trichomes

Figure 7-42 *Stinging trichomes on leaf of* Urtica, *the stinging nettle; magnification 10×.*

214 *plant organization*

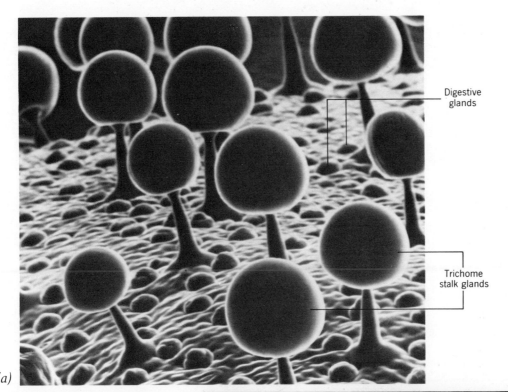

(a)

Digestive
glands

Trichome
stalk glands

*Figure 7-43 The insec-
tivorous plant Pinguicula.
Trichomes form stalk
glands, which secrete a
sticky mucilage that traps
insects. Magnification
150×. (b) An ant is trapped
by the plant. Digestive
glands on the leaf surface
secrete enzymes to digest it.*

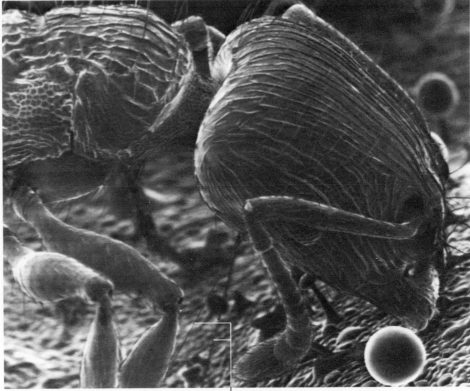

(b)

Mucilage strands bind
insect to surface

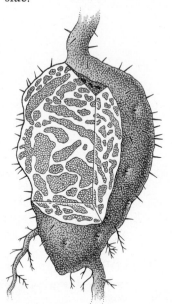

Figure 7-44 The stem of Myrmecodia, *showing the chambers in which ants reside.*

(a)

(b)

(c)

(d)

Figure 7-45 The chemicals produced by the milkweed plant are poisonous to most animals, but the monarch butterfly can tolerate them. It feeds on milkweed as a caterpillar and retains the compounds, becoming poisonous itself to predators such as the blue jay. Blue jays quickly learn to avoid monarch butterflies.

and plant and animal cooperate to protect a plant against other browsers. A number of plants, for instance, produce a sugary exudate that attracts specific insects who live in association with the plant. One such plant, *Myrmecodia*, develops a hollow tuberous stem in which ants take up residence (Figure 7-44). The ants inflict painful stings on any animal that attempts to eat or otherwise damage the plant.

Plants also protect themselves from animal damage by producing unpalatable or poisonous compounds. Most animals simply learn to avoid bitter and unpalatable or poisonous plants. The poisons are not 100-percent effective, however, because some animals develop a tolerance for these protective chemical agents. This development is bad for the plant, but very good for the animal. Not only

does the animal then have a source of food, but if the poison remains intact, the browsing animal itself may become unpalatable or poisonous to its predators.

This relation has recently been studied in the Monarch butterfly. The juvenile stage—the caterpillar—feeds on milkweed and related plants. Milkweeds contain various amounts of cardiac glycosides, compounds that are poisonous to most animals. However, if the Monarch caterpillar eats a milkweed plant containing them, the adult butterfly also contains them. Figure 7-45 shows the reaction of a blue jay after consuming such a butterfly. The jay recovers after vomiting but subsequently will refuse Monarch butterflies.

Man's relationship to the protective chemical agents produced by plants is somewhat different from that of other animals. In many cases man has learned to remove the poison in order to use the plant as food. There are many examples. The root of the taro plant, a staple food crop in the tropics, has to be beaten, soaked, and allowed to ferment in order to remove crystals of poisonous oxalic acid. Raw olives are processed to remove bitterness. Acorns have been used for flour, but only after extended soaking to render them palatable. In other cases, man uses the poison itself. Many plant compounds are extremely potent drugs. Humans have learned to use these compounds for curing diseases, for religious purposes, and for pleasure.

Recommended Reading

Bell, C. R., *Plant Variation and Classification*. Belmont, Cal.: Wadsworth (1967).

A discussion of natural variation in vascular plants, in relation to plant identification and classification.

Salisbury, F. B., and R. V. Parke, *Vascular Plants: Form and Function* (2nd ed.). Belmont, Cal.: Wadsworth (1970).

A modern treatment of plant physiology in conjunction with the structure of higher plants.

Sutcliffe, James, *Plants and Water*. London: Edward Arnold (1968).

A brief paperback monograph on the movement of water in plants and its control. It includes an excellent discussion of osmosis.

8 plants and man

Green plants support all the life on our planet. Animals are directly and continually dependent upon plants for oxygen, food, and shelter, and man is no exception. This relation is obvious, yet it is easily ignored by urban populations. For most of us, food, clothing, and shelter are found in the yellow pages or the local shopping center. We often behave as though these resources are inexhaustible. We assume that an increased demand will simply increase the available supply. Of course, this assumption is quite wrong: our plant resources have very real limits. When the corn crop fails and supplies of livestock feed are limited (as happened, for example, in the southern corn blight attack of 1970), meat prices increase, and the number of people who can afford meat decreases.

The consumer's dependence on plant products tends to be obscured by sophisticated preparation and packaging of plant products. Cotton means sheets and permanent-press shirts, not plants. Few people realize that synthetic fibers, too, are made from plant products—rayon uses wood cellulose, and nylon is made from coal or petroleum, which themselves are of plant origin. The timber in houses is evident during construction but often concealed in the finished product. Aside from these primary uses—food, oxygen, clothing, and shelter—plants provide a long list of other materials only slightly less important, such as medicines, drugs, insecticides, dyes, spices, soaps, and waxes. We will first consider plants as food, with emphasis on food production and world food requirements.

Opium poppy

219

Plants as Food

Humans were hunters and gatherers for the first two or three million years of their existence; some still are. Agriculture began comparatively recently, only about ten thousand years ago. The river valleys in Mesopotamia were the original site of agriculture in western civilization. The name Mesopotamia means "between the rivers," and it refers to its location between the Tigris and Euphrates rivers in Iraq, which drain into the Persian Gulf.

Mesopotamian agriculture was based on wheat and barley. Presumably wild wheat and barley grew naturally in the area. At first, existing patches of wild wheat were harvested. But soon it was discovered that if some grain fell on the ground, grain plants would reappear the following year. The next step was to scatter the grain on specially prepared plots, remove competing plants, and protect the ripening seed from birds and neighbors. (An alternate theory of early agriculture argues that the first domesticated plants were used for drugs and dyes, and that the cereal grains were brought into production only secondarily.)

Wild cereal grains were probably improved rather quickly by the early farmers. By modern standards, wild grains are not very good crops. Grain drops from the seed heads and is lost during gathering; at harvest the small grain heads return only three or four times the amount of grain that was originally used for seed. The farmers were able to increase their harvests by selecting rugged plants that bear large seed heads.

At the same time that food plants changed, man was changing as well. Before the development of agriculture, humans had been organized into small bands that existed by hunting and gathering. But agriculture permitted the support of larger populations by producing a stable and abundant food supply. The human population probably increased 50- to 100-fold as agriculture developed and spread throughout the world. Hunting was quickly reduced to a supplementary food source, because survival was possible on grain alone. To provide variety, other wild plants (such as olives, dates, figs, and grapes) were domesticated as well, but cereal grains remained the major crops.

Hunting and a nomadic existence were eventually abandoned for the security of a fixed abode and a predictable food supply. But at the same time there were also negative costs. Man became bound to soil, seed, and season. He could never return to the old way of life, since hunting and gathering were not productive enough to feed the new, larger population.

Extensive social changes followed in the wake of early field agriculture. Crops, and hence the new societies, were restricted to rich and wet river-delta lands because rainfall was undependable during the growing season. River water was diverted into channels to irrigate low-lying fields. The risk of flood and disaster was great, so dams and elaborate irrigating channels were constructed to control flooding and extend the growing area. Work in the fields and the maintenance of dams and dikes and channels became vitally necessary. Therefore, social machinery had to evolve to insure cooperation from the whole community.

For the first time there was actually a food surplus; it was no longer necessary for every member of the community to devote most of his time to the hunt for nourishment. The grain surplus produced by the farmers now supported people in other full-time occupations: a ruler, a military establishment to protect the food supply from raiders and to enforce community labor, a priesthood to coordinate the technical and religious aspects of the whole effort. But, though life in such a society was secure, the work was heavy and dull.

When Egypt was at this stage of development, the pharaohs controlled peasant life through military and religious organizations. The whole population was effectively enslaved. However, at the same time, the fact that some individuals were freed from the fields, coupled with complex management requirements, created a civilization. The sciences prospered. Astronomy (for measuring time and predicting the seasons), geometry and mathematics (for surveying, tallying, and taxation), writing (for keeping records and general administration), and architecture (for the creation of aqueducts, granaries, palaces, and temples) were brought into existence.

Further social changes accompanied these scientific developments. Concepts of land tenure, property, and inheritance, and a code of law all developed at this stage. They are still recognizable in our modern culture. In fact, the word "rival" has the same root as "river"; rivals were "those who share the river water."

The Mesopotamian cultures eventually declined. Repeated barbarian invasion of the wealthy delta was a major factor. However, the failure of the irrigation system was equally important. As populations increased, trees were removed from hillsides and used for timber. Grass replaced the trees, but it was removed by grazing goats. Soil washed from the hillsides into the great rivers. Silt settled out near the dams and along the irrigation channels, so dredging and maintenance became more and more costly. Restoration of the irrigation system after invasion became increasingly difficult and, finally, impossible. The foothills were denuded and barren, and the rivers were laden with silt; the aqueducts were in ruins and the irrigation channels gone. The region has still not recovered. What were once the fertile valleys of Mesopotamia are so desolate today that it is difficult to believe that they once supported millions of people; that their crops and the wealth of their people were the envy of the Western world.

This pattern has occurred over and over again: a civilization arises and develops into a large and complex society, supported by irrigation for the production of cereal grains; the food-production system eventually succumbs to salt and silt; and the civilization declines. A major culture went through this cycle along the Indus River in what is now Pakistan. The New World also has its extinct river-valley cultures. A thousand miles of ancient irrigation channels have been found in Arizona and New Mexico. The origins of the Aztec culture have been traced to agricultural villages in the Tehuacan Valley south of Mexico City, where by 7000 B.C., maize was already being raised. The Mayan and Inca civilizations developed further south.

All these agricultural systems were based on the crops that are obtained by planting seed. Another form of agriculture, based on crops that require individual handling, arose independently. Root crops, such as potatoes, sweet potatoes, and cassava, supported civilizations in the New World; in the Old World the equivalent crops were a potato-like plant called taro, and rice. (Rice is actually a cereal grass, not a root crop, but it is handled in much the same way; seedlings have to be planted by hand in rice paddies.) We know less about early societies based on root-crop agriculture.

The influence of agriculture is obviously not restricted to the origins of civilizations; crop plants have had a continuing importance in human affairs. For example, agricultural needs encouraged slavery in the Roman Empire and serfdom throughout Europe. The institution of slavery in the southern United States was at first related to the value of tobacco crops. As tobacco prices declined in the late 1700's, slavery might have disappeared. However,

the invention of the cotton gin made cotton a profitable crop, and it, too, was based on slave labor. The survival of slavery then became a key issue of the Civil War.

Other examples of the way agriculture can shape a society come from Latin America. Many countries in this region have been referred to as "banana republics," suggesting that their social and political development has been controlled by their agriculture. These countries began to export bananas (and other crops) to the United States at the end of the nineteenth century. This activity was a direct result of the financing and economic development provided by large United States corporations. The accompanying changes in roads, communications, transportation, and employment profoundly altered their entire social and political structures.

History is certainly more than an account of agricultural changes. However, a history focusing on agriculture would be recognizable and interesting, and as illuminating as histories based on the actions of political or intellectual figures. The development of an agricultural system always involves a mutual adjustment between crop plant and society. The plants are selected and modified by the cultivators. In turn, the social systems and agricultural practices of humans are modified and constrained by plant and soil and season.

This interaction produces very basic differences between societies, differences which are often overlooked. If a proposal to improve primitive tropical agriculture is based on techniques developed in Iowa, the results may be disastrous. Traditional agricultural practices throughout the world survive for the very good reason that they work. Various improvements in production are possible, but drastic changes usually produce unanticipated and negative results.

crop plants

Since many of the characteristics of a society are influenced by the characteristics of its crops, we will briefly outline the biological nature of the major crops.

Cereal Grains

Cereal grains are the most important food source for man. More than two thirds of all the world's arable land is devoted to their cultivation. They provide half of man's food supply directly, and indirectly they contribute to the production of beef, milk and dairy products, poultry, and other animal foods.

Wheat, rice, and corn are the most important grain plants; the others, such as barley, rye, millet, oats, and sorghum play a relatively minor role in the overall food supply. All the cereal grains are grasses. The grain itself is technically a single-seeded fruit. Grains are concentrated foods, composed of about ten percent protein, five to ten percent lipid, and the rest carbohydrate. The grain itself comprises about a quarter of the weight of the entire plant. It is easily stored and requires no special drying or handling.

Cereal grasses can be adapted to many different climates and agricultural practices. Wheat, *Triticum aestivum*, is grown on prairie land and can survive cold and dry conditions. Rice, *Oryza sativa*, is usually grown on flood plains or areas where the land can be worked into ponds. Corn, more properly called maize, or *Zea mays*, grows best in warm, humid conditions. Most wheat and rice is consumed directly as human food. Maize is more commonly used as animal feed; 90 percent of the corn grown in the United States is used for fattening hogs and cattle. Maize is also processed for corn starch and oil.

All three cereals provide poor-quality protein for human nutrition. They are all particularly deficient in the amino acids lysine, tryptophan, and threonine, which are essential in the human diet. All of these plants concentrate protein in the outer layer of the grain (see Figure 8-1). For this reason processing, such as the milling of corn, reduces the protein content of the grain even more.

Major improvements in grain crops will probably come from genetic modification of their protein content through plant breeding. A particular gene in corn that increases the amount of lysine and tryptophan has already been isolated. Modern breeding techniques are now being applied to corn and rice and wheat with spectacular results. However, the "green revolution," which has greatly increased the world's food production, is based on new wheat and rice varieties that still have the amino acid deficiencies typical of these grains.

Root Crops

The category of **root crops** includes underground stems (bulbs, tubers, corms) as well as roots in a

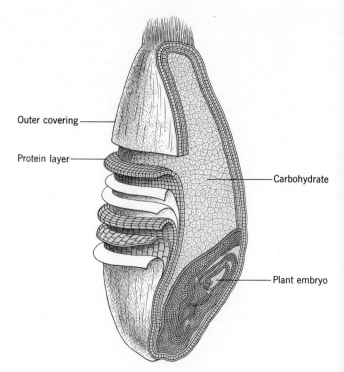

Figure 8-1 *The structure of a wheat grain.*

Labels: Outer covering; Protein layer; Carbohydrate; Plant embryo

strict botanical sense. Potatoes, sweet potatoes, cassava, and beets are the most common root crops. The yearly tonnage of root crops rivals that of the cereal grains, but their contribution to the world food supply is considerably less because they are starchy and low in protein. They also have a high water content, which makes them subject to spoilage. For these reasons they are less desirable than cereals as staple crops. On the other hand, they can be grown in climates and soils that are unsuitable for cereals.

The potato is the most important root crop in temperate climates. Its food storage organ is an underground stem called a **tuber;** the "eyes" are buds. Potatoes are usually grown from chunks of potato containing buds, rather than from seeds. This procedure keeps the quality of the plants constant and also assures the faster development of the new plant. But it also allows virus disease agents to build up in the tuber from generation to generation. The potato originated in South America as a highland crop in areas too cold for cereal grains. It was introduced into Europe in the sixteenth century by returning explorers, and it rapidly became a staple crop in cooler regions of Europe. By 1840, potatoes were almost the sole food for most of the eight and a half million people in Ireland. Two successive crop failures in the 1840's resulted in the great potato famine.

Cassava is the staple root crop of tropical areas. In America it is consumed in the form of tapioca. Its swollen roots resemble sweet potatoes; they are rich in starch and contain little protein. But cassava leaves are a relatively good protein source. Cassava, however, also contains considerable quantities of cyanide, which must be removed. It is remarkable that such a poisonous plant was ever domesticated in the first place. Sweet varieties now exist which do not contain much cyanide.

Vegetables and Fruits

A large group of plants is consumed as fresh vegetables. Vegetables are usually prized for their flavor and are used as food supplements rather than as staple foods. They are important sources of minerals and vitamins, and they relieve the monotony of cereals or potatoes or beans. They also supply essential amino acids that are sparse in basic foodstuffs. Some vegetables are roots, such as carrots, onions, parsnips, and radishes; others are stems or leaves, such as asparagus, cabbage, celery, cauliflower, lettuce, and spinach. Tomato, eggplant, and squash are actually fleshy fruits.

All these plants are rather perishable and usually must be consumed near the growing area. Fresh-frozen vegetables are widely available in countries like the United States. But freezing and distribution are expensive processes; they are simply not possible in undeveloped countries. Most populations depend on vegetables grown locally or collected wild. Primitive populations achieve a balanced and nutritious diet from local plants by trial and error, even in forbidding environments.

Fleshy fruits such as apples, oranges, plums, grapes, bananas, and pineapples are sometimes rich in sugar and vitamins (for instance, vitamin C in citrus fruits), and they also provide variety in the diet. They are quite perishable and rather watery but can be preserved in several ways: by drying (turning grapes into raisins or plums into prunes), by cooking in a sugary syrup (making jams and jellies), or by canning or freezing. Wine-making is also a good way to preserve the caloric content of fruit. In fact, because of their use in wine, the tonnage of grapes harvested each year exceeds that of any other fruit.

Most fruit for American markets is picked green. Varieties are selected primarily for appearance and for their keeping and shipping qualities. Taste and texture are secondary considerations. We simply never have the opportunity to sample most high quality, tree-ripened fruits, because they are too perishable.

Legumes

Legumes—peas, beans, soybeans, and peanuts —belong to the plant family Leguminosae. Legume seeds are borne in pods (see Figure 8-2). Most legumes contain large amounts of nitrogen, because they form symbiotic associations with the nitrogen-fixing bacteria *Rhizobium*. The bacteria lie in root nodules; they convert atmospheric nitrogen (N_2) to nitrate (NO_3^-) or other nitrogenous compounds that can be utilized by the plant. Some free-living soil bacteria and blue-green algae also fix nitrogen, but the process is faster in legume nodules. For this reason, protein-rich legume crops can be grown without the use of nitrogen fertilizers. In fact, legume crops are often alternated with cereal crops to enrich the soil; the legume seeds are harvested, but the root nodules and other nitrogen-rich parts of the plants are plowed back into the soil to supply nitrogen for the next crop.

Even though legumes provide less than ten percent of our direct food intake, they are very important as a protein source; the seeds of legumes are about 25-percent protein. Furthermore, legume protein contains adequate amounts of all the amino acids essential for human nutrition. Legume protein is therefore equivalent to milk or eggs. Of course, meat is a richer source of protein, but it is also more expensive to produce and is simply not available to most of the human population. Beans, peas, and their relatives are vital protein sources for most people of the world.

Almost any legume seed is also suitable as food for domestic animals. The entire plant is a rich food source for herbivores. Legumes such as clover and alfalfa are extremely important forage crops.

Pasturage

Grasses, legumes, and other plants are important forage for animals. They therefore are vital for man, because the animals themselves are an important source of high-quality protein. Almost 15,000 species of grasses and legumes are useful as forage. These plants do well on land that cannot support intensive agriculture, so pasturage based on grass

and legume plants extends the capacity of the earth to support human population. The world's pasture land comprises about ten million square miles, whereas only about six million square miles are farmed for crops of one sort and another. (For comparison, the area of the continental United States is about three million square miles.) In the United States, cattle are raised on forage and fattened for market on grain. Grain feeding improves the quality of the meat, but it is certainly not necessary.

Utilization of pasturage takes many other forms. Forage crops may be used to produce milk and dairy products, rather than meat. Some societies, such as the Dodo tribe in Uganda, obtain protein from cattle by drinking their blood as well as by consuming milk and meat (see Figure 8-3). This practice supplies salt (which is scarce) as well as protein. Other groups in Africa "crop" wild game for meat instead of removing the natural fauna and replacing it with domestic animals. This method proves to be quite efficient. The sacred cattle in India are used even more efficiently. They are not eaten as meat. Instead, they provide milk, and their dung is used for fertilizer or dried and used for fuel. The cattle are also vitally important as a source of power—in plowing, for instance. And most of the forage they consume cannot be used by man for other purposes.

(a)

Figure 8-2 Legumes. (a) A mature pea pod. (b) A green pod cut open to show developing seeds.

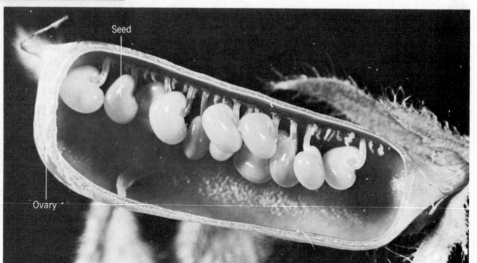

(b)

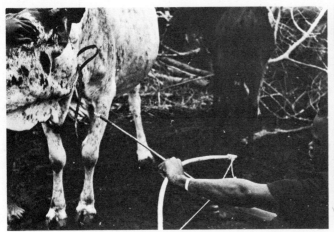

Figure 8-3 Cattle blood is harvested by the Dodo tribe of Uganda. The blood is an important source of protein and salt. (a) The animal's jugular vein is pierced by an arrow, and (b) the blood is collected in a bowl.

food production

When man moved from food gathering to agriculture he altered the natural environment to increase food productivity. The increased productivity is achieved in various ways, but all of them represent an addition of energy to the natural system.

Agriculture and Energy

The amounts of energy added to an agricultural system are usually small compared to the energy of the sunlight used for photosynthesis. Still, the additions can be very significant, because they may re-channel and control larger flows of natural energy. Let us consider an undisturbed river valley. Some of the native plants provide food for man. Wild wheat, berries, nuts, and tubers are present. They would presumably be there whether or not man is present. This whole system exists with no human control or intervention, and sunlight is the sole energy input. No single species of plant dominates the system; many different species live in the same area. Thus, a particular plant disease or grazing animal is unlikely to affect more than a few plant species at one time. The overall plant community will be constant and stable.

The diversity and stability of the plants provides for a corresponding diversity of burrowing and browsing animals and microorganisms. This diversity, too, creates stability. For example, a varied community of soil organisms ensures that critical metabolic reactions will continue even if conditions change. A reservoir of potential predators is always present to prevent any one animal species from becoming overabundant. The same diversity and complexity provides for the creation and maintenance of soil fertility, the aeration of the soil, and the trapping of occasional rainwater. For all these reasons, the most complicated natural systems are the most stable and well balanced. Any such interacting system of organisms and environment is called an **ecosystem.**

If we destroy the original ecosystem by removing the natural plants and replacing them with a stand of wheat, many things change. Sunlight energy is still available to the field, but the various systems that maintain the fertility and structure of the soil and prevent epidemic disease are gone; we have removed them. We can simply use the field until its nutrients are exhausted, disease organisms accumulate, and it no longer supports a crop. Alternatively, we can provide for each of these natural functions externally.

Soil structure and fertility can be maintained by plowing, crop rotation, and fertilizers. Our crop can be protected from other organisms with scarecrows, weeding, and other direct methods. We can use pesticides and herbicides to kill competitors chemically. We can also breed plants that are resistant to predation and disease agents. These are standard agricultural techniques. Agricultural systems are unnatural and unbalanced ecosystems. They have to be maintained by such external inputs; the farmer must replace many of the things he removed when he cleared the field. All of these procedures supply energy to the system.

Energy is supplied in many forms. Work energy is supplied by manual labor with a digging stick and a hoe, or by plowing with domestic animals or a tractor. Fertilizers and herbicides and tractors and the other material things used in agriculture also represent energy inputs, because energy is required to manufacture them. The yield in a food production system is directly related to the external energy input; a larger energy input produces a larger yield (see Figure 8-4). Interestingly, the external inputs are about the same as the increased yield obtained.

As supplementary energy is used more and more to maintain soil structure and fertility, the result is a greater food yield. However, a second result is the decreased stability of the system. Diverse ecosystems, ones that contain many species and a variety of habitats, are quite stable; but the converse is also true: an ecosystem that contains only one crop is very unstable. This fact makes agricultural systems very susceptible to wind, weather, insects, and plant pathogens. Farming is always a gamble because the ecosystem is so fragile.

If the energy supplements to a food-production system are interrupted, production promptly drops, and the diversity of the ecosystem increases. However, the area will probably not return to its original undisturbed condition. Soil structure and fertility, leaf cover and shade, the water-holding capacity of the land, and other factors will have been changed by the introduction of the energy supplements. Abandoned farm land will be replaced by a more complicated ecosystem, but it may be less productive than the original one. The ecosystem in the Mesopotamian delta and the surrounding hills was diverse and productive even before it was brought under intensive cultivation. Energy supplements such as plowing, the construction of dikes and waterways, the clearing of land, and grazing increased food production until the area supported millions of people. But when the energy supplements stopped, food production dropped, and the ecosystem has never regained its original productivity.

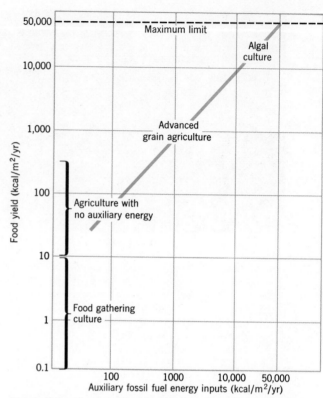

Figure 8-4 *The greater the amount of energy that is put into an agricultural system (in the form of hand labor, irrigation, cultivating and harvesting machines. fertilizers and pesticides, or other inputs), the greater the food yield.*

Human Food Requirements

An individual's food energy requirement is usually expressed as kilocalories[1] per day. The actual requirement depends on the size and physical activity of the individual. If we can estimate the total number of humans and their average size and activity, then we can also estimate the total amount of food required by the human population. Since size is closely related to age, we can use census information about the number of people in various age cat-

[1] The **kilocalorie** is a metric unit of heat. It represents 1,000 **calories**, and each calorie is defined as the amount of heat required to raise the temperature of one gram of water by 1° C. In conventional dietary discussions, however, the prefix "kilo-" is usually dropped; the "calories" listed on "calorie counters" are actually kilocalories. To avoid confusion, we will adhere to scientific usage throughout.

egories to get an estimate of size as well as numbers. A normal day's activity is arbitrarily defined as eight hours of sleep or bed rest and sixteen hours of light activity with occasional periods of heavier physical work. This pattern of activity can be converted into a requirement for kilocalories of food. These data are then fed into a computer to obtain the total amount of food required each day by the human population. Such a calculation was made by the President's Science Advisory Council in a report on world food. Each person required 2,354 kilocalories per day in 1965 (the date is necessary because the age distribution—and therefore the size—of individuals in the world population is not constant). For that same year, the Food and Agriculture Organization of the United Nations estimated the retail availability of food at 2,420 kilocalories per person per day. However, a tenth of this food is lost between the retail market and actual consumption; hence only 2,178 kilocalories are actually consumed.

The numbers still agree rather closely, but this correspondence does not actually mean that there is sufficient food for all the people in the world. There is no reason to think that food is equally distributed among countries of the world, or among individuals within a country. In fact we know that food distribution is very unequal (see Figure 8-5). A depressingly large fraction of the world population is **undernourished**—unable to secure an adequate caloric intake. Roughly half the people in the world are constantly hungry and semistarved; they survive only by drastically curtailing their physical activity.

If we focus our attention on the requirements for specific elements in the diet, the situation is even worse. There is a considerable gap between getting enough calories for survival and having an adequate diet. Energy can be supplied by protein, carbohydrate, or fat, singly or in combination. An adequate diet is one that not only contains enough overall energy from these foods, but also contains the proper amounts of specific proteins, fats, vitamins, and minerals. These compounds are needed for specific functions in the body, yet they cannot be synthesized by metabolic reactions. When the required

Figure 8-5 Food distribution in the world. The number of kilocalories consumed per person per day in various regions of the world, as estimated by the Food and Agriculture Organization of the United Nations.

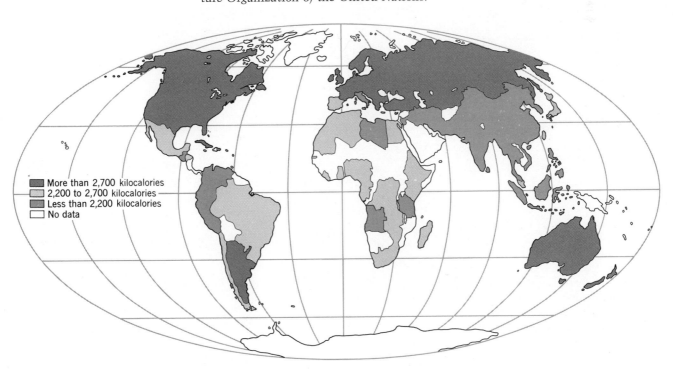

More than 2,700 kilocalories
2,200 to 2,700 kilocalories
Less than 2,200 kilocalories
No data

molecules are absent or in short supply, the body develops abnormally and many functions are impaired. The world's protein supplies are particularily poor. At least half the world's population is **malnourished,** or deprived of one or more essential elements of the diet.

Protein deficiency is the most widespread form of malnutrition. It is especially damaging to young children and pregnant or lactating (milk-producing) women. On a worldwide basis, protein deficiency affects the growth of seven out of ten children under the age of six. Protein deficiency is particularly tragic because its effects are not reversible, even if the diet is improved later in life. The brain is the most rapidly growing organ in the body during the first three years of life. But if protein intake is inadequate, brain growth simply slows down, because the necessary raw material is absent. If protein becomes available later this lost growth is not restored. About 350 million children—or ten percent of the world's population—are victims of protein deficiency.

If undernutrition and malnutrition are so widespread, then death by starvation must be very common. However, it is almost never reported as such. A starving individual is very likely to die of diarrhea, measles, or even a head cold. His death will be attributed to the disease, and the underlying cause will be ignored. Figure 8-6 is a record of weight increase in an Indian child. Each square below the record of weight reports an illness of some sort, mostly gastrointestinal or respiratory. These diseases are trivial challenges to a well-nourished child but potentially deadly to the under-nourished and malnourished infant.

These facts are reflected in various statistics. In the United States, about twenty children of every 1,000 born alive die before the age of one year. The comparable number in undeveloped countries is as high as 250 and typically is between 100 and 150. In several countries, fewer than half the children born alive reach school age. These numbers partly reflect differences in public health and medical care, but they also result from undernutrition and malnutrition of these children. Pregnant or lactating women and the aged are in a similar situation; their deaths are often attributed to dysentery, exposure, or childbirth complications, rather than to simple starvation.

Any estimates of deaths by starvation must

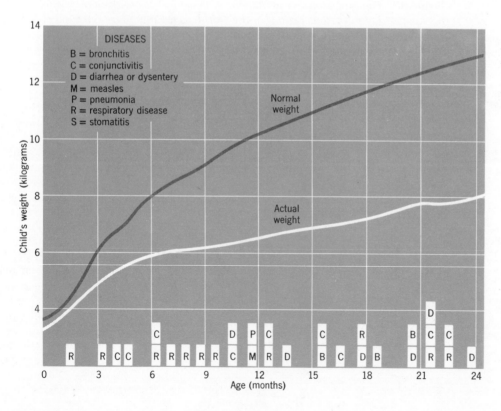

Figure 8-6 Weight gain and health record of an undernourished child. Normal weight gain is shown for reference.

therefore be quite indirect, but a good estimate is that 20 to 30 million deaths a year are attributable to inadequate food. Yet many people reject the idea that famine is constant in the world. They reserve the term "famine" for specific localized food shortages associated with wars or crop failures: World War II is designated as responsible for famines in Bengal, India, and in Honan Province in China that killed millions in 1943; the Biafran War (1967–1969) produced a famine in which more than a million people died. Such drastic events stretch back through recorded history. Yet right now, even in a time of relative peace, food is scarce and badly distributed. Almost one percent of the world's population dies of starvation each year. We tend to overlook this toll for several reasons: because it is not localized, because it is disguised as death by disease, and because it is simply so hard to believe.

We must do several things if we are to alleviate world food problems. We must improve food distribution and production for the current population, and we must try to deal with the requirements of a growing population. The human population is now growing at two percent per year. Two percent may seem like a modest increase, but it means that the world's population will double every 35 years. And as the population doubles, so does our requirement for food. In 70 years the world's food requirement will be four times its present level. At a later time (Chapter 19) we will discuss the details of human population growth. But regardless of the future, we need more food right now; there is already a scarcity among people who are now alive. What are the prospects for further increases in food production?

Control of Food Production

There are many estimates and opinions about the limits of food production on earth, and hence of the limit of the human population. Some people feel that we have already reached the practical limit of food production and are systematically destroying arable land by farming it too intensively. At the other extreme are individuals who see no limit because they believe that science and technology can keep pace indefinitely with the demands of a growing population. In this section we will look at some of the factors that limit and control food production right now. Then we will be prepared to evaluate some of the proposals for changing these limits and increasing food production.

The quantity of food that can be produced by any agricultural system depends on a complex of interacting factors. The nature of the crop plant itself is one factor, of course. But a host of others must be weighed if we are to understand food production. The subject is complicated by the fact that agriculture is part of a whole social system. Therefore, political and social factors are fully as important as the characteristics of the plants themselves in determining the limits of food production.

One way to investigate food production in various systems is simply to measure the amounts of crops grown now. If we express production in terms of kilocalories harvested per acre or per square mile, we can make direct comparisons of the energy yields of different foodstuffs. However, such figures completely ignore important qualitative aspects of the crops, such as their protein content. One can easily improve kilocalorie production per acre, for instance, by planting sugar beets in place of soybeans, but the loss in protein might well be more important than the gain in total energy. Bearing this fact in mind, though, it is still useful to consider the productivity of various crops in terms of energy units.

The yields from rice, wheat, and potatoes under intensive modern agriculture in the United States are all about the same: about 3.2 billion kilocalories per square mile per year. If consumption rates are 2,354 kilocalories per person per day, then a square mile of wheat or rice or potatoes can feed about 3,700 people for a year. Table 8-1 lists similar calculations for these and other crops grown under modern conditions.

These figures do not necessarily represent the upper limits of production, but because they are based on the most modern techniques, they are higher than the yield for much of the world. The cattle-production figures, for example, are based on feeding rich hay and sugar beets to animals that have been bred for maximum productivity. The worldwide average figure would be much lower, because cattle usually browse on sparse vegetation in areas unsuitable for tillage. Nevertheless, the table shows the expected result: it is most efficient to consume plant products directly. If plants are fed to domestic animals and the animals are then used for food, the carrying capacity of the land decreases. When plants are used to support animals, only a fraction of the kilocalories produced by photosynthesis appears as usable milk or meat or eggs. The most obvious conclusion is that we could feed more people by increasing our direct consumption of vegetation.

However, this is a very simplistic approach to the

Table 8-1

Agricultural Yield per Square Mile[1]

Crop	Feeds (number of people)
Rice	3,700
Corn	3,700
Potatoes	3,700
Wheat	1,800
Prunes	1,450
Milk	1,450
Soybeans	1,300
Beef	780
Pork	560
Eggs	75

[1] A square mile of most crops contains enough calories to feed about 3,000 people per year. However, this same amount rarely contains enough protein. Sugar beets, for instance, have a high caloric yield but so little protein that the diet must be supplemented with protein-rich foods. Legumes such as soybeans are exceptions; their protein content is very high.

problem, because it ignores the cultural and political aspects of food production systems. If you consider the number of people per square mile actually supported by various kinds of agricultural systems, cultural influences become apparent. Hunting and gathering cultures, in the Congo and Amazon basins, for example, support only about one person per square mile. At the other extreme, intensive agriculture using human labor and draft animals can support over 600 persons per square mile. This method is now practised in some areas of India. In 1970, United States agriculture supported about 120 persons per square mile. It is the supplementary energy that is supplied through agriculture that accounts for the increased carrying capacity of both the Indian and the American systems.

American productivity is only slightly greater than Indian productivity in terms of yield per square mile, yet Indian agriculture supports five times as many people. Obviously, the standards of consumption in the two societies must be very different. However, the differences extend well beyond patterns of consumption. In each system, agriculture and food production are integral parts of the whole culture.

In the United States, the farm population is only 2.5 percent of the total population. Yet the relations of the rest of the population with agriculture are very direct. Many people are involved in food-related occupations such as food packaging, merchandising, and transportation. Industries provide the energy inputs for farming by manufacturing tractors, fuels, and pesticides. Industries also send manufactured goods back to the farm population. But the relation between agriculture and society is even more fundamental. In the United States, agriculture is supported by the whole apparatus of urban life, based on the manufacture and consumption of goods and the provision of services.

This kind of supporting urban cultural apparatus is poorly developed in a country like India. About 80 percent of the population of India is rural and is engaged directly in farming or in farm-related work. In such a country, our agricultural system would be inappropriate and, in fact, disastrous. In a sudden transition, hundreds of millions of people would be driven away from the countryside and into cities that are unprepared to support them. Furthermore, the productivity figures suggest that an industrialized agriculture like ours could support only about a fifth of the present population of India.

The corresponding transition—moving 80 percent of the United States population to farming activities—might be less of a disaster, but it would also cause a dramatic change in our whole culture. We must therefore avoid oversimplification in discussions of food production. We cannot alleviate world hunger by forcing a diet of rice and potatoes on the rest of the world, because this approach ignores the reality of the complex relations between agriculture and man. There are no simple solutions. Any major change in food production carries with it changes in the general culture.

The economic relations between various societies are just as important as internal cultural factors. The hunting and gathering tribes that survive in rainforests are still relatively isolated, but all the other food production systems of the world are interrelated. Agricultural imports and exports are a major form of world trade. Large acreages of prime agricultural land in undeveloped countries are used for export crops such as sugar, coffee, and bananas. These crops are sold abroad, but the income rarely goes to improve the living standards of the population that produces them. Furthermore, land on which export crops are grown is unavailable for the production of local food crops. The high standard of food consumption in developed countries is maintained at the expense of land and labor and a low standard of consumption in the undeveloped countries.

The flow of protein in the world is another crucial international aspect of food production. Basically, protein moves to the developed countries from the undeveloped countries. There is some flow of protein in the other direction, but it usually takes the form of cereal grains, which are of low quality. The protein imported by the developed countries often takes the form of fish meal and legumes. Peru, for instance, has this kind of relationship with the United States and Europe. The developed countries have helped to expand the Peruvian fishing industry in the last two decades. The Peruvian catch now exceeds the total volume of all the European fisheries, and all of it is exported. Two thirds goes to Europe, slightly less than one third to the United States, and the rest to Japan. Only a small amount arrives as fish for direct human consumption. The rest is in the form of fish meal, which is used primarily as a protein-rich feed for animals such as pigs and chickens. Thus, imported protein is used for the economical production of pork and chicken in developed countries. Huge amounts of protein are involved: South America's fish exports exceed its total meat production.

The United States plays an important role in these relations because of its great wealth, but it is not as dependent on such subsidies as some other countries are. Denmark leads the world in protein imports per person. The imported protein is used to raise pigs which are then processed, canned, and exported as ham and bacon. This industry is an important part of the Danish economy. The original protein comes from South America and Africa, but it does not return there. Very few African farmers will ever taste canned ham.

It seems immoral that countries in which protein deficiency is the greatest are major exporters of high quality protein, and that this protein is used to maintain the high living standards of Western European countries and other importers. This trade relationship is not widely known, and it is just one facet of the development and exploitation of the rest of the world by the West. No doubt it will change with time, but the change will be difficult. The present relationships have a great deal of inertia, and schemes for modification of worldwide food distribution receive little support in the developed countries. Most approaches to food problems have focused not on improving food distribution, but rather on increasing production.

Increasing Food Production

There are many proposals for improving and expanding the world's food supply. We will comment briefly on some of the major suggestions.

New Land

One third of the earth's 52 million square miles of land surface is already used for food production. And of this agricultural land, one third is tilled for crops and two thirds are pasture land. It would seem that the remaining two thirds of the earth's total land surface is available for exploitation. However, the vast majority of this remaining land simply cannot be used. About a fifth of it is covered with ice and snow. And mountainous regions, inaccessible forest land, desert, and wasteland account for most of the rest. A generous estimate is that cultivated land or pasturage might be increased as much as ten percent without disastrous short-term damage; such an increase would represent an additional three percent of the earth's total land area. But this new land would be quite poor, because all the good agricultural land is already under cultivation.

Schemes for increasing agricultural acreage, therefore, usually propose the cultivation of large land areas that experts now consider unusable. As an example, some people think that the forests of Africa and Central and South America are a vast untapped reservoir of land, capable of supporting billions of people. This forest land certainly does support complex ecosystems with lush vegetation. But efforts to introduce intensive agriculture have been unsuccessful.

Tropical ecosystems are quite easy to disrupt and destroy. Most of the nutrients in these forest communities are held in the leafy vegetation, and the decay of leaf litter normally returns them to the soil. However, the nutrients in the soil are likely to be leached away by rain, because tropical soil is very porous. As long as there is a leafy forest canopy the flow of rainwater is retarded. Nutrients can be recaptured and used to maintain the structure of the forest. However, at the same time, the canopy also keeps light away from the forest floor, so the ecosystem does not produce much food for man.

The food productivity of the tropical forests can be temporarily increased by certain agricultural techniques, but this productivity cannot be maintained. **Slash and burn** agriculture, for example, is widely used in tropical areas. A portion of the forest is cut to allow sunlight to reach the ground, and burned to return nutrients to the soil. Root crops can then be grown for a few years. However, the soil fertility rapidly decreases, since nutrients are leached away by heavy rains and are no longer renewed by leaf litter. After a few crops the first strip is abandoned, another area is cut and burned, and the original area is allowed to reforest. Refores-

Figure 8-7 *Slash and burn agriculture in the jungle. A strip of land is cleared by felling and burning the trees; it is then used for crops until the soil nutrients are exhausted.*

tation takes a decade or more, depending on the nature of the soil and the climate (see Figure 8-7).

In many areas of the tropics it is possible to establish a rotation system that allows the same area to be cut and burned every fifteen or twenty years. Thus, eight or ten strips farmed in rotation can constitute a self-restoring food production system. But in other areas, soil conditions do not permit the reuse of an area. Once the soil is exposed and the minerals are leached away, it bakes to a bricklike consistency and will not support plant growth. We simply do not know how to accomplish artificially all the maintenance functions that are carried out by the natural ecosystem in the tropics. We may eventually develop a suitable agricultural system for such areas, but little research is currently directed toward this goal. Tropical forests simply cannot be treated as a usable land reserve at present, nor are they likely to become available in the near future.

Other categories of unusable land must also be discarded as candidates for immediate cultivation. Desert and wasteland cover one fifth of the earth's surface. Portions of the desert are now being recovered and brought under irrigated cultivation. However, our recent track record is not encouraging: we are losing more land through erosion and destruction of the fertility of existing lands than we are gaining through irrigation. In the last half century,

man's agricultural activities have *doubled* the area of desert and wasteland; the net balance is in favor of deserts. In addition, we are steadily degrading the quality of most land now under cultivation.

Perhaps these trends can be reversed. But unfortunately, few governments are supporting research in this field. Nor is it easy to communicate with largely illiterate populations about the need for better agricultural techniques. For these reasons, we will do well just to hold our present land area under food production and maintain its fertility.

Food from the Sea
The area of the earth's oceans is more than twice the area of its land. The oceans provide only one percent of our total food, but that one percent is rich in high-quality protein. Any increase in food from the sea would thus be very significant in terms of protein production. Unfortunately, large increases are highly improbable.

Most of the food in the oceans originates through the photosynthesis of minute drifting algae, the phytoplankton. Minute animals feed on these unicellular plants and are in turn preyed upon by larger animals. Humans enter this food chain later, by eating large marine animals three to four steps removed from the primary photosynthetic production. By that point, most of the photosynthetic energy trapped by the phytoplankton has already been dissipated by the other animals in the food chain. The seafood used by man is rich in protein, but it represents a very small part of the total amount of food produced by photosynthesis in the oceans. We can therefore try to increase the yield of food from the sea in several ways:

1. We can improve our fishing techniques or attempt to exploit hitherto untapped fishing grounds.
2. We can begin to harvest smaller animals or plants directly from the plankton.
3. We can try to farm the waters instead of just hunting and gathering their natural populations.

The first two proposals have little chance of success. The only new fishery we might tap is the krill. These animals are small, shrimplike crustaceans that once fed large whale populations. Schools of krill stay below 400 meters during the day. At night they migrate toward the surface but become scattered. The technical problems of catching large numbers of these animals have not been solved.

Many other fish resources are already being cropped very heavily; any increases might endanger the existing populations. The cropping of whales is about at an end. The best estimates show that even more intensive harvesting of present fishing grounds, coupled with the exploitation of new grounds, cannot do more than double our present harvest.

Cropping smaller marine animals or harvesting phytoplankton directly is impractical because it requires tremendous expenditures of energy. These organisms are dispersed through very large volumes of water, and they constitute only a minute fraction of the total mass. Therefore, very large volumes of sea water must be filtered in order to obtain any significant return. There is no immediate prospect of harvesting plankton to increase oceanic food production.

The most likely addition to our food supply, therefore, comes through **aquaculture,** or farming the seas. Food production can be increased by fertilizing bays and inlets or by cultivating kelp beds to increase natural fish production. Ponds can be stocked with particularly productive fish species (see Figure 8-8). Though the development of aquaculture is slow at present, it may eventually create important additions to our protein supply. Still, even a large increase in seafood will not represent a major addition to our total caloric intake.

Figure 8-8 Aquaculture. Farmers drawing their nets as they harvest the Tilapia fish in breeding ponds in the Central African Republic.

Microorganisms

Two methods for the culture of microorganisms for human food are being studied. One is the growth of microorganisms on organic wastes. Petroleum wastes, the fibrous by-products of sugar cane, paper, spoiled fruit and vegetables, sawdust, and other wastes all can serve as food for specific microorganisms. The microorganisms themselves can then be harvested as protein. The protein content of the harvested bacteria and yeasts is high, and their amino acid composition is comparable to that of plant proteins. The product is suitable food for both humans and livestock. Pilot projects are now testing the feasibility of combining industrial processing with food production, by growing microorganisms on waste substrates.

It is difficult to assess the potential importance of such processes. The growth substrate is cheap and may even have a negative cost—that is, the waste would have been difficult to dispose of any other way. But despite this fact, current production methods are not as economical as those for extracting protein from soybeans or sesame or other plant crops. The process requires a significant capital investment for the construction of a culture apparatus, and it requires power to supply oxygen and maintain a constant environment for the growth of the microorganisms. Factories for producing such **single-cell protein,** as the material is called, can probably be established in industrialized countries, where large quantities of both wastes and power are readily available. But such projects would be much less feasible in undeveloped countries, where the costs of producing it would simply not be competitive with those for traditional agricultural products.

A second, related form of single-cell food production is the mass culture of algae. Under optimum conditions, single-celled algae such as *Chlorella* are capable of converting 50 percent of the incident light energy into organic carbon (see Figure 8-9). But several decades of investigation and large-scale attempts at the production of algal protein have not yielded a usable process. Again, the reasons are mostly economic: even though the substrate costs may be negligible, processing and harvesting are expensive.

Leaf Protein

Many plants are rich in protein, but animals such as man simply cannot remove protein from the tough plant cells by normal digestive processes. However, a simple process can be used to extract

Figure 8-9   Mass culture of algae for food. In this Japanese pilot project, algae are grown in shallow outdoor tanks. Revolving booms keep the water in motion.

protein from the foliage of many varieties of trees and brush. The leaves are shredded, and their juice is squeezed out and heated to coagulate the protein. The final product is edible but not very palatable. Nonetheless, it could be a vital protein source or supplement in primitive societies, where it is needed most.

Unlike the culture of microorganisms, the recovery of leaf protein requires no equipment that is beyond the reach of primitive societies. Surprisingly, the work required to obtain leaf protein is only slightly greater than the effort of growing traditional crops. The major obstacle is communicating with primitive peoples and convincing them that the whole procedure can be worth the effort.

Synthetic Food

Science fiction writers often predict that future man will be sustained by synthetic foods from chemical factories or computerized automats. Chemists can synthesize a variety of organic compounds, and many of them are perfectly good foodstuffs. However, large-scale synthesis would be hopelessly impractical. First, it would require an enormous capital investment for the necessary industrial apparatus. Second, even synthetic food requires large amounts of raw materials: mainly hydrogen, oxygen, carbon, and nitrogen. Hydrogen and oxygen are abundant, but concentrated sources of carbon must come from fossil fuel, and plant tissue. These resources are limited. Alternatively, carbon would have to be extracted from minerals or air and then reduced to organic compounds. This process requires huge amounts of energy. The same is true for the extraction of nitrogen from the atmosphere. And the processing itself would be wasteful. The synthesis of carbohydrates and fats in the laboratory requires 500 times as much water as does their synthesis by plants.

Synthetic food production is simply too inefficient to compete with the productivity of green plants. The energy requirements are enormous, and other valuable resources would still have to be consumed. Synthetic chemicals are invaluable as sources of vitamins and other compounds that are scarce in the natural world. But chemistry simply cannot provide an economical supply of major foods.

The Green Revolution

Improvements of existing crop plants have so far been the best way to increase food production. Crop yields in the United States have risen steadily. For example, American corn production has doubled at the same time that the acreage planted has actually decreased. These improvements are due to a number of factors: improved plant varieties; increased use of fertilizers, pesticides, and herbicides; and many changes in methods of cultivation and harvesting.

The success of American agriculture depends on research and development by the United States Department of Agriculture, colleges, universities, and private firms such as seed producers and chemical industries. Their work is quickly transmitted to a literate farm population through county agents of the Department of Agriculture as well as by secondary means, such as 4-H Clubs and private merchandising. The agricultural system is a complicated and flexible apparatus that responds to local as well as general problems. It contains a large cadre of trained experts and many channels of communication to individual growers. Most developed countries have similar supporting networks. But such an establishment is often lacking or quite inadequate in undeveloped countries. It is important to recognize that improvements in a food production system are dependent on the existence of an apparatus for research, development, communication, and education.

The improvements generated by such a system can be exported to undeveloped countries with great success, but only if the recipient countries put together their own supporting apparatus to make them work. The **green revolution** is a general term for the success of various highly productive varieties of wheat and rice in undeveloped regions

Figure 8-10 N. E. Borlaug inspecting dwarf wheat in an experimental field. Dr. Borlaug received the Nobel Prize for his work in developing and introducing these strains.

of the world. Short-stemmed, disease-resistant wheat was developed in Mexico under the sponsorship of the Rockefeller Foundation (see Figure 8-10). The new plants produced much more grain than native plants. They were exported to many parts of India and Pakistan, where they showed similar improvements in yield. The new wheat increased the yield so much in Mexico that the country changed from an importer to an exporter of cereal grains. India and Pakistan obtained bumper crops in the late 1960's from the new strains. High-yield rice varieties developed at the Philippine Rice Institute have also been very successful. The Philippines recently grew enough rice for their own needs for the first time in decades. These spectacular results produced great optimism about the prospects for increasing world food supplies.

Despite their obvious successes, however, the miracle seeds are not an instant solution to world food shortages. Mexican dwarf wheat does well only if it receives adequate fertilizer and water. If the seed is planted without irrigation and without fertilizer it actually produces less grain than native varieties, which are better adapted to local conditions. The governments of undeveloped countries that adopt the green revolution must therefore educate their farm populations to use the new seed properly; they must also provide fertilizer and arrange for irrigation. Their farmers are typically living at a subsistence level, so the government must supply money for fertilizer and equipment, must guarantee to purchase the crop, and must somehow induce the farmers to try something new.

Some countries have accomplished all these things with great speed. But in other regions, governmental efforts have created bitter social upheaval. The green revolution's chances for success depend heavily on the existing social structure. If the landholding system promotes equality, and the new seed is carefully distributed, production increases and the whole community prospers. But if inequalities of distribution exist, the wealthier farmers are often the only ones able to take advantage of the new techniques. The improved yields are not shared by their poorer neighbors. The wealthier farmers are able to expand their holdings, and subsistence farmers are driven off the land. But the urban centers are unable to absorb these untrained people. The result is rioting, crop burning, and great bitterness. In many areas, the social problems fostered by the green revolution have been as formidable as the food supply shortages they alleviate. As noted above, agriculture is part of the total culture of a people; you can't change one without influencing the other.

In spite of all these problems, the green revolution should be an abrupt step forward in adding to the world food supply. It offers a respite of fifteen or twenty years during which our food supply can keep pace with increasing world population. It also offers at least an opportunity for undeveloped countries to catch up with the developed world. If the developed world were to provide economic assistance (loans, equipment, technical expertise) and were to agree to curtail deliberately their own production of basic foodstuffs, then the undeveloped countries could grow and export food for profit, accumulate capital, and strengthen their economies. These conditions, in fact, are a very simplified description of the Indicative World Plan, developed by the Food and Agriculture Organization of the United Nations. But the plan was unanimously rejected by the developed countries in 1970. This action is no surprise. As we noted earlier, the international aspects of food production have a great

deal of inertia and are part of international relations in the same way that local agriculture is part of local cultures. Nonetheless, rejection of this plan was a great tragedy and a lost opportunity.

Minimizing Crop Losses

About one third of our food is eaten by other organisms or destroyed by disease or spoilage. Reducing this toll would be equivalent to increasing our food production. In the United States, food losses in the field and in storage cost seven billion dollars a year. The losses would be even greater if we did not spend about three billion dollars per year on control measures and storage and handling facilities. Effective programs to reduce crop losses require a sophisticated technology. Success with pesticides and herbicides requires literate and educated growers, chemical industries, a distribution system, and a supporting research and development system. Storage facilities must be equally sophisticated. Human labor can be substituted for technology to some extent—caterpillars can be killed and weeds pulled by hand. But crop losses in the undeveloped world are still larger than those in developed regions, despite more intensive use of human labor.

Reducing crop losses also requires a continuing program of research in plant breeding and control agents of various sorts. An intensive agriculture is inherently unstable, because it represents an extremely simplified ecosystem. Tremendous fields of a single plant crop are susceptible to large populations of pests and predators. Even crops that are bred to be disease-resistant are endangered, because disease agents mutate to more virulent and invasive forms. New control methods and new resistant plant strains must be developed on a continuing basis. Developed countries will continue to reduce crop losses, but their achievements will have little impact on the total world food supply.

Limits of Food Production

Plant production cannot increase indefinitely. The amount of radiant energy that strikes the earth's surface provides an absolute upper limit to productivity. But that limit is obviously not achievable, anyhow. And a more realistic estimate is difficult to make.

A Russian botanist, A. A. Nichiporovich, has made a careful and optimistic projection of the maximum productivity we can achieve in the forseeable future. He estimates that crop-plant production can be increased to 7.5 times its present level,

Table 8-2

The Limits of Food Production

	Present Levels	Possible Levels
Area Under Cultivation	1.5×10^7 km²	3×10^7 km²
Efficiency of Photosynthesis[1]	0.5%	2%
Total Yield	5×10^{12} kg	40×10^{12} kg
Nitrogen (additional)[2]	4×10^9 kg	450×10^9 kg
Irrigation Water	0.5×10^{12} m³	3×10^{12} m³
Minerals (additional)[2]	3×10^{10} kg	300×10^{12} kg

[1] The conversion of available radiation to plant biomass.
[2] Supplied by fertilizer.

but only if certain conditions are met. The world's cultivated land area must be doubled, and crop plants must become four times more efficient in their utilization of sunlight. These achievements would require tremendous increases in irrigation and even more astonishing increases in the amount of nitrogen and other minerals cycling in the system (see Table 8-2).

Nichiporovich describes this estimate as the most optimistic figure for the maximum attainable yield. He does not even discuss the likelihood of reaching this limit, but he does make it clear that it would be an incredible achievement. For instance, he assumes that we can double agricultural land. To do so, we would have to create a productive agriculture in the dry and the rainy tropics, recover and sustain the productivity of deserts, and improve the fertility of marginal lands. We would have to increase irrigation by a factor of six, and much of the land that is easily irrigated is already under irrigation; a six-fold expansion of that area would be a heroic task. There may not even be sufficient water to permit this increase. In fact, increasing the amount of irrigation water to Nichiporovich's goal of 3×10^{12} cubic meters per year is beyond heroism: it is simply unrealistic. That figure is equivalent to ten percent of the total annual drainage by all the rivers of the world. Most hydrological experts would agree that there is no chance of capturing and utilizing one tenth of the world's fresh water flow for irrigation.

The postulated increases in quantities of nitrogen and other minerals would also be nearly impossible to achieve. A potential source of nitrogen is the atmosphere, which contains 4×10^{15} tons of N_2 gas. As yet, however, no cheap extraction process is even remotely possible. Supplies of other minerals such as potassium and phosphate are more limited.

Other problems will definitely arise. The earth's land area is finite, and food production competes with other uses of it. Forests (for timber, paper, wilderness, watersheds, and the prevention of erosion) and developed land (for residential and industrial use) both compete with farmland. Converting forests to tilled land on a large scale introduces major weather changes and encourages erosion. In many areas of the world, the ratio of woodland to tilled land is already dangerously low. During World War II, Japan began clearing forests to increase its wartime food production. But this campaign resulted in serious erosion and a decline in the water table, which threatened already productive land. The land clearing had to be stopped; reforestation was immediately undertaken to restore stability to the ecosystem.

Urban settlements may also compete with the most fertile and productive agricultural lands. In California, for instance, three million acres had already been transferred from agricultural to residential use by 1960. Extrapolating this rate to the year 2020, fully half of the state's agricultural land will have been taken over. Since California now provides 43 percent of the United States' production of fruits and vegetables, this change would have a significant effect on the whole country's food supply.

The human population and its food requirements are already pressing against the world's resources of land and water and minerals. Our choices are now quite limited. We can no longer "open" new land as our needs increase. Land must now be used for food production *instead of* for forest, or for urban communities *instead of* for food. These uses are mutually exclusive.

Secondary Plant Products

The direct use of plants as food is not the only important relation between plants and man. Plants contribute a host of other products. Plant fibers give us important materials like timber, paper, and cotton. Chemicals from plants are used in such diverse products as waxes, solvents, lacquers, paint, and plastics. In this section, we will discuss yet another category of organic compounds known collectively as secondary plant products.

Secondary plant products are not directly related to the structure of the plant, nor are they part of the major metabolic pathways, the ones concerned with photosynthesis or respiration or growth. The secondary products are chemically diverse; they include such major categories of compounds as quinones, terpenes, sterols, alkaloids, tannins, and oils. Specific products that are of particular interest to man include caffeine, nicotine, quinine, reserpine, strychnine, morphine, atropine, lysergic acid, ephedrine, theobromine, and many others.

The plant produces its secondary products as part of its chemical survival strategy. Many chemicals serve to make plant foliage taste bitter, or act as extremely effective poisons; these devices protect the plant against browsing animals. Plants can also produce herbicidal agents that inhibit the growth or germination of other competing plants. Some plants synthesize insecticides as well. But not all the secondary plant products are defensive. Plants may produce oils and other aromatic substances that attract insects to pollinate flowers.

Plants have evolved specific chemical agents to enhance their chances for survival and reproduction. However, at the same time, herbivores and competing plants have evolved. Herbivores may develop resistance to plant poisons or may even come to use them for their own protection against predators. Man himself has learned to use secondary plant products for many purposes. Our use of plant compounds for flavorings and scents, drugs, and pesticides is an integral part of our culture.

beverages

We have already discussed the use of plants in beer and wine making (Chapter 5). Almost any plant carbohydrate can be similarly fermented to produce an alcoholic beverage. In addition, various plants are used to produce non-alcoholic beverages as well, such as tea, coffee, and chocolate (see Figure 8-11).

Tea is an extract of leaves from an evergreen plant, *Thea sinensis*. The young leaves of the plant contain many soluble materials, but the most important ingredients are caffeine, theophylline, and xanthine. These compounds are all mild stimulants. Tea increases the level of glucose and fatty acids in the blood, increases the heart rate, and stimulates the central nervous system, among other effects. Mild doses of caffeine relieve fatigue and boredom and improve concentration; larger doses cause tremor and failure of eye–hand coordination.

Tea drinking is an old practice in the Far East. It was introduced and popularized in the West through the activities of the Dutch and English. Most tea is still grown in the Orient. Tea leaves are usually crushed after gathering and allowed to ferment to a brownish-black color and rich aroma; green tea is tea that has not been allowed to ferment. The quality of tea depends more on the locality where it is grown than on variations in plant stocks. In fact, an expert tea taster can pinpoint the locale where a particular tea was grown. Teas offer a great variety of tastes and aromas to less educated palates as well.

Coffee is made from beans, or seeds, in the berries of *Coffea* trees. The beans are extracted from the berries, peeled, roasted, and finally ground up. Wild coffee trees were discovered in Ethiopia. Coffee drinking reached the Arabic world in the thirteenth century. From there it spread to Europe and reached America about 1660. Most coffee is now grown in South America; Brazil alone accounts for almost half of the world's production of coffee.

Cacao seeds, the source of cocoa and chocolate, were first domesticated in South America. In processing, the seeds of cacao trees are first fermented and then roasted. Next, fats and oils are pressed out of them to produce **cocoa butter.** Cocoa butter is a rich food source and the critical ingredient in chocolate. The residues of the pressed and roasted seeds, by the way, contain another xanthine drug, theobromine. This compound is similar to caffeine and is used in cola drinks.

Figure 8-11 Sources of plant beverages. (a) Coffee tree berries. (b) Cacao tree. (c) Tea plants.

spices

Spices add flavor and scent to foods. They are consumed in such small amounts that they make little contribution to our energy requirements, but they have many other effects. We have already suggested that a low level of personal hygiene in our ancestors, as well as difficulties in preserving meat, provided a strong impetus for the heavy use of both perfumes and spices. Another reason for the popularity of spices is that they contain essential oils and agents with drug-like effects. Camphor, found in various spices, is a cerebral and respiratory stimulant; menthol has local effects on sensory nerves, causing tingling and a cool sensation; some kinds of pepper contain chemicals that produce numbness and salivation; cloves contain eugenol, which is still used in dentistry as a mild antiseptic. These effects may have contributed to the great demand for spices.

Spices have been quite important historically. The chief reason is that most spice plants grow only in the Orient. Spices were already valuable items of trade in Greek and Roman times. The overland trade between Europe and the Far East was monopolized by the Arabs for millenia, and their monopoly was not broken until the great age of exploration in the fifteenth and sixteenth centuries. Columbus was searching for a western route to the Indies (the "Spice Islands") when he discovered America.

drugs

Plants provide man with purgatives, cathartics, poisons, hallucinogens, sleeping potions, pain relievers, and a wide range of other medicines. In 1820, the first edition of the United States Pharmacopoeia (catalogue of drugs) listed about two hundred medicines of plant origin. But during the next century and a half, over three fourths of these original plant products were dropped from the pharmacopoeia, while only about twenty were added. The number of plants in the pharmacopoeia decreased for two reasons. First, some plants proved to be inactive and useless as medicines. But even more important, many of the plants have been replaced by synthetic drugs.

As the first step in the preparation of a synthetic drug, the active compounds that give the plant its medicinal properties have to be isolated and identified. They can then be synthesized by a chemist. Quite typically, molecules that are closely related to the actual plant compound are more effective than the natural plant compound itself, or are more desirable because of their reduced side effects. Even small changes in structure, such as replacing a hydrogen atom (–H) with a methyl group (–CH$_3$) or a hydroxyl group (–OH), can make a big difference. The synthetic compound will also be superior because its concentration is known, permitting careful control of the dosage—the concentration of natural compounds varies from plant to plant. The source of a particular drug often becomes a matter of economics; if it is cheaper to synthesize the drug from abundant precursor compounds such as other plants or petroleum, the synthetic drug will replace the original drug plant.

Plants are gradually becoming less important as a bulk source of drugs. However, they are irreplaceable as a reservoir of new ones. Many valuable drugs were first discovered in traditional plant remedies. For instance, willow bark and oil of wintergreen were common remedies to relieve pain and control fever, long before a chemical investigation of the plants revealed that they contained large quantities of methyl salicylate. The discovery of the active ingredient led to the production of a related molecule, aspirin, or **acetylsalicylic acid.** Aspirin is the most commonly used drug at the present time in the United States.

Foxglove was a folk remedy for epilepsy for a long time. It was finally introduced into medicine in the eighteenth century by an English physician, William Withering, who obtained his sample of foxglove from the garden of the "Witch Woman of Shropshire." The plant was first used for all sorts of ailments, though usually without success. It was finally discovered that its active agent, **digitalis,** is a specific heart stimulant. In a similar manner, **reserpine,** a useful tranquilizing drug, came from snakeroot (*Rauwolfia*), a shrub used in Indian folk medicine.

It is still profitable to investigate the cultural use of plant drugs. Drug companies and other organizations continue to sponsor safaris to collect possible drug-producing plants. Plant drugs continue to be important even though the number of medicines obtained directly from plant sources is declining.

Opium

Opium is collected from the flower capsules of the opium poppy, *Papaver somniferum*, after the petals have been shed. The capsules are lanced, and the milky exudate is allowed to flow overnight. The next morning the coagulated drops of this crude opium are collected. Opium contains a complicated mixture of active agents; the most important are **morphine** (10–20%), **codeine** (1–2%), and **thebaine** (0.5–1.5%). All three compounds are closely related chemically. Morphine is converted to **heroin** by the addition of two acetyl groups (–COCH$_3$) to the molecule. **Methadone,** now used in the treatment of heroin addiction, is a synthetic compound. In diagrams it does not appear to be very similar to morphine (see Figure 8-12); however, a three-dimensional model reveals that the two compounds have very similar structural features.

Opium is one of the most effective analgesic, or pain-relieving, drugs. It raises the pain threshold and has a soothing effect that removes fear as well as pain. The medicinal use of opium was first mentioned six millenia ago in Sumerian writings. At first, opium was taken by mouth in wine or mixed with other drugs and spices in potions. In the nineteenth century it was most commonly smoked. In that case the active ingredients are vaporized and absorbed through the lung.

The initial response to taking opium is euphoria, followed by relaxation and lethargy. **Tolerance** to the drug is developed very rapidly; in other words, larger and larger doses are required to obtain the same effects. A typical therapeutic dosage of morphine sulfate for an adult is five to twenty milligrams. But after tolerance has been developed, more than one hundred times as much may be consumed daily.

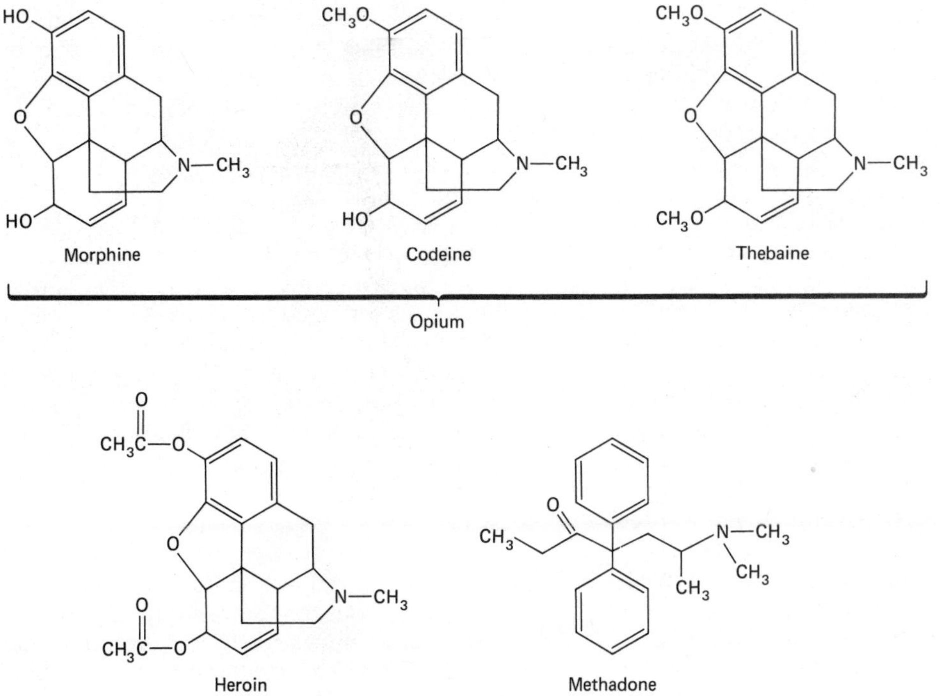

Figure 8-12 Natural and synthetic opiate drugs.

Opium tolerance is accompanied by physical dependence; many bodily functions are disrupted when the drug is absent. Withdrawal from morphine produces its first symptoms about twelve hours after the last dose. Yawning, watering eyes, and perspiration are followed by restlessness and muscle twitching, and then cramps, chills, and hot and cold flashes. The withdrawal symptoms reach a peak two to three days after the last dose—nausea, vomiting, diarrhea, and rapid weight loss. The withdrawal symptoms decrease after about a month, but six months may be necessary for the full recovery of normal physiology and behavior.

Many individuals in earlier cultures probably became physically dependent upon opium, but Western society has only recently considered dependence a problem. Two events in the nineteenth century stimulated the use of opium. The first was the introduction of the hypodermic needle, which permitted direct injection of the drug. The second was the isolation of morphine from crude opium. Before the development of morphine, users had been somewhat protected from addiction by the impurity of the opium preparations. The drug's negative effects, including addiction, were attributed to the impurities rather than to the drug agents themselves. It was felt, therefore, that the purification of

opium would eliminate these harmful effects. As a result, morphine was very widely used during the Civil War. Soldiers were taught to dose themselves with hypodermic syringes, and morphine preparations were freely used for any painful condition. The result was widespread addiction among veterans (termed "soldier's disease"). Morphine use was also common in the civilian population. Morphine syrups and elixirs were touted as cures for all ills—including opium addiction.

Heroin was first synthesized from morphine at the end of the nineteenth century. At first it, too, was considered non-addictive, and it was freely used in an attempt to cure physical dependence on morphine. This unfortunate sequence of events has been repeated several times. Methadone is currently being used to cure heroin addiction. However, it is at least recognized that methadone is addictive, and one may hope that this same lesson will not be repeated yet again.

Heroin is the most commonly used opiate among addicts in the United States. Since heroin is several times as potent as morphine per unit weight, and since its synthesis from morphine is very easy, it is quite profitable for illegal importers to refine raw opium to heroin. This cost factor is probably the main reason that heroin replaced morphine and raw opium on the illegal market. About 200 tons of opium per year are imported into the United States for medical purposes. The illegal trade in opiates is larger. At the beginning of 1970, there were 65,000 known heroin addicts. But since this figure represents only those known by the F.B.I. and other law enforcement sources, it should probably be increased by a factor of at least five or ten. There are probably several hundred thousand addicts in New York City alone. In 1972, for the 15- to 35-year-old age group in that city, drug abuse was the leading cause of death, through overdosing, infection, and reactions to adulterants and additives.

Most addicts are malnourished and debilitated. But these developments are not primary responses to the drug. An addict who receives doses just high enough to avoid withdrawal symptoms has normal body weight; blood pressure is somewhat elevated and respiration is depressed, but the addict's behavior is not grossly changed. Infection from the hypodermic needle is a much greater health risk. Many heroin addicts are in poor condition because they have personality problems that date from before their drug use: anxiety, psychosexual immaturity, and feelings of futility are common. These difficulties are often aggravated by social conditions. Most addicts live in slum areas of large cities. Returning veterans were recent additions to the group. As the addict develops a tolerance for heroin, larger doses are required. The habit becomes more expensive, and the addict must devote more energy to maintaining his drug supply. The exorbitant cost to the user is a direct result of the illegal nature of the drug.

Various approaches to the addiction problem are now being tried, but none with any great success. Some rehabilitation programs substitute methadone for other opiates to avoid withdrawal symptoms in an addict. Methadone may be less debilitating than heroin. But it is not clear whether this apparent difference is a real physiological one or just a reflection of the social context in which methadone is supplied. In any case, withdrawal symptoms are less violent with methadone; it is possible to cure the physical dependence of heroin addicts by hospitalizing them, shifting them to methadone, and then slowly reducing the dosage.

Ultimately, however, almost all of the treated addicts return to heroin when they are released. This result is quite predictable. Any personality problems most likely remain, and the ex-addict usually returns to the same social context that produced addiction in the first place. Self-help programs such as Addicts Anonymous and addict communities such as Synanon attempt to approach this problem by changing the addict's life style and by indoctrinating members in socially acceptable values.

Addiction must be interpreted in its social context. The legalization of opiates would eliminate many of the present health problems, but at the same time it might risk an increase in the number of individuals physically dependent on drugs. The medical uses of opiates are important enough so that it would not be possible simply to destroy all sources of the drugs and ban their synthesis, even if such a ban could be enforced. The problem of addiction is complicated and there is no simple solution.

Tobacco

Tobacco (*Nicotiana*) was first used by North American Indians. Tobacco leaves contain **nicotine**, a drug that acts on the nervous system. Smoking produces tremor, a slight increase in heart rate, and various effects on the nervous system. The amount of nicotine absorbed in smoking is small—only one or two milligrams from a typical cigarette. Larger doses are extremely toxic. Nicotine is an important drug for experimental analysis of the nervous system, but it has no therapeutic uses.

Tobacco is difficult to grow. The young plants are extremely sensitive to a variety of diseases and plant predators. They must be started in fumigated, protected greenhouse areas. Later the plants are set out in fields. The leaves are harvested carefully by hand throughout the growing season. They are cured by drying and then processed.

The introduction of the manufactured cigarette early in the twentieth century brought a tremendous increase in the incidence of smoking in the population. At the present time, about 45 percent of the adult male population and 30 percent of the adult female population of the United States are smokers. There has been a corresponding increase in the economic importance of the tobacco crop. The value of the United States tobacco crop is about six percent of the value of all harvested crops. To this value must be added the income generated by taxation of tobacco products. In all, the tobacco industry represents an annual economic activity of seven to eight billion dollars.

Smokers exhibit a tolerance for larger and larger doses. The physical dependence is minor, but withdrawal is often very upsetting psychologically. The use of tobacco has many harmful effects and is a substantial health hazard. This conclusion is supported by many retrospective statistical studies which compare the rates of disease and death in smokers and non-smokers. There is a clear correlation between smoking and a host of serious conditions: lung cancer, coronary disease, arteriosclerosis, pulmonary emphysema, birth defects, and premature death (see Figure 8-13).

Such correlations do not necessarily establish a causal relationship between smoking and these health hazards, of course,—a proneness to lung cancer and a proneness to smoking might both be due to yet a third characteristic of the individual. But this argument is not very convincing for several reasons. First, increases in the incidence of lung cancer (once a rare condition) correlate well with the increased incidence of smoking in this century. Second, tobacco smoke contains residues that are clearly **carcinogenic** (cancer-producing) in experimental animals. And third, cancer is correlated with smoking in a quantitative way. That is, the risk of cancer increases with the amount one smokes. The effect is reversible, because the risk of cancer also decreases when one stops smoking.

However, we still don't know precisely how smoking causes these harmful effects. Other effects of smoking become clear only when the habit is broken: energy, sexual potency and libido, and general enjoyment of life all seem to increase.

The cumulative effects of smoking as a health hazard are considerable. Smoking is now our major preventable cause of death. About fourteen percent of all the deaths that occur in the United States are due to smoking, in one way or another. And our economic losses through work days lost because of smoking far outweigh the economic importance of the tobacco industry itself. For these reasons, tobacco use is actually a far greater health hazard and economic problem in our society than addiction to opiates.

pesticides and herbicides

Many secondary plant products kill plant-eating insects or repel their attacks. Others inhibit

Figure 8-13 *Tobacco smoking and health.* (a) *Smoking is correlated with an increased death rate.* (b) *The death rate is linked to the number of cigarettes smoked per day.* (c) *The death rate decreases after cessation of smoking.*

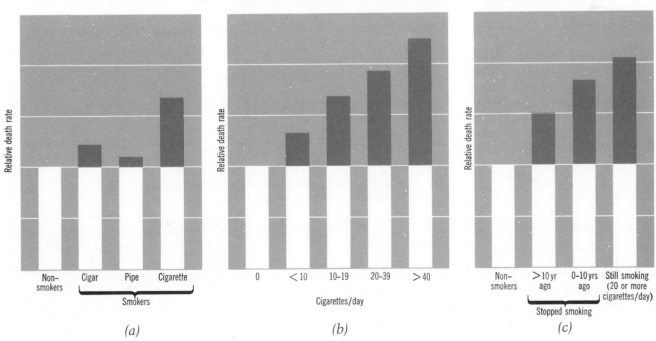

the growth of competing plants. In both cases the plant products can be used directly or modified by man and then used to control the environment.

Pesticides

Pyrethrins, nicotine, and rotenone are examples of natural plant products that are widely used by man as insecticides. Pyrethrum, produced by several species of chrysanthemums, is particularly useful because it is an insecticide with little or no toxic effects on warm-blooded animals. Organic gardeners recommend that chrysanthemums be included in plantings in order to control insect pests.

Plants produce other substances which discourage insect pests in subtler ways. Some plants, for instance, synthesize compounds that interfere with insect development. Insect development is normally controlled by hormones. The process requires periodic molting, or shedding, of the old external skeleton and the laying down of a new one. A hormone called **ecdysone** (from *ecdysis*, the Greek word for molting) must be present in the insect or molting will not occur, and the insect will not advance to its next stage of maturation. A second hormone, the **juvenile hormone,** moderates the effect of ecdysone. It inhibits the advance to the next stage; if the juvenile hormone is present a caterpillar will simply become a larger caterpillar. Both hormones must be in balance for normal development.

Many plants protect themselves by synthesizing these insect hormones or analogous compounds with the same activity. The compound pronasterone, for instance, is an analogue of ecdysone. It is vastly more potent than the normal insect hormone, and it accelerates the molting cycle and upsets development. It prevents the normal development of the *Cecropia* moth, even at concentrations as low as one part per billion in the diet. To date, about 30 ecdysone-like compounds (including ecdysone itself) have been found in 60 species of plants. The compounds are highly concentrated in plants. A ton of stems from a common fern would contain about ten kilograms of ecdysone, while a ton of insects might yield only 25 milligrams. Any plant that contains such huge amounts of hormone is well protected from insects.

Plants produce analogues of the juvenile hormone as well. This discovery was made at the Harvard laboratory of an insect physiologist, Carroll Williams. A Czechoslovakian biologist, Karel Slama, came to work in the laboratory and brought along stocks of a European bug. But when he tried to raise them for use in experiments, they failed to develop into adults. They simply continued to molt into abnormally large juvenile forms (see Figure 8-14). The investigators recognized that the juvenile hormone must be present somewhere in the system. They soon discovered that the paper towels on the bottoms of the insect containers contained juvenile hormone. They went on to find that many samples of American and Canadian paper contained the hormone, though it was absent from European

Figure 8-14 The normal nymph of a linden bug (left) becomes a giant nymph when treated with juvenile hormone (center), instead of molting to a normal adult (right).

paper. The trees used for pulp in North America (balsam fir, hemlock, tamarack) have an analogue of the juvenile hormone.

There are several significant differences between the action of juvenile hormone analogues and that of ecdysone analogues. First, the juvenile agents are specific for particular groups of insects, whereas ecdysone appears to be the molting hormone for all arthropods—insects, spiders, and crustaceans alike. Second, juvenile agents are absorbed through the insect's exoskeleton, while ecdysone must be injected. Third, the juvenile hormone prevents the normal development and hatching of insect eggs. These characteristics suggest that juvenile hormone analogues might be used to control specific pest insects, while leaving neutral or beneficial insects unharmed.

The specificity of particular juvenile hormone analogues is still unknown, and we don't know how many of these agents are naturally produced by plants. However, we may also develop specific synthetic agents. Alternatively, we might use nonspecific juvenile hormone analogues on particular insects in another way. The hormone agent could be applied to male insects of a particular species raised in the laboratory. Then, when these males mated with local females, the hormone-active material would be transferred to the female, and her eggs would be prevented from hatching. In this way, it might be possible to attack one particular insect species, even though the hormone agent is not itself specific. These exciting possibilities for new insecticides are being developed, but they are not yet ready for broad use in agriculture. We are, for the moment, still heavily dependent on insecticides such as DDT. And unfortunately, many insects have become resistant to DDT through many years of repeated exposure to it.

The development of resistance to the standard insecticides has made the exploitation of insect growth regulators even more critically important. Many people believe that this form of regulation will be extremely successful, because they are convinced that an insect cannot defend itself against its own hormones. But this conclusion is probably overly optimistic. It will probably be more difficult for insect pests to accomodate to such insecticides, but not impossible. The evolution of both plants and animals proceeds in parallel because they are dependent on one another. So far, animals have always been able to cope with the chemical defenses that plants have evolved. It is unlikely that insects will find it impossible to cope with this new development.

DDT

DDT, dichloro-diphenyl-trichlorethane, was first synthesized in the nineteenth century. It was rediscovered as a potent contact poison for insects in Switzerland in 1939 and put into immediate use to protect the Swiss potato crop against insect predators. In 1942, DDT was sent to the United States, where it was employed in killing the insect transmitters of typhus, malaria, and yellow fever.

The chemistry of DDT bears no relation to insect growth regulators. Instead, it is a highly effective toxin which is absorbed through the insect's cuticle, or outer covering. Some insect populations, however, quickly develop resistance to DDT. In that case a new insecticide must be used. The same process, of course, may then be repeated with the new agent, and eventually with all our insecticides, until ultimately we are back to a flyswatter.

Herbicides

The development and use of herbicides by man, which we touched on in Chapter 7, is entirely analogous to the development of insecticides by plants. The most effective herbicides are compounds that interfere with the normal growth of weeds but do not affect the crop plants themselves. The herbicide 2,4-D (2,4 dichlorophenoxyacetic acid) is an analogue of the natural plant growth regulator, auxin (indole-3-acetic acid). When 2,4-D is applied to broad-leafed plants, it stimulates excessive growth, and the plants die. Plants that belong to the grass family are not affected (see Figure 8-15). Plant growth regulators have been quite successful as herbicides. Weed pests have shown no sign of

Figure 8-15 Effects of the herbicide 2,4-D. The tract on the right was sprayed with herbicide; the tract on the left was not.

Figure 8-16 Compounds given off by sagebrush inhibit the growth of competing plants. The zone A-B is completely barren; partial inhibition occurs in zone B-C.

developing resistance to the auxin analogues.

Plants themselves employ a different sort of chemical defense against other plants. They release substances that inhibit the germination or development of other plants. A common sagebrush in southern California is an example. It is usually found isolated from other plants by a ring of bare ground (see Figure 8-16). Other grasses and herbs, which might compete with it for moisture and nutrients and sunlight, have been inhibited by chemicals synthesized by the sagebrush. Its leaves release camphor and other volatile compounds. These substances adhere to soil particles and prevent the germination of other plants.

Plants did not evolve simply to be food for animals. Man has carefully selected and modified existing plants to use as food sources, and his own culture has changed in the process. The secondary chemical compounds that plants synthesize are part of a complicated tactical game. They must repel animal pests, injurious microorganisms, and plant competitors; yet they attract other animals to act as pollinators, or microorganisms to live as symbionts. Man again uses these plant products for his own purposes. The plant's chemical strategy for survival provides us with profitable hints for controlling and modifying our own behavior and environment.

Recommended Reading

Baker, Herbert, *Plants and Civilization* (2nd ed.). Belmont, Cal.: Wadsworth (1970).

Discusses the origins of crop plants and drugs, and miscellaneous uses of plants and plant products.

Borgstrom, Georg, *The Hungry Planet: The Modern World at the Edge of Famine.* New York: Collier Books (1967).

A discussion of food production, water and land, resources, and their relation to human population growth.

Brown, Lester R., *Seeds of Change.* New York: Praeger (1970).

A discussion of the "green revolution," the development and introduction of new varieties of wheat and rice into various regions of the world.

The Yearbook of Agriculture, published annually by the United States Department of Agriculture, United States Government Printing Office, Washington, D.C.

Each year a different major topic is discussed. The books are inexpensive, beautifully illustrated, and authoritative.

9 animal organization I: nutrition

Photosynthesis provides plants with the reduced carbon compounds necessary for their growth and survival. Animals also need reduced carbon for energy and growth. Many aspects of a plant's structure are influenced by its requirements for photosynthesis, and similarly, the feeding patterns of an animal are extremely important in determining its whole organization. The speed and grace and complicated behavior of many animals can usually be interpreted as devices that enable the animal to acquire food. The sluggish and sedentary habits of other animals are related to their problems of food acquisition. Food is a primary animal requirement, and thus it makes a good starting point for our discussion of animal organization.

Food must supply adequate amounts of all the compounds that constitute an animal's cells and cell products. These substances include proteins, fats, carbohydrates, nucleic acids, minerals, various trace organic compounds that we know as vitamins, and, of course, sufficient water. But an animal's organic requirements are somewhat elastic, since animals can convert many types of compounds to others.

Fat and protein and carbohydrate are generally interconvertible. First they are broken down into their monomers. Then, the carbon chains of fatty acids and amino acids can be used to synthesize sugar molecules, and vice versa. Of course, the conversion of sugar to amino acids can occur only if some nitrogen is added to the sugar molecule. The nitrogen ultimately comes from the breakdown of other amino acids. Thus, sugar or fats are no substitute for protein in the diet.

Animals can also manufacture particular fats, sugars, or amino acids from

Pieter Breughel, "The Large Fishes Devouring the Small Fishes"

249

other compounds in their diets. But not all animals can carry out all the interconversions. Often an animal may not have the enzymes that are necessary to carry out a particular metabolic reaction; and when a necessary compound cannot be synthesized from other compounds, it must be obtained from the diet. Thus, the specific dietary requirements of a particular organism depend on its ability to convert organic compounds from one form to another. Some animals require many and varied foods from the environment. Others get along on an astonishingly narrow diet. Food must also provide an adequate supply of chemical potential energy, in a form that can be metabolized, and it must supply adequate amounts of compounds that act as precursors for growth and the repair of tissues.

feeding patterns

A food source may be plant or animal, live or dead. Animals may consume food in large chunks (nuts or berries or artichokes or cows), or as small particles (microscopic plankton or flakes of decaying matter), or as liquid (blood or plant juices). Almost everything is used as food by something. Often, to us, the material seems hopelessly unsuitable as food. Some marine animals live by consuming the insulation of submarine cables. A small species of worm survives by eating the felt mats used in German taverns.

Filter-Feeders

Animals that feed on the fine particulate food in lakes and oceans provide a good starting point for a discussion of animal nutrition and organization. These organisms, called **filter-feeders**, are less familiar than land animals, and their specializations are quite different from those of cats and cows. Nonetheless, they are biologically and economically important.

Some animals from almost every major phylum (that is, from almost every major subdivision of the animal kingdom) feed on thinly distributed microscopic plants and particles by filtering food particles out of a stream of water. The particles are then sorted, collected, and passed to the mouth. The necessary elements for pumping, filtering, and sorting are derived from a variety of structures. The water flow that is required may be produced by

cilia or flagella, by muscular movements, or by natural current flow. Water may move past a stationary animal, or the animal may move through the water. The filter may consist of stiff bristles, cilia, or a mucus sheet.

The whole organization of a filter-feeding animal is adapted to the necessities of food collection. Anatomical features that do not contribute to this process may be reduced or lost altogether. Such structural modifications can be quite drastic. Molluscs—the animal group containing clams, snails, and their relatives—provide a good example. The clam's ancestor was a snail-like creature that crawled about on a muscular foot and browsed on algae. It had a well-defined head and tentacles, a small pair of gills, and a protective shell (see Figure 9-1).

The modern clam is very different from its ancestor, because it has become specialized for filter-feeding. The gill, which functioned in the ancestral molluscs as an oxygen and carbon dioxide exchanger, is a water pump and filter in modern clams. The whole animal is enclosed in a mantle of flesh that directs water circulation through the gill. The muscular foot is reduced or lost altogether, so the animal can no longer move along the substratum. In addition, clams have lost the head and tentacles that were features of the original mollusc organization. Adaptations to a particulate food source have totally transformed clams into efficient filterers; other features have been suppressed and subordinated to this end.

The clam's apparatus for pumping water, filtering it, and sorting the particles is constructed of cilia, which sweep particles along the surface. The gills are bars of tissue lying side by side, forming a flat sheet. Each bar, or **gill filament**, is separated from its neighbor by a space through which water can flow. The cilia lining these spaces, called the **lateral cilia**, pump water between the bars. At the front corners of each bar are long **latero-frontal** cilia that beat in the opposite direction (though not strongly enough to change the direction of water flow). As suspended particles approach the space between any two bars, they are caught by the latero-frontal cilia and are pushed, against the flow of water, onto the front surface of the gill filament. **Frontal cilia** on the surface of each bar then carry particles toward a food groove at the base of the gill sheet (see Figure 9-2). Particles are sorted in folds of the overall gill structure or on special sorting surfaces. Large and heavy particles are carried away from the mouth and rejected. Smaller particles are carried to the mouth and consumed.

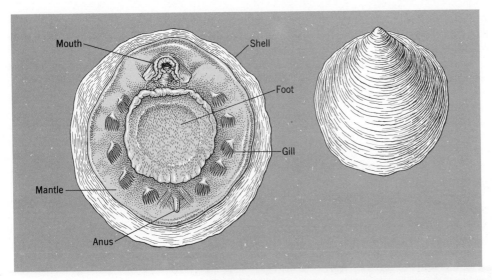

Figure 9-1 *This primitive mollusc,* Neopilina, *is thought to resemble the ancestor of all modern clams, snails, and related animals.*

Figure 9-2 *Filter feeding in clams. (a) A clam with one shell valve, mantle, and half of the outer left gill removed. Food particles carried in the water are collected on the gills and moved to the palps and the mouth by tracts of cilia. (b) Gill filaments. Water is pumped through the water pores by lateral cilia. Suspended paritcles are filtered out by the latero-frontal cilia and are carried along the front of the gill bar to the food groove by frontal cilia. They then pass to the palps and the mouth.*

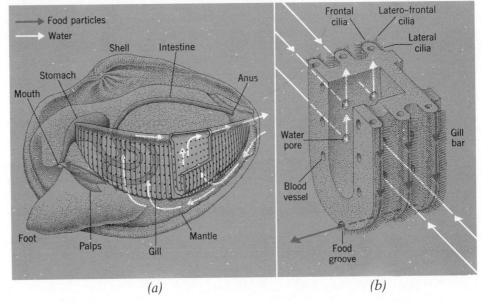

What is the origin of such elaborate ciliary tracts? They can scarcely have sprung full-blown into existence. The answer comes from observing modern animals such as snails. Some snails use the gill for gas exchange, and others use the gill for food-collection. In respiration, as in filter-feeding, lateral cilia pump water between the gill sheets to facilitate gas exchange. But particles in the water tend to settle on the gills and foul them, hindering gas exchange. Animals with respiratory gills have therefore developed latero-frontal cilia to strain fine particles out of the water before they reach the gas exchange surface; the particles are conveyed to the exterior by various rejection tracts. All of the elements of the filter-feeding apparatus are derived directly from these anti-fouling devices of the respiratory gill. The first step in the evolution of a filter feeder occurred when the animal began to eat material from the rejection tracts instead of just expelling it.

Clams also seem to have another way to collect particles from the water. They secrete a sheet of mucus that covers the whole surface of the gill. The mucus itself appears to behave as a filter, collecting suspended particles. This sheet is eventually rolled up, particles and all, and conveyed to the mouth. The mucus-feeding mechanism may occur in quiet water, and in waters where the suspended particles are very fine and rejection of large particles is not a problem. But in rougher water large particles are abundant, so the ciliary mechanism is needed to sort out and reject the larger particles.

An animal that strains food from the sea must balance the energy it expends in water filtration against the energy it obtains from its food. This problem is basically the same as the one encountered when man attempts to extract minerals from the sea. A cubic mile of sea water contains several thousand dollars worth of gold. However, no one has ever devised an economical way to get this gold out of solution. The cost is always greater than the value of the metal recovered. Several dozen schemes have been published and patented, but they simply aren't economical.

Some early experiments on filter-feeding made it look as though the animal was expending more energy than it was gaining as food. First, the volume of water filtered by a mussel was measured, as well as the concentration of minute plants and bacteria in the water. These figures were then used to calculate the total amount of organic matter available to it—about four grams per year. Yet the average weight gain for a mussel in one year turned out to be slightly more than four grams. In other words, these particular clams appeared to be converting organic matter in their food to protoplasm with an efficiency of more than 100 percent. This is clearly nonsense. Food is not only a source of raw material for the synthesis of new protoplasm, but it must also supply energy for all of the activities of the animal. Clams may be rather quiet, but they certainly require energy for their activity, particularly for pumping water during filter feeding.

By measuring the rate of oxygen consumption in the mussel, we can estimate that it must obtain at least 0.05 mg of organic material for each liter of sea water it filters, just to stay alive. If it gains weight, reproduces, or loses organic matter in any way, its input of food must be correspondingly greater. Off the coast of California, the phytoplankton in the sea water represent from 0.0 to 0.06 mg of dry organic material per liter. In other words, there is just barely enough phytoplankton to sustain a filter-feeding animal at favorable times and insufficient food most of the time. However, there is a second kind of organic matter in the sea, **detritus.** Detritus is non-living organic material covered with bacteria. It is so abundant that it outweighs the phytoplankton by a factor of ten; the

logical conclusion to draw is that filter-feeding animals also use this non-living matter for their food source.

We still know very little about particulate detritus in the open ocean, but we know it is an important food source for animals in other environments. A salt marsh is one such area. When plant tissues die or are shredded by wave action, they begin to decay. Each particle becomes a tiny community that supports a population of bacteria and other microorganisms. These microorganisms can use the cellulose and other plant carbohydrates as an energy supply to synthesize their own protein. The detritus particles are collected by various animals, such as clams, snails, and worms. The protein portion of the particle is digested, and the cellulose fraction is passed through the gut to be released as fecal pellets, or solid waste. The fecal material becomes detritus again as organic nitrogen reappears in the form of bacterial protein. Detritus is actually recycled through the community until all the plant cellulose has been converted to microbial protein and all the protein is consumed by detritus feeders (see Figure 9-3).

Detritus particles are much more attractive than fresh plants as food because they are more digestible. Most animals are unable to digest plant tissues directly. Plant biomass is composed mainly of cellulose and similar carbohydrates that are digested by animals slowly or not at all. However, bacteria and other microorganisms can grow on plant material, and these microorganisms are thoroughly digestible. In fact, the smaller and more decayed the particles are, and the higher the microorganism content, the more satisfactory they are as an animal food source.

Herbivores

Animals that eat fresh plants are called **herbivores.** Many kinds of animals fit into this category: various insects, worms, snails, fish, birds, and mammals. Some of the herbivores digest plant tissues directly; but many of them—like the detritus feeders—require the assistance of microorganisms.

Ruminants—a category of mammals that includes deer, sheep, and cattle—are herbivores that have very elaborate stomachs containing microorganisms. The cow's stomach is divided into four regions (see Figure 9-4). The first area, the **rumen,** receives grass and other forage after it has been crushed by the grinding teeth. Portions of this fi-

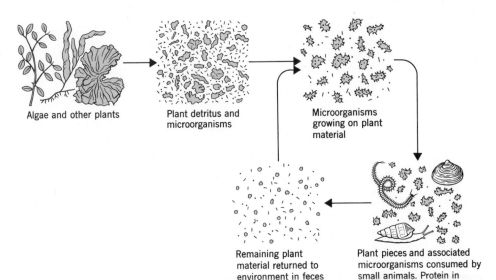

Algae and other plants

Plant detritus and microorganisms

Microorganisms growing on plant material

Remaining plant material returned to environment in feces

Plant pieces and associated microorganisms consumed by small animals. Protein in microorganisms is digested

Figure 9-3 Microorganisms grow on pieces of plant tissue, producing detritus. The microorganisms convert the organic molecules in the plant tissue to protein and other compounds digestible by small animals.

brous material in the rumen are periodically regurgitated and reduced to a semi liquid state. In other words, the cow "chews its cud." The rumen contains bacteria and protozoa capable of digesting cellulose and other plant compounds. The cow is simply maintaining a culture of microorganisms in the rumen. Forage material is broken down by the chemical action of the microorganisms, aided by the mechanical crushing provided by the cow's teeth. In effect, cattle carry their own salt marsh communities around with them.

Since the rumen contains no oxygen, the microorganisms carry on fermentation—they oxidize the plant material only to various organic acids, not all the way to carbon dioxide and water. The acids produced (acetic, propionic, and butyric) are absorbed through the walls of the rumen and the next section, the **reticulum,** to become the main energy source for the ruminants. Periodically, the culture liquid in the reticulum passes into the posterior regions of the stomach, the **omasum** and the **abomasum.** There, the cow's enzymes digest the microorganisms themselves in order to supply the cow with digestible protein.

If the contents of the rumen and reticulum become too acid, the microorganisms will be unable to function. The cow reduces the acidity by producing large amounts of saliva rich in sodium bicarbonate, a weak base. The saliva neutralizes the stomach acidity, but this reaction also produces several liters of carbon dioxide per minute. Hence the animal belches continually. Herbivores in

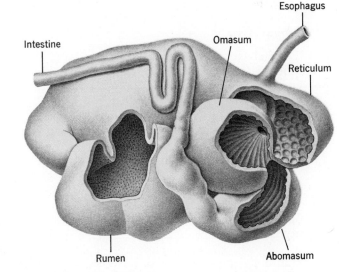

Esophagus

Intestine

Omasum

Reticulum

Rumen

Abomasum

Figure 9-4 The stomach of a ruminant has four compartments. Anaerobic digestion of plant material by microorganisms occurs in the rumen and the reticulum. The resulting organic acids are absorbed into the animal's blood. Microorganisms are themselves digested in the omasum and abomasum, providing protein for the ruminant.

general have long intestines, and food is retained in their guts for a long time. The ruminant's digestive tract may represent as much as one seventh of its total weight.

Ruminants can be quite efficient in obtaining the chemical energy from a forage diet. In cattle, about half the total energy in the diet becomes available for their growth and metabolism. But other herbivores are simply not very efficient. The reason that caterpillars are so destructive to vegetation, for example, is that they utilize very little of the material that they crop. Consequently, they have to devour huge amounts of vegetable matter in order to survive.

Animals' solutions to the problem of obtaining nourishment from plants are diverse. But most animals are able to feed on plant material only with the assistance of microorganisms.

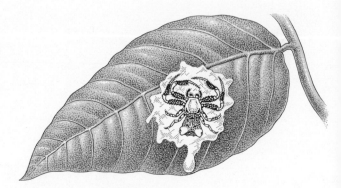

Figure 9-5 Ornithoscatoides dicipiens *camoflagues itself to resemble a bird dropping.*

Carnivores

The richest and most suitable source of food for an animal is another animal. Though animals vary in composition, they are generally equally suitable as food. Their differences in composition are mainly due to variations in the structure of their individual polymers (particularly the proteins and lipids). However, all the polymers are easily broken down into monomers, and the monomers are the same.

Carnivores are animals that consume other living animals. Carnivores are probably descended from **scavengers,** animals that feed on waste material or on dead animals or plants. Some scavengers probably developed specializations that permitted them to attack living organisms. The carnivorous mammals and birds of prey fit this picture very well. Cats and their relatives provide a good illustration of the transition from scavengers to hunting carnivores. Many relatives of the cats, such as hyenas, dogs, and bears, are wholly or partly scavengers, and their structures are quite unspecialized. The modifications that transform a scavenger into an effective hunting carnivore involve almost all of its structure. Cats are probably the most specialized of carnivores. Their brains are relatively large and their behavior patterns are complicated. The large head has well-developed eyes, long sensitive hairs, and an extensive nasal passage with a large surface area for detecting odors. Their jaw muscles are massive and their lower jaws are hinged for biting and cutting movements. Their premolars and first molars are modified into cutting

teeth. Much of the skeleton is specialized for leaping and rapid movement. The long tail aids in balance.

Not all carnivores are specialized for active hunting. Many of them have protective coloration or camouflage that allows them to seize unwary wanderers. Spiders have a wide range of structural and behavioral patterns, and their number includes both active and passive carnivores. Hunting spiders actively search for prey. At the other extreme are the trappers, such as *Ornithoscatoides dicipiens.* This tropical spider spins a flat, white web on the surface of a leaf and positions itself off-center on the web, so that spider plus web are disguised as a bird dropping (Figure 9-5). Web-weaving garden spiders are more familiar examples of carnivores who are trappers rather than hunters. To paraphrase Milton, they are also served who stand and wait.

Symbiosis

Some of the most remarkable feeding patterns are found in plants and animals that live in intimate association with each other. Such associations are extremely common, and a whole terminology has been introduced to describe the various relationships. The term **symbiosis** has now come to refer to the whole category of intimate associations between organisms, but it was originally coined to describe lichens.

Lichens are actually fungi and algae living entwined together in a single structure (see Figure 9-6). The fungus provides a tough structural framework in which the algae are embedded, while the algae,

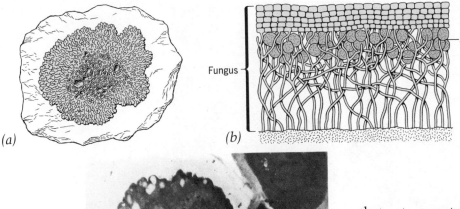

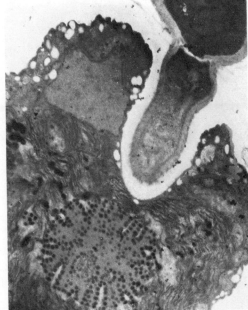

(a)

(b)

(c)

Figure 9-6 Lichens. (a) A lichen growing on a rock. (b) Cross section of a lichen. (c) In this micrograph, a single algal cell partially surrounds a fungal cell. Presumably this arrangement facilitates the exchange of materials.

Fungus

Algae

through photosynthesis, provide food for both organisms. Both organisms benefit from the association, but each one can still be cultured separately. In culture, the fungus loses its characteristic "lichen" form, and the algae appear to be very similar to free-living relatives.

Symbiosis is not always mutually beneficial. One partner may not be helped by the relationship or may even suffer. If both partners benefit, we speak of the symbiosis as **mutualism.** But if one partner benefits to the detriment of the other, we call the relation **parasitism.** An intermediate condition, in which one organism benefits and the other neither benefits nor is harmed, is called **commensalism.** The larger of the two a sociates is usually termed the **host,** while the smaller one is the **symbiont.**

The total number of animals and microorganisms that act as mutuals, parasites, and commensals is probably greater than the number of free-living forms, the ones not involved in such associations. A field collection of insects will probably contain more parasites than herbivores and carnivores combined. The role of symbiotic microorganisms in ruminant digestion is one of many important symbioses between mammals and microorganisms. And microorganisms in the human gut synthesize vitamins for their human hosts. In fact, animals without microbes, ones raised in germ-free environments, tend to be delicate and runty.

Although the various terms for symbiotic relationships are clearly defined, it is usually difficult to establish the precise nature of a particular interaction. The association between the freshwater animal *Hydra* and its green symbiotic algae has been studied extensively, but it is still only partially understood. The composite organism is the green hydra, *Chlorohydra.* Its color results from the presence of green unicellular algae inside the cells that line the hydra's digestive cavity (see Figure 9-7). Each animal cell may contain as many as 30 algal cells. When the algae are isolated and cultured alone, they turn out to be a species of *Chlorella* quite similar to a free-living *Chlorella* species. The hydra part of *Chlorohydra* also can live quite well when freed of its symbiotic algae.

Relations between hydra and algae have been approached in several ways. The early work consisted of simple growth and feeding experiments. It was found that when the external food supply is reduced, the green hydra can continue to grow at feeding levels that would retard growth in hydra without algae. But this ability exists only if the green hydra are kept in the light—a fact which suggests that the algae contribute to the nutrition of the hydra. Later experiments with radioactive isotopes support the same conclusion. When green hydra are exposed to radioactive bicarbonate ions

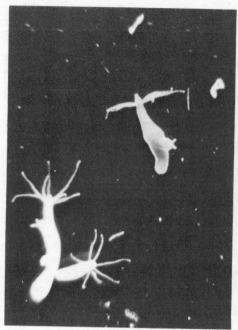 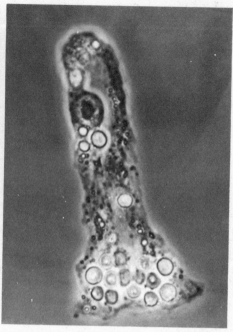

Figure 9-7 (a) Two Chlorohydra. (b) The symbiotic algae are clearly visible in this cell from the host animal's digestive cavity.

(HCO₃⁻), the algae incorporate the radioactivity into organic compounds during photosynthesis. Radioactivity also appears in the animal portions of the organism, suggesting that organic products of photosynthesis are transferred from plant symbiont to animal host.

These and other lines of evidence show that the algal symbiont provides reduced carbon energy to its host. Yet the analysis of the whole relationship is still incomplete. What does the alga receive in return? Protection? Carbon dioxide? Nitrogen? Perhaps all of these and other things as well. Then, too, the algae may provide other things for the host besides food.

Symbiotic associations between photosynthetic cells and non-photosynthetic organisms are quite common. Many protozoa contain algal symbionts. Coral animals and sea anenomes also have algal symbionts. The most spectacular example is the giant clam, *Tridacna*, which may be familiar to you through movies about Polynesia; its empty shell is often shown clamping over the ankle of an unwary diver. In life, *Tridacna* does attain very great size, but it is essentially a shallow-water organism. And it is so sluggish that it would take a lot of hard work and commitment to jam your foot between its valves. Thick fleshy tissue protrudes from the edges of the *Tridacna* shell, brilliantly colored by

symbiotic algae (see Figure 9-8). A detailed examination of the clam, and a comparison with related forms, shows that its whole structure has been radically reorganized to accomodate to its partnership with the alga. These modifications range from microscopic changes (it has transparent spheres in its surface tissues which focus light on the algae in deeper layers within it) to a total reorientation of its feeding and filtering currents.

Such relations are virtually impossible to untangle completely, but it is likely that symbiotic relations between animal hosts and symbiotic plants always involve nutrition on the part of one or both members of the association, and that they are cases of mutualism or at least commensalism.

The relation between parasite and host probably always has a nutritional component. Animal parasites, like the carnivores, probably descended from scavengers. This relationship can be illustrated by examining insects that are attracted to dead or decaying animals. Many insects feed on dead animals. Several species of flies deposit their eggs on freshly-killed meat. Then, as decay begins, various moths and beetles are attracted to the carcass. Other beetles appear later when tissues begin to liquefy. Finally, still other insects attack the remains of the carcass as it dries. (This sequence is predictable enough to be used as a means of assessing the state of decay of a carcass, in order to provide an estimate of time of death.) These same scavenging insects occasionally act as parasites by attacking live animals. Flesh flies that lay their eggs on freshly killed meat can and do lay eggs in wounds and sores on living animals. Similarly, scavenging insects adapted to feed on the dried skin, hoofs, and hair of dead animals may attack these areas while the animal is still alive. The behavior of these insects suggests that animal parasites probably evolved from scavenging ancestors that resemble species alive today.

In the case of parasitic relationships, nutritional considerations are often obscured by side effects that damage the host. Many parasites have deleterious effects on their hosts, ranging from minor annoyances to complete debilitation and death. In fact, it is quite difficult to decide where parasitism ends and predation begins. Insects such as body lice and fleas are usually considered parasites. They may live with their host continually or may hop on and off for periodic feeding. But since the symbi-

onts take only a small fraction of the concentrated organic material of the host, while the host's substance is continually being renewed, they are in effect living off interest rather than spending their capital. Having located a host, they permit him to function but subject him to minor taxation in the form of a periodic blood meal.

On the other hand, parasitic insects such as wasps can be lethal to their hosts. Some adult wasps, for instance, lay their eggs in the body of a caterpillar. As the caterpillar feeds and grows, its tissues are systematically consumed by the developing parasites. A brood of parasitic wasps ultimately emerges from the cocoon instead of the original moth or butterfly host. This relationship is very close to predation. Hunting wasps, which are considered predators, paralyze other insects such as locusts and lay their eggs directly on them. As the eggs hatch they have a supply of fresh food for their development. In fact, the major difference between parasitic and hunting wasps is perhaps simply that the parasitic wasp makes more efficient use of its

Figure 9-8 The giant clam, Tridacna. The edge of the mantle is thick and brilliantly colored because of algal symbionts in the clam's tissues.

Parasites of Man

The general definition of parasitism—an association wherein one partner benefits and the other is harmed—actually includes all kinds of disease agents. But in common usage, the term parasite is reserved for larger organisms: protozoa, various kinds of worms, insects, and other animals. Viruses and bacteria are usually excluded. The parasites that infest man are usually viewed with horror by middle-class Americans. Though they are considered extremely unpleasant, parasites are also thought to be rare and unimportant. This is not the case, however. Parasites are both common and extremely important in all countries, including the United States. In the southern United States, the effects of hookworm infestation were responsible for the apparent laziness and shiftlessness that led to contempt for the "poor white trash."

The parasite burden is still related to poverty and malnutrition in our country, and its effects are not yet fully recognized. The problem does not receive much public attention, but not because it does not exist. We are misled by two circumstances. First, parasitic worms are less common in temporate climates than in the tropics. Second, middle-class standards of hygiene are sufficiently high to interrupt the transmission pathways of many parasites.

In 1947 a parasitologist made a careful estimate of the number of worm parasite infestations in the human population. He concluded that there are more infestations than there are people in the world, because many individuals carry more than one parasitic worm species. The prevalence of worm parasites has not decreased in the last 25 years. In fact, some parasites are gaining ground alarmingly. **Schistosomiasis** (also called **bilharzia**) is probably the most important human disease caused by a worm parasite. The parasite lives alternately in a snail host and a human host. Larvae are liberated from the snails into fresh water. When a human comes into contact with the contaminated water, the larvae penetrate through the skin, migrate to blood vessels near the intestine or the bladder, and develop into adult worms. The adult worms then lay eggs that emerge with urine or feces, are ingested by the snail host, and develop into larvae. The cycle is completed. Schistosomiasis produces diarrhea and bloody urine, liver damage, lethargy, and fever. It is common in areas where cropland is irrigated, where the snail hosts are present, and where standards of hygiene allow completion of the life cycle.

Schistosomiasis has been blamed for the retarded development of several countries, such as Egypt. Ten million Egyptians are chronically infested. And, in fact, the problem is now worsening, because the construction of the Aswan Dam increased the extent of irrigation and encouraged the spread of the snail host. Schistosomiasis also was probably responsible for the failure of mainland China to invade Taiwan when the Nationalist Chinese fled to that island in 1949. An invasion was planned, but tens of thousands of troops were infested with schistosomiasis, and the invasion was postponed. During the resulting delay, the United States posted ships to the Formosa Strait and the invasion was cancelled.

The influence of parasitic disease on human history is difficult to assess, because historical events can always be attributed to multiple causes. However, there is no doubt that parasites have played a major and sometimes dominant role in social and political affairs, and that they will continue to do so.

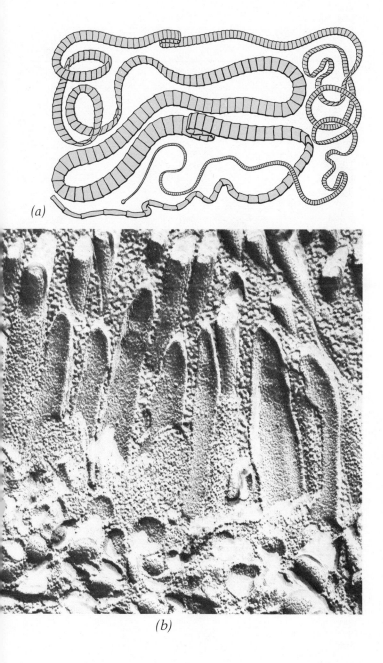

(a)

(b)

prey, since the prey can continue to gather food to feed a larger brood of developing parasites.

Parasites are very highly evolved in relation to their specialized environments. There are several parasite groups whose digestive tracts have disappeared altogether. To these animals, food is available in the form of blood or pre digested food in the host gut. It can simply be absorbed directly across the body wall. The external body wall of a tapeworm has extensive folds in the external covering, which produce fine tubes of protoplasm to increase the area for absorption. These parasites have no trace of a digestive tract at any stage in their development. They also lack sensory equipment or any elaborate nervous system (see Figure 9-9). Such structural characteristics sometimes cause them to be described as **degenerate** organisms, meaning that they have lost many features essential to free-living forms. But "degenerate" is an unfortunate term because of its pejorative connotations. Rather, tapeworms and other parasites should be recognized as animals that are extremely highly adapted to their specialized food sources and environments.

Food and Structure

The specializations of a tapeworm's body wall that facilitate absorption, the compartments in a ruminant's stomach that permit its survival on a plant diet, and the structural characteristics in cats that contribute to their efficiency as carnivores, are all good examples of the relation between food source and structure. The nature of the food source influences the structure and the behavior of the animal.

Insects provide many interesting examples of this principle. Insects are the most successful animals, in terms of both numbers of species and numbers of individuals. Part of their success is due to their ability to exploit many different food sources. This exploitation depends on evolutionary modifications of a basic set of appendages to form different kinds of mouthparts. Three pairs of appendages are involved, and they have dramatically different shapes and functions in different insects. The same appendages have been modified for tearing and masticating, for sucking nectar, for biting and piercing, for the production of external enzymes and for sucking up the digested food, and on and on (see Figure 9-10).

Changes in mammalian teeth likewise are adaptations to a wide range of food sources. All of the mammalian patterns of dentition are derived from a

Figure 9-9 Tapeworms. (a) A tapeworm may be several feet in length. Having no digestive tract, it absorbs food across the body wall. (b) Electron micrograph of a tapeworm body wall, showing tiny projections that increase the surface area for food absorption. Magnification 43,500×.

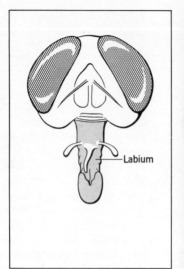

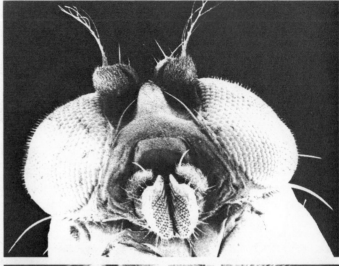

(a)

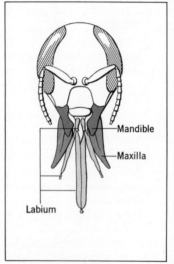

Figure 9-10 Insect mouthparts. (a) The fruitfly sponges up liquids with the labium. (b The honeybee has chewing mandibles; the maxilla and labium fit together to form a tube for sucking nectar. (c) The mandibles and maxilla of a mosquito form a sharp, hollow needle; the labium acts as a sheath.

Labium

Mandible

Maxilla

(b

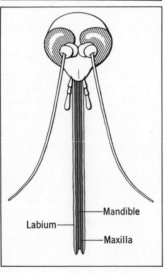

Mandible

Labium

Maxilla

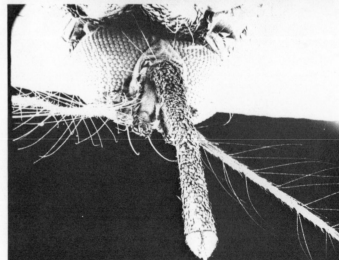

(c

basic ancestral pattern comprising four types of teeth. **Incisors** (I), at the front of the mouth, are specialized for cutting. **Canines** (C), just behind the incisors, rip and tear. **Premolars** (P) are specialized for crushing and grinding or sometimes for cutting. **Molars** (M) are usually grinding teeth. Primitive mammals possessed a dental formula of 3I, 1C, 4P, and 3M on each side of both the upper and lower jaws (Figure 9-11). Modern mammals show reductions in the numbers of some of these different kinds of teeth, depending on their food sources. Horses, for example, retain the three incisors for cropping grass, but the canines and the first premolars are absent, leaving a gap between the incisors and the remaining premolars and molars. The molars have broad grinding surfaces. They continue to grow throughout the life of the animal to compensate for wear.

Rabbits and rodents lack canine teeth and some of the premolars. Their large incisors are used for gnawing. The hard enamel layer on the incisors is restricted to the front surface, so that as the tooth wears it keeps a sharp, chisel-shaped edge. These teeth also grow throughout life to compensate for wear. An elephant gathers forage with its trunk, which is actually an elongated nose fused with the upper lip. Its tusks are derived from incisors in the upper jaw and have nothing to do with eating. Lower incisors, canines, and premolars are absent in adult elephants. The molars develop in sequence and are used one at a time throughout the life of the elephant; each one is enormous. Human dentition, like many other structural features of man, is relatively unspecialized. All four categories of teeth are represented, though their total number is reduced.

The variety of feeding methods and associated specializations is nearly inexhaustible. Many carnivorous animals such as snakes simply swallow their prey whole. The process can take an hour or more, so some snakes have their **tracheae** (the passages leading to their lungs) modified as snorkels so they can continue to breathe while swallowing oversized rats or mice. Snake jaws are specially hinged, enabling them to engulf animals larger in diameter than they are. One snake feeds on intact eggs. The egg passes down the gut until it comes to a set of special backbone extensions that neatly break the shell. Other animals that swallow food whole have muscular regions in the stomach or esophagus that grind food mechanically while enzymes digest it chemically. Birds use a region of the gut called the **gizzard** for grinding. They swallow small stones or other objects, including such things

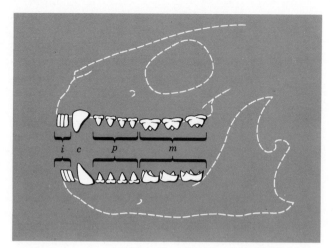

Figure 9-11 *Dentition of a hypothetical ancestral mammal.*

as nails, and deposit them in the gizzard to increase its efficiency as a grinding mill.

Other modifications of the gut facilitate food storage—eating is not a frequent activity for all animals. A leech may take a blood meal ten times its own weight, extract the water, and store the rest to be used as a food source for many months.

Animals do many other things, but they must eat. Many aspects of their musculature and sensory equipment, internal structures and organs, and even their external form and color are related to the problems of obtaining food.

digestion

Once obtained, food is broken down and distributed to the animal's cells. There it is used as raw material for growth and repair, or as an energy source. The mechanical and chemical breakdown of food particles is called **digestion.** The details of digestion depend on the size and the chemical composition of the food source. Large food particles are broken down chemically in several ways. **Mastication,** or chewing, is the most familiar means of breaking up large chunks of food, but some animals have muscular regions and hard ridges in their digestive tracts which accomplish the same end. Even in animals that do masticate, mechanical churning by muscular contractions of the stomach completes the process of mechanical decomposition.

The chemical events in digestion involve the reduction of large polymers to their constituent monomers: protein to amino acids, carbohydrate to sugars, lipid to glycerol and fatty acids. (Table 9-1 lists the human digestive enzymes, their sources, and their actions.) The monomers are then absorbed and distributed to the tissues. In small, simply-organized animals, digestion may occur in the individual cells. Food particles are engulfed and digested within the cells, and the products of digestion are then distributed to other tissues. The more common arrangement is extracellular digestion, which takes place in the **lumen,** or central cavity, of the digestive tract.

An animal with a simple structure may have only one opening to its digestive tract, through which food is ingested and wastes are ejected. But most animals have a complete tubular gut with a mouth at one end and an anus at the other. Regions specialized for food reception, storage, digestion, absorption, and excretion are found along the length of the gut. The human digestive tract is a good example, since it is nonspecialized and therefore is representative of the general mammalian pattern.

In the human digestive system, food is first received by the **stomach,** which initiates digestion. Digestion is then completed in the **small intestine,** which is divided into three major regions: the **duodenum,** the **jejunum,** and the **ileum.** The absorption of digested foodstuffs takes place in the small intestine. The remaining food then passes into the **large intestine,** which functions mainly in water resorption and the compaction of feces. The feces are then voided through the **rectum** (see Figure 9-12). The human digestive tract is about 30 feet long, of which twenty are the small intestine. Compared to those of other mammals, the human small intestine is intermediate in length. In carnivores the gut tends to be only two or three times the length of the body; in herbivores the gut may be twenty times the body length.

Structurally, our gut is a long tube with the same basic cellular structure throughout. The inner layer of epithelium, or lining of the lumen, is called the **mucosa.** Its cells function in secretion or absorption or both. Beneath the mucosa is the **submucosa,** a layer containing connective tissue, blood vessels, nerves, and muscle fibers. Next are two layers of smooth muscle. The first is **circular muscle,** oriented around the long axis of the gut; and the second is **longitudinal muscle,** oriented parallel to it. Finally, the whole is covered with a layer of flattened epithelial cells called the **serosal layer** (see Figure 9-13). Various ducts that supply the lumen with digestive juices (such as saliva, pancreatic

Table 9-1

Human Digestive Enzymes

	Acts on	Produces
Produced by Salivary Glands		
Salivary amylase	Starch	Maltose
Produced by Stomach		
Gastric lipase	Fats	Fatty acids and glycerol
Pepsin	Protein	Peptides
Produced by Pancreas		
Trypsin	Protein	Peptides
Chymotrypsin	Protein	Peptides
Carboxypeptidase	Peptides	Amino acids
Pancreatic amylase	Starch	Maltose
Pancreatic lipase	Fats	Fatty acids and glycerol
Produced by Intestine		
Aminopeptidase	Peptides	Amino acids
Dipeptidases	Dipeptides	Amino acids
Sucrase	Sucrose	Glucose and fructose
Maltase	Maltose	Glucose
Lactase	Lactose	Glucose and galactose

Figure 9-12 The human digestive tract.

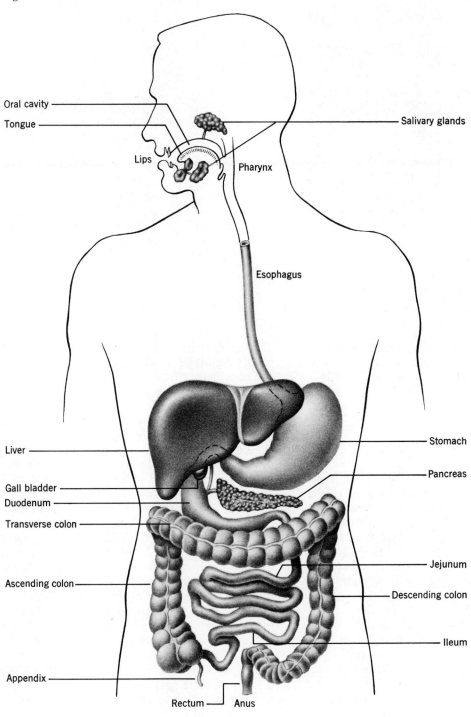

Oral cavity

Tongue

Lips

Pharynx

Salivary glands

Esophagus

Liver

Gall bladder

Duodenum

Transverse colon

Ascending colon

Appendix

Rectum

Anus

Stomach

Pancreas

Jejunum

Descending colon

Ileum

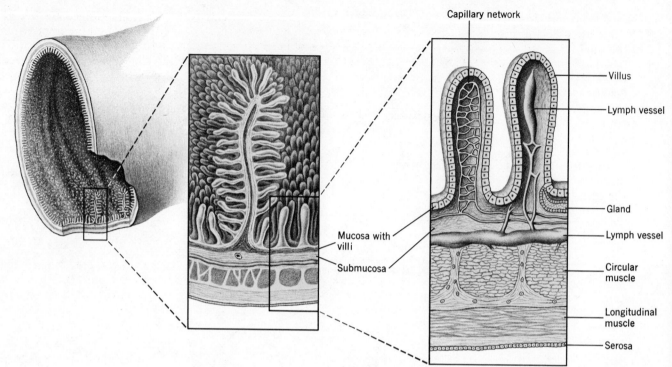

Labels on figure: Capillary network, Villus, Lymph vessel, Mucosa with villi, Submucosa, Gland, Lymph vessel, Circular muscle, Longitudinal muscle, Serosa

Figure 9-13 Structure of the mammalian small intestine.

juice, and bile) enter along the length of the gut. In addition, small multicellular glands whose secretion also enters the gut lumen are found in folds of the mucosa.

Movement of Food

Swallowing is initiated voluntarily, when food mixed with saliva is thrown against the back of the throat. A series of reflex contractions then passes the food mass to the opening of the **esophagus,** the tube leading to the stomach. The trachea (leading to the lungs) and the digestive tract cross over at this point, so that the trachea must be closed off briefly during the swallowing process. It takes only a few seconds for food to pass into the stomach.

The stomach entrance is normally closed by a contracted band of smooth muscle, called a **sphincter.** When the wave of contraction in the esophagus arrives, the sphincter relaxes and permits the food to enter. (These waves of contraction of circular muscle, which move along the gut from the mouth to the anus, are termed **peristalsis.**

They occur throughout the length of the digestive tract.) Stomach contractions occur more or less continuously; they have been compared to the beating of the heart. Peristaltic waves are initiated or enhanced by various factors, including nervous and chemical influences, the amount of food in the stomach, and the consistency of it. Waves churn the stomach contents and gradually move the liquid portion, called **chyme,** toward the small intestine. The larger food particles are mixed with acid and digestive enzymes, and they are gradually reduced to chyme. Most of the food taken in a meal passes into the small intestine in about four hours.

The entrance to the small intestine is guarded by another sphincter. This valve also opens automatically, upon the arrival of a wave of peristalsis from the stomach. However, fluid chyme is not propelled into the small intestine by a peristaltic wave. Instead, it is carried onward by a small difference in pressure between the stomach and the intestine.

The rate of food transfer from the stomach to the small intestine is under careful control. As food enters the small intestine, the resulting stretching of the intestine sets up a reflex that inhibits stomach contraction. In this way the entry of further food from the stomach is slowed. Stomach contractions are also inhibited by a hormone secreted by the small intestine, enterogastrone. Both of these mechanisms are of the "feedback" type. In other words, the result of the process tends to prevent its continuing. Other nervous reflexes in the gut act in the opposite way and are described as "feedahead" controls. For instance, the presence of food in the stomach promotes peristalsis in the small intestine and the colon.

Internal reflex control, of both the feedback and feedahead varieties, coordinates activity throughout the gut. There is also external nervous control of the general level of activity of the gut. The gut is supplied with two kinds of nerve fibers. One type, called parasympathetic, promotes stomach contractions, the secretion of gastric juice, and blood circulation to the mucosa. The others, the sympathetic nerves, have inhibitory effects on these same functions. Nervous discharge in the sympathetic nerves is characteristic of emotional states, accounting for the well-known sensitivity of digestion to fear, rage, anxiety, or other strong feelings. (The nervous system is the subject of Chapter 11.)

The small intestine has another pattern of muscle activity, in addition to peristalsis, described as **segmentation.** Various bands, or segments, of the circular muscle contract, giving the small intestine the appearance of a chain of sausages. As the first contractions relax, new ones replace them in different positions along the tube. In this way, the contents of the small intestine are mixed with digestive juices, and the mixture is brought into contact with the mucosa for the absorption of food molecules. This process continues as the food passes through the three divisions of the small intestine, from the duodenum to the jejunum to the ileum.

The small intestine and the large intestine, or **colon,** are joined by a third sphincter. Again, waves of peristaltic contraction (in the ileum) open the sphincter. The large intestine also undergoes segmentation and peristalsis, but it is more sluggish. Three or four times a day, large waves of peristaltic contraction transfer the contents of the colon toward the rectum. A massive transfer of fecal material into the lower portion of the colon is usually triggered by the morning meal. Defecation is produced by peristalsis of the colon and rectum, assisted by contraction of the abdominal muscles.

Feces are often described—incorrectly—as simply the undigested portion of food. This widespread misconception probably results from observations that seeds and other indigestible material pass through the digestive tract unchanged and are eliminated in the feces. But in fact, the composition of feces is surprisingly independent of diet, even when one compares a rice diet with one of meat. Roughly one third of all feces by weight consists of bacteria. Residues of various secretions of the digestive tract, and of cellulose and other indigestible material, are also present. But, though the composition of feces is relatively constant, their quantity does depend on diet. It varies from seven or eight grams per day when one is fasting to several hundred grams on a diet high in vegetables.

One's frequency of defecation depends partly on the volume of feces. In spite of the suggestions of laxative manufacturers, there is no particular reason why defecation should occur on a daily basis. Likewise, there is no basis for the fear that toxins are absorbed from feces that remain in the colon (though one famous nineteenth-century biologist attributed the entire process of aging to this cause).

Digestive Enzymes

The release of digestive enzymes into the gut coincides with the presence of food to be digested. This synchronization is accomplished by both nervous and hormonal control. The Russian physiologist Ivan Pavlov established the existence of nervous control of these secretions. Everyone is familiar with his classic experiments on the conditioning of salivation in dogs. Normally, the presentation of food produces an increased flow of saliva. By repeatedly ringing a bell before presenting food to the dogs, Pavlov succeeded in substituting the sound of the bell for the normal stimulus to salivation. He demonstrated a similar reflex in the production of gastric juice. The sight of palatable food produces a gastric secretion, which Pavlov called "appetite juice."

These reflex secretions must be controlled by the nervous system, at least in part. However, the control of gastric secretion is accomplished by chemical means as well. The chemical aspects can be demonstrated in an experimental animal prepared so that food can be introduced directly into its stomach and samples of the stomach contents withdrawn. As we already noted, if appetizing food is presented to the animal some gastric secretion occurs. But this secretion will continue only if certain foods, such as meat or partly digested protein, are present in the stomach. If, instead, dry bread or some inert and indigestible material is placed in the stomach unknown to the animal, no further secretion occurs. Some chemical component of the partly digested food acts as a chemical stimulus for the release of a hormone, **gastrin,** into the blood stream. The gastrin in turn stimulates the continuing production of gastric juice. Gastric secretion is a self-reinforcing, or positive-feedback, process; it continues as long as partly digested products remain in the stomach.

Several digestive enzymes come from the **pancreas,** a carrot-shaped gland located near the stomach. Digestive enzymes from the pancreas—**pancreatic juice**—move through a duct that empties into the duodenum. The secretion of pancreatic juice is also under hormonal control. Pavlov had found that the presence of acid in the duodenum produced the secretion of pancreatic juice, but he interpreted this mechanism to be a reflex. However, in 1902, W. M. Bayliss and E. H. Starling showed that pancreatic secretion was independent of nervous reflex. They completely severed the nerve connections between a segment of the small intestine and the rest of the body, while leaving the circulatory connections intact. When they placed acid in the intestine, secretion occurred as usual. Acid injected directly into the bloodstream of the animal did not produce pancreatic juice. But when they extracted the intestinal mucosa with acid and injected the extract into the bloodstream, the pancreas again was stimulated. Bayliss and Starling concluded that the acid caused the liberation of a chemical messenger from the intestinal lining, which passed to the pancreas via the bloodstream and stimulated secretion. The chemical messenger, which they called **secretin,** was the first hormone ever described (see Figure 9-14).

Even though details are lacking, it is clear enough that the presence of an enzyme at the right time as food passes through the gut is ensured by both nervous and chemical coordination. The quantity of digestive juices is enormous. The body produces roughly half a liter of gastric juice in digesting a meal, and its daily production of pancreatic juice may be almost a liter. The pancreatic juice is two percent protein, representing several grams a day. Much of the protein in these digestive secretions is eventually recovered through the digestion of the enzymes themselves, but some protein is lost to the feces.

One of the interesting characteristics of the digestive tract is its failure to digest itself. After all, it is pouring out large quantities of strong acid much of the time, and the digestive enzymes are

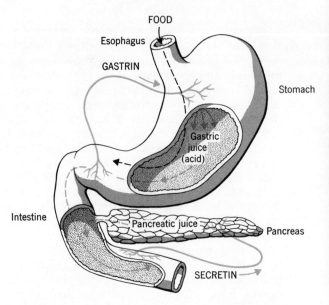

Figure 9-14 *Gastrin and secretin are two hormones that regulate secretion of digestive juices. Partly digested food in the stomach stimulates the release of gastrin from the stomach into the blood stream; gastrin stimulates the secretion of gastric juice by various stomach glands. The entry of food from the stomach causes secretin to be liberated from the small intestine into the bloodstream; secretin causes the secretion of pancreatic juice.*

designed to break down animal tissues. There are several reasons why the digestive tract is able to survive its own rough treatment. For one thing, the mucosa is protected by a layer of mucus secreted throughout the digestive tract. Also, the digestive enzymes usually become active only after entering the lumen of the gut. Gastric juice, for example, contains a mixture of secretions from at least four different kinds of glands in the stomach lining. The mixture includes hydrochloric acid and the enzyme pepsin. Pepsin, a protein-digesting enzyme, is not active until it is mixed with the acid. Since the enzyme and the acid are produced in separate glands, the pepsin and the acid are not mixed until the secretions enter the stomach. Other enzymes behave in similar fashion. Yet, despite all these protective measures, self-digestion sometimes occurs. Gastric and intestinal ulcers are the result. Acidic substances such as aspirin can also damage the stomach lining (see Figure 9-15). And even in the normal digestive tract, the lining of the small intestine has to be completely replaced by new cells every three days.

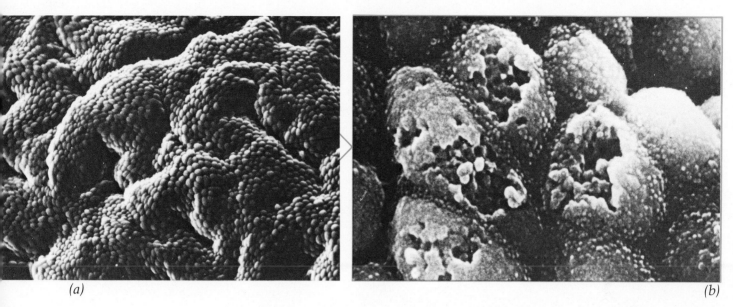

(a)   *(b)*

Figure 9-15 The effect of aspirin on the stomach mucosa. (a) A scanning electron micrograph of intact mucosa; magnification 215×. (b) Damaged mucosal cells; magnification 5,000×. Such damage to the mucosa often produces bleeding into the stomach lumen, though it is not usually serious.

Absorption of Food

Little absorption of food occurs in the stomach itself (with notable exceptions, such as alcohol). The major site of food absorption is the small intestine. Its lining has a large surface area, to increase the rate at which food molecules move from the lumen into the blood stream. This increase in area is achieved through elaborate folding of the intestinal wall. The mucosa is folded into thousands of small, finger-like extensions called **villi.** In addition, the surfaces of the individual mucosal cells are covered with small protoplasmic extensions called **microvilli.** These myriad extensions provide a tremendous increase in absorptive surface.

The small intestine absorbs several liters of water each day. Only about twenty percent of it is water that came directly from eating and drinking. The bulk of it, about 80 percent, comes from saliva, stomach juice, bile, and pancreatic juice. The movement of water into and out of the intestinal lumen is controlled by osmotic concentration. As food molecules or inorganic ions move out of the gut, water follows. The movement of sodium ion (Na^+) is particularly important. Sodium is actively transported out of the gut lumen into the bloodstream and the fluid between the mucosal cells.

The movement of sodium lowers the osmotic concentration in the intestinal lumen. As a result, there is a corresponding movement of water out of the gut.

Conversely, if the osmotic concentration in the lumen is raised, the water movement out of the gut slows down. "Epsom salts" — magnesium sulfate ($MgSO_4$) — is an effective laxative because the magnesium and sulfate ions are not easily absorbed by the intestinal mucosa; as a result, the osmotic concentration is raised and the gut contents retain water. Diarrhea is also caused by high osmotic concentration. When the intestine is irritated and inflamed, peristaltic contractions speed up and move food through the intestine so rapidly that some food molecules cannot be absorbed. As a result, the osmotic concentration of the mixture stays high, water is retained, and diarrhea results.

Much of the organic material in the intestine, as well as the water, is derived from digestive secretions. A considerable fraction of the amino acids in the intestine come from the body itself. The rest are the breakdown products of ingested food. As a result, the amino acid composition of the intestinal contents is quite stable. In fact, some parasites of the digestive tract depend on this constant spectrum of amino acids for survival. The sugar composition in the gut, however, is quite variable and depends entirely on sugar intake. Since sugar is a strict requirement for some tapeworms, it is possible to "starve out" tapeworm parasites by eliminating carbohydrate from the diet.

Some food molecules, particularly lipids, pass into the bloodstream passively simply because they are more concentrated in the gut lumen. But the movement of sugars and amino acids depends almost entirely on active transport. This transport requires energy, because it occurs against a concentration gradient. The molecules are more concentrated in the cells lining the gut than in the lumen itself.

The movement of amino acids and sugars across the membrane is aided by sodium diffusion. Their transfer from the lumen to the cell is rapid when there is a high concentration of sodium ions in the lumen. The amino acid or sugar attaches to a protein binding site on the membrane more tightly if a sodium ion binds to the site along with it. The system is described by the term **co-transport.** The effect is the transport of an amino acid-sodium complex or a sugar-sodium complex into the cell. This transport occurs against a concentration gradient for the amino acid, but in the proper direction for sodium diffusion. Thus, the transport step itself does not require energy. But the sodium must be moved out of the cell again to maintain a low internal concentration, relative to the high concentration in the lumen. This step requires metabolic energy.

Fats are prepared for digestion and absorption in several steps. First they are emulsified (dispersed into droplets) by the action of **bile salts** and **bile acids.** These compounds are produced by the liver, stored in the **gall bladder,** and emptied into the small intestine. The fat droplets are then further broken up by the action of fat-digesting enzymes, or **lipases.**

Tiny droplets of partly-digested fats enter the cells passively since they can pass through the lipid component of the cell membrane. Once in the cell, the droplets recombine to form larger fat drops, which pass out of the cell into the surrounding fluid. Since the drops are too large to enter blood capillaries they are collected in lymph ducts. They are later transferred into the large veins of the bloodstream (see Figure 9-16).

The small intestine also can absorb some large protein molecules intact. This ability allows suckling young to receive antibodies from their mothers' milk before they themselves have developed immunological competence. The protein hor-mone insulin is also absorbed intact across the gut. When it enters the bloodstream it causes a reduction of the level of blood sugar. The only reason insulin is not usually taken orally in the treatment of diabetes is that we cannot predict just how much of a given dose will be digested and how much will be absorbed. But on the negative side, the absorption of large molecules that act as antigens is responsible for some food allergies.

human nutrition

Food must provide an adequate source of metabolic energy. Any of the major categories of foods can be oxidized for energy—carbohydrate, fat, or protein. Food must also provide the necessary raw materials for the growth and replacement of cells and cell products. All the molecules in the body must either be obtained in the diet directly or synthesized from other compounds in the diet. Some of these molecules are required in large quantities because they are used for structural material in muscle, bone, cell walls, membranes and so forth. Other molecules, the **vitamins,** are required in small amounts because they are involved in the activities of enzymes in the body. They direct chemical reactions but do not participate in them, so they can be used repeatedly.

Detailed information on human dietary requirements is of tremendous value, of course, but unfortunately our information is meager. Much of what we know comes from experiments with rats and cats and rabbits. For obvious reasons, it is difficult to carry on controlled feeding experiments that produce gross dietary deficiencies in man. The experimental techniques that are routinely used with other animals depend on being able to alter the animal's diet drastically, to look for changes in its ability to function, and then to sacrifice it and examine all its tissues. These animal experiments are very important, but they do not allow us to specify an adequate diet for man in qualitative or quantitative terms. And even if we had methods for determining precise human dietary requirements, individual variation is so great that a given result would still apply only to that particular individual, at that time, and at a given level of activity.

Because of these difficulties, published dietary standards usually refer to "allowances," rather than requirements. Table 9-2 reproduces the daily dietary allowances recommended by the Food and Nutrition Board of the National Academy of Sci-

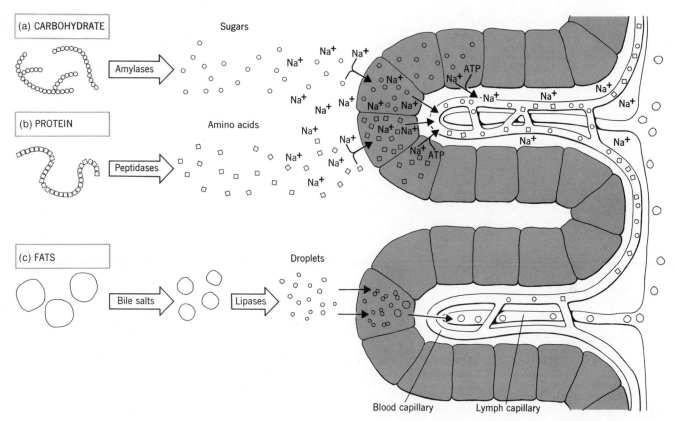

Figure 9-16 Food digestion and absorption. (a) Carbohydrates are digested to simple sugars, which enter the mucosal cells by means of cotransport. From there the sugars enter the blood stream. Sodium is actively transported out of the cells into the tissue fluid and blood stream. (b) Proteins are digested to amino acids and transported into mucosal cells in a similar way. (c) Drops of fat are emulsified and broken up by the action of bile salts and then are digested by lipases into minute droplets, 30 to 100 A in diameter. The droplets can enter the mucosa, since they are soluble in the lipids of the cell membrane. The droplets are later reformed into fat drops, which enter the lymph circulation.

ences–National Research Council. The table is full of estimates and must be hedged with many disclaimers. For example, the vitamin requirements are obtained by taking the smallest amount that will prevent overt deficiency symptoms and then multiplying it by a reasonable safety factor. In fact, larger amounts might be desirable. Massive increase in vitamin C intake may prove to be effective in preventing respiratory infections, for example. But in each case, professional advice or personal study of the issue is needed to decide whether or not a supplement is desirable. A table of allowances must not be interpreted as a rigid guide for

personal diet. The Recommended Dietary Allowances are described by the NAS–NRC as "goals toward which to aim in planning food supplies for groups of people."

Protein Requirements

Protein deficiency is probably the most important disease in the world today. Protein is a major dietary requirement for two reasons. First, our bodies require a large amount of nitrogen because many of the molecules in our body contain it: mainly our

Table 9-2

Recommended Daily Dietary Allowances[1]

From the Food and Nutrition Board, National Academy of Sciences–National Research Council.
Allowances are intended for persons normally active in a temperature climate.

	Age[2]	Kilo calories[3]	Protein (gm)	Calcium (gm)	Iron (mg)	Vit. A (I.U.)	Vit. B$_1$ (mg)	Vit. B$_2$ (mg)	Niacin Equiv.[4] (mg)	Vit. C (mg)	Vit. D[5] (I.U.)
Men[6]	18–35	2,900	70	0.8	10	5,000	1.2	1.7	19	70	
	35–55	2,600	70	0.8	10	5,000	1.0	1.6	17	70	
	55–75	2,200	70	0.8	10	5,000	0.9	1.3	15	70	
Women[7]	18–35	2,100	58	0.8	15	5,000	0.8	1.3	14	70	
	35–55	1,900	58	0.8	15	5,000	0.8	1.2	13	70	
	55–75	1,600	58	0.8	10	5,000	0.8	1.2	13	70	
Pregnant (2nd and 3rd trimesters)		+200	+20	+0.5	+5	+1,000	+0.2	+0.3	+3	+30	400
Lactating		+1,000	+40	+0.5	+5	+3,000	+0.4	+0.6	+7	+30	400

[1] The allowance levels are intended to cover individual variations among most normal persons as they live in the United States under usual environmental stresses. The recommended allowances can be attained with a variety of common foods, providing other nutrients for which human requirements have been well defined.

[2] Entries on lines for the age range 18–35 years represent the 25-year age. All other entries represent allowances for the midpoint of the specified age periods.

[3] Calorie adjustments are calculated for weight and age.

[4] Niacin equivalents include dietary sources of the preformed vitamin and the precursor, trytophan; 60 mg tryptophan represent 1 mg niacin.

[5] Vitamin D requirements for most categories are undetermined.

[6] Weight 70 kg., height 175 cm.

[7] Weight 58 kg., height 163 cm.

protein molecules. Thus, the best way to get nitrogen for our own bodily proteins is to consume someone else's protein. But if nitrogen as such were the only problem, we would still be able to get nitrogen from other kinds of nitrogen-containing molecules. The second reason we need protein in our diet is that we cannot make all of the amino acids for our own proteins from other compounds, even if we do have enough nitrogen. We can synthesize only about half of the necessary amino acids. Eleven amino acids are essential in human nutrition because we lack the enzymes for making them out of other molecules. These amino acids must be obtained from dietary protein.

Various sources of protein can thus be graded and evaluated in terms of their essential amino acid content. The Food and Agriculture Organization of the United Nations accepts eggs as a superior "reference" protein. Other protein sources are then compared to them in terms of amino acid content and digestibility. Assigning eggs a value of 100 units, other protein sources rank as follows: fish (83), meat (82), milk (75), potatoes (71), vegetables and fruits (70), cereal grains (56), and peanuts (48). Nonetheless, a diet based exclusively on any one protein source will usually be inadequate. Wheat protein, for example, is usually deficient in the amino acid lysine; rice lacks two amino acids, lysine and methionine; corn lacks yet another, tryptophan.

It is important to know what the minimum protein requirement is since protein is much more rare in the world than fat and carbohydrate. But some of the problems associated with trying to designate an adequate level of protein intake are quite complex.

One approach is to keep track of nitrogen intake in food and its output in the urine and feces. Most of the required nitrogen in the diet is acquired in protein; and any nitrogen that is excreted represents a breakdown and loss of protein, which must be replaced. Measurement of nitrogen output permits us to assign a level of intake just adequate to balance the output. As a rough average, a 70-kilogram man (about 154 pounds) loses ten to twelve grams of nitrogen per day. This figure is the amount of nitrogen found in 60 to 70 grams of protein. Some men can maintain normal weight and body function with a lower intake of protein, in some cases with half the amount indicated. In these cases, their nitrogen loss is correspondingly smaller. If excess protein is ingested, more nitrogen is lost. Apparently 60 to 70 grams has been accepted as an adequate estimate of the daily protein requirement, with a safety factor built into the recommendation.

Another approach to establishing protein requirements is to look at the nitrogen losses that occur during voluntary starvation. Presumably nitrogen losses at such times represent a minimum and could therefore be used to define the minimum quantity of protein required for survival. However, when such experiments are actually done, it is found that nitrogen losses continue to decrease for some time after food intake ceases. This decrease is accompanied by weight loss, a lower metabolic rate, and, eventually, by impairment of various functions. It is thus difficult to decide when an adequate maintenance minimum has been reached. A better approach would be to starve the subject first, and then increase his protein ration until nitrogen balance has been achieved and he feels fit and in vigorous health. Unfortunately, this approach is rarely used.

At present, populations in undeveloped countries survive on protein and calorie intakes below the levels that are required to maintain optimum health and growth. They suffer from very high infant mortality rates and have low growth rates, small adult size, decreased longevity, and poor resistance to disease. Yet man's ability to survive on inadequate diets is spectacular. The average diet for prisoners at Dachau concentration camp in the poorest months of 1945 contained only fifteen to twenty grams of protein and less than 600 calories per day. Adults became extremely emaciated and lethargic, but many survived and recovered.

Protein deficiency in infants has more serious and irreversible consequences. In the first year or two or life the brain grows at a rate of one to two milligrams per minute. If protein intake is inadequate to support the requirements of this protein synthesis, growth is permanently depressed. Seventy percent of the children in the world today risk incomplete brain development because of inadequate protein intake. Protein malnutrition is, of course, often complicated by other dietary deficiencies. A precise estimate of the number of people suffering from undernutrition and malnutrition is therefore not possible. We simply don't have straightforward tests that permit us to say that one person is undernourished and another is not. However, according to the best estimate we can make, half the population of the world is undernourished, or malnourished, or both.

Consumption of protein in amounts beyond the normal requirements has no particular value. At best, the excess protein merely provides energy, which can equally well be supplied by fat or carbohydrate. A high-protein diet offers no advantage to people engaged in heavy labor or competitive athletics, despite the popular view to the contrary. Such people usually do have protein intakes higher than the suggested minimum, simply because they require more calories. Since most food (other than fat and refined carbohydrate) contains at least some protein, their protein intake automatically rises as their food consumption increases. But when their protein intake is deliberately kept low in spite of a high caloric intake, their work performance is not impaired in any way.

Food Intake

Several physiological mechanisms influence food intake. Hunger pangs, which stimulate the appetite, may be caused by waves of peristaltic contraction in the stomach. This connection was demonstrated by having a subject swallow a balloon with a tube attached to it; the tube passed to a recorder. Stomach contractions were then recorded by measuring any changes in pressure on the balloon. With the recording device out of their range of vision, the subjects were asked to report when they felt hunger pangs. A high correlation was established between hunger pangs and measured contractions. Unfortunately, this observation does not go very far in helping to understand the regulation of food intake. Quite typically, one will eat before the onset of hunger pangs. Furthermore, patients who have had their stomachs surgically removed still feel hungry.

Observations on experimental animals have revealed the presence of brain centers that control food intake. In rats, the removal of a small area on the floor of the hypothalamus, a region of the brain, produces a threefold increase in food intake. The animals become obese. If regions on the lateral walls of the hypothalamus are destroyed, the animals stop eating. They will die of starvation unless they are force-fed. The first region is called the **satiety center,** since it signals when sufficient food has been ingested. The second region is the **feeding center,** since it signals the rat to begin feeding. The destruction of both centers leads to a reduction in eating and to weight loss.

Presumably the two centers interact and produce the signals that start or end the complicated behavior sequences involved in food-seeking and eating. Levels of blood sugar may provide the actual input to the brain. A decrease in blood sugar would favor the activity of the feeding center and inhibit the satiety center. A high level of blood sugar would have the reverse effect. Support for this idea is based on recordings of electrical activity from these regions of the brain, but the issue is not yet settled.

Habit and experience must also play an important role in the control of food intake. One normally ceases eating long before changes in blood chemistry have provided a complete assessment of the food acquired. The expectation of satiation, based on past experience, must therefore play a considerable role in the control of food intake.

The attractiveness and palatability of food also play a large role in human food consumption. An abnormal sensitivity to these factors may contribute to overeating and human obesity. In one experiment, obese individuals and a normal control group were provided with a limitless quantity of tasteless but nutritionally adequate liquid diet food. All the participants found the food bland and mildly distasteful. A control group reached normal levels of caloric intake quickly and maintained constant weight. The obese individuals decreased their caloric intake dramatically and lost weight. (Their weight loss was promptly regained at the end of the

Health Foods

Food selection is often influenced by fads or by claims about the nutritional superiority of "organically" grown or "natural" foods. Such fads are often based on misinformation. In spite of the claims of the local health food store, for instance, price is the only demonstrable difference between synthetic vitamin C and the vitamin C obtained from rose hips. The molecules themselves are, of course, identical. Likewise, "special protein concentrates" are likely to consist of wheat germ, soybean meal, dried eggs or milk solids, and so forth; the only difference between the original foods and the "concentrated" product is the appearance of the protein and the price. Similarly, there is little evidence for a difference in the food content of fruit and vegetables grown with "natural" fertilizers and those grown with manufactured fertilizers. In fact, much of our progress in agriculture has depended on our increasing ability to overcome "natural" deficiencies of soils—or even to avoid poisons that are "naturally" present in some soils. Obviously, a substance that is naturally present is not necessarily desirable, unless one accepts a rather mystical or fatalistic view of the relations between the organism and its environment.

Some food fads can be quite dangerous. Vegetarian diets, for example, can be perfectly safe, but only if the plants are carefully selected to provide the proper protein intake. Most plants have very little protein, and often the protein they do contain lacks one or more essential amino acids. The vegetarian must therefore be careful to consume a combination of plants that supplies the proper kind and quantity of protein. Legumes (soybeans and peas and beans) and nuts are particularly rich in protein. Meat-eaters, by contrast, have little problem. Most meats and dairy products supply adequate protein.

On the other hand, food fads and interest in health foods do focus attention on diet and may lead to an increased understanding of the difficulties and principles involved in nutrition. For example, the proponents of health foods have focused attention on the fact that we really know relatively little about vitamin requirements—both in terms of which ones we need and how much to take for optimum health. As a result, what information is available has been widely disseminated to the public. Also, food regimes that encourage the use of plant proteins are beneficial, because they take some pressure off the limited supplies of meat and fish protein in the world.

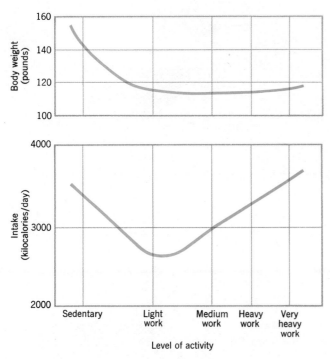

Figure 9-17 *Body weight and caloric intake as a function of physical activity. The data are for male Indian workers engaged in various occupations.*

Table 9-3

Weight Loss of Rabbit Organs and Tissues During Starvation

	Percent of Normal Weight
Blood	91
Heart	72
Liver	52
Muscles	82
Bones	99
Skin	95
Subcutaneous Tissues	45

experiment when they were again given access to the bakery and candy shop.)

There is usually a strong relationship between the individual's level of activity and the weight he maintains. Figure 9-17 shows the relationship between average weight, food intake, and physical work. In general, increasing activity is correlated with lower weight. Moderately active people have the lowest food intake. Very sedentary individuals eat more and are therefore heavier. Very active people also eat more, but this extra intake just balances the increased work they do. Normal individuals modify their weight by diet and exercise within limits set by their general body configuration.

The usual correlation between high activity and low weight sometimes breaks down, however. The most spectacular example is one Daniel Lambert, who suffered from progressive obesity. Lambert led an active and athletic life even though his weight increased steadily. At 500 pounds he worked as a police constable and played rugby on the side. He was 40 years old and still active when he died in 1809—weighing 739 pounds. His condition was exceptional because corpulence is usually self-limiting. Normally the affected individual will sooner or later strike a balance between a high caloric intake and the increasing energy output required to maintain the increasing weight. Lambert was probably obese because of some metabolic or regulatory defect, not because of inactivity. The great emphasis on weight-loss diets in the United States is partly due to a decrease in routine physical activity, which produces a tendency toward obesity, and partly because of a cultural emphasis on asthenic standards of beauty.

Starvation resulting from inadequate food intake is a more significant problem than obesity for human populations. Normal men die from starvation in time of famine when they lose 25 to 30 percent of their body weight. The figure is slightly higher for women. Weight loss in wasting diseases may reach more dramatic levels. Over half of the normal weight can be lost in some cases, presumably because such individuals are in hospitals where they receive supportive care. The internal organs show considerable weight change during starvation. In spite of the folklore stating that vital organs are spared during starvation, a decrease in heart weight is always found (see Table 9-3). Considerable water is lost in the first days of starvation, but after a week, most of the weight loss is due to fat (see Figure 9-18). Extreme starvation results in the virtual disappearance of fat.

metabolic rates

Discussions of the quantitative aspects of food intake and food requirements are usually framed in terms of the calorie content of food and the caloric expenditures involved in daily activities. In discus-

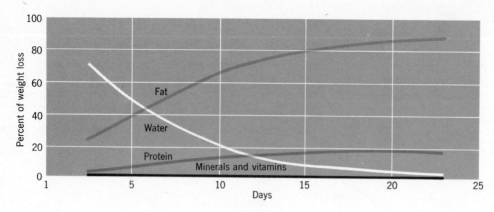

Figure 9-18 Weight loss due to loss of water, fat, protein, and minerals during starvation.

sions of metabolic rates, the term "calorie" can actually be considered to be interchangeable with "oxygen consumption," "carbon dioxide production," or "heat production." To say that an animal requires a certain number of calories per day is the same as saying that it has oxidized a proportional amount of organic material by consuming oxygen and producing carbon dioxide and heat. If we know what kind of food the animal is oxidizing, we can equate all of these factors.

For simplicity, assume that the organism is burning sugar. The chemical equation for this oxidation is:

$$C_6H_{12}O_6 + 6O_2 = 6H_2O + 6CO_2 + \text{heat}.$$

In other words, for every six molecules of oxygen consumed, one molecule of sugar is burned and six molecules of carbon dioxide and water are produced, plus heat. We can calculate how many molecules of oxygen are present in one milliliter of air, divide that number by six, and convert the resulting number of sugar molecules to milligrams. Thus, for each milliliter of oxygen consumed, we find that an animal must burn about 1.25 milligrams of sugar. For protein and fat, the figures are about 1.0 and 0.5 milligrams per milliliter of oxygen, respectively.

Given this information, we can now simply measure the oxygen consumption of an organism and then translate it into food requirements. These requirements are usually given in terms of a standard unit, the kilocalorie. (The "calories" in calorie charts and cookbooks and books on dietetics are actually kilocalories.) The kilocalories in food are determined by burning a measured weight of it and measuring the heat produced.

It may seem odd that the entire caloric content of food should be expressed as heat. After all, we know that the metabolic machinery in biological systems converts roughly half the energy produced by the oxidation of glucose into phosphate bond energy. Why should we now conclude that it all appears as heat? In fact, the conclusions are completely compatible. The rest of the energy will also appear as heat, since the phosphate bond energy is used by the organism for driving synthetic reactions, for muscle contraction, for the transmission of nerve impulses, and so forth.

Metabolic rate—that is, the rate of oxygen consumption—depends on an individual's activity, size, and general organization. In turn, the metabolic rate indicates the minimum food consumption necessary for maintaining the integrity of the animal.

The variations in metabolic rate caused by activity can be enormous. Table 9-4 presents the oxygen consumption and metabolic rate associated with each of several activities for an average man. The minimum expenditure (during sleep or rest) and the maximum exertion (swimming the breast stroke at three miles per hour) differ by a factor of 80. The maximum exertion can be sustained only for very short periods of time, partly because not enough oxygen is delivered to the tissues to sustain oxidation. A trained athlete can maintain a tenfold increase of metabolism for about an hour. However, energy expenditure per·day is probably a more meaningful statistic. A 155-pound lumberjack may have a total energy output of more than 6,000 kilocalories per day. The same man, with complete bed rest, would expend about 1,000 kilocalories. A man doing light work uses about 3,250 kilocalories.

Similar relations between metabolic rate and activity hold for other kinds of animals. The most

Table 9-4

Metabolic Rates During Various Activities

	Kilocalories per Minute
Sleeping	1.2
Sitting, at rest	1.7
Sitting, at lecture	3.0
Walking (2.5 mph)	3.8
Cycling (10 mph)	5.7
Walking briskly (4.5 mph)	7.8
Skiing (on level, 3 mph)	9.0
Jogging (6 mph)	11.6
Rowing (975 strokes/hr)	12.6
Wrestling	13.0
Running	
10 mph	20.6
15 mph	43.7
Swimming	
Butterfly	
1 mph	12.0
2 mph	29.0
3 mph	75.0
Breast Stroke	
1 mph	7.0
2 mph	29.0
3 mph	97.0

Figure 9-19 Exchanges of respiratory gases are measured in a parakeet flying in a wind tunnel. The bird's oxygen consumption is used as a measure of its energy expenditure during flight.

extensive measurements have been made on birds. There has been a great deal of curiosity about the energy cost of flight, and considerable ingenuity has been expended in devising methods to study it. The most convincing experiments have been done on birds trained to fly in a large wind tunnel. Parakeets and other birds were placed in the apparatus with masks clipped over their nostrils (see Figure 9-19). The results showed that the energy cost of flight exceeds the bird's resting metabolism by a factor of ten or more. This ratio is roughly comparable to that which can be sustained by man.

Such measurements allow us to ask whether birds are really capable of the nonstop, long-distance migratory flights they appear to make. Some of these flights are quite remarkable. The golden plover takes off from Newfoundland and flies south along the east coast of the United States until it reaches Cuba. In Cuba, it simply verifies its position and then takes off again across the Caribbean nonstop to South America. This feat would be comparable to a person's running from New York to Chicago, turning left, and then trotting on to St. Louis.

When calculations are performed to determine how far the birds should be able to fly, based on their measured metabolic rates and their amounts of stored fat, it sometimes turns out that they fly farther than we expect. Our measurements of metabolic rate and total energy expenditure come out right for the flight of hummingbirds across the Gulf of Mexico. But similar calculations for the long migratory flights of other birds give them a cruising range of 1,000 to 1,500 miles. Yet they undertake and survive journeys almost twice that long. Obviously, we do not fully understand what the birds are doing. Perhaps they depend on favorable winds and glide patterns to become more efficient. In any case, flying animals have no competition from walkers for long-distance travel.

Because of the great differences in metabolic rate that can be induced by different activities, animals are usually compared on the basis of resting metabolism, or **basal metabolism.** Figure 3-38 plots the basal metabolic rates for mammals covering a 200-fold weight range: from a two-kilogram rabbit to a 440-kilogram horse. As we expect, larger animals consume more oxygen and require more food. However, larger animals also have lower metabolic rates per unit of weight.

An unexpected relation between size and food requirements appears when we consider the efficiency of food utilization. Assume that one steer

Figure 9-20 The conversion of food to flesh is fairly constant despite differences in animal size. Smaller animals consume food more rapidly than larger ones because their metabolic rates are higher, but they also grow more rapidly.

	1 steer	300 rabbits
1 ton of hay consumed (in days)	120	30
Metabolism (kilocalories/day)	20,000	80,000
Growth rate (lbs/day)	2	8
Total weight gain (lbs)	240	240

and an equivalent weight of rabbits (about 300 of them) are each fed a ton of hay. The metabolic rates for rabbits are substantially higher because of their small size. As a result, the ton of hay will be consumed by the rabbits in only 30 days, while the same quantity feeds the steer for four months. However, the efficiency of conversion of hay to animal flesh is the same in both cases. Both experiments produce a gain of 240 pounds in the animals' weights (see Figure 9-20). The faster metabolic rate in rabbits is precisely balanced by their greater growth rate. In other words, a constant fraction of food is available for growth, regardless of the animal's metabolic rate. Similarly, a Jersey cow gives more milk per pound than does the larger Holstein. But again the figures balance out, because the Jersey must consume more food per unit weight in order to sustain her higher level of metabolism.

An animal's efficiency of food utilization in growth does, of course, vary with its age. Young animals convert about 35 percent of their food to tissue. But the figure drops off sharply with increasing age, becoming zero for animals at maturity.

Many aspects of animal organization assure a continuing supply of raw material for cellular activities related to growth, repair, and synthesis of protoplasm. We have looked at many of them here. Animals also have to maintain appropriate internal environments for their cells so that these activities can be carried out. Oxygen and other materials must be supplied to each metabolizing cell, and wastes must be removed. The chemical composition of the cellular environment has to be carefully regulated to facilitate these activities.

Recommended Reading

Jennings, J. B., *Feeding, Digestion and Assimilation in Animals*. New York: Pergamon Press (1965).

A general introduction to the study of digestion and assimilation.

Rosebury, Theodor, *Life on Man*. New York: Viking (1969).

Man's symbiotic microorganisms, internal and external.

Schmidt-Nielsen, Knut, *How Animals Work*. New York and London: Cambridge University Press (1972).

The book reports a series of lectures presented at Cambridge University in 1971. The principle focus is on energy and body size.

ISIS

RHINOCERVS

Nach Christsecburt. 1513. Jar Adi. May. hat man dem großmechtigsten König Emanuel von Portugall/gen Lysabona aus India pracht/ain solch lebendig Thier. Das nennen sie Rhinocerus. Das ist hie mit all seiner gestalt Abconterfect. Es hat ein farb wie ein gespreckelte schildkrot/vnd ist von dicken schalen vberlegt sehr fest/vnd ist in der gröfs als der Helffandt/aber niderdrer von baynen vnd sehr wehrhafftig es hat ein scharffstarck Zorn vorn auff der Nasen/das bes gundt es zu wetzen wo es bey staynen ist/das da ein Sieg Thir ist/des Helffanden Todtfeyndt. Der Helffandt förchts fast vbel/den wo es Jhn ankompt/so laufft Jhm das Thir mit dem kopff zwischen die fordern bayn/vnd reist den Helffanten vnten am bauch auff/vnd er würget Jhn/des mag er sich nicht erwehren. dann das Thier ist also gewapnet/das Jhm der Helffandt nichts Thun kan. Sie sagen auch/das der Rhinocerus/Schnell/fraydig vnd Lustig sey.

10 animal organization II: the internal environment

The environment of an individual cell within a large animal is usually quite different from the outside environment encountered by the animal as a whole. Muscle cells never feel wind, and the liver cell has no contact with the sun. Such cells experience only the internal environment provided by what amounts to a private internal pond, whose characteristics are carefully maintained by the activity of the whole animal. Each constituent of this internal pond is regulated within limits that permit the normal functioning of the internal cells. This regulated state is termed **homeostasis.**

Cellular environments have a characteristic composition of salts, dissolved gases, and organic materials; their osmotic concentration, acidity, and temperature are also regulated as well. The regulation of the internal environment is accomplished in part by physiological mechanisms: for instance, by changes in blood circulation, respiration, or metabolism. In part, it is accomplished by modifications of the animal's gross behavior: the selection of a proper habitat, or the control of drinking and eating. As Claude Bernard, a famous French physiologist, summed it up, the maintenance of a constant internal environment is the condition of a free life.

Regulation is never perfect, however, and the internal environment is not completely constant in any animal. In fact, if it were constant the organism could not function, because the only way information can be transferred from the gross environment to the functioning tissues is by small internal changes in the cellular environment. Changes in the external environment cause internal changes. The internal changes coordinate cellular activities and produce behavior designed to restore the original condition. Nevertheless, Ber-

Woodcut by Albrecht Dürer

279

nard's insight is substantially correct. Changes can carry information only if they stand out against the background of a regulated environment.

The surface of an organism represents an **interface,** or dividing line, between external and internal environments. In most organisms this interface is composed of an inert layer, which covers the living cells and protects them from direct exposure to the external environment. Organisms that dwell in the ocean are the only exceptions. Starfish and sea urchins, for example, have naked external layers of cells. They do not maintain major chemical differences between their internal and external environments, and their body fluid has an inorganic composition very similar to that of the surrounding sea water. Some soft-bodied marine animals such as worms, snails, and sea anemones are covered by elaborate cuticles that provide mechanical protection. But their surface cells are still in direct chemical contact with the environment; inorganic ions and the organic molecules dissolved in sea water are transported directly across the boundary layers of their bodies. Such transport may occur so rapidly that much of the organic material they need as food is obtained by this method. This sort of transport is the sole source of nutrition for animals that have no digestive tracts.

Nevertheless, the most successful animals in the ocean—that is, the most abundant and diverse animal groups—are sharply isolated from direct contact with the surrounding water. The crustaceans—crabs and lobsters and shrimps—are covered by external skeletons that are impermeable to nearly all dissolved compounds. Fish and marine animals also have protective skins, composed of layers of dead cells impregnated with tough materials. The living cells are deep beneath the actual interface. Living cells become even more heavily protected as we move from marine animals to the groups that inhabit brackish and fresh water, and finally to terrestrial animals. The boundary between self and surroundings becomes sharper and more necessary in fresh water and on land because these environments are progressively less favorable for cellular function.

Complete withdrawal from the environment is impossible for an organism, of course, because exchanges must occur. Gases must enter and leave. Food must be obtained and wastes eliminated. The entry and exit of water has to be regulated. In short, animals must remain open systems. Nonetheless, the areas where exchanges occur need to be limited and controlled. Some animals' limitations on exchanges can be spectacular. Certain desert snails, for example, lie in direct sunlight most of the year. They emerge from their shells perhaps only once a year to browse on the algae that spring up after a rain. After a few hours of activity, they retire to their shells to wait for another rain. But even with this regime, between periods of activity the animal has to maintain its private pond within bounds that permit cellular functioning.

The methods evolved for restricting and regulating exchanges with the environment vary, of course, within and between major animal groups. Two groups, insects and land vertebrates (particularly mammals), are unusually good at regulating exchanges and maintaining a constant internal environment. As a result, these animals are found in all kinds of habitats.

Insects

By several criteria, insects are the most successful of all the land animals. About 750,000 different species are alive today, more species than in any other animal or plant group. The number of individuals in each species is also immense. Insects cover the land from the arctic to the tropics and dominate all land surfaces, along with man. But the long-range survival prospects of insects are considerably better than those of man. This is due in part to their great fecundity and in part to the adaptability of their structure. But another major reason for the success of insects is the superior construction of their skin. This interface offers good protection from the environment while permitting controlled exchanges with it.

interface

Insect skin is not just a surface covering. It also functions as a skeleton, so it is called an **exoskeleton** ("outside skeleton"). The exoskeleton is a multi layered structure produced by secretions of a layer of subsurface cells called the **epidermis.** The exoskeleton is composed of an outer **epicuticle** and an inner **procuticle**, each a multilayered structure. The thin epicuticle contains protein, wax, and a hard, shellac-like cement for mechanical protection. The procuticle is much thicker and is composed of protein and a carbohydrate derivative called **chitin** (Figure 10-1).

Regions of the procuticle are hardened into rigid plates that are separated by flexible areas. The insect is in effect completely enclosed in a hinged box. When the animal moves, muscles flex the hinged regions between adjacent skeletal plates. Though the exoskeleton provides protection, it is also confining. Hence an insect can increase in size only by discarding its old exoskeleton (through molting) and secreting a new one. In addition, there is an upper limit to the size of an exoskeleton. As the insect becomes larger the exoskeleton must get bigger, and it must also be thicker to support the increasing weight. At some point it just becomes too large and heavy for the insect to carry around. Really large insects have never evolved, probably because of limitations imposed by the exoskeleton.

The outermost layer of the epicuticle is an extremely thin wax layer that waterproofs the animal, preventing rapid and fatal dehydration. Insects are quite susceptible to such rapid water loss, because they are very small and have large surface areas per unit of their weight. A simple experiment demonstrates the importance of the superficial wax layer. Placed in dry air, a dead insect loses water quite slowly below a certain critical temperature. But above that temperature, the water loss accelerates, and the insect quickly dries out. The critical temperature corresponds to the point at which the waxy layer melts. The melting point of the wax is related to the insect's normal habitat; cockroaches have a wax with a low melting point, and insects exposed to the sun have high-melting-point waxes.

regulated exchanges

Exchanges between an animal and its environment are controlled at the interface. The materials to be exchanged must be transported to and from the interface, and must be distributed within the organism.

Circulation

Insect blood is responsible for the distribution of food to tissues and for the removal of wastes. The food is in the form of dissolved amino acids, protein, and a sugar composed of two glucose units. (Most insects burn sugar to provide energy, but there are some interesting exceptions. Butterflies feed only on nectar, yet they convert this sugar into fat before using it; their flight muscles are incapable of oxidizing sugar. Perhaps the reason is that fat provides more calories per gram than sugar, and is therefore a more compact and efficient energy source for long flights.)

The nitrogen-containing wastes from cellular metabolism enter the blood and are transported to excretory organs called **Malpighian tubules.** These

Figure 10-1 Insect exoskeleton is composed of a waterproof epicuticle and a thicker, multilayered procuticle; the exoskeleton is secreted by underlying epidermal cells.

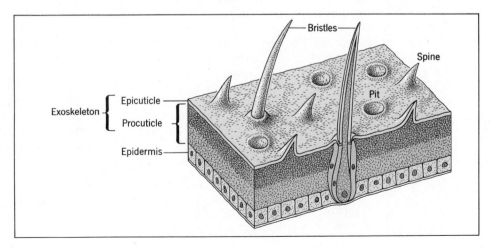

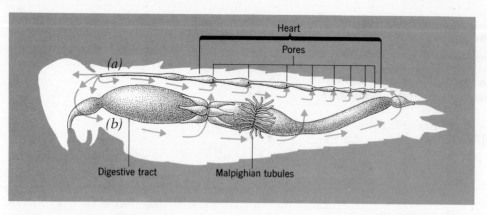

Figure 10-2 *Internal anatomy of a typical insect.* (a) *The circulatory system consists of a long, tubular heart. Blood is drawn in through pores in the heart wall and pumped anteriorly and posteriorly through the tissue spaces.* (b) *The digestive and excretory systems. Wastes are absorbed by the Malpighian tubules and are passed into the digestive tract for elimination.*

organs consist of blind sacs off the central digestive tract. In them, the wastes are extracted from the blood stream and converted into uric acid for excretion. The uric acid then passes through the remainder of the digestive tract (see Figure 10-2). This mechanism is thought to be related to the insect's need for water conservation. Uric acid is quite insoluble in water; it can be accumulated as solid crystals and voided occasionally, with only very little water loss.

Blood in insects does not flow through a well-defined system of vessels. Instead, it flows in spaces between the cells and between the tissues. The insect has a tubular heart lying in a large blood-filled space along its upper surface. The heart is pierced with holes, and the holes are protected by valves. Blood from the surrounding space enters the heart through the valves. Heart contraction then drives the blood forward, where it flows into spaces between various tissues (see Figure 10-1). Even though the blood is not confined in vessels, the insect can develop considerable blood pressure, because the whole organism is enclosed by the body wall. Insect blood pressure is quite variable, however, and can even be negative; in that case, a drop of fluid placed on the surface of an insect will be sucked into any cuts in the exoskeleton. Circulation is sluggish, compared to that of mammals, even in active insects. An insect may take half an hour to circulate all of its blood, while man and other mammals require little more than a minute.

Respiration

In most animals, gases are dissolved at an exchange surface such as gills or lungs, and then are carried to and from tissue in solution, usually by blood. But in insects, gas exchange is completely independent of the circulatory system. Surface pores called **spiracles** are connected with various tissues of the body by a system of branching tubes called the **tracheal system** (see Figure 10-3). Tracheal tubes are lined with both epicuticle and procuticle.

The tracheal tubes supply air to the insect's tissues in gaseous form. Oxygen is withdrawn from the air, and carbon dioxide diffuses into it. The spiracles open and close under control of the nervous system. A high carbon dioxide content inside the tracheae will cause the spiracles to open. (Fumigation is therefore more effective if carbon dioxide is first put into the atmosphere to keep the spiracles open, before hydrogen cyanide is added.) This mechanism is comparable to that for gas exchange in the leaves of green plants. The fine tips in the tracheal system are equivalent to the gas space surrounding the mesophyll in leaves. Both insects and leaves are covered with waxy cuticles to conserve water. And the spiracles are analogous to leaf stomata, since both are important devices for avoiding water loss. The pupa of a butterfly (that is, the dormant stage between the caterpillar and the adult butterfly) metabolizes for periods as long as several months without feeding or drinking. A

(a)

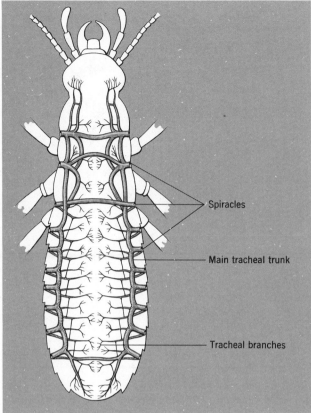

Spiracles

Main tracheal trunk

Tracheal branches

(b)

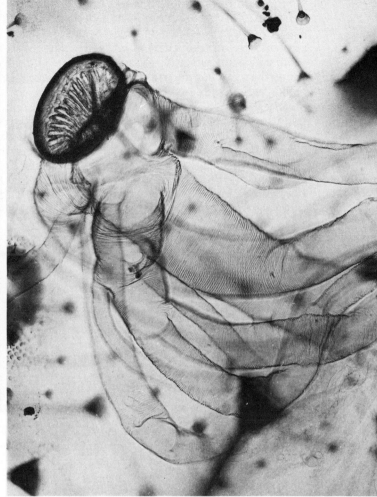

Figure 10-3 The tracheal system of an insect. (a) The air-filled tracheal tubes open to the exterior through the spiracles; the finest branches of the system terminate in or on individual cells. (b) A spiracle and attached tracheal tubes in a butterfly. Magnification 115×.

pupa can be kept in completely dry air and still survive, because its spiracles are tightly closed except for brief periods when the carbon dioxide concentration rises above the threshold level. The spiracles then open briefly for discharge of carbon dioxide and water vapor and for oxygen intake. The water loss that occurs through this mechanism is so low that water of oxidation, produced in the metabolism of fats and carbohydrates, is sufficient to balance the loss. However, if carbon dioxide is added to the air, the pupa quickly dries out and dies.

Gas exchange may also be aided by a pumping action of muscles in the thorax and abdomen. The flight muscles in butterflies, for example, help to pump air. Some aquatic diving insects use quite a

different mechanism to obtain oxygen. When the insect enters the water, it carries with it a layer of air over portions of its surface. The oxygen in the air layer is consumed fairly quickly, but more oxygen diffuses into the trapped air from the surrounding water. This mechanism is called a **gas gill** (see Figure 10-4). Work is proceeding on a comparable gas gill for human divers that would extract oxygen from solution in the surrounding water.

Water Balance

Water conservation is a dominant problem for many insects. Yet they have adapted to conditions where liquid water is almost totally unavailable. Spiracles are an important mechanism for restricting water loss. In addition, insects may actually be able to transport water into their bodies against a concentration gradient. Some studies show that insects such as fleas can actually gain weight when kept in air that is damp but not completely saturated with water. The observations seem to be sound, but the insect would have to expend tremendous amounts of energy to move water from the air into solution. A more recent study demonstrates that when water is placed in the rectum of an insect, it moves into the blood against a concentration gradient. Again, active transport of water seems to occur, but it is difficult to understand just how it can be done. If this is true, it is a unique and remarkable feat among organisms.

Figure 10-4 *The water beetle,* Dytiscus, *uses a gas bubble for respiration under water.*

Mammals

In a typical dictionary, **mammals** are defined as "a class of animals comprising man and all other animals that nourish their young with milk from mammary glands, and that are usually more or less covered with hair." Both of the distinguishing characteristics, mammary glands and hair, are derived from the mammal's interface—the skin.

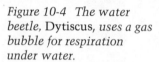

interface

In mammals, skin produces nourishment for the young; protects, conceals, or camouflages; supplies information about temperature and touch; helps regulate temperature; and is used to signal aggressive or courtship behavior. Mammalian skin is not an exoskeleton, but it can produce bony armor (as in armadillos), horns (as in the rhinoceros), hooves, claws, and fingernails. The skin is much more than a simple boundary layer. It is an unusually complicated organ system (see Figure 10-5).

Skin is divided into two layers, an outer **epidermis** and an inner **dermis.** The epidermis is produced by a deep basal layer of continuously dividing cells. As new cells are produced, the older

ones are displaced toward the surface in layers. As the epidermal cells age they secrete a tough protein, **keratin,** which eventually surrounds the cells and kills them. Dead, keratinized cells are finally pushed to the surface and sloughed off (see Figure 10-6).

Differences in the thickness of skin—for example, between the skin on the sole of your foot and the skin covering your ankle—result from both genetic and environmental forces. Genetic factors produce somewhat thickened skin on the sole of the foot of a newborn baby. But additional toughening depends on the environment. People who habitually go about barefooted develop thick and horny skin on the soles of their feet. Mechanical stimulation of the skin accelerates cell division in the basal cell layer of the epidermis.

Hair is derived from keratinized epidermal cells. Sunken regions of epidermis form **hair follicles,** the glandular sacs that generate the hair. The follicle produces keratinized cells, which are then formed into hollow tubes. These proteinaceous tubes are the hairs. When the tube is round in cross-section the hair is straight. When it has a flattened cross-section the hair curls.

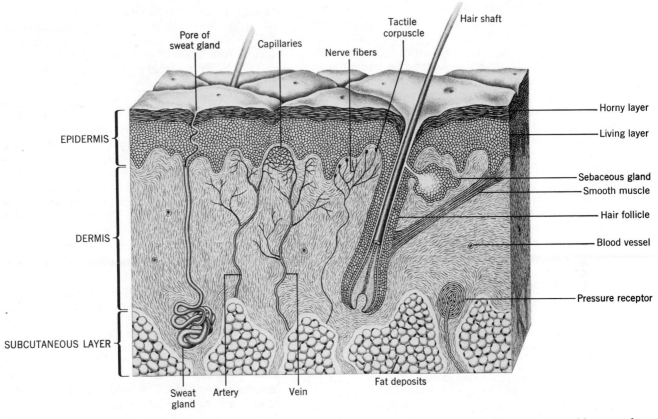

EPIDERMIS

DERMIS

SUBCUTANEOUS LAYER

Pore of sweat gland

Capillaries

Nerve fibers

Tactile corpuscle

Hair shaft

Horny layer

Living layer

Sebaceous gland

Smooth muscle

Hair follicle

Blood vessel

Pressure receptor

Sweat gland

Artery

Vein

Fat deposits

Figure 10-5 *The structure of human skin.*

Figure 10-6 *Scanning electron micrograph of human skin. Keratinized surface cells are in the process of being sloughed off. Magnification 1,500×.*

Hair is an important insulator because it prevents the movement of air on the animal's surface. Fur traps an insulating air layer that restricts exchange of heat across the interface. Small muscle fibers attached to each follicle can erect the hairs to increase the thickness of the layer of air trapped in the pelt. Glands associated with each hair follicle secrete **sebum.** This substance is a mixture of fatty and waxy materials that keeps hair and skin oily, flexible, and waterproof.

The basal layer of the epidermis holds other secretory cells, including sweat glands, the mammary glands, and various scent glands. The underlying dermis has rich blood circulation and contains many cell and tissue types. Specialized receptors for heat and cold, touch and pain, are located there. The dermis binds the epidermis to muscle and other tissues, and, through the circulatory system, brings raw materials to the epidermal glands that produce milk, sebum, and sweat.

regulated exchanges

Regulated exchanges in mammals are somewhat different from those in insects, because mammals have different surface/volume relationships. An insect has a large surface relative to its volume. As a result, the mass of its internal tissues is quite close to an exchange surface; exchanges of vital substances are easy. As we have seen, rapid exchange is desirable for such substances as respiratory gases, but not for others such as water, which need to be conserved.

Mammals have a different set of problems, because they are much larger and have less exchange surface relative to their volume. The loss of vital substances to the environment is a less serious problem for them; but at the same time it is also more difficult to supply their internal tissues with required substances. Thus, mammals need to have more elaborate apparatus for transporting materials from exchange surfaces to the internal cells and tissues.

Circulation

Blood circulation is much more elaborate and efficient in mammals than it is in insects. The heart, blood vessels, and blood provide for the rapid distribution of food, wastes, heat, ions, organic compounds, and gases. (In insects, you recall, gas ex-

change is independent of circulation.) The subject of circulation is important in a personal sense, too, because over half the deaths in the United States result from heart or circulatory failure.

The human circulatory system rapidly moves fluid and dissolved substances throughout the body. It is a closed system; that is, the circulatory fluid is contained in a system of blood vessels. Obviously, the circulatory system is not completely closed, however. The blood must exchange substances with both the external and the internal environments. The circulatory system carries on exchanges with the outside through the lungs (gas exchange), the intestine (acquisition of food molecules and water), and the kidney (removal of metabolic wastes). Blood circulation maintains the stability of the internal environment by transporting gases, food, and wastes. These substances are absorbed across the walls of the blood vessels, into the fluid surrounding the cells and tissues; tissue fluid, in turn, conducts exchanges with the cells.

Heart and Blood Vessels

The blood vessels of the circulatory system are given various names, according to their size and their position in relation to the heart. Vessels leading away from the heart are **arteries.** Arteries branch and divide, becoming smaller **arterioles.** Arterioles branch into a fine network of minute vessels called **capillaries.** The capillaries permeate the tissues, and they are the area where most of the exchanges with the tissue fluid take place (see Figure 10-7). As blood leaves the capillaries, it moves into blood vessels that lead back toward the heart,

Figure 10-7 A capillary bed in the muscle tissue of a living rat. Magnification 55×.

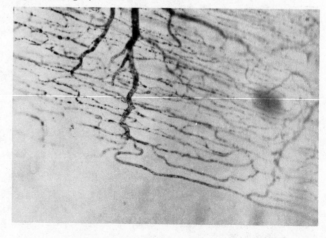

called **veins.** Capillaries first join small veins, or **venules,** and the venules then coalesce into the larger veins, which travel to the heart.

Fluid is circulated by the heart. The human heart has a left side and a right side, each a separate pump. One pump passes blood to the lungs, while a second delivers it to the rest of the body. Each of the two pumps has two sections. The **atrium** (from the Latin for "room" or "chamber") receives blood from major veins and moves it on into the second section. This muscular chamber, the **ventricle** (from the Latin for "cavity"), contracts to force blood back into an artery (see Figure 10-8).

More specifically, blood from the body enters the atrium on the right side of the heart through a large vein called the **vena cava.** The right atrium moves the blood into the right ventricle. The contraction of the right ventricle then delivers the blood to arteries that supply the lungs, called the **pulmonary arteries.** The blood passes through capillaries in the lungs and is returned to the left side of the heart by means of **pulmonary veins.** It moves through the left atrium to the left ventricle and is pumped into the **aorta.** The aorta is a huge artery that supplies blood to all parts of the body except the lungs. The blood circulates through capillaries in various parts of the body and is finally returned to the vena cava, completing the cycle (Figure 10-9).

The entire circulatory system is under fairly high pressure. The contraction of the heart, called **systole,** forces blood into the arteries under pressure. The arteries themselves have elastic, muscular walls that help keep arterial blood under positive pressure. The blood is under pressure even during the rest phase, or **diastole,** when the heart is filling with blood. Blood-pressure measurements are given in millimeters of mercury and are presented as two numbers, **systolic pressure** and **diastolic pressure.** A blood pressure of 120/80 means that during heart contraction (systole), the pressure in the large arteries would support a column of mercury 120 mm high, while the pressure during diastole drops to 80 mm. Blood pressure normally increases with age. 120/80 is average for a healthy 20-year-old.

Blood pressure remains at about the same levels throughout the arterial system, but there is an abrupt drop at the smaller arteries and arterioles. About three quarters of the arterial pressure is lost by the time blood reaches the capillary bed. The delivery of blood to the tissues can be changed rapidly and efficiently by simply changing the diameter of the small arteries. Increasing their diameter allows blood to be driven into them. Constricting or reducing their diameter channels blood else-where in the system. In this way, very rapid changes in circulation can be accomplished in response to the needs of different tissues. Changes in the diameter of blood vessels are produced by both nervous and chemical influences.

As blood passes through the capillaries, its pressure is very low. This pressure is sufficient to return blood from the feet to the heart when you are lying down, but not when you are standing. The return of blood to the heart therefore must be assisted by contraction of the muscles of the body. The veins are provided with valves, so that whenever blood is squeezed forward it will not fall back when the pressure of muscle contraction is released. But if you reduce muscle activity by standing rigidly at attention, the return of blood to the heart is greatly reduced; blood flow to the brain decreases, and you may faint.

The patterns of blood supply in some regions of the body are quite complicated. Two capillary beds rather than one may lie between artery and vein. Blood leaving the first capillary system is carried to the second by **portal veins.** Such a portal system, for example, lies between the capillary beds in the small intestine and those in the liver. These veins allow food molecules absorbed in the blood to be carried directly to the liver. The liver is the regulator of the sugar and amino acid levels in blood. If it were to receive blood by means of an artery, rather than by the portal route, newly acquired food would reach it only gradually, by a circuitous route, and regulation of blood chemistry would be correspondingly slower.

The length of time a blood cell spends in the capillary bed is quite short compared to the total time required for the cell to pass through the system. In fact, it takes roughly a minute for a blood cell to make a complete circuit of the system, and less than a second of this minute is spent in the capillaries. Yet, insofar as exchanges are concerned, this short time is the only significant part of the circulation. In its brief passage through a capillary, the blood delivers food to the tissue fluid, collects waste, and exchanges oxygen and carbon dioxide. The remaining 99 percent of the time the blood is simply being shifted around from capillary to capillary. It is fair to say that the heart and large vessels exist only to maintain capillary circulation.

The control and regulation of circulation is a complex subject. The rate at which blood is pumped by the heart can be altered about threefold by various nervous and chemical influences. These influences modify the rate of the heartbeat, the amount of blood delivered to the ventricles, and the

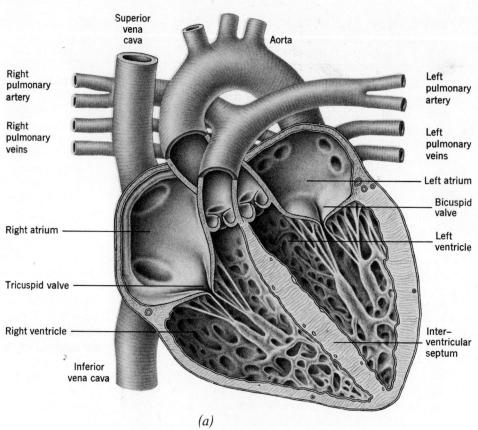

Superior
vena
cava

Aorta

Right
pulmonary
artery

Left
pulmonary
artery

Right
pulmonary
veins

Left
pulmonary
veins

Left atrium

Bicuspid
valve

Right atrium

Left
ventricle

Tricuspid valve

Right ventricle

Inter–
ventricular
septum

Inferior
vena cava

(a)

Figure 10-8 The human heart. (a) *A longitudinal section of the heart, showing the four chambers and the major veins and arteries.* (b) *Blood flow in the heart.*

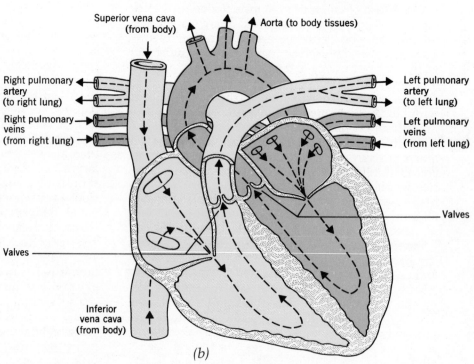

Superior vena cava
(from body)

Aorta (to body tissues)

Right pulmonary
artery
(to right lung)

Left pulmonary
artery
(to left lung)

Right pulmonary
veins
(from right lung)

Left pulmonary
veins
(from left lung)

Valves

Valves

Inferior
vena cava
(from body)

(b)

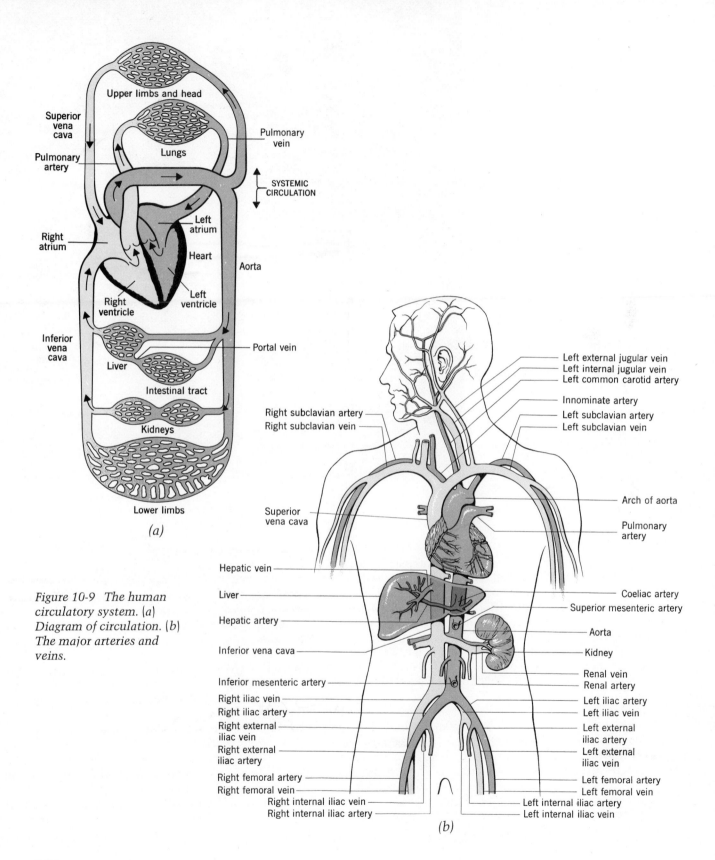

Figure 10-9 The human
circulatory system. (a)
Diagram of circulation. (b)
The major arteries and
veins.

(a)

Upper limbs and head

Superior vena cava

Pulmonary artery

Pulmonary vein

Lungs

SYSTEMIC CIRCULATION

Right atrium

Left atrium

Heart

Aorta

Right ventricle

Left ventricle

Inferior vena cava

Portal vein

Liver

Intestinal tract

Kidneys

Lower limbs

(b)

Right subclavian artery
Right subclavian vein

Left external jugular vein
Left internal jugular vein
Left common carotid artery

Innominate artery
Left subclavian artery
Left subclavian vein

Superior vena cava

Arch of aorta

Pulmonary artery

Hepatic vein

Liver

Hepatic artery

Coeliac artery
Superior mesenteric artery

Inferior vena cava

Aorta

Kidney

Renal vein
Renal artery

Inferior mesenteric artery
Right iliac vein
Right iliac artery

Left iliac artery
Left iliac vein

Right external iliac vein

Left external iliac artery

Right external iliac artery

Left external iliac vein

Right femoral artery
Right femoral vein

Left femoral artery
Left femoral vein

Right internal iliac vein
Right internal iliac artery

Left internal iliac artery
Left internal iliac vein

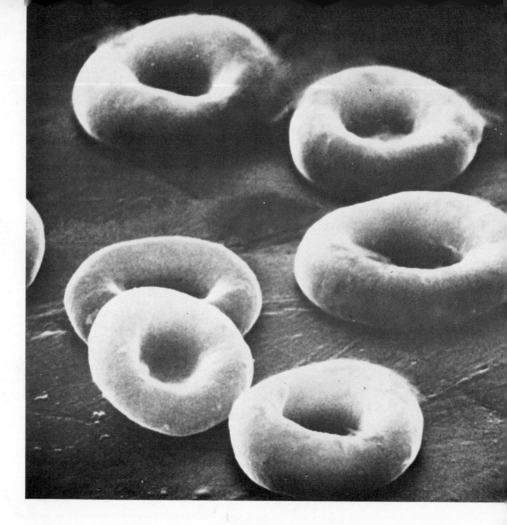

Figure 10-10 Human red blood cells, magnification 10,000×.

completeness of ventricular contraction. The blood delivered to any particular tissue is subject to further nervous and chemical control; these rates can vary by factors of twenty or more.

Overall, in a normal lifetime, the heart is expected to contract three billion times and to pump 40 million gallons of blood. The work it does is equivalent to raising your body to a height of 1,500 miles, and all this work is accomplished by an organ which is only five percent of the body weight.

Blood and Plasma

The blood of mammals is half cells and half fluid. Most of the blood cells are **red cells,** or **erythrocytes,** which transport carbon dioxide and oxygen in respiration (Figure 10-10). The red blood cells contain a red pigment called **hemoglobin,** which is involved in the transport of gases. Erythrocytes are produced by tissue in the bone marrow and are shed into the bloodstream at a rate of two million per second. A mature erythrocyte has no nucleus and it begins to break down after two to three months. Aged erythrocytes are collected and broken down by the **spleen,** an organ located behind the stomach.

Less than one percent of the cells are **white cells,** or **leukocytes,** which are concerned with defense against invasion. The lymphoid cells that produce antibodies are one kind of leukocyte; other leukocytes are the phagocytes that ingest invading bacteria and other foreign particles. In addition the blood contains a third type of cell, the **platelets.** Platelets are tiny cells that contain one of the substances needed for blood clotting.

The fluid portion of blood, or **plasma,** is a complex solution of inorganic compounds, small organic molecules, and proteins (see Table 10-1). The total osmotic concentration of plasma is about one third that of sea water. Inorganic solutes account for most of its osmotic properties, although their total weight amounts to only about one gram per liter of blood (or ten percent of the total solute weight). The reason is that the osmotic concentration depends on the number of solute particles rather than on their weight. The inorganic ions are quite small but are present in large numbers. Amino acids and glucose in the plasma are reservoirs of raw material and energy for cell activities. Urea and lactate are present as waste products.

Amino acids and glucose may simply diffuse out of the capillary into the neighboring tissues, and urea and lactate can diffuse back from tissues to plasma in the same way. However, some of the plasma fluid itself (without the proteins) flows out of the capillaries into the surrounding tissue fluid at the arterial end of the capillary bed and returns to the circulatory system at the venous end. This flow creates a slow circulation between the blood plasma and the tissue fluid surrounding the cells. About a liter of fluid per hour is exchanged between plasma and tissue fluid in this way.

Some protein does leak out of the capillaries, along with smaller molecules. If this leakage were not rectified, protein would accumulate until the system broke down. Therefore, protein and other macromolecules that leak into the tissue fluid are returned to blood by another system of vessels, the **lymphatic system** (see Figure 10-11). Tissue fluid containing these macromolecules (now called **lymph fluid,** or simply **lymph**) is collected in **lymph vessels** and returned to the venous blood circulation at a rate of about 100 milliliters per hour. If the lymph flow is stopped, fluid accumulates in the tissues. **Elephantiasis**—a spectacular disease in which the lower limbs or scrotum become enormously swollen—is caused by a parasite that blocks the lymph ducts draining the affected area.

Plasma proteins serve many functions. The antibodies, discussed in Chapter 4, are found in the globulin fraction of plasma proteins. Clotting is one of the most important functions of plasma protein. The clotting mechanism is necessary to minimize blood loss from any cuts or breaks in the system. Yet at the same time it must allow blood to remain fluid when the circulatory system is intact. Both conditions are vital. The plight of the individual whose blood does not clot is pitiable; every minor accident is a crisis. On the other hand, blood clots within the blood vessels can be lethal.

Current hypotheses state that clotting is initiated by some factor in blood that responds to contact with a foreign surface. Thus, when blood touches anything but the normal endothelial lining of the blood vessel, clotting begins. Internal blood clots, called **thromboses,** are formed on irregularities on the blood vessel lining. Clotting is also initiated when blood contacts substances contained in brain and lung tissues. This mechanism helps to control internal bleeding.

Clotting begins with a sequence of reactions between various substances dissolved in the blood plasma and a compound that is released by the platelet cells. Eventually, an enzyme called **thrombin** is produced. Thrombin attacks a soluble protein, **fibrinogen,** dissolved in the plasma. The enzyme splits off small peptides from the ends of

Table 10-1

Major Constituents of Blood Plasma
Units are milligrams per liter.

Inorganic		Organic	
Negative Ions		Monomers	
Bicarbonate	25–30 mg/l	Urea	200–300
Chloride	100–110	Amino Acids	400–600
Phosphate	2	Glucose	650–900
Sulfate	1	Lactate	50–150
		Cholesterol	400–800
Positive Ions		Polymers	
Calcium	5	Polysaccharides	1,500–2,000
Magnesium	2	Lipids	4,500–6,500
Potassium	4–5	Protein	60,000–80,000
Sodium	135–150		

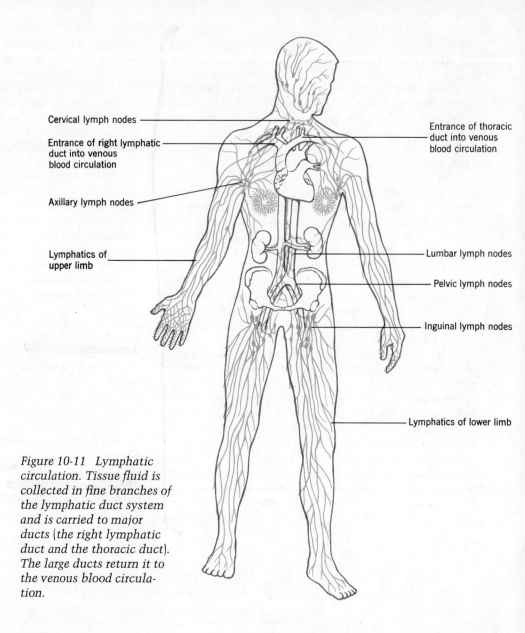

Figure 10-11 Lymphatic circulation. Tissue fluid is collected in fine branches of the lymphatic duct system and is carried to major ducts (the right lymphatic duct and the thoracic duct). The large ducts return it to the venous blood circulation.

Labels on figure:
Cervical lymph nodes
Entrance of right lymphatic duct into venous blood circulation
Axillary lymph nodes
Lymphatics of upper limb
Entrance of thoracic duct into venous blood circulation
Lumbar lymph nodes
Pelvic lymph nodes
Inguinal lymph nodes
Lymphatics of lower limb

the fibrinogen molecule. The remaining central strand is **fibrin**, a rather insoluble protein. Fibrin molecules polymerize to produce fiber-like strands; these strands, in turn, are bound together by cross-linkages. They form a netlike clot across the wound (Figure 10-12).

Solute Regulation

Each solute in the blood plasma is regulated so that its concentration stays within tolerable limits. Since plasma (without the protein) bathes the cells

and tissues, regulation of the solutes in it is required for maintaining a constant internal environment. Different factors control the levels of sodium, potassium, lactate, amino acids, sugars, and other components; but many of the control mechanisms have common characteristics, and they tend to interact. Blood glucose, therefore, provides a good example of solute control.

Control of Blood Sugar
Glucose from the blood moves into cells, where it is used as an energy source. The circulating levels

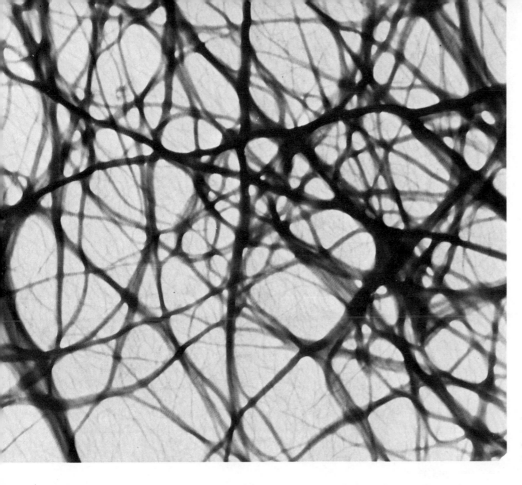

Figure 10-12 *Strands of fibrin provide the framework of a blood clot. Magnification 21,000×.*

Circulation and Exercise

Circulatory disease, particularly heart failure, is correlated with obesity and lack of exercise. Blood is supplied to the heart itself by way of several arteries that branch off the main aorta. Collectively these vessels are termed the **coronary arteries.** If their blood flow is interrupted, heart failure and death result. Coronary occlusion (blockage) or insufficiency account for almost a quarter of all the deaths in the United States each year.

The role of exercise in the prevention of coronary disease has been clearly demonstrated in an experiment with dogs. Surgery was performed on the dogs to change their coronary circulation. A major coronary vessel was partially tied off, restricting but not blocking the blood flow. The animals still had normal heartbeats and were all perfectly healthy after recovery from the operation. The dogs were then divided into two groups. One group was exercised on a treadmill several times a day, while the other group received no exercise. After about two months, surgery was again performed on all the dogs. This time the coronary vessel was completely tied off. The dogs that had been exercised daily were unaffected by the complete closure of the coronary branch. However, the second group showed profound changes in their heartbeats and symptoms of coronary insufficiency. Apparently, exercise had produced alternate pathways of blood flow. These pathways were necessary to support the increased circulation to the heart that was required during periods of greater activity. This so-called **collateral circulation** was so well developed by exercise that complete closure of a major coronary vessel did not affect the resting animal.

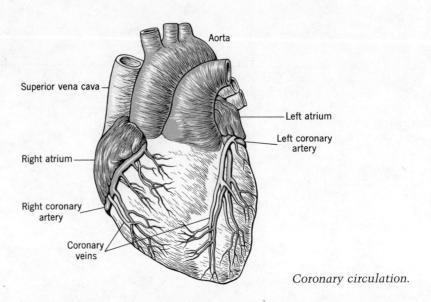

Coronary circulation.

The effects of partial blockage of heart circulation through surgery are quite comparable to those of **arteriosclerosis.** This disease is caused by deposits of fatty material or calcium on the walls of arteries. Many people in the United States are in its early stages as they approach middle age, though they do not yet show its symptoms. Eventually, however, the deposits may begin to block the coronary arteries, causing damage to the heart muscle. If individuals with arteriosclerosis exercise properly, they will develop collateral circulation. These collateral pathways permit the individual to survive a blockage of any one pathway.

Various statistical studies support this hypothesis. For example, death from arteriosclerotic heart disease was studied in groups of railroad workmen between the ages of 60 and 64. The death rates were 10.4 per thousand for inactive clerks, 6.7 for switchmen, and only 4.2 for active section hands. Many other factors may predispose an individual to coronary disease: cigarette smoking, high-fat diets, or hypertension (high blood pressure). Nonetheless, the relation between physical inactivity and susceptibility to coronary disease is quite clear.

of glucose in human blood are normally 650 to 900 milligrams of glucose per liter. This concentration of sugar provides sufficient energy for about half an hour of normal activity. When the concentration rises above 900 milligrams per liter (as after eating, for example), some of the excess glucose is transported into the liver. There it is converted into glycogen and stored, serving as an energy reserve.

Normally, the blood sugar level is under very tight hormonal control. The pancreas, in addition to producing digestive enzymes, synthesizes the sugar-regulatory hormones insulin and glucagon. Islands of hormone secretory cells are scattered throughout the normal glandular portion of the pancreas. The secretory regions contain **alpha cells,** which produce glucagon, and **beta cells,** which produce insulin.

Insulin was discovered in 1922 and was identified as the factor that relieved the symptoms of diabetes mellitus. "Diabetes" comes from a Greek word for "siphon," referring to the excessive water consumption and urination that characterize this condition. "Mellitus" means "sweet" and refers to the fact that the urine of a diabetic is high in sugar. In diabetes, the glucose level of the blood becomes so high that some glucose is excreted by the kidney. The administration of insulin brings diabetic symptoms under control. It has permitted victims to lead more or less normal lives.

Insulin produces a dramatic fall in the blood

sugar concentration in three ways. First, it facilitates the movement of glucose from the blood into the cells. Second, it increases the production of glycogen. Finally, it inhibits the synthesis of new glucose in the liver by acting on the enzymes involved in carbohydrate metabolism.

Normally, any increase in blood glucose stimulates the secretion of insulin, which in turn lowers the blood sugar; this mechanism constitutes negative feedback. The release of insulin is also promoted by stimuli from other control systems. For example, hormones concerned with the control of digestion, such as gastrin from the stomach and secretin from the small intestine, interact with glucose regulation. The presence of any of these hormones in the blood is a stimulus for insulin release. This mechanism is a type of feedahead regulation: it primes the liver to deal immediately with an influx of blood glucose from a new meal.

In general, insulin regulates the upper limit of blood glucose concentration. **Glucagon,** on the other hand, controls the lower limit. This hormone is secreted by the alpha cells of the pancreas. It was first noticed as a contaminant of insulin. Injections of glucagon-contaminated insulin produced an initial increase in the blood sugar level, only later followed by the expected reduction. A decrease in blood sugar is the stimulus for release of glucagon into the bloodstream. Glucagon causes the liver to release sugar into the bloodstream by stimulating the splitting of glycogen (see Figure 10-13).

Glucose is also regulated by some very complex behavioral sequences. A drop in blood glucose, for instance, stimulates hunger, through its effect on receptors in the central nervous system. Hunger, in turn, activates a larger and more unpredictable feedback loop, initiating the complicated behavior associated with food-seeking and eating. Eating ultimately restores blood glucose to an acceptable level.

Kidney Function

The mammalian kidney is another organ involved in maintaining the composition of solutes in the internal environment. The kidney regulates the osmotic concentration of inorganic ions and organic compounds in blood plasma, and also the blood's acidity. The kidney is also the pathway for the elimination of waste products, such as urea and ammonia.

Human kidneys are two massive organs, weighing about 300 grams each. One kidney contains about a million small functional units called **nephrons.** The nephron is a small tubule, one end

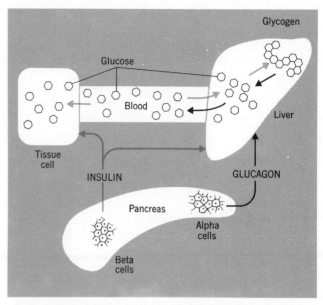

Figure 10-13 *The control of blood sugar by hormones. Insulin, secreted by the beta cells of the pancreas, lowers blood sugar by facilitating glucose entry into cells and by promoting glycogen synthesis in the liver. Glucagon, produced by the alpha cells of the pancreas, raises blood sugar by increasing the breakdown of liver glycogen.*

of which receives fluid from the bloodstream. Then, as this fluid moves through the nephron, some solutes are absorbed out of the tubule and returned to the general circulation, while others are secreted into the tubule from surrounding capillaries. Water is also resorbed from the fluid, and the remainder drains into tubules that pass out of the kidney to the bladder as urine.

Blood fluid is transferred from the circulatory system to the nephron through a ball of capillaries called the **glomerulus.** The arterial blood pressure forces about twenty percent of the blood fluid through the porous walls of this knot of thin-walled vessels. Proteins and blood cells are filtered out so that they remain in the general circulation. The filtered plasma, now called **filtrate,** is collected in **Bowman's capsule,** a curved sac at one end of the nephron, surrounding the glomerulus. At this point the filtrate is equivalent to blood plasma without its normal proteins. Throughout the kidneys as a whole, about 125 milliliters of filtrate per minute are produced in this way. But over 99 percent of this volume of liquid will be resorbed in the process of urine formation, so that urine production itself is only about one milliliter per minute.

From Bowman's capsule the filtrate flows through the nephron itself, which is divided into three regions. The one closest to Bowman's capsule is called the **proximal convoluted tubule** ("proximal" meaning "near"). It is followed by the **"loop of Henle"** a long U-shaped region that bends back upon itself. The final region is the **distal convoluted tubule** ("distal" meaning "away from"). The distal convoluted tubule of the nephron empties into a **collecting tubule.** The collecting tubules, in turn, drain into the **ureter,** which leads to the bladder. The glomeruli and Bowman's capsules and the proximal and distal convoluted tubules all lie at the surface, or **cortex,** of the kidney. The loops of Henle and the collecting tubules dip into the center of the kidney, the **medulla.** Capillaries leading from the glomerulus surround each nephron. This arrangement facilitates the exchanges of solutes between the filtrate in the nephron and the blood plasma (see Figure 10-14).

The proximal convoluted tubule is where most of the solutes from the filtrate are resorbed back into the blood plasma. The secretion of some solutes from plasma into the filtrate also occurs there. Both of these processes involve the movement of particular solutes against concentration gradients, a process that requires metabolic energy.

Some resorption systems can remove all of a particular solute from the filtrate. For example, all of the glucose will be removed unless the blood sugar exceeds normal values by a considerable margin. Amino acids are also completely resorbed; none are lost in the urine. Other transport systems remove most but not all of a solute; sodium is an example. Much of the sodium ion in kidney filtrate is transported back to the blood, but three to five grams per day pass into the urine. As a result, human beings have strict salt requirements and must consume several grams of sodium chloride per day in order to live. This unavoidable loss of sodium has had considerable consequences in human affairs. The importance of salt is apparent from such expressions as "worth his salt." The word "salary" is derived from the same root as "salt" and refers to wages that were paid in the form of salt bricks. Historically, personal liberty and independence have flourished in areas where salt is abundant. Where salt supplies are restricted (for example, where it must be mined), the individual or group that controls the salt supply has absolute power over the local population.

As solutes are resorbed from the filtrate in the proximal tubule, water moves with them back into the blood plasma, keeping the osmotic forces in balance. By the time the filtrate enters the loop of Henle, about 90 percent of the fluid volume of the original filtrate has passed back into the capillaries.

In the distal convoluted tubule the filtrate undergoes further changes in composition. Inorganic ions are actively extruded from the filtrate or actively secreted into it from blood. This region of the tubule also regulates the overall acidity of the blood. Ammonia (synthesized from amino acids or other precursor compounds) combines with hydrogen ion to form ammonium ions:

$$NH_3 + H^+ \rightleftharpoons NH_4^+.$$

The ammonium ions are then secreted into the filtrate to be voided in the urine. The net effect of this process is the removal of one hydrogen ion. In other words, it causes a decrease in acidity. Relatively constant acidity, as we have seen, is vital for maintaining the tertiary structure of protein molecules.

Water Regulation

The kidney also functions in the regulation of the water content of the whole body. Urine can be more or less concentrated than blood and plasma, depending on the quantity of water in the body at the moment. If water is low, the kidney produces a concentrated urine, and wastes are eliminated without excessive water loss. When the body contains too much water, a dilute urine results, allowing the elimination of excess water with a minimal loss of solutes.

Urine concentration is controlled by the collecting tubules. The filtrate that enters the collecting tubule in the kidney's cortex is relatively dilute, because most of the solutes have been removed. As the filtrate moves down the tubule, it passes through the medulla, where the osmotic concentration of surrounding fluid is high. If the walls of the collecting tubules are permeable to water, water moves out of the filtrate into the surrounding fluid and the urine becomes more concentrated. On the other hand, if the collecting ducts are impermeable to water, urine will remain dilute (that is, somewhat less concentrated than blood). Thus, the kidneys can either conserve water or eliminate it by changing the permeability of the collecting ducts. Their permeability is controlled by a hormone called **anti-diuretic hormone,** ADH. ADH is released in greater or lesser amounts de-

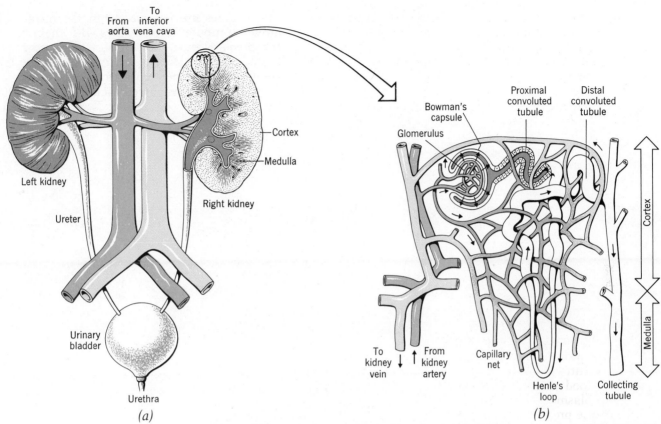

Figure 10-14 Kidney excretion in humans. (a) The kidneys, ureters and bladder. (b) An individual nephron with its associated capillary net.

pending on the water concentration in the blood.

Urine can be concentrated only if the osmotic concentration of the medulla is high. This high concentration results from the activity of the central part of the nephron, the loop of Henle. The limbs of the loop are U-shaped, so that flow along descending and ascending arms occurs in opposite directions. The *ascending* limb is impermeable to water but not to sodium. Sodium, is actively extruded from the filtrate in the ascending limb into the surrounding medullary fluid. Since water cannot move out of the ascending limb along with the sodium, the surrounding fluid becomes more concentrated than the filtrate. But the *descending* limb is quite permeable to both sodium and water, so the high concentration of the surrounding fluid causes sodium to enter the filtrate in the descending loop and water to leave it. As the filtrate moves from the cortex toward the tip of the loop, it becomes more and more concentrated. The osmotic concentration

at the tip is three or four times that of the entering filtrate. As the filtrate ascends away from the tip, back into the cortex, it becomes less concentrated again because of sodium extrusion (see Figure 10-15).

The U-shaped loop is called a **counter-current** system, because the flows in the two arms move in opposite directions. It is a very efficient way to create a high osmotic concentration, because sodium is always actively transported from the filtrate to the fluid against a small concentration gradient. The osmotic concentration of filtrate inside the ascending arm is always just a little less than it is in the surrounding fluid. This is the case even though there is a large concentration difference between the fluid around the tip (in the medulla) and the fluid around the ends of the tubule (in the cortex). Compare this system to a situation in which sodium was simply being pumped out of a straight tubule. In that case, the concentration inside would get lower and lower while the outside

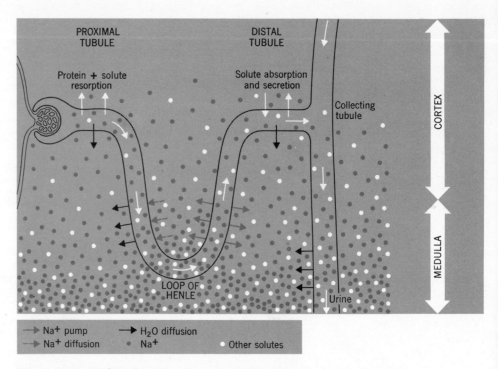

Figure 10-15 Kidney function. Proteins and solutes are removed at the proximal convoluted tubule, and other materials are secreted into the filtrate. The loop of Henle creates a gradient of osmotic concentration from cortex to medulla, through the active extrusion of sodium from the ascending limb of the loop. Further solute extrusion and secretion into the filtrate occurs at the distal convoluted tubule. The final filtrate enters the collecting duct at the cortex, passes through the medulla, and flows into the ureter as urine.

concentration increased. As a result, sodium would soon have to be transported against a very large gradient. The larger the gradient, the more the energy that would be required to oppose the forces of diffusion.

In summary, the nephron performs two functions. The proximal and distal regions regulate the composition of solutes in the urine by selectively transporting solutes in and out of the filtrate. The central part, the loop of Henle, helps to control the overall concentration of the urine by raising the osmotic concentration of fluid in the medulla. This process allows water to move out of the filtrate, leaving concentrated urine.

In humans, the osmotic concentration of urine ranges from one fourth to four times the concentration of blood. Though the maximum is slightly greater than that of sea water, the ocean is still too salty for humans to drink. Urine and sea water are similar in osmotic concentration, but their compositions are very different. Many of the inorganic salts in sea water are not abundant in urine and cannot be concentrated effectively by the kidneys. Consequently, the body must contribute water to the ingested sea water before all of the salts can be excreted. Drinking sea water actually leads to a net loss of water.

Mammals adapted to dry conditions are able to produce extremely concentrated urine. They have longer loops of Henle, enabling them to maintain greater concentration differences between the cortex and the medulla of the kidney. The kangaroo rat, an extremely efficient water conserver, can produce urine six times as concentrated as sea water—about eighteen times as concentrated as its own blood plasma. Most of the time it does not drink at all. It can obtain enough water from oxidizing its food.

Respiration

Oxygen must be procured from the environment and carried to the respiring cells of the body. Carbon dioxide must be transferred from the cells to the surroundings. In most animals, with the notable exception of insects, such exchanges between environment and cell are mediated by the blood.

Gas Exchange Surfaces

Gas exchange between the environment and the blood often occurs at a specialized surface: a gill or a lung. In other animals, gas exchange simply occurs across the whole body surface. Some water-living salamanders depend entirely on gas exchange across the skin. Other amphibians may use both mechanisms. A frog has lungs for gas exchange on land, but it obtains oxygen across its skin when submerged. All gas-exchange structures are richly vascularized; that is, they are well supplied with blood.

Gills occur in many forms. The beautiful plumes of marine slugs are actually gills (Figure 10-16). The gill of the sea cucumber represents an internal extension of the posterior gut. This tree-like internal gill is alternately filled and emptied as water is pumped in and out of the anus. In fish, the gill consists of slit-like openings in the anterior, or front, portion of the gut, which communicate with the environment.

Aquatic animals obtain fresh supplies of oxygen by moving water across their gills. This flow is usually one-way: water enters the mouth and pharynx, sweeps over the gills, and returns to the environment (Figure 10-17). Inactive fish that dwell in rocks or sand or mud produce a flow with a muscular pump. More active fish depend on swimming movements to drive water across their gills. Mackerel and tuna will suffocate if they stop swimming.

The water flow across the gill and the blood flow through it are in opposite directions, ensuring an efficient transfer of oxygen from water to blood. This device is another counter-current system, analogous to the loop of Henle in the kidney. It depends only on diffusion—active transport is not involved. But the principle is the same. Because the flows run in opposite directions, there is a small gradient at every point along the exchange surface—at any point there is always slightly more oxygen in the water than in the blood. As a result, oxygen diffuses from the water into the blood across the entire surface of the gill. But if the two flows were parallel, much less oxygen would be absorbed. In fact,

(a)

(b)

Figure 10-16   The gills of sea slugs are elaborate external structures. (a) Tridachia crespata. (b) Glaucus.

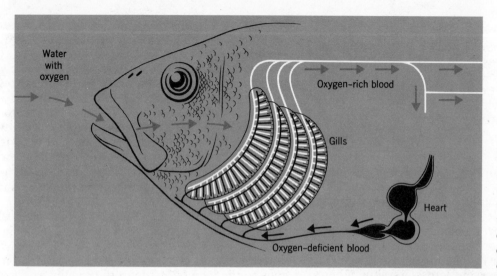

Figure 10-17 *Fish gills.
Water is drawn in through
the mouth and exits
through the gill openings in
the pharynx; the gills provide a surface for exchange
of respiratory gases and
other materials with the
environment.*

transfer would cease when the oxygen content of
the blood equalled that of the water; only half the
available gas would be transferred (see Figure
10-18).

Using counter-current exchange, many aquatic
animals are so efficient that they withdraw three
quarters or more of the oxygen from the water as it
passes over the gills. By contrast, land animals extract less than one quarter of the available oxygen
from the air. But in terms of energy expenditure,
aquatic animals do poorly. They must spend a fifth
of their total metabolic energy to provide a fresh
gas supply, while land animals require only about
1/50 of theirs. This difference results from the high
viscosity of water, which makes it hard to move
across the gill, and also from its low oxygen content. Air is approximately 21 percent oxygen; therefore, every liter of air contains 210 milliliters of oxygen. By contrast, a liter of water contains only
about 6.5 milliliters of oxygen at 20° C and about
10 milliliters at 0° C. At best this amount is still
only about five percent of the amount of oxygen in
the same volume of air. As a result, the oxygen consumption of an aquatic animal is usually lower
than that of a land dweller.

The lungs of terrestrial animals are always internal structures. This location serves to protect
delicate tissues. It also restricts evaporation from
the moist exchange surface and minimizes the loss
of water: lung surfaces must be kept moist so that
gas can move into blood in dissolved form.

The main air passage leading to the lungs from

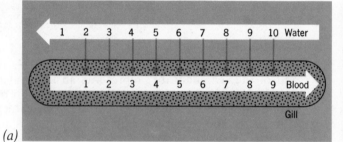

(a)

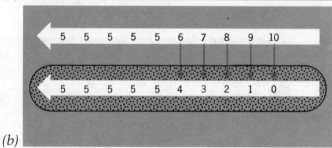

(b)

Figure 10-18 *Counter-current exchange. (a) The counter flow of
two streams (in this case blood
and water) provides a gradient
for exchange along the entire
length of the exchange surface.
Virtually all the oxygen in the
water can be transferred to the
blood. (b) Parallel flow would be
much less effective, because the
gradient, though initially steep,
disappears after only a fraction of
the oxygen has been transferred.*

the outside is the **trachea,** or windpipe. The trachea branches to form two tubes, the right and left **bronchi,** which lead to the right and left lungs. Within the lungs, the bronchi branch repeatedly to form small **bronchioles.** The bronchioles finally terminate in minute air sacs. Each air sac consists of a group of interconnected spheres called **alveoli.** The alveoli are surrounded by capillaries, and they are the site of gas exchange (see Figure 10-19). Lungs enclose a modest volume of air—about six liters in man. Nevertheless, through this branching, they have a tremendous surface area for gas exchange—about 600 to 800 square feet.

Recently, experiments have been performed in which land animals, including man, breathe water. The water must be kept under high oxygen gas pressure to increase the amount of dissolved oxygen. It must also contain salts of the proper concentration so that lung tissue will not go into osmotic shock. Laboratory mice have survived for days under water, but their lungs are often damaged by the extraordinary effort involved in moving water in and out (see Figure 10-20). The procedure has also been successfully carried out on a man. So it is possible for humans to breathe water, but only under very special conditions. The prospects for diving and breathing water directly, instead of depending on an air hose or SCUBA, are not very encouraging.

Oxygen Exchange

No matter how efficient the exchange surface is, the amount of oxygen delivered to tissues depends on how much oxygen the blood can carry. If blood were simply a salt solution, oxygen transfer to cells would be very ineffective. A liter of water in equilibrium with air at 37° C (body temperature) contains about five milliliters of oxygen. But human blood can carry about 200 ml of oxygen per liter—40 times as much. The carrier of most of the oxygen in blood is hemoglobin.

Hemoglobin is composed of a protein linked to a heme group. The heme is structurally similar to chlorophyll but lacks a side chain (Figure 10-21). It is this portion of the molecule that associates with an oxygen molecule.

$$\text{Hemoglobin} + O_2 \rightleftharpoons \text{Hemoglobin } O_2$$

The hemoglobin in human blood is all contained in the erythrocytes; in fact, these cells are about one-third hemoglobin. If the same quantity of hemoglobin were dissolved directly in the plasma, the consequences would be disastrous. The osmotic concentration of blood would be so high that water

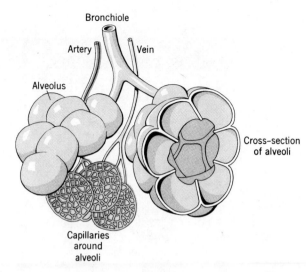

Figure 10-19 *Bronchioles terminate in clusters of tiny sacs, or alveoli. Gas is exchanged between the alveolus and the surrounding capillaries.*

would tend to move out of the tissues and into the blood. Despite such difficulties, some animals (such as earthworms) do have blood with dissolved hemoglobin. But in their systems less hemoglobin is present and the protein molecule is very large, to minimize osmotic problems. Such blood carries less oxygen than blood in which the hemoglobin is contained in cells.

The actual quantity of oxygen in blood may be considerably less than the maximum possible. It is lowered when oxygen in the lungs drops or when tissues consume oxygen at rapid rates. The availability of oxygen at a given time is expressed as a **partial pressure,** that is, as the portion of the total air pressure that it represents. Air has a total pressure of 760 millimeters of mercury at sea level, and it is 21-percent oxygen. The partial pressure of oxygen is therefore 760×0.21, or about 160 millimeters. If the total air pressure is reduced, the partial pressure of oxygen decreases accordingly.

The amount of oxygen gas dissolved in water at any time depends on the partial pressure of oxygen in the gas phase above the water. So, when the partial pressure of oxygen gas increases, the oxygen concentration of the adjacent water mass increases proportionally. However, the relation between the partial pressure of oxygen and the oxygenation of hemoglobin is quite different. Single heme units carry (or "load") oxygen quite efficiently, even at

Figure 10-20 *A mouse breathing water.*

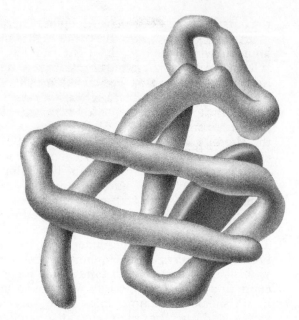

Figure 10-21 *A model of one protein chain of a hemoglobin molecule. The entire molecule consists of four similar chains. The heme group is shown in color.*

or gill, where carbon dioxide is lower. This effect increases the efficiency of oxygen delivery to the tissues. Fully oxygenated hemoglobin is bright red; deoxygenated hemoglobin is purplish red. Thus, the level of oxygenation can be determined quite precisely by measuring the light absorption of a blood sample.

Carbon Dioxide Exchange

We have focused on the transport of oxygen to cells. But the removal of carbon dioxide from the tissues is equally important. As carbon dioxide dissolves in water, it combines with a water molecule to form carbonic acid, and carbonic acid separates into bicarbonate ion and hydrogen ion:

$$CO_2 \underset{H_2O}{\rightleftharpoons} CO_2 + H_2O \rightleftharpoons H_2CO_3 \rightleftharpoons HCO_3^- + H^+$$

| (carbon dioxide gas in air) | (carbon dioxide gas dissolved in water) | (carbonic acid) | (bicarbonate ion + hydrogen ion) |

Thus, carbon dioxide can be present in blood as dissolved gas, as carbonic acid, or as bicarbonate ion. Most of the carbon dioxide in blood is carried as bicarbonate, formed from carbonic acid. But the reaction of carbon dioxide with water to produce

low partial pressures. As oxygen becomes more plentiful the heme unit quickly becomes fully saturated. The hemoglobin of animals that live in oxygen-poor water has a very high affinity for oxygen. The pigment may be saturated by oxygen at partial pressures of less than five millimeters of mercury. The blood of air-breathing animals, by contrast, has a lower affinity for oxygen. In man, hemoglobin is saturated at 90 millimeters of mercury.

Hemoglobin loading is modified by several factors, particularly carbon dioxide. A high carbon dioxide concentration decreases hemoglobin's affinity for oxygen. For this reason, oxygen tends to be "unloaded" at the tissues, where carbon dioxide levels are high. Oxygenation is favored at the lung

carbonic acid, and the reverse reaction, are both comparatively slow. The interconversion of carbon dioxide and carbonic acid is catalyzed by an enzyme in red blood cells, **carbonic anhydrase.** Without the enzyme, the reaction would proceed so slowly that blood would have to remain in the capillary bed a long time for carbon dioxide exchanges to be completed.

The major pathway for gas exchange proceeds as follows: at the tissues, carbon dioxide diffuses from the cells into the tissue fluid, then to the capillaries, and, finally, into the red blood cells. The red blood cells, in turn, transfer oxygen from hemoglobin to the tissues. The exchanges of both carbon dioxide and oxygen are linked. Carbon dioxide is converted into carbonic acid, which in turn becomes bicarbonate ion and hydrogen ion. The hydrogen ion combines with hemoglobin, as the oxygen is released:

$$H_2CO_3 \rightleftharpoons HCO_3^- + H^+$$

$$Hb{\cdot}O_2 + H^+ \rightleftharpoons Hb{\cdot}H + O_2$$

Since these reactions occur simultaneously, there is no change in acidity (see Figure 10-22).

All of these reactions are reversed at the lung. Carbonic anhydrase in the red cells catalyzes the production of carbon dioxide from carbonic acid, so carbon dioxide begins to diffuse into the alveolus. At the same time, hemoglobin is being reoxygenated, providing a source of hydrogen ions for the formation of carbonic acid from bicarbonate ion.

Breathing

The varying demands of body tissues to obtain oxygen and get rid of carbon dioxide can be accommodated to some extent without much change in the ventilation of the lungs. Under normal circumstances, about a quarter of the oxygen carried in the blood is delivered to the tissues. When oxygen demand increases in a particular area of the body, as much as 80 percent of the oxygen can be unloaded in those tissues without much change in circulation or lung ventilation. However, the depth and pace of lung ventilation—and the heart rate—do respond to overall increases in oxygen consumption and carbon dioxide production.

Up to a point, a person's lungs can be filled and emptied voluntarily. But several nerve centers in the brain can override the voluntary control of breathing. These centers are sensitive to the carbon dioxide content, acidity, and oxygen content of the

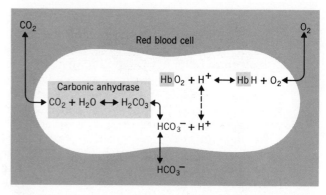

Figure 10-22 *Oxygen and carbon dioxide exchanges between red blood cells and tissue fluid are interrelated. In the tissues, oxygen is delivered and carbon dioxide is acquired. As hemoglobin is deoxygenated, it combines with the hydrogen ions that are produced as carbon dioxide moves into solution. As a result, large volumes of carbon dioxide can be carried with little change in the acidity of the blood. These reactions are reversed in the lungs: there, carbon dioxide is delivered and oxygen is acquired. Carbonic anhydrase in the red blood cells accelerates the reversible reaction between carbon dioxide and water by which carbonic acid is formed.*

blood. Because of this mechanism, the average person can hold his breath for only about a minute. Even experienced pearl divers rarely exceed this time. In fact, the control system for breathing is quite complicated. Like many other important body functions, one feedback system modifies the behavior of others. The most interesting aspects of breathing control are encountered in extreme environments—that is, under low and high pressure conditions.

High Altitude

As altitude increases, atmospheric pressure decreases. At 21,000 feet the pressure has dropped to half that at sea level. As the external pressure drops, the partial pressure of each air gas drops at the same rate, so there is no qualitative change in the air itself. Its composition is constant—there is just less of it. Also, the total pressure in the lung is the same as the surrounding pressure. Nevertheless, the composition of the gas in the lungs does change when the total pressure decreases.

The gas in the alveoli is always saturated with water vapor. Furthermore, the carbon dioxide level in the lungs is kept high by body metabolism. As the total air pressure drops outside and the partial pressures of oxygen and nitrogen go down, the carbon dioxide and water levels within the lungs remain high. Above 50,000 feet the alveoli will contain only carbon dioxide and water, even when pure oxygen is available. Thus, at 40,000 feet or higher, oxygen must be supplied to the lungs under pressure.

Actually, breathing responses become inadequate long before one reaches such heights. As the external pressure falls, the amount of oxygen in the lungs goes down, and less oxygen is delivered to tissues. But the control of breathing responds more strongly to the level of carbon dioxide in the blood than to the oxygen level. When the carbon dioxide level in the blood goes up, the breathing rate increases. But at high altitudes the carbon dioxide level in the lung, and therefore in the blood, doesn't change. Therefore, the breathing rate does not change. Finally, as the oxygen content of the blood continues to fall, the low oxygen level will override the carbon dioxide control. This occurrence is only an emergency measure, however; it does not permit normal activity when you climb rapidly to 10,000 feet.

If you remain in the mountains your body will gradually adapt to the change. The respiratory centers become less sensitive to carbon dioxide, and breathing rates can increase even though the blood carbon dioxide becomes low. As a result, normal activity is possible in a day or two at intermediate heights of 5,000 to 10,000 feet. People who live and work year-round at heights of 10,000 to 15,000 feet develop a more extensive set of adaptations. Their carbon-dioxide sensitivity reappears, but it is adjusted to tolerate a lower level. The number of red cells increases, as does each cell's hemoglobin content; chest volume increases; and mild high blood pressure may develop.

At extreme heights all adaptive changes become inadequate. A Himalayan expedition occupied a hut at 19,000 feet for several months. All members of the party lost weight. At the end of the winter, their physical condition and adaptation to heights were worse than those of newcomer colleagues. The limit for climbing without oxygen breathing apparatus is around 28,000 feet. Several Everest expeditions reached this altitude without oxygen. None could establish a higher camp.

Animals differ in their ability to adjust to altitude. Humans are considerably more tolerant than horses, for instance. In 1799, when ballooning was popular, one man elected to go up astride a horse.

The horse was in considerable distress during the trip and began bleeding from the nose and ears, but the man suffered no discomfort.

In aircraft flight, the reduction of the oxygen partial pressure produces a graded series of psychological and physiological problems. Above 10,000 feet supplementary oxygen must be supplied. And above 40,000 feet even pure oxygen must be supplied under pressure. These conditions reverse the normal work of breathing; that is, gas must be forced out of the lung against a higher pressure. Such breathing is very tiring. The alternative approach is to pressurize the cabin to mimic lower altitudes.

Diving

Fully oxygenated blood contains enough oxygen to support the oxygen consumption of the resting body for two to three minutes, or for less than one minute of slow swimming. If the air in the lung is full of gaseous oxygen as well, these figures are doubled. However, the oxygen supply itself does not usually control breath-holding. As we mentioned earlier, the voluntary inhibition of breathing is overridden after about a minute by an increase in blood carbon dioxide.

By **hyperventilating** (that is, by breathing deeply and rapidly) before diving, sport divers can increase their active time without breathing to two or three minutes. Hyperventilation reduces carbon dioxide in the alveoli and in the blood, so more time passes before the carbon dioxide concentration rises to an intolerable level. The record for breath-holding under water is thirteen minutes and 45 seconds. The record-holder hyperventilated with oxygen to wash carbon dioxide out of the lung and then lay supine on the bottom of a swimming pool. Hyperventilation should not be used by divers, however, because it is very dangerous: the oxygen may be used up before enough carbon dioxide has accumulated to signal the need to breathe. If that happens, the diver quietly passes out and dies. Hyperventilation also reduces the acidity of the blood. This effect can cause severe muscle cramps and, again, unconsciousness.

Depth records for free diving are about 240 feet. The limitation in this activity is pain from lung compression. For each 33 feet of descent the pressure increases by one atmosphere. Most people cannot tolerate more than four atmospheres (at a depth of about 130 feet). At that pressure, the lungs are compressed to about a fourth of their maximum volume.

Seals can reach a depth of 1,400 feet. Whales can descend to 3,000 feet. Diving mammals have several adaptations that help them function underwater. For one thing, they have rib structures that allow for complete lung collapse. A second adaptation is a change in their patterns of circulation after they submerge. In animals that are good divers, such as seals or ducks, these alterations are very striking. The heart rate decreases. The blood supply to the body muscles is reduced, and more blood is channeled to the nervous system. This procedure produces an **oxygen debt** in the muscles; that is, in the absence of oxygen, organic compounds are not completely oxidized by the muscle cells. Instead, they are converted to intermediate products by fermentation. Later, oxygen must be supplied to complete the oxidation of these compounds to carbon dioxide and water. The "debt" must be repaid when oxygen becomes available, and before the concentration of intermediate compounds such as lactic acid becomes intolerable.

Whaling is possible because whales build up an oxygen debt. When a whale sounds, it can stay down for a long time, but it accumulates an oxygen debt. If it is forced to dive again and again to escape a whale hunter, its time below the surface becomes progressively less, because the animal cannot carry an oxygen debt for extended periods of time. Eventually, it can be approached and harpooned.

Caisson workers, hardhat divers, and SCUBA divers breathe air or special gas mixtures that are at the same pressure as the surrounding water. The diver who set a SCUBA depth record of 437 feet was breathing air at twelve atmospheres. When special gas mixtures are used, such as oxygen plus helium, SCUBA divers can survive at 1,000 feet.

Divers breathing gases at depths encounter two basic difficulties. The first stems from the increased solubility of gases at high pressure. If an animal breathes in a high-pressure atmosphere, increased quantities of all the gases present will gradually dissolve in its blood. Then, when external pressure falls as the animal surfaces, there may not be sufficient time for the excess dissolved gases to leave the blood by crossing alveolar surfaces back to the environment. Instead, the gas may come out of solution as tiny bubbles, which are carried along in the circulatory system. Eventually the bubbles become lodged in arterioles, where they block the blood flow, causing "the bends." The precise symptoms of this condition depend on where the bubbles are formed, how large they are, and where they lodge. "The bends" can be fatal.

In order to prevent bubble formation, surfacing divers may have to spend time at intermediate depths (in other words, at intermediate pressures) to allow the dissolved gases to escape. In this way, their blood can equilibrate gradually with lowered pressures without having bubbles form. At 200 feet, for example, 30 minutes of working time requires more than an hour of subsequent decompression, with stops at 40, 30, 20, and ten feet. One way to get around this limitation is to maintain the divers in a submerged, pressurized diving chamber between work periods. In that case, decompression to atmospheric pressure at the surface needs to be done only once, at the end of a longer period. If the diver can come "home" to a pressure of four or five atmospheres, his actual working time is greatly increased. But such divers still face the second danger from high-pressure gases, toxicity.

Regular SCUBA dives are probably too short to create problems with oxygen gas toxicity, or **oxygen poisoning.** However, long exposure to oxygen at pressures of three times normal produces lung damage and nervous system attacks that are rather like epileptic seizures. We do not understand the reason for oxygen poisoning, but the symptoms can be avoided by using special gas mixtures with only three or four percent oxygen, rather than the 21 percent found in air. High pressures still provide ample partial pressure for the transfer of oxygen to the blood.

Another form of gas toxicity, **nitrogen narcosis,** occurs during short dives. The so-called "inert" nitrogen gas is actually toxic at high pressure. The symptoms of nitrogen narcosis are comparable to those of alcohol intoxication or exposure to nitrous oxide ("laughing gas"). The diver feels uneasy, or may behave in foolish or irresponsible ways that are potentially lethal. The effects show up as changes in brain wave patterns in rather shallow water, but they become troublesome only at pressures of four or five atmospheres. At still greater depths, it becomes almost impossible to function. Helium must be substituted for nitrogen in the air mixture for deep dives.

Human existence under high or low pressures requires constant attention to an artificial support system. Technology has not succeeded in creating foolproof environmental aids. We have alluded to only some of the major problems. Humans can exist below the sea for months, but the cost is constant alertness and attention to detail. We are still a long way from constructing underwater cities or establishing orbiting communities.

Temperature Regulation

Mammals and birds are sometimes described as **warm-blooded** animals. This term is a bit misleading, because their body temperatures may sometimes be cooler than that of their environment. What the term does mean is that they maintain their body temperature at a relatively constant level. For any warm-blooded animal, there is some temperature below which it must warm itself by producing additional metabolic heat to prevent a drop in body temperature. This temperature is called the **critical temperature.** When the environment is warmer than the body, on the other hand, the cooling effect of evaporating water must be utilized to keep the body temperature from rising.

Between these extremes there is an intermediate temperature range in which the hairy coats of mammals and the feathers of birds can be extremely efficient insulators. Fur's effectiveness as a barrier to heat flow can be modified by varying the depth of the air layer trapped among the hairs. Air conducts heat rather poorly. The erection of the hairs, therefore, traps a static layer of air and increases the insulation. "Gooseflesh" in humans is the result of the contraction of small muscles at the base of hair follicles, which erect the hairs. But humans' hairy coats are too thin to make this device effective. Arctic animals, however, have such excellent insulating pelts that their normal metabolism can maintain their body temperature when the external air is −30° C, or even lower. White foxes can survive in the lowest temperature recorded on earth, about −65°, by generating only a 25 percent increase in their metabolic heat.

As might be expected, the insulating capacities of the fur of tropical mammals is much lower. Humans are functionally tropical animals and become chilled below air temperatures of 27° C. There is little difference in people's tolerance to cold between summer and winter, or between cold and warm climates.

Changes in subdermal blood circulation—circulation beneath the skin—also regulate heat loss. Water (and therefore blood) is a good heat conductor. If blood is routed rapidly and continuously through the skin, heat will quickly be transferred to a colder environment and the body will be cooled. On the other hand, if circulation to the skin is minimal, heat flow is reduced and warmth is conserved. This mechanism explains our familiar responses to heat and cold. The skin is flushed in the former case and pale in the latter. However, if we remain in the cold, our skin may become

flushed again as heat is returned to the skin to keep it from freezing.

A layer of subdermal tissue is also used to restrict heat loss. Such tissue is extremely important to aquatic mammals, whose heat loss is greater because water is such a good heat conductor; critical temperatures for land animals are always higher in water. In water, humans actually begin to shiver at 33° C, a temperature that seems uncomfortably warm at first. Heat loss cannot be retarded by wet fur, either, since its insulating properties are impaired when the air spaces are lost. Seals and whales are therefore insulated by a thick layer of fat below the epidermis. Some heat is still lost from the blood that maintains these fat cells and the overlying epidermal cells, but this loss is minimized by a special pattern of circulation. The arteries carrying blood to the surface are surrounded by the veins returning blood to the central tissues. This close proximity of artery and vein permits heat exchange between the warm blood moving toward the skin and the chilled blood returning from it. In this way, heat is returned to the central body tissues before it has an opportunity to reach the surface and be lost. This is yet another example of a counter-current system, analogous to the systems that facilitate ion transport in kidneys and oxygen exchange in gills. Conversely, when it is required, heat loss may be aided by increasing the circulation of blood to special areas of skin. The ears of jackrabbits are used as surfaces for heat loss. And, in fact, the rabbits in the southern regions of the United States have longer ears than do their northern cousins (see Figure 10-23).

But any device that depends on insulation or circulation for cooling simply doesn't work when environmental temperatures are higher than body temperatures. Under these conditions, in fact, low insulation or increased circulation will facilitate the transfer of heat from the environment to the animal. Yet the animal continues to metabolize and generate heat, and if this heat is not transferred back to the environment its body temperature will rise. The evaporation of water accomplishes the transfer of heat to warm environments.

Every gram of water that goes from liquid to vapor state requires about 575 calories of heat to make the transition. The evaporation of water uses heat energy from the surface where it occurs, and the surface itself is therefore cooled. The air that leaves the lungs is saturated with water vapor. It represents a continual and unavoidable loss of both water and heat. But sweating increases the amount of water evaporation from the body surface, in re-

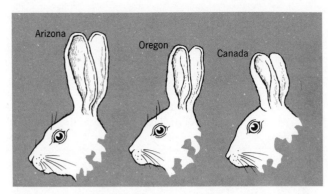

Figure 10-23 *Hares from warmer climates have longer ears to facilitate heat loss.*

sponse to heat stress. Sweat is produced at a low rate, about twenty milliliters per hour in man, in the absence of a heat load. But more than a liter per hour would be produced by the same man walking naked in the desert sun. This figure is close to the maximum rate of production. Hard physical work produces the same increase in sweat production, but suitable insulating clothing can drop the figure by 20 to 25 percent. The effectiveness of sweating in promoting heat loss to a hot environment is really quite dramatic. A man can survive in a dry oven for 30 minutes or more at air temperatures of 250° F, provided he has adequate water and is insulated from the oven walls. He can sit and watch his dinner cook.

Sweat is a more dilute solution than the blood and tissue fluid from which it is derived, but it does contain some salts, particularly sodium chloride. This lost salt must be replaced, along with the evaporated water. Salt tablets should be issued to individuals under heat stress. Commercial "thirst quenchers," which are simply flavored salt solutions, are also satisfactory.

Environmental situations that combine high temperature and high humidity are dangerous because they retard evaporation. Sweating is copious, but ineffective. Heat cannot be lost without evaporation. Heavy work under such circumstances is simply not possible for extended periods.

The size of an animal is an important factor in its temperature regulation. A small animal has a large surface area per unit weight. Under cold stress, the surface of such an animal is so large that insulation and high metabolic rate combined cannot maintain its body temperature. This is the reason that there are no marine or arctic mammals as small as mice. And at the other extreme, under heat stress, small animals don't have enough water to spend on temperature regulation. Small desert mammals such as kangaroo rats and desert mice avoid heat stress by staying in relatively cool burrows during the day. They are active at night. When a kangaroo rat is forced out of its burrow into the desert sun, it salivates on its fur to cool itself (it lacks sweat glands), but it still dies in minutes.

Homeostatic mechanisms such as the ones we have examined provide a constant internal environment in which all the body's systems can function smoothly. The nervous system, too, depends on a constant internal environment for efficient function, but it also aids in maintaining it. Sensory endings collect information from the environment; the information is processed; and, if necessary, a response may be ordered—muscles contract or hormones are released. Information about external temperature may induce a change in an animal's insulating layer or cause it to move to a different environment. The work of the nervous system in obtaining information from the environment and directing responses to it may, in fact, be interpreted as an extension of homeostasis.

Recommended Reading

Langley, L. L., *Homeostasis.* New York: Van Nostrand Reinhold (1965).

An introduction to the concept of homeostasis and regulatory mechanisms.

Smith, Homer W., *From Fish to Philosopher.* New York: Doubleday (1959).

The physiology of excretion and the evolution of the kidney in relation to vertebrate evolution.

Wigglesworth, V. B., *The Life of Insects.* New York: World (1964).

A famous insect physiologist presents a discussion of the largest and most diverse group of organisms.

11 animal organization III: behavior

Animal behavior—that is, the sum of all the obvious and varied movements we associate with being an animal—can be regarded in a sense as just another aspect of homeostasis. An animal moves about to obtain food, which is necessary for maintaining its homeostatic systems, or to avoid environmental situations that threaten homeostasis. But in another sense homeostasis could be interpreted as subordinate to behavior: it maintains the physical integrity of the animal, and thereby "permits" behavior. Homeostatic mechanisms support, create, and modify behavior. The converse is also true.

Biology assumes that animal behavior is ultimately the result of the activity of cells and tissues in the nervous system. The nervous system receives information from the environment, evaluates it, and causes appropriate responses to it. The sum of these responses is behavior. Animal behavior is often a direct response to information received from both the external and the internal environments, and it is designed to relieve real or potential stresses on the internal environment. For example, the nervous system may direct the animal to reduce its body temperature by panting, sweating, moving to a cooler place, or removing clothing. Other forms of behavior, such as learning, memory, and instinct, bear a more distant and complex relation to environmental influences, but ultimately they are all directed toward the same end—the preservation of the integrity of the organism.

The nervous system is composed of nerve cells, or **neurons.** Neurons form a network throughout the entire body, but all of them are connected in one way or another to the massive collection of nerve cells that makes up the brain and spinal cord.

M. C. Escher, "Encounter"

309

Parts of the nervous system can be studied directly because they will continue to function even when isolated from the rest of the organism. Nerve cells and muscle cells, sensory receptors and reflexes have all been exhaustively studied as specific, isolated units. And as a result of this type of work, certain kinds of simple behavior can be described in considerable detail. Yet in the whole organism these units do not occur in isolation. They interact in complex ways, and the functions that are observed in the isolated units become highly modified in the organism. Some aspects of behavior can be described with confidence despite these difficulties; others are poorly understood.

Components of the Nervous System

The nervous system receives many kinds of information. Receptor organs such as eyes, ears, and nose provide information about the external environment. Specific cells receive other environmental inputs, such as touch, pain, heat, or cold. The internal environment also channels information to the nervous system by way of nerve cells that are sensitive to specific types of chemical imbalances, such as high carbon dioxide concentrations or fluctuations in blood sugar. The information received by all of these receptor cells is converted to electrical signals that can move along the length of a neuron. The signals may also be passed on to adjacent neurons and may eventually reach the brain.

The brain and spinal cord together are usually called the **central nervous system** (CNS). In some ways the CNS resembles an information processing center. Incoming information may be screened out and ignored, or it may lead to several kinds of behavior: a rapid and simple behavioral response, or **reflex;** a complex "emotional" response (fear, rage, or pleasure, for instance); or a very complicated behavioral sequence such as food gathering, migration, reproduction, or fighting. Incoming information may also be stored, in the processes we call "learning" and "memory." It may be related to other information ("thinking"). And it may be used either immediately or later to direct additional behavior. The CNS directs behavior by signalling particular muscles to contract, or by signalling special secretory cells to release hormones.

The CNS sends signals and communicates with the rest of the body by means of nerves in the **peripheral nervous system.** One part of the peripheral system, the **somatic nervous system,** is a direct link between the CNS and various skeletal muscles. The other part of the peripheral nervous system, the **autonomic nervous system** (ANS), sends signals to the various glands or smooth muscles that regulate the internal environment. Both the ANS and the somatic nervous system accomplish complex functions by regulating muscle contraction and secretion. The operation of the nervous system, including inputs, processing and integration, and outputs, is diagrammed in Figure 11-1.

the neuron

The basic unit of the nervous system is the nerve cell, or **neuron.** Each neuron consists of a cell body containing the nucleus, and various long extensions, or nerve fibers. These fibers are of two types. The shorter, branched ones are called **dendrites.** The single long, sheathed fiber is the **axon** (Figure 11-2).

Though many neurons are small, some can be immense. A nerve extending from the spinal cord to a particular muscle cell, for instance, may have an axon five to twenty microns in diameter and up to a meter in length. If the cell were enlarged until its body was the size of a baseball, its dendrites would take up all the volume of a small room. The axon would be about a centimeter in diameter, and would extend a mile or more before reaching the muscle cell.

The neuron receives a signal from the environment or from another nerve cell, and it conducts the signal in the form of an electrical impulse along its length. Finally, it passes the signal to the next cell in the chain—a muscle cell, a secretory cell, or another neuron. Incoming signals enter the neuron by way of the dendrites, move through the cell body, and move out again along the length of the axon.

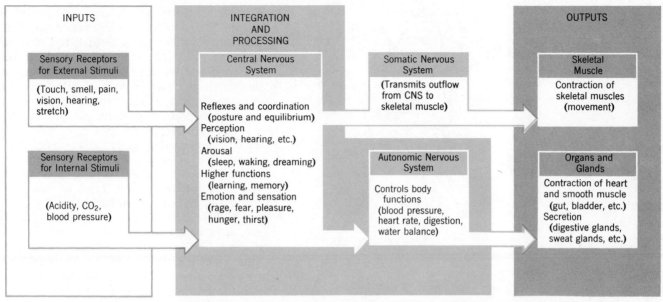

Figure 11-1 The organization of the nervous system. Information enters through receptors from the external and internal environments. The information is processed and integrated by the central nervous system (the brain and spinal cord). The autonomic nervous system controls the internal environment, thereby participating in integration and processing. The output to skeletal muscle and to glands and internal organs flows by way of the somatic nervous system via motor nerves, and by means of the autonomic nervous system.

the nerve impulse

Information is conducted along the axon in the form of discrete electrical disturbances called **impulses.** An incoming signal is coded into one or several impulses. All the information flow in the nervous system is carried in this pulse code.

Membrane Potential

The energy for the nerve impulse is electrical energy, provided by the neuron itself. But the electrical energy in nerves is somewhat different from more familiar electrical currents. Household electricity depends on the movement of electrical charges carried by electrons. Electrons are particularly useful because there are many materials that can be used as insulators to contain the electrons

and direct their movement. But most living tissues are very poor insulators. Electrons can move freely through the watery matrix around the cells, so there is no way to control their movement and convey information. Electrons, therefore, are not appropriate for electrical conduction in biological systems.

Neurons make use of electrical charges carried by ions. Ions are vastly larger than electrons and, as we have seen, can be controlled by cell membranes in various ways. Ultimately, the activity of the nervous system depends on the neuron's ability to maintain an unequal distribution of inorganic ions on each side of its cell membrane.

Sodium (Na^+) and chloride (Cl^-) ions are much less concentrated inside the cell than outside. Potassium (K^+) is distributed in the opposite way: its concentration is much lower outside than inside. This steady-state distribution of ions characterizes most cells, not merely neurons. It occurs because sodium is actively transported out of cells, in a process requiring metabolic energy; potassium ions are moved inward in the same way. Typical

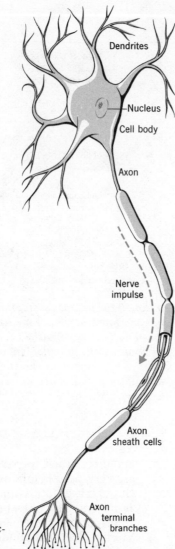

Dendrites

Nucleus

Cell body

Axon

Nerve
impulse

Axon
sheath cells

Axon
terminal
branches

Figure 11-2 A neuron.
The sheath cells as-
sociated with some
axons contain fatty
material and act as elec-
trical insulators.

The concentrations of potassium chloride inside and outside are unequal, however. As a result, the potassium ions begin to diffuse out through the membrane. (The chlorides would, too, of course, if the membrane were permeable to them.) As the potassium ions move outside, they carry their positive charges along with them. The outside of the membrane becomes positive and the inside negative (since it has lost some positive ions). But, since positive and negative charges attract each other, the positive ions at the same time tend to be electrically attracted back toward the negatively-charged side of the system. In short, the ions tend to move out because of diffusion, but they tend to move back because of electrical forces. The system comes into balance when the electrical attraction that develops just balances the force of diffusion.

This electrical attraction, or potential, is the same as a voltage applied across the membrane. If the solutions on both sides of the membrane were now joined by an electrical connection, current would flow. The current would reduce the potential difference across the membrane, of course. But as the electrical attraction diminished, more potassium ions would be able to diffuse across the membrane. The process thus would continue until all the potassium was equally distributed (see Figure 11-3).

The system we have described is called an electrical concentration cell, a kind of battery. Its voltage depends on the concentration ratios of the ions. As current flows, the battery gradually runs down, until the ion concentrations are finally equal on both sides of the membrane. At that point the battery is dead.

The situation in a neuron is analogous. Potassium is much more concentrated inside the neuron than outside, and the cell membrane is permeable to it. The outside of the cell membrane should therefore be positive with respect to the inside. In other words, neurons should behave like concentration cells. And in fact, knowing the ratio of potassium concentrations, we can predict the voltage across the membrane using an equation derived from physics. A ratio of 38 potassium ions outside to every one inside should produce a voltage of slightly less than a tenth of a volt. We actually do measure a voltage of 0.09 volts. Also, if our analogy with a battery is accurate, adding potassium to the external medium should decrease the voltage. It does.

Still, the concentration cell analogy provides only a partial description of a neuron. Sodium ions are also distributed unequally, but in the opposite

ion ratios, inside to outside, are: sodium ions, 1/12; chloride ions, 1/30; and potassium ions, 38/1.

The unequal distribution of ions on the two sides of a membrane produces an electrical potential, or voltage. Consider for a moment a model system with a membrane permeable *only* to potassium (K^+). We place a highly concentrated solution of potassium chloride (KCl) inside the membrane, and a more dilute solution of potassium chloride outside. Since KCl dissociates to K^+ and Cl^- ions, there is now a positive K^+ ion for every negative Cl^- ion on each side of the membrane. At this point the system is still electrically neutral.

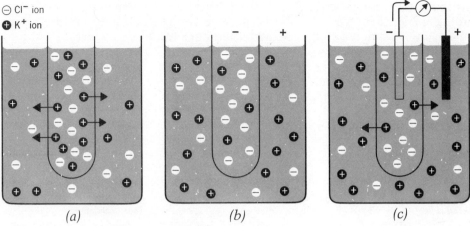

⊖ Cl⁻ ion
● K⁺ ion

(a)   (b)   (c)

Figure 11-3 An electrical concentration cell. (a) A concentrated solution of KCl inside is separated from a less concentrated solution of KCl outside by a membrane permeable only to K⁺ ions. (b) As K⁺ ions diffuse out across the membrane, the outside becomes positive with respect to the inside. When the electrical charge balances the force of diffusion, the net movement of K⁺ ions ceases. (c) If the two compartments are connected electrically, current flows. The current flow permits the diffusion of more K⁺ ions, until the K⁺ concentrations are equal inside and outside. The amount of charge, (that is, the voltage) depends on the ratio of K⁺ ions on the two sides of the membrane.

direction, and the membrane is slightly permeable to sodium. We would expect the sodium, therefore, to produce a voltage in the opposite direction of 0.06 volts, the outside being negative with respect to the inside. Yet this voltage is not found in the resting cell. Although the cell membrane is indeed permeable to both potassium and sodium, it is normally much more permeable to potassium. The electrical potential due to potassium predominates, and the resting cell behaves most like a potassium concentration cell. In other words, the cell has a "potassium membrane." Under certain conditions, however, the sodium permeability increases dramatically and surpasses potassium permeability. At that point the cell has a "sodium membrane." It becomes negative outside with respect to the inside. In this way, rapid changes in voltage accompany changes in the membrane's permeability to different ions.

The nerve impulse depends upon just such a rapid change in membrane permeability. The impulse is a temporary state of excitation, during which the membrane's sodium permeability increases while its potassium permeability remains unchanged. This change brings a drastic alteration of the voltage across the membrane—but it does not necessarily involve the transfer of large numbers of ions. The ions need only *begin* to flow

in order to establish the voltage difference. The overall ionic ratios of sodium, potassium, and chlorine are hardly disturbed by the transition from the resting to the excited state.

In the living cell, we can measure changes in permeability by several methods. The inward and outward movement of ions can be measured directly with chemical techniques, but it is quite difficult. The quantities of material are minute, and the whole process is over in a few thousandths of a second. On the other hand, these same few milliseconds represent a long time for electronic instruments. For this reason, most studies of the nervous system have been based on electronic measurement of potential changes across the membrane, rather than on direct observation of ion movement or permeability change.

The most important electronic instrument in a neurophysiology laboratory is an oscilloscope. Changes in potential near a cell membrane are first detected by electrodes. The electrodes then send a signal to the oscilloscope, where it is displayed on a screen like that of a television set. The incoming signal deflects a single beam of electrons that is sweeping across the screen. The sweep rate is controlled by the experimenter and is used as a measure of time, while the vertical position of the beam indicates the voltage of the signal. The oscilloscope draws a graph of potential change with time. Figure 11-4 is a recording of a single nerve impulse as it passes a particular point in a nerve cell.

In spite of our electronic sophistication, however, we really know very little about how these dramatic changes in permeability come about. One theory holds that pores in the membrane restrict and control the flow of ions. Rapid changes in the size or the number of pores could cause changes in permeability.

Initiation of Nerve Impulses

Under normal conditions, the permeability changes that carry a nerve impulse begin near the junction of the axon and the cell body of the neuron. In most nerve cells, the impulse begins when the membrane is exposed to a chemical agent released by a neighboring neuron. But nerve impulses can also be initiated directly in response to signals from the environment. Special receptor neurons perform this function.

Sensory Receptors
Receptor is the general term for a sensory structure. Each type of receptor is very sensitive to a particular form of energy. Eyes are responsive to light, ears and the various touch and pain receptors respond to mechanical deformation, and so forth. Sensory receptors are selective converters, or transducers, of energy. They receive information from the internal or external environment in one form of energy (heat or light, for instance) and transduce it into another form, the nerve impulse.

Receptors can actually be stimulated by several forms of energy, even though they are most responsive to only one kind. Your eye is most sensitive to light, but it also responds to touch. If you press your finger against the side of your eye (or get punched) you will see light. The photoreceptors are responding to mechanical energy. Similarly, the chemoreceptors that control the blood sugar level can be stimulated electrically to produce a rise or fall in blood sugar. But once initiated, any one nerve impulse is like another. An impulse moving down the optic nerve will be interpreted as "light" regardless of how it was actually initiated. In fact, if we could perform an operation to trade the connections of the optic nerve and the auditory nerve, we would "see" thunder and "hear" lightning. But because the eye is much more sensitive to light than

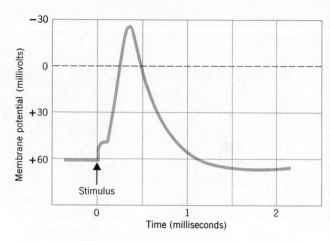

Figure 11-4 *Electrical recording of a nerve impulse. As the impulse passes down the axon, the change in voltage across the membrane is measured by electrodes placed on the surface of the membrane and inside the cell. The potential changes from +60 millivolts (outside positive) to −30 millivolts (outside negative) in a fraction of a millisecond.*

to any other kind of energy, a signal along the optic nerve usually corresponds to the presence of real light in the physical world.

Receptors have been studied in a number of ways. The basic problem is understanding the relationship between environmental signals and the resulting chains of nerve impulses. One method is to measure the output of nerve impulses from the receptor in response to controlled energy inputs from the environment. Usually, light or sound is sent to a single receptor neuron in an eye or an ear. The output of the nerve cell is then followed with an oscilloscope. The electron beam is set to sweep across the oscilloscope tube quite slowly, so that each nerve impulse appears as a vertical streak upward. A repetitive series of impulses from a single nerve appears as a series of vertical pips; sometimes the screen is photographed so that the pips can be counted and measured. The impulses may also be fed into a loudspeaker. A trained ear can tell a great deal from the sputtering, popping, or humming that results. (This technique makes conversation among neurophysiologists a bit strange to the outsider. The results of experimental procedures are likely to be imitated for the listener by a series of tongue clicks, "popopopopop," "brrrt," and the like.)

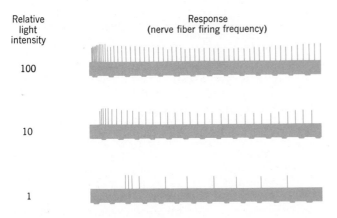

Relative light intensity

100

10

1

Response
(nerve fiber firing frequency)

Figure 11-5 The conversion of light energy to nerve impulses by a single sensory nerve fiber in the eye of the horseshoe crab. The relationship between light intensity and the frequency of nerve impulses is logarithmic. As a result, the eye can discriminate a very wide range of light intensities.

Such experiments have shown that the frequency of nerve impulses is related to the input energy logarithmically. That is, if one flash of light produces nerve impulses at a rate of ten per second, then another flash ten times as strong (in terms of energy) might produce twenty impulses per second. A flash a hundred times the original energy input will give an output of 30 per second (see Figure 11-5). This logarithmic relation means that the receptor will be sensitive to any changes in input level that are larger than a certain constant percentage—regardless of the overall energy level of the input.

This formula agrees reasonably well with experimental observations. It has been found that the smallest difference in weight or in light intensity that one can detect is always a constant fraction of the initial input. In other words, you can detect a three-percent difference in weight, whether you are judging sheets of paper or buckets of water.

If the input remains the same for an extended period of time, the output of nerve impulses is found to decline: the neuron **accommodates** to the stimulus. Different types of receptors accommodate at different rates. Touch receptors at the bases of hairs accommodate very rapidly; temperature receptors in the tongue accommodate slowly, if at all. Figure 11-6 illustrates the time course of accommodation for two different receptors. Any description of the input-output relations in sensory receptors

must include both the initial response to an input and the time course of accommodation.

A second approach to the study of receptors asks what happens between input and output. That is, what events follow the input of energy from the environment, and what leads to the production of nerve impulses? A great deal of ingenuity has been expended in this field. At the moment, though we still cannot trace the complete chain of events from input to output, we do have a good deal of information. Some of the important events are well understood.

The generation of an impulse is always preceded by a slow change in electrical potential at a localized region of the receptor membrane. These so-called **generator potentials** have been recorded from various types of receptors. Generator potentials can also be produced artificially, by placing an electrode on an axon and applying a voltage to decrease the electrical potential across the membrane. When the decrease gets below a specific threshold value, an impulse is generated. Furthermore, if the external voltage is applied continuously, a train of impulses is produced. The frequency of the impulses is related to the strength of the voltage.

How is the generator potential itself normally produced? It must be different for each kind of receptor, since each type is sensitive to a different form of input energy. At present the process is best understood for light receptors. The chemical changes that occur when light strikes the eye are analogous to those that occur in photosynthesis. In both cases light must be absorbed and converted to another form of energy before it can be utilized by the organism. In photosynthesis, chlorophyll and the accessory pigments absorb light. In vision, light

Figure 11-6 The frequency of nerve impulses generated by a receptor decreases, or accommodates, in response to a continuing stimulus. The rate of accommodation differs among various receptors.

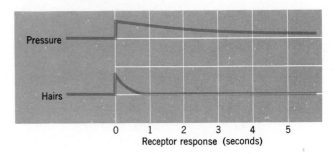

Pressure

Hairs

0   1   2   3   4   5
Receptor response (seconds)

is absorbed by a pigment called **visual purple,** or **rhodopsin.**

When rhodopsin is extracted from the eyes of animals in the dark or in dim red light, it is purplish-red. When the pigment is then exposed to light, it bleaches and changes to a pale yellow solution. The same color change can be observed in a whole, isolated eye. If a light pattern is cast on the **retina,** the sensory layer, of a dark-adapted eye, the image of the pattern becomes visible on the retina. The image is formed by the bleaching of the visual purple where the light has struck (see Figure 11-7).

Rhodopsin is found in sensory cells in the retina called **rods.** The action spectrum of these cells is just like the absorption spectrum of rhodopsin. This correspondence indicates that rod vision depends on light absorption by rhodopsin. The rods account for most of our vision at low light intensities. Rod vision is black-and-white. Color vision comes from another set of sensory cells, the **cones,** which are effective only at higher light levels. Color vision requires different pigments, which are sensitive to blue, red, and green light. (Color-blindness occurs when a person is not able to synthesize all three color pigments because of a genetic defect; a lack of the red-sensitive pigment is the most common form.) The mechanism of color vision is more complicated than that of rod vision, but similar to it in principle. A look at rod vision and rhodopsin will make the mechanism clear.

Rhodopsin consists of a molecule of **retinal**

bound to a molecule of the protein **opsin.** Retinal is synthesized from a Vitamin A molecule, and this fact accounts for the well-known connection between good night vision and Vitamin A, supplied by carrots and some other foods. Retinal is a carbon ring to which are attached some methyl $(-CH_3)$ groups and a side chain with four double bonds between the carbon atoms. There is also a pair of methyl groups along the side chain. Theoretically, retinal can exist in many shapes, because the side chain can fold in several different ways. However, most of these shapes are unstable. The molecule normally exists in only two shapes: one with the side chain continuing straight away from the ring, and one with it turned downward 90 degrees (see Figure 11-8). When light is absorbed by the molecule, the configuration changes from the angular form to the straight form. The retinal no longer fits tightly to the protein surface, and the opsin is split off. Bleaching has occurred. An enzyme found in epithelial cells surrounding the rods eventually returns the retinal molecule to its angular form, so it can reassociate with opsin to form rhodopsin again. Dark adaptation—the gradual increase in visual sensitivity that enables one to adjust to low light—represents the resynthesis of rhodopsin from retinal and opsin (see Figure 11-9).

The change in retinal configuration involves only small amounts of energy, and it cannot produce the generator potential directly. However, the energy is amplified by the bleaching steps that follow the change in retinal configuration; the splitting of opsin from retinal is the final one of these steps. The energy released by this sequence is considerably larger than the original change, but it is still not enough. The generator potential is probably produced, therefore, by an association between rhodopsin molecules and membranes in the rod cells. Rhodopsin is located on a series of flattened membranous sacs in the light-sensitive region of the cell. One can speculate that changes in the configuration of opsin, set in motion by light, eventually produce larger electrical changes in the membrane system.

This account of sensory input leaves a number of questions unanswered. How is rhodopsin associated with the membrane sacs, for instance? How can the rods continue to generate impulses when a considerable portion of the rhodopsin is bleached? But however incomplete, our description illustrates some important principles. Very small energy inputs can activate a sensory system. In fact, a rod can respond to a single quantum of light. The initial change in the retinal is very small, but this

Figure 11-7   The image formed by bleached and unbleached visual purple was first observed by a German scientist in 1878. He exposed the eye of a rabbit to a barred window and then examined the retina. This is his drawing of the image.

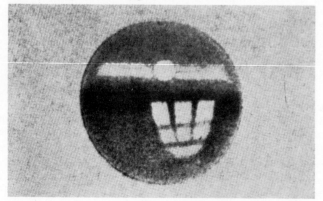

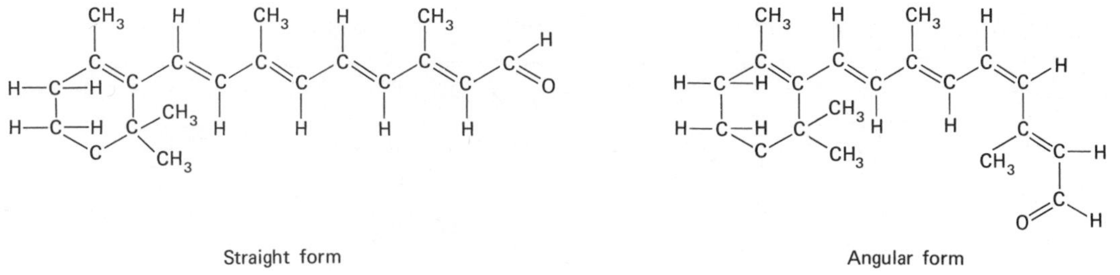

Straight form

Angular form

Figure 11-8 The two forms of retinal. Only the angular form binds with the protein opsin.

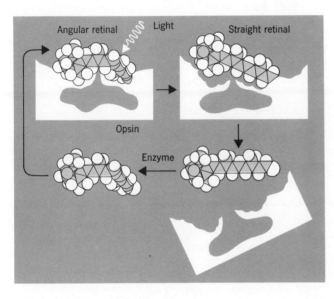

Figure 11-9 Vision depends on the bleaching of rhodopsin. Light is absorbed by a molecule of retinal, changing it from the angular to the straight form, so that it dissociates from opsin. This event bleaches the visual pigment and initiates a chain of events that generates nerve impulses in the optic nerve. The released retinal is reconverted by an enzyme to the angular form. It can then reassociate with opsin to reconstitute the visual pigment.

small initial event is amplified first on a molecular level (the bleaching steps) and then on a cellular level (the changes in the membrane system) to produce the final nerve impulse. An input of radiant energy results in a change in the electrical potential of the generating region of the sensory cell.

In other sensory systems the input may be mechanical, as in touch and hearing, or chemical, as in smell and taste. The intermediate steps are different but the final output is always the same: a generator potential and a train of impulses.

Synaptic Transmission

In most neurons, an impulse is triggered by the activity of a neighboring neuron rather than by a sensory input. The functional connection between any two neurons is called a **synapse** (Greek for "junction"). By definition, an impulse is transmitted from the **pre-synaptic cell** to the **post-synaptic cell.**

The outer membranes of the pre- and post-synaptic cells lie very close to each other. Tiny globules, or vesicles, are located just inside the membrane of the pre-synaptic cell. These pre-synaptic vesicles are membrane-bounded globules, and the chemical they contain is known as the **transmitter.** This compound is an agent that can modify the permeability of the post-synaptic membrane.

Nerve impulses are transmitted at the synapse by the release of the transmitter chemical. When a nerve impulse arrives at the pre-synaptic membrane, transmitter is simultaneously released from hundreds of vesicles in the pre-synaptic cell. The transmitter then diffuses across the narrow gap between the cells, the **synaptic cleft,** to the post-synaptic membrane (see Figure 11-10).

Acetylcholine is the chemical transmitter at many synapses. When a nerve impulse reaches the synaptic vesicles, a few thousand molecules of acetylcholine are released from each one. They diffuse

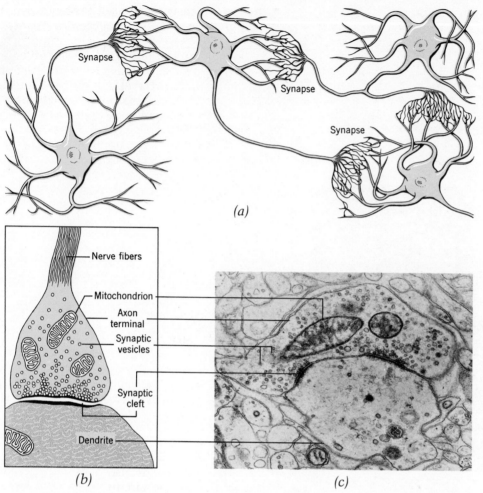

(a)

Nerve fibers

Mitochondrion

Axon terminal

Synaptic vesicles

Synaptic cleft

Dendrite

(b)

(c)

Figure 11-10 *The synapse.*
(a) A drawing of several
nerve cells, showing syn-
apses between axon
branches and dendrites. (b)
The presynaptic ending of
the axon contains mi-
tochondria and numerous
vesicles of the chemical
transmitter. (c) An electron
micrograph of a synapse.

across the synaptic cleft to the membrane of the post-synaptic cell. Since the cleft is only 200 or 300 Angstroms wide, this diffusion is almost instantaneous. Acetylcholine causes an increase in the post-synaptic cell membrane's permeabilities, and these permeability changes in turn cause the potential difference across the membrane to decrease. This drop initiates a nerve impulse in the post-synaptic cell.

Nerve impulse transmission is a one-way process. The vesicles exist only on the pre-synaptic side of the synapse, and thus a nerve impulse can be transmitted in only one direction. An impulse may arrive at the post-synaptic membrane from elsewhere in the cell, but there are no vesicles or transmitter agents there, so it cannot be transmitted.

The acetylcholine liberated by a nerve impulse cannot remain in the system very long. It must be

eliminated, or the post-synaptic cell would produce a continuous chain of nerve impulses. An enzyme called **acetylcholine-esterase** in the post-synaptic membrane splits the acetylcholine molecule and inactivates it. In some synapses, acetylcholine is destroyed rather slowly, and these cells consequently fire repeatedly in response to a single incoming impulse. Drugs that interfere with the destruction of the transmitter agent have a similar effect. Acetylcholine is called an excitatory agent since it tends to produce an impulse across the synapse.

Some transmitters are inhibitory agents: they actually decrease the probability that a post-synaptic cell will fire. An inhibitory transmitter may work by increasing the permeability of the post-synaptic membrane to chloride ions. This "chloride membrane" has an inside potential that is negative with respect to the outside. If inhibitory and excitatory transmitter agents arrive at a synapse at about the same time, the effects will counterbalance, and the membrane voltage will not decrease enough to fire the cell. The action of inhibitory transmitters should not be confused with that of acetylcholine-esterase. Acetylcholine-esterase allows the post-synaptic cell to fire, but keeps it from firing repeatedly. In the case of inhibition, the post-synaptic cell does not fire at all. As we shall see, inhibitory influences are an important part of nervous system function.

Conduction of the Nerve Impulse

We have now seen how a nerve impulse originates and how it is transmitted from cell to cell. The conduction of the nerve impulse down the axon has also been studied extensively. The usual means is to produce a nerve impulse artificially with an electrode, and then to examine it as it moves down the axon.

When a negative electrode is placed near the surface membrane of an axon, current flows from the negative electrode toward the positively-charged membrane surface. This current reduces the membrane potential at the electrode. That portion of the membrane becomes slightly less positive than the neighboring regions of the membrane—it is, in effect, negative with respect to them. At the same time, the membrane becomes more permeable to sodium. It remains in this state as long as the outside voltage is applied through the negative electrode (Figure 11-11a).

If we increase the strength of the current flow through the negative electrode, the membrane potential is finally reduced to the **threshold** point. At that level the change in membrane potential becomes an explosive discharge; it is no longer under our control. The electrical charge on the membrane reverses polarity, becoming negative on the outside with respect to the interior. The membrane is now **depolarized.** Its sodium permeability increases drastically. The threshold electrical stimulus has produced a nerve impulse, also called an **action potential** (Figure 11-11b).

This local disturbance can stimulate adjacent regions of the neuron. Since the active region has reversed polarity, it is negative with respect to an adjacent, unexcited region. Current again flows, and the potential of the inactive region drops. The neighboring area reaches the threshold, and it also becomes active. In this way, excitation is conducted to successive regions of the neuron, until all the membrane eventually becomes involved in sequence (Figure 11-11c). The nerve impulse is conducted from one end of the cell to the other.

If the negative electrode is placed near the middle of an axon, impulses travel in both directions to the ends of the cell. The same thing happens normally. Nerve impulses arise near the junction between the cell body and the axon and then sweep over the cell in both directions. As the impulse travels down the cell, the recovery process begins at its trailing edge. Current flows between active and inactive regions at the trailing edge as well as at the leading edge (Figure 11-11 d-e). For a time, the recovering membrane cannot be stimulated anew by a negative electrode; in this state it is said to be **refractory.** If it were not so, the entire membrane might remain depolarized. The neuron would be unable to conduct subsequent impulses. Instead, however, the nerve impulse is a discrete wave of negative polarization. It sweeps down the length of the cell, lasting only a few thousandths of a second in any region. The permeability changes associated with the nerve impulse do cause some Na^+ ions to enter and K^+ to leave the neuron. These ions are eventually restored to the opposite side of the membrane by the membrane's active transport system.

The rates at which nerve impulses move are determined by local current flow. Anything that increases the electrical conductance of the medium surrounding the cell also increases the speed of the nerve impulse. Current also must flow inside the cell to complete the circuit. Thus, the larger the diameter of the cell, the easier the pathway for electrical flow and the more rapid the speed of the impulse. Nerve impulses in a typical **motor** nerve (that is, a nerve that sends impulses to a muscle) may travel twenty meters per second. In other words, assuming that the active state lasts 5/1000 of a second, about ten centimeters of the nerve will be in the active state at any given time as the impulse passes down the length of the cell.

The character of the impulse is entirely determined by the membrane properties of the cell, including its size and conduction rate. It doesn't matter how the impulse is generated—an electrode or a blowtorch will produce an identical one. The constant character of the nerve impulse is described by the term **all-or-none.** Many analogies have been used to convey this idea. For example, the conduction of the impulse is compared to the burning of a train of gunpowder: it burns in the same way and at the same rate whether it is ignited by a match or by a bolt of lightning.

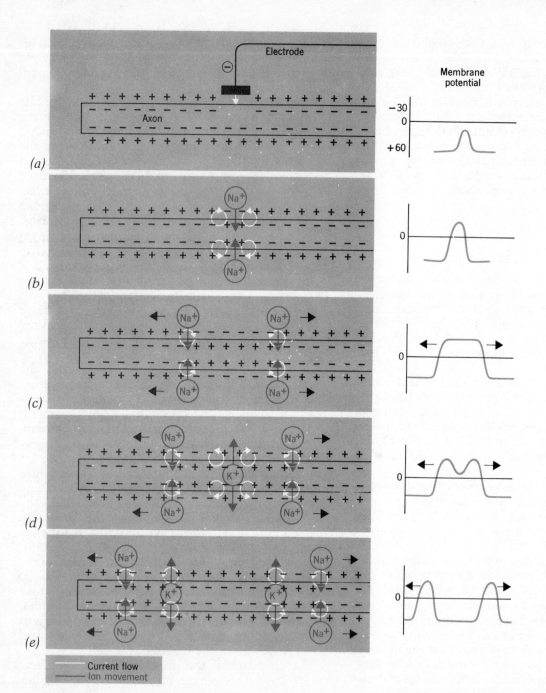

Figure 11-11 The nerve impulse. (a) An impulse can be initiated in an axon by reducing the voltage across the membrane. A negative electrode is placed against the surface. (b) As the voltage across the membrane decreases, the membrane's permeability to Na^+ increases. When the voltage reaches the threshold level, the increase in Na^+ permeability occurs with explosive speed. The membrane potential is changed so that the outside is negative with respect to the inside; this region is now excited and is in the active state. (c) The excited region of the axon membrane itself behaves like the negative electrode, producing a current flow and exciting adjacent regions of the axon. (d) The state of excitation spreads in both directions. The explosive increase in Na^+ permeability is self-limiting, and the original low permeability is soon restored at the trailing edge of the impulse. The restoration is accompanied by a temporary increase in the permeability of the membrane to K^+ ions. (e) The impulse travels in both directions from the point of initiation.

These discoveries gave rise to several theories about how the nervous system works. One proposal compared the nervous system to a digital computer. The all-or-none character of the neuron response—the way it either fires or does not—is similar to the behavior of tubes and transistors in a computer. The basic computer circuits are either closed or open, on or off. If neurons must also be either "on" or "off," then why not think of the nervous system as a digital computer?

Following this reasoning, scientists have created elaborate digital circuits in an attempt to mimic the functions of the nervous system. The model systems contain a code for the intensity of inputs, complicated but stable interconnections to a central digital array of "neurons," and a final coded output of nerve impulses that represent the instructions to appropriate muscles. These analyses were stimulated by the rapid development of digital computer technology during World War II. Unfortunately, they have not been very successful. Digital analogies do work when applied to simple reflex units isolated from the rest of the organism. But they cannot account for the behavior of the organism as a whole. Intensity coding and all-or-none behavior are characteristic of individual neurons, but other aspects of the transmission and control of nerve impulses are more complex.

Some synapses are simple relays. Whenever the pre-synaptic cell fires, an impulse is generated in the post-synaptic cell. But in other junctions, an incoming impulse produces a whole chain of impulses in the post-synaptic cell. Sometimes only a rapid train of input impulses can cause the post-synaptic cell to fire. In other cases, input activity from several pre-synaptic cells is required for transmission of the nerve impulse. The behavior of a particular synapse depends on the numbers of the excitatory and inhibitory endings on the post-synaptic cells, the positioning of these endings, and the rate at which transmitters are inactivated. These features are not necessarily stable. The behavior of a particular junction may depend on whether it has recently been active, the general condition of the body, and other factors. In principle, we could predict the behavior of such a junction if we had sufficient information. But in fact, the behavior of a synapse is predictable in only a few instances. We cannot assume that synapses are like the simple and constant digital circuits in a computer.

The synapses between neurons are also complicated structurally. The axon of a single cell may branch, contributing pre-synaptic endings to many synapses. Alternatively, a single neuron may receive pre-synaptic endings from many neurons. A single neuron may excite many other neurons, and input stimuli from many cells may converge on a single output neuron. In fact, every neuron in an organism is probably connected, at least indirectly, to every other neuron. These interconnections mean that a single incoming stimulus would mobilize the entire nervous system if the original impulse were not inhibited at some synapses.

The vital importance of such inhibition can be illustrated with a simple experiment. **Strychnine** is a drug that interferes with the action of the inhibitory transmitters. If a small amount of it is injected into a frog, there is no effect as long as the animal is protected from input stimuli. But if the animal is lightly stimulated with just a puff of air, all its major body muscles go into violent contraction. The animal becomes rigid and finally dies. The inhibition of its nervous activity has failed, its nervous system fires massively, and all coordination is lost. Many drugs that produce dramatic changes in emotional state, perception, and orientation act by modifying the action of transmitter agents.

muscle contraction

Neurons receive inputs from the environment either through sensory receptors or through adjacent neurons. In either case, the input is coded in the form of one or more nerve impulses and is conducted along the axon. The major *output* from the nervous system is muscle activity. In fact, many forms of animal behavior are simply patterns of muscle contraction.

When a nerve impulse arrives at a muscle, the state of excitation is transmitted to the muscle cell at a special junction called the **neuromuscular junction** (Figure 11-12). The mechanism is comparable to that of transmission at a synapse. The excited muscle cell generates its own impulse, which sweeps along its length just like a nerve impulse. The changes in sodium and potassium permeability are also similar. The muscle action potential then activates the contractile apparatus, and the muscle cell shortens. Muscle contraction is incredibly fast; less than a tenth of a second may elapse between the arrival of the impulse at the neuromuscular junction and the completion of a contraction.

A muscle is constructed of several different tissues and there are several types of muscle. **Skeletal**

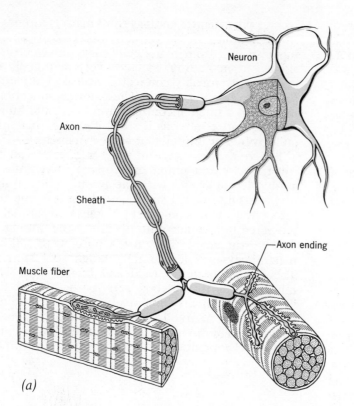

Neuron

Axon

Sheath

Muscle fiber

Axon ending

(a)

muscles, sometimes called **voluntary muscles,** are attached to the skeleton and are responsible for movements of the whole body. Muscle cells in skeletal muscle are arranged in bundles bound together by connective tissue; a single muscle is composed of many bundles. Connective tissue surrounds the whole muscle and also forms **tendons,** tough cords that connect the muscles to bones. The tendons are attached across joints in the skeleton, so that a shortening of the muscle produces motion at the joint. Muscles often come in sets arranged on opposite sides of a joint; the contraction of one set of muscles produces motion in one direction and also lengthens the opposing set (its **antagonists**). Contraction by the antagonists produces motion in the opposite direction (Figure 11-13).

Another type of muscle, **smooth muscle,** is formed by spindle-shaped muscle cells that occur in sheets. Smooth muscle is found in layers in the digestive tract, the uterus, the urinary bladder, arteries, and elsewhere. Smooth muscle contracts slowly. A third type of muscle, **heart muscle,** is formed by a network of branching fibers. But all

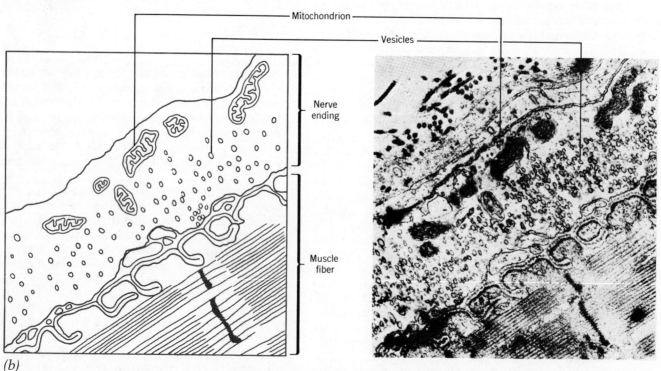

Mitochondrion

Vesicles

Nerve ending

Muscle fiber

(b)

Figure 11-12 The neuromuscular junction. (a) A motor neuron and its neuromuscular junctions. (b) Electron micrograph of a neuromuscular junction.

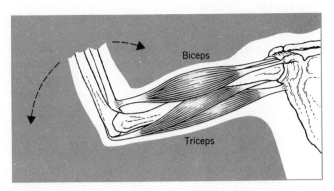

Figure 11-13 (left) *Antagonist muscles. In the forearm, flexion is produced by contraction of the biceps, extension by contraction of the triceps.*

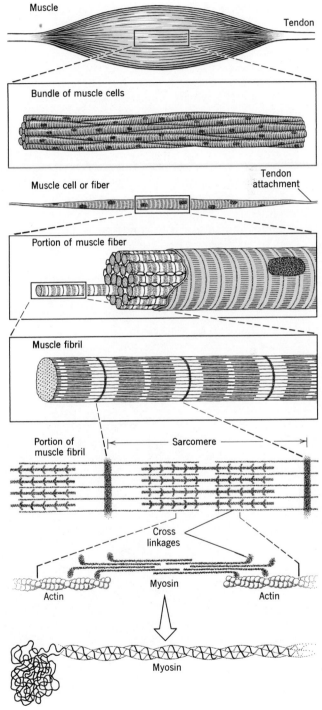

muscles are comparable in overall function: they all contract, thereby exerting a force, or tension. When relaxed they can be passively lengthened.

Individual muscle cells are called **muscle fibers.** A single muscle fiber in a skeletal muscle contains many nuclei and appears to be banded along its length. Minute banded strands, or **fibrils,** run lengthwise down the cell. A closer look with an electron microscope reveals that the apparent cross-banding is produced by a regular arrangement of light and dark areas on the parallel strands of muscle filaments within the fibril. The filaments are made of protein, either **actin** or **myosin.** Pairs of actin and myosin filaments are fitted together like the teeth of a zipper. They are not continuous throughout the length of a fibril but are arranged in short functional units called **sarcomeres.** The basic scheme bears repeating: muscles are composed of muscle cells (or fibers), muscle cells are composed of muscle fibrils, and muscle fibrils are composed of protein filaments (see Figure 11-14).

Figure 11-14 *Structure of skeletal muscle. A whole muscle contains bundles of muscle cells, or fibers. A single muscle fiber is a multinucleate cell that contains many muscle fibrils. A single muscle fibril is composed of parallel protein filaments. Filaments are composed of the proteins actin and myosin and are organized into repeating units called sarcomeres.*

The contraction of the muscle is produced by a sliding movement of the actin and myosin filaments to "close the zipper." Parallel filaments are linked by chemical bonds that hold them in place. In contraction, the old chemical linkages are broken, the filaments slide to a new, more compact position, and chemical bonds are reformed in the new locations. This process shortens the muscle fibrils and the muscle cell that contains them, and therefore the whole muscle contracts. Actually, muscle contraction takes vastly longer to explain than to accomplish. Just think of the rate at which actin and myosin filaments must be coupling and uncoupling in the legs of a gazelle, or the beating wings of a hummingbird.

The energy for muscle contraction is supplied by ATP. The actin filaments associate very tightly with ATP. Myosin, in addition to forming the filaments that interlock with actin, also acts as a powerful ATPase; that is, it splits ATP to release phosphate bond energy. This energy is used to break and reform the bonds between the actin and myosin filaments.

Muscle contraction is actually controlled by calcium ions; myosin acts as ATPase only in their presence. In the resting state, muscle cells contain little free calcium. But calcium is localized in a network of fine tubes, called the **sarcoplasmic reticulum,** which penetrates throughout the muscle cell. When a muscle action potential sweeps along the cell, it is conducted from the cell surface to the inside through another system of tubules, the **T tubules** (Figure 11-15). This action potential causes the movement of calcium ions from the sarcoplasmic reticulum into the region of the protein filaments. Myosin can then split the ATP to release the energy required for muscle contraction.

Two kinds of muscle contraction occur in animals. If one or both ends of the muscle cell can move, the cell will shorten. But if both its ends are held in fixed positions, its pull (or **tension**) increases, though its length does not change. The former kind of contraction is **isotonic contraction** (*iso* means "the same," and *tonic* means "tension") because muscle tension remains constant during the change in length. The latter type of contraction is **isometric contraction** (*metric* meaning "distance"). Isometric exercises have become quite popular because they are easy to do and require no special equipment. Both kinds of contraction are brief events, followed by relaxation.

Physiologically, muscle relaxation depends on the inactivation of myosin ATPase through the return of calcium ions to the sarcoplasmic reticulum.

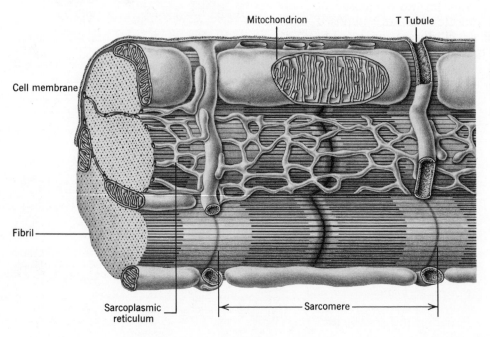

Figure 11-15 *The sarcoplasmic reticulum contains calcium ions, which flow to the protein filaments when a action potential sweeps down the system of T tubules.*

Mitochondrion   T Tubule

Cell membrane

Fibril

Sarcoplasmic reticulum

Sarcomere

The withdrawal of calcium also requires energy, since the ions have to be moved against a concentration gradient. After the muscle cell relaxes, it can be passively stretched. In fact, muscle can return to its rest length only if an outside force is applied. This force may be supplied by the contraction of an antagonist muscle, for instance, or by gravity.

When a nerve impulse stimulates a single muscle cell, the resulting contraction is called a **twitch.** Whole muscles may also twitch. This process can be demonstrated with the nerves and muscles of the leg of a frog. The large sciatic nerve is often used to stimulate the calf muscle. The sciatic nerve contains hundreds of separate axons leading to thousands of muscle cells. If a single nerve impulse is sent down the leg's sciatic nerve, it produces simultaneous impulses in all the motor axons. The impulses cause a twitch — that is, the simultaneous contraction of many muscle cells. However, neither man nor frog moves by twitching. The simultaneous contraction of all the muscle cells would not allow for coordinated movements. Muscle contraction has to be controlled.

The strength of a muscle contraction depends on the number of impulses arriving at the muscle cells at any time. Each motor axon branches to supply several muscle cells. If only a single axon is active the muscle exerts only a small force. If greater muscle tension is required, more axons come into play and more muscle cells are activated. The frog muscle behaves in this way: as the strength of a stimulation of the sciatic nerve is increased, the size of the action potential of the whole nerve increases as more individual axons are stimulated. As a result, the strength of the twitch increases (see Figure 11-16).

The coordination of muscle contraction depends on the frequency of the nerve impulses arriving at the muscle. A single impulse produces one twitch contraction. When the muscle is stimulated by a train of low-frequency nerve impulses, a new twitch is initiated before the first has completely decayed. The result is a new tension increase superimposed on the end of the first contraction. New twitch contractions will be generated as long as new impulses arrive from the nerve. If the nerve impulses arrive even faster, the individual twitches can no longer be discerned, so that muscle tension increases smoothly. This occurrence is called a **tetanic contraction;** its tension is much greater than the tension produced by a single twitch contraction. Increasing the frequency of the inputs along the motor nerve converts a twitch into a smooth con-

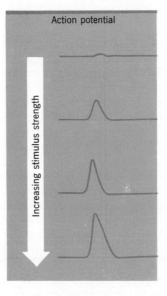

Figure 11-16 (left) As the strength of the stimulus is increased, the action potential in the nerve increases, because more axons are conducting impulses. The axons stimulate more muscle fibers, and the strength of the muscle contraction is increased.

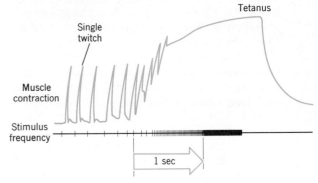

Figure 11-17 At slow rates of nerve stimulation, a series of muscle twitches is produced. As the frequency of stimulation is increased, the contractile mechanism is reactivated before relaxation can occur. Finally, a smooth, sustained contraction, called a tetanic contraction is attained.

traction of the muscle (see Figure 11-17).

Muscle contraction can be adjusted in strength by recruiting additional cells in the muscle and by modifying the frequency of impulses arriving from motor axons. The action and control of other kinds of muscle follow the same basic principles, but with differences related to the specific function of the muscle. For example, it is impossible to produce tetanic contraction in heart muscle. This arrangement has obvious adaptive value, since a continuously contracted heart can't do much pumping.

The initiation of a nerve impulse (through sensory input or synaptic transmission), the conduction of the impulse, and the stimulation of muscle contraction together make up the sequence known as the **reflex.** The reflex is the basic behavioral sequence in the nervous system. The simplest reflex is produced by a **reflex arc,** consisting of the following elements: a sensory receptor nerve cell, which has a junction with a second nerve cell; the second, or post-junction nerve cell; and a muscle cell (Figure 11-18). With modifications, the reflex arc is the basic physiological unit of all reflexes.

When the doctor taps below your knee with a rubber mallet, he is testing a reflex called the "knee jerk." When the mallet strikes a tendon attached to the lower end of the large thigh muscle, the muscle is slightly stretched. The quick stretch causes **stretch receptor** cells to generate a train of nerve impulses. These impulses flow along axons to your spinal cord, where they synapse with motor nerves. The impulses cross the synapse to the motor nerves and are conducted back to the muscle. The impulses cause the muscle to contract, and your foot swings forward (Figure 11-19). Such a reflex is called a **postural reflex** because it helps to maintain a particular posture by preventing changes in muscle length. If you are standing and begin to sway slightly, postural reflexes will cause the appropriate muscles to contract and check your movement.

The knee-jerk reflex arc is a single-junction reflex. That is, the axon of the stretch receptor synapses directly with the motor nerve that leads back to the muscle. However, the overall response to the doctor's mallet is actually much broader. The stretch receptor axons branch and synapse with other nerve cells as well. Some branches activate other muscles in your leg, which assist in the foot-swinging movement. Other axon branches activate intermediate neurons that produce inhibitory impulses. These fibers keep antagonist muscles on the back of your thigh relaxed while the opposing set contracts. Thus, the initiation of the simple knee jerk reflex involves changes in all the muscles of the lower body. Still other branches send impulses along your spinal cord toward the brain. This last circuit may involve higher regions of the central nervous system.

Voluntary movements may also use reflex pathways. In their case, however, the stretch receptors are activated by a different pathway. Stretch

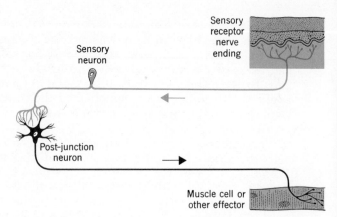

Figure 11-18 A simple reflex arc. The receptor synapses directly with a motor neuron that stimulates a muscle fiber.

receptors contain muscle-cell elements internally. In voluntary movement, a set of nerves activates these internal elements and causes them to contract. The internal contraction causes the stretch receptors to increase their firing rate, so the muscle contracts by way of the reflex (Figure 11-20). Voluntary movement is usually produced in this way.

This system seems unnecessarily indirect and complicated until we consider what would happen if muscle contraction were activated directly. Normally, the stretch receptors are firing more or less continuously to maintain postural muscle tone. If the muscle were to contract directly, however, the firing rate of the receptors would suddenly and drastically drop, as the muscle contraction removed the stretch. Thus, until the receptors could adjust their length to the new posture, muscle tension would also drop sharply. Movements would be rough and uncoordinated. But by routing voluntary movements through the reflex arc, muscle tension is maintained, and movement is smooth and controlled.

Reflex behavior, represented by the knee-jerk reflex, is called **spinal behavior,** because the neural circuits loop through the spinal cord and may be independent of the brain. In one sense, these nerve–muscle circuits have little adaptability or variability. Responses such as the knee jerk are stereotyped. But in another sense, spinal behavior is astonishingly complex. Fine adjustments in muscle tension and coordination, such as those involved in walking, are controlled almost entirely at the spinal level. Complicated behavior can occur in

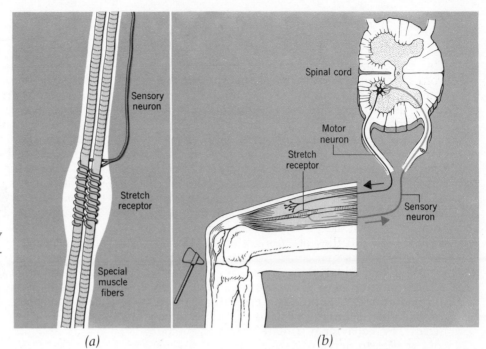

Figure 11-19 Postural reflexes. (a) Stretch receptors embedded in muscles control movement. (b) When they are stretched (in this case by tapping the muscle tendon with a mallet), they initiate a burst of impulses. The impulses travel to the spinal cord and activate motor neurons that in turn activate the contraction of muscle fibers.

(a)

(b)

the complete absence of higher-level inputs. If the brain of a frog is removed or destroyed, leaving the spinal cord intact, the animal will be quiescent. But if a small piece of filter paper soaked in some irritating chemical is placed on its skin, the frog will begin to make coordinated efforts to scratch away the paper. These movements continue until the source of irritation is removed.

Such autonomous behavior in a brainless animal is quite consistent with the computer model of behavior. A sensory input enters a circuit, is processed, and is transformed into an appropriate muscular output. At this level, the elements involved in the behavior, such as receptors and motor neurons, can be identified and isolated. Collectively these isolated functions account for the complexity of the behavior of the brainless frog.

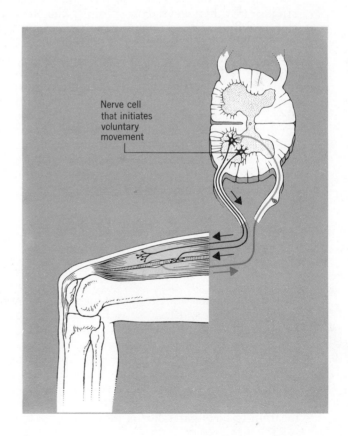

Figure 11-20 Voluntary movement of skeletal muscle. Motor nerves from the spinal cord initiate contraction in special contractile fibers in the stretch receptors. This contraction produces an increase in firing by the sensory portion of the receptor, and the muscle fibers are activated through the reflex arc.

Organization of the Nervous System

The preceeding description of neurons is drawn from experiments with isolated neurons and with small groups of neurons. These studies have allowed us to discuss the nature and transmission of the nerve impulse, and to understand the mechanics of simple interactions between neurons and muscles. Certain kinds of behavior, such as postural reflexes, are well explained by this kind of information. And the individual neurons almost certainly display similar behavior in the whole animal. Yet the nervous systems of most animals are so complex that it has thus far been impossible to trace the overall flow of information, or to understand the factors that control this flow, at least in terms of individual neuron activity. The normal cellular environment of the nervous system must have additional chemical and electrical influences on individual neurons that have not yet been recognized. The behavior of isolated neurons undoubtedly applies to nervous system function in general terms, but it is only part of the picture.

Complex behavior involving the brain is usually studied in other ways. For example, the information flow out of the brain of an intact animal may be traced through the electrical activity of large groups, or **tracts,** of nerve fibers, rather than single axons. In other cases, brain function is studied by analyzing its electrical activity. Surgical removal or electrical stimulation of portions of the brain are also important tools. In still other studies, the animal is treated as a "black box"; that is, the inputs and outputs of its nervous system are measured but no attempt is made to trace the course of the intervening nervous activity. This technique is common in studies of learning and memory.

cephalization

The organization of the nervous system reflects the fact that most animals move. An animal's orientation during movement is not random. Whether it glides, swims, flies, or walks, one surface is oriented downward (the **ventral** surface) while the other faces upward (the **dorsal** surface). During locomotion, the head is carried foremost, at the **anterior** end, while the other regions of the body trail behind toward the **posterior** end. Anterior, posterior, dorsam, and ventram are terms that describe the location of body parts during movement.

The organization of the nervous system likewise reflects the nature of the animal's motion. An animal is more interested in where it is going than where it has been. Its sense organs and neurons therefore are concentrated in the regions that first encounter the environment, the anterior end. As animals evolved and became more and more complex in their structure and activities, the concentration of neurons in the anterior end became more pronounced. This trend is called **cephalization** —literally, "head formation."

Inactive animals such as clams and sea anemones are exceptions to this rule. Their nervous systems are diffuse and netlike. They lack any concentration of neurons that could be called a brain. As a famous neurophysiologist once put it, motor behavior is "the cradle of the mind."

The Central Nervous System

Most animals have a central nervous system; that is, they have concentrations of neurons in the head and bundles of axons that extend toward the posterior end. In man and related animals, the central nervous system consists of a hollow, segmented tube of nervous tissue. The tube extends from the head, where it has expanded into a brain, back through the body, where it becomes the spinal cord. The spinal cord is enclosed in hollow bone segments called **vertebrae,** which together comprise the **vertebral column.** Any animal having a vertebral column is termed a **vertebrate.** Each vertebra in the spinal column represents a segment of the nervous system. In the brain, the segments are somewhat obscured because the walls of the tube have expanded and folded over on each other as cephalization evolved. The segmented nature of the nervous system reflects the overall organization of vertebrates: other vertebrate organ systems as well are segmented.

Spinal Cord Structure

Each segment in the spinal cord receives and gives off a pair of **spinal nerves.** Each spinal nerve divides to form a **dorsal root** and a **ventral root** just as it enters the spinal cord. The dorsal root contains sensory axons that send impulses to the spinal cord, and the ventral root contains motor nerve fibers that activate muscles.

The spinal cord has a central core of grey matter and, around it, an outer layer of white matter. **Grey matter** consists largely of the cell bodies of motor neurons that synapse directly with receptors (stretch receptors, pain receptors, and so forth). In each segment the cell bodies of the receptors themselves lie outside the spinal cord in a cluster called the **dorsal root ganglion.** (A ganglion is any mass of nerve cells outside the brain and spinal cord.) Grey matter also contains other neurons that make connections between sensory inputs and motor outputs in spinal reflex behavior.

White matter is composed of nerve cell axons that run to and from the brain. The white color is due to a fatty insulation around the axons. Axons carrying incoming sensory information, and axons from intermediate reflex neurons in grey matter, ascend toward the brain by way of the white matter. Descending axons leave the white matter to synapse with the motor and intermediate neurons of the grey matter. Thus, the cells and nerve fibers in the spinal cord provide ascending information about reflex events, and they also provide a channel for central information to descend and modify local, spinal responses (see Figure 11-21).

Brain Structure

The human brain contains ten billion neurons and even more interneuron connections. These cells are organized into anatomical and functional groups, but the arrangement is complicated, and the task of trying to sort out pathways is formidable. A simple account of brain structure describes three regions, ascending from the spinal cord toward the forehead: **hindbrain, midbrain,** and **forebrain.**

The hindbrain consists of two segments, the medulla oblongata and the cerebellum. The **medulla oblongata** is the junction of the spinal cord and the brain. It has the same tubular organization as the spinal cord, but the nervous tissue has become thickened into lateral columns and the roof of the tube is a thin layer of non-nervous tissue. The neurons in the medulla are concerned with temperature regulation, respiration, and other homeostatic mechanisms. Above the medulla lies the **cerebellum.** It is a conspicuous folded mass of nervous tissue that controls and coordinates posture and balance. It receives inputs of nerves that ascend the spinal cord from receptors for stretch, position, and pain; from the inner ear it receives nerve signals concerned with balance.

The midbrain has two main areas. The dorsal region, the **corpora quadrigemini,** receives sensory

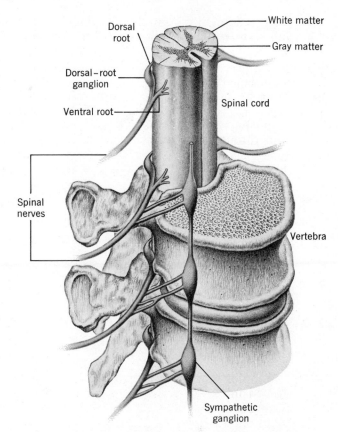

Figure 11-21 *The spinal cord and vertebrae. Spinal nerves are arranged segmentally, and they connect to the spinal cord between the vertebrae. Sensory axons enter through the dorsal root, and motor axons exit through the ventral root of the spinal nerves. The sympathetic ganglia contain cell bodies of neurons belonging to the autonomic nervous system.*

flow from the spinal cord and coordinates it with inputs from the eye and ear. The ventral region, or **tegmentum,** is involved with the coordination of movement.

The forebrain is divided into two major regions: the **diencephalon,** which lies centrally and deep in the forebrain, and the cerebral hemispheres, also called the **cerebral cortex.** The floor of the diencephalon is the **hypothalamus.** It receives sensory and motor inputs from the viscera; electrical stimulation of particular centers in the hypothalamus produces eating, drinking, salivation, urination, defecation, vomiting, or even more complicated responses such as defensive or escape movements.

The midbrain is quite small in higher vertebrates, and brain structure is dominated by the large cerebral cortex (see Figure 11-22).

Brain Function

Neurons in the brain are organized into clusters of nerve cell bodies, called **nuclei.** Nerve nuclei communicate with one another and with the spinal cord through bundles of axons called **tracts.** The functions of particular nuclei in the brain can be studied in several ways. Tissue destruction and electrical stimulation are two techniques that are commonly used. Neurons can be destroyed surgically, by cutting out regions of the brain, or they can be killed by heat. In the latter case, an electrode is placed in the brain and its tip is heated to "cook" the cells. The feeding and satiety centers in the hypothalamus were identified and localized with

(a)

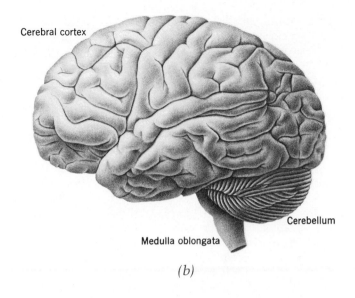

Cerebral cortex

Cerebellum

Medulla oblongata

(b)

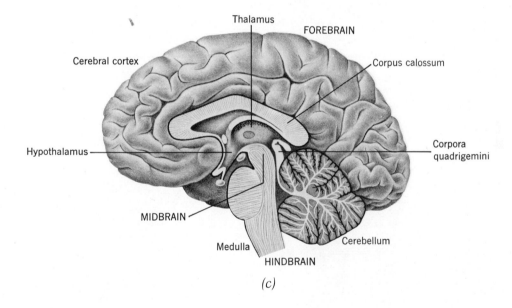

Thalamus

FOREBRAIN

Cerebral cortex

Corpus calossum

Hypothalamus

Corpora
quadrigemini

MIDBRAIN

Medulla

Cerebellum

HINDBRAIN

(c)

Figure 11-22   The vertebrate brain. (a) The evolution of the major groups of vertebrates involved tremendous changes in the relative sizes of the forebrain, midbrain, and hindbrain. The most striking development is the enormous expansion of the cerebrum. (b) The intact human brain shows a continuation of this development. (c) The major regions of the human brain are shown in longitudinal section.

these techniques. The destruction of specific nuclei in the hypothalamus produces a cessation of eating. The destruction of other areas produces excessive eating and obesity. The first nucleus was called the "feeding center," because it seemed to initiate the organized behavioral acts concerned with food intake. Its function was later confirmed with electrical stimulation by electrodes inserted in the nucleus. Electrical stimulation caused a quiescent animal to begin eating. Such a feeding center would also, presumably, show considerable nervous activity when an animal is eating, and this prediction was also verified.

The function of particular nerve tracts in the brain can be studied in several ways. One common method again depends on electrical stimulation. An animal is anaesthetized to reduce its nervous activity to a minimum. Then a particular tract is located and stimulated electrically. Recording electrodes are moved along the tract, tracing the course of electrical activity. An alternative procedure depends on the fact that when an axon is cut, its fatty sheath—as well as the axon itself—degenerates. A degenerating axon and sheath are easily located in cross sections of the brain because they strongly absorb certain chemical dyes. Thus, the individual axons in a tract can be followed by locating particular stained, degenerating axons.

Assiduous application of all these techniques has produced a tremendous body of descriptive information about the location of various centers and about their interconnections. These approaches have produced some particularly interesting information about the cerebral cortex. Sensory inputs, such as those from the eye or ear, have been traced electrically along nerve tracts to specific areas in the cerebral cortex. These regions, collectively known as the **sensory cortex**, have been located and mapped in detail. In addition, there have been many experiments in which the cortex of a lightly anaesthetized animal has been stimulated directly at various points to produce motor responses. These regions belong to the **motor cortex** (see Figure 11-23). But when all such results are combined, large regions of the cerebral hemispheres still remain unidentified. In the anaesthetized animal

Figure 11-23 The sensory and motor portions of the cerebral cortex, shown in a cross section. The parts of the body of the cartoon homunculus correspond to the relevant areas of the cortex. Note the relative prominence of the face and hands.

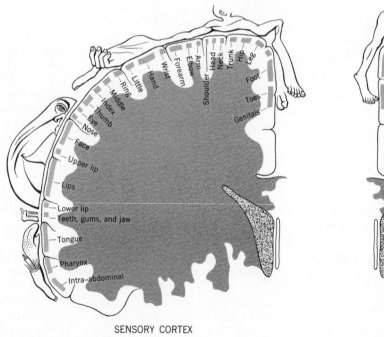

SENSORY CORTEX

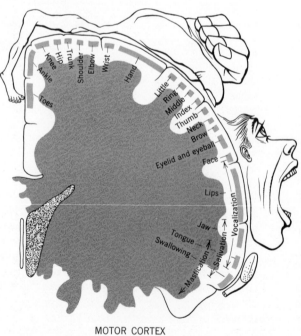

MOTOR CORTEX

these regions do not respond to sensory inputs; direct stimulation of them produces no consistent responses in the animal. These so-called "silent" areas of the cortex appear to have no simple connections to input signals or to output messages. Yet, they are not really silent, or inactive; electrical activity can be detected in these areas in the normal animal.

The stimulation of some areas in the brain can produce predictable and rigidly defined behavior. Electrical stimulation of the feeding center will produce food intake even in a satiated animal; stimulation of the satiety center will stop a hungry animal almost in mid-bite. Stimulation of the centers for erection and copulation produces performances beyond adolescent fantasy. Stimulation of a "pleasure" center in rats also produces extreme behavior: a rat would rather press a lever that delivers stimulation to its pleasure center than press a lever that delivers food. In fact, the animal will starve rather than interrupt brain stimulation. A spectacular example of such stereotyped responses was demonstrated by a neurophysiologist, J. M. R. Delgado. He implanted electrodes in the brain of a fighting bull. Delgado then stood in the ring while the bull charged, armed only with a remote control apparatus that stimulated the electrodes. When the charging bull bore down on him he pushed the button. The bull stopped short, became amiable, and wandered away.

Such rigid and predictable behavior is spectacular, but it is highly unnatural and not at all representative of normal brain function. Normal inputs from the environment to the whole animal rarely produce mechanical and stereotyped responses—at least not in higher animals. Neither mice nor men are reflex machines. As the poet puts it, "If sex were all, then every trembling hand could make us squeak, like dolls, the wished-for words." An adequate analysis of brain function must include the factors that modify and alter rigid responses to make them reasonably appropriate to environmental circumstances. Any theory of brain function must allow for real or apparent spontaneity. We do not behave as reflex machines, and we must account for the so-called "higher functions," such as choice, plasticity, spontaneity, learning, and thought.

The first modern account of brain function attributed such higher functions to the cerebral cortex. As we saw, an anaesthetized animal shows clear pathways or tracts ascending into various regions of the brain. It was assumed that sensory input from the body (such as pain, touch, or posi-

tion) was fed into the brain through the tracts of the spinal cord, while sensory information from the head (from the eye, ear, or nose, for instance) entered the brain directly through cranial nerves. Presumably this input all passed to the cerebral cortex, where it was processed, transformed, and ultimately re-emitted as signals to muscles. It was believed that the whole process of integration and coordination occurred in the cerebral cortex. These assumptions were natural enough, for two reasons. First, the cortex is more highly developed in man than in any other animal, so it was tempting to associate it with complex higher nervous functions. Second, major regions of the cortex—the silent areas—had no obvious function. Thus, they seemed likely to be the centers of such capacities as thought and memory and choice.

This account, persuasive as it was, was nevertheless quite wrong. It is now known that the input relays along the spinal tracts and in nerves from the eyes, nose, and ears, are not mechanical and automatic. In fact, most of the information moving up the spinal cord toward the central nervous system is screened out before it ever reaches the brain. This fact has been clearly demonstrated by several experiments. For instance, it has been found that in quiescent animals, impulses generated at lower levels of the spinal cord simply disappear at higher relay stations. Only in anaesthetized animals, whose central nervous system is incapable of normal activity, does the spinal cord relay all information passively.

Inputs from the ear are screened in the same way. This process was observed in a cat that was prepared so that impulses along the tract leading from the auditory nerve to the brain could be recorded. When a click was sounded in the cat's presence, a burst of nerve impulses was picked up on the nerve tract. But if the clicking continued, nerve activity decreased and finally disappeared; auditory information was no longer getting through to the brain. The initial response to clicks also disappeared if a mouse was presented to the cat. These results agree with common experience. You will usually accommodate to a repeated signal until finally you simply don't hear it. And when something "interesting" happens in the environment, you focus attention on it and screen out other inputs.

Such experiments show that the brain does not simply accept and process a mass of sensory inflow. Rather, it selectively inhibits inputs. Every input channel that has been studied possesses some kind of central inhibitory pathway. Perhaps the question "How does the brain process so much information?" is inappropriate. A better question to ask is, "How does any information get through to the brain?"

Certain sensory inputs not only get through inhibitory relays, but they actually pass through focusing systems that activate and stimulate brain responses. One important activating system is a mass of loosely organized nerve cell bodies in the central hindbrain and midbrain, called the **reticular formation.** It gets its name because it is netlike, lacking discrete nuclei. Sensory relay columns that pass from the spinal cord to the brain contribute some nerve fibers to the reticular formation, while other fibers go directly to the brain. The sensory impulses that do enter the reticular formation eventually arrive at the cortex, where they produce electrical changes and responsive behavior. The animal is aroused. Electrical stimulation of the activating system of the reticular formation will arouse a sleeping animal. But if this area is destroyed, the animal behaves as if it is permanently asleep, no matter what sensory inputs are used. The destruction of the reticular formation does not actually prevent sensory inflow from reaching the brain by the direct route. Recording electrodes in the sensory cortex verify that direct connections do function. But nothing happens behaviorally. Apparently, the brain ignores the input unless it is aroused.

Brain function does not rely on passive circuitry to feed all inputs directly into the controlling cortex. Instead, many inputs are screened out at the periphery of the nervous system. Others are focused and projected up to the cortex.

the peripheral nervous system

Behavior is actually caused by the activity of muscle and gland cells, in response to signals from the nervous system. These cells are called **effectors,** a general name for cells whose activity produces an effect on the internal or external environment. The CNS communicates with the effectors by means of the **peripheral nervous system** (PNS), which includes all the neurons and nerve fibers that lie outside the CNS.

We have already discussed part of the peripheral nervous system: namely, the motor fibers that control one kind of effector, the skeletal muscles.

These nerve fibers, along with fibers from sensory receptors, are a subdivision of the peripheral nervous system called the **somatic nervous system.** Sensory fibers and motor fibers in the somatic system synapse in the spinal cord, forming reflex arcs. In these reflexes, the body of the nerve cell that activates the skeletal muscle lies in the spinal cord. Only the cell's axon, running from the spinal cord to the muscle, is in the PNS.

Another group of effectors is activated by a separate part of the PNS, the **autonomic nervous system** (ANS). The ANS controls smooth muscles, and glands such as the pancreas, tear ducts, and salivary glands. The sensory receptors that are concerned with regulating the smooth muscle and gland effectors also form reflex arcs in the spinal cord, much like the reflex arcs involved in the control of skeletal muscles. But in the ANS, the nerves leading from the spinal cord to the effectors synapse with cells in a series of interconnected ganglia that lie outside the CNS. The neurons in these peripheral ganglia send fibers to the effectors.

The peripheral ganglia also receive inputs from the CNS through the spinal nerves, but there is no direct flow from the CNS to the effectors: it must always pass through the peripheral ganglia of the ANS. In summary, the CNS exerts direct control over skeletal muscle through the somatic nervous system. But smooth muscle and glands are controlled only indirectly, by way of the peripheral ganglia of the ANS (see Figure 11-24).

the autonomic nervous system

Physically, the autonomic nervous system itself is divided into two sections, the **sympathetic** and the **parasympathetic** divisions. These systems differ in the location of their peripheral ganglia and in the connections between their ganglia and spinal cord. In the sympathetic division, the ganglia lie on either side of the spinal cord, at some distance from the final effectors. Sympathetic ganglia receive connections from central portions of the spinal cord. The parasympathetic division receives outflow from nerve fibers at the base of the spinal cord, the sacrum, and from nerve fibers from the hindbrain. Its external ganglia are in or near the effector organs (see Figure 11-25, page 337).

These anatomical divisions, sympathetic and parasympathetic, are approximately equivalent to

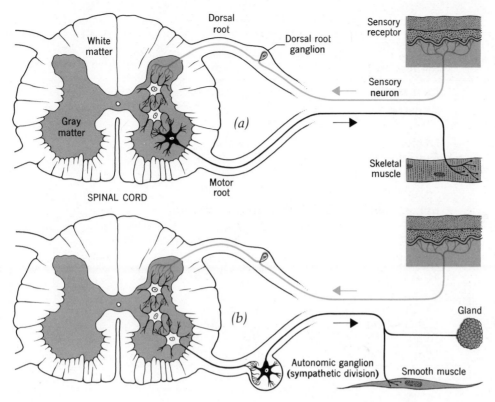

White matter

Dorsal root

Dorsal root ganglion

Gray matter

Sensory receptor

Sensory neuron

(a)

SPINAL CORD

Motor root

Skeletal muscle

(b)

Autonomic ganglion (sympathetic division)

Gland

Smooth muscle

Figure 11-24 The somatic and autonomic nervous systems. (a) Motor outflow from the somatic nervous system proceeds directly to skeletal muscle. (b) Motor outflow from the autonomic system makes a synapse with another neuron before proceeding to the effector. The sympathetic ganglia are located close to the spinal cord; the parasympathetic ganglia lie in or near the effectors.

two separate sets of coordinated functions. Broadly speaking, the sympathetic division is **ergotropic,** or "directed toward work." It supports vigorous physical activity and prepares the animal for an emergency. The parasympathetic division is **trophotropic,** "directed toward nourishment." It is concerned with regulating the internal environment in order to maintain homeostasis.

Stimulation of the sympathetic division produces a group of ergotropic responses, all of which prepare the animal for vigorous activity. The heart rate is increased, blood sugar rises, the pupil of the eye is dilated, and the eye accommodates for distant vision. Other functions that are not critical for strenuous activity are simultaneously inhibited. Sphincters contract, restricting activity in the digestive tract and bladder; the blood supply to the gut is restricted; and peristalsis slows down. Stimulation of the parasympathetic, or trophotropic, system has the reverse effect. The heart rate slows, digestive secretion and peristalsis are promoted, and nutritive and maintenance functions are supported (see Table 11-1).

These opposite responses are possible because many visceral organs receive two sets of nerves, one from the sympathetic and one from the parasympathetic division. In some cases their action is equivalent to antagonism in skeletal muscles. If you stimulate one nerve to contract your biceps, your arm will flex; stimulate another to contract the triceps, and your arm will straighten out. In the same way, one set of autonomic nerve fibers will dilate the pupil of your eye; an alternate one will produce contraction. Since two opposing sets of muscles control the size of the pupil, the arrangement is similar to that of skeletal muscle control. Some sets of autonomic fibers, however, can cause antagonistic actions in the same muscle. The heart, for example, receives two kinds of autonomic nerves. Stimulation of one set increases heart rate, and stimulation of the other decreases it. In this case, autonomic nerve impulses decrease the excitability of heart tissue by inhibiting the transmission of impulses from nerve to muscle. Some organs, however, receive inputs from one system only. Major glands of the digestive tract receive only parasympathetic signals; blood vessels in skin and muscle have only sympathetic inputs. The

Table 11-1

Actions of Autonomic Nerves in Humans

Organ or Function	Sympathetic	Parasympathetic
Heartrate	Increase	Decrease
Blood Vessels	Constriction	Dilation
Gut		
Peristalsis	Decrease	Increase
Sphincters	Contraction	Relaxation
Digestive glands	(Not supplied)	Secretion
Liver	Glycogen breakdown	Flow of bile
Eye	Iris dilation	Iris contraction
Skin		
Hair muscles	Contraction	(Not supplied)
Sweat glands	Secretion	(Not supplied)
Adrenal medulla	Secretion	(Not supplied)
Urogenital system		
Bladder sphincter	Contraction	Relaxation
Vas deferens, prostate	Contraction	(Not supplied)
Uterus (non-pregnant)	Relaxation	(Not supplied)
Uterus (pregnant)	Contraction	(Not supplied)

activity of these organs is controlled by increasing or decreasing a resting level of nervous activity.

Nerve impulses are normally present in both the sympathetic and parasympathetic systems. One system or the other may predominate at any given time, producing changes in the condition of the body. But only under emergency conditions is one system or the other totally silent.

Chemical Control of the ANS

Autonomic nerve fibers activate organs in the usual way, through nerve impulses that arrive at neuromuscular junctions. But effectors in the sympathetic division of the ANS may also be activated by hormones from the **adrenal medulla.** The adrenal medulla is the central portion of the adrenal glands, so-called because they lie close to the kidneys (*ad* meaning "near," and *renes*, "kidneys") (Figure 11-26). The adrenal medulla receives a direct nerve supply from the spinal cord. Its cells secrete products into the bloodstream, to be carried to responsive tissues and organs throughout the body.

One hormone product of the adrenal medulla was identified in the late nineteenth century and called **adrenalin. Epinephrine** is the modern name for the same compound, and the one more commonly used. A second hormone of the adrenal medulla was discovered in 1950. This compound, called **noradrenalin** or **norepinephrine,** is chemically identical with adrenaline except that it lacks a methyl group ($-CH_3$). It is also similar in function. Both these hormones stimulate effectors in the same way that the chemical transmitter agents from sympathetic nerve fibers do. The only difference is that the adrenal medulla liberates its hormones directly into the bloodstream, while the nerve fibers liberate their transmitter only at specific neuromuscular junctions.

These transmitter agents are destroyed very slowly, so modifications in membrane permeability, and therefore in the active state of the target cell, are long-lived. In skeletal muscle, the twitch response to a single stimulus is over in a tenth of a second. But a twitch response in smooth muscle stimulated by the ANS may last for a minute. This time is sufficient for some of the transmitter agent to diffuse from the neuromuscular junction. Thus, transmitter agent from the nerve fibers as well as from the adrenal medulla can activate many organs

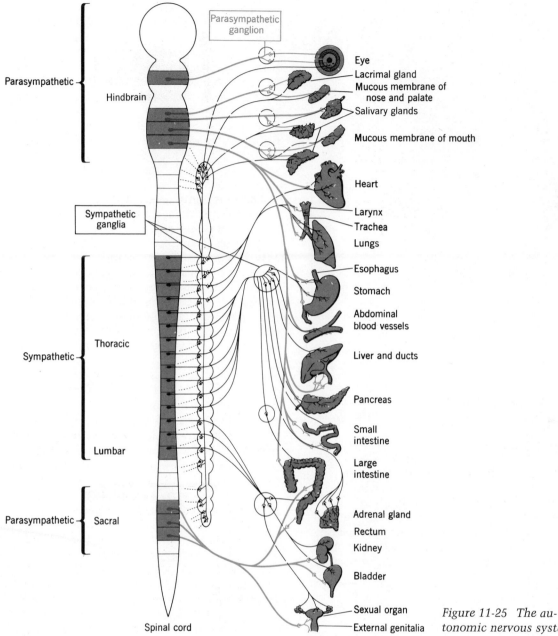

Parasympathetic

Hindbrain

Parasympathetic
ganglion

Eye

Lacrimal gland

Mucous membrane of
nose and palate

Salivary glands

Mucous membrane of mouth

Heart

Sympathetic
ganglia

Larynx

Trachea

Lungs

Esophagus

Stomach

Abdominal
blood vessels

Liver and ducts

Sympathetic

Thoracic

Pancreas

Small
intestine

Lumbar

Large
intestine

Parasympathetic

Sacral

Adrenal gland

Rectum

Kidney

Bladder

Spinal cord

Sexual organ

External genitalia

Figure 11-25 The au-
tonomic nervous system.

in the sympathetic division simultaneously.

The activation of the sympathetic system by transmitter from nerve fibers was first demonstrated in a simple experiment with dogs. Struggling, exercise, or excitement produces an increase in a dog's heart rate of about 100 beats per minute. If the autonomic nerve supply to the dog's heart is cut, the increase in heart rate is still obtained. If the

adrenal medulla is removed, the heart rate again increases, but more slowly. But if all the sympathetic ganglia are removed, the response is finally abolished. This experiment strongly suggests that something like epinephrine or norepinephrine is released by sympathetic nerve fibers (both at the heart and in the sympathetic ganglia), as well as by the cells of the adrenal medulla. It now appears that

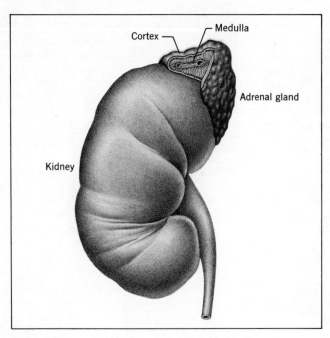

Figure 11-26   Location of the adrenal glands.

epinephrine and norepinephrine are the transmitter agents throughout the sympathetic division of the autonomic system.

The parasympathetic division, by contrast, uses the same transmitter as does skeletal muscle: acetylcholine. There is still a difference in its action in the two systems, however, since acetylcholine is destroyed very slowly in parasympathetic junctions. There is also a difference in the response of the effectors: the permeability changes produced by acetylcholine in some heart muscle cells are quite different from those it produces at neuromuscular junctions in skeletal muscle.

Central Control of the ANS

Both divisions of the ANS are strongly influenced by higher centers in the brain. Homeostatic mechanisms and emergency responses both depend on central control. If the nervous pathways from the brain are interrupted, blood pressure drops, temperature regulation fails, the bladder and gut are paralyzed, and sexual responses such as erection and ejaculation disappear. The animal is incapable of rage or escape responses.

An important higher center for coordination of

the autonomic system is the hypothalamus, on the floor of the forebrain. Stimulation of appropriate regions of the hypothalamus can affect such diverse conditions as water balance, body temperature, food intake, and digestion, and can produce emotional responses of fear, rage, or pleasure. For example, the hypothalamus controls acid secretion in the stomach through two autonomic pathways. One is a direct nervous pathway, and the other is hormonal, using the adrenal hormones. Stimulation of the hypothalamus activates both pathways, and prolonged stimulation produces excess acid secretion and ulcers.

The autonomic system is also influenced by the general level of activity in both the central and somatic nervous systems. Gastric ulcers are related to prolonged anxiety or emotional stress, presumably because of hypothalamic activity. Ulcers are sometimes regarded as psychosomatic disorders, but they might better be called hypothalamic disorders. High levels of nervous activity in the somatic system, produced by muscle tension, may stimulate the entire autonomic system. This effect can be counteracted by relaxation therapy. The relaxation of skeletal muscles reduces the inputs from muscle receptors to the brain. This reduction, in turn, decreases the ergotropic output from the hypothalamus and other brain areas. Both "physical" and "emotional" tensions are relieved. In fact, the physical and emotional effects of sympathetic nervous activity are so closely related that it is impossible to be thoroughly relaxed and enraged at the same time.

Disciplines such as meditation and yoga no doubt produce changes in ergotropic-trophotropic balances as well. Some clinical procedures in psychotherapy are simply devices to modify and to train autonomic responses. One therapeutic procedure, for example, encourages the expression of rage to help overcome fear. Fear and anxiety on the one hand, and rage and aggression on the other, have strong ergotropic components. However, the balance between epinephrine and norepinephrine secretion is quite different in the two cases, epinephrine being predominantly associated with fear and anxiety. There is evidence that this balance can be therapeutically modified. As one scientist remarked, "psychology is like a tapeworm that keeps losing segments to physiology." We should hastily remark that a good many segments remain. Still, recent advances in our knowledge of the biochemical basis of emotion are extremely interesting.

The autonomic system has been described in many terms. In part it is an emergency system:

ergotropic discharge prepares the body for "flight or fight." But as everyone knows from personal experience, the emotional changes that accompany this state of readiness are not necessarily helpful, and may even be counterproductive. So why are they present? Such emotions may simply be necessary side effects of a system that can produce emergency responses to threatening environmental conditions, but that must also provide for adjustments in homeostatic mechanisms.

Higher Functions

An animal is not a simple reflex machine. Environmental information is selectively received, processed, stored, and integrated with other stored information. Behavioral responses to this information can be direct or delayed, simple or complex; they can be modified by new information. We have already discussed some simpler aspects of information processing. For example, sensory inputs can be screened out or reinforced, depending on activity in the reticular formation and in other parallel pathways. Sensory information undergoes processing that results in a systematic distortion of the input. All of this processing assures that only "important" information reaches the brain.

Importance, of course, depends on the particular organism. Frogs are interested in flies; the visual system of the frog is organized to facilitate the recognition and capture of flies. The frog "knows" whether or not an object is a fly before the input ever reaches its brain. The visual system itself makes the decision. In higher mammals such as cats or primates, special areas of the brain cortex are involved in sensory processing.

perception

In tracing the processing of sensory inputs we will begin with a simple example of visual detection, or **perception,** of light patterns in the horseshoe crab. Significant processing and distortion occur at the eye itself. The eyes of the horseshoe crab are sheets of receptors, each one of which responds to light directly in front of it. The crab's world is therefore transformed into a pattern of dark and light spots. Such an eye is called a **mosaic eye,** because the image it produces is comparable to a picture formed with hexagonal tiles of various shades of gray.

The processing of visual information occurs in receptors in the following way. Suppose we choose several receptors in a group. When light strikes one receptor, it generates a train of nerve impulses

whose frequency depends on the brightness of the light. If a second light is then directed on several neighboring receptors, the frequency of firing by the first receptor decreases. Each receptor unit in the eye of the horseshoe crab inhibits the activity of adjacent units (see Figure 11-27).

These interactions distort the actual environmental input by emphasizing contours or discontinuities in the visual field. A very simple experiment illustrates the distortion effect. If we take a

Figure 11-27 Lateral inhibition in the eye of the horseshoe crab. (a) A light shining on a single receptor produces a train of impulses. (b) The illumination of adjacent receptors reduces the frequency of firing in the original receptor.

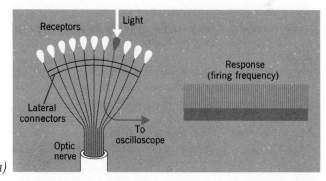

(a)

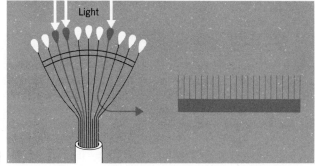

(b)

sheet of smoked glass or cellophane and overlap a second sheet over part of it, the area where the two sheets overlap is twice as dark as the single sheet. An instrument now plots the frequency of firing of a single receptor unit in the crab eye as it scans this pattern. As the metered receptor approaches the dark region, its firing frequency increases, because the receptor is released from inhibition by darkened receptors ahead. Upon entering the dark region, the receptor's firing is depressed, because it is inhibited by trailing units that are still illuminated. Instead of seeing the true physical pattern, therefore, the horseshoe crab sees a light and a dark, region separated by a dark line. This distortion of sensory input automatically emphasizes borders and edges in the environment (see Figure 11-28).

The sort of optical "illusion" we have described also occurs in human vision. But in man the analysis—or distortion—of the input signal becomes more complicated. Visual receptors, the rods and cones in the retina, are connected to a relay system of neurons that lead to the **visual cortex** on the surface of the brain. Some of the connections are arranged so that various cells in the visual cortex, called "simple" cells, receive inputs from groups of receptors in the retina. Each group covers a small elliptical area, called a **receptor field.**

Some regions within the receptor field are excitatory and some are inhibitory. When light falls on an excitatory region, the simple cell in the cortex fires. But when light also falls on an inhibitory region, the simple cell's firing is inhibited. The excitatory and inhibitory regions are arranged in distinct patterns in the receptor field. Some light patterns inhibit the firing of the cell in the cortex, and others stimulate it. For example, one elliptical receptor field responds strongly to a vertical line of light and weakly to lines in other orientations (Figure 11-29). The reason is that a vertical line of light falls on excitatory regions and not on the neighboring inhibitory regions. Other receptor fields will fire only in response to a line that is oriented and moving in a particular direction. These fields have inhibitory elements clustered on only one side of the excitatory line.

"Complex" and "hypercomplex" units in the cortex receive inputs from "simple" cells. Information originally provided by retinal receptors is analyzed further at this level. In humans as well as in horseshoe crabs, recognition of contours and of their patterns of movement is an important part of perception.

In higher animals, such contour perception develops only with experience. If animals do not

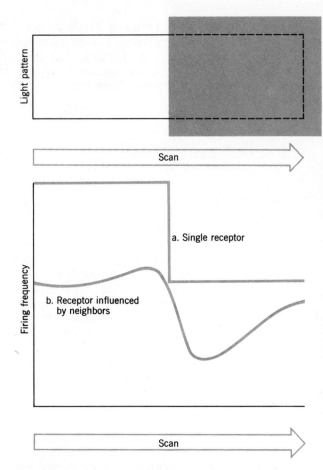

Figure 11-28 *Perception in the horseshoe crab.* (a) *If only one receptor is exposed to a light pattern that shows a sharp transition from light to dark, the receptor's firing response is directly proportional to the actual brightness of the pattern.* (b) *If neighboring receptors are exposed as well, lateral inhibition produces the bottom curve. This response emphasizes the edge between the light and dark areas.*

receive patterned visual input from birth onward, they fail to develop the "simple" and "complex" units in a normal way. This fact probably accounts for the behavior of blind individuals whose vision is surgically restored. At first they cannot recognize even very simple geometrical figures. They must "learn" to see. Other observations suggest that continual sensory input may actually be a requirement for the proper functioning of the analytic elements. If a sighted individual is deprived of normal visual input for a time with eye shields, he is disoriented when input is restored. In one experiment an indi-

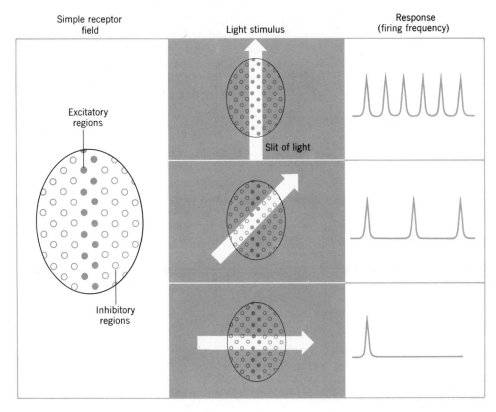

Simple receptor field

Excitatory regions

Inhibitory regions

Light stimulus

Slit of light

Response (firing frequency)

Figure 11-29 Visual receptor fields in mammals. The receptor field responds most strongly to a vertical line of light because of its arrangement of excitatory and inhibitory regions. Each region is a small group of rods and cones on the retina.

vidual was fitted with translucent eye shades that transmitted diffuse light but filtered out any light patterns. After wearing the shades for only a few hours, the subject experienced disorientation when the shades were removed. Walls appeared to tilt, corners were not square, the floor was too close or too far away.

A simple demonstration will illustrate the complexity of visual perception. If you look at a lamp and then press the side of your eye with your finger, the lamp's image will jump. Next, look at the lamp for a few seconds and then close your eyes; you will see an after image of the lamp. But if you now press the side of your eye with your finger, the image does not jump as you might expect. This suggests that the after image is more than a simple continuing response of your retina. Rather, it represents an image of the world which is constructed centrally in the brain. It does not respond to manipulation of the receptor itself—the pressing of your eye.

This demonstration is an example of **reafferenz,** a concept that regards perception as an indirect, secondary inspection of an abstraction of the original sensory input. Energy is first transduced into a series of nerve impulses, the impulses are then interpreted centrally in the brain, and finally they are reprojected for inspection and interpretation. Unusual inputs to a normal individual may be distorted and transformed until they fit into a previously-established pattern. This transformation may happen at the receptor itself, or at other points during relay to the brain.

Adelbert Ames used this characteristic to create optical illusions with oddly shaped rooms. He arranged distance cues so that the rooms appear normal when viewed through a peephole. However, when people are included in the room their size relations are all wrong. One person may appear nearly three times as tall as another (see Figure 11-30). Yet even after the illusion has been explained, the observer cannot reject this impossible situation and reevaluate the angles of the walls and floors; the room remains square, and the sizes remain distorted.

Interpretation takes place in the cortex, and it is subject to revision and modification based on experience. That is why we can learn to interpret unfa-

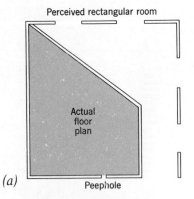

(a)

(b)

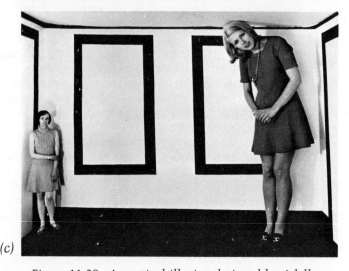

(c)

Figure 11-30 *An optical illusion designed by Adelbert Ames.* (a) *The actual shape of the room is quite irregular.* (b) *Yet when it is viewed from one specific angle it appears rectangular.* (c) *The room still appears rectangular even when human figures stand in it; the size of the figures is distorted by the room.*

miliar inputs. The learning process may be slow, however, and subject to error. In many ways, perception is an act, not a passive event. The active participation of the organism is an intrinsic part of its perception of the environment.

In summary, then, the organism is not a passive transducer for translation of energy input into motor output. Its attention can be withheld or focused. Its receptor inputs are transformed and distorted to fit functional requirements. The active distortion of inputs, and the reprojection of the transformed inputs, may produce an indirect and inaccurate view of the world. It is impossible to "just look at what's there."

learning and memory

Learning and memory are brain functions that require more complex information processing. **Memory** can be viewed as the part of learning that involves information storage and retrieval. **Learning** can be defined as changes in behavior that result from encounters with the environment. In learning, information from the environment is processed, integrated with previously stored information, and stored in a modified form. The modified information produces changes in behavior. The capacity to learn from experience is one of the most important adaptive characteristics of animals. We cannot yet give a complete account of information coding, sorting, integrating, and retrieving, but we know quite a bit about it.

Conditioning

Learning processes can be carefully examined only if the investigator has a way to control and measure the amount of learning that occurs. Most studies use one of two procedures to measure learning: **classical,** or **Pavlovian, conditioning,** and **instrumental conditioning.** The term "conditioning" simply refers to some kind of training or learning.

In classical conditioning, two stimuli, or signals, are used. One is called an **unconditioned stimulus** because it is one that already produces a particular response in an untrained animal. For example, the presentation of food produces salivation in dogs. Pavlov took this unconditioned stimulus as his starting point. Then the experimenter chooses a second stimulus which at first does not produce this response; Pavlov used the ringing of a bell. A training trial or learning period follows, consisting of successive presentation of the two stimuli. In this case the experimenter rings the bell just before presenting food to the dog. After several training trials, the bell alone produces salivation. The ringing of the bell has become a **conditioned stimulus,** and a **conditioned response** has been established. If the experimenter continues to present only the conditioned stimulus (the ringing of the bell), without the unconditioned stimulus (food), the salivation response drops off again and finally disappears. The number of learning trials required to produce a conditioned response is a measure of an animal's learning speed. A great deal of ingenuity has been used in classical conditioning, but it can be used only to study learning behavior that has an existing stimulus–response pattern, such as salivation or muscle reflexes. In such experiments the animal is usually passive.

Instrumental conditioning requires the animal to make an active response. A particular response may be encouraged by rewarding the animal in some way. For example, an animal may be given food or water when it reaches a specific region in a maze. This method is called **positive reinforcement.** Alternatively, the animal can be subjected to an unpleasant stimulus—an electric shock or a loud noise, for instance—to teach it to avoid certain areas of a cage or maze. This method is **negative reinforcement.**

Instrumental conditioning is more flexible than classical conditioning. Active behavior can be shaped by reinforcement in remarkable ways. Suppose that you want to teach a dog to climb a wall (a skeptical journalist did in fact propose this task). The experimenter begins with a hungry dog and allows him to move around a walled area. When he moves toward a particular wall, the experimenter rewards him with a small food pellet. The animal quickly learns to move and jump toward the wall to obtain more food. As stronger and stronger effort is demanded to obtain the reward, the dog is leaping several feet up the wall within ten or fifteen minutes.

The **Skinner box,** invented by B. F. Skinner, is a mechanical application of the same principles; it is widely used in instrumental conditioning. The Skinner box contains a lever. When it is pressed by an animal the event is recorded, and a reinforcement is automatically dispensed in the form of food, water, electrical stimulation, or some other

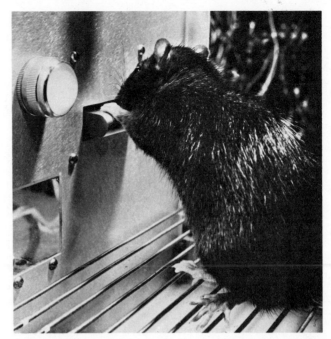

Figure 11-31 *A rat in a Skinner box. Pressing the bar delivers food or another reward, according to a preset schedule of reinforcement.*

reward (see Figure 11-31). The experimenter can adjust the lever mechanism so that a lever press delivers food every time, every third time, on a random schedule, or on whatever basis he wishes. Through such experiments, instrumental conditioning has produced a large body of information about how learning is related to reinforcement. Conditioning techniques allow us to identify and measure learning.

Psychologists often use classical and instrumental conditioning to study learning speed, for example, or to compare the learning speeds of various tasks. But they are not concerned with the changes going on inside the organism. In most psychological investigations the animal is treated as simply a "black box," an entity of unknown structure whose abilities are revealed during the experimental sessions. The box receives certain inputs; it generates certain outputs. The outputs are measured, and the relation between the two is analyzed. (The phrase "black box" actually refers to a procedure in which a student is given a box with some terminals and is asked to deduce what electronic components are inside.) But there is no reason to use only a black-box approach to learning, even though opening the box may be difficult and confusing. Biologists are trying to understand the components within the black box. Their goal is to find connections between the characteristics of the learning process, as discovered by psychology, and the known biological properties of the organism. In other words, how do the inputs produce the outputs?

Most theories of learning assume that changes in behavior are accompanied by some physical change in the nervous system. Learning often occurs slowly; the ability to perform a task may improve gradually with repeated trials. This delay probably reflects the gradual establishment of a physical change in the nervous system. The general term for this change is **engram** (derived from "engrave"). Most biological studies of learning are aimed at answering the following questions about the engram: what is the physical nature of the engram? how rapidly is the engram formed and stored? where is the engram stored in the animal?

What is the Engram?

We know that functional capacities of cells depend on their chemical structure, and that most cellular activities are directed by proteins. This role naturally makes proteins suitable candidates for the engram. Furthermore, the synthesis of proteins is in many ways analogous to the learning process. Learning is a response to the environment, and protein synthesis also responds to environmental influences, in the sense that the nucleic acids that direct protein synthesis can be turned on and off. Once the nucleic acids are turned on, the protein product is synthesized. Also, learned behavior persists for various periods, just as the lifetime of protein in a cell is variable. And finally, we know that proteins are associated with cell membranes and can modify their properties. Presumably, newly synthesized proteins could change the behavior of neurons dramatically.

These analogies are merely suggestive, but they have provided a working hypothesis for investigations of the physical nature of the engram. The likelihood that protein is crucial is supported by the general relationship between protein synthesis and the development of the nervous system: it is known, for instance, that brain development is irreversibly stunted by protein insufficiency during infancy and early childhood.

Many biologists have attempted to study the relations between protein synthesis and learning more directly. No simple summary of this work is possible, because the results are complex and often ambiguous. But some brief examples will illustrate the present state of our knowledge.

Inhibition of Protein Synthesis

Certain chemical agents inhibit or stop protein synthesis. When these agents are injected into the brain of an animal, the animal cannot learn new tasks, though it still recalls older patterns of learned behavior. A number of drugs that interfere with protein synthesis have been employed, but the experimenters have not been able to prove that interference with memory is actually caused by the interruption of protein synthesis and not by other effects of the drugs.

Passive Transfer of Learning in Planarians

Planarians are simple worms that eat meat, including other planarians. One group of planarians was classically conditioned to learn a task, and then ground up and fed to a group of untrained planarians. This second group then learned the same task faster than a control group permitted to ingest untrained planarians. RNA extracted from trained animals and injected into untrained worms also improved their performance. These observations have been confirmed several times, but some workers deny that planarians can be classically conditioned at all. Many factors seem to influence the rate of learning, and some of them are difficult to control. At present, we must reserve judgment concerning the meaning of these studies.

Passive Transfer of Learning in Rats

Rats were trained in a variety of tasks, brain extracts were prepared from them, and the extracts were injected into untrained rats to see if learning was facilitated. Several laboratories conducted these experiments, and about half had positive results. A comparison of the results is difficult, however, because different methods were used. The passive transfer of learning in higher animals has not yet been studied in any standard and repeatable manner.

This discussion will be out of date very shortly. The relationships between learning and protein synthesis are so tantalizing that many laboratories are pursuing them, even though the work is technically difficult.

How Rapidly is the Engram Formed and Stored?

When does learning become memory? Memory storage has been studied extensively with a one-trial form of instrumental conditioning. An animal is taught a very simple task in a single learning trial. For example, a rat is placed on a low platform. As he attempts to step off, he receives a mild foot-shock. In subsequent trials, the rat will stay on the platform for a longer time. A mouse placed on a brightly lit platform will quickly enter a hole in the wall behind the platform, where he receives a foot-shock. Both rat and mouse learn to stay on the platform. The amount of learning that occurs in a single trial is measured by recording the time it takes the animal to leave the platform. The longer the time, or **latency,** for the response, the more learning has occurred in the single trial.

A typical experiment compares the behavior of several groups of mice. Each mouse receives a single learning trial on the platform apparatus. Twenty-four hours later, they receive another trial. The mice will hesitate for a longer time on the second trial—that is, they have an increased latency—if there is no other treatment. These mice form a control group. Other groups of mice may receive additional experimental treatment before the second trial. Their latency times are then compared with those of the control group to see if the experimental treatment has any effect on learning.

One group, for example, is anaesthetized with ether immediately after the first trial. This treatment has no apparent effect; their latency is the same as that of the control group. Another group receives an electrical shock strong enough to produce muscular convulsions one minute after the learning trial. When the shocked animals are tested 24 hours later, their latency is the same as that for an *untrained* group. In other words, no learning has occurred. Yet the convulsions themselves are not necessary to prevent learning. When anaesthetized mice are given such brain stimulation, the anaesthesia prevents overt muscular convulsions, yet these animals still show no evidence of having learned the task. Brain stimulation, not convulsion, has produced **retrograde amnesia**—a task that was learned prior to the shock has been forgotten. In other words, brain shock has interfered with the storage of an engram.

Instrumental conditioning and electroconvulsive shock can be used to study the time course of memory storage. That is, how rapidly is learned behavior stored in memory? Convulsive shock right after the training trial produces amnesia. When learning trial and shock are separated by an hour, retrograde amnesia still occurs. This result shows that memory storage is still proceeding even after one hour. But a longer separation of trial and shock, such as 24 hours, produces no amnesia.

The learning process itself, however, appears to be complete after one hour. This timing is shown by the performance of animals on a second trial. At first, as the time between the initial learning and the second trial is increased, their performance steadily improves (in other words, the latency lengthens). But it does not improve beyond an hour; at that point, in other words, learning is complete. This experiment demonstrates that the establishment of behavior or learning, and the storage of learned behavior in the memory, are separate processes. The former process is called **short-term memory;** the slower storage process is **long-term memory.**

Long-term memory can be disturbed in several ways, in addition to convulsive shock. Potassium chloride (which interferes with the ionic basis of the nerve impulses), sedative drugs such as barbiturates, and inhibitors of protein synthesis all may prevent the establishment of long-term memory. Humans who have suffered head injuries have retrograde amnesia and cannot remember events that occurred prior to the injury. As we have seen, long-term memory is susceptible to disturbance for a long time after the learning trial.

By the same token, it should also be possible to facilitate the establishment of long-term memory. Several drugs in fact do improve long-term memory. A small dose of strychnine sulphate, for example, improves memory if it is administered shortly after the learning trial.

Where is the Engram Stored? Where is Memory?

Brain size in mammals increases as one moves from primitive mammals toward the primates, the order to which man belongs. The cerebral cortex in particular increases dramatically (see Figure 11-22). In man, the cerebral cortex dwarfs more ancient brain areas. At the same time, the "mind," particularly with reference to learning ability and memory, is also highly developed in man. Furthermore, much of the cerebral cortex has no clearly assignable function. Neurophysiologists have shown that large areas of the cortex are "silent," that is, they are not clearly associated with sensory or motor functions. Thus, it was tempting to assume that the highly developed mind of man is in the highly developed cortex.

These arguments are not sufficiently convincing, however, and many efforts have been made to localize "mind"—and particularly memory—more precisely. The psychologist K. S. Lashley conducted a famous series of investigations "in search of the engram." He trained animals to run mazes and then surgically removed portions of their cortices. After each animal recovered, he retested its ability to run the maze to see if the removal of the cortex had caused a loss of memory. The animals proved to be surprisingly adaptable. When the brain damage actually interfered with some physical movements involved in the task, they would substitute other movements and still reach the goal. In other words, if the animal could not make a left turn, it would substitute a 270-degree right turn. (This result makes it impossible to think of learning as only a fixation of particular neuromuscular patterns.)

The removal of "silent" portions of the cortex did not produce any specific change in behavior. However, the animals didn't learn or remember quite as well. Lashley suggested that the memory and learning ability of a rat is related to the quantity of cortex it has, rather than to any specific areas. The engram had not been localized.

More recent experiments help explain why Lashley and others had difficulty in locating the engram. New work shows that a task learned by one half, or **hemisphere,** of the brain is transferred to the other half. Such a transference was shown in experiments in which animals were given tasks that depend on visual discrimination. The optic nerve sends nerve fibers to the visual cortex in both halves of the brain. At one point, called the **optic chiasma,** fibers from each eye cross over and go to the opposite half of the cortex. If the optic chiasma is cut, a visual input from one eye can reach only half of the brain directly. However, the brain hemispheres themselves communicate through a large C-shaped tract called the **corpus callosum.** As a result, if an animal with a severed optic chiasma is taught a task with one eye covered, it still performs it perfectly well with the other eye covered. Visual input from the uncovered eye reaches one half of the cortex, and then the information is passed to the other half through the corpus callosum. Input from the other eye to the second cortical hemisphere produces identical behavior.

When the same experiment is performed with an animal with a split brain—that is, one in which both the optic chiasma and the corpus callosum are se-

vered—the task is again readily learned. But this time, when the first eye is covered and the animal is forced to use the other one, it behaves like an untrained control animal. The engram has been stored in only one hemisphere (see Figure 11-32). Under normal conditions, the engram is stored in at least two locations.

Recent work on engram localization uses electrical recordings from single nerve cells in the brain. Learning, defined as changes in electrical activity, was found to occur in several brain regions when a stimulus was paired with presentation of food. Such "learning points" were found in the cortex but were also located in at least three other regions of the brain. In several areas no "learning points" could be found.

These experiments show that we should no longer refer to a single engram. Instead, we should look for many. If memory is stored in many locations, then Lashley's results make more sense. The removal of a particular portion of the cortex would impair memory in general, since it eliminates sites for engram storage. However, specific memories of learned behavior are stored in other areas as well and would not be lost.

Range of Learning

Learning is usually equated with conscious behavior. In reality, this attitude is simply cultural prejudice, reflecting our emphasis on rationality, free will, and other qualities attributed to the human mind. Learning is actually a more general phenomenon. It can occur at the reflex level, for example, without any intervention of the brain; the synapses in the spinal cord can change their response patterns after repeated stimulation. There is also evidence that learning can occur in the autonomic nervous system.

The name "autonomic" reflects the earlier conviction that this portion of the nervous system was isolated from environmental influences. In fact, however, the ANS does respond to the environment, and many learned autonomic responses are known. The autonomic system regulates smooth muscle and glandular secretion. It modifies heart rate, changes blood circulation by contracting the arteries, controls gut contractions, and is generally concerned with homeostatic function. Yet animals can learn to control some of these processes.

An animal can be trained to increase or decrease its heart rate, for example, to obtain a reward or to avoid punishment. Still, it is difficult to prove that

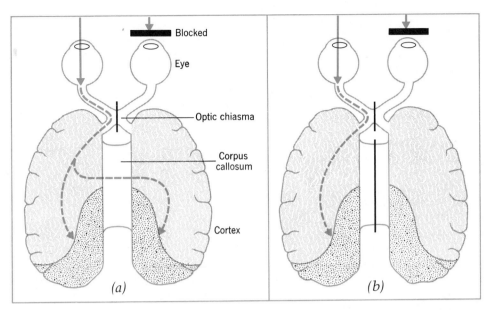

Figure 11-32 *Learning in a split-brain animal. (a) When the optic chiasma is severed, information obtained from one eye is still stored in both brain hemispheres because it can pass across the corpus callosum. (b) When both the optic chiasma and the corpus callosum are severed, a task learned using one eye is stored in only one hemisphere. If the animal is presented with the task and forced to use the other eye, it behaves like an untrained animal.*

the ANS itself produced this behavior. We have to be sure that the animal is not really using skeletal muscles controlled directly by the CNS to produce the desired effect. For example, skeletal muscle tension can initiate a chain of physiological reactions that lead to an increase in heart rate. To minimize difficulties with skeletal muscle, therefore, the rats are anaesthetized with curare (arrowhead poison). Curare produces paralysis by blocking the passage of nerve impulses across neuromuscular junctions to skeletal muscle. While they are paralyzed the rats cannot breathe, so they must be kept alive with artificial respiration. Nor can they eat or drink. Therefore, direct brain stimulation in the pleasure center, rather than food of drink, is used as a reward. As a further complication, the anticipation of the reward or punishment might itself produce the effect in question. The anticipation of an electric brain stimulation, for example, might cause the heart rate to increase. To insure that the stimulation does not produce the effects directly, therefore, some groups of animals are trained to increase their heart rates while alternate groups are

trained to decrease them.

The results of such experiments are spectacular. Learning proceeds rapidly. In fact, anaesthetized rats learn faster than normal animals. The paralysis of skeletal muscle not only prevents false responses, but it also reduces the "noise level," the number of unrelated inputs. As a result, the animal is not distracted from its task. Training follows the lines of instrumental conditioning. The heart rate is monitored, and each time an increase (or decrease) in rate is noted, the animal is rewarded. Subsequently, larger and larger changes in rate are required before the reward is provided. After only a 90-minute training session, changes in heart rate of plus or minus twenty percent of normal are obtained.

Several autonomic responses have been conditioned in this way. They include change in heart rate (with no change in blood pressure), change in blood pressure (with no change in heart rate), and changes in salivation, intestinal contraction, and rate of urine production. Animals were even conditioned to dilate the blood vessels in their left or right ears, but not both (see Figure 11-33). These accomplishments are all very specific learned responses.

Some evidence suggests that humans can also learn autonomic responses, but it is not yet conclusive. It seems likely that humans would be just as capable as rats in this regard. The occurrence of such learning at the autonomic level may have many important consequences for human behavior. Psychosomatic problems in some individuals may actually be learned responses. A stressful situation, such as going to school or giving a speech, may produce a variety of autonomic symptoms such as pallor or an upset stomach. If the individual is rewarded for such behavior by escaping the stress—by being permitted to remain at home "ill"—he may learn this response and incorporate it into his repertoire. This behavior never becomes conscious. But in fact such psychosomatic symptoms can become very debilitating, and the individual may consciously wish to be rid of them.

Even certain kinds of homeostatic regulation may also represent learned behavior. We may learn to control changes in blood sugar levels, body water content, or other aspects of the internal environment that need to be carefully regulated. This idea may seem outlandish, but it does account for the fact that many homeostatic responses develop slowly. Temperature regulation in young mammals, for instance, is quite poor. It would be interesting to know if the ability to regulate temperature improves at the same rate in animals kept at a constant temperature as in animals exposed to a fluctuating environment.

The ability of humans to learn at the autonomic level has not been proved unambiguously. The control of heart rate, changes in circulation, and brain waves can be learned, but in humans there remains the possibility that these changes may actually be

Figure 11-33 An organism can learn to control many "involuntary," autonomic responses. This graph shows the progress of a rat trained to dilate the capillaries in one ear and not the other.

accomplished through skeletal responses. Human volunteers cannot readily be subjected to deep paralysis in order to eliminate the possibility. For practical purposes, anyhow, the distinction would not be important as long as the effects were the same.

Nevertheless, the distinction is important to our analysis of organisms and the way they interact with their environments. As we have repeatedly emphasized, the organism interacts with the environment as a whole unit, and the ANS is an integral part of that interaction. From this perspective, the idea that the autonomic nervous system can show learned responses is not surprising. In fact, learning may be a basic property of all cells. Animals such as amoebae, which lack nervous systems, can still learn. The elaboration of the nervous system in higher animals merely provides more ways for an organism to interact. But the overall aim of the interaction, the preservation of the integrity of the organism, is the same in both amoeba and man. From this viewpoint, consciousness and rationality are simply valuable adjuncts to other forms of behavior. They increase the ability of the organism to interact with its environment.

Varieties of Consciousness

The more complex forms of animal behavior—perception, learning, memory—all occur in an animal that is awake. An animal that is alert receives and responds to sensory input. But animals normally experience many other states of arousal. In addition, humans have learned to alter their own consciousness with drugs.

sleep

Sleep is a state characterized by muscular relaxation, decreased responsiveness to environmental inputs, lowered heart rate, blood pressure, and respiration, and loss of consciousness. We spend about a third of our lives asleep, yet we know astonishingly little about this state. We are quite ignorant of its function and really cannot improve much on Shakespeare's judgment that it knits up the ravelled sleeve of care. Yet we do know some things. Sleep includes a variety of arousal states, just as wakefulness ranges from playing football to meditation. Sleep is necessary. Animals deteriorate and finally die when deprived of sleep. Sleep is more than the absence of wakefulness: it is a highly organized and complex state.

Most of our information about sleep comes from a technique called **electroencephalography.** An electroencephalograph (EEG) is a device that records changes in brain waves. In humans, recordings are made from electrodes placed externally on the skull. In animal experimentation, electrodes are sometimes attached or implanted in the skull. **Brain waves** are changes in electrical potential of a few microvolts, which result from the coordinated and rhythmic firing of nerve cells in the brain cortex. EEG records show the frequency and size of these rhythmic electrical potentials.

The frequency and size characteristics of an EEG can be correlated with observed or reported behavior. The EEG of an individual concentrating on a task is characterized by high-frequency, spiky electrical activity (15 to 50 peaks per second). In an individual who is awake but resting, this pattern is replaced by **alpha waves** (8 to 12 peaks per second).

The onset of sleep is accompanied by an orderly sequence of changes in the EEG. First, the alpha waves flatten out. Then they are gradually replaced by **delta waves,** slow, large waves with a frequency of about one per second. Arbitrary divisions in this sequence, designated Stages I through IV, are defined according to the "depth" of sleep. An individual in Stage I sleep is easily awakened, but a Stage-IV sleeper is aroused only with difficulty (see Figure 11-34).

A sleeping individual does not stay at Stage IV from the time he enters it until he awakens. Instead, several times during the course of a seven- or eight-hour sleep, totalling about 90 minutes in all, his EEG takes on the appearance of Stage I sleep and the rhythmic potential changes become small. However, during these periods, a new pattern of spiky potentials emerges. These spiky brain wave patterns can be correlated with rapid movements of the eye. For this reason this state is called **REM (rapid-eye-movement) sleep.** Though the overall EEG is characteristic of rather light Stage I sleep, an individual in REM sleep is difficult to arouse. Rapid eye movements never occur in Stage I at the onset of a sleep period. Only after "deep" sleep (Stage IV) has been achieved—then and only then—does the

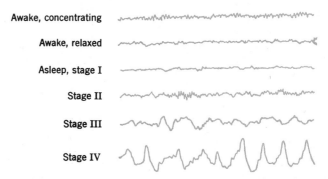

Figure 11-34 *EEG records of brain waves during various states of arousal. An alert state is characterized by rapid, spiky brain waves. These waves are replaced by slower alpha waves during relaxation or the early stages of sleep. Deep sleep is characterized by large, slow delta waves.*

EEG shift toward an apparent Stage I to produce episodes of REM sleep.

The REM stage is strongly correlated with dreaming. An individual awakened from REM sleep reports that he has been dreaming; awakened from Stage IV he does not. This observation made it possible to investigate the effects of "dream deprivation." Every time a group of sleepers entered the REM state they were awakened. Control subjects were awakened from "deep" sleep an equal number of times. The controls showed no ill effects. But the subjects deprived of dream time became increasingly irritable and unable to carry on normal activities. As the experiment proceeded they entered the REM state more and more frequently and had to be awakened more often to prevent dreaming. When the experiment was terminated, the subjects showed a rebound; that is, they spent considerably more time in REM sleep than the normal 90 minutes, until they had made up the loss.

These initial observations were made about twenty years ago. Prior to these discoveries, sleep was considered a comatose state like the one that could be produced artificially by destroying parts of the reticular formation. But we can no longer think of sleep as simply the absence of arousal. Sleep is complex and organized, and it involves considerable activity in selected portions of the neuromuscular system. Other physiological changes besides eye movements accompany REM sleep, including alterations in the heart rate, blood pressure, blood chemistry, muscle tone, and muscle activity. REM sleep in man, for example, is characterized by extreme relaxation of the neck muscles.

Vitalists and Mechanists

Our discussion assumes that all animal behavior, from knee-jerk reflex to learning, results from the activity of the nervous system, and that such activity is reducible to physical and chemical laws. However, many people object to the idea that animal behavior—and human behavior in particular—is reducible to physical principles. They feel that to describe behavior in physical terms is to demystify and belittle man's nature. "Vitalists" maintain that life contains a vital element that cannot possibly be explained by physical science. They concede that the remainder of the physical world, including other "physical" aspects of man, is describable and predictable, but human behavior is an exception. The vitalist accepts physical explanations for most of his world, but exempts his own behavior.

Individuals with an alternative view willingly accept physical explanations of human behavior. These people are "mechanists." Mechanists feel there is more than enough complexity and uncertainty in a detailed physical description of behavior to provide beauty and wonder. But they also believe that the world is orderly, that prediction is possible, and that the human animal is part of the physical world.

Of course, the question of whether or not man and his universe are totally reducible to this kind of mechanistic analysis cannot ever be resolved. No matter how crude the current physical and chemical account of the world, the mechanist can always say that his account is true "in principle." Yet no matter how sophisticated his account, an opponent can always postulate yet another level of analytical resolution, below and beyond the laws of physics, where a nonphysical and perhaps supernatural influence becomes important.

REM sleep occurs in mammals and birds but not in reptiles or lower vertebrates. In fact, the occurrence of REM sleep in animals has been used to settle arguments about whether some animals sleep at all. For example, horses sleep standing up unless they are in familiar surroundings; goats keep their eyes open at all times. Nevertheless, REM patterns can be identified in these animals, so presumably they do sleep. Their sleep must be accompanied by other kinds of organized muscular activity, however, such as the open eyes in goats and the continued activity of the postural muscles in horses.

Requirements for REM sleep depend on age and on the kind of animal. Young animals show a high percentage of REM; newborn kittens have almost continuous REM sleep patterns. This requirement is reduced to about twenty percent of total sleep time in adult cats. Adult humans also spend about twenty percent of their sleep in the REM state, though their total REM time may be lower because they spend less overall time asleep; cats may sleep about 60 percent of the time.

The total daily sleep requirement is another important subject for investigation. We know how much people *do* sleep, but we have no idea how much sleep a normal human adult actually needs. Babies are awake only a few hours a day. But adults vary tremendously in their sleep habits. Some individuals get by on four hours of sleep a day, while others seem to require nine or ten. There is some evidence that the first group sleeps more efficiently by going into REM very quickly. The efficient sleeper may also show some kinds of physical and mental superiority compared to long sleepers. Unfortunately, it is almost impossible to separate out the relevant social, psychological, and physiological variables that contribute to these differences.

The biochemical and hormonal basis of sleep has also been studied, again with the aid of EEG. In one study, water extracts from brains of sleeping animals were injected into the bloodstreams or brains of other animals. The injections induced sleep. Transmitter agents in the brain have been specifically implicated in the control of sleep, because the levels of some of the agents change in the sleeping animal. Drugs that induce sleep may act by affecting these transmitter levels.

hallucination

Dreams, fantasy, and hallucination are all closely related forms of consciousness. Hallucination is roughly distinguished from fantasy in that hallucinations are experienced subjectively as "real," and may be unpleasant. They are personal and are less under control than fantasies. However, we have already seen how our perception of the physical world can differ from our sensory inputs. This discrepancy suggests that dreams, fantasy, and hallucinations cannot be clearly separated.

Despite these difficulties, a hallucination can be defined as a private perception believed to be "real." Hallucinations can be produced by a tremendous variety of means: brain stimulation, sensory overloading, sensory deprivation, hunger, fatigue, loss of sleep, drugs and toxic agents, zen or yoga or meditation, or simply trying to hallucinate. The probability that one procedure will produce hallucinations in an individual depends on his personal expectations and on society's attitude toward the procedure and toward hallucinations in general. Such cultural aspects of hallucination and related states of consciousness more properly belong to the fields of psychology, sociology, and anthropology. Nevertheless, we will trespass enough to provide a context for biological studies of hallucination.

The biologist wishes to identify the particular activity of the nervous system that is responsible for hallucinations. No single physiological change characterizes all hallucinations, but several different changes are associated with various hallucinatory states. The yogi master, for example, believes that the world is illusory, so he attempts to achieve a meditative state that blocks out the external world. In this state, his EEG shows alpha activity that cannot be easily disrupted by sensory input. Placing his hand in ice water will probably not distract him. The zen master, on the other hand, tries to place himself in the world. He believes that full responsiveness to the external world is desirable. He also shows alpha activity in meditation, but his alpha pattern is easily interrupted. Furthermore, he does not show the gradual accommodation to repeated stimuli that occurs in normal individuals.

One theory which explains many experimental observations suggests that hallucination, fantasy, and dreams actually are all forms of memory storage and retrieval. Dreams and hallucinations both may be memories that are fragmented, distorted, and recombined. But since dreaming occurs only during deep REM sleep, the dreaming individual is protected from possible harmful side effects of waking hallucinations, such as anxiety and dangerous isolation from reality. If this theory is correct, hallucinations represent a failure of normal control over memory retrieval. One piece of supporting evidence is the fact that many hallucinogenic drugs inhibit REM sleep and interfere with the establishment of short-term memory.

A major approach to the biological basis of hallucination is to study the chemical and physiological characteristics of hallucinogenic drugs. These drugs can be divided into classes according to their effects on biochemical processes in the nervous system.

marijuana, LSD, and other hallucinogens

Marijuana is one of several names for the common hemp plant of the genus *Cannabis*. It has been used as a drug for many centuries and in many societies. Some cultures, like India, tolerate the use of marijuana while prohibiting other drugs such as alcohol. Marijuana is illegal in the United States but is still widely used. Because trade in it is prohibited there are no reliable figures about its use, but it is estimated that five to ten million people in the United States smoke marijuana regularly; a much larger number have tried it. This widespread use has created considerable concern about its composition and effects.

The active component of marijuana is a compound called tetrahydrocannabinol (THC). In addition, several related compounds have been found in hemp resin, but their hallucinogenic activity is still in doubt. THC is found in the resin of the flowers, leaves, and seeds of female *Cannabis* plants; male plants have less of it. Dried marijuana leaves look like grass (hence the popular name). The leaves are usually smoked, but they may also be boiled in water to make a drink (hence "pot" or "tea"), or they may be eaten. Hashish consists of the dried resin alone and is a more concentrated source of THC.

THC produces an increase in pulse rate, dilation of the pupil, and reddening of the eye. Marijuana users report an increase in sensory experience: music seems richer, colors are brighter, food tastes better, and there is a general feeling of well-being. The user's time sense may be distorted. Long periods spent in meditative contemplation may seem brief, or brief intervals may be expanded so that the world seems to slow down. Some work suggests that short-term memory is impaired by THC. But the same results would occur if the subject were simply not paying attention to the task. It

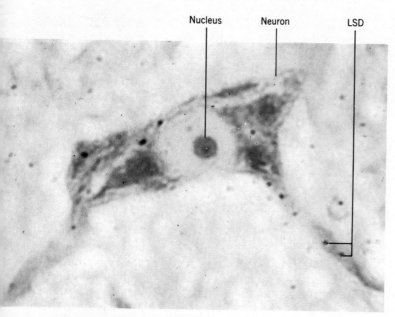

Nucleus Neuron LSD

Figure 11-35 Radioactive LSD injected into a rat appeared in its brain within fifteen minutes. Brain sections were covered with photographic film, and radioactive particles from the LSD exposed silver grains in the film. In this way, the radioactive molecules show up as black dots on the developed film. LSD appears to be associated with serotonin-containing neurons.

is difficult to distinguish actual memory impairment from the effects of inattention.

The mechanism of THC action is hard to study, because responses are variable. Often an individual will not experience changes in consciousness the first time he uses marijuana. His performance of intellectual and perceptual tasks will be somewhat impaired, though. Experienced users report a "high," a sense of well-being and happiness, but they show little or no functional impairment. In one study, experienced users were given a placebo (that is, an inert substance) and told it was marijuana. They experienced the same perceptual and psychic changes usually attributed to marijuana. In fact, some of them felt the placebo was stronger than authentic marijuana. This experiment suggests that the hallucinogenic action of THC is partly a learned response; the effect of smoking marijuana is strongly conditioned by experience, by expectation, and by the social circumstances surrounding its use. This variety of individual and

learned responses to marijuana has produced much confusion in studies of the mechanism of THC action.

THC is a comparatively weak drug. A typical dose is 100–300 micrograms per kilogram of body weight; lethal doses are so high that they have not been established. LSD (lysergic acid diethylamide) is quite a different matter. It is one of the strongest drugs known and is active in doses of only 0.5 micrograms per kilogram. LSD was first synthesized in 1938, and its hallucinogenic properties were discovered five years later.

Subjective responses to LSD are similar to those reported by THC users. LSD affects the pulse rate and dilation of the pupils, and it decreases sensory input. Actually, both THC and LSD depress sensory input, even though their users report increased sensory experience. EEG recordings from an LSD user show high-frequency EEG activity similar to that which characterizes an aroused and alert condition. This abnormally high arousal state reconciles the two contradictory findings about sensory input: the threshold for sensory input is higher, but the inputs themselves generate greater than normal arousal responses. Thus the subject experiences elevated sensory input.

LSD may work by changing the levels of transmitter agents in the central nervous system. It has been found that LSD molecules become associated with neurons in the brain that secrete serotonin (see Figure 11-35), and that levels of serotonin in the brain are elevated during LSD use. Serotonin is probably a transmitter agent, comparable in its effects to acetylcholine in the synaptic junctions in the brain.

Many other drugs as well act on transmitter agents. Like LSD, psilocybin, mescaline, and yohimbine all affect serotonin levels. Atropine and scopolamine block the transmission of nerve impulses at junctions where acetylcholine is the normal transmitter agent. Opium and its derivatives—codeine, morphine, and heroin—depress serotonin levels and also affect junctions where acetylcholine or norepinephrine act as transmitters. Barbiturates modify both norepinephrine and acetylcholine junctions. Amphetamines (Benzedrine, Dexedrine, "speed") block the normal breakdown of serotonin, norepinephrine, and epinephrine, so that nerve impulse transmission is stimulated rather than blocked.

The interaction between drugs and chemical transmitters forms a consistent pattern. Still, we cannot yet be sure that a drug's hallucinogenic properties are directly related to this interaction.

Drug Use

Drug use is common in most cultures and is certainly prevalent in western society. Most people use consciousness-modifying drugs routinely. Caffeine in coffee, tea, and cola; nicotine in tobacco; and ethyl alcohol in beer, wine, and liquor are acceptable drugs to most people in our culture. There are more prescriptions for the tranquilizer chlordiazepoxide (Librium) than for any other single drug. Obviously any drugs that stimulate pleasant emotional states will be popular.

At the same time, however, individuals in every culture have been concerned about widespread drug use, and drug abuse has always been an emotionally-charged subject. There are sound reasons for concern: drugs are potent and effective, and the dangers of drug abuse are very real. The personal suffering of the alcoholic or the heroin addict and the cost to his society are great. Nonetheless, it is often difficult to discern any rational policy in cultural attitudes toward drugs.

One of the common concerns is the addictive character of some drugs. Physical addiction is usually distinguished from psychological dependence on a drug in question. According to the nature of the withdrawal symptoms, physical addiction produces measurable physical distress—such as nausea, sweating, or changes in blood pressure. Psychological addiction produces emotional stress, rather than physical symptoms. But such a clear distinction is difficult to maintain. Emotional and physical distress are closely interrelated. We have already seen how emotional stress affects body functions controlled by the autonomic nervous system. The term "addictive potential" is often used to describe the tendency of a drug to cause withdrawal symptoms, without specifying the nature of the withdrawal response. Some drugs, like heroin, have considerable addictive potential, while others, such as marijuana, have little. A drug such as nicotine may be very difficult for some individuals to give up and easy for others. Of course, addictive potential is not limited to drugs. Comic books, television, sex, and horse racing have high addictive potential. Yet not many people would consider all of these things pernicious. A second concern about drug use centers around potential damage to an individual's health and reduction of his ability to contribute to society.

Muscle damage caused by alcohol. Volunteers were given about one cup of alcohol per day for 28 days. At the end of that time, a sample of skeletal muscle showed serious deterioration. (a) Wide spaces, filled with lipid drops and glycogen granules, appeared between the muscle fibrils. (b) The muscle regained its normal appearance after the volunteers went for six months without alcohol. Magnifications 10,000×.

Muscle fibril

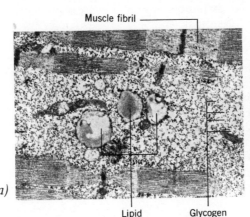

(a)

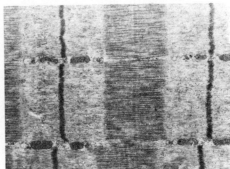

(b)

Lipid drops   Glycogen granules

In terms of addictive potential and damage to the individual and society, alcohol is the most serious drug problem in the United States. The alcohol user experiences a feeling of expansiveness and well-being. But alcohol is a general nervous system depressant. Low doses depress the brain's ability to handle multiple information inputs. Alcohol is widely used as a social lubricant, since the emotional tone of a gathering is often more important than the intellectual exchange. Nevertheless, higher doses impair motor behavior and finally produce unconsciousness. Alcohol can seriously impair health. Muscle tissue may be severely damaged. Alcoholics suffer from liver deterioration, and they may become malnourished as well because of disinterest in solid food.

There are five to six million alcohol addicts in the United States. In economic terms, alcoholism costs billions of dollars in lost working time and impaired efficiency. Half our auto accidents involve drinking drivers. The family and associates of the alcoholic suffer in ways that have been told and retold to exhaustion.

The personal and social costs of widespread use of the tranquilizer Librium are less obvious. The drug reduces anxiety but produces lethargy, loss of coordination, and sometimes hostility. It has considerable addictive potential. The use of Librium is medically justifiable in only one out of twenty cases. Yet routine use of this unnecessary and dangerous medication is supported by a complicated set of customs and perhaps by social pathology.

The use of illegal drugs attracts much more attention than the misuse of socially acceptable ones. Heroin is the most significant illegal drug. It has a fairly high addictive potential and disastrous effects on personal health of the addict. At the beginning of 1971 there were 65,000 known heroin addicts in the United States. All the other "hard" drugs add only an additional 3,000 addicts. This figure should be multiplied by at least five or ten since it includes only the addicts known to law enforcement agencies. Nonetheless, this number is still only a small fraction of the number of alcohol and Librium addicts. Most heroin addicts are outcasts. They are concentrated in slum and ghetto areas of large cities; typically they are young, poor, or members of minority groups.

Society has a punitive attitude toward some drugs but considers others socially acceptable. Acceptance or intolerance is often unrelated to the actual danger of the drug. A drug is accepted if and only if it has a social function (cocktail parties) or a ritual role (for religious use, for instance). Our society is committed to many kinds of drug use, yet it is intolerant of a minority that uses drugs for which we have no ritual and no community role.

Other factors may be important. For instance, two drugs with identical effects on chemical transmitters may be distributed very differently in the body. One drug may easily enter the brain; another may be restricted to the peripheral regions of the nervous system. Some drugs are broken down and metabolized rapidly; others persist. Certain drugs affect other important biochemical systems in addition to transmitters; for example, barbiturates interfere with ATP formation. And finally, some drugs that do not seem to affect transmitter agents may also be hallucinogenic. All psychoactive drugs probably modify activity of the nervous system in some way, but that statement in itself tells us very little. We do not have enough information about the nervous system to establish a clear relation between a drug's biochemical effects and the associated changes in consciousness.

the internal world

We have discussed how man can exist in a variety of conscious states, each of which involves a slightly different kind of sensory input and a different perception of the environment. But in general, most sensory inputs are discarded by the nervous system. Selected inputs are often distorted to emphasize environmental features of particular significance. Thus the organism deals with a perceived environment which is necessarily different from the "real" environment. And, through its continuing interaction with its environment, the animal reshapes its internal view of the world. The wanderings of reverie, imagination, dreams, and visions may actually be attempts to incorporate new sensory inputs into our internal world.

The fact that REM sleep (and hence dreaming?) occurs in many mammals and birds suggests that such reworking of the perceived world is not limited to man. This is not surprising, since animals other than man can also learn and remember. The existence of learning behavior means that the animal must have some kind of functional internal world.

Just as the chemical composition of an organism is related to but different from its environment, behavior of that organism is based on a world related to but different from its environment. The internal world does not need to be a simple or relatively accurate picture of the world. The only constraint is that the internal world must organize behavior so that the animal can survive and reproduce in the external world.

Recommended Reading

Davis, David E., *Integral Animal Behavior.* New York: Macmillan (1966).

A discussion of animal behavior with emphasis on the survival of the individual and the species.

Klopfer, Peter H., and Jack P. Hailman (eds.), *Function and Evolution of Behavior.* Reading, Mass.: Addison-Wesley (1972).

A collection of research papers and discussions of the role of instinct in animal behavior.

Teitelbaum, Philip, *Physiological Psychology: Fundamental Principles.* Englewood Cliffs, N.J.: Prentice-Hall (1967).

A neurophysiological approach to human behavior, including discussions of emotion, sleep and wakefulness, and motivation.

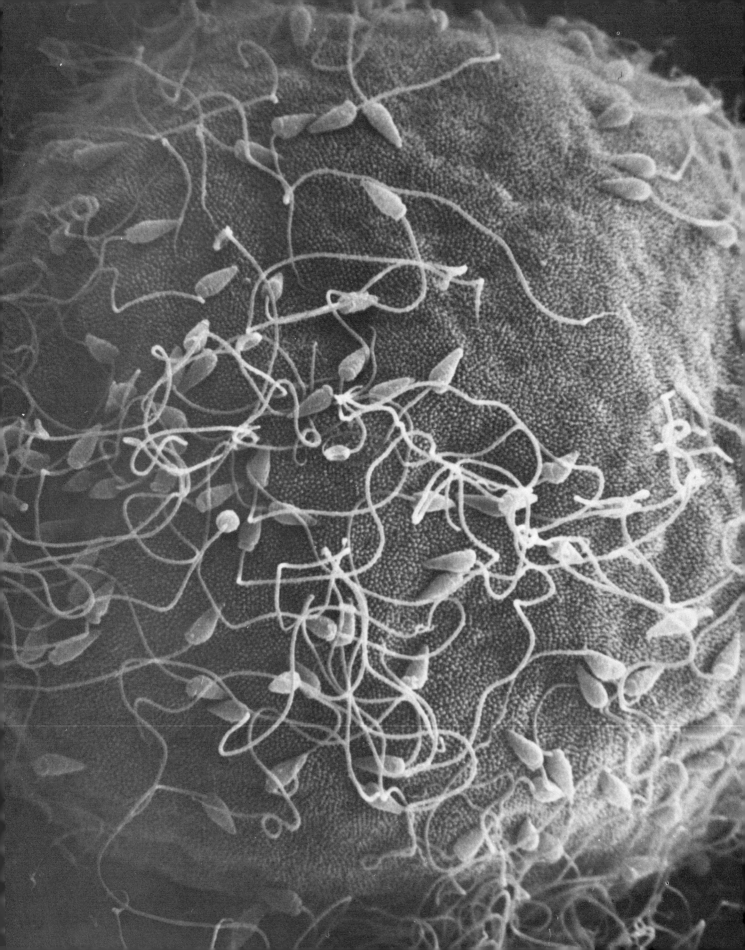

12 reproduction

We have so far discussed the organization of plants and animals in terms of their cell chemistry and cell structure. Throughout these chapters we have ignored the question of how organisms grow and reproduce. Basically, the growth of organisms results from an increase in the size of their cells, an increase in the number of their cells, or both. Reproduction—the process by which an organism produces more of its own kind—also results from the production of new cells. Obviously, then, cell division forms the basis for both growth and reproduction. The fundamental role of cells and cell division finally became apparent in the middle nineteenth century. Before that time, the relation between the structure of living tissue and the materials of the physical world was unknown, and understanding growth and reproduction was viewed as a philosophical problem rather than a scientific one.

Cell Division

Cell division is an orderly sequence of events in which a single cell divides to form two "daughter cells." In the process, each daughter cell is provided with a complete set of instructions (DNA) and sufficient organelles to enable it to carry out metabolic reactions. Cell division in higher animals and plants involves a sequence of three events: (1) DNA replication, (2) mitosis, and (3) cytokinesis. The replication of nuclear DNA provides two full sets of DNA. **Mitosis** is the formation of two separate nuclei, each of them supplied with a full complement of DNA. **Cytokinesis** is the division of the whole cell.

Fertilization of a sea-urchin egg

357

DNA replication

In plant and animal cells, most of the DNA is located in the nuclei. Under a light microscope the nucleus of a non-dividing cell appears granular and shows no particular organization of its DNA content. We know that DNA is present, however, since the nucleus can be stained by dyes that react specifically with DNA. Under the electron microscope, strands of DNA become visible.

When cells are stained for DNA, we can use optical techniques to measure the amount of DNA in each nucleus. We transmit light through the tissue and measure the amount absorbed by the nucleus. Such light measurements show that the DNA content of nuclei is always identical to, or a simple unit multiple of, the DNA content of other cells in the organism. That is, if the nucleus of a typical body cell (liver, muscle, leaf, root) is used as a standard, other cells will have the same amount or half or twice as much DNA. The multiples may be higher in some tissues, but they are always whole numbers. This evidence supports the idea that cells have one or two (rarely three or more) full sets of DNA instructions. No cells have one and a half or two and a half sets of instructions. Each full set of DNA instructions is called a **genome.**

Each genome is packaged into one or more chromosomes. In bacterial cells, all the DNA is packaged into one chromosome that is located in the middle of the cell. The cells of more complex, eucaryotic organisms have a genome distributed among many chromosomes, and the chromosomes are all contained within the nucleus. Cells that have two sets of DNA instructions have two identical sets of chromosomes. Cells with a single genome are called **haploid;** those with two are **diploid;** and those with three or more are termed **polyploid.**

During normal cell division, the entire DNA content of the cell doubles, even though more than one genome may already be present. The DNA is then equally distributed to the two identical daughter cells. This doubling of the DNA occurs through the semi conservative replication of nuclear DNA. In semi conservative replication, the two strands of each existing DNA molecule unwind, and each one acts as the template for the synthesis of a new, complementary chain. The duplicated DNA molecules are then packaged into identical, duplicate chromosomes. This mechanism is a crucial feature of cell division because it assures that the DNA in

the chromosomes of daughter cells will be identical to the parental DNA. It provides continuity of the DNA throughout growth, from egg cell to adult, and from generation to generation. During the development of a multicellular organism, millions or billions of cells may be produced from a single precursor cell. But each cell in the organism has the same set of genetic instructions, that is, the same genome.

The classical experiment demonstrating semiconservative chromosome duplication was performed by J. Herbert Taylor about fifteen years ago. He supplied thymine, labelled with radioactive hydrogen atoms, to the dividing cells of a higher plant. Taylor reasoned that this thymine would be incorporated into any DNA being synthesized. He then replaced the labelled thymine with regular, unlabelled thymine and allowed the cells to grow and divide normally. All the chromosomes in the first generation daughter cells were labelled. However, when the cells were permitted to go through a second division cycle, only half the chromosomes were labelled. Taylor explained his results as follows: during DNA synthesis and duplication, each DNA molecule separates into its constituent strands. Labelled thymine is incorporated into these newly synthesized duplicates; thus, one strand of each DNA molecule is radioactive. Each molecule is then incorporated into a chromosome, so that all the chromosomes are radioactive at the end of one division cycle. After a period of growth, DNA molecules in the daughter cell chromosomes separate for another duplication. This time, duplication occurs in the presence of unlabelled thymine, so that the new strands are unlabelled. Thus, only half of the newly duplicated chromosomes are radioactive at the end of two division cycles (see Figure 12-1). This result strongly supports the contention that DNA in chromosomes is duplicated by means of semiconservative replication, and that each daughter cell receives an identical set of DNA.

DNA duplication occurs with incredible speed at the molecular level. Each molecule of DNA in the nucleus of a human cell contains about ten billion nucleotide pairs. The DNA double helix makes a complete turn with every ten pairs. Thus, a billion turns must be unwound into strands for duplication to occur, and then they must be rewound. Yet human cells duplicate their DNA in a few hours. Even if the DNA in all 46 chromosomes of the human cell started duplicating at the same time, each DNA helix must still unscrew and rescrew at a rate of a thousand turns per second. If replication were to begin at several points on the same chromosome rather than just at one end, the winding rate would be less. However, it seems unlikely that it does so, since the 46 strands do retain their continuity and integrity throughout duplication.

DNA replication is an astonishing process with respect to precision and accuracy as well as speed. If the wrong nucleotide is inserted into a new DNA chain, an error will be made when that DNA segment is used to direct protein synthesis. The wrong amino acid (or no amino acid at all) will ultimately be placed in the new protein molecules. The actual error rates for DNA duplication can therefore be measured by looking for changes in proteins in daughter cells. These rates turn out to be remarkably low. The transmission of DNA-coded instructions from one cell generation to the next is virtually error-free.

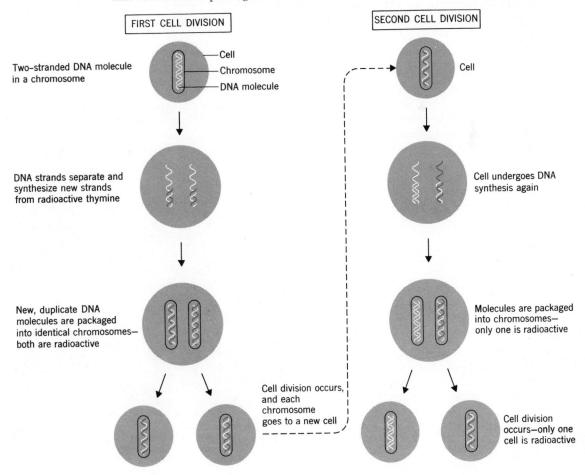

Figure 12-1 Taylor's experiment on chromosome duplication. A cell is exposed to radioactive thymine during the time DNA synthesis is occurring and then is returned to a solution of non-radioactive thymine for a second division. At the end of the first division all the chromosomes in the two daughter cells are labelled but at the end of the second division cycle only half the chromosomes are radioactive. These results are consistent with the hypothesis that each strand of the double helix molecule serves as a template for the synthesis of a new, complementary strand, and that each helix is packaged in a chromosome.

FIRST CELL DIVISION

SECOND CELL DIVISION

Two-stranded DNA molecule in a chromosome

Cell
Chromosome
DNA molecule

Cell

DNA strands separate and synthesize new strands from radioactive thymine

Cell undergoes DNA synthesis again

New, duplicate DNA molecules are packaged into identical chromosomes—both are radioactive

Molecules are packaged into chromosomes—only one is radioactive

Cell division occurs, and each chromosome goes to a new cell

Cell division occurs—only one cell is radioactive

Mitosis is the term for the formation of two identical nuclei in the parent cell. It begins after the cell's DNA has been duplicated and packaged into duplicate chromosomes.

The process of mitosis is conventionally divided into four steps, or phases. The first phase is called **prophase.** During this period the chromosomes themselves begin to coil tightly into thick strands. These strands gradually become visible in the light microscope. The entire nucleus begins to look like a ball of yarn. The coiling and thickening continue until the chromosomes become short and thick. The two duplicate chromosomes are connected to each other at one point called a **centromere,** and each one is called a **chromatid** at this stage. The membrane around the nucleus usually disappears at this time.

Prophase is followed by **metaphase.** In metaphase, the thickened chromosomes, each consisting of two chromatids, become oriented at the center of the cell; all the chromosomes lie in one plane, called the **metaphase plate** (Figure 12-2).

At the third phase of mitosis, **anaphase,** the two chromatids of each chromosome separate and move to opposite ends of the cell. Then, during the last stage, **telophase,** the chromosomes go through a sequence that reverses the events of prophase: membranes form around the two separate sets of chromosomes, creating two new nuclei. The individual chromosomes elongate and disappear again, producing a non-mitotic, or **interphase** nucleus. These phases are presented in Figure 12-3.

The orientation and movement of chromosomes during late prophase, metaphase, and anaphase are caused by the **spindle** apparatus. This apparatus is an arrangement of microtubules radiating from two poles at opposite ends of the dividing cell. It forms during mitosis and comes to occupy a large part of the cell. Some microtubules are attached to the centromere joining each chromatid pair. Others run from one pole of the spindle structure to the other, or radiate out from each pole to form a starlike structure. The formation of the spindle apparatus requires a tremendous amount of protein: fully fifteen percent of the cell's total protein is used.

Figure 12-2 The metaphase stage of mitosis in an animal cell. The chromatids are lined up at the center of the cell to form a "plate." Duplicate chromatids are joined by centromeres attached to microtubules of the spindle apparatus.

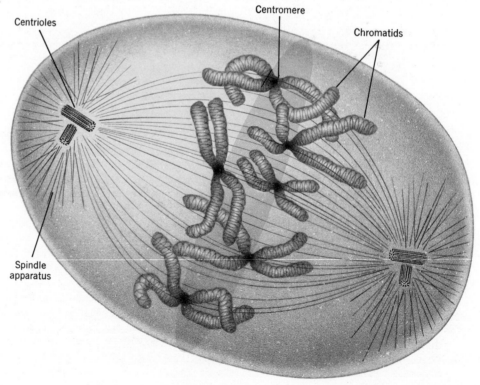

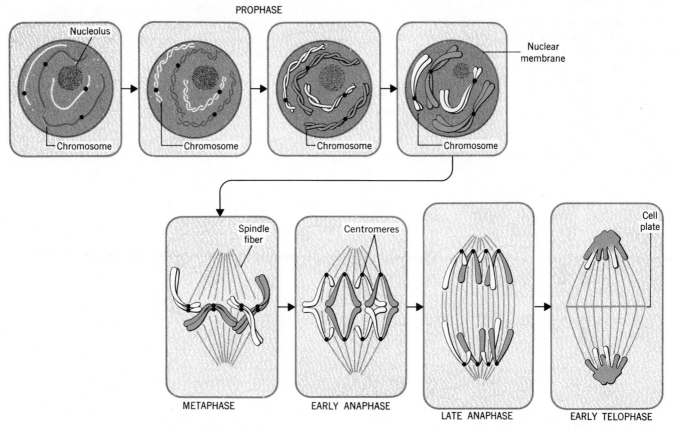

PROPHASE

Nucleolus

Chromosome

Nuclear membrane

Chromosome

Chromosome

Chromosome

Spindle fiber

Centromeres

Cell plate

METAPHASE

EARLY ANAPHASE

LATE ANAPHASE

EARLY TELOPHASE

Figure 12-3 Mitosis in a plant cell.

The chromosomes move along the spindle tubules during anaphase toward the spindle poles. The movement of the chromosomes may involve the contraction of spindle fibers; alternatively, the chromosomes may be pushed to the spindle poles on the ends of rapidly growing fibers, rather than pulled. The exact mechanism of their movement is still unknown.

In most animal cells, each spindle pole contains a **centriole** (see Figure 12-2). Centrioles are structurally similar to short pieces of cilia or flagella. At telophase, each nucleus is reformed adjacent to a centriole. The centriole than divides in preparation for the next mitotic cycle. However, as yet, we don't understand the function of the centriole in mitosis. The cells of higher plants undergo mitosis, but they do not have centrioles. Though this organelle may be important for mitosis in some organisms, it is obviously not necessary in all.

cytokinesis

Mitosis is usually (but not always) accompanied by the division of the whole cell into two portions. This last process is called cytokinesis. In a typical animal cell, cytokinesis occurs during telophase in the mitotic cycle. The outer edge of the cell starts to fold inward at an area midway between the spindle poles. The furrowing and folding of the cell membrane proceed until the cell body is pinched in two.

In plant cells, which have rigid walls, the mechanism is different. A structure called the **cell plate** appears in the middle of the spindle apparatus. It is formed from dictyosome vesicles that deposit various substances along the middle of the cell. This plate extends laterally until it divides the cell into two portions. The remnants of the cell plate in each resulting daughter cell become sites for the deposition of cellulose fibrils to form a new wall between the cells (Figure 12-4).

Cytokinesis can be very complicated in certain kinds of dividing cells. In some cases the cell body may be divided into very unequal fractions. The cell division that leads to the production of an egg, for example, is followed by unequal cytokinesis. One of the new nuclei is retained in the egg, while the other is confined in a tiny cell which then degenerates.

Unequal cytokinesis is partly responsible for the development and differentiation of specialized cells in multicellular organisms. Mitosis assures that all the nuclei in an organism are at least potentially identical, because each cell nucleus contains a full set of the DNA instructions for protein synthesis. In theory, then, the nucleus of any bean plant cell might direct the activity of any other cell in the plant: a root cell, a leaf cell, or a phloem cell. Similarly, the nucleus of a human liver cell ought to be able to direct the activity of a neuron. But cells are, in fact, different. Roots differ from leaves,

Figure 12-4 Formation of the cell plate in a plant cell. (a) The cell plate is laid down by vesicles that aggregate along a plane dividing the daughter cells. Magnification 25,000×. (b) A cell wall composed of cellulose fibers forms on the cell plate. The cellulose fibers are laid down parallel to an aggregation of microtubules. Magnification 50,000×.

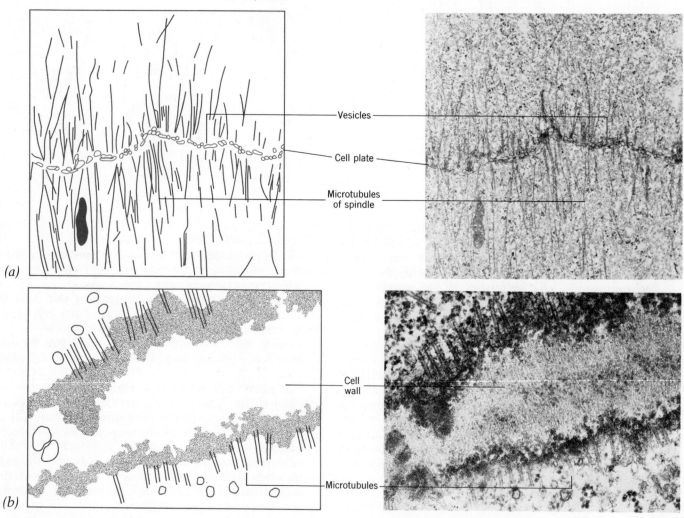

(a)

Vesicles

Cell plate

Microtubules of spindle

(b)

Cell wall

Microtubules

and livers from brains. Presumably, such differences result from interactions between the nucleus and the cytoplasm. The cytoplasm may activate or suppress portions of the DNA strands in particular cells. Differences between the cytoplasmic activity of cells result partly from environmental influences and partly from unequal distribution of cytoplasm in dividing cells. Through the unequal distribution of the cytoplasm of the dividing cell, differences in the daughter cells can be achieved and sustained. As a result, a single cell can give rise to a tremendous variety of cell types.

growth

Cell division is a vital part of growth as well as of reproduction. The growth of organisms is brought about by an increase in both the size of cells and the number of cells. In unicellular organisms, growth is simply an increase in cell size. The growth of a single cell usually occurs at two points in the cell division cycle. One period of growth occurs in a new daughter cell immediately after division. The parent cells do not double their size before dividing, so the daughter cells start out at a size smaller than normal. During this first period of growth, they achieve their characteristic size. This period is the most variable phase of the growth cycle and may be directly controlled by environmental factors.

The initial growth period is followed by a period of DNA synthesis, during which the DNA strands are duplicated. At this point the cell grows again to greater than its normal size. The enlarged cell then undergoes mitosis. After mitosis, the cycle begins again with a new initial growth period for each daughter cell. All the phases of the cell division cycle, except the initial growth period, are constant in length for a particular type of cell (see Figure 12-5).

The cytoplasm must have a specific volume, relative to the size of the nucleus, before cell division can occur. Therefore, the initial growth period must be long enough to ensure that the critical cytoplasm/nucleus volume ratio is reached. Cell division can be stopped by preventing this volume increase. We can keep the cell from growing either by periodically cutting away portions of the cytoplasm or by simply starving it. In both cases, cell division will be delayed indefinitely.

Growth in multicellular organisms is also aided by an increase in the size of preexisting cells, but an

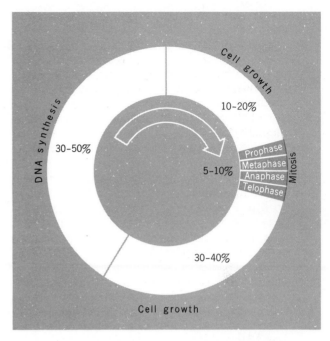

Figure 12-5 *The cell growth cycle is divided into periods of DNA synthesis, growth, mitosis, and a second growth phase. Mitosis lasts a small fraction of the total cycle.*

increase in the number of cells is more important. Each new individual usually starts as a single cell. But mature adults have a tremendous number of cells, all derived from the original one by mitosis and cytokinesis. Man has about 100 trillion (10^{14}) cells. (This number is really too large to grasp. What does it mean to have as many cells in your body as the number of *pennies* in the American gross national product?) The division of the original single egg cell can produce all these cells in a comparatively short time; in fact, if none of the cells were to die, the 10^{14} cells could be produced in only 47 division cycles. But, in fact, the fertilized egg does not simply divide repeatedly through 47 cycles to produce the 10^{14} cells in your body. Patterns of cell division are extremely variable.

Some animal tissues divide throughout life. About three million red blood cells are produced each second by blood-forming tissue in bone marrow. The epithelial cells that cover your body surface and line your digestive tract are continuously lost and renewed by cell division. Skin is completely replaced every two weeks. Every day about two percent of your cell population is lost and replaced.

Other animal cells retain the capacity to divide, but do so only on demand. For example, if a portion of your liver is removed surgically, the cells of the remaining portion begin to divide rapidly. When the original mass is restored, cell division turns off again. In other animal tissues, cell division is completed before birth, and no more cells are ever produced. Skeletal muscle cells and nerve cells in man behave this way. The cells may still enlarge, however, after birth. For example, a weightlifter enlarges his muscles by producing additional fibrils within individual muscle cells. Nerve cells get larger by producing new synaptic connections with neighboring nerve cells. However, the number of cells is fixed early in development in these tissues.

The capacity for cell division appears to be related to the nature of the cell. Generally, cells that synthesize and *retain* specialized structures lose their capacity to divide. In this category are muscle fibers, and neurons. But cells that *extrude* synthesized products (such as hormones, or digestive enzymes) can continue to divide; in this class are liver cells and gland cells. As yet we don't understand why this correlation exists, but further study of cell division may provide an explanation.

Growth in plants is quite different from growth in animals. A plant grows throughout its lifetime—it has no fixed adult size and shape, as an animal does. Continued growth is possible because new cells are continually produced by meristems, and these cells continually differentiate into the specialized tissues of leaf, root, and stem. All the living cells in a higher plant, even a long-lived plant like a tree, are young. Even differentiated tissues are constantly renewed by further division of the meristematic tissues. Thus, no living cell in a sequoia is more than a few years old, even though the tree itself may have been growing for a millenium.

Plants, unlike animals, depend on cell enlargement for a considerable amount of their growth. Most of the height increase of a young stem, for example, is due to enlargement of newly-produced cells. But plant cell walls are rigid; molecules in cell walls are stabilized by calcium ions, which form cross-linkages between cellulose chains. During growth, the cross-linkages must be broken so the cell wall can expand. Plant hormones have precisely this effect in higher plants. It is not clear exactly how they work—they may withdraw or inactivate the calcium. In any case, once the bonds have been broken, the long cellulose molecules are stretched and aligned along the direction of growth. The bonds are then reestablished in new positions. After growth has concluded, secondary thickening of the wall may occur through the deposit of additional material on the inside of the existing wall.

No individual cell is immortal. It ultimately must perish, divide, or fuse with another cell. The fact that some cells retain their capacity for division confers immortality on the species, or on a particular cell line, but not on the individual. Your body cells are descended in an unbroken line from those of your ancestors, but most of your own body tissue is a dead end. An old man has old muscle and nerve cells, and when these cells no longer function, he will die. Neither are higher plants immortal, even though they are constantly renewed by cell division. Plants accumulate a mass of dead tissues that eventually interfere with the functioning of the living cells and cause death.

Growth involves the production of new cells, but reproduction is the formation of new individuals. There are two basic kinds of reproduction. **Asexual** (or non-sexual) **reproduction** is defined as the production of new individuals from the genetic material of one parent cell. **Sexual reproduction** is the production of offspring from the genetic material of two parent cells.

Asexual Reproduction

Most unicellular organisms reproduce asexually through simple cell division. In eucaryotic unicells, cell division proceeds in the usual way, by mitosis and cytokinesis. It is a little different in organisms that have procaryotic cells: the blue-green algae and bacteria. These organisms divide by a simple process of **binary fission:** the cell simply pinches in half. No mitotic spindle is present (Figure 12-6).

Unicellular fungi, the yeasts, have another form of asexual reproduction, called **budding.** A protuberance forms on the mother cell during mitosis. This protuberance, or **bud,** enlarges and finally breaks off to form a new yeast cell (Figure 12-7).

A third method of asexual reproduction is the

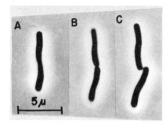

Figure 12-6 *Bacteria divide by a process of binary fission. They simply elongate and divide in half. Magnification 2,500×.*

production of **asexual spores.** This method is common in algae and fungi. Spores are single cells that develop into new individuals under favorable environmental conditions. In unicells, spores are formed inside the cell wall. In multicellular organisms, spores are formed in special structures called **sporangia** (Figure 12-8). Spore formation begins with repeated mitoses within a cell, to produce several nuclei. These nuclei are enclosed in cell walls, becoming spores. Many asexual spores are **zoospores,** propelled by flagella. Others are nonmotile and depend on air or water currents for dispersal (see Figure 12-9).

Asexual reproduction in higher plants depends on the activity of meristematic tissues. Many grasses will put out new roots and leaves along parts of the stem that lie on the surface of the ground. This type of asexual reproduction is used in gardening. If the necessary meristematic tissues can be separated from the parent, a new individual will be produced. If you place a willow twig in water, for instance, meristematic cells in the cambium of the stem will produce new root meristems. When the cutting has "rooted" in this manner, it can be planted.

Even multicellular animals can reproduce asexually. The pond animal, hydra, is a good example. In the hydra, dividing cells are found along the column of the adult individual. Cells are lost primarily

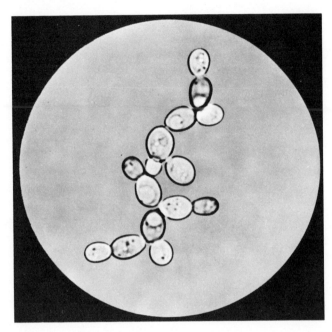

Figure 12-7 (above) *Yeast cells with buds.*

Figure 12-8 (below) *The sporangia of slime molds are specialized structures that produce asexual spores.* (a) Physarum (b) Stemenitis.

(a) (b)

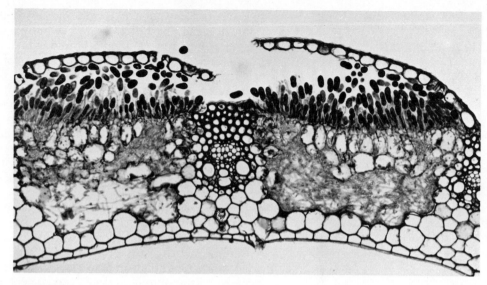

Figure 12-9 (*left*) *A cross section of a wheat leaf, showing the nonmotile spores of the wheat rust fungus (Puccinia graminis) inside a sporangium. Magnification 100×.*

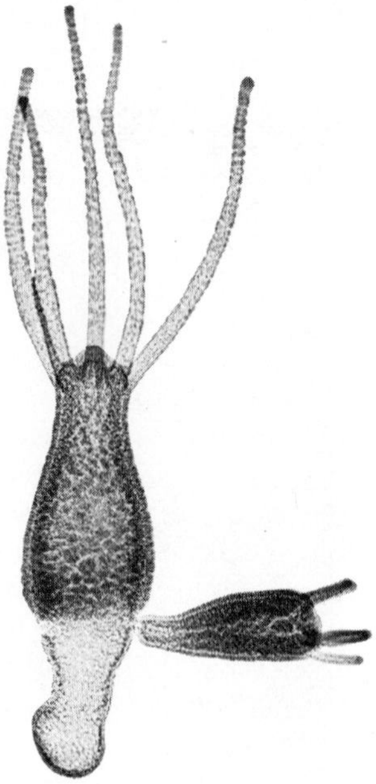

at the tentacles and the base of the animal. The lost cells are replaced by cells migrating from the column. If excess food is available, extra cells migrate and accumulate at the side of the column, producing a bud. The bud grows and produces its own tentacles while still attached to the parent hydra. Finally it separates from its parent and becomes a free-living hydra (Figure 12-10).

Figure 12-10 (*right*) *A hydra with an attached bud. (Courtesy Carolina Biological Supply Company.)*

Sexual Reproduction

Sexual reproduction is any process in which progeny receive DNA from two parent cells. Sexual reproduction is vastly more complicated than asexual reproduction, simply because the DNA from each parent must somehow be brought together. As a result, sexual reproduction utilizes many of the organism's resources; it is often costly to the individual participant and may even be lethal. Males often are casualties. In the praying mantis, for example, the female eats the male while he is mounted and copulating. Her behavior seems ungracious, but it is actually necessary. The brain of the male contains a center that inhibits copulatory movements and sperm transfer; it must be destroyed for successful copulation. In most animals the females survive copulation, but even a female may not survive long after laying her eggs. In some animals neither parent survives the repro-

ductive process. A species of marine worm liberates gametes to the surrounding sea water by splitting open. The gametes then fuse, and young larval worms develop in the water. Obviously, the adults who provided the gametes do not survive.

Sexual reproduction is also costly in terms of energy and tissue requirements. In higher plants, the quantity of pollen that is produced is immense compared to the amount that actually functions in sexual reproduction. In animals, elaborate, energy-consuming courtship behavior is common.

Since sexual reproduction is costly and may be dangerous to the individual, why does sex exist at all? Since many organisms reproduce asexually, sex is obviously not necessary for reproduction. Asexually-reproducing organisms have survived perfectly well; they maintain themselves in the face of competition from organisms that do reproduce sexually. Yet sexual reproduction is widespread in spite of the costs. Perhaps a better question is, what are the advantages of sexual reproduction?

The answer is that a population of organisms that reproduce sexually has a better chance to survive in some kinds of changing environments. Let us first consider an asexual population. In asexual reproduction, the daughter cells have precisely the same DNA complement as the parent cell. The reproductive process itself does not produce new characteristics or abilities in the population. Changes in DNA are produced only through a process of mutation. A **mutation** is any change in a gene that results in a permanent change in some characteristic of the individual. Mutation alters the order or content of the base pairs in a DNA strand. Such changes in the DNA code will of course change the RNA, and hence the resulting proteins.

A large population of rapidly growing microorganisms will always be producing a few mutant individuals with new proteins, and therefore with new characteristics. Should the environment change, one of the mutant cells might be the only one able to survive under new conditions. All the other organisms, those with identical DNA instructions, might simply be wiped out. Through random mutations, then, asexual populations of bacteria and other microorganisms are prepared for changed conditions. Given enough individuals and the randomness of the mutation process, there might be a few individuals who can survive in a new environment. A mutant disease agent may survive because it is resistant to penicillin while the rest of the population is not. A mutant microorganism may be able to survive extreme heat, thrive in an acidic soil, or metabolize a new food source. However, the population must be very large for this process to succeed, because mutation is really quite a rare event. Furthermore, mutational changes in the DNA strand are random. Mutations usually impair cell function rather than improve it. Asexual reproduction and mutation can provide adaptability, but only for small, simply-organized forms with a high reproductive capacity—ones that can reproduce rapidly to produce large populations.

Sexual reproduction increases adaptability and variability in a population. Two parents, rather than one, are contributing genetic material to each daughter organism. The progeny therefore represent a recombination of two sets of DNA. The new individual will be different from each parent. Thus the reproductive process itself provides variability in a population, because each new individual has a new combination of genes.

moneran sexual reproduction

Several kinds of sexual reproduction have been discovered in bacteria. In each case, a fragment of DNA chromosome from one bacterial cell is introduced into a second cell. That fragment is incorporated into the whole chromosome of the recipient cell by a process of chromosome breakage and recombination. This process is called **crossing over.** The incorporated fragment usually does not increase the size of the chromosome; rather, it replaces an equivalent section of it.

The chromosomal fragment can be transferred to the recipient cell in three ways, via conjugation, transformation, or transduction. **Conjugation** is a common form of sexual reproduction in bacteria; it involves the direct transfer of DNA between two living cells. A cell defined as male has a small segment called an **F factor** (fertility factor) on a circular chromosome. Females lack this fragment. Conjugation begins with the formation of a bridge between male and female. The male chromosome begins to replicate, and the newly duplicated section immediately moves across the bridge (see Figure 12-11). Conjugation may cease long before the entire chromosome is replicated and transferred, so that only part of the F fragment is transferred to the female. Simple mechanical agitation is often the reason. Crossing over then takes place; a segment of the female's chromosome breaks off and is replaced by the injected fragment.

In **transformation,** DNA fragments released to the environment by dead bacteria enter living cells. This DNA is incorporated into the chromosome of the living cell by crossing over. The live bacteria become "transformed," since they may acquire some of the characteristics of the dead cells. In this way, for example, a harmless bacterium can be transformed into the dangerous pathogen responsible for pneumonia (Figure 12-12). The harmless strain incorporates a DNA section from the dead cells, and this segment directs the synthesis of a carbohydrate capsule around the cell (Figure 12-13). The capsule makes the microorganism extremely pathogenic, presumably because the capsule protects the bacterium against attack by white blood cells.

The **transduction** of genetic material from one bacterium to another requires the participation of virus particles. When a bacteriophage virus infects a bacterial cell, it uses the synthetic machinery of

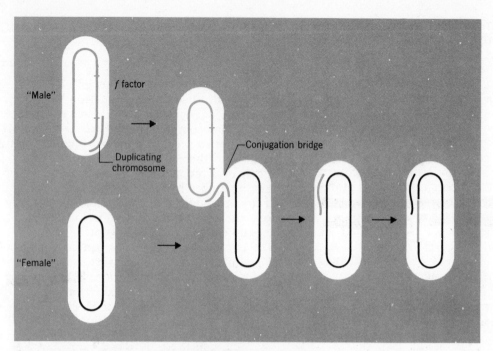

Figure 12-11 Conjugation in bacteria. A DNA fragment from a male cell is passed to a female cell across a conjugation bridge. Conjugation occurs as the male chromosome is being duplicated.

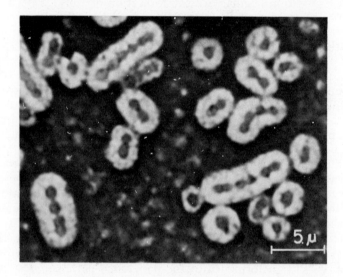

Figure 12-12 (above) Streptococcus pneumoniae, a virulent disease agent. A capsule is clearly evident around the cells in the picture. These organisms were obtained from spinal fluid of an individual with a fatal case of meningitis. Magnification 3,000×.

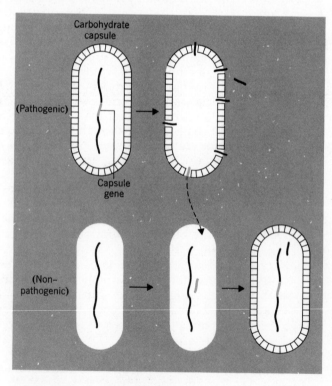

Figure 12-13 Transformation. A fragment of DNA from a dead cell is taken up and incorporated into the genome of a live cell.

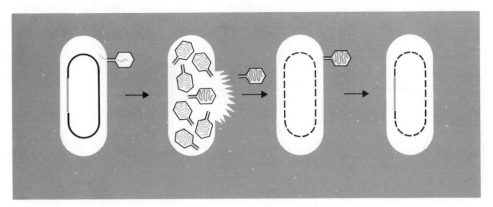

Figure 12-14 Transduction.
Phage virus particles carry
a fragment of bacterial
DNA from one cell to an-
other.

the host cell to produce many copies of the viral coat proteins and viral chromosomes. As the viral particles are produced inside the host cell, a portion of the host chromosome may accidentally be incorporated into a viral particle. After the host cell is lysed, the faulty viral particle may then infect another bacterium and inject the original bacterial DNA (Figure 12-14).

Transformation and transduction are not usually discussed as forms of sexual reproduction, since the mode of DNA transfer is unusual. Nevertheless, the result is identical to those of other, more common forms of sexual reproduction; in all these cases a single cell ends up with chromosomal material from two cells. Sexual reproduction in bacteria is also unusual because the progeny do not receive a complete set of DNA from both parent cells.

In all other organisms, sexual reproduction is accomplished by fusion of cells each of which contains a whole genome. The resulting cell gets one complete genome from each parent cell. Thus, the cell that is the immediate product of sexual reproduction has twice as much DNA as the parent cell. The DNA content must therefore be halved again at some point, or each generation would contain greater and greater quantities of DNA. The reduction of DNA content occurs as part of a cell division process called **meiosis** or **reduction division.**

meiosis

Meiosis in man is fairly typical of the process in all animals. Each human body cell is diploid; that is, it contains two genomes. Each genome is contained in 23 chromosomes, so there are 46 chromosomes in all. The chromosomes can be arranged into 23 pairs. The members of a pair are similar because both influence the same characteristics. The chromosomes in a pair are said to be **homologous** to each other; they are nonhomologous to other chromosomes (see Figure 12-15).

Meiosis, like mitosis, is initiated by a synthesis phase, in which the DNA in the chromosome is duplicated, forming two identical chromatids. At this point, the homologous chromosomes come together to form 23 pairs. Since each chromosome in the pair is split into two chromatids, the unit contains four chromatids in all, and is called a **tetrad.** As the next step, the chromatids from each homologous chromosome wind tightly around each other, so that each tetrad seems to have two rather than four strands. The two strands then cross over by breaking and rejoining at various points. As a result, DNA segments are traded among the chromatids (Figure 12-16). Two nuclear divisions now occur. The first separates the intertwined homologues, and the second separates the chromatids in each homologue (Figure 12-17).

In both meiosis and mitosis, the cell begins with two sets of DNA in 46 chromosomes—it is diploid. During DNA synthesis every chromosome is duplicated. In mitosis, the synthesis phase is followed by one nuclear division; two nuclei result, and each nucleus has 46 chromosomes. In meiosis, synthesis is followed by two nuclear divisions, giving four nuclei. Each nucleus has only 23 chromosomes. That is, it has only one set of DNA; it is haploid. Meiosis halves the chromosome number and the DNA content of the cell. The fusion of two nuclei resulting from meiosis will not merely restore the chromosome number and DNA content of the cell, but it will reestablish 23 homologous chromosome pairs as well.

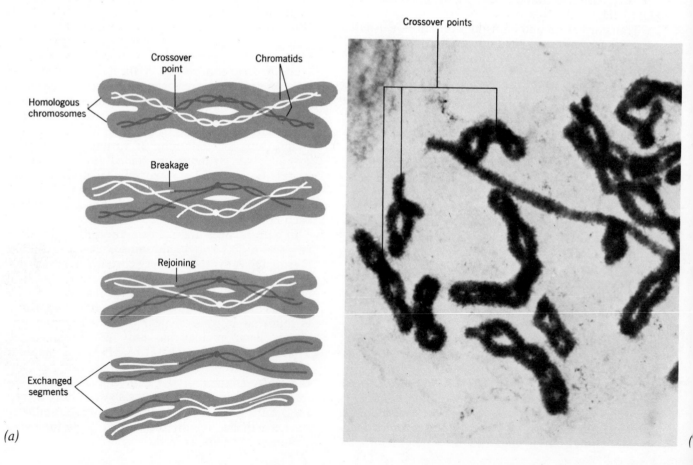

Male					Female				
1	2	3	4	5	1	2	3	4	5
6	7	8	9	10	6	7	8	9	10
11	12	13	14	15	11	12	13	14	15
16	17	18	19	20	16	17	18	19	20
21	22	X Y			21	22	X		

Figure 12-15 (left) The chromosomes of a normal human male and female, arranged in homologous pairs.

Figure 12-16 (below) Crossing over in meiosis. (a) The chromatids of homologous chromosomes break and rejoin, exchanging genetic material. (b) Crossover points in chromosomes of an amphibian during sperm formation. Magnification 1,500×.

Crossover points

Crossover point

Chromatids

Homologous chromosomes

Breakage

Rejoining

Exchanged segments

(a)

Figure 12-17 Meiosis in plant cells.

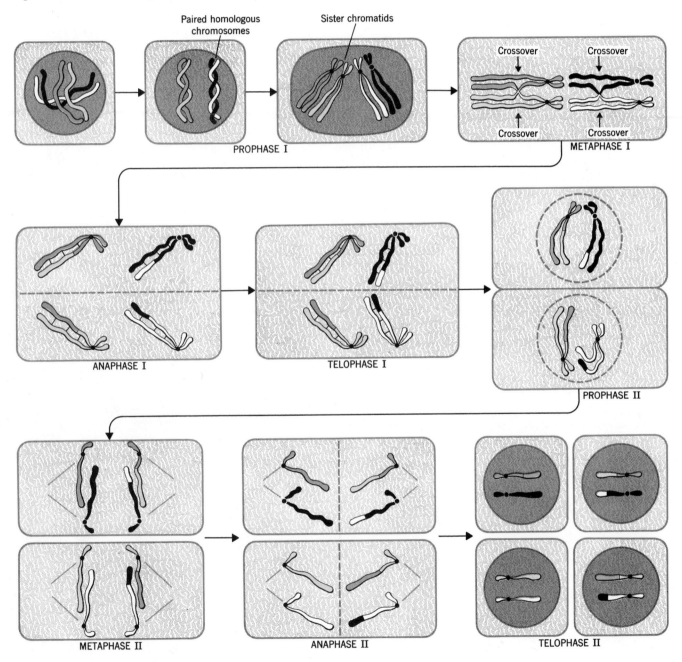

Paired homologous chromosomes

Sister chromatids

Crossover   Crossover

Crossover   Crossover

PROPHASE I

METAPHASE I

ANAPHASE I

TELOPHASE I

PROPHASE II

METAPHASE II

ANAPHASE II

TELOPHASE II

Sexual reproduction in animals is usually an alternation between the diploid and haploid condition. Diploid cells undergo meiosis to produce haploid cells, and haploid cells—called **gametes**—fuse to produce a diploid cell. This new diploid cell is the **zygote.** A multicellular individual arises by mitosis from this zygote.

Patterns of Sexuality

Most of the animals that are familiar to us are divided into two sexes. They are all either female or male, depending on their chromosome complements. The body cells of most animals are diploid, so each cell contains two sets of chromosomes arranged in homologous pairs. One pair is the **sex chromosomes,** which contain the DNA instructions that determine the individual's sex, as well as some other characteristics. All the other homologous pairs of chromosomes are called **somatic chromosomes.** They control the rest of the individual's characteristics. The number of somatic chromosomes depends on the particular species; in humans, there are 22 pairs of somatic chromosomes. Typically, the sex chromosomes consist of two X chromosomes in the female and one X and

one Y in the male. But this scheme is not invariable: in frogs, for example, the XX combination belongs to the male.

Though most animals are either male or female, this pattern is by no means the only one. Many animals are **hermaphrodites** (named for the god Hermes and the goddess Aphrodite). A hermaphrodite possesses both types of sex organs. Earthworms and some garden snails are in this category; they produce eggs and sperm at the same time, but they do not fertilize themselves. Instead, individuals copulate in pairs, so that each receives sperm from the other and the eggs of both become fertilized. In other words, cross-fertilization occurs (see Figure 12-18).

In other hermaphrodites, eggs and sperm are produced at the same time, and self-fertilization will occur if no partner is around when the gametes are ripe. This scheme is common in parasites such as tapeworms, which live in the digestive tracts of animals. Even when only one adult tapeworm manages to find a suitable host, it can still reproduce.

Still other hermaphroditic animals produce both eggs and sperm, but not at the same time. An individual may be a functional male for a time and then become a functional female. Molluscs (clams and snails) often have this kind of reproductive pattern. The slipper limpets (*Crepidula fornicata*) are marine snails that cluster together into crescent-shaped stacks (see Figure 12-19). The large snails at the bottom are females, the small ones at the top

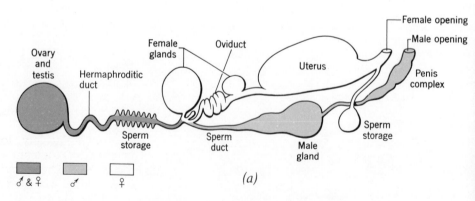

(a)

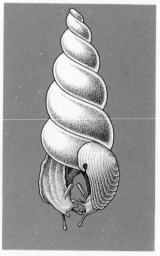

(b)

Figure 12-18 *Hermaphroditic animals.* (a) *The reproductive tract of a hermaphroditic freshwater snail.* (b) *Cross-fertilization in the garden slug. During copulation, each individual receives sperm from the other partner.*

Figure 12-19 Sexual reproduction in the slipper limpet, Crepidula fornicata. *The individuals are found in stacks with the largest females on the bottom and the smallest males on top. Several clusters are shown.*

are males. When the largest (and therefore the oldest) females die, the males above them increase in size and become females. New, small males then settle on the top of the commune. The same thing happens in some clams: the small clams are males, then most of them change into females as they grow. They may even change back; the same individual may function as male, female, and male in alternation.

Gametogenesis

Animals reproduce sexually by means of haploid gametes, eggs and sperm. Typically, these gametes are formed in clearly-defined organs called **gonads.** The male gonad is the **testis,** and the female gonad is the **ovary.**

Gametogenesis is the formation of gametes through meiosis. Special cells in the gonad, called **germinal cells,** ultimately give rise to the gametes through meiotic division. In the testes, **spermatogonia** enlarge to become **primary spermatocytes,** which undergo meiosis. The first meiotic division produces **secondary spermatocytes.** The second meiotic division then produces **spermatids.** The spermatids develop into sperm without further nuclear changes. Similarly, in the ovaries, germinal cells called **oogonia** enlarge to **primary oocytes.** The

primary oocyte produces a **secondary oocyte** and then an **ovum,** or egg, by meiosis (see Figure 12-20).

Gametogenesis begins in the gonad, but it may not be completed until after the gametes are shed from the testis or ovary. In human egg cells the first meiotic division occurs early in development, while the female is still an embryo. But the second meiotic division does not occur until the egg has been released from the ovary and is fertilized. Human sperm are also immature when they leave the testis, and they must go through a period of "ripening" in ducts associated with the male reproductive tract.

Eggs contain food reserves that supply energy for early development of the zygote. Eggs are usually non-motile. Sperm are motile, containing only enough food to allow them to reach the egg. For these reasons, there are differences in the way the two types of gametes are formed, though the basic principle is the same.

Sperm

When sperm cells differentiate from spermatids, they become little more than haploid nuclei with flagella. A typical sperm cell from an animal has three regions: a head, a midpiece, and a tail (see Figure 12-21). The head contains the haploid nucleus and is covered by a caplike organelle, the **acrosome.** The acrosome develops from dictyosomes in the spermatid. It facilitates the penetration of the egg membrane at fertilization. There is a lot of variation in shape of the sperm head. In humans it is a flattened oval, but in other animals it may be elongated, sickle-shaped, or spherical. However, the similarities between sperm cells outweigh any differences.

The midpiece of the sperm contains a spiral array of mitochondria that provide energy for the flagellar tail. Two centrioles are also present. One enters the egg, along with the nucleus, and participates in the first division of the zygote. The other controls tail movement. A few groups of animals (for example, crayfish) have sperm that are non-motile, but they are exceptions (Figure 12-22).

Sperm are able to survive for relatively long periods, even though they have little cytoplasm and few organelles to support metabolism. They survive because they remain in a quiescent state until an environmental factor causes their activation. At that point, flagellar movement begins. Depending on the animal, activation can be triggered by mixing with other secretions of the male reproductive tract to form **semen,** by exposure to

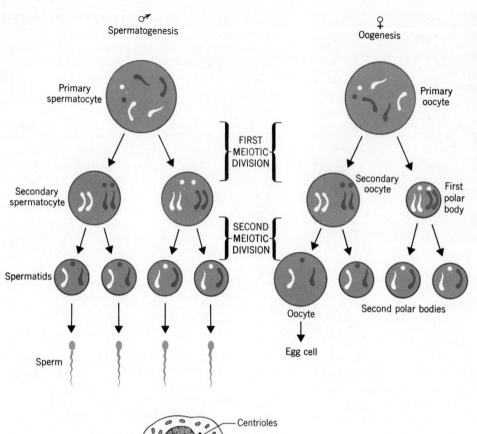

Figure 12-20 Spermatogenesis and oogenesis. Diploid primary spermatocytes and oocytes form haploid sperm and eggs through meiosis. A primary spermatocyte produces four sperm, but only one egg results from the meiosis of a primary oocyte. The other nuclei are sequestered into tiny polar bodies, which soon disintegrate. The first polar body may or may not divide.

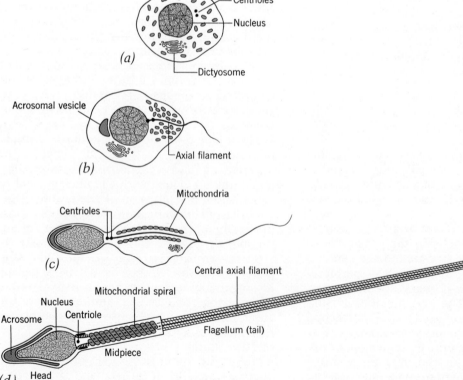

Figure 12-21 The development of sperm from spermatids. The head of the sperm cell is formed from the nucleus, and the acrosome is produced by dictyosomes. An axial filament is formed by microtubules in the region of the centrioles. The mitochondria assume a spiral arrangement around the filament to form the sperm cell's midpiece, and other fibers associate with the filament to form the flagellum.

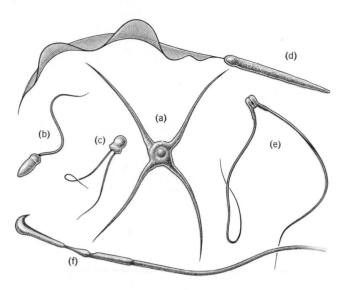

Figure 12-22 *Various types of sperm. (a) Crayfish. (b) Sea urchin. (c) Toadfish. (d) Toad. (e) Opossum. (f) Guinea pig.*

surrounding water, by entry into the female reproductive tract, or by other events. But until they are activated, sperm maintain a low metabolic rate and draw nourishment from other cells or from the surrounding fluid in a storage organ.

Hundreds or thousands of quiescent sperm are often packaged into larger units called **spermatophores.** The whole spermatophore is transferred to the female at copulation, and the sperm are then activated "on demand." Queen bees, for example, copulate with drones only once and store the sperm, meting them out and activating them during their entire reproductive life.

Sperm can be maintained in a quiescent state artificially for later use. Semen from bulls is kept cold in glycerin and then used for artificial insemination. In this way a bull with particularly desirable genetic traits can sire whole herds of cattle. Another major economic use of artificial insemination is in apiculture (growing honey bees). Honey bees mate only in flight, and it is difficult to retrieve the fertilized queen so she can get down to work. It is easier to keep track of the female if she is artificially inseminated and placed in a new hive rather than permitted to mate normally. Human sperm can also be kept in cold glycerin with no apparent harmful effect. Such a human sperm bank might have several functions. A man undergoing vasectomy could store some sperm just in case he

decides to have another child. (Vasectomy is an operation that produces sterilization by cutting the duct that conveys sperm from the testis; it is not always reversible.) Others have suggested that sperm of selected men might be collected and saved for controlled human breeding to upgrade the genetic potential of the human population. But this scheme has some obvious social difficulties, which don't require discussion here. (It is not necessarily a "male chauvinist" proposal, though, since it depends on the fact that we know how to preserve sperm but not eggs.)

The testis, or male gonad, contains various kinds of cells in addition to spermatogonia. Typically, testes have cells for the support and nourishment of developing sperm and cells that secrete hormones. The overall form of the gonad varies tremendously. In some cases, the testis (and the ovary) is much reduced during the non-breeding season and becomes prominent only at the onset of sexual reproduction. In other cases, the male may consist of nothing but testis and survives as a sac-like parasite of the female. This is the case in some organisms that might have trouble finding a free-living mate, such as deep sea fishes and sedentary worms.

Eggs

Eggs cover a much greater range of size than sperm. The size of an egg is related to the amount of nourishment that it must supply to the developing embryo before outside nutrition becomes available. The mature egg of humans and other mammals is small, only about 100 microns in diameter. It need not be larger, because it develops inside the mother; nourishment is provided by her tissues. At the other extreme, the yolk of an ostrich egg cell is about twenty centimeters in diameter. It is about a billion times larger than a human egg. Most of this tremendous volume is stored food (fat and protein), which is used to support the development of the chick until it is ready to hatch. Obviously, such eggs are rich food sources for other animals as well.

The general organization of eggs varies as well as the yolk content. Human eggs have a central nucleus and a small amount of yolk, evenly distributed throughout the protoplasm of the egg cell. Yolky birds' eggs consist of a small disc of protoplasm containing the nucleus, and located on the surface of the yolk. Insect eggs have still a different organization. In insect eggs, the yolk is in the center, surrounded by a layer of protoplasm (Figure 12-23).

In humans and most mammals, egg development begins in **follicles.** A follicle is simply a pocket of

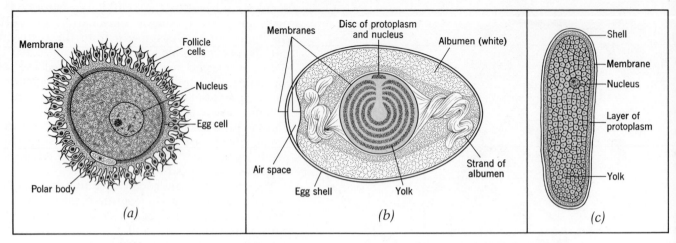

Figure 12-23   *Various types of egg.* (a) *Mammal.* (b) *Bird.* (c) *Insect.*

the ovary that is lined with cells. Follicle cells adjacent to the egg cell provide nourishment and material for its growth. Only one mature egg develops from each oogonium. The first meiotic division of the primary oocyte divides the cytoplasm very unequally. One nucleus becomes the secondary oocyte, which will divide again to form the egg. The other nucleus is pinched off from the main body of the developing egg into a minute portion of cytoplasm. This cell is called the **first polar body.** It plays no further role in the development of the egg, but it may divide once more before it degenerates. The secondary oocyte then undergoes a second meiotic division. This division produces the ovum and another daughter nucleus, which becomes the **second polar body.** The second polar body also degenerates (see Figure 12-20). The release of the developing egg from the follicle is called **ovulation.** The nuclear events are the same in oogenesis and spermatogenesis. But in spermatogenesis, all four nuclei produced develop into sperm, while in oogenesis, only one nucleus becomes an egg.

A human female produces only a few hundred viable eggs during her reproductive life. And her potential fecundity is actually much lower than this figure, because a pregnant woman does not produce mature eggs. The record for childbearing by one woman is 69 (27 confinements, many multiple births). By contrast, a codfish may produce a million eggs, all fertilized and all capable of development into a new adult. Some kinds of tapeworms may produce hundreds of thousands of fertile eggs a day, over a life span of years. The number of fertilized eggs an animal produces is essentially a measure of the probability that they will survive. Mammals have small numbers of eggs but a high survival rate. A tapeworm egg has a very low chance of survival, since it must leave one host and find another before it can develop into an adult.

Fertilization

Fertilization is the fusion of a haploid sperm nucleus and the haploid egg nucleus. This fusion produces a zygote with diploid chromosome number. The zygote then develops into the new individual.

Fertilization is really a complicated process. In mammals, the sperm prepares for fertilization as it moves up the female reproductive tract. As the sperm nears the egg, it secretes an enzyme to dissolve surrounding mucus and speed its passage through the female reproductive tract. The actual penetration of sperm into the egg follows many patterns. It has been most thoroughly studied in the sea urchin, a relative of the starfish. When a sea urchin sperm contacts the egg, the acrosomal cap on the sperm produces a stiff, elongated **acrosomal filament,** which touches the egg. Then the egg cytoplasm bulges up around the filament, and a clear area appears below. This area forms the **fertilization cone,** through which the sperm enters (see Figure 12-24). The entry of additional sperm at this point is somehow prevented. Usually only one sperm penetrates the egg membrane, even though

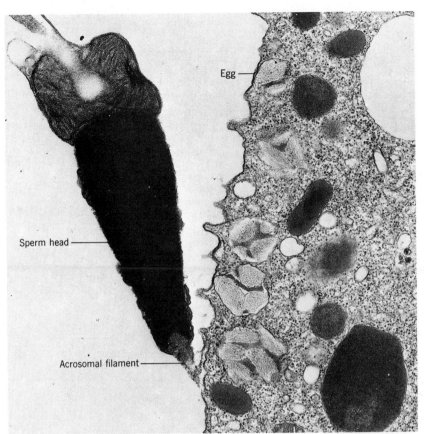

(a)

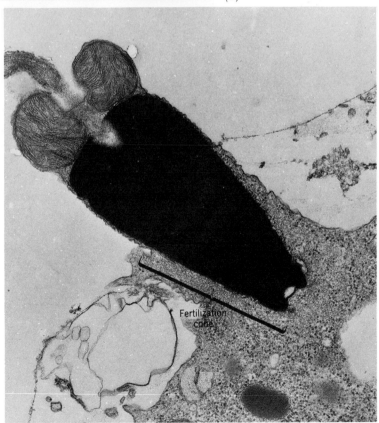

(b)

Egg

Sperm head

Acrosomal filament

Fertilization cone

Figure 12-24 Fertilization. (a)
Upon contact with the egg, the
sperm produces an acrosomal filament, which penetrates the egg
membrane. Magnification
15,000×. (b) The cytoplasm
bulges up around the filament,
forming a funnel-shaped path
called the fertilization cone,
through which the sperm head
enters. Magnification 23,000×.

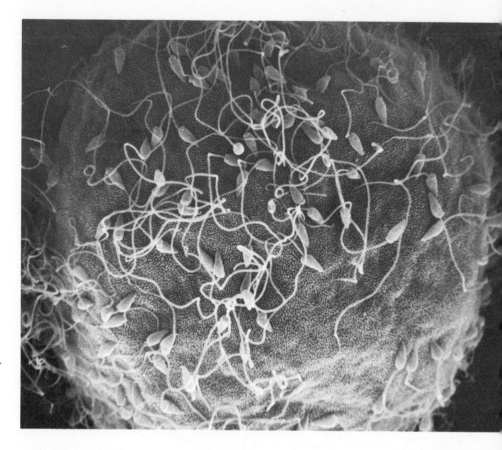

Figure 12-25 Scanning electron micrograph shows sperm attached to the membrane of a sea urchin egg a few seconds after they were mixed together. Only one sperm will penetrate and fertilize the egg. Magnification 3,000×.

millions of sperm cells may be present swarming around the egg (Figure 12-25). A membrane may form around the egg after fertilization to act as a barrier; the entry of sperm might also be prevented by droplets of mucus in the egg cytoplasm that liberate their contents just beneath the membrane surface.

After the entry of the sperm, the oocyte is said to be "activated;" it then completes the second meiotic division to form the egg nucleus, now called the **female pronucleus.** The haploid sperm nucleus (now called the **male pronucleus**) enlarges and migrates toward the center of the egg. The female pronucleus also migrates to the center. The membranes that surround the male and female pronuclei disappear, and fusion occurs, producing a zygote. Mitotic cell division commences.

Activation of the egg usually involves sperm penetration. However, egg division can be started artificially in several ways. Treatment with poisons, salt solutions, heat, or cold, or simply pricking the surface of the egg with a needle, are effective methods in some animals. If artificially activated

eggs are permitted to develop, normal adults may result. This process is called **parthenogenesis** (Greek for "virgin birth"). The largest animal to have been successfully produced by parthenogenesis is a rabbit. Rabbit oocytes were collected after they were shed from the ovary. They were then placed in a tissue-culture medium. This procedure activated some of the oocytes. The activated oocytes were then implanted in the uterus of another rabbit. Two of them developed normally. The rabbits turned out to have diploid cells, even though they developed from a haploid egg cell. The diploid condition must have been restored early in their development, presumably by a mitotic duplication of chromosomes that was not accompanied by cytokinesis. For this reason, both of the parthenogenic rabbits had to be females. The haploid set of chromosomes in an egg always contains one X chromosome. Therefore, duplication of the X chromosome to restore the diploid condition produced two X chromosomes.

Some of the insects and their relatives undergo parthenogenesis as a normal part of their life cycle.

As we shall see, all male bees develop this way. In this case, the diploid condition is not restored; all the male bee's cells are haploid.

For successful fertilization, the eggs and sperm must be brought together in a proper liquid medium. The gametes must mature simultaneously and then be brought into contact. Close cooperation between the participants may be required. In fact, much of animal behavior that is not directly related to food-catching concerns reproduction. The mechanics of fertilization tend to be relatively simple in aquatic animals and rather elaborate in animals that dwell in dry, terrestrial environments. The reason is simple: aquatic organisms may simply shed their eggs and sperm into the surrounding water, but land animals must themselves provide the proper liquid environment for fertilization.

Many marine animals liberate their eggs and sperm into the surrounding sea water, and fertilization occurs externally. But the production and release of gametes by different individuals must still be coordinated so that eggs and sperm are present in the water at the same time. The release may be triggered by changes in daylength, by a temperature change, by the presence of other gametes in the water, or by a combination of such factors. These timing devices are often very precise. The palolo worm, a marine animal, responds to a combination of daylength and phase of the moon that occurs only one day a year. The worms swarm to the sur-face of the sea to shed their gametes, which are carried in specialized segments of their bodies. These segments break off from the adult worms, swarm around, and split open to liberate the gametes. At this time, it is claimed, the worms are so dense that an oar placed vertically in the water will stand upright for a time. Another kind of marine worm will luminesce, or glow, during swarming in order to attract other worms to the same place.

Some aquatic animals release only their sperm into the surrounding water. The eggs are retained by the females. Sponges, for example, release sperm which are collected by the females. So-called **collar cells** collect the sperm cells just as they normally collect food. However, instead of digesting the sperm, the collar cell loses its flagellum and becomes amoeba-like. It then migrates over to an egg and releases the sperm, and fertilization occurs (Figure 12-26). The retention of the egg in this manner may be advantageous to the animal, because eggs require more energy for synthesis than sperm, particularly if they are large and yolky. Fewer eggs are wasted if they are retained and protected.

The storage of sperm and their direct transfer to the female is called **internal fertilization.** Internal fertilization has evolved independently in many groups of animals, and many methods exist. Presumably internal fertilization does improve the chances for a successful union of egg and sperm nuclei. In land animals, internal fertilization is not just advantageous, it is a necessity. Amphibi-

Figure 12-26 Fertilization in sponges. (a, b) A collar cell engulfs a sperm cell. (c) The collar cell migrates to the vicinity of an egg and transfers the sperm to the egg. (d) The sperm and egg nuclei fuse, producing a zygote.

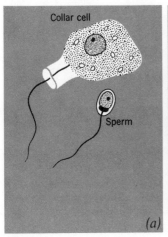
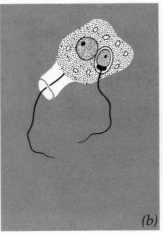
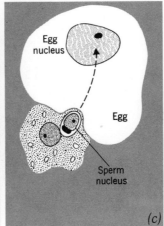
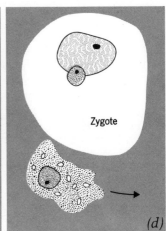

ans—frogs, toads, salamanders—still return to water to breed, and some of them use external fertilization. But reptiles, birds, mammals, insects, and other terrestrial animals must use internal fertilization to provide a liquid environment for gamete fusion.

In vertebrates, unlike other animal groups, there is an intimate relation between the urinary and the reproductive systems—in other words, they use much of the same apparatus. Presumably this relation is a result of their gradual adaptation from the sea to fresh water to land. As the ancestors of the present-day land vertebrates gradually moved upriver, they began to encounter osmotic problems. They could not survive unless they evolved ways to keep the entry of fresh water from diluting their body fluids. The development of the kidney's glomerular filtration units was one response to this environmental pressure; another was to minimize the number of body openings through which water could enter. Hence, it can be argued, it was more efficient for the male to use one duct both to convey urine to the exterior and to convey sperm to the female. This account is speculative, but it does explain the intimate relationship between urinary and genital tracts found in the vertebrates and not in other major animal groups. Whatever the origin of this relation, many human individuals and even whole societies still confuse elimination and reproduction. The human urinary and reproductive tract is typical of the mammalian pattern; it is diagrammed in Figure 12-27. A detailed account of human reproduction is provided in Chapter 14.

Hormonal Control of Reproduction

Oogenesis and spermatogenesis are controlled by hormonal systems. These systems respond to environmental signals conveyed by way of the nervous system. Such control assures the simultaneous maturation of eggs and sperm. In most animals, **neurosecretory cells** provide the link between the external environment and the gonads. As their name implies, neurosecretory cells have character-

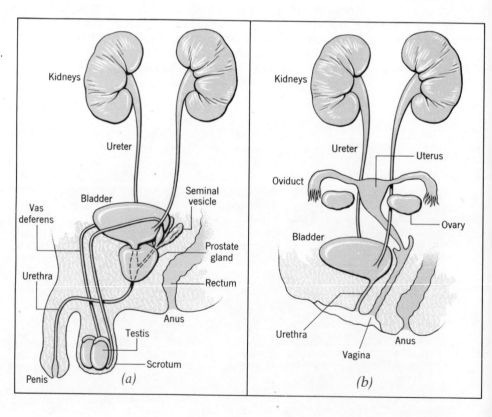

Figure 12-27 The human urinary and reproductive tract. (a) Male. (b) Female.

istics of typical nerve cells, but they also can secrete substances. Suitable staining reveals secretory granules in such cells. The electron microscope shows that these granules are actually membrane-bounded droplets. The droplets move to the end of the neurosecretory axon, where they are discharged into a capillary.

In vertebrates, groups of neurosecretory cells concerned with reproduction are located in the hypothalamus in the brain. Most of the axons from these cell groups end up at the junction of the pituitary and the hypothalamus. There the axons secrete products called **releasing factors** into a capillary bed; the releasing factors are then carried to the anterior portion of the pituitary, where they stimulate the release of hormones. The pituitary hormones then pass into the general circulation.

The anterior pituitary produces several hormones; each is controlled by its own releasing factor. Two of these hormones are called **gonadotropins,** because they act on the gonads. Gonadotropins stimulate oogenesis and ovulation in the female, and spermatogenesis in the male. Since the gonadotropins are controlled by releasing factors from the hypothalamus, we should be able to modify their effects by stimulating or by destroying portions of the hypothalamus. In fact, stimulating these areas can cause ovulation, and tissue destruction may prevent it.

There is other evidence to show that the nervous system controls the secretion of pituitary hormones. In rats, changes in female reproductive physiology are periodic; in other words, ovulation and the associated changes in the reproductive tract occur in cycles. The male has no cycle. But if the anterior pituitary of a female rat is replaced by the anterior pituitary from a male, her reproductive cycle will continue. The male pituitary now produces the proper hormones in the right amounts and at the right times to support a normal female reproductive cycle. The reason is that the glandular activity of the anterior pituitary is controlled by the female's own releasing factors.

The hormones produced by the anterior pituitary stimulate the gonads to secrete the so-called sex hormones. The ovary secretes two kinds of female sex hormones, **estrogens** and **progestogens.** The testis secretes several male sex hormones called **androgens.** Sex hormone secretion sets in motion a feedback system: the sex hormones act on the hypothalamus, and the hypothalamus in turn inhibits the secretion of the pituitary hormones (see Figure 12-28).

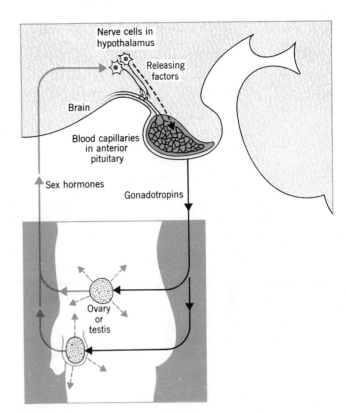

Figure 12-28 *Hormonal control of reproduction. Releasing factors from brain neurons stimulate the anterior pituitary to secrete gonadotropins. The gonadotropins control the development of the gonads, the maturation of gametes, and the secretion of sex hormones. The sex hormones—estrogens and progestogens in the female, androgens in the male—in turn influence the production of releasing factors, forming a feedback loop.*

Reproductive Behavior

Successful fertilization requires behavioral cooperation between the male and the female. This requirement becomes overriding in species that depend on internal fertilization. Rape is rare in the animal kingdom; more often, copulation is prolonged and enthusiastic. Nonetheless, reproductive behavior can be very aggressive. A male crayfish, for example, may simply tire of courtship, throw the female on her back, and place his spermatophore in a special external storage pouch near her reproductive tract. A remarkable instance of aggressive courtship occurs in one kind of spider. The male traps an immature female and keeps her prisoner in a silken tent he spins. When she matures into an adult (and is thus fertilizable), he tears open the tent, fertilizes her, and dashes off. If he were to wait, her new exoskeleton would harden; not only could she then defend herself, but she would probably eat him.

The life of male spiders is often risky. The male is usually much smaller than the female, just about snack size. He must therefore find a way to persuade the female to temporarily abandon thoughts of food and think about sex. The male usually approaches the female with considerable caution until she shows some interest in sexual behavior. A web-weaving male spider may stand at the edge of the web and agitate it rhythmically so that the female will not mistake him for a trapped fly. A male hunting spider may have brightly-colored appendages for identifying himself to a female of the same species (the process is called **species recognition**). As he walks around in search of a female, he stops every few steps and waves his appendages so he will not be mistaken for food. The procedure is complicated by the fact that both he and his potential mate have poor eyesight. Males of another species of spider initiate sexual behavior by catching flies. The male then attaches the fly to a hollow ball of silk and presents the package to the female. He quickly fertilizes her while she is occupied with his present. Sometimes, overanxious males can't find flies and will substitute dried insect husks. In such cases they are promptly eaten by the female. In all of these cases, the male is expendable once fertilization has occurred. From an evolutionary viewpoint, if he can end his life by providing food for the female, so much the better.

Other patterns of sexual behavior provide for species recognition and for the controlled release of eggs and sperm. The courtship of the stickleback, a small freshwater fish, is a good example. As the breeding season approaches, the male constructs a tubelike nest by cementing small stones and other material with a sticky secretion from his kidneys. He awaits a female and defends the area around his nest from other males. When a female swims by, elaborate courtship behavior begins. The male starts with a "zigzag dance." If the female is ready to breed, she swims straight toward him. He then leads her to his nest. He points out the opening with his head, and she enters. While she is in the nest he stands on his head and touches her tail, inducing her to lay eggs. She then leaves, and he immediately enters and sheds sperm to fertilize the eggs. The male then settles down to the task of fan-

ning water through the nest until the eggs have developed and hatched (Figure 12-29). Each step in the process depends on specific interactions between male and female. If the sequence breaks down at any point the whole enterprise is dropped. The individual signals are very simple: for example, the male stickleback recognizes other males because of their red bellies. But the whole sequence is so improbable that only two sticklebacks can carry it to completion. This mechanism guarantees that only sticklebacks will mate with sticklebacks.

Reproductive behavior in birds is notoriously complicated. It often involves elaborate courtship. Pairing may last for life. Threatening behavior toward others of the same species establishes territory for the breeding pairs and insures a good food supply for the developing chicks. Nest building, egg laying, hatching, and the subsequent care of the young are all rich fields for descriptive natural history.

life cycles

Birth, and copulation, and death.
That's all the facts when you come to brass tacks;
Birth, and copulation, and death.

— T. S. Eliot

Individual human beings do go through a process of development from birth to death. But the species as a whole goes through a life cycle. The life cycle is the alternation of diploid individuals with the haploid gametes they produce. The fusion of the gametes makes the process cyclic by producing new diploid individuals. A life cycle includes only the period from birth to copulation. The survival and evolution of the species depends on its operation. But the post-reproductive individual—one in the period between copulation and death—is outside the loop. Such an organism is a pointless tag end from an evolutionary point of view, because it can no longer create new individuals or pass its genes on to another generation.

Many different life cycles exist. They are usually described in terms of the number of genomes per cell, and in terms of the occurrence of mitosis, meiosis, and fusion of gametes. Asexual reproduction involves only mitosis; fusion and meiosis do not occur. But sexual life cycles involve a **haploid generation** (that is, individuals with haploid cells), the fusion of the haploid cells to create a **diploid**

Figure 12-29 *Courtship behavior in the stickleback.*

generation (individuals with diploid cells), and the production of haploid cells again from the diploid individual by meiosis. This cyclic process is called the **alternation of generations** (see Figure 12-30). Either generation or both may be structurally complicated and conspicuous in a particular species. But in many cases one generation is dominant, and the other is small or present only briefly and inconspicuously.

The life cycle of man and many other animals is dominated by the diploid generation. The diploid phase starts with a fertilized egg and continues until the mature organism produces gametes. This portion of the life cycle is obviously the most conspicuous and complicated. The haploid phase consists simply of the gametes produced by meiosis. These gametes do not normally divide, nor do they have any conspicuous independent life. They exist only in a carefully maintained environment within the diploid individual. It is usually assumed that the dominance of the diploid phase is typical of animal life cycles, but in fact, the life cycles of many lower animals have not been carefully stud-

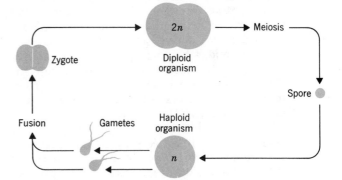

Figure 12-30 *Many organisms have life cycles in which a diploid stage (2n chromosomes) alternates with a haploid stage (n chromosomes). The diploid organism produces a haploid form by meiosis. The fusion of haploid gametes restores the diploid condition. Either stage or both may be conspicuous or absent in different organisms.*

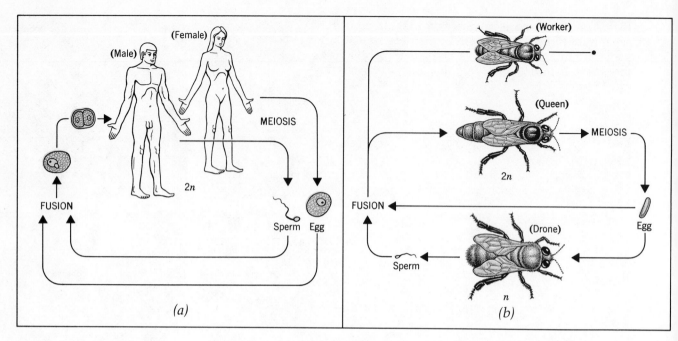

Figure 12-31 Animal life cycles. (a) In humans, the diploid organism is the only free-living form. (b) In bees, both haploid and diploid adults are present.

ied. In animals such as bees, the cycle is known to be much more complicated. Both worker bees and queen bees are females, and both have a diploid chromosome number. The differences between them are caused by diet. Drone bees, however, are haploid males derived from unfertilized eggs. The drone provides haploid sperm during mating, but his sperm are derived from mitosis, not meiosis. Thus, drones are conspicuous and free-living animal representatives of the haploid phase of the life cycle (Figure 12-31).

Plants exhibit a much greater variety of life cycles than animals. Any cycle of sexual reproduction must at some point involve meiosis and a subsequent fusion of gametes. But in plants, these events occur at various places in the life cycle, depending on the species. Many plants do have a simple alternation of haploid and diploid generations. Some plants, however, may have as many as four distinctive plant forms and may produce several kinds of asexual spores, as well as sexual gametes. Reproductive patterns in plants—particularly in algae—become so complicated that many botanists prefer to speak of "life histories" rather than "life cycles."

plant sexual reproduction

A plant that produces gametes is called a **gametophyte** (*phyton* is the Greek word for "plant"). A gametophyte is itself haploid. It produces gametes by mitosis in specialized cells or structures called **gametangia.** Fusion of gametes produces a zygote that develops into a diploid plant called a **sporophyte.** The sporophyte produces haploid spores by meiosis in structures called **sporangia.** (Since they arise by meiosis, they are called **meiospores.**) The spores develop into gametophytes. This terminology is complicated by the fact that some gametophytes and sporophytes can also produce asexual spores: diploid asexual spores may be liberated from the diploid sporophyte; haploid asexual spores may be formed by the haploid gametophyte.

Probably all plants have some kind of alternation between haploid and diploid phases. But this basic similarity is often obscured by the dominance of one phase over another, or by the presence of several forms of asexual reproduction. Since plant life histories are extremely varied, we shall consider separately three basic groups in the plant kingdom: aquatic plants (the algae); simple land plants (the mosses and ferns); and complex terrestrial plants (seed plants).

Algae

Life histories may take many forms in the algae. One is like a typical animal life cycle. It occurs in some green algae that are common in the ocean. *Caulerpa*, for example, produces fields of small, green plants in tropical waters. The plant is diploid and produces gametes through meiosis. The haploid gametes fuse to give a diploid zygote, which develops into a diploid adult. As in animals, the haploid, gametophyte stage consists only of the gametes.

Another pattern is seen in the unicellular green alga *Chlamydomonas*. The free-living individuals of this genus are haploid. Each cell may produce one or more haploid gametes by mitosis. The gametes in many species of *Chlamydomonas* look identical, but they behave like two different sexes. As soon as two gametes fuse to form a zygote, the zygote undergoes meiosis, giving rise to four haploid individuals. The zygote is the only diploid cell in the life history (see Figure 12-32).

A third type of life history is found in algae with both haploid and diploid plants. Kelp plants grow in this way. The kelp sporophyte is large and conspicuous. It may reach 200 feet and weigh over one ton. The sporophyte produces meiospores in sporangia on special blades at the base of the plant. These spores are released into the water; they settle to the bottom and develop into male and female gametophytes. The kelp gametophyte is a microscopic filamentous plant that produces eggs or sperm. When the egg is fertilized, it grows into the giant sporophyte (Figure 12-33).

These contrasting life cycles all result from varying the timing of meiosis and fusion. When fusion follows meiosis immediately, the life cycle is like that of animals, and the dominant phase is diploid (illustrated by *Caulerpa*). When meiosis follows fusion immediately, the haploid phase is dominant (*Chlamydomonas*). The separation of meiosis and fusion leads to an alternation of well-defined haploid and diploid generations (kelp).

Other algal life cycles can become much more complex. A red alga, for example, may have four separate plants in its life history; they include

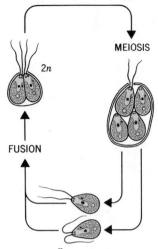

Figure 12-32 *The uni-cellular green alga* Chlamydomonas, *has a life cycle in which the haploid generation predominates.*

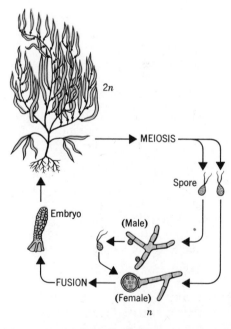

Figure 12-33 *The brown alga* Macrocystis *has both haploid and diploid free-living individuals. The diploid plant is large and complex, and the haploid plant is small and inconscipuous.*

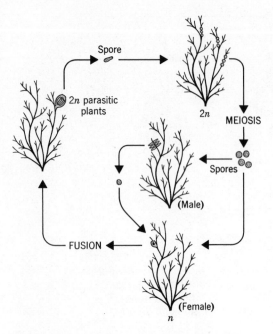

Figure 12-34 *The red alga* Polysiphonia *has haploid plants and two kinds of diploid plants. One grows parasitically on the female haploid plant, and the other is a large, free-living plant.*

separate male and female haploid plants and two kinds of diploid sporophytes. One sporophyte is a free-living form; the other develops on the female plants (Figure 12-34).

In many cases we simply don't know the sequence of stages in a particular life cycle. In other cases, certain stages have never actually been seen and must be supplied by guesswork. But our information gets better as techniques for culture of organisms in the laboratory improve.

Mosses and Ferns

The Bryophyta—the mosses and their relatives —depend on water for reproduction. In the bryophyte life cycle, the haploid gametophyte is the dominant form. The gametophyte begins as a meiospore. The spore germinates in moist soil and forms a branching, filamentous structure that eventu-

ally grows into a leafy gametophyte. This gametophyte is the conspicuous moss plant that can be found in damp, sheltered places (Figure 12-35).

The gametophytes produce eggs or sperm, or sometimes both on the same plant. If the soil and the plant are wet enough, the sperm can swim to the egg in the female gametangium. The fertilized egg then develops into a diploid sporophyte while still attached to the gametophyte. In fact, the sporophyte is permanently dependent on the gametophyte for its water and nourishment. Eventually, the sporophyte elongates above the leafy gametophyte and produces a capsule containing meiospores. As the capsule dries out, it opens and the haploid spores are liberated to begin the cycle again (Figure 12-36).

The ferns and their relatives have an alternation of generations in which the diploid sporophyte gen-

Figure 12-35 *A moss. The leafy gametophyte is the dominant plant. As sporophytes develop they remain attached to, and dependent on, the gametophyte.*

Sporophyte

Leafy gametophytes

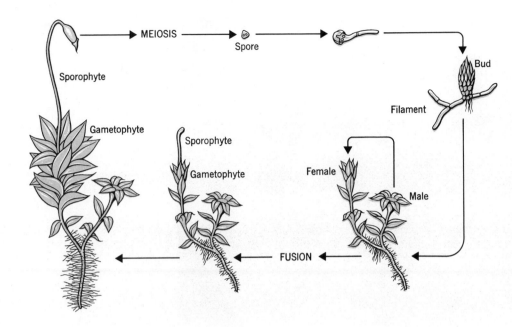

Figure 12-36 *Life history of a moss. The spore germinates and forms a filamentous mass of cells that differentiates into the leafy gametophyte plant. The gametophytes produce eggs and sperm, and the fertilized egg grows into an attached sporophyte. The sporophyte produces a spore capsule, within which spores are produced by meiosis, completing the cycle.*

eration is the dominant form. The sporophyte plant forms the large and familiar leafy structure, while the gametophyte is just a small, prostrate sheet of cells. The leaves of fern sporophytes bear meiospores on their lower surfaces. The spores are shed and germinate into small gametophyte plants called **prothallia.** The prothallia are flat, green, and heart-shaped; they are short-lived and quickly produce their eggs and sperm. As in mosses, the sperm are free-swimming, and water is required for fertilization. This requirement is an important one, because it limits the distribution of ferns to moist, damp places. Again, the egg is retained on the gametophyte. The fertilized egg produces a sporophyte attached to the prothallium (Figure 12-37).

Seed Plants

The life cycles of seed plants—gymnosperms and angiosperms—show the same sort of alternation of generations that is found in algae, mosses, and ferns. But in seed plants, the sporophyte generation is now completely dominant. The gametophyte is reduced to a few cells and a completely dependent existence: the male gametophyte is a pollen grain; the female is a small group of cells growing within the tissues of the sporophyte, within a cone or a flower. A key feature of seed plants is their ability to achieve fertilization without water. This ability has allowed them to dominate land habitats. The male gametes, encased in pollen grains, are carried directly to the female gametophyte by various means—wind, or insects and other animals. Fertilization can occur without external water.

Gymnosperms

The life cycle of a pine is typical of gymnosperms. The pine tree represents the sporophyte phase of the life cycle. Its meiospores are produced in male and female pine cones. The male cones consist of a whorl of small scales. The male meiospores, called **microspores,** are formed by meiosis in specialized structures on the lower surface of each scale. Each of the resulting four haploid spores then divides several times to produce a pollen grain. The mature pollen grain contains two functional haploid nuclei, one of which will fuse with the egg, as well as other nuclei involved in maintaining its own structure. The pollen grain is actually an immature male gametophyte plant; its development is completed in the female cone.

Female cones are a whorl of larger woody scales. **Megaspores,** the female meiospores, are formed by meiosis in specialized structures called **ovules** on the upper surfaces of the scales. Only one of each set of four haploid megaspores becomes a female gametophyte; that megaspore divides about a dozen times to form the female gametophyte. The female gametophyte plant develops within the ovule, and

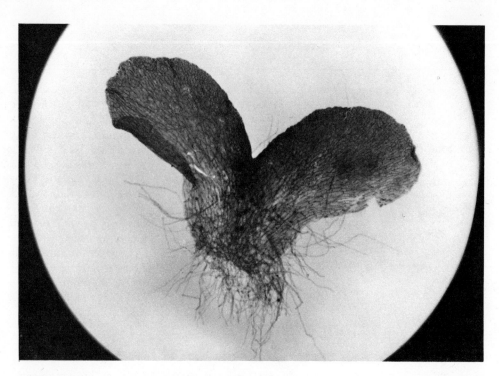

each one may contain several eggs. Nourishment for the gametophyte is provided by surrounding sporophyte tissue. In pines, it may take more than a year for the female gametophyte to mature.

Male cones are small (about a centimeter in length) and are borne in clusters on lower branches of the tree. The female cones are the larger and more familiar pine cones borne in higher branches. This arrangement promotes fertilization between different trees rather than between male and female cones on the same tree. Pollen is carried by the wind to fertilize the egg in the female gametophyte cone. Since air rarely moves straight up, it is more likely that pollen will move to the female cones on neighboring trees.

Fertilization begins when pollen drifts between the scales of the female cone and is trapped in a sticky liquid. As this liquid dries, the pollen is drawn toward an opening at the end of the ovule called the **micropyle.** At this point the pollen (that is, the male gametophyte) begins to grow again. It sends out a **pollen tube,** a long extension that moves into the micropyle and toward the female gametophyte. This process, too, takes about a year. During this time, the male gametophyte is nourished by the ovule tissue. Finally, one of the haploid sperm nuclei moves from the pollen tube into the female gametophyte and fertilizes an egg nucleus.

After fertilization, the new zygote begins to develop into an embryo sporophyte within the seed. Another year elapses before the seed has matured to the point where it is capable of germinating and producing a new pine tree (see Figure 12-38).

Angiosperms

The life cycles of angiosperms resemble those of gymnosperms. The gametophyte generation is brief and is parasitic on the sporophyte generation. Both the female and the male gametophyte have only a few cells.

The flower is the reproductive organ of the angiosperms. The flower is actually a terminal shoot bearing four whorls of modified leaves. The four sets of leaf elements are attached to the end of the stem, called the **receptacle.** The modified leaves of the lowest whorl are called **sepals;** next are the **petals,** then the **stamens,** and finally the **pistil.** Sepals are often green and resemble foliage leaves. Flower petals are usually large and brightly colored. A ring of stamens is usually found within the petals and surrounding the pistil. Stamens and pistils are very unleaflike, but studies of their growth and

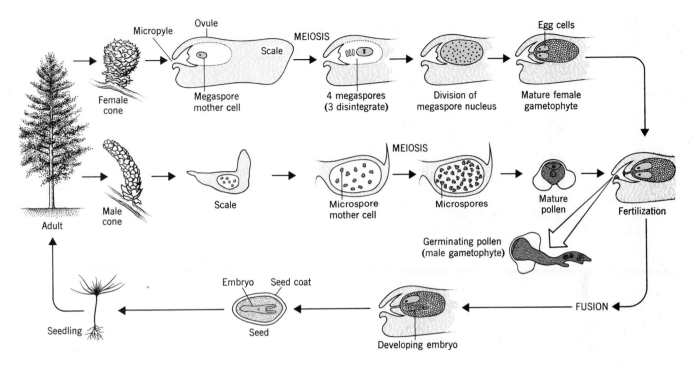

Figure 12-38 The life history of a gymnosperm, a pine tree.

development reveal a similarity to leaf structure. The leaf elements that comprise the pistil are usually fused together to form a vase-shaped structure (see Figure 12-39).

Stamens are the site of pollen formation. Male meiospores are produced at the tip of the stamen in a structure called the **anther.** Eggs are produced from female meiospores in the pistil. A pistil has three basic parts: a **stigma,** a **style,** and an **ovary.** The stigma, at the tip of the pistil, is the specialized region for reception of pollen. Fertilization begins when pollen is transferred to the stigma. The pollen grain sends out a pollen tube that moves down the style to the ovary, and in through the micropyle to the egg cell. One of the haploid nuclei in the pollen grain occupies a position at the tip of the pollen tube. This nucleus is called the **tube nucleus.** The other haploid nucleus divides into two **sperm nuclei,** which lie behind the tube nucleus. When the pollen tube reaches the micropyle, the sperm nuclei are discharged into the female gametophyte. One of them enters the egg cell and fuses with the egg nucleus to restore the diploid condition. The other sperm nucleus associates with the two haploid nuclei in the central cell of the female gametophyte. That cell is now triploid—it has three full sets of chromosomes. The triploid

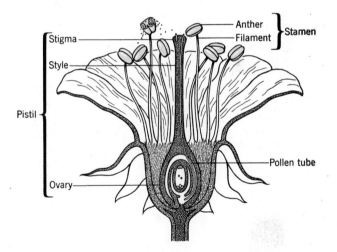

Figure 12-39 Flower structure.

nucleus divides repeatedly, and the resulting cells synthesize food for the developing embryo sporophyte. This food is called **endosperm.** Thus there are two separate nuclear fusion events in angiosperm fertilization: the restoration of the diploid condition in the zygote, and the production of triploid endosperm to service it. The process is termed **double fertilization.**

Directly after fertilization, the zygote begins to develop into an embryo sporophyte. Distinct leaves called **cotyledons,** a shoot, and a root soon become detectable. The sporophyte is surrounded by endosperm and is covered with several layers of tough floral tissues, which form a seed coat (Figure 12-40).

Thus, seeds are simply embryo sporophytes, packaged with a food source and covered with a protective layer. The food supplies energy for the growth of the embryo when it germinates. This food must support growth until the plant is large enough to expose its leaves to sunlight and photosynthesize. Sepals, petals, and stamens are usually lost from the parent plant as seed development proceeds. The ovary enlarges and develops layers of tissue that cover and protect the seed. These tissues are the **fruit.** In everyday usage "fruit" refers to seeds covered by fleshy and edible portions of the ovary. However, technical use of the term includes all ovarian tissue and seeds, whether they are edible or not.

Figure 12-40   The life history of an angiosperm.

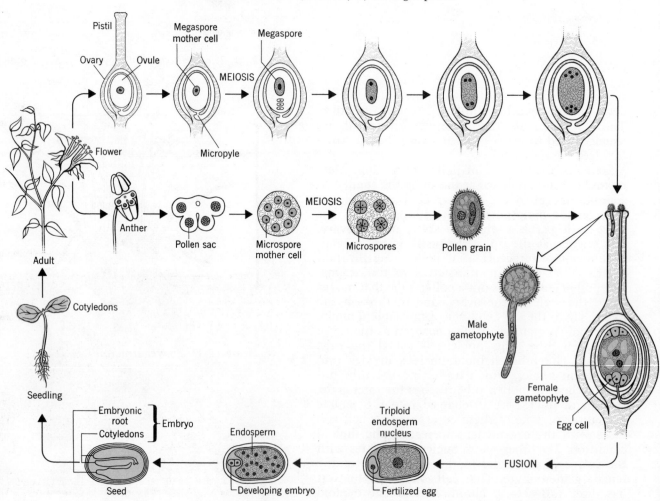

Pollination

Seed plants came to dominate the land because they devised ways to achieve fertilization in the absence of water. Sperm is packaged into pollen grains and is transferred by various means from the anther to a pistil for fertilization.

The pollen from many gymnosperms and from some angiosperms is windborne. Successful pollination by windborne pollen depends on several factors. First, huge quantities of pollen must be produced. Second, the release of pollen has to be timed to match the receptivity of the egg-bearing structures. And finally, various devices must be used to promote **cross-fertilization** (pollination between different individuals) rather than **self-fertilization.** Pines, for example, use all of these methods to promote fertilization. In the spring, so much pine pollen is produced that it forms visible yellow clouds in the sky. The maturation of the male and female cones is timed by daylength. Cross-fertilization is promoted by the position of the cones (the males on the lower branches, the females above).

Analogous devices are used by wind-pollinated angiosperms. The style may be branched or elongated to increase the probability of wind pollination ("corn silk" is actually such a style). And there are several ways in which self-fertilization is prevented. The male and female flowers may be situated on different plants or on different parts of the same plant. Alternatively, the pollen on a flower may mature before the pistil does.

Pollen from flowering plants is often carried from one flower to another by birds, mammals, and many kinds of insects. Everyone is familiar with the role of honey bees in the pollination of flowers. The bees collect nectar and protein-rich pollen produced by the flower and thus obtain food. In return, some of the pollen collected by the bee is inadvertently transferred to another flower and deposited on its stigma, so that fertilization is accomplished. In this way honey bees serve as pollinators for a wide variety of flowering plants.

The flower attracts its insect pollinator by scent or by sight. Flower color is an important factor. Bees are sensitive to light in the ultraviolet region. **Nectar guides**—lines along the petals which lead to the nectar—may be invisible to our eye but well-defined in the ultraviolet light visible to bees (Figure 12-41). Many scents are used to attract pollinators, and some are not very pleasant to the human senses. Some flowers give off an odor of rotten meat to attract flies.

Once the pollinator has been attracted by scent or vision, further inducement is offered to make sure that pollen is transferred between insect and flower. In the case of the honey bee, the flower offers food in the form of nectar and pollen, directs the bee to it by nectar guides, and delivers as promised. In other cases, food may be supplied in the form of pseudo-pollen—grains of waxy material rich in carbohydrate but poor in protein; the real pollen is conserved and becomes attached to the pollinator while it collects the pseudo-pollen.

Enticements other than food may also be employed. The flower may mimic the prey of the pollinator, invite attack, and attach its pollen to the attacker. Orchids may mimic the female of a male

Figure 12-41 Nectar guides in the silverweed (Potentilla). (a) To the human eye the flower is yellow. (b) If the flower is photographed with ultraviolet light, which is visible to insects, a dark pattern is apparent at the center of the flower.

(a)

(b)

pollinator and invite copulation. One orchid attracts its bee pollinator by scent and then supplies it with an intoxicating drug. When the bee lands on the flower it scratches a petal, which exudes a drop of the drug. The bee presses a special pad on its leg against the droplet, the fluid is absorbed, and the bee becomes groggy, uncoordinated. At this point it staggers into the flower. Any pollen the bee may be carrying is removed, and a fresh packet of pollen is attached in anticipation of the bee's next "trip" through another flower. The complete list of floral enticements includes food, violence, sex, and drugs.

Other flowering plants have developed even more complicated relations with insect pollinators. The fig and its wasp pollinator provide a classical example. The fig plant produces two kinds of flowers. The female wasp lays its eggs in one kind, a cup-shaped mass of flowers. As the fruit develops, the wasp eggs hatch into larvae, feed on the fig, become adults, and mate. This fruit has no seeds and is inedible to humans. Female wasps can enter the other kind of flower, but the structure of the flower is such that she cannot deposit her eggs. However, her visit does pollinate the fig, producing the normal seeded—and edible—fruit. The fig fruits without seeds are simply hatcheries for the pollinating wasp. If a single branch bearing the sterile and inedible fruit is grafted on to a normal fig tree, it will provide enough wasps to pollinate the edible fruit on all the other branches.

Pollination in Orchids

The orchids are a very large and diverse family of plants. In fact, about ten per cent of all angiosperms belong to the orchid family. Orchid flowers and their insect pollinators have evolved together to produce some remarkable interactions. An orchid flower has many special features. One of the petals forms the **lip,** which acts as a convenient and attractive landing platform for the pollinator. The sexual parts of the flower—stamens, anthers, style, and stigma—are fused together to form the **column.** Orchid pollen is bound together in masses called **pollinia.** The pollinia are supplied with a sticky protrusion, called a **viscidium,** which attaches the pollen to the insect. Pollinia are located in a pocket at the tip of the column, underneath a hinged cover called an **anther cap** (see Figure A, a-d).

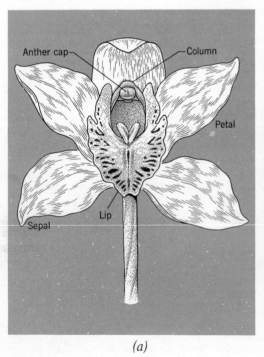

(a)

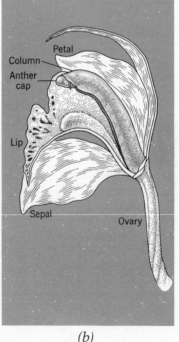

(b)

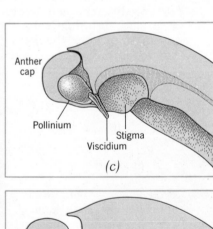

(c)

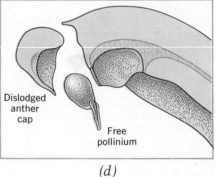

(d)

Orchids belonging to the genus *Cattleya* attract pollinating insects by scent. The insect forces its way between the lip and the column as it attempts to reach the source of the attractant odor. The pathway is so narrow that the insect must back out once it has entered. As it does so, it brushes against the viscidium and pushes back the anther cap. Pollinia are cemented to its back (see Figure A, e-g). On a visit to another flower, the same sequence of events occurs. But this time, as the insect backs out, the attached pollinia are forced into the sticky cavity of the stigma and thereby removed from the insect. The insect then receives another set of pollinia as it passes the tip of the column. Pollinia brought by the insect become attached to the stigma, and a new set is affixed to the insect to be carried to another flower. *Cattleya* with a pollinator is shown in Figure B.

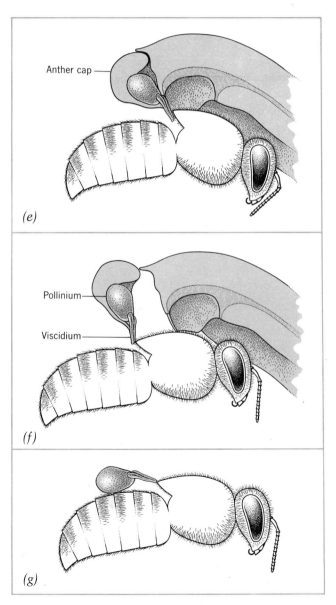

(e)

(f)

(g)

Figure A Structure of an orchid flower. (a, b) Front and side views of the whole flower. The petals and sepals are similar; in this orchid, one of the petals has developed into a showy structure called the lip, which serves as a landing platform for pollinators. The reproductive parts of the flower (filaments, anthers, style, and stigma) are fused into the column. (c, d) An enlarged lateral view of the tip of the column. Pollen grains are contained in the pollinium, which is covered by the anther cap and has a sticky extension, the viscidium. The stigma lies just below the pollinium and anther cap and is coated with a sticky material. (e, f, g) Attachment of the pollinium to the pollinator. The column lies above the lip, forming a narrow tube through which the pollinator must enter. When the pollinating bee leaves, it must back out. In the process, the viscidium attaches the pollinium to its back; the anther cap is dislodged. When the bee enters another flower, the pollinium is pressed into the pocket-like stigma and retained, thus fertilizing the flower.

Another orchid genus, *Coryanthes,* literally traps the pollinator. The lip of its flowers forms a bucket that is filled with water by a pair of special glands. Pollinating bees are attracted by the scent of the orchid; the insect then tumbles into the bucket while it scratches around in search of the source of the scent. It can escape only by moving through a tunnel that brings it to the column. Here, pollinia are attached to the insect and deposited on the stigma (Figures C and D).

Figure B Cattleya *and a pollinating bee.*

(B)

Figure C (below) The pathway of a pollinating bee through the flower of Coryanthes. *The insect drops into the bucket and can escape only by brushing against the column.*

(D)

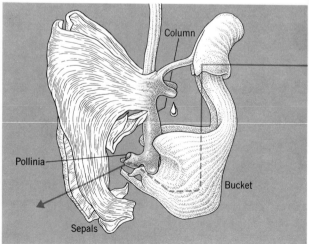

(C)

Figure D (above) Coryanthes *with a pollinator emerging.*

Catasetum has an even more complex arrangement for pollination. It possesses both male and female flowers. Male flowers are much more numerous than female flowers, so the pollinating bee will probably first be attracted to the male flower—again by scent. As it scratches around for the scent (Figure E) its movements set off a trigger. The trigger releases the pollinia, and the viscidium is attached to the bee's back (Figures F and G). Once the pollinia is released, the male flower rapidly loses it attractant odor. The bee eventually lands on a female flower and pollinates it (Figure H).

The orchid *Oncidium* uses yet another device. Its insect pollinator is a very aggressive bee that drives other insects from its territory. Flowers of *Oncidium* mimic an antagonist, causing the bee to attack them. The bee does not land on the flower, but merely strikes it, so that the pollen becomes at-

Figure E The male flower of Catesetum *and a pollinator.*

Figure F The male flower of Catesetum *and the pollinator after the trigger has been tripped, releasing the pollinium.*

Figure G Diagram of the male flower of Catesetum, *showing the position of the pollinium before and after the trigger has been tripped.*

(E)

(F)

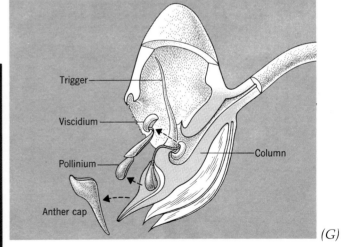

(G)

(H)

Figure H (left) The female flower of Catesetum *and a pollinator with attached pollinium.*

tached to its head (Figures I and J). The pollinia bend downward slightly after they are cemented to the front of the bee's head, so that they press into the stigma when the bee attacks a second flower (Figure K). The bees must be accurate to within a millimeter to effect the exchange, yet they rarely miss. *Oncidium* flowers do not really look much like insects, yet something about their color or shape causes the bee to attack (Figure L).

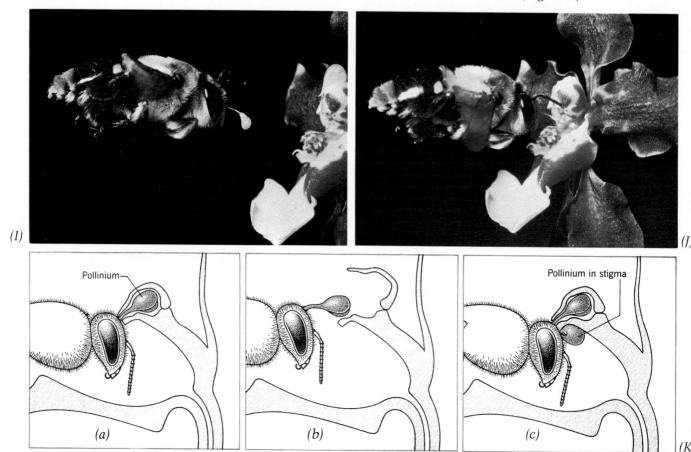

(I)   *(J)*

(a)   *(b)*   *(c)*   *(K)*

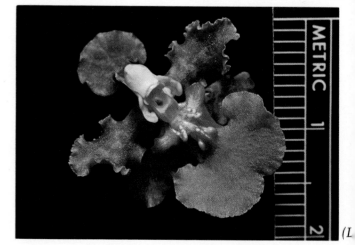

(L)

Figure I (top left) A pollinator approaching the flower of Oncidium.

Figure J (top right) The bee strikes the flower of Oncidium but does not land.

Figure K (center) Pollination of Oncidium. (a) When the bee strikes the flower, the pollinium is attached to its head and the anther cap dislodged. (b) The weight of the pollinium causes it to bend downward. (c) When the bee strikes another flower, the attached pollinium is forced into the stigma. A new pollinium is acquired.

Figure L (bottom right) Front view of a flower of Oncidium (the stigma is marked by a small pebble). The resemblance to an intruding insect is not obvious to the human eye.

The genus *Ophrys* attracts male insects by mimicking females (Figures M and N). The flowers are pollinated as males attempt to copulate with them. The insect's copulatory apparatus is extruded, but no sperm are ejaculated, so the process is called "pseudocopulation" (Figures O and P). The removal and deposition of pollinia occur during pseudocopulation.

(M)

(N)

Figure M   *The flower of* Ophrys insectifera.
Figure N   *The flower of* Ophrys speculum.
Figure O   *A male wasp of the genus* Gorytes *attempting to copulate with* Ophrys insectifera.
Figure P   *A male wasp of the genus* Andrena *attempting to copulate with* Ophrys speculum. *The pollinium is attached to the tip of the wasp's abdomen.*

(O)

(P)

environmental control of reproduction

The control and timing of reproduction are linked to the environment in many ways. Food, oxygen supply, various aspects of the chemical environment, seasonal changes, and other factors may all initiate or inhibit reproduction.

Unicellular Organisms

Life cycles of unicellular organisms can respond to rapid changes in environmental conditions. A bacterium or alga or protozoan divides rapidly during favorable conditions and when food is plentiful. This type of growth can be studied in culture by placing a few cells in a flask of growth medium that contains food and oxygen and is maintained at a suitable temperature.

At first, the cells adapt to their new surroundings. The period of adaptation is called the **lag phase.** Then they begin to grow and divide. The organisms go through their division cycles at a rate that depends on food supply, temperature, and other factors. The population enters a growth phase called the **log phase.** "Log" refers to the fact that there is a linear relation between the logarithm of the number of cells and the time. Suppose we start with one cell, which goes through one division cycle in twenty minutes. After each twenty-minute period the cell number doubles; that is, we have two, four, then eight, and then sixteen cells. The numbers quickly become very large. After ten division cycles, we have 1,024 cells. After twenty cycles, more than a million cells are present in the culture flask. When growth is rapid like this, it is convenient to plot the logarithm of cell numbers rather than cell number itself. This plot yields a straight line.

The cell numbers quickly become astronomically large. If a bacterial culture were able to stay in log growth for just a few months, and if all the cells survived, they would have a mass equal to that of the earth. But no population can stay in log phase growth for an indefinite period. The human population is now growing logarithmically at two percent per year; this rate represents a doubling time of about 35 years. If this rate could be maintained, all the mass of the universe would be converted to human flesh in about 6,000 years. Obviously, such unlimited increase cannot occur, with either bacteria or man. In cell culture, therefore, the log phase is followed by a **stationary phase.** At that point the cells either cease to divide or divide less often. Food scarcity or oxygen deficiency may slow growth, or toxic waste products of metabolism may accumulate in the medium. The stationary phase is soon followed by a **death phase.** This period starts when some essential substance is completely exhausted or when toxic products become lethal. The phases of growth in culture are diagrammed in Figure 12-42.

The growth of organisms in culture is a highly artificial procedure. Still, it provides a fair model of many normal situations encountered by unicellular organisms. Microorganisms are opportunists. They respond to favorable environments with rapid reproduction, until millions of individuals are produced. Then, when food is exhausted or when local conditions become unfavorable for continued growth, most of the organisms die off. However, a few enter into a resting stage. They survive in an inactive state until conditions permit the resumption of active metabolism and reproduction.

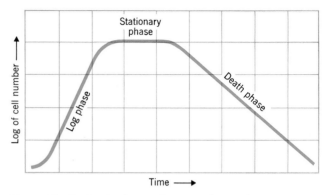

Figure 12-42 Typical growth curve shows changes in the number of organisms in a limited environment such as a culture flask.

Animal Reproduction

Many small animals (hydra, water fleas, insects) reproduce asexually throughout favorable environmental periods. This process allows a rapid increase in numbers, much like that found in unicells. In the water flea, *Daphnia,* sexual reproduction is a response to unfavorable conditions. When the environment becomes permissive, asexual reproduction resumes and a large population is produced (Figure 12-43).

Most animals only reproduce sexually, and timing signals from the environment often initiate breeding behavior. In mammals, the physiological basis of such signals has been worked out. The environmental control of reproduction often works through hormonal systems that are triggered by light. The main interaction is between the pituitary gland, the gonad, and the pineal gland. The **pineal gland** is an outgrowth of nerve tissue from the roof of the midbrain Figure 12-44).

These hormonal interactions have been demonstrated in rats. If one ovary is removed from a female rat, the pituitary produces more gonadotropin, so the other ovary increases in size and egg production continues. If the rat is kept in the dark after the operation, this doesn't happen. However, if the pineal gland is removed, the ovary enlarges even in darkness. Apparently the pineal gland produces an inhibitor substance when the animal is kept in the dark, but not when it is kept in the light; the inhibitor prevents release of gonadotropin. Thus, the removal of the pineal removes the inhibitor. Compensatory growth of the ovary can then occur. The conclusion is that the pineal gland

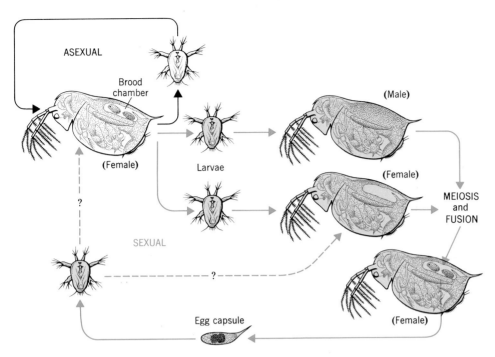

Figure 12-43 *The life cycle of the water flea,* Daphnia. *In favorable environmental conditions, females reproduce asexually; diploid eggs develop directly into larvae. Unfavorable conditions initiate sexual reproduction. The fertilized eggs are a resistant stage that can survive low temperature or drying.*

itself responds to the amount of light in the environment.

We know that light reception is important in normal reproductive cycles. For example, puberty will be delayed in rats if they are kept in the dark. Evidence from other animals suggests that the light signal is directly perceived by the pineal gland. In some reptiles, the pineal gland is a clearly developed photoreceptor; it is even called the "third median eye." Experiments on ducks, too, show that light is received by something other than the eye. Ducks will respond to a single light flash by initiating gonad enlargement. But even if the retinas of both of a duck's eyes are removed, the response still occurs. In fact, if light is conducted directly to the brain through a quartz rod, the response is still obtained. Light is clearly an important environmental signal that controls reproduction. But the light signal is received not by the eyes but by various cells in the nervous system that control secretion of hormones. Many details need to be worked out, however. For instance, what kind of environmental light signal is required? How is this signal fed into the nervous system? Which neurosecretory cells participate in a given response?

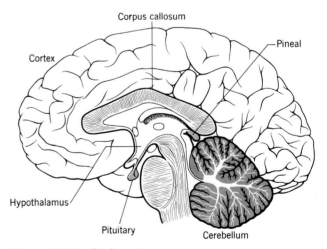

Figure 12-44 *The location of the pineal gland in the human brain.*

Higher Plant Reproduction

The induction of flowering and sexual reproduction in higher plants has been carefully studied. As a result, we can now describe the environmental events that cause the plant to shift from vegetative, non-reproductive growth to flowering.

Flowering and the germination of seeds are often seasonal. They are timed to the annual climatic events that will maximize the chances of reproductive success. Seeds germinate at different times. Some sprout immediately; others require a period of dormancy or exposure to low temperature before they begin development. The details depend on seasonal weather changes in the particular environment.

Plants also flower at different times. Some blossom in early spring, some in mid-summer, and others in the fall. The length of the daily light period is often used as a timing signal. Summer days are long; winter days are short. The seasonal variation in daylength is a good environmental signal for two reasons. First, it is stable and predictable, while other signals are undependable. Temperature can fluctuate unpredictably, for instance, and rains may be unseasonable. Second, daylength provides a precise sort of signal that can be used throughout the year, not merely at one season. Some plants respond to long days, others require short days.

How does the plant respond to light at a chemical level in order to initiate flowering? One theory suggested that light absorbed by photosynthetic pigments controls flowering. Photosynthesis during long days would increase the plant's store of reduced carbon and might stimulate flowering. However, it would take altogether another scheme to account for plants that respond to shortening days. But it seems highly unlikely that the entirely separate mechanisms, operating on different principles, should have arisen. A more satisfactory explanation, therefore, is that flowering (and several other responses, including seed germination) depends on light absorption by another pigment, specialized for the purpose, called **phytochrome.**

The first clue to the existence of phytochrome came from studies of the germination of lettuce seeds. Some types of lettuce seed germinate after exposure to light. Plant physiologists used light of various wavelengths to determine the action spectrum for seed germination. They found that red light was most effective in initiating germination, and that germination could be reversed by exposure to infrared light. If seeds were exposed to an alternating series of red and infrared lights, the seeds germinated or failed to do so depending on what they were last exposed to. The same thing happened with other responses, such as flowering, the expansion of leaves, and the formation of pigments (such as the red pigment of apple skins). These induction and reversal effects showed that phytochrome could take two different forms. In one form, called **phytochrome red,** or Pr, it absorbed light in the red region of the spectrum. In the other form (**phytochrome far red,** or Pfr) it absorbed light in the infrared. Daylight probably converts Pr to Pfr. Pfr is then reconverted to Pr metabolically in the dark, or when exposed to infrared light. The phytochrome system, therefore, allows the plant to distinguish between day and night.

Phytochrome has finally been isolated, but we still do not know precisely how it acts to produce flowering. One attractive hypothesis is that phytochrome somehow controls the synthesis of a flowering hormone called **florigen.** But this is still speculative, and our information is still far from satisfactory, especially considering the economic importance of flowering and plant reproduction.

In any case we know that many reproductive processes in many organisms are controlled by environmental signals, even though the details are poorly understood.

Recommended Reading

Corner, E. J. H., *The Life of Plants*. New York: World (1968).

Includes extensive discussion of the reproduction of algae, land plants, and the fungi, in the context of their physiological and structural responses to the challenge of new environments.

Mitchinson, J. M., *The Biology of the Cell Cycle*. New York and London: Cambridge University Press (1971).

A detailed monograph on the cycle of cell growth and division, in organisms ranging from bacteria to mammals. The main emphasis is on patterns of biochemical synthesis and their control.

Saunders, John W., Jr., *Animal Morphogenesis*. New York: Macmillan (1968).

Includes a discussion of gametogenesis and fertilization, as well as an account of animal development.

13 development

Living things are characterized by ceaseless change. In all organisms, dynamic exchanges proceed continuously at the molecular level. We are all aware of change on a longer time scale. We and our friends mature and age, pets grow old and die. Birth, childhood, senility, and death are universal concerns. Change through education and through personal endeavor is also familiar; we learn or diet or swim or run or become alcoholic. These changes are all part of development. Some changes are more or less reversible, others are progressive and linear in time. We can gain and lose weight repeatedly, but we cannot grow younger, despite heroic efforts to give that impression to the world.

mechanisms of development

Underlying this ceaseless and universal change — at all levels — is a constant interaction between the organism's genetic instructions and its environment. Genetic instructions work by directing the synthesis of specific proteins in particular cells. The environment works at various levels. Cytoplasm interacts with the genome to determine which fraction of the instructions becomes activated at any given time. The cell itself is influenced by its surrounding chemical environment; that environment may also interact with the genome. In multicellular organisms, the environment of each individual cell is moderated by the activities of other cells and by the whole array of

403

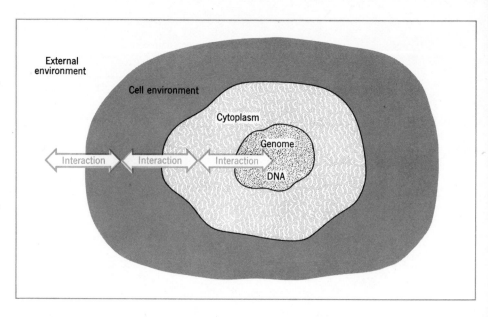

Figure 13-1 The genome of an organism (its nuclear DNA) interacts with the environment throughout the organism's development. The immediate environment of the genome is the cell cytoplasm, which itself interacts with the cellular environment (tissue fluid, other cells). The cellular environment in turn interacts with, and responds to, changes in the external environment.

homeostatic mechanisms. The cellular environment, in turn, responds to changes in the external environment. Every one of these levels interacts with every other level, either directly or indirectly (see Figure 13-1). A change in any one of the interacting elements will alter the morphology and physiology of the organism. The sum of these changes is what we know as development.

Developmental biology is concerned with any and all of the changes that take place in the organism, at all levels of this balanced interaction, from genome to environment. It is concerned with the sequential changes that occur over an organism's life history, with the repair of injury and the regeneration of lost parts, and with diseases caused by defects in normal developmental processes. The effects of environment and genome are often difficult to detect directly. But their action is usually reflected in aspects of the organism's development: changes in cell number, and the differentiation and movement of cells.

Increase in Cell Number

Most multicellular organisms go through single-celled stages at some time in their life cycles. They develop multicellular structures through cell division. Increases in cell number are simply one expression of the interaction between genetic in-

structions and the organism's internal and external environments.

In some organisms, the cytoplasm surrounding a fertilized egg during its early development is an obvious environmental influence on the egg's division. The fertilized eggs of some invertebrate animals have a complicated pattern of cell division called **spiral cleavage** (Figure 13-2). This pattern is already established in the cytoplasm while the egg cell is being produced by the mother. Special cells called **nurse cells** surround the oocyte during its development. The positioning of the nurse cells around the egg controls the structure of cytoplasm within the egg. And, in turn, the cytoplasm's structure determines the location and orientation of future spiral cleavage. In this case, cytoplasmic information interacts with the dividing nucleus, resulting in a predictable pattern of increase in cell number.

The next environmental level is the area surrounding the whole cell, the cellular environment. This environment includes tissue fluids, other cells, and cell products. The cellular environment influences cell division in many developing systems. The meristematic tissues in higher plants retain the capacity to divide, but in fact only one of the two daughter cells from each meristematic cell division has the ability to divide (see Figure 13-3). The "decision" as to which of two daughter cells remains meristematic is controlled by the cellular environment.

Figure 13-2 *Spiral cleavage in a mollusc egg. After the first four cells are formed, division becomes unequal. Note that the four small cells at the 8-cell stage are not directly above the large cells, but rather lie in the furrows between them.*

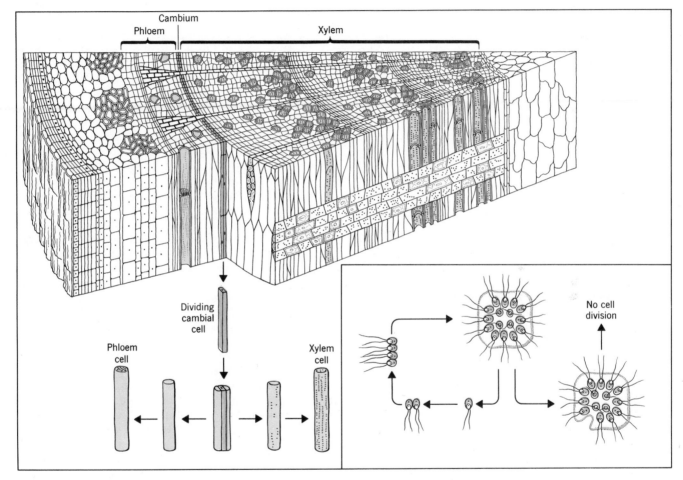

Figure 13-3 *Cells produced by the vascular cambium of higher plants become xylem cells on one side and phloem cells on the other, and the cambial cells themselves continue to divide. Presumably the differences between the three types of cells reflect differences in the cellular environment.*

Figure 13-4 *A single cell isolated from a colony of the green alga* Gonium *divides and reconstitutes the colony. The colony from which it was removed does not replace the missing cell; the presence of the other cells inhibits division.*

405 *mechanisms of development*

Another example of the interaction between the cellular environment and cell division is found in the colonial alga *Gonium*. This alga normally exists in groups of sixteen cells arranged in a flat plate. One cell, if isolated, will divide four times to reconstitute the colony of sixteen. But if a whole colony loses just a few cells, the missing cells will not be replaced (see Figure 13-4). The constant presence of neighboring cells inhibits cell division.

Increases in cell number can also clearly be influenced by the external environment in some cases. The production of red blood cells in mammals is a good example. Ascent to high altitudes, loss of blood, and other external factors all will increase the rate of red blood cell production.

Unfortunately, our information about the various factors that influence cell division is only descriptive. The actual pathways from environment to genome, which ultimately control the rates of cell division, are not yet understood.

Cell Differentiation

A second kind of developmental change is **differentiation:** the production of specialized cells. A multicellular organism has many kinds of cells, which have different functional roles. Like changes in cell number, differentiation results from interactions among the genome, cytoplasm, cell environment, and external environment. As one or more of these elements change, cells may respond with a particular pattern of differentiation.

Cell differentiation presents an interesting problem. By means of cell division, a single cell generates all the cells in the organism. Yet the offspring of that original cell become nerve cells or liver cells, phloem cells or root cells. One might guess that cell differentiation results from an unequal distribution of the genome to different cells. But the evidence is overwhelming that each daughter cell receives a complete set of DNA instructions, identical to that of the parent cell. First of all, we know that all diploid cells in an organism contain a constant amount of DNA. But the most spectacular proof comes from the work of F. C. Steward. In 1958, Steward reported that he was able to grow a complete carrot plant from a single cell taken from a carrot root. Although the cell was already a root cell, its nucleus must have contained all the necessary genetic instructions to guide the formation of the whole plant.

Related experiments in animals have tested the competence of nuclei to direct development. The nucleus of a frog egg is removed by microsurgery. It is then replaced by a nucleus derived from another frog cell, one at a later stage of development. Often the replacement nucleus is able to direct the development of the egg into a normal frog. However, as donor nuclei are taken from later and later stages of development, the percentage of success decreases, as differentiation occurs.

The differentiation process allows only part of the genome to be active in protein synthesis at any given time. Any active portion of the DNA genome will be synthesizing mRNA; parts of the genome that are inactive produce no mRNA. Therefore, one can predict that differentiating cells will have identical sets of DNA but different kinds of mRNA. In fact, hybridization experiments do reveal differences in mRNA composition between differentiated cells. The mRNA from any one kind of cell matches up with only a portion of the total DNA.

In procaryotic organisms, the activation of particular portions of the genome is governed by **repressor** molecules. These molecules can prevent mRNA synthesis, at least in some cases. In eucaryotic organisms, the control of various parts of the genome is more complicated. The histone proteins associated with DNA chains are part of the control mechanism, but detailed information about how they work is still lacking.

The activation of portions of the genome can be directly observed, however, in certain insect tissues that have giant **polytene chromosomes.** These chromosomes consist of hundreds of parallel strands of DNA. Particular genes can be identified on the DNA strands by their chemical staining properties. Active synthesis of mRNA by any one of these genes shows up as a visible enlargement, or "puff," on the chromosome (Figure 13-5). DNA's from different larval tissues in the insect have different puffing patterns. This evidence supports the idea that specific parts of the genome control differentiation in each type of cell and tissue.

The activation of genetic information in the egg is obviously influenced by the cytoplasm. In both plants and animals, mature eggs often show regional differences in their cytoplasm. The mature sea urchin egg, for example, has an unequal distribution of yolk granules, oil droplets, pigment granules, and clear regions along an imaginary axis in its cytoplasm. One end of the axis, called the **animal pole,** has relatively little yolk. The other end, the **vegetal pole,** is quite yolky. This axial organization is preserved during the zygote's early divisions and throughout the organism's early development. The unequal distribution of materials in the egg's cytoplasm has a considerable influence

Hybridization is the term for the formation of hydrogen bonds between complementary strands of DNA, or of DNA and RNA. The amount of bonding, or hybridization, naturally depends on the number of complementary base sequences in the strands. DNA bonds very tightly to its complementary strand; RNA hybridizes most strongly with the DNA that directed its synthesis.

The fraction of any cell's total DNA that is actively synthesizing protein can be determined through hybridization. The DNA of first bound to solid particles packed in a glass column. Then, strands of radioactive RNA from various kinds of cells in the same organism are added. The RNA will hybridize to whatever portion of the DNA directed its synthesis. The amount of radioactivity that remains on the column provides a measure of the amount of hybridization that has occurred: non-hybridized RNA simply washes through the column.

Samples of mRNA from various organs—for instance, liver, brain, and kidney—can be added to the column one at a time. Liver mRNA will become bonded to some of the DNA but not all. A later addition of mRNA from kidney or brain tissues will produce additional bonding with other sections of the DNA, reflected in additional radioactivity. This evidence indicates that only a fraction of the total DNA is active in any one type of cell. If all the DNA were active, then the RNA from any one type of cell would hybridize with all the DNA; the addition of RNA from other cells would not produce additional bonding.

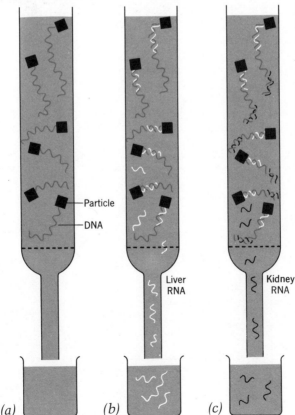

Particle

DNA

Liver RNA

Kidney RNA

Figure A   DNA-RNA hybridization. (a) DNA from mouse cells is attached to particles in a column. (b) RNA from mouse liver cells is added and allowed to bond with the DNA. Excess liver RNA is washed away. (c) When RNA from mouse kidney is added to the column, additional bonding with DNA occurs. Different cell types have different RNA's, which bond with different parts of the DNA.

(a)   (b)   (c)

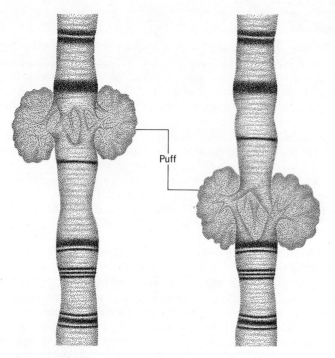

Figure 13-5 Giant chromosomes of an insect. The same chromosome from two kinds of cells shows different "puff" patterns, where different portions of the chromosome have uncoiled and are engaged in RNA synthesis.

Puff

on the sea urchin's early differentiation (see Figure 13–6, facing page).

We can demonstrate the extent of the cytoplasm's influence with an experiment. Unfertilized sea urchin eggs are cut into sections, and the resulting egg fragments—those that contain nuclei—are then fertilized by sperm. If the cuts are made along the animal–vegetal axis, each half can develop into a normal larva, only half the usual size. But if unfertilized eggs are divided across the axis, so that one half receives all of the vegetal pole and the other all of the animal pole, neither half develops normally. The animal half eventually produces a hollow sphere of ciliated cells, and the vegetal half develops into a skeletal rod embedded in a mass of cells with a partial digestive tract (see Figure 13-7, below).

The sea urchin egg contains a gradient of materials in its cytoplasm, and this gradient is established before it is fertilized. Experimental procedures that disturb the balance of cytoplasm in various regions along the animal–vegetal axis interfere with normal cell differentiation. Yet the developing embryo can accomodate to the loss of a large amount of cytoplasm, provided only that the overall balance of cytoplasmic materials along the main axis of development is undisturbed.

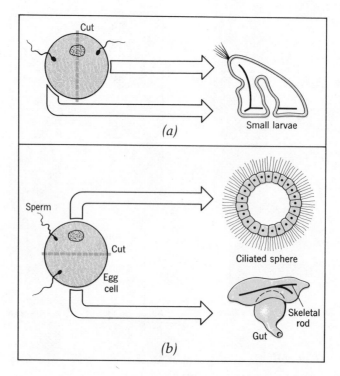

Cut

Small larvae

(a)

Sperm

Cut

Egg cell

Ciliated sphere

Skeletal rod

Gut

(b)

Figure 13-7 The influence of cytoplasm on development. (a) If a sea urchin egg is cut parallel to the animal–vegetal axis, development is normal, and diminutive larvae result. (b) If the egg is cut across the axis, the animal half produces a ciliated sphere, and the vegetal half produces an abnormal larva.

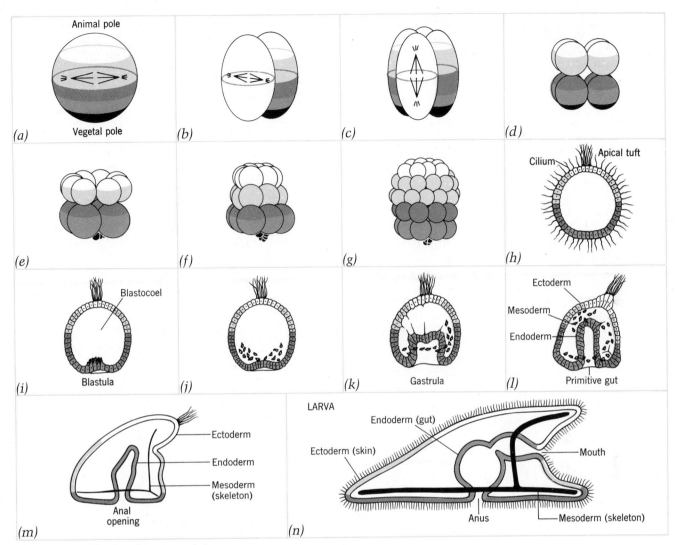

Figure 13-6 *The development of the sea urchin egg. The egg divides repeatedly to produce a hollow ball of cells, the blastula. One side of the blastula invaginates, forming two layers. Cells then migrate between the two layers to establish a three-layered tube-within-a-tube organization, called the gastrula. The cytoplasmic contents of the egg are distributed unequally in subsequent divisions as shown by the different bands of color. Specific portions of the cytoplasm are sequestered in different regions of the embryo as it develops.*

This capacity to accommodate to the removal of cytoplasm is eventually lost, as cells become **determined** with respect to their subsequent fate. A determined cell is one that has lost its plasticity; it can differentiate into only one kind of cell. When small groups of cells are removed from an embryo that is already determined, the embryo will develop defects in later stages. The defects correspond to the specific cells that were removed. All systems are capable of some degree of accommodation during early development. But they ultimately become determined and lose their plasticity.

Forces outside the cytoplasm control differentiation after cell division has begun. At this point, the cellular environment becomes important. The developmental fate of a cell or group of cells is often directed or influenced by a neighboring mass of cells. This kind of interaction is called **induction.** Inductive action may be reciprocal: that is, each cell group may restrict and direct the differentiation of the other cell group.

Many inductive interactions occur during an organism's development. Epithelial cells—to take just one example—are induced to form a wide range of differentiated structures. In birds, for instance, the epithelium produces feathers, scales, and beak, and the cornea and lens of the eye. The differentiation of these structures and cell products is induced by the underlying cells. Epithelial cells that would have produced feathers differentiate to scales if they are transplanted to the developing leg, or to cornea or lens if transplanted to the eye.

The reciprocal effect of induction is best shown by the development of the leg in a chick embryo. The outer epidermal cells interact with the underlying cells, and induction is mutual. If the epidermal portion is removed, the limb fails to develop: If it is replaced by the epidermal layer from another limb, normal development is restored. Clearly, the epidermal layer is necessary for the normal development of the underlying tissues. But the underlying cells, in turn, control and induce the development of the surface cells. In one experiment, chick epidermis was grafted over the exposed underlying tissue of the limb bud of a duck. The result was a webbed foot like a duck's rather than a chicken foot (see Figure 13-8).

Presumably, differentiation is controlled by something transferred or exchanged between cells. Some experiments suggest that specific chemical compounds diffuse from one cell mass to another to direct differentiation. These is some evidence that RNA is transferred during inductive interactions. However, no single compound can be responsible for all the observed interactions.

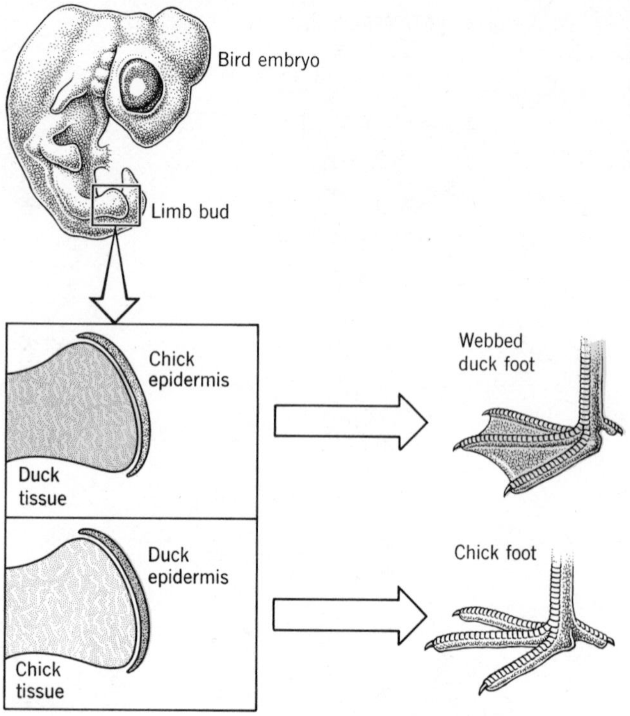

Figure 13-8 *Induction. The presence of epidermis is necessary for the development of the embryo's limb buds. However, the cells of the underlying tissue control the details of the epidermal cells' differentiation.*

Cell Movements

Animal development depends on cell movement, as well as on differentiation and division. The orderly movement of individual cells and groups of cells changes the basic form of the developing organism. Higher plant cells cannot move individually, because of their heavy cell walls, but the plant achieves changes in the relative position of cells through the orderly production of new cells and their subsequent enlargement and change in shape. Such relative growth processes also occur in animal development, and often go along with cell movements.

As animal development begins, the fertilized egg divides to form a hollow sphere of cells. The sphere is then transformed into a three-layered cup of cells which eventually produces the basic tube-within-a-tube plan of higher animals. The internal tube represents a primitive gut. The transformation process is called **gastrulation,** which means "gut formation."

Gastrulation is accomplished by a combination

of mass movement of cells and relative growth. The details vary in different animals. In the sea urchin, long, arm like extensions of cells stretch across the cavity of the sphere to the opposite wall, attach, and contract, pulling cells into the interior. This movement is accompanied by continued cell division and changes in cell shape (see Figure 13-6). Gastrulation in vertebrates is more complex. But the result is the same: the conversion of a one-layered structure to a multi-layered one, by a combination of movements of cell sheets and continued relative growth.

Similar movements accomplish changes later in development. For example, the formation of the central nervous system in vertebrates depends on cell movement. After gastrulation has occurred, some of the surface cells thicken to form a plate. This plate then rolls in at the edges, forming the **neural tube,** which will eventually differentiate into brain and spinal cord (Figure 13-9). Similarly, other sheets of cells move into position to form muscle, glands, and other tissues and organs.

Cells also move individually in the developing organism. In vertebrates, cells above the neural tube migrate outward to establish nerve ganglia. Other cells from the same source migrate outward and differentiate into pigment cells. Primordial germ cells, which eventually will form eggs and sperm, originate outside the gonad and then migrate into position. Cell migration follows pathways along blood vessels or nerve tracts.

The movements of all such cells, except those

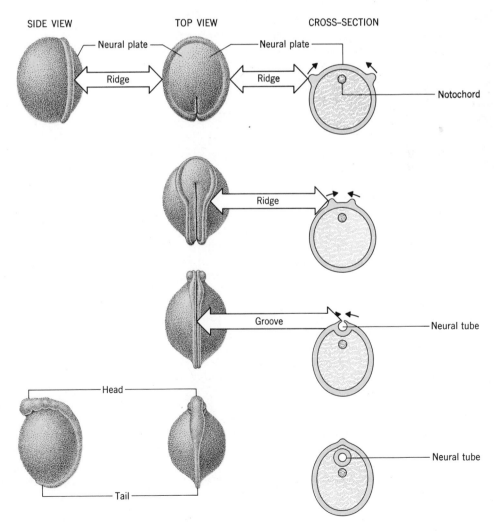

Figure 13-9 The development of the neural tube in an amphibian.

SIDE VIEW TOP VIEW CROSS–SECTION

Neural plate — Ridge — Neural plate — Ridge — Notochord

Ridge

Groove — Neural tube

Head — Tail — Neural tube

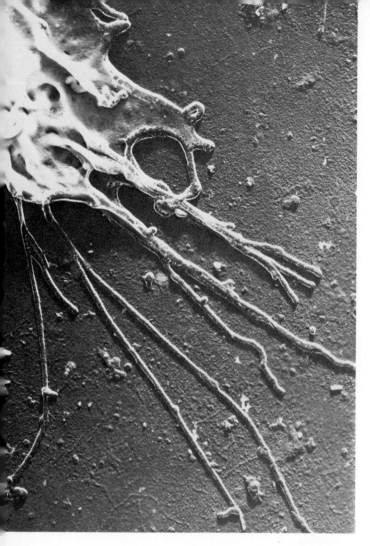

Figure 13-10 Mammalian cells grown in culture produce long extensions called microspikes. Similar extensions permit the movement of cells and cell sheets during development. Magnification 11,000×.

Figure 13-11 Cell aggregation. When liver cells are mixed with cartilage cells, they aggregate into an organized mass with liver cells surrounding the cartilage.

0.1 mm

with cilia or flagella, obviously are achieved by the attachment of cell extensions to neighboring cells or to other surfaces in the environment (Figure 13-10). The characteristics of the cell surfaces are therefore very important in directing orderly cell movements. Surface characteristics vary considerably. If we mix cartilage cells and liver cells, for example, they will arrange themselves so that the cartilage cells are on the interior, and the liver cells are on the outside (see Figure 13-11). Apparently, there are various kinds of cohesive forces between cells that cause them to aggregate in specific patterns to form tissues. These cohesive forces probably depend on differences in cell surfaces. Further studies of surface properties may help explain how cells move during development.

The basic processes of development—cell division, cell differentiation, and cell movements—all reflect the various interactions between the genome and its surroundings. Some of these in-

teractions take place in sequence. First, the genome of the parent directs the differentiation of the egg's cytoplasm. Then, as fertilization occurs and development begins, the cytoplasm of the egg determines further differentiation. The genome of the zygote then becomes involved, as various genes are activated in sequence. Development proceeds as an interaction between genome and all levels of the environment.

These processes are the general mechanisms of development. They will serve as the background for a more detailed description of normal development in a few selected organisms.

plant development

Higher plants have a basic bipolar organization, consisting of a root and a shoot. That is to say, the plant maintains dividing tissues at the top of the stem (the apical meristem) and at the tip of the root (the root meristem).

The Embryo

The bipolar pattern is established early in the development of the plant embryo. The female gametophyte is already a polarized structure. It consists of an oocyte flanked by a pair of haploid cells at the micropylar end and three haploid cells at the other end. The oocyte's central endosperm contains two haploid nuclei. At fertilization, the two sperm nuclei enter the female gametophyte by way of the micropyle. One sperm nucleus fuses with the egg cell to create the diploid zygote, while the other sperm nucleus fuses with the two nuclei of the endosperm to produce triploid endosperm.

The zygote divides to form a chain of cells. The cells at the micropylar end of the chain form a stalk, called a **suspensor.** The cell at the other end divides to produce an embryo. The suspensor elongates and pushes the embryo into the endosperm tissue. The endosperm nourishes the embryo. The embryo forms a globular cell mass, which then flattens into one or two seed leaves, called

cotyledons, and an embryonic root. At this stage, conducting tissue differentiates between the shoot apex and the root apex. The basic bipolar organization of the mature plant is now established (Figure 13-12).

The seed consists of the plant embryo and the surrounding seed coats. The seed coats protect the embryo, and they often assist in spreading seeds to new locations. Some seeds are distributed by the wind, for instance; they have feathery or wing like seed coats to increase their air resistance. Other seed coats are specialized for attachment to the fur or feathers of animals. The fruit around the ripe seed may also participate in dissemination. For example, some fleshy fruits are consumed and digested by animals, but the seeds pass through the digestive tract and are thereby distributed to new

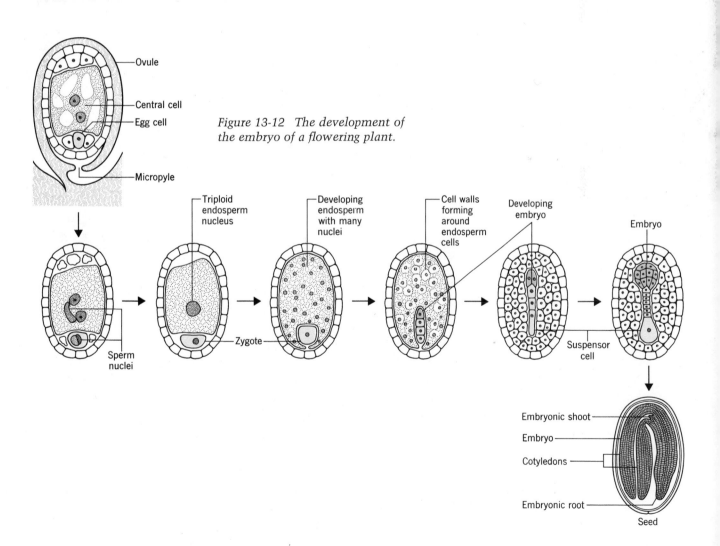

Figure 13-12 The development of the embryo of a flowering plant.

Figure 13-13   Seed dispersal. Seeds dispersed by wind include (a) a garden flower, Clematis; *(b) dandelion; (c) Coulter pine. Seeds may be attached to other organisms for dispersal, as in (d)* Geranium; *(e) foxtail, and (f) burr clover. Seeds may be dispersed by being cast out when the fruit dries and splits open, as in (g) vetch, and (h) California poppy.*

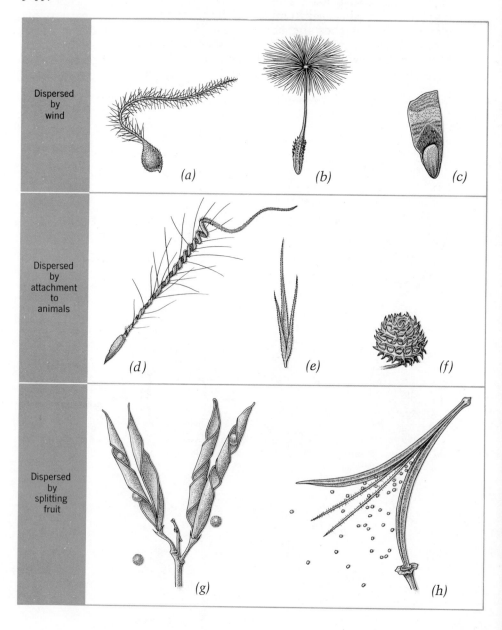

Dispersed by wind

(a)

(b)

(c)

Dispersed by attachment to animals

(d)

(e)

(f)

Dispersed by splitting fruit

(g)

(h)

locations. In other cases, the fruit ovary may dry out and split open, casting its seeds away from the parent plant (see Figure 13-13).

The development of the early plant embryo involves the various mechanisms we have already discussed. We know that inductive mechanisms operate in higher plants. In one experiment, a shoot tip was grafted next to a mass of undifferentiated cells in tissue culture. The graft caused xylem cells to differentiate adjacent to the shoot tip. The same result was obtained even when the graft was separated from the root tissue by a membrane (Figure 13-14). Nevertheless, direct evidence about the details of plant development is difficult to obtain. Plant embryos are well protected by sporophyte tissues, and most current experimental techniques are based on the isolation or destruction of portions of the embryo.

The development of the zygote in the seed is relatively independent of environmental influences. The female gametophyte and young embryo are completely dependent on the adult sporophyte. The zygote undergoes relatively little growth in the seed, so gas exchange, food supply, and waste disposal are not significant problems. Seeds sometimes represent a resting stage in the life cycle, permitting the plant to survive unfavorable environmental conditions. During these periods of dormancy, the seed's metabolism may be reduced so much that it is undetectable.

Germination

Germination, and the subsequent development of the embryo into an independent sporophyte, depends upon more direct interactions with the environment. Water, of course, is necessary for continued development. But seed coats may be impermeable to water; they must therefore be broken or scratched or decayed before water can enter and development can proceed. And even if water is available, there may be other physical requirements as well. In general, dormancy is brought to an end by physical factors that signal the onset of a favorable environment for growth. The seeds of annual plants in temperate climates often require exposure to low temperatures before they can germinate. In that way, the renewal of active growth coincides with the spring of the year. The seeds of other plants may lie inert for years on a forest floor before their dormancy is finally broken by a flash fire. The fire signals the destruction of the forest canopy,

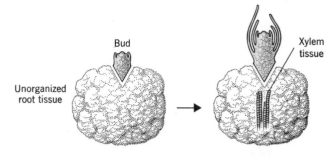

Figure 13-14 Induction in plant development. Contact with a bud induces the formation of xylem elements in undifferentiated root cells grown in tissue culture.

which normally shades the forest floor and prevents the growth of young plants.

Growth begins as the germinating seed rapidly absorbs water. The seed coats are split, the embryo's metabolism increases, and the shoot and root begin to grow. At first, energy is derived from stored organic material in the cotyledons or endosperm; later it comes from photosynthesis.

The pattern of the new plant's growth depends on the interaction of genetic instructions (to the meristems and to the cells that secrete growth-regulating hormones) with environmental influences. For example, auxins and other growth regulators probably control the spacing and orientation of the leaves produced by the apical meristem. Auxins from the apical meristem are also used to control patterns of shoot branching. This mechanism can be shown in various ways. If the apical meristem is removed from a plant, lateral buds become meristems and grow into new shoots. But this effect can be prevented by applying auxin to the cut tip. Grasses have persistent intercalary meristems in the main stem between the leaves. Cutting the apical meristem of the grass blade activates these intercalary meristems, so the grass continues to elongate. In both these cases, auxins from the apical meristem inhibit the growth of other meristems. When the auxin is removed, through the cutting of the apical meristem, the other meristems begin active growth.

Plant life cycles typically are divided into stages of juvenile growth, maturity, and senescence. The juvenile stage is characterized by rapid growth and a characteristic form. The juvenile plant may have leaves of a specific juvenile shape (Figure 13-15), or

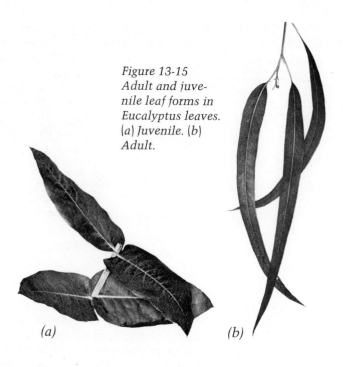

Figure 13-15
Adult and juvenile leaf forms in Eucalyptus leaves.
(a) Juvenile. (b) Adult.

(a)   (b)

it may produce structures such as thorns, which disappear in the mature plant. The period of juvenile growth may last a very long time. The common ivy plant, for instance, with a vine-like growth, is a juvenile form that lasts for years. At maturity, the plant becomes erect and bushlike, and the leaves change shape. The mature form, with flowers and seed, is rarely seen.

Plant Senescence

Flowering and sexual reproduction often signal the onset of senescence. Many plants senesce and die shortly after they flower. In such cases, senescence can sometimes be retarded simply be preventing flowering or by removing flower buds. Senescence is an active process; it is not simply deterioration. The senescent phase of the plant life cycle involves developmental changes in cell permeability, cell organelles, and chemical composition. Leaves and floral parts synthesize enzymes that break down protein and carbohydrate polymers into their constituent monomers. The monomers are then transported to other parts of the plant. Senescing plant organs also synthesize compounds that promote **abscission,** the loss of leaves or flowers through the

destruction of tissues at their bases.

Many of the changes in later plant development are associated with a group of growth regulators called the **kinins.** Like other plant growth regulators, the kinins have a variety of effects. They stimulate cell enlargement, counteract auxin by activating the growth of lateral buds, promote cell division, and initiate seed germination. Kinins also increase protein synthesis and stimulate the uptake of amino acids, sugars, and nucleotides into the cells. Kinins are particularly concentrated in the tissues of the fruit and seed, where they appear to promote flowering and senescence by stimulating the uptake of nutrients into the rapidly developing tissues. These nutrients are supplied by the breakdown of polymers in senescent parts of the plant. Nevertheless, the action of the kinins does not explain the onset of senescence in other, non-flowering parts of the plant.

The life spans of various plant species are very different. Grasses such as wheat have sharply-limited life spans. Death ensues immediately after sexual reproduction, so all the wheat plants in a wheat field die at the same time. Other plants, such as fruit trees, flower and produce fruit repeatedly, but they still have well-defined life spans. Still other plants never seem to die of senescence. The giant sequoia grows slowly throughout its life. The sequoia tree usually dies by simply being blown over; sequoia root systems are rather shallow, and at some point the roots are simply no longer capable of anchoring the massive tree. There are no changes in the sequoia that might be interpreted as senescence.

Plants may escape senescence by **vegetative reproduction.** A shoot system may grow up each year from a persistent underground corm or bulb and then die back. The bulb itself shows no sign of senescence. Potatoes are normally propagated vegetatively, by planting portions of the underground stem that contain buds, or "eyes." Potato stocks do lose vigor over the years, but probably only because the stems accumulate viruses, not because of senescence. Some potato stocks have been maintained for more than a century with no sign of declining vigor. These observations support the idea that senescence and death can be deferred by the presence of an active meristematic region. As we shall see, there is some evidence that juvenile tissues also defer senescence in animals. But true vegetative reproduction—that is, the ability to produce a whole new organism from a fragment of the old organism—is found in few animals.

animal development

The study of animal development has produced a wealth of descriptive information. The development of the vertebrate embryo is of most immediate interest to us, of course. Since all vertebrates have similar body plans, the principles of vertebrate development can be adequately illustrated in any number of organisms. Additional details of human development will be presented in the next chapter.

Vertebrates

The frog and the chick have both been studied extensively. A typical pattern of development is found in the frog. The first cell divisions in the zygote produce a spherical mass of cells of different sizes. The smaller cells are clustered around one end of the sphere, the animal pole. The cells at the opposite end, the vegetal pole, are larger and divide more slowly (Figure 13-16). At the 32-cell stage, a small cavity appears in the center of the cell mass. This cavity is called the **blastocoel.** The developing organism now becomes a hollow ball of cells; this stage is called the **blastula** (from the Greek word for "sprout").

The blastula develops into a more complex tubular structure, the **gastrula,** by means of gastrulation. Gastrulation in the frog proceeds by a combination of cell growth and cell movement. Cells on one side of the blastula turn in and move into the blastocoel through an opening called the **blastopore.** The moving cells are dividing rapidly as they move, so a new layer of cells forms below the surface cells. At the same time, the small surface cells at the animal pole are also dividing and moving over the larger cells at the vegetal pole. Eventually, both of these dividing cell layers—the surface layer and the new layer just beneath it—grow over and around the larger cells. The result is a tubular structure with a three-layered cross section. The cavity of the tube is a primitive gut (Figure 13-17).

The three layers of cells at this stage are collectively called **germ layers.** Individually they are the **endoderm** ("inner skin"), the **mesoderm** ("middle skin"), and the **ectoderm** ("outer skin"). Each of these germ layers will later develop into specific tissues and organs. The ectoderm gives rise to the skin and its derivatives (hair, scales, sweat glands, and so forth); it also is the source of the nervous system. The mesoderm layer produces the skeleton, muscles, circulatory system, urogenital system, and most of the cartilage and connective tissue. The endoderm produces the digestive tract and related organs: liver, lungs, pancreas, thyroid, and others.

The Development of an Organ System: The Urogenital System

The development of the urinary organs illustrates how development is tied to the organism's evolutionary history. Kidney development in vertebrates is based on various modifications of a primitive kidney structure. In this primitive structure, the excretory units, or nephrons, are arranged in segments. This primitive kidney is now only hypothetical; it occurs in no living vertebrate. But its structure was deduced from studying kidney development in many animals. Different portions of the pattern appear in different animals and at different stages of their development.

The nephrons develop in sequence from head to tail. The anterior nephrons constitute the **pronephros,** or "head kidney." The ducts draining these units join to form two **pronephric ducts,** which run to the rear, where they empty into a bladder or some other structure. The pronephros is a functional kidney in the larvae of vertebrates such as amphibians.

The pronephros is later replaced by functional units that develop farther back in the **mesonephros,** or "middle kidney." Mesonephric units make connections with the original pronephric ducts. In birds and mammals, the mesonephros is the functional kidney during embryonic development, while the pronephros itself shows little development. But the pronephric duct does develop. (It is called the Wolffian duct.) The mesonephros drains into the Wolffian ducts. In adult mammals and birds, the mesonephros is superseded by a still more posterior set of nephrons, the **metanephros** ("rear kidney").

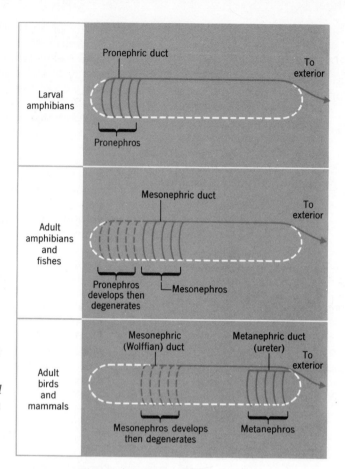

Figure A Kidney development. Different portions of a primitive kidney—pronephros, mesonephros, and metanephros—are found in various animals. No living vertebrate possesses all three portions together.

The metanephric units are also joined to the Wolffian duct (Figure A).

The closely associated development of the gonad further complicates this sequence. The gonad develops from mesoderm tissue adjacent to the kidney. Primordial germ cells, which eventually will develop into ova and sperm, migrate from elsewhere in the embryo into the developing gonad. At this point the male and female gonads are identical. If the outer layer of the gonad tissue continues to develop and predominates over the inner layer, the gonad becomes an ovary. If the inner layer predominates, it becomes a testis. The gonad develops separate duct systems for ova and sperm. In male mammals, sperm pass through the Wolffian duct. In females, the Wolffian ducts degenerate, and a separate pair of ducts, called the Mullerian ducts, arise alongside the Wolffian duct (Figure B).

The embryonic gonad is a bipotential structure: that is, it can become either an ovary or a testis. The embryo also contains both female and male duct systems. This developmental bipotentiality can sometimes override the genetic basis of sex determination. In amphibians, for example, genetic females can be made to develop as functional males by exposing them as embryos to high temperatures. This bipotentiality may persist into adult life in some animals. The domestic hen has only one functional ovary; the other gonad does not develop. If the functioning ovary is removed, however, the right gonad may develop as a testis, and the animal then becomes a fully functional rooster. Thus, the development of the urogenital system may extend well into adult life.

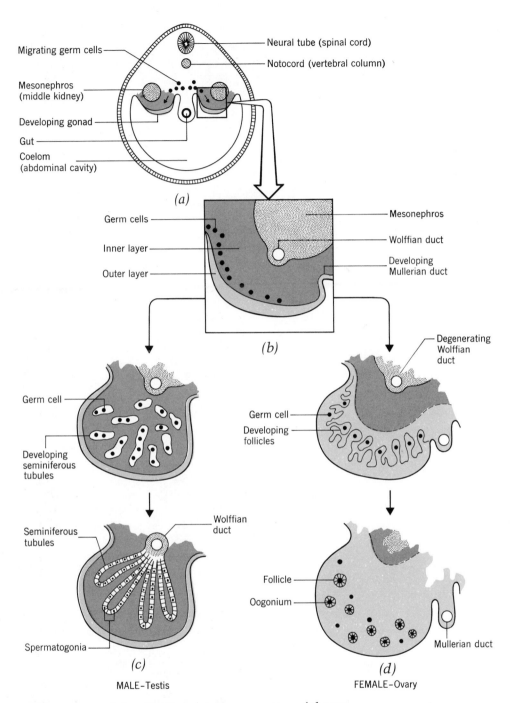

Figure B Gonad development. (a) A cross section of the embryo in the region of the mesonephros. (b) The developing gonad: germ cells migrate in. (c) In development of the testis, the inner cell layer predominates, and the germ cells become spermatogonia in seminiferous tubules. (d) In the development of the ovaries, the outer cell layer predominates, and the germ cells form oogonia in follicles.

FERTILIZED EGG

Animal pole

Vegetal pole

Blastocoel

BLASTULA

Head region

Tail region

Figure 13-16 (above) Cleavage and the forma- tion of the blastula in an amphibian egg.

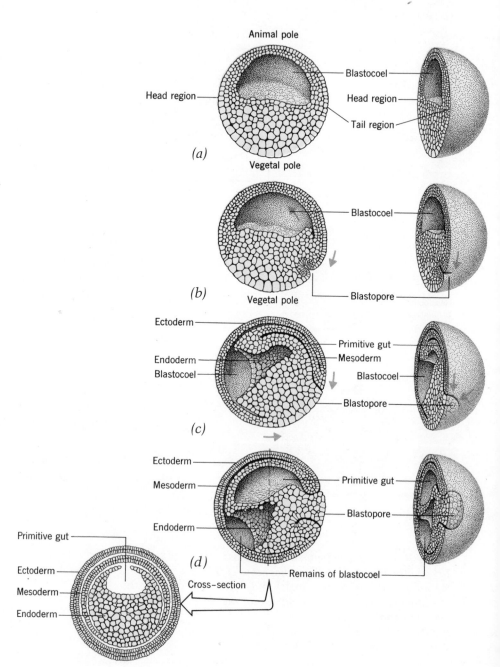

Animal pole

Head region — Blastocoel

Head region

Tail region

(a)

Vegetal pole

Blastocoel

(b)

Vegetal pole

Blastopore

Ectoderm

Endoderm
Blastocoel

Primitive gut
Mesoderm

Blastocoel

Blastopore

(c)

Ectoderm

Mesoderm

Primitive gut

Endoderm

Blastopore

Primitive gut

Remains of blastocoel

(d)

Ectoderm

Mesoderm

Endoderm

Primitive gut

Cross–section

Figure 13-17 (right) Gastru- lation in amphibians. (a) The late blastula stage. (b) Gastrulation begins when surface cells move inward to form a blastopore. (c) Cell division accompanies the continued inward movement of surface cells. The surface layer (ec- toderm) and the newly created layer just below it (the mesoderm) gradually surround the larger cells from the vegetal pole (the endoderm). A cavity, the primitive gut, appears between the endoderm and the mesoderm. (d) The gas- trula stage. The three germ layers (ectoderm, meso- derm, endoderm) are now well-established.

The development of the salamander embryo is illustrated in Figure 13-18. All the vertebrates and most other animals develop from this same sort of tubular organization, formed by gastrulation. But the details of development vary widely.

Development is not neatly divided into phases. As the basic tubular organization is being laid down in one region of the organism, differentiation and organ formation have already begun elsewhere. The embryo is actually a four-dimensional entity—that is, it changes not only in length, breadth, and depth, but also through time. Each major organ has a specific temporal pattern of differentiation; each pattern is coordinated into an overall time sequence of embryonic development.

Patterns of development are complicated by the fact that organisms have evolutionary histories: modern organisms are descended from more primitive organisms. As a result, many organs or structures characteristic of more primitive vertebrates may appear early in the embryo's development. These structures may later disappear altogether, or may be converted to other structures that have totally unrelated functions in the modern organism. A satisfactory account of development, therefore, requires information about the evolutionary relationships of organisms. Otherwise, many aspects of development make no sense.

The developing animal embryo is, of course, an open system. Food must be obtained to provide energy for growth and synthesis. Oxygen must be supplied to the cells during metabolism. Wastes must be eliminated. The simplest way to deal with these requirements is to minimize them. If the embryo develops quickly and remains small, it can maintain exchanges with the environment across its surfaces; it does not require special structures.

Among the vertebrates, many fish and amphibians have this reproductive strategy. They release small zygotes into the surrounding water. Each zygote is provided with yolk, to supply food for its embryonic development. Gas exchange and the elimination of nitrogenous wastes can occur across the surface because the embryo is small. The embryo eventually hatches into a small **larva,** a feeding stage that is quite different from the adult in appearance. The larva feeds to supplement the original food supply while it develops into an adult. This pattern of development has one serious disadvantage: the embryo is relatively unprotected and has a small chance for survival. To compensate for this low survival probability, therefore, the adults have to produce large numbers of eggs.

An alternate reproductive strategy appears again and again in the vertebrates. In this pattern, the embryo is allowed to achieve a larger size and a more elaborate organization before it enters postembryonic life, so it can cope with a more rigorous environment. Individuals therefore have a greater chance to survive, and fewer zygotes need to be produced. Embryonic development is most elaborate in the terrestrial vertebrates: the reptiles, birds, and mammals. These organisms must be able to survive life on land by the time they are born or hatched.

The development of a large, complex embryo presents several problems of its own. The food supply must be increased, either by supplying more yolk or by keeping the embryo in contact with a food supply from the parent. Arrangements for gas exchange and the disposal of wastes must also be present. These requirements are met by specialized structures located next to the developing embryo.

Birds and mammals offer good examples of the development of a complex embryo. The zygote of a typical bird begins as a small plaque of protoplasm lying on the surface of the yolk, enclosed by the eggshell. The embryo obtains food from an immense mass of yolk; gas exchange and waste disposal are accomplished by membranes that surround the embryo.

These membranes are formed from two sheets of cells that grow out from the developing zygote. The external sheet is an extension of the embryonic ectoderm and mesoderm. It folds over the developing embryo and fuses to form two sacs: an internal **amnion,** and an external **chorion.** The second, internal sheet of cells is an extension of the endoderm and mesoderm. This sheet also produces two sacs: the **yolk sac,** which surrounds the yolk, and the **allantois** (see Figure 13-19).

The amnion is filled with a liquid, the **amniotic fluid,** which encloses and protects the embryo. The yolk sac remains connected with the embryo's digestive tract. Food is transferred from the yolk sac to the embryo by a system of blood vessels that is continuous with the circulatory system of the developing bird. The cavity of the allantois receives the nitrogenous wastes produced during development. The wastes are in the form of uric acid crystals, which are left behind when the bird hatches. The membranes of the allantois and the chorion function in respiration. They contain extensive sets of blood vessels connected to the embryonic circulatory system. The membranes lie just beneath the surface of the shell to facilitate the exchange of oxygen and carbon dioxide with the environment.

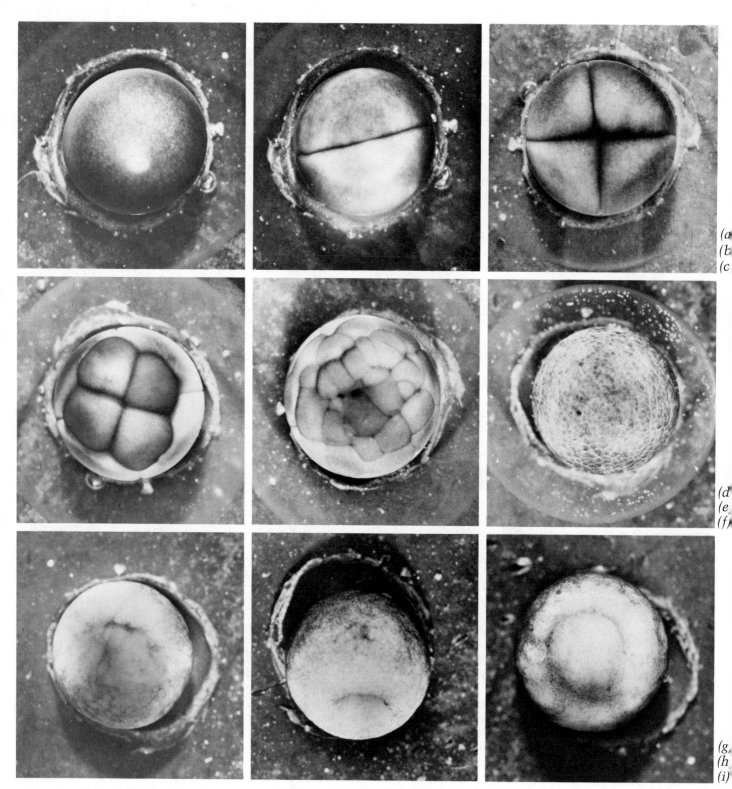

(a)
(b)
(c)

(d)
(e)
(f)

(g)
(h)
(i)

Figure 13-18 The development of a salamander. This series of photographs
illustrates (a–d) cleavage; (e–f) the formation of the blastula; (g–j) gastrulation;

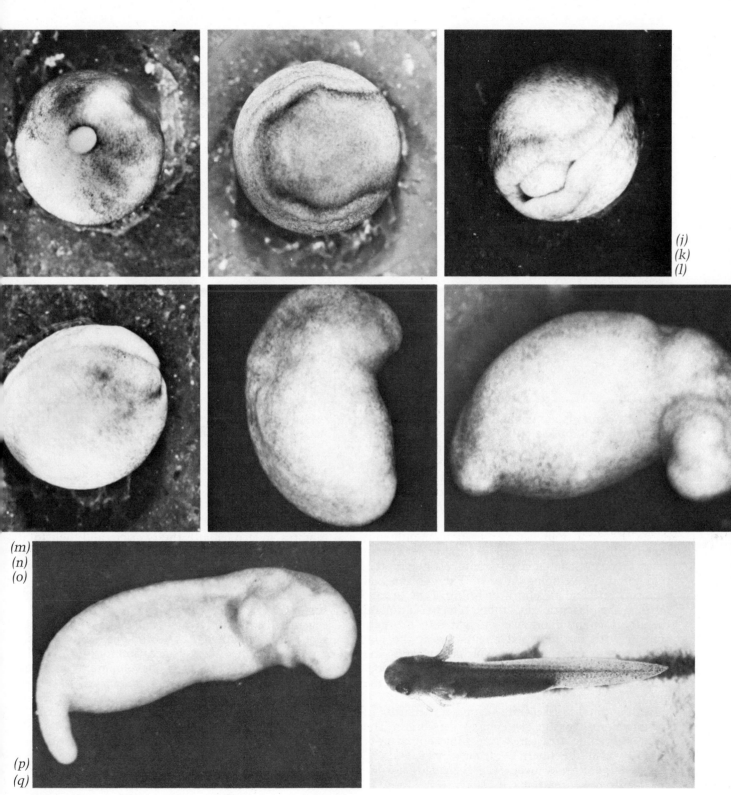

(j)
(k)
(l)

(m)
(n)
(o)

(p)
(q)

(k–m) the formation of the neural tube; and (n–q) the subsequent development of the larva.

423   *animal development*

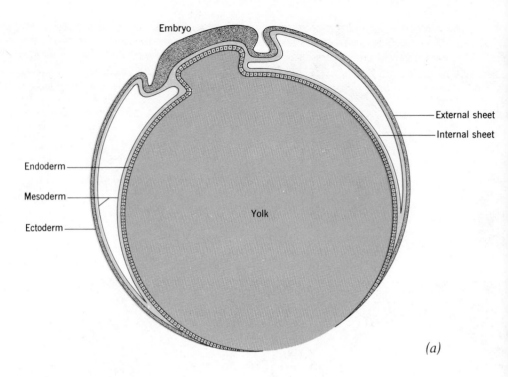

Embryo

External sheet

Internal sheet

Endoderm

Mesoderm

Ectoderm

Yolk

(a)

Most mammals are **viviparous;** that is, they bear their young alive. The only exceptions are the few egg-laying mammals such as the platypus. The embryos of viviparous mammals depend on the tissues of the mother for nutrition, gas exchange, and excretion. These exchanges take place through a structure called the **placenta.** The mammal embryo develops from a plate of protoplasm, and it produces the same four extra-embryonic membranes that the bird embryo does. The amnion again provides a controlled liquid environment and protection for the developing embryo. The chorion and allantois fuse and attach to the lining of the uterus, forming the placenta. The embryonic circulation on the allantois and chorion is brought into intimate contact with the maternal circulation at the placenta, permitting an exchange of food, wastes, and respiratory gases. Figure 13-20 shows the development of these extra-embryonic membranes in the pig.

Birth or hatching is a time of sudden transition in the physiology of the developing vertebrate. The extra-embryonic structures stop functioning quite abruptly, and the internal organs—lungs, kidney, digestive tract—take over. These changes are accompanied by alterations in blood circulation to support the embryo's new physiological organization. Many birds and mammals are relatively help-less at birth, and post-embryonic parental care is necessary for their survival.

Insects are by far the most important class of invertebrates, judged by both their variety and their total number. Since adult insects are more complex than many other invertebrates, their development is correspondingly involved.

The adult insect tends to lay its eggs in moist surroundings, since the development of the insect egg usually requires water. However, a typical insect egg is surrounded by various protective coverings, and it can survive desiccation for long periods. In some cases the egg's development may be triggered by some environmental factor, to assure its hatching in an appropriate season; this mechanism is comparable to seed dormancy in plants.

Many insect eggs have structures that control exchanges with the environment. Water enters through specialized areas in the shell. Respiratory gas exchange occurs across small areas of the protective shell, which function like the spiracles of adult insects. This mechanism limits the loss of water while permitting oxygen and carbon dioxide

Figure 13-19 Formation of the extra-embryonic membranes in the development of a bird. (a) (left) Two sheets of cells grow out from the developing embryo. The outer sheet is composed of ectoderm and mesoderm; the internal sheet is composed of mesoderm and endoderm. (b, c) (below) The external sheet folds over the embryo and joins to produce two sacs, the chorion and the amnion. The internal sheet also forms two sacs, the allantois and the yolk sac.

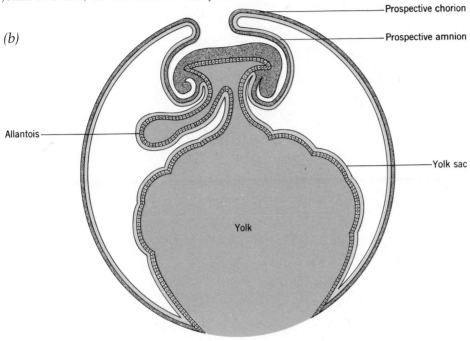

(b)

Prospective chorion

Prospective amnion

Allantois

Yolk sac

Yolk

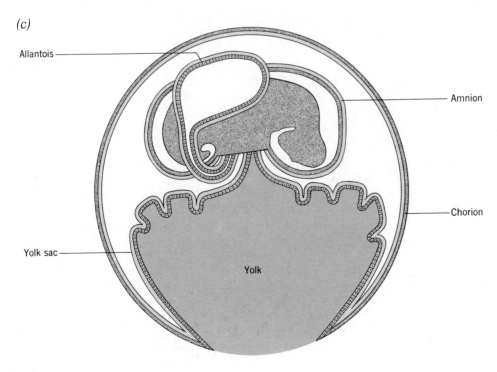

(c)

Allantois

Amnion

Chorion

Yolk sac

Yolk

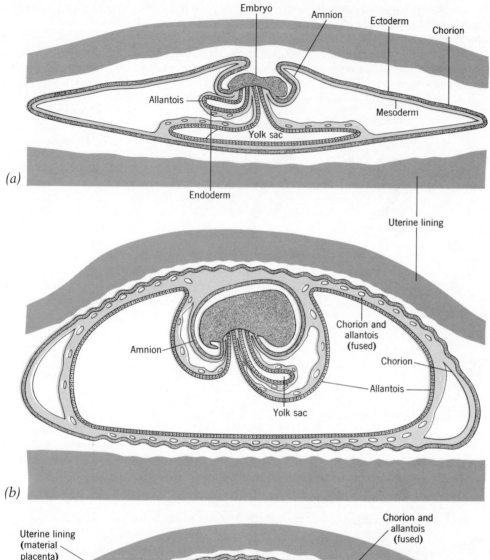

Figure 13-20 The formation of the extra-embryonic membranes in the development of the pig. (a) The early stages of formation are similar to those in birds. (b) Later in development, the chorion and allantois fuse and contribute to the formation of the placenta. (c) The amnion and the allantois also fuse around the embryo.

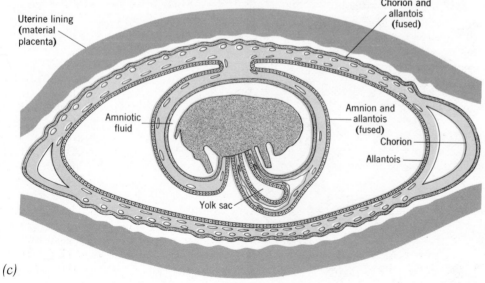

to be exchanged. The nitrogenous wastes that accumulate during development are stored in the egg in the form of insoluble compounds.

Insects that lay their eggs directly in water have other adaptations to provide for gas exchange. The water scorpion, for instance, inserts its eggs into the leaves of water lilies. The egg shell has long, tube-like extensions that reach above the water surface and permit gas exchange. Mosquito eggs are laid in clusters or rafts provided with floats to keep them at the water surface.

The insect zygote nucleus is surrounded by a mass of yolk and then a layer of protoplasm. The development of the zygote begins with a series of nuclear divisions that produce several hundred nuclei. Most of the nuclei migrate to the surface layer of protoplasm surrounding the yolk mass, where they are enveloped with cytoplasm. Adjacent nuclei are then separated by membranes. The result is a single layer of cells surrounding the yolk.

The insect body begins as a disc of cells on one surface of the egg. The primary germ layers are established by a simple process of gastrulation, whereby the germ band thickens and folds over to give two layers. The external layer is ectoderm, and the internal layer is both mesoderm and endoderm. While gastrulation is proceeding, the developing embryo sinks inward, and the sheet of cells that surrounds the egg folds over the embryo. The folds coalesce to form two protective extra-embryonic membranes, called the amnion and the serosa. This procedure is similar to the production of the extra-embryonic membranes in vertebrates (see Figure 13-21).

The germ layers begin to show a segmental organization immediately after gastrulation. At first the segments are similar from head to tail, and each segment bears a pair of rudimentary appendages. But the appendages of the posterior segments fail to develop, the anterior appendages eventually grow into the standard insect pattern: one pair of antennae, three pairs of modified appendages that serve as mouthparts, and three pairs of legs (Figure 13-22).

The edges of the segmented, two-layered embryo eventually fold toward each other. They grow together, giving the usual tubular organization, with ectoderm surrounding the mesoderm and endoderm. Ectoderm produces the body wall, the exoskeleton and appendages, the tracheal system, glands, and the nervous system. The mesoderm differentiates into heart, gonads, muscles, and part of the digestive system. The endoderm appears to contribute only to a segment of the digestive tract. The embryo finally hatches from the egg as a fully-functioning juvenile form.

After hatching, the insect's development to adult form must proceed stepwise, because the exoskeleton limits the amount of growth that can occur at any one stage. The insect exoskeleton is a complex cuticle. It consists of a series of hardened plates secreted by an underlying epidermis. The plates are joined by flexible regions that permit movement. The outermost layer of the cuticle waterproofs the insect, limiting water loss. However, this outer layer cannot be extended. As a result, if the insect is to increase in size, it must periodically discard the old exoskeleton and replace it with a new and larger one. This process is called **molting, or ecdysis.** As a molt approaches, the underlying epidermal layer detaches from the exoskeleton and enlarges by cell division and cell growth. This new, larger epidermis is folded under the old exoskeleton. The epidermis then secretes the inner layers of a new exoskeleton inside the old one. The animal sheds its old exoskeleton and emerges in its new and roomier covering. The plates harden, and the outer, waterproofing layer is added as a final step.

Size increases may be gradual or abrupt. Soft-bodied insects such as caterpillars grow between molts by gradually adding to their exoskeletons until the outer layer of cuticle is fully stretched out. The caterpillars of some silkworm moths may increase 5,000 times in volume in only five molts. But other insects have more extensive networks of hardened plates and can grow very little between molts. These insects take in water or air after a molt to expand the folded and enlarged cuticle. The skeletal plates are then hardened, and the growth of the internal organs proceeds until the expanded armor is again fully occupied. The whole process is then repeated.

Molting not only allows the insect to increase in size but it also permits the whole external form of the insect to change. When the epidermis is freed from the old cuticle just before a molt, the organism can secrete a new cuticle with a radically different morphology. Such changes in form are called **metamorphoses.** Primitive wingless insects, such as the common silverfish, show little metamorphosis when they molt. The young look just like the adults. Molting simply increases their size and allows the development of the reproductive organs.

In other insects, there are two major patterns of metamorphosis. The first is a gradual assumption of adult form through a series of molts. The immature forms are called **nymphs** (in terrestrial insects) or

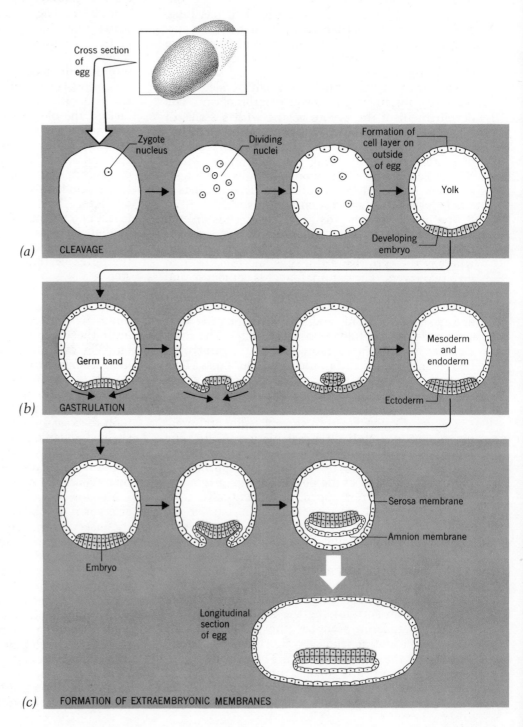

Figure 13-21 The early development of a typical insect.

Cross section of egg

(a) CLEAVAGE

Zygote nucleus

Dividing nuclei

Formation of cell layer on outside of egg

Yolk

Developing embryo

(b) GASTRULATION

Germ band

Mesoderm and endoderm

Ectoderm

(c) FORMATION OF EXTRAEMBRYONIC MEMBRANES

Embryo

Serosa membrane

Amnion membrane

Longitudinal section of egg

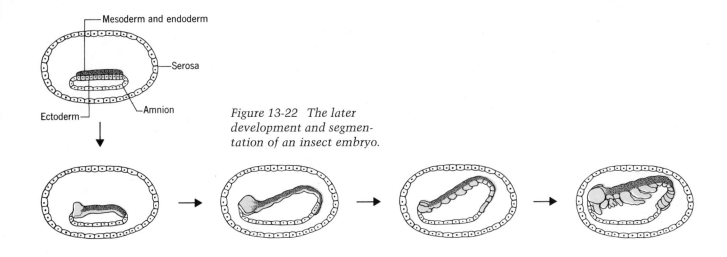

Figure 13-22 *The later development and segmentation of an insect embryo.*

naiads (in aquatic insects). The young are similar to the adults, but they lack wings; cockroaches and dragonflies are examples. As cockroach nymphs increase in size, wings appear as pads. These pads become increasingly prominent, until the adult form is ultimately achieved. Dragonfly naiads live in fresh water. They have gills and other structures that are not found in the adult. However, their metamorphosis is still gradual.

The second form of metamorphosis, complete metamorphosis, is more dramatic. The insect hatches as a larva. The larva is primarily a feeding stage that acquires raw material for development. The wings and other adult structures, such as legs, are developed internally through a series of larval molts. A new stage, the pupa, appears in the last molt before the adult form is achieved. The pupa then undergoes a major reorganization of the body form, and the insect finally emerges as an adult. Butterflies are the most familiar insects that show complete metamorphosis. The butterfly egg hatches a small caterpillar. The caterpillar feeds on vegetation and goes through a series of larval molts, increasing in size each time. Throughout the larval stages, wings develop as pads of tissues under the cuticle; they are not yet visible. The last larval stage molts to a pupa, which has visible adult wings and legs. The pupa is immobile; it is a stage of active synthesis and reorganization, in which the larval form is destroyed and the adult form is achieved. The adult butterfly eventually emerges from the pupa. The butterfly is totally different in form and habit from the immature stages. Gradual and complete metamorphosis are illustrated in Fig-

ure 13-23. The butterfly is shown in Figure 13-24.

Molting is coordinated and controlled by several hormones that are responsive to environmental inputs. Molt is initiated by a brain hormone, produced by neurosecretory cells and liberated into the blood. The brain hormone itself activates the prothoracic glands and stimulates them to produce the molting hormone, ecdysone. Minute quantities of ecdysone initiate the synthesis of a new cuticle, the resorption of the old cuticle, and molting.

The neurosecretory cells are activated by events in the internal or external environment. The nature of the activation depends on the insect. Blood-sucking bugs molt shortly after they take a blood meal. Apparently, ingested blood stretches the midgut, activating neurosecretory cells that in turn activate the prothoracic glands. The pupa of the silkworm moth molts after exposure to low temperature. Silkworm pupae maintained at room temperature will never molt. However, if a pupa is chilled and rewarmed, the brain hormone is secreted and development proceeds. In nature, the silkworm pupa spends its winters in a resting state; chilling in the winter and rewarming in the spring are necessary for the completion of its development.

Another hormone, the juvenile hormone, can modify the effects of ecdysone. It causes the epidermis to continue to secrete a larval cuticle. The form of the insect is determined by the balance between ecdysone and juvenile hormone. A high proportion of juvenile hormone stimulates the larval portion of the genome to direct the secretion of a juvenile cuticle. A decrease in the amount of

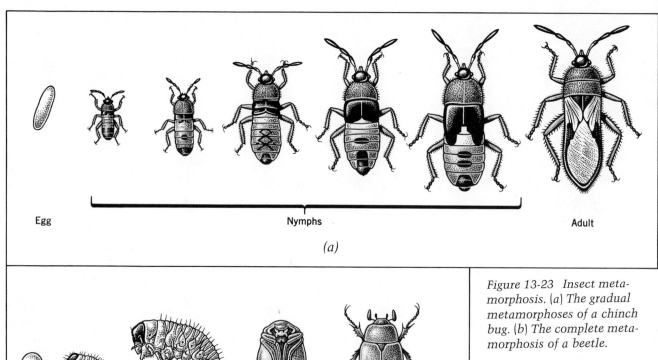

Egg | Nymphs | Adult

(a)

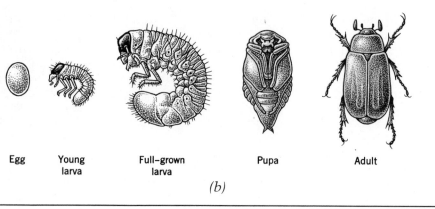

Egg | Young larva | Full-grown larva | Pupa | Adult

(b)

Figure 13-23 Insect metamorphosis. (a) The gradual metamorphoses of a chinch bug. (b) The complete metamorphosis of a beetle.

juvenile hormone results in the activation of the pupal and then the adult genome. (see Figure 13-24).

If the hormone balance is disrupted, the insect's development is abnormal. An insect will undergo additional molts if excess juvenile hormone is present. The result will be an abnormally large nymph. Conversely, if the tissue that produces juvenile hormone is removed, the insect will metamorphose prematurely into a diminutive adult.

Ecdysone appears to be the molting hormone not merely for insects but for all other arthropods as well. Juvenile hormones are more specific; the juvenile hormone active in one insect group may have no influence on development in other insects.

Patterns of development in vertebrates and insects have many similarities. For instance, some in-

sects have developed a kind of placentation: their young, like those of vertebrates, are highly developed and are hatched in small numbers. On the other hand, some vertebrates, like insects, undergo metamorphosis. Tadpoles, for instance, metamorphose into adult frogs by resorbing their gills, growing lungs, rearranging their digestive tracts and other internal organs, resorbing their tails, and growing limbs. These changes are all under hormonal control. For that matter, human development, from newborn to adult, is characterized by significant morphological and physiological changes. Although these changes are more gradual than the transition from tadpole to frog, they are also under hormonal control and equally deserve the name "metamorphosis."

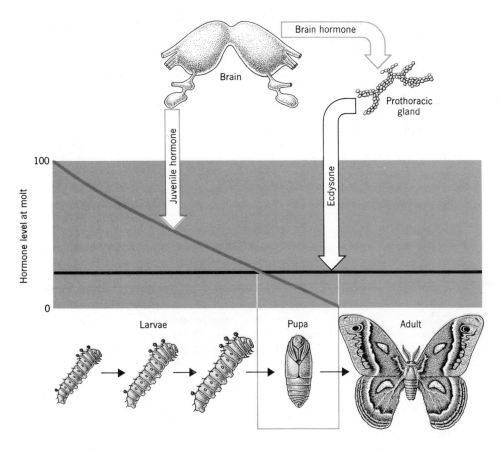

Figure 13-24 Hormonal control of insect metamorphosis. Ecdysone, secreted by the prothoracic glands in response to brain hormone, causes molting. Juvenile hormone, secreted by the brain, prevents the insect from advancing to the next stage of development when it molts. During early development, juvenile hormone levels are high, so the larva molts into another, larger larva. As the level of juvenile hormone drops, the larva metamorphoses into a pupa and then into an adult.

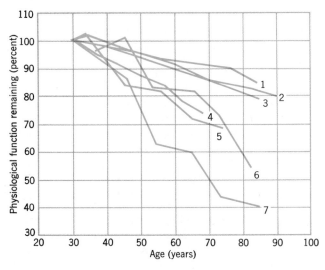

1 **Nerve** conduction velocity 5 Cardiac output (at rest)
2 Basal metabolic rate 6 Filtration rate of kidney
3 Body water content 7 Maximum breathing capacity (voluntary)
4 Work rate

Figure 13-25 Effects of aging in man. Many physiological functions decline steadily after age 30.

Aging and Death in Animals

Aging is part of the normal developmental life history of most animals. Cell aging, or senescence, first shows up early in development as irreversible differentiation. The later development of most animals is characterized by the aging, senescence, and death of the whole organism. Aging in mammals is associated with a general decline in physiological function. In man, most measures of vitality or general health peak at about age 30 and decline thereafter (Figure 13-25).

Various theories have been advanced to explain again in animals. Cell death has been suggested as a cause of aging. Short-lived cells, such as red blood cells or epithelial cells, are produced throughout the lifetime of the organism. But long-lived cells are lost and not replaced. Brain cells are lost throughout life. Most of the loss of brain cells, however, occurs between birth and maturity. And the total cell loss caused by aging is small compared to the deficit that can be tolerated. Mammals can survive with one lung, one kidney, one third of the liver, and no limbs, and they can function without large areas of the brain. Clearly, aging is not the result of a simple decrease in cell number, or in the number of functional units in various tissues.

Many hypotheses about aging focus on specific changes in cells and cell products. If cellular changes that are responsible for aging can be identified, we can investigate ways to modify them and possibly retard the aging process itself. A good example is the investigation of changes in collagen. **Collagen** is a major protein constituent of mammals; it represents about one fourth of our total protein content and is found in all the tissues of the body. Collagen maintains the form and elastic properties of the organs and tissues. It is also found in the connective tissue that surrounds muscles and internal organs, connects them to each other, and suspends them in body cavities. Collagen literally holds the body together.

Collagen protein is secreted by cells in the connective tissues. While an animal is growing, its tissues contain considerable amounts of collagen in soluble form. By maturation, most of the collagen has aggregated, forming insoluble fibrils that surround the cells in the connective tissue (Figure 13-26). As the animal ages, the collagen fibrils become more strongly cross-linked and even more insoluble. This effect may be caused by an enzyme-catalyzed chemical reaction. Aging collagen also associates with other materials that gradually accumulate in the extracellular spaces of tissues.

These changes in collagen can account for many of the specific changes that are associated with aging. Increased cross-linkage between collagen molecules, and an accumulation of calcium and cholesterol, are both associated with a decrease in the flexibility of collagen fibrils. This stiffening of the collagen must interfere with various physiological activities. The contraction of heart muscle becomes more difficult because of the stiffness of the connective tissues. Blood vessels become less elastic and less able to absorb pressure changes in the circulatory system. Aging collagen in connective tissue matrix may restrict the passive diffusion of materials such as food and oxygen to organs and tissues; the removal of wastes would also be impaired. The deterioration of collagen may explain why arteriosclerosis (hardening and degeneration of artery walls) and hypertension (high blood pressure) are two of the most common causes of death in aging humans.

Some observations, however, disagree with the collagen hypothesis. If changes in collagen cause aging, then preventing these changes should retard it. Researchers have treated rats and mice with chemical agents that do just that: retard the formation of cross-linkages in collagen. The treatment did not slow the aging process. The theory of aging based on collagen changes is still unproven.

In mammals, aging can be retarded by restricting their food intake to slow the growth rate. Underfed rats survive much longer than their better-fed brothers (Figure 13-27). We don't understand why aging is retarded in these circumstances, but it does suggest that the aging of an animal is related more to its stage of development than to time—if development slows down, so does aging. This relation is quite apparent in cold-blooded vertebrates such as fish. Bass that live in northern waters grow much more slowly than southern populations of the same species, and they live twice as long.

Some populations do not face the problem of aging at all. Because of the rigors of their environments, they have few old or senescent individuals—the animals die before they have a chance to grow old. The planktonic algae that filter feeders consume are an example. Such organisms simply face a constant probability of being killed, and this probability is independent of their age. Many other populations, however, do contain aging individuals. In such populations, older individuals are more likely to die than younger ones. Human populations show both of these characteristics. That is, they contain some individuals whose deaths are independent of aging, and many whose deaths are related to it. Death in a population is often described by a mortality curve. The early mortality of infants and children is not related to senescence; it reflects accidents, disease, and inadequate nutrition and hygiene. In intermediate ages, some deaths due to accidents and disease are still independent of age; but other causes of death in middle age, such as cancer, are age-related. Senescence and aging become strongly correlated with deaths in the population only at relatively advanced ages (see Figure 13-28).

Thus, though human death is obviously related to aging and senescence, it is not a simple or strict

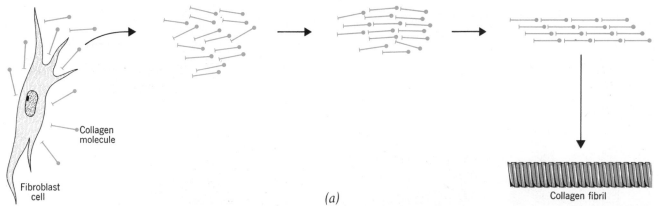

(a)

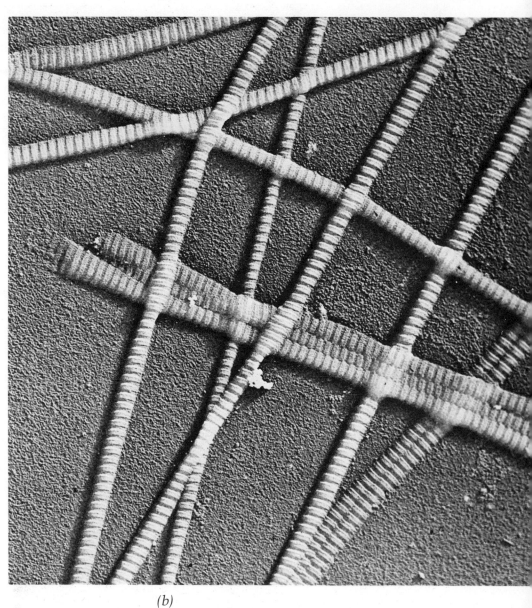

Figure 13-26 Collagen fibrils. (a) Collagen molecules are secreted in soluble form by connective tissue cells. They polymerize and cross-bond to produce fibrils. (b) Electron micrograph of collagen fibrils from mammalian skin. Magnification 38,000×

Collagen
molecule

Fibroblast
cell

Collagen fibril

(b)

433 *animal development*

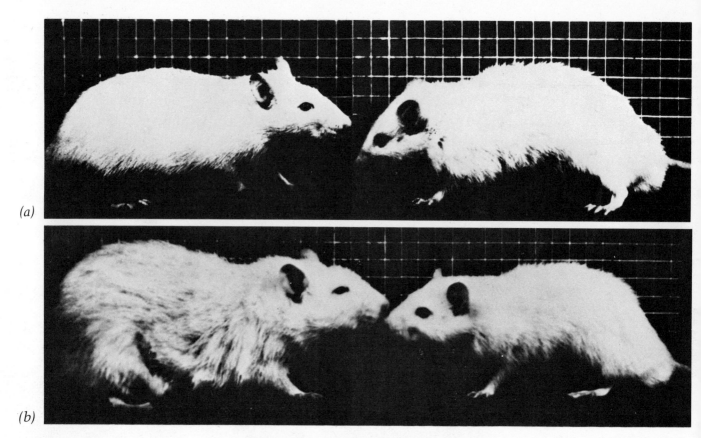

(a)

(b)

Figure 13-27 *Aging in rats. (a) An underfed 24-month-old rat (right) looks similar to a normally-fed three-month-old rat (left). (b) By comparison, a normally-fed 24-month-old rat (left) has reached old age compared to an underfed rat of the same age.*

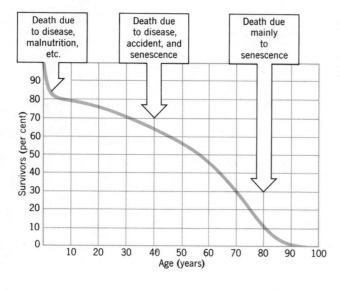

Figure 13-28 *Human mortality curve. The curve represents the United States population about 1900.*

relationship. Aging is a heritable characteristic in humans, but people do not automatically die on their 70th birthday. Aging must interact with environmental forces to produce death. Presumably, environmental stresses that threaten the functional integrity of the human organism occur at random times. But we know that many physiological functions decline with age. Thus, the probability that an individual will succumb to an environmental challenge increases with age.

regeneration

All plants and animals have some ability to survive injury by regenerating lost or damaged tissues. But this capacity is extremely variable. Simple organisms may be very limited in their repair abilities, and some rather complicated organisms do very well. Various salamanders can regenerate entire limbs. Still other organisms have a spectacular capacity to restore their normal form and function by regeneration. Sponges can reform from completely dispersed individual cells. A new hydra can arise from a small fragment of parent tissue; the same is true of some flatworms. Higher plants respond to the loss of the shoot tip by activating new meristems in lateral buds. Plants also repair and replace other lost parts, and sometimes they can build new sporophyte plants from cuttings or fragments of tissue. In fact, regeneration may serve as a normal mode of asexual reproduction in some organisms.

Mechanisms of repair and regeneration depend on the usual developmental processes: cell division, cell differentiation, and cell movement. The study of regenerative processes is therefore in the province of developmental biology. For the developmental biologist, regeneration represents a system of cell division and differentiation that can be initiated to suit the convenience of the investigator, and analyzed at leisure. As such, it should yield considerable information about mechanisms of development. But for the most part, this promise has not been fulfilled, even though a great deal of descriptive information has been accumulated.

Much of our present information is interesting and tantalizing. We know, for example, that the regeneration of a salamander limb after amputation will occur only if a critical mass of nervous tissue is present in the regenerating stump. The regenerative capacity is lost if the leg nerve is removed, and it is restored by the insertion of nerve tissue from another source, such as other nerves, brain tissue, or even embryonic nerve cells. Furthermore, we have learned that amphibians that do not normally regenerate limbs can be induced to do so by the implantation of additional nerve tissue in the stump. It is assumed, therefore, that some material is transferred from the nerve to the regenerating tissue, but we do not really understand the basis of these effects.

Similar experiments with mammals are more difficult, but it has been possible to obtain regrowth and some regeneration in the hind limb of the opposum (Figure 13-29). A newborn opposum is a

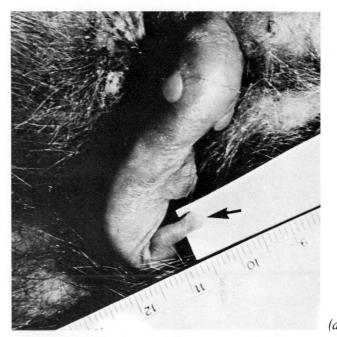

(a)

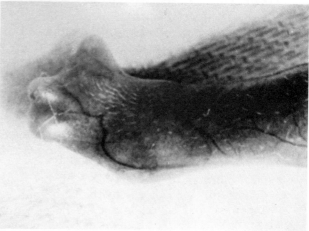

(b)

Figure 13-29 *Regeneration of the hind limb of a newborn opossum. (a) The stump of a foot amputated at birth shows little regeneration after five weeks. (b) Implantation of nervous tissue before amputation results in significant regeneration.*

good experimental system for two reasons: it is born in a very immature state (about thirteen days after conception), and it has no functional immune system. Thus it is possible to work with a rapidly developing limb and to implant nervous tissue without risk of transplant rejection. Both features are important to the success of this experiment. Attempts to produce the regeneration of a more mature limb failed.

Such experiments should help us to understand the biological basis of regeneration. It is even possible that, given such knowledge, we will eventually be able to induce the regeneration of mammalian limbs and other structures whose loss or damage is now irreversible. But we are still very far from understanding the mechanism of regeneration, let alone achieving the ultimate practical aim.

cancer

From a biological point of view, cancer is an example of a developmental defect. "Cancer" is the term for a whole class of diseases, but all of them are produced by cells that escape the controls limiting cell division in normal development. In cancer, normal cells, whose cell division is regulated, become transformed into cells whose growth is uncontrolled.

Cancer damages the organism by destroying normal tissue in two ways. First, **tumors**—masses of cancer cells—prevent the normal functioning of the body. Second, rapidly growing cancer cells starve normal tissues by using up necessary nutrients. Any kind of cell in the body may become cancerous, and cancer is usually described on the basis of the cells that make it up: liver, bone, nerve, or blood. Cancer cells sometimes become invasive; that is, dividing cancer cells invade adjacent tissues or are carried by the blood or lymph to more distant sites. The spread of cancer cells is called **metastasis.** Cancer cells that are highly invasive and that set up secondary tumors are described as **malignant,** while relatively non-invasive cancer cells are **benign;** but the distinction between benign and malignant is relative, and all intermediate conditions can occur.

The cancer that is seen by the medical practitioner is already in an advanced stage. Tumors are visible or palpable, and abnormal cells can be detected (see Figure 13-30). Cancer is treated by removing the tumorous growths with surgery or radiation therapy or chemotherapy (treatment with drugs). Most chemotherapy uses chemicals that selectively poison dividing cells. Unfortunately, however, the poisons now being used are equally effective on cancer cells and on normal dividing tissues such as blood or epithelium. Furthermore, both chemotherapy and radiation treatment depress the immune system of the patient and thereby impair the body's normal defenses against the spread of cancer cells. Such treatment also makes the individual susceptible to other diseases.

Modern cancer research focuses on the processes that transform a normal cell into a cancer cell. One important discovery is that cancer cells undergo changes in the surface properties of their outer membranes. Cultures of normal cells will divide and grow rapidly until the surface of the culture dish is covered with a single layer of cells. But then cell division slows or ceases; this process is termed **contact inhibition** of cell division. Transformed cells, however, have lost contact inhibition; they continue to divide, and they pile up, forming a mass several cell layers thick.

The changed surface properties of cancer cells make them susceptible to attack by the immune system. Antibodies recognize the cancer cell's surface as an antigen. As a result, normal cells that are transformed to cancer cells may be kept in check by immune responses. Immune responses occasionally fail, however, and cancer results.

The loss of contact inhibition gives researchers a way to detect transformed cancer cells and single them out from cultures of normal cells; cancer cells reveal themselves by forming multilayered masses in culture. It then becomes possible to search for the agent or the combination of factors that produces the transformation.

The best-known cancer agents are viruses. When certain viruses are added to a tissue culture, some or all of the cells begin to divide, change their surface properties, and become cancer cells. Viruses are known to be associated with some cancers of mammals and other vertebrates. Current thinking (or more properly, hoping) is that such viruses are in fact responsible for the most dangerous cancers. If they are, we have a promising approach to understanding the basic transformation from normal cell to cancer cell, because we already know something about the invasion of cells by viruses.

The normal life cycle of a virus begins when a virus invades a cell. Genetic information from the virus directs the synthesis of enzymes, which themselves direct the synthesis and replication of structural proteins and nucleic acids for the virus. New virus particles are assembled from these com-

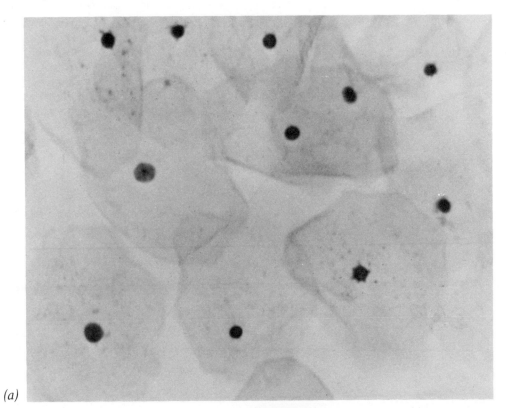

(a)

Figure 13-30 Normal (a)
and cancerous (b) cells from
the uterine cervix. Mag-
nifications 550×.

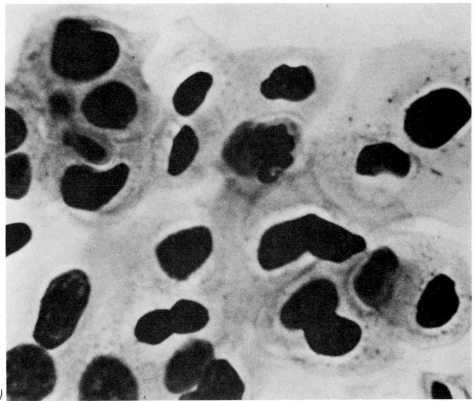

(b)

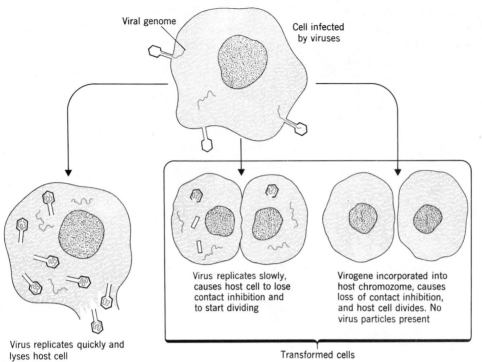

Viral genome

Cell infected by viruses

Virus replicates quickly and lyses host cell

Virus replicates slowly, causes host cell to lose contact inhibition and to start dividing

Virogene incorporated into host chromozome, causes loss of contact inhibition, and host cell divides. No virus particles present

Transformed cells

Figure 13-32 (facing page) A virogene is a piece of DNA or RNA from the virus genome inserted into the host genome. DNA can be inserted directly. RNA must first direct the synthesis of complementary DNA, which is inserted.

Figure 13-31 (left) A normal cell may be destroyed by the process of virus replication. But it may be transformed into a cancer cell instead, in either of the ways indicated.

pounds inside the host cell; they are finally released, causing the death of the invaded cell. However, in some cells, only the first step occurs. That is, the host cell's synthetic apparatus is turned on, presumably by a protein coded on the virus genome, but it does not synthesize duplicates of the virus. Instead, the host cell synthesizes its own components, and either begins to divide or increases its rate of cell division. This development is accompanied by a change in surface properties. The cell has been transformed into a cancer cell (see Figure 13-31).

If this viral hypothesis is correct, we should be able to understand the transforming event by isolating the protein that is synthesized early in the virus life cycle. Unfortunately, most cancers do not seem to be produced directly by viruses. Furthermore, even when certain cancers have been associated with viruses, it is often impossible to demonstrate that the virus is present in all tumors of that type of cancer; in some cases one can even exclude the possibility that there was any exposure to the virus agent when the transformation occurred.

This evidence has cast doubt on the importance of viruses in the transformation process. However, it has recently been found that some mammalian cells carry portions of virus genome in their own chromosomes. Viral DNA may be inserted into the chromosomal DNA of the host cell; in this case, the viral DNA is called a **virogene.** It is duplicated during normal host cell division and passed on to the daughter cells. The virogene is repressed, so that its genetic information does not synthesize whole viruses. However, the existence of virogenes has been proven by DNA hybridization experiments.

The idea that viral DNA is incorporated into the chromosomal DNA of a host cell is not too surprising—this process is comparable to transduction (described in Chapter 12). However, it has also been discovered that viral genomes composed of RNA can become part of the host chromosome. The RNA directs the synthesis of complementary DNA, and the DNA is then inserted into the host genome. This discovery is exciting in general terms, since it violates the principle that information flow in the cell is always from DNA to RNA. In the context of cancer biology it is exciting, because it explains the presence of virogenes in host chromosomes that are complementary to the viral RNA genome. Thus, virogenes exist for both RNA and DNA viruses (Figure 13-32).

These facts suggest that a portion of the virogene

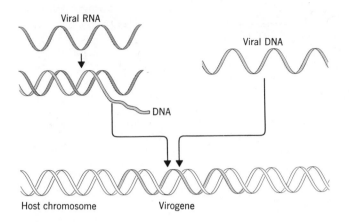

Viral RNA

Viral DNA

DNA

Host chromosome   Virogene

may transform normal cells to cancer cells even when a whole virus is absent. It would be necessary only to activate the initial portion of the viral genome in the host chromosome to cause the transformation. The genome would then synthesize a protein that activates nucleic acid replication and cell division in the host cell.

This mechanism explains why such a great variety of agents can produce cancer. The agents could act in various nonspecific ways to activate virogenes already present in cells. The virogene theory would also explain the heritable character of some cancers. That is, cancer is transmitted as a

virogene from one cell generation to the next, and from parent to child. The virogene may remain inactive throughout the life of the individual. But the appropriate combination of environmental factors, external or internal, may activate the virogenes so that the infected cells are transformed into cancer cells. The virogene theory is a promising hypothesis. But even if it is verified, it may not be the only or even the most important cause of cancer. Many cell types are affected by many kinds of cancers.

Cancer research is now receiving a great deal of attention in the United States. In 1974, the National Institutes of Health received $500 million for cancer research. This is $74 million more than the amount for 1973. At the same time, support for most other kinds of biological research was reduced. Most biologists believe that it is futile to attempt this kind of "moonshot" approach to cancer biology while our information about normal development remains so meager. Cancer is, after all, a process that modifies the normal controls of development. Past experience has shown that, in the long run, a balanced approach to research is most productive. A balanced program is one that encourages research on specific problems—such as cancer—but that *also* supports studies of basic biological processes. This is the only way to be sure that our medical practices are based on a solid understanding of cellular mechanisms.

Recommended Reading

Ebert, James, and Ian M. Sussex, *Interacting Systems in Development* (2nd ed.). New York: Holt, Rinehart, and Winston (1970).

The regulation of development by interactions at all levels of organization—molecules, organelles, cells, and tissues.

Kohn, Robert R., *Principles of Mammalian Aging.* Englewood Cliffs, N.J.: Prentice-Hall (1971).

A discussion of aging at the molecular, cellular, and organismic levels, with particular attention to mammals and man.

Torrey, John G., *Development in Flowering Plants.* New York: Macmillan (1967).

A summary of the development of flowering plants from single cells to adult organisms.

Trinkaus, J. P., *Cells into Organs: The Forces that Shape the Embryo.* Englewood Cliffs, N.J.: Prentice-Hall (1969).

A discussion of the mechanisms of animal development, with particular emphasis on the movement of cells and cell sheets.

14 human reproduction and development

The cellular aspects of human reproduction follow the general principles discussed in the preceding two chapters. Human gamete formation, fertilization, and development are not unique or even particularly unusual. Social and behavioral factors are an important part of human reproduction. But these factors are not unique to humans, either. Other animals have elaborate mating behavior and supporting social organizations. In fact, the details of reproduction are better known in many other organisms than they are in man simply because other organisms are easier to study. Nonetheless, a separate discussion of human reproduction is easily justified because of the tremendous personal and social importance of the subject.

Biological information about human reproduction, and about human ecology (the subject of Chapter 19), is particularly fundamental to the social sciences. All the social sciences concern themselves in one way or another with the behavior and ecology of man, and they all deal with reproduction and development. Even though each field has its own set of principles and methods of investigation, the work of the anthropologist or the sociologist ought to be consistent with the work of the biologist. The relationship between biology and the social sciences is much the same as the relationship between physics and biology: the biologist is not a physicist, but biology is based on physical principles. Consequently the biologist must make his work consistent with those principles; and at the same time, the physicist often provides insight into biological problems. Similarly, the sociologist must work within the constraints of biological information.

Albrecht Durer, "The Rape of a Young Woman"

441

Human reproduction depends on a complex system of organs and ducts, called the **genitalia.** The genitalia inside the body include the gonads, which are the source of eggs and sperm; various ducts for conveying the gametes away from the gonads; glands for nourishing the gametes; and organs for sperm storage (in males) and the growth of the embryo (in females). The genitalia on the outside of the body are mainly concerned with sexual arousal and copulation. They insure that eggs and sperm are safely brought together so that fertilization will occur. These external organs also support a general level of sexual activity that is an integral part of human behavior.

The Male Genital System

The internal genitalia in the male are composed of testes, ducts, and glands (Figure 14-1). Each testis is a mass of long, coiled **seminiferous tubules.** Sperm are produced by the cells that line the tubules.

Sperm pass from the seminiferous tubules into a coiled duct called the **epididymis,** which is attached to the surface of the testis. Sperm are stored there while they mature and await ejaculation. A major sperm duct, the **vas deferens,** transports sperm from the epididymis to the exterior during ejaculation. Peristaltic waves of contraction in the walls of the epididymis and vas deferens move the sperm along. The vas deferens leaves the testis and loops over the **ureter,** a duct that drains urine from the kidney to the bladder. The vas itself empties into the **urethra,** which drains the bladder. Three glands, the **seminal vesicle, prostate,** and **bulbourethral gland,** secrete various fluids into the duct as the sperm move along. This mixture is **semen.**

The external genitalia are the scrotum and the penis (Figure 14-2). The **scrotum** is a sac of skin and smooth muscle that contains the testes. The **penis** is a mass of erectile tissue that surrounds the urethra. A cylinder of erectile tissue is located on each side of the urethra, and a third tube of erectile tissue surrounds it. The tissue enlarges and becomes firm when engorged with blood, causing erection. The central, tubular mass of erectile tissue at the tip of the penis forms the **glans.** Skin on the penis forms a fold over the glans called the **prepuce,** or foreskin.

Figure 14-1   The internal and external genitalia of the male.

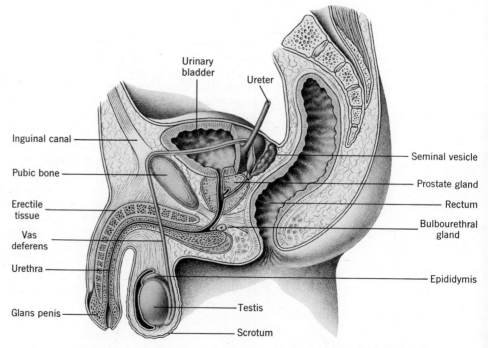

Urinary bladder

Ureter

Inguinal canal

Pubic bone

Erectile tissue

Vas deferens

Urethra

Glans penis

Seminal vesicle

Prostate gland

Rectum

Bulbourethral gland

Epididymis

Testis

Scrotum

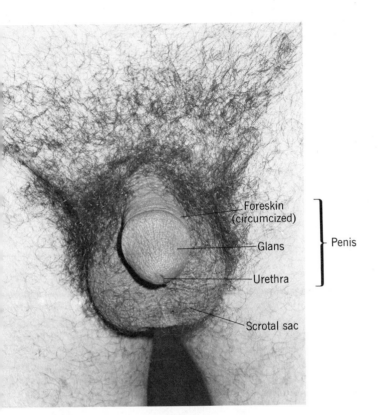

Figure 14-2 *The male external genitalia.*

Foreskin
(circumcized)

Glans

Urethra

Penis

Scrotal sac

The Female Genital System

The female internal genitalia are the ovaries, which produce eggs; the ducts and organs that transport the egg; and the organs concerned with the development of the embryo (Figure 14-3). The **ovary** is a bean-shaped gonad located next to the oviduct. Eggs are shed from the ovary as secondary oocytes—they have already undergone the first meiotic division. The egg is trapped by the funnel-shaped end of the oviduct and moved along by cilia and by contractions of smooth muscle. The oviducts lead to the **uterus,** a thick-walled organ that houses the developing embryo. The lining of the uterus is called the **endometrium;** it undergoes dramatic changes during the reproductive cycle. The endometrium is covered by a thick coat of smooth muscles.

The base of the uterus is the **cervix,** which opens into the muscular **vagina.** The external opening of the vagina is surrounded by a membrane called the **hymen,** or maidenhead. The hymen is perforated at the center, so that the vaginal canal is in communication with the exterior. Often the opening is too small to admit the erect penis. Hence it may be torn during initial copulation, persisting as shreds of tissue around the opening of the vaginal canal. However, the hymen can be ruptured in many

Figure 14-3 *The internal and external genitalia of the female.*

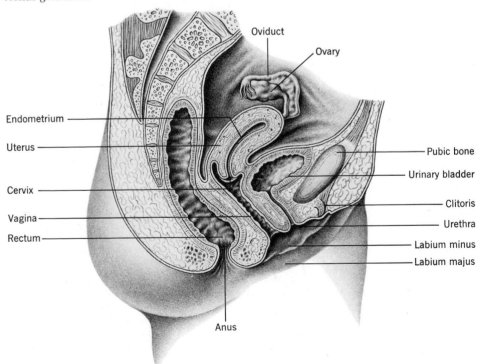

Oviduct

Ovary

Endometrium

Uterus

Cervix

Vagina

Rectum

Pubic bone

Urinary bladder

Clitoris

Urethra

Labium minus

Labium majus

Anus

other ways—by vaginal examination, masturbation, or various accidents. In other cases, the normal opening of the hymen is sufficiently elastic to stretch and admit the penis without tearing. Obviously, there is little correlation between an intact hymen and virginity. The external opening of the urethra is close to the vaginal opening. The only function of the urethra is to empty the bladder; it does not participate in the transfer of gametes to the exterior as it does in the male.

The female external genitalia include the clitoris, the labia minora, and the labia majora (Figure 14-4). The **clitoris** is a homologue of the penis in the male. That is to say, both organs have the same embryonic origin and the same basic structure. Like the penis, the clitoris has a pair of cylindrical masses of erectile tissue, and it terminates in an enlarged glans. (The third tube of erectile tissue, which surrounds the urethra in the penis, is not present in the female.)

Figure 14-4   The female external genitalia.

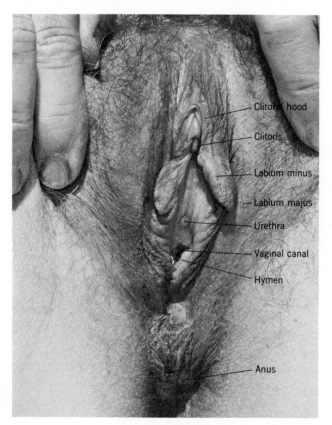

Clitoral hood
Clitoris
Labium minus
Labium majus
Urethra
Vaginal canal
Hymen
Anus

The clitoris is unique in that it receives stimuli and focuses the sexual response without playing any direct functional role in gamete transfer and fertilization. Apparently the clitoris is not necessary for sexual responsiveness, however, since its surgical removal does not necessarily interfere with orgasm. The **labia minora** and **majora** are lips or folds of skin that surround and enclose the urinary and vaginal openings and the clitoris.

human sexual response

The human sexual response has three basic components: arousal, orgasm, and copulation. Sexual arousal involves physiological changes in the whole body. Arousal often leads to a set of physiological responses and sensations called **orgasm.** Arousal may lead as well to **copulation**—that is, the insertion of the penis into the vagina. However, in humans, sexual arousal, copulation, and orgasm are often separated. Human sexual responses may lead directly to reproduction, but frequently they do not. Human sexuality is much more than a reproductive act—it is an important part of human behavior and social structure.

Arousal and Orgasm

Sexual arousal is a complicated phenomenon. Various psychogenic factors (fantasy, visual and auditory inputs), as well as the stimulation and manipulation of the genital area by touch, all contribute to arousal. Thus, arousal behavior is extremely variable; it depends on the individual, the situation, social convention, and custom.

Self-arousal and achievement of orgasm (**masturbation**) is an extremely common practice not only in humans but in other primates as well; it is virtually universal among individuals in our society. It has no known ill effects. Some individuals can achieve orgasm entirely by erotic thoughts. However, a combination of erotic fantasy and mechanical stimulation of the genital area is more common.

Between partners, arousal behavior often involves gentle stroking and mutual manipulation of the genitals. But this practice is hardly universal. In some cultures, painful scratching and biting replace genital manipulation. Heterosexual and homosexual patterns of arousal and achievement of orgasm show great variety, as does their social acceptability. "Normal" arousal behavior simply cannot be defined.

The physiological response to arousal is much more standardized, however. That is to say, the effects of arousal and orgasm on the genitalia are much the same in all individuals, regardless of how they are produced. Arousal in women is accompanied by secretion of mucus in the barrel of the vagina. The labia majora flatten and move laterally to expose the vaginal opening. The labia minora engorge with blood, become swollen, and change from pink to a bright red color. The swelling in the labia minora forces apart the labia majora; it also increases the effective length of the vagina. The clitoris becomes engorged with blood and increases in diameter. The inner two thirds of the vagina expands, and the vaginal canal lengthens (Figure 14-5). Rotation and elevation of the uterus enlarges the inner two thirds of the vaginal canal still further.

During orgasm, the first third of the vaginal canal contracts rhythmically several times. Deeper areas of the vagina widen further and elongate, and the uterus contracts. A woman may return to an intermediate state of arousal soon after orgasm, and she may achieve subsequent orgasms without returning to a non-aroused condition.

A distinction is sometimes made between a "clitoral" and a "vaginal" orgasm. There is no physio-logical basis for such a distinction. The orgasm itself is the same, whether the clitoris or the vagina is stimulated. In fact, mechanical stimulation of the vagina nearly always agitates the clitoris as well. This distinction refers only to the nature of arousal and to the more psychological aspects of orgasm.

The initial response to sexual stimulation in men is the erection of the penis (Figure 14-6). Erection is apparently controlled by the parasympathetic division of the autonomic nervous system. (In rats, at least, erection is directed by a parasympathetic control center in the hypothalamus. Stimulation of specific neurons in the hypothalamus results in erection and ejaculation.) The erectile tissue in the penis contains sinuses divided into many compartments. Blood flow through them is controlled by the diameter of the arterioles leading to the sinuses, and by valves in the veins that drain the sinuses. The penis becomes engorged with blood during erection, because the flow to the penis is increased by the enlarged diameter of the arterioles, while flow away from it is simultaneously restricted by valves in the veins. Erection is a comparatively rapid response, and it may lead quickly to orgasm. But erection may also be sustained for long periods, depending on the intention of the individual and the intensity of sexual stimulation. Other features of male sexual arousal are the contraction of smooth muscle in the scrotal sacs and the elevation and enlargement of the testes.

Orgasm accompanies the ejaculation of semen. It

Figure 14-5 Changes in the female genitalia during arousal (a) and orgasm (b).

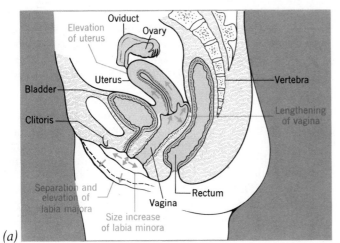

(a)

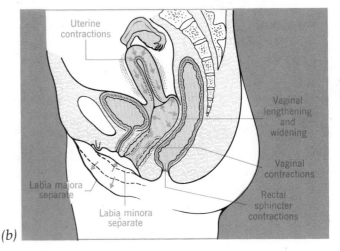

(b)

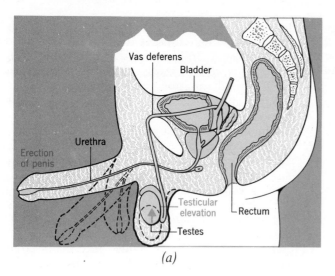

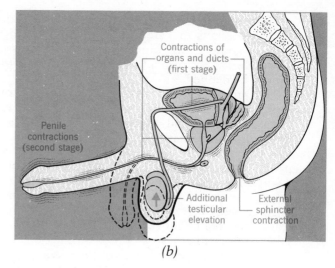

Figure 14-6   Changes in the male genitalia during arousal (a) and orgasm (b) .

occurs in two phases (see Figure 14-6). In the first phase, the contraction of the smooth muscle of the various ducts and glands delivers semen to the urethra. The second phase is the forceful contraction of skeletal muscles, which compresses the urethra and expels the semen. Immediately after orgasm, blood drains from the penis, and the male quickly returns to a non-aroused state.

This account has focused on changes in the genital area. But arousal and orgasm are responses of the whole body, in both men and women. The localized genital effects are accompanied by changes in heart rate, blood pressure, the distribution of the blood supply, and the contraction of major muscles. These general components of arousal and orgasm are somewhat more variable than the local genital events. Psychological aspects and subjective experience are, of course, also extremely variable. Even though the genital aspects are quite predictable, the overall nature of arousal and orgasm is influenced by various factors. These factors include the type of sexual stimulation (copulation, masturbation, oral stimulation, homosexual or heterosexual interaction), psychological influences (anticipation, emotional tension, guilt), and whole-body responses.

The frequency of orgasm produced by all forms of sexual activity can range from total abstinence to several orgasms per day. The average frequency depends on age and sex. In most Western cultures studied, the individual's frequency of orgasm reaches a maximum in the early twenties.

Copulation

Copulation is the insertion of the penis into the vagina to effect the transfer of sperm by ejaculation. Attempts to insert the erect penis into the vagina will be painful in the absence of sexual arousal in the woman. Arousal, however, causes the enlargement and lubrication of the vagina so that it can contain the erect penis.

A great deal of mythology surrounds the question of size of vagina and penis. The size of neither organ is correlated with general body stature. That is, a large man does not necessarily have a large penis; a small woman does not usually have a small vagina. The vagina is a muscular sac that conforms to penis size during copulation. It will accommodate either a large or a small penis and can essentially be distended indefinitely if sexual arousal is adequate. The flaccid (non-erect) penis shows considerable variation in size; however, the erect penis shows less variation. That is to say, a small flaccid penis tends to enlarge relatively more than a larger flaccid penis.

Ejaculation deposits a pool of semen in the vagina just below the cervix (Figure 14-7). The secretions in the vagina are normally quite acid. However, the mucus secretion that accompanies sexual arousal reduces the acidity, creating a more favorable environment for the survival of the sperm. The sperm migrate through the cervix and the uterus and into the oviducts. Fertilization occurs if an oocyte is present in the oviducts.

Several facts suggest that copulation is a very im-

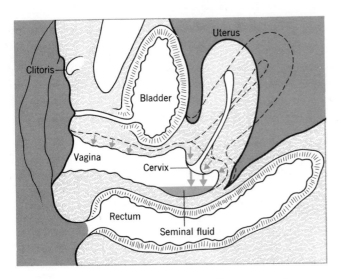

Figure 14-7 The female genitalia after orgasm. The cervix dips down into the pool of semen.

portant part of human behavior—that it is more than a simple reproductive act. First, orgasm in the woman is unnecessary for fertilization. Second, human beings' usual frequency of copulation is greater than that which is required for fertilization. And third, copulatory behavior continues when it is obviously non functional—for example, during pregnancy or after menopause, when ovulation has ceased. Copulation is more or less continuous throughout adult life. Women and men are always potentially active. Obviously, therefore, human copulation is not concerned with maximizing the probability of fertilization. This fact is somewhat surprising, because many other animals have mechanisms that coordinate ovulation and copulation to ensure fertilization. But the nature of human sexuality suggests that copulation serves a broad function in human society. Copulation in man and other primates is actually a form of communication, related to aggressiveness and affection.

sexual behavior

Sexual behavior is firmly embedded in the culture of each human group and is heavily influenced by education, tradition, and other social forces. These forces are in turn related to the religious, political, and economic features of each particular society.

The Gusii people provide a particularly good ex-ample of the influence of culture on sexual behavior, because their view is quite different from the usual Western attitude. The Gusii are an agricultural people living in the highlands of Kenya. Their culture is characterized by great antagonism between men and women, and all their sexual relations border on rape. In fact, it is difficult to distinguish rape from seduction or even from marital sex. Women greatly resent copulation. Men view copulation as a means of hurting and inflicting pain on the women. They even seem to be sexually excited by expressions of pain or rage. At marriage the bride arms herself with charms and potions to render the groom impotent. She resists copulation with all the force and guile at her command. For the groom, the wedding night is a battle. His reputation depends on the number of times he forces copulation; six is considered the minimum for respectability. If his bride is unable to walk the next day, the groom's reputation is greatly enhanced. Subsequent married life is just as tumultuous. Young married men are expected to demand copulation at least twice each night against their wives' continued resistance. But even if the man begins to tire there is no easy escape from this pattern, because wives complain if they are consistently ignored sexually. Copulation among the Gusii is an aggressive act that establishes the dominance of a husband over his wife.

Such an aggressive pattern is not accepted in most Western countries, but elements of it crop up in countries such as the United States. Sources as diverse as the sociological studies of Kinsey and the fiction of Mickey Spillane indicate that extremely aggressive male behavior is quite acceptable in some groups.

The United States has many different views of proper and acceptable sexual behavior. Various groups hold different views, depending on their level of economic success, their ethnic and religious backgrounds, and other influences. Many of these viewpoints are mutually exclusive. That is, if an individual adheres to one code of sexual behavior, he necessarily violates basic tenets of other codes. Most individuals in our society are presented with a variety of patterns of sexual behavior from which to choose. The entire range of opinion concerning sexual behavior is presented through major communication channels, such as public education, radio and television, films, and literature. Special interest groups push their own codes of sexual behavior. This variety of competing viewpoints generates a great deal of social and political tension, and the individual often ends up in a state of confusion and bewilderment.

The view that men and women should share equally in sexual experience has developed steadily in western countries since the advent of the woman suffrage movement. It is part of a more general position which says that men and women should be educated comparably, should participate equally in the full range of economic and political activity, and should share equally in the responsibilities of child-rearing. The double standard of sexual conduct, which condones premarital sexual activity in men and forbids it in women, is specifically rejected; men and women are encouraged to expect and seek equal satisfaction in sexual relations. The achievement of equal satisfaction usually depends on the ability of sexual partners to communicate with each other about the effectiveness of mutual stimulation, their levels of arousal, and other such factors. And in fact, mutual satisfaction rarely comes automatically. Sexual behavior, particularly with respect to achieving satisfactory orgasm, is very much a learned behavior pattern; it depends on personal experience as well as on cultural and social influences.

Sexual equality is not a dominant and uncontested view in any society at present, but it is most widespread in Sweden, where it is supported by appropriate laws. In the United States, the concept of equality in sexual behavior is receiving more and more attention. Adherents of other sexual codes are concerned about this development, however, largely because sexual equality has tended to encourage increased permissiveness for women, rather than increased restrictions on male sexual activity. Such people fear that increased permissiveness will weaken the social arrangements that favor the formation of stable family units. They favor a restrictive code of sexual behavior for both men and women, one that maintains a tight connection between sexual activity, reproduction, and child-rearing.

Human sexual behavior is so complex that it is almost impossible to disentangle its biological from its social determinants. That is to say, it is difficult to decide which aspects of human sexual behavior are controlled by our physiology, and which are controlled by our cultural environment. One problem is that accurate descriptive information is hard to obtain. Most of our information about sexual behavior comes from interviews and questionnaires. Direct neutral observation of human behavior and of physiological responses during copulation is a very recent development, and still a very limited one. Direct observation provides the basis for a description of arousal and orgasm, but it

is still not extensive enough to support much theorizing.

The confusion between biological and social factors in sexual behavior is very obvious in investigations of sexual activity in children. Some societies encourage regular heterosexual intercourse in children from the age of eight. They believe that frequent copulation before puberty is necessary for children to mature and to become fertile. By contrast, most western countries have strong cultural sanctions against such behavior. These sanctions are so strong that Sigmund Freud was led to believe that children from ages six to twelve go through a biological latency period, during which time they have no interest in sex. Obviously, questionnaire responses about sexual behavior in children from members of such diverse cultures would produce very different information.

Even simple questions about the biological basis of sexual activity are still in doubt, because reliable information is lacking. For example, does **castration** (the surgical removal of the testes) affect **libido** (sexual desire)? The consensus of opinion is that libido is reduced by castration, and that administration of testosterone, the hormone normally secreted by the testes, may counteract this effect. This view is still debated, however, simply because "libido" is difficult to define and measure. Definitions such as "true sexual desire" or "lust" are rather vague. If libido is equivalent to the frequency of sexual activity, there is a clear decline after castration. But if libido is defined by sexual interest and erotic fantasy, castration does not seem to affect it at all.

Libido in other animals can be measured by several criteria: the frequency with which the male mounts the female or inserts the penis into the vagina, or the willingness of a male to accept pain to reach a female (or vice versa). Such animal experiments have produced some interesting information about castration, and about the control of sexual behavior by hormones. Specific receptor sites for male and female sex hormones have been located in the brains of several mammals. These receptor sites are groups of nerve cells that accumulate hormone molecules. Apparently the accumulated hormones direct some parts of sexual behavior. When small pellets of hormone are implanted near the receptors in the brain of a castrated animal, the animal will mount a female. This effect is a specific behavioral one and has no effect on the animal's genital tract.

In female rats, the stimulation of receptor sites produces sexual receptivity to the male. Receptivity is normally cyclic because of the pattern of hormone secretion in the female rat. A cycle of receptivity is also found in female rhesus monkeys, and this cycle is correlated with the hormones secreted during the menstrual cycle. Receptors analogous to those in the rat are probably responsible. A similar cycle has been reported in women, but it is very inconspicuous and easily overriden by social and cultural factors.

Our inability to describe the strictly biological aspects of human behavior is partly due to a lack of direct information. We must rely heavily on studies of animals. Yet the fact that this difficulty exists at all reinforces our earlier conclusion: human sexual behavior is very plastic—it does not have rigid biological limits.

reproduction

Reproduction and development in humans follow a typical mammalian pattern. Gametes mature, the egg is fertilized, and the zygote is implanted in the uterus, where the embryo continues to develop. After birth, human development is usually divided into several phases: infancy, childhood, puberty and adolescence, maturity, and senescence.

The Menstrual Cycle

In women, the maturation of the egg is coordinated with the preparation of the uterus for receiving and nourishing a fertilized egg. Both of these events are controlled by cyclical changes in hormones. The overall sequence of events is called the **menstrual cycle.** Within the overall menstrual cycle, the process of egg maturation is called the **ovarian cycle,** while the changes in the uterus make up the **uterine cycle.**

The ovarian cycle produces and releases an oocyte every 26 to 28 days. The growth and maturation of the oocyte in the follicles of the ovary take about two weeks of that time (Figure 14-8). When the oocyte is mature, **ovulation** occurs: the oocyte is liberated into the oviduct. In the next two weeks, the cells lining the empty follicle form a mass known as the **corpus luteum** (literally, "yellow body"), which is involved in hormone secretion.

At the same time, the uterine cycle is preparing the lining of the uterus to receive and nourish the developing embryo in the event that the egg should be fertilized. The uterine cycle begins with an increase in the thickness of the endometrium lining the uterus. This thickening is correlated with the

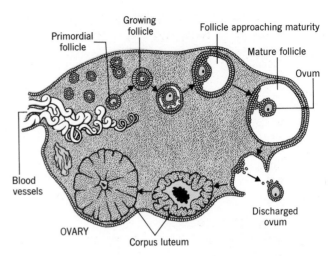

Figure 14-8 *The development of the egg and follicle in the ovary. The developmental stages are shown in clockwise sequence.*

maturation of the oocyte in the ovarian cycle. In the second half of the uterine cycle, glands in the endometrium produce a secretion of glycogen and mucus. If fertilization does not occur, the uterine lining is sloughed off and the cycle repeats (Figure 14-9). The periodic discharge of uterine tissue and blood through the vagina is the most obvious feature of the menstrual cycle. This process is **menstruation.** Menstruation is usually taken as the beginning of the 28-day cycle; ovulation, then, occurs near its midpoint, about two weeks after the beginning of menstruation.

Fertilization takes place in the oviduct shortly after ovulation. Several days are required for the fertilized egg to pass along the oviduct to the uterus, so it undergoes a few divisions by the time it reaches the uterus. At this point in the cycle the endometrium has become thick and vascular, so that it can nourish the developing embryo.

The menstrual cycle is regulated by a hormonal control system that works through the hypothalamus and pituitary in the brain. The anterior pituitary gland produces two gonadotropins, hormones that act on the gonads. They are, respectively, **follicle-stimulating hormone** (FSH) and **luteinizing hormone** (LH). FSH stimulates the growth and maturation of an oocyte in the ovarian follicle. LH causes the development of the corpus luteum in the empty follicle after ovulation. An increase in the secretion of both LH and FSH immediately precedes and, in fact, causes ovulation. **Prolactin** is a

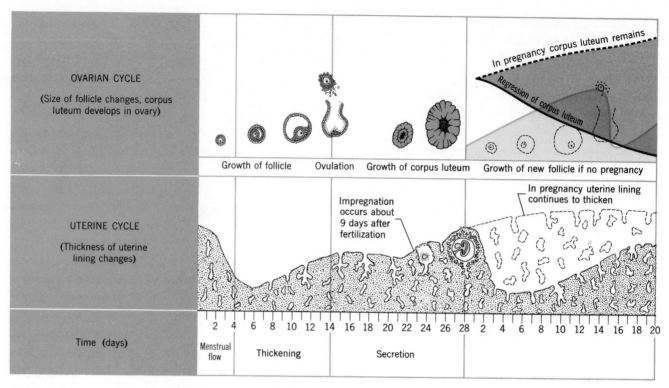

OVARIAN CYCLE				In pregnancy corpus luteum remains	
(Size of follicle changes, corpus luteum develops in ovary)				Regression of corpus luteum	
	Growth of follicle	Ovulation	Growth of corpus luteum	Growth of new follicle if no pregnancy	

Growth of follicle · Ovulation · Growth of corpus luteum · Growth of new follicle if no pregnancy

UTERINE CYCLE

(Thickness of uterine lining changes)

Impregnation occurs about 9 days after fertilization

In pregnancy uterine lining continues to thicken

Time (days)

2 4 6 8 10 12 14 16 18 20 22 24 26 28 2 4 6 8 10 12 14 16 18 20

Menstrual flow	Thickening	Secretion

Figure 14-9 *The human menstrual cycle consists of regular and sequential changes in the ovary (the ovarian cycle) and in the uterus (the uterine cycle).*

third hormone secreted by the anterior pituitary. Its major function is to stimulate milk secretion during late pregnancy and nursing.

The release of the hormones from the anterior pituitary is controlled by other hormones, produced by nerve cells in hypothalamic centers in the brain. These hormones are the **FSH, releasing factor,** the **LH releasing factor,** and the **prolactin inhibiting factor.** All three factors are transported along the nerve axons to the point where the hypothalamus joins the pituitary (Figure 14-10). There they are secreted into the bloodstream. They are carried to the anterior pituitary, where they stimulate FSH and LH production and inhibit the secretion of prolactin. The hypothalamus has two centers that control the production of the releasing factors. One center called the **tonic center,** is responsible for a continuing production of low levels of FSH and LH. The other center is responsible for the surge of FSH and LH that occurs before ovulation; it is called the **cyclic center.**

The ovary itself produces two types of sex hormones: **estrogen** and **progestogen.** These terms de-

scribe two classes of steroid compounds secreted by the ovary. Estrogen hormones are secreted by various cells in the ovary immediately after menstruation. Estrogen directs the repair and regrowth of the endometrium after menstruation. Cell division increases, glycogen is deposited, and the blood supply is stimulated. Progestogen becomes important after ovulation. It continues to work on the uterus, preparing the endometrium for the implantation and maintenance of the developing embryo. The blood supply to the endometrium increases further, and glandular mucus secretions increase markedly. Circulating levels of the pituitary and ovarian hormones are diagrammed in Figure 14-11.

Pituitary hormones, the releasing factors from the hypothalamus, and the ovarian hormones all interact to control the menstrual cycle (see Figure 14-10). At the beginning of the cycle, FSH from the pituitary stimulates an ovarian follicle to produce an egg. The ovary then begins to secrete estrogen, which in turn directs the growth of the endometrium. At mid-cycle, an increase in progestogen (of unknown origin) probably activates the cyclic

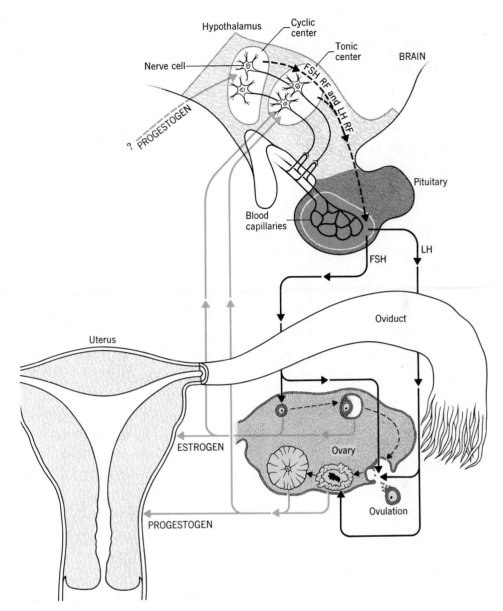

Figure 14-10 Hormonal control of human reproduction. Nerve cells in the hypothalamus produce
releasing factors (FSH RF amd LH RF), which control the production of the gonadotropins (FSH and
LH) by the pituitary. FSH and LH act on the ovary
promoting the growth of the egg and the follicle,
ovulation, and the development of the corpus luteum. The follicle and the corpus luteum produce
the sex hormones, estrogen and progestogen. These
hormones act on the uterine lining, preparing it to
receive a fertilized egg. They also form a feedback
loop by influencing the production of releasing
factors by the hypothalamus.

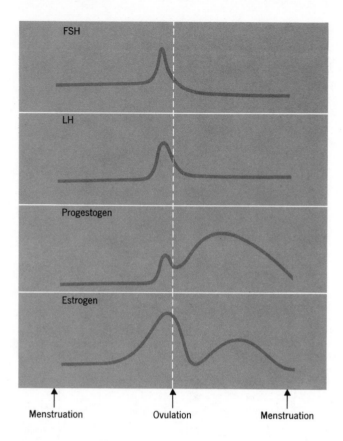

FSH

LH

Progestogen

Estrogen

Menstruation   Ovulation   Menstruation

Figure 14-11 Changes in circulating hormone levels during the menstrual cycle.

ample, the menstrual cycle of a girl living in a college dormitory tends to become synchronized with those of her close friends or roommates. It has been suggested that, in this case, odor is the critical sensory influence. Humans do produce secretions with musklike smells, and such an odor may synchronize menstrual cycles. A woman's emotional state also modifies her menstrual cycle. There are several probable pathways: under stress conditions, for example, progestogen is produced by the adrenal glands.

The menstrual cycle involves more than just the cyclic changes of the ovary and the uterus. The associated hormonal events are complex. Estrogen from the ovary does much more than control development of the endometrium: it affects the genital tract, the breasts, the brain, glands, and many physiological functions. Estrogen has a direct effect on a woman's water balance, for example. The high estrogen levels that occur before ovulation and before menstruation cause the body to retain sodium chloride. This increased salt content produces swelling, thirst, and measurable weight gain. The excess water is then lost at the onset of menstruation.

Fertilization and Implantation

When an oocyte is liberated from the ovary into the oviduct, it has already completed the first meiotic division. The second does not occur until a sperm penetrates the egg membrane. The unfertilized oocyte remains viable for 24 to 48 hours after ovulation. Sperm move from the rear of the vagina through the uterus, reaching the oviduct in a few hours. They retain their capacity for fertilization for two to three days. Contact between egg and sperm is random, however—there is no chemical attraction between them. Copulation and ovulation must be closely timed if fertilization is to occur. (In other animals, such as rabbits, copulation triggers the surge of LH that brings on ovulation, thus increasing the probability of fertilization.)

At fertilization, the sperm head enters the oocyte, and the second meiotic division occurs in the egg. The egg and sperm nuclei then fuse to restore the diploid chromosome number. The division of the zygote begins immediately in the oviduct.

The developing embryo takes three or four days to move down the oviduct to the uterus. By the time it reaches the uterus, it is a mass of 16 to 32 cells (Figure 14-12). Continued cell division produces a blastocyst, composed of a wall of cells surrounding a fluid-filled cavity and a mass of

center in the hypothalamus to produce releasing factors. The releasing factors then cause a surge of LH and FSH and produce ovulation. LH then initiates the development of the corpus luteum, and the corpus luteum itself begins to secrete progestogen. As the corpus luteum reaches the end of its development, the progestogen level declines. This drop in progestogen level causes the sloughing of the endometrium and the menstrual flow. The circulating levels of the ovarian hormones, estrogen and progestogen, provide a feedback system to inhibit or facilitate the secretion of releasing factors. In this way they control the production of the anterior pituitary hormones.

The menstrual cycles of various animals are influenced by sensory inputs. Vision, smell, touch, and hearing all can modify ovulation cycles and their associated sexual behavior. This correlation is not unexpected, since the nerve cells that produce the releasing factors are open to sensory influences, as well as to stimulation by estrogen and progestogen. The same may be true for humans. For ex-

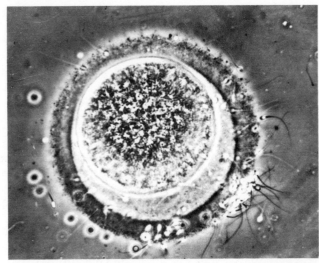

(a)

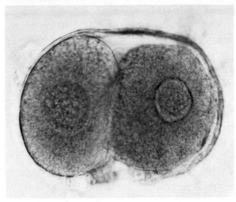

(b)

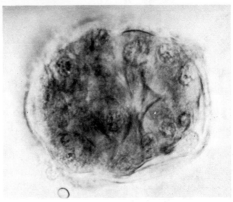

(c)

Figure 14-12 Fertilization and the early development of a human zygote. (a) A human egg with its surrounding layer of supporting cells, being penetrated by sperm. (b) The two-cell stage. (c) The 32-cell stage. Magnifications about 500×.

inner cells. The blastocyst becomes implanted in the endometrium about a week after fertilization. The external wall of the blastocyst grows a layer of cells that attaches to the uterine wall. Under this layer the lining of the uterus is eroded away, so the blastocyst sinks into the cavity. The external cell layer then covers the whole blastocyst (Figure 14-13). Later on, the embryo becomes attached to the uterine wall by means of the **placenta,** which forms from uterine tissue and the external cell layer of the blastula. Later stages of embryonic development follow the general mammalian pattern.

Birth Control

Birth-control practices rely on four basic methods. The first is to prevent contact between viable sperm and a viable oocyte. The second is to interfere with the production of viable eggs or sperm.

The third is to prevent the implantation of the dividing zygote. And finally, the developing embryo may be aborted, that is, removed or expelled from the uterus.

Methods for preventing the union of a viable egg and sperm vary in complexity and reliability. The most effective methods are surgical interruption of either the woman's oviduct (an operation called **tubal ligation**) or of the man's vas deferens (**vasectomy**). Viable eggs and sperm are still produced in the gonads, but they simply disintegrate. The person's hormonal balance and libido are unchanged. Tubal ligation used to be a major surgical procedure, because a large incision was made in the abdomen to reach the oviducts. However, newer and simpler procedures are now used. The surgeon makes a small incision in the abdomen or the wall of the vagina. He inserts a special instrument through the incision which allows him to locate the oviducts. He then cuts the oviducts and seals

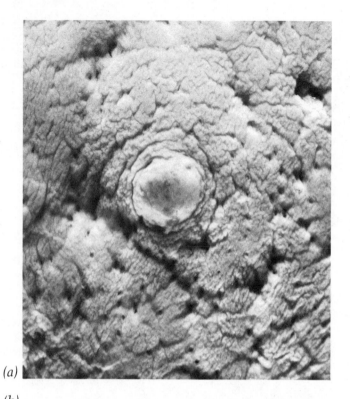

(a)

(b)

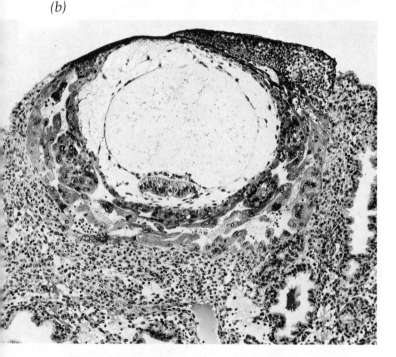

Figure 14-13 Implantation. (a) A blastocyst after twelve days of development implanted in the lining of the uterus. (b) A section through the implanted blastocyst; magnification 500×.

the cut ends by tying them or cauterizing them with electric current. The procedure may take only fifteen minutes. Vasectomy is even simpler. A small incision is made at the base of the scrotum, and the vasa deferentia are located, tied, and cut (Figure 14-14).

Vasectomy is sometimes reversible, but the surgery is tedious and difficult. Success rates measured in terms of the production of viable sperm range from twenty to 80 percent. But the success rate measured by the ability to cause pregnancy is much lower: only five to ten percent. The reason may be that the original vasectomy cuts the nerve supply that controls the contractions of the vas deferens. The contractions transport sperm from the epididymus during ejaculation. If they are reduced because of nerve damage, fewer sperm are ejaculated, and the chance of conception is reduced.

The union of viable sperm and egg can also be prevented by various mechanical and chemical barriers. A **condom,** or rubber sheath, can be placed over the penis; a **diaphragm** filled with spermicidal jelly can be inserted in the vagina over the cervix; and spermicidal foams or jellies can sometimes be used alone in the vagina. These methods are all reasonably reliable if used carefully. Failures often result from improper use. For example, a condom must be used throughout copulation and must be carefully withdrawn with the penis after ejaculation to avoid loss of sperm. A diaphragm should always be used together with spermicidal jelly. And it should be kept in place for several hours after copulation to insure that all the sperm in the vagina are dead before it is removed. (Sperm survive for several days in the oviduct, but not in the vagina. The vaginal secretions are very acidic, and therefore harmful to sperm.) When spermicidal foams are used by themselves, they must be freshly applied each time the penis is reinserted.

The **rhythm method** of contraception attempts to prevent the union of egg and sperm by restricting copulation to "safe" times, when no egg is present in the woman's genital tract. But this method is extremely unreliable. It can work only if the woman's body temperature is measured daily, and if the resulting record is carefully interpreted. Body temperature rises about one degree following ovulation (Figure 14-15). Copulation must be restricted to the period that begins after this temperature rise and ends with the onset of menstruation; copulation must be avoided altogether when the temperature record does not show a clear rise. Very few couples are patient and sophisticated enough to employ this method successfully.

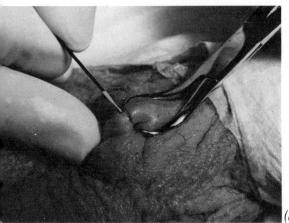

(a)

(b)

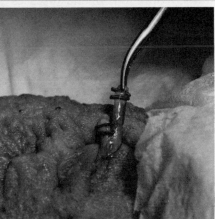

(c)

Figure 14-14 *Vasectomy.* (a) *A small incision is made in the scrotal sac.* (b) *The vas deferens is located and exposed.* (c) *Metal clips are attached to close the cut ends of the vas.*

inactivating them. But there are many difficulties with this technique. The reaction must be very specific. The antibodies must react with an antigen found only in sperm, or undesirable side effects are likely. Furthermore, this immunological technique must be easily reversible, or it offers no advantage over the easy and entirely reliable vasectomy. At present these problems are unsolved.

Another approach to sperm inactivation is to find a male pill, a compound that can be ingested by a man and that inactivates sperm. It should be specific in its action, free of side effects, and reversible. No such agent has been developed, but there are some promising candidates. The search for a male pill is a difficult task, however, because it cannot be easily tested on humans. There is no way to be sure the sperm have been inactivated except by copulating and attempting to cause pregnancy. If the sperm are only partially damaged but not incompletely inactivated, fertilization may occur anyway,

Coitus interruptus, or withdrawal of the penis from the vagina before ejaculation, is a very ancient contraceptive method. It is extremely unreliable, however, because sperm may pass along the urethra into the vagina before orgasm.

Techniques that prevent the development of viable gametes are the most successful ones for women. The best known and most effective method is the use of an oral contraceptive pill. "The pill" is a mixture of steroids that usually acts by preventing the surge of LH, and thereby preventing ovulation. Some women using the pill still undergo ovulatory cycles, but they are still protected against conception.

Many attempts are currently being made to inhibit the production of viable sperm, but none has met with much success as yet. One approach is to immunize the male against his own sperm. In other words, the male would produce antibodies that react with antigens on his own sperm, thus

Figure 14-15 *Changes in body temperature during the menstrual cycle.*

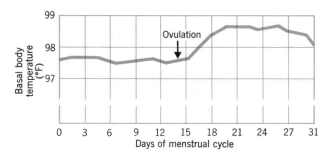

Oral Contraceptives

The pill is widely used as a contraceptive in the United States. Many brands are on the market, and all of them use synthetic steroids with activity similar to that of estrogen and progestogen. Synthetic compounds are more effective and have fewer side effects than the natural steroids. Synthetics are also metabolized more slowly, so they remain in the body longer. A balanced mixture of estrogen and progestogen is taken daily for 20 to 22 days, followed by a withdrawal period of six to eight days. Normal menstruation occurs at this time.

Taken as directed, the pill is an extremely reliable contraceptive. Side effects have been observed, such as nausea and headache, depression, weight change, discharge of small amounts of blood, and similar unpleasant but minor discomforts. But these effects are usually transient and disappear a few months after pill consumption is begun. The only medically significant side effect is an increase in the risk of death due to thromboembolic diseases (diseases that impair circulation by causing the formation of clots in blood vessels). However, this effect is still insignificant compared to the risk of death from legal abortion, pregnancy, or childbirth (see Table 14-1). Thus, it is more sensible to risk the side effects of the pill than to risk unwanted pregnancy.

and a genetically deformed embryo could result. By contrast, the female pill is easily tested by taking temperature readings to verify the suppression of ovulation. It is not necessary to risk the conception of an abnormal embryo.

Another major contraceptive method works by preventing the implantation of the embryo. This method is the **intrauterine device** (IUD). The term "device" covers a wide variety of foreign objects that have the same effect when introduced into the uterus (Figure 14-16). Loops, rings, spirals, or threads of foreign material in the uterus all will interfere with implantation. An IUD probably works

by stimulating white blood cells to invade the lining of the uterus and to enter the uterine cavity. This invasion is part of the normal uterine cycle, though its function is not understood: invasion occurs immediately after ovulation and then disappears. But if an IUD is present, the white blood cells remain and create an environment unfavorable to the survival of the embryo. The white blood cells are thought to liberate enzymes that attack the blastula. The effectiveness of an IUD depends on its total surface area and on the material used to make it. A large surface area enhances the effectiveness, and metals are more effective than plastics.

Table 14-1

Risk of Death[1]

Age	Death from Thromboembolism		Death from Complications of:			
	Non-users of Oral Contraceptives	Users of Oral Contraceptives	Abortion	Pregnancy	Delivery	All Risks of Pregnancy[2]
20–34	2	15	56	75	71	228
35–44	5	39	104	138	265	576

[1] Yearly deaths per million users or non-users of oral contraceptives, or per million abortions, pregnancies, or deliveries.
[2] Pregnancy, delivery, and the period immediately following delivery.

The IUD is the only contraceptive technique that acts indefinitely and that requires no attention. It is also reversible. Unfortunately, there are side effects, such as intermenstrual bleeding and cramps; these effects are particularly serious in women who have not borne children. A new IUD design looks very promising, however. It is a small T-shaped device that is wrapped with copper and zinc wire. The metal wire increases its contraceptive effectiveness, but the side effects are minimal (see Figure 14-16). The metal probably stimulates the white-cell invasion.

Abortion is the expulsion or removal of the embryo (and placenta) before it becomes independently viable. Abortion can be accomplished mechanically or chemically in a great variety of ways. Almost all the techniques are extremely dangerous in unskilled hands. Mechanical procedures can easily puncture the uterus, and most of the effective chemical agents are extremely toxic. For this reason, illegal abortion is usually very risky. However, when abortion is undertaken by a competent doctor with supporting medical personnel and facilities, the risk is very low; it is considerably lower than the risk of childbirth (see Table 14-1). The standard abortion procedure is called **dilation and curettage,** or simply **D-and-C.** The cervix is dilated to permit the introduction of metal instruments, and the lining of the uterus is scraped clean. Alternatively, a suction tube may be used to remove the embryo. Either procedure can be used until the twelfth week of pregnancy. At later stages of pregnancy, other procedures must be used that involve hospitalization and greater discomfort. If a pregnancy is to be terminated by abortion, the decision should be made promptly.

The most promising candidates for inducing abortion with chemicals are the **prostaglandins.** They were first observed in semen, and they originate in the prostate gland, hence their name. The prostaglandins are actually a family of chemically-related compounds found in many tissues. They have a wide spectrum of physiological effects, but basically they act by modifying metabolic control systems. Two kinds of prostaglandins, when given in therapeutic doses, can induce abortion in early stages of pregnancy by stimulating uterine contractions. However, they must be administered intravenously for several hours. A D-and-C is faster and often involves less discomfort. Recently, prostaglandins administered vaginally have successfully terminated pregnancy. It should be possible, then, to develop a self-administered preparation that can be used at the end of each menstrual cycle, that is,

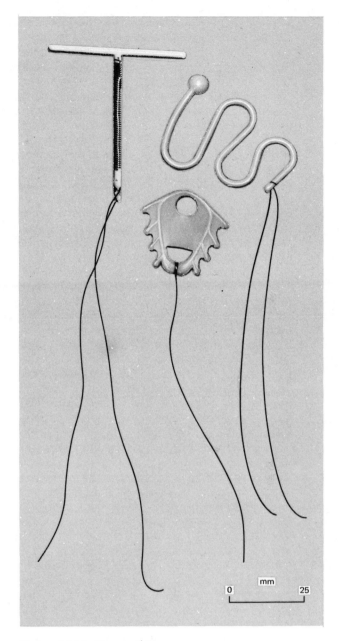

Figure 14-16 Types of intrauterine devices. Left to right: the copper T, the Dalkon shield, and the Lippes loop.

28 days after the onset of the last menstruation. This method would ensure that the uterus is in a nonpregnant state once a month. Such prostaglandin "postcontraceptives" will probably become available in the near future.

The effectiveness of each contraceptive method has been measured in statistical studies of large populations of users and in detailed studies of small, selected groups. The results are usually expressed as the number of pregnancies for each 100 woman-years of use—that is, 100 women using it for one year apiece. Table 14-2 gives representative figures from recent studies. The measured effectiveness of a contraceptive method in a test population will always be less than its potential effectiveness, because some individuals fail to use the method properly or conscientiously.

Effectiveness is obviously an important consideration in choosing a contraceptive technique, but other factors are also important. For example, a couple who are sexually active but whose planned family is complete might find vasectomy or tubal ligation most satisfactory. On the other hand, the simplicity and effectiveness of the pill recommend it to sexually active couples who still want children. Individuals who are less active might reasonably accept a less effective contraceptive with no known side effects, such as a condom or diaphragm. These methods, of course, would expose the woman to a small but non-negligible chance of pregnancy. But she could plan to use postcontraceptive methods such as abortion if pregnancy occurred. In this case, the woman is accepting the potential risk from abortion. However, there is only a small chance that she will need one, so her overall health risk in using a diaphragm (or condom) with abortion as a back up is actually lower than the risk of side effects from the pill. Personal circumstances should be considered along with the effectiveness and availability of particular methods when one chooses a specific contraceptive strategy.

Anyone who is aware of the facts about contraception and who chooses one of the first seven methods listed in Table 14-2 should not be surprised by a pregnancy. The risk may be low in some cases, but it still is certainly real.

Fetal Development and Pregnancy

The developing embryo first acquires a generalized mammalian appearance and then gradually takes on specific human characteristics (see Figure 14-17).

Table 14-2

Risk of Pregnancy

Contraceptive Method	Pregnancies per 100 Woman-Years
None	50–115
Douche	31
Rhythm (by number of days after menstruation)	24
Coitus interruptus	18
Condom	14
Diaphragm	12
IUD	1.5–3.4
Rhythm (by body temperature)	1.0
Vasectomy	0.15
Pill	0.07
Tubal ligation	0.04

The normal **gestation period**—that is, the duration of pregnancy—is about 40 weeks, counting from the last menstruation. Variations of two to three weeks in either direction are common. The embryo may survive delivery as early as the 28th week with intensive hospital care, but if it is removed from the uterus much earlier its lungs are insufficiently developed for survival. The average weight of a newborn infant is about eight pounds, plus two pounds for the placenta and amniotic fluid.

Women are affected by the embryo in two basic ways during pregnancy. First, the embryo puts a significant physical and physiological load on the mother. It develops, in effect, as a parasite, depending on the pregnant woman for nutrition, gas exchanges, and elimination of wastes. These metabolic demands become very significant as the embryo enlarges.

The second major way the developing embryo affects the mother is through hormone production. The embryo and the placenta are major new sources of hormones. These hormones abridge the normal menstrual cycle and produce major changes in the pregnant woman. The two kinds of changes, physiological and hormonal, overlap considerably. Some of the physiological effects that help the mother meet the demands of the developing embryo are in fact caused by changes in hormones. Rising estrogen levels cause the maternal blood volume to increase by about 40 percent during pregnancy, for instance. This change increases blood circulation to the placenta. The pregnant woman

Abortion and Public Health

Open discussion of contraception and abortion, and provision of up-to-date information about them, are much less common than they should be if we place a high priority on personal and public health. Part of the reason is that such discussion is discouraged by people who fear sexual permissiveness. In addition, it is sometimes opposed by groups that consider contraception or the termination of pregnancy offensive and immoral. Clear discussion is also hampered by the vagueness that is intrinsic in distinctions between contraception, abortion, and infanticide. Should the moment of conception be defined as the instant when fertilization occurs, or as the time the dividing zygote is implanted in the uterus? Is the IUD a contraceptive or an abortive technique? At what point does the embryo become an infant? Biological information is not really very helpful here, because development and reproduction are really continuous and cyclical processes. The whole organism follows one cycle of development:

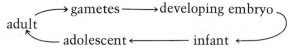

The reproductive tissue follows another cycle:

Obviously, both these cycles represent continuous processes. The stages and arrows are entirely arbitrary and could be rewritten in any number of ways. It is also obvious that either cycle can be interrupted at any point. The distinction between acceptable and unacceptable interruptions of the life cycle is made on social—not biological—grounds.

Abortion is very much a problem of health, and the public-health impact of legal abortion is well documented. The New York State abortion law is a good example. It passed the state legislature by a margin of one vote and went into effect on July 1, 1970. It allowed abortion at the request of a pregnant woman for the first 24 weeks of pregnancy. Within the first year, the number of legal abortions increased significantly. The total maternal death rate and infant death rates in New York dropped to new low levels. The overall death rate from legal abortion during the first year was 4.7 deaths per 100,000 women. This rate is less than half the death rate associated with live births. The evidence for the overall improvement of public health is clear. However, the opposing social sentiment was so strong that both houses of the New York State Legislature voted to repeal the law in May 1972. (The repeal was vetoed by the Governor.) Obviously, public health considerations do not necessarily play a dominant role in political or legal decisions about these sensitive and emotionally-charged matters.

Two important features of the abortion problem are often ignored. First, even when prevailing sentiment makes it illegal, abortion is not eliminated from the society—instead, it merely is no longer equally available to all groups. Safe, medically-supervised abortion is always available to women who have sufficient leisure and income to travel to a country where it is socially sanctioned. Thus, legislation or social pressure that prevents free access to medically safe abortion discriminates against the young and the poor. (A study of the New York situation showed that the law had in fact

Estimated Number of
Induced Abortions, 1967

Country	Abortions per 100 Live Births
Argentina	100
Austria	150–200
Belgium	150–300
Chile	40–80
Finland	40
France	50–150
Italy	70–100
United Kingdom	10–20
United States	10–60
Uruguay	300
West Germany	100–300

made abortion services available to low-income and minority groups.) Even stringent punishment of both the abortionist and the woman undergoing the operation has been ineffective in eliminating abortion. In countries where legal abortion is difficult or impossible to obtain, the frequency of illegal abortion is very high. In fact, it may equal or exceed the number of births (see table). Obviously, the motivation to terminate pregnancy is strong enough so that large numbers of people choose to risk death and social punishment. A second important point is that groups to whom abortion is offensive can, of course, choose not to avail themselves of such services.

Unfortunately, these facts don't have much impact on the political and social process, probably because most people believe that their own moral and religious beliefs should apply to everyone. The increased infant and maternal mortality that results from restricted access to abortion is often viewed as just punishment of the immoral. Thus, political and medical authorities must attempt to protect public health with minimum offense to the groups who find contraception offensive in general, and abortion particularly unacceptable.

also reduces her rate of glucose utilization, loses muscle tone, and increases fat deposition. Progestogen produces all these effects by inhibiting the production of high energy phosphate bonds (ATP) in the mother. This cutback favors the development of the embryo by reducing maternal metabolic demands.

Progestogen and estrogen increase dramatically during pregnancy. Progestogen secretion is stimulated by another hormone, called **chorionic gonadotropin,** from the chorionic membrane around the embryo. Chorionic gonadotropin works like LH in that it maintains the corpus luteum and stimulates it to produce progestogen. It is also similar to LH in another way: it will induce ovulation. This characteristic is the basis of pregnancy tests. If a woman is pregnant, chorionic gonadotropin appears in her urine. When a small amount of urine is injected into a female rat or rabbit, the animal will ovulate. The ovaries of the test animal may show detectable changes in as little as two hours. Another pregnancy test depends on the fact that a male frog or toad will expel sperm two hours after being injected with the urine of a pregnant woman. Chorionic gonadotropin appears in the urine just before the first missed menstrual period and reaches a peak during the next month. The placenta itself soon begins to make progestogen, however, and gradually takes over from the corpus luteum.

The estrogen increase during pregnancy is caused by both the placenta and the embryo. Levels of estrogen in late pregnancy provide a practical index of the embryo's health, because the estrogen level increases only when the embryonic circulation is normal and when the adrenal glands of the embryo show normal activity. The normal levels of

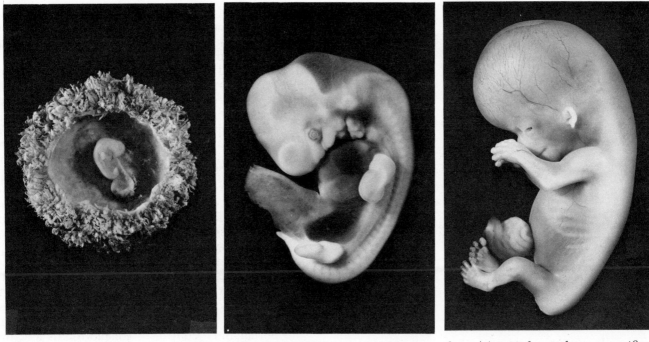

Figure 14-17 The development of a human embryo. (a) A 28-day embryo; magnification 2×. (b) A 34-day embryo; magnification 5×. (c) A 56-day embryo; magnification 2×.

chorionic gonadotropin, estrogen, and progestogen during pregnancy are shown in Figure 14-18.

Estrogen and progestogen perform many functions during pregnancy. One major function is to maintain the rich circulation and secretory activity of the uterine lining so that it nourishes and supports the embryo. Progestogen and estrogen also promote breast development during pregnancy. They stimulate the growth of milk-secreting tissue and of the duct system that conveys milk to the nipples. Breast development may depend on other hormones as well. Pregnant rats achieve full breast development only if they stimulate their own breasts by licking them. This discovery suggests that the licking reflex influences the hypothalamus to reduce its production of prolactin inhibiting factor which acts on the pituitary. Breast manipulation in late pregnancy is advised for women in order to encourage breast development.

Another function of estrogen is to cause further development of the muscular layer of the uterus to prepare it for labor. Small local contractions of the uterus occur throughout pregnancy. They increase slowly in frequency and intensity after the seventh month. But large contractions of the developing muscle layer are at this time inhibited by progestogen. During the last month of pregnancy, the head of the embryo moves into the mother's pelvis.

The onset of muscle contractions in labor is a response to declining progestogen levels and to the secretion of **oxytocin,** a hormone from the posterior pituitary. The frequency and strength of the contractions increase and they may be extremely painful. Uterine contractions start near the oviducts and progress downward. The cervix gradually dilates until it is large enough to permit passage of the baby. Contractions are usually supplemented by the mother, who "bears down" by inhaling, holding her breath, and contracting her abdominal muscles. In some animals, birth is rapid and not very demanding for the female. But in humans the close fit between the diameter of the baby's head and the opening of the pelvis makes birth a longer process. Women vary tremendously with respect to the duration and pain of labor; it can be prolonged, painful, and exhausting. Psychological preparation can help reduce the trauma.

The mother should be examined periodically by a doctor during pregnancy so that any possible delivery complications can be anticipated. Competent medical assistance at birth facilitates delivery and greatly reduces maternal and infant mortality rates.

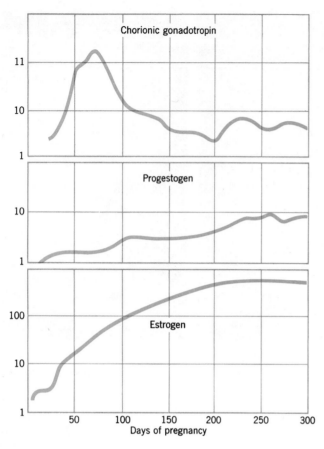

Figure 14-18 *Production of hormones during pregnancy.*

istics and the gradual differentiation of the embryo into an adult male or female. The chromosomal basis of sex determination depends on a single difference in the set of chromosomes.

Diploid cells in humans have a total of 46 chromosomes. Of these, 45 are similar in men and women; one is different. Each diploid cell in women has two X chromosomes, while each diploid cell in men has one X and one Y. This single difference initiates a train of events during development which eventually produces an adult man or woman. At early stages, the embryo is bipotential—that is, it is capable of developing in either a feminine or a masculine direction. But at the end of normal development the sexuality of a man or woman shows up in many aspects of his or her physiology, structure, and behavior.

Sexuality is determined by the nature of the gonad (testis or ovary), duct system (oviducts and uterus, or sperm ducts), and the associated glands; by sex hormone production (estrogen and progestogen or androgens); by differentiation of the hypothalamus (control of gonadotropins and interaction with gonadal hormones); by development and maintenance of secondary sex characteristics (breast development, body hair); and by sexual behavior and sex roles in society. This is an impressively long list. Some of these features of sexual differentiation are quite reliable, and variations in them are rare; the reproductive organs are usually reliably male or female. But other features of sexual differentiation are quite variable and are not necessarily controlled by the chromosomes. They include the distribution and amount of body hair, the size of shoulders and hips, muscle development, physical and intellectual aggressiveness, and so forth. These features are neither all "male" nor all "female"; they show considerable variation in individual development along a continuous scale, caused by differences in nutrition, hormone level, cultural forces, and a host of environmental influences. Sexuality in an adult is the sum of all these features. Each constituent has a so-called normal range. That is, the sexual differentiation of a particular feature is considered normal if it is structurally and physiologically possible for the individual to play the male or female role in reproduction successfully. But this definition is satisfactory only if you add to it a very large component of social and cultural influence. If society considers an individual to be outside the normal range of variation, it will apply social pressures that make it difficult for the person to reproduce, even if he or she is physically capable of doing so.

Most features of human sexual differentiation are

This statement seems obvious, but prenatal medical care is not standard procedure in major social and economic groups in the United States. Most children are born under medical supervision, but medical care and examination throughout pregnancy is standard only among the more affluent. Lack of prenatal medical care no doubt contributes to the rather high infant mortality rates that are characteristic of low-income areas. In fact, many countries have better infant and maternal survival rates and better health than the United States (see Figure 14-19.

sex determination

One of the most interesting aspects of human development is the acquisition of sex character-

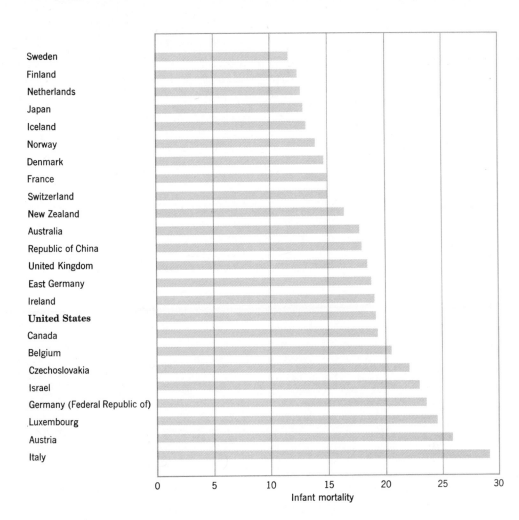

Figure 14-19 *Infant mortality in various developed countries in 1972. Infant mortality is defined as the number of children dying before one year of age, per 1,000 live births.*

poorly understood. Information about development in the embryo is based on experimental analysis of other species and on observations of normal and abnormal human development. Our information about puberty is so incomplete that we do not even understand what triggers it. And the development of sex roles after puberty presents the same problems as other forms of sexual behavior, because it is impossible to disentangle cultural influences from physiological differences between men and women. Yet the personal and social importance of sexual differentiation makes it extremely difficult to resist seizing upon explanations that are still tentative and poorly documented.

Embryonic Differentiation

The gonad of the early embryo begins as a ridge of tissue adjacent to the developing kidney. The cells that eventually become eggs or sperm originate elsewhere and migrate into the developing gonad. Although the sex of the embryo has been determined by the chromosome complement of its cells, ovaries and testes are indistinguishable at this stage of development (see Figure 14-20). The genital ducts begin to differentiate from the embryo's Wolffian or Mullerian duct systems. The male ducts that convey sperm from the testes to the exterior are derived from the Wolffian duct, which drains the middle portion of the kidney. The oviducts and uterus are derived from a separate duct, the Mullerian duct, which is independent of kidney development. In the male embryo, the Wolffian duct system develops and the Mullerian duct degenerates; in the female embryo, the Mullerian duct is retained while the Wolffian system degenerates.

Later in development, both ovaries and testes descend, or move down in the body. The ovaries

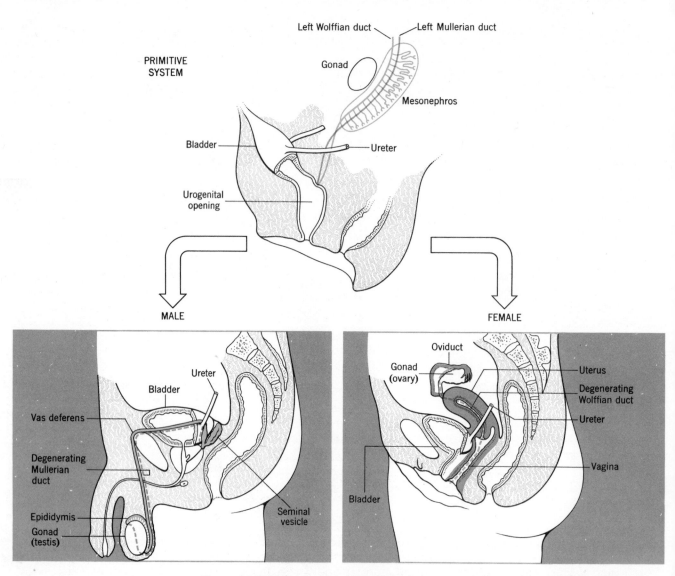

Figure 14-20 *The development of the human reproductive tract. The ducts that convey gametes to the exterior develop from the Wolffian duct in the male and from the Mullerian duct in the female.*

remain in the pelvic cavity; the testes descend below the pubic bones into the scrotal sacs by moving down through openings at the base of the pelvic cavity. The openings are called inguinal canals. After the testes have descended, the muscles around each inguinal canal close around the vas deferens; this point is always a weak spot in the muscular wall of the abdomen, and it can rupture to produce an **inguinal hernia.**

The development of the gonad into ovary or testis depends on the chromosome complement of the individual; hormonal influences do not seem to

be involved. However, the subsequent development of ducts and of the external genitalia in males does depend on hormones from the gonad. This conclusion is based on two kinds of evidence. The first comes from animal experimentation. If the embryonic testis is removed before it begins to secrete hormones, the animal develops the female duct system (oviducts and uterus) and female external genitals. Apparently the masculine pattern of development is caused by androgens from the testis. Two androgens are involved: one, **testosterone,** promotes development of the male ducts and external genitals; a second, still unidentified, promotes the degeneration of the Mullerian duct system.

The hormonal basis of male development is also supported by observations of human embryos. The embryonic testis shows clear signs of secretory activity at the same time that the male duct system and external genitals begin development. Ovaries develop much more slowly than testes, and the ovary shows no secretory activity. Embryonic hormones do not seem to be required for the development of the Mullerian ducts into the female genital ducts, or for the development of the female external genitals. A developing embryo will acquire female sex organs unless male hormones appear to redirect the process—even if the embryo has a male chromosome complement. Conversely, a female embryo will acquire masculinized external genitals if testosterone is present during its development. (This sometimes happens when testosterone is produced by a tumor of the mother's adrenal glands.)

Another aspect of early sexual differentiation occurs in rats and other mammals, and possibly in humans. Apparently the cyclic center in the hypothalamus, which produces the periodic surges of LH and FSH in the female menstrual cycle, is inactivated by testosterone. This relation can be shown by removing the testes from a newborn male rat, to prevent testosterone secretion. If we allow the animal to mature and then implant an ovary in it, the ovary will behave normally and ovulation will occur periodically. On the other hand, if a newborn female rat receives a single injection of testosterone, she becomes permanently sterile. Her ovary develops normally but she is unable to ovulate. Since the male castrate with an implanted ovary is capable of ovulating, apparently he has developed a hypothalamic center that can interact with the ovary. The hypothalamus of the normal male rat lacks the capability, presumably because testosterone is present at the critical developmental period.

The evidence for this kind of development pattern in humans is very tentative. However, it is known that human males do lack the cyclic center for gonadotropin release. In this case, too, the female pattern develops without hormonal support unless testosterone is present to interfere with it.

Childhood and Puberty

Hormone secretion by the testis virtually ceases at birth; the ovary of a newborn is also quiescent. During childhood the hypothalamus–pituitary–gonad hormonal system operates at a very low level; animal experimentation suggests that the brain keeps the hypothalamus inactive. For example, when lesions are made in the nerve tracts that lead to the hypothalamus in experimental animals, the animals show precocious sexual development. In addition, human children with abnormalities in the hypothalamus, in the nerve fibers leading to the hypothalamus, or in the pituitary, also show sexual development long before they reach adolescence.

A definite change in hormone balance develops at puberty. **Puberty** is a period characterized by a surge of growth, and by visible changes in sexual differentiation. The most obvious changes are the growth of pubic hair, breast development, and increase in penis size. The internal sexual organs also grow, and levels of gonadotropins and gonadal hormones increase dramatically. At this point the brain releases the inhibition and the hypothalamus begins to function. The hypothalamus then activates the whole hormonal system.

In boys, the activation of the hypothalamus at puberty starts the pituitary producing LH and FSH. LH stimulates the interstitial cells of the testis to produce testosterone. FSH stimulates spermatogenesis in the testis. Both LH and FSH are controlled by releasing factors from the hypothalamus, just as in women. Androgens from the gonad, mainly testosterone, stimulate the growth and development of the penis, testes, prostate, and seminal vesicles (Figure 14-21). Testosterone produces abundant hair on the pubis, armpits, chest and face; the voice becomes lower because of enlargement of the larynx and thickening of the vocal cords. These changes are termed **secondary sex characteristics** because they are not involved in the primary sexual function, the production and transfer of gametes. Other secondary sex characteristics such as muscular strength and patterns of fat deposition are related to testosterone secretion.

In women, the activation of hormonal systems is associated with the onset of the menstrual cycle at puberty. The time of the onset of puberty has changed considerably in the last century, particularly in women. In the mid-nineteenth century, menstruation began at ages sixteen to seventeen; now the average age is thirteen to fourteen in western industrial societies. Menstruation still begins later in girls of less developed countries, possibly because of poor nutrition.

During puberty, increased estrogen secretion causes the development of the gonads, uterus, vagina, and associated glands (Figure 14-22). Female secondary sex characteristics appear; breasts de-

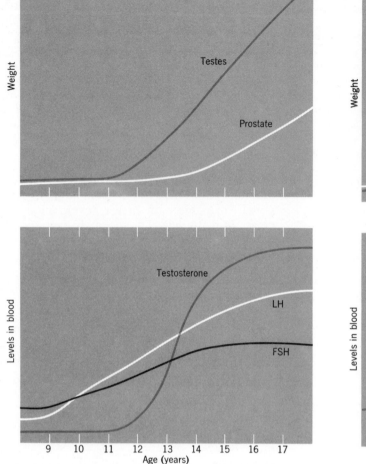

Figure 14-21 *Changes associated with puberty in boys.*

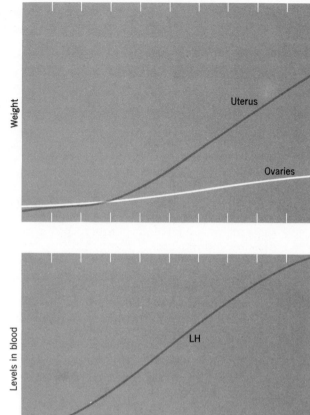

Figure 14-22 *Changes associated with puberty in girls.*

velop, and the skin becomes thicker as a result of an increase in its water content and subcutaneous fat. Estrogen also causes bone growth to cease, ending the period of increase in height. The cyclic production of progestogen also commences at the onset of menstruation but it is much less important than estrogen in establishing secondary sex characteristics.

Both men and women secrete both testosterone and estogen. Testosterone in women comes from the ovary and from other tissues that convert steroids from the adrenal glands to testosterone. In men, small amounts of testosterone are converted to estrogen. Therefore, the hormone-related sex differences in adults depend on the circulating levels of hormones rather than on any absolute differences in hormone makeup. Testosterone levels

are twenty times higher in men than in women, and estrogen is much more concentrated in women than in men; but both types of hormones are present in both sexes.

Adult Sex Roles

There is a great deal of individual variation in the sexual characteristics of adults. The circulating sex hormones and the physical changes that establish the masculine or feminine adult form can vary considerably. Human form is described by a variety of terms; hypomasculine, masculine, and hypermasculine; or hypofeminine, feminine and hyperfeminine. These classifications are based on observed muscular development, genital size, hair distribution, and body shape.

There is also variation in sexual characteristics between races and cultures. Some body differences are clearly genetic. Oriental males, for example, have sparse facial hair regardless of their levels of testosterone. Other differences are clearly environmental. Balinese men are typically slim and lightly muscled, but Balinese men who work as dock laborers develop heavy musculature. Different body configurations are associated with femininity and masculinity in each culture.

Some differences among adult humans result from failures in sexual differentiation. In women, abnormally high levels of androgens may cause the clitoris to enlarge into a penis-like organ. Alternatively, the penis of a genetic male remains very small if his testes fail to descend and to secrete testosterone. Individuals of either sex whose sexual differentiation is ambiguous are termed **androgynous.**

The great majority of humans assume the behavioral and reproductive role that matches their chromosome complement and their sexual differentiation. That is to say, they look like, behave like, and reproduce like men or women. However, what constitutes proper behavior for a man or a woman is determined largely by cultural influences. The process of induction into a culturally defined sexual role begins very early. (Actually, it often begins at the moment of birth—many hospitals wrap boys in blue blankets and girls in pink ones.) Many details of male or female roles depend on the specific culture. Thus, sexual behavior and sexual roles are certainly learned behavior in part. For example, in some cultures, traditional Western sex roles are reversed: women are aggressive, perform the major tasks, and are economically dominant; men are cast in submissive and nurturant roles.

The remarkable influence of culture on sex roles is dramatically illustrated by studies of androgynous individuals. Apparently whether an androgynous child adopts a male or a female sexual role is determined almost entirely by the parental de-cision concerning that role. The pattern of the child's rearing is more important than its chromosomal composition or the details of its external genitalia. An individual who is raised as a woman successfully adopts the appropriate female cultural and sexual role. This development is possible despite rather dramatic morphological discrepancies; for example, such a woman may have a penis and scrotal sacs. But the same androgynous individual rasied as a male would successfully adopt the male role.

Physical differences between the sexes also determine people's sexual behavior and sex roles. Since men tend to be larger and potentially stronger than women, they are usually assigned the more physically demanding tasks in a culture. Part of the female role usually reflects the physiological requirements of pregnancy and nursing. However, the less dependent a society is on heavy subsistence labor, the less stringent these limitations become.

Hormonal differences may control yet other aspects of sexual behavior. Testosterone is strongly associated with aggression in most of the vertebrates that have been studied. Fighting behavior in male animals begins at the time of puberty and testosterone secretion. In animals that breed seasonally, fighting is correlated with annual cycles of testosterone production. Castration reduces aggressive behavior, and the administration of testosterone restores it. Aggression-related behavior, such as marking a territory with scent from special glands, or defending a territory against other males, is also influenced by testosterone levels. These effects have been described for many different kinds of animals, including reptiles, birds, and several mammal groups as well as the primates. Thus it is a reasonable hypothesis that the aggressive role of men in most cultures is an expression of high levels of testosterone. On the other hand, an important characteristic of primates is the tremendous range of behavior found within and among different primate species. Human behavior is particularly plastic. And in any case, aggression in modern society is so indirect and complicated that the hypothesis remains largely untested.

The Range of Human Somatotype

Somatotype is the term for the size, shape, and proportion of the body. The existing variation in human somatotype is impressive. Men of the pygmy Bush tribes of South Africa average less than five feet in height and weigh less than 90 pounds. Other ethnic groups average more than six feet in height and 175 pounds in weight. The average differences between men and women also depend on ethnic origin. Among the Klamath Indians on the northwest coast of the United States, women are two inches shorter than men on the average; the difference is eight inches in an African tribe from the eastern Sudan. Women in a population of south Russian Jews, though shorter, outweigh the men. Body proportion in various populations also covers a tremendous range.

Classification of somatotype is difficult, but William Sheldon has developed a system that is reasonably dependable. It is not influenced by the individual's state of health or by weight changes. Also, it is objective enough to give the same result when applied by different people. An individual is classified by photographing him in standard poses and then taking measurements

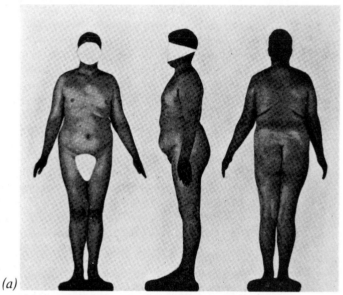

(a)

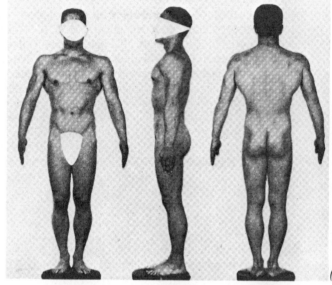

(b)

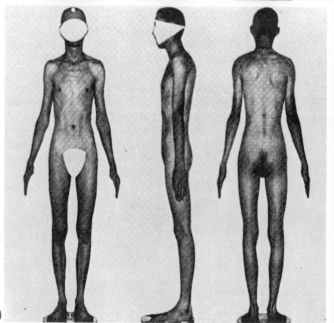

(c)

Body forms. Extreme somatotypes in men. (a) Endomorph. (b) Mesomorph. (c) Ectomorph.

from the photographs. Each measurement is expressed as a fraction of height and then subjected to mathematical analysis. The final classification is given as three numbers that rate the individual on a scale of 1–7 for each of three basic conditions: endomorphy, mesomorphy, and ectomorphy.

Endomorphy is a somatotype with a porpoise-like appearance. A **mesomorph** looks like Superman. The **ectomorph** looks emaciated. Most men and women occupy an intermediate position along all three scales; women are usually more endomorphic than men. Sheldon's somatotype rating is particularly useful because it can be correlated with other characteristics such as weight gain with age, longevity, and susceptibility to different diseases.

Any classification of somatotype shows that there is a wide range of adult morphology, and that morphology is, to a considerable extent, genetically controlled. An environmental regime of diet and exercise will never convert a slim distance runner into a football lineman, or vice versa. Similarly, few women can ever achieve the appearance of a high-fashion model or the mammalian opulence and callipygous dimensions of a pinup girl. Appearance is only partially under the control of the individual.

This statement may seem too obvious to merit discussion, but it is not obvious to society as a whole. Our culture is full of fads and fashions dictating that one particular somatotype is the proper one. Worse yet, it is implied or even stated that failure to conform to this idealized body type somehow constitutes a moral lapse. The modification of the individual's appearance—through extreme diets, chemicals, constrictive undergarments, and even surgery—is a major industry. And many of these attempts to change somatotype involve personal pain and danger to health.

Recommended Reading

Llewellyn-Jones, Derek, *Everywoman and Her Body*. New York: Taplinger (1971).

A discussion of female reproductive physiology from adolescence through menopause, including a detailed account of the events of pregnancy and birth; clearly written and profusely illustrated.

Masters, W. H., and V. E. Johnson, *Human Sexual Response*. Boston: Little, Brown (1966).

A detailed presentation of the morphology and physiology of arousal and orgasm in men and women, based on an extensive series of observations.

Planned Parenthood Federation of America, *The New York Abortion Story*. Washington, D.C.: Planned Parenthood Federation (1972).

The medical and political aspects of the New York State abortion controversy. Planned Parenthood is a non profit organization that offers information and counselling and a variety of services in the family planning area. It maintains clinics in most major cities.

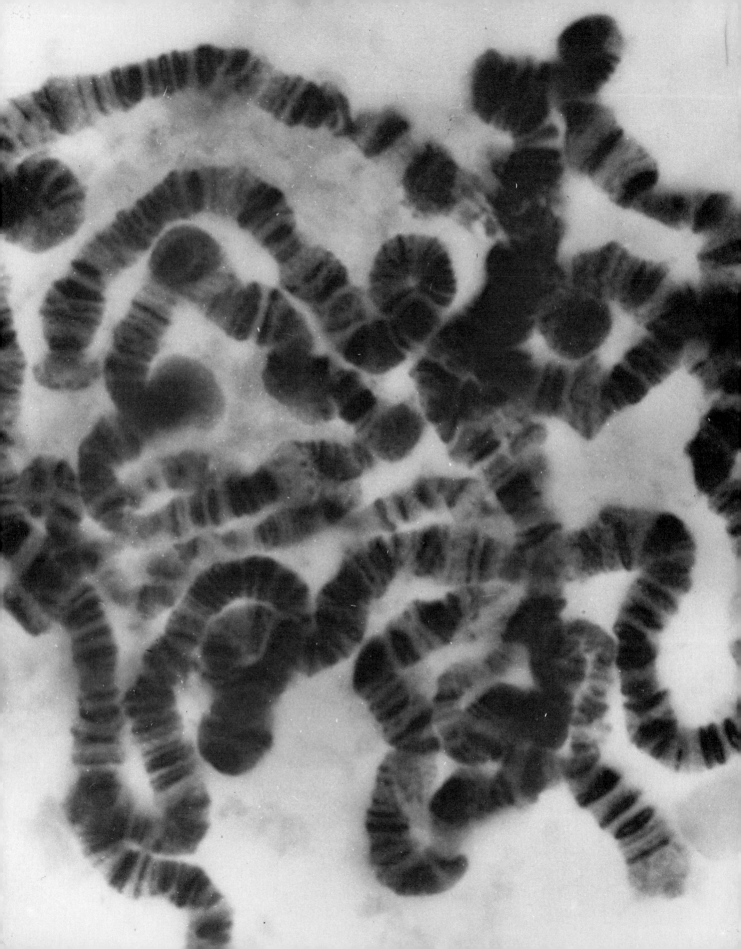

15 inheritance

Offspring resemble their parents; but offspring also differ from their parents. Sex determination is the most obvious case. After all, daughters are different from their fathers. The progeny of a male and female cannot resemble both parents in all respects. But offspring differ from their parents in many subtle ways as well.

The similarities and differences between parent and offspring have been a major preoccupation of western thought. Some thinkers believe that physical and behavioral differences between people are based entirely on inherited, or genetic, characteristics. This theory is sometimes used to support a rather Calvinistic view of the world: through a random distribution of genes (it says) individuals are predestined to be successes or failures. This attitude reinforces the view that poverty is associated with unfitness, and that charity and sympathy just retard the elimination of the poor and the sick.

An analogous view, based on the same belief in the genetic determination of human behavior, is the "beast in man" account of human nature. Man is described as an aggressive hunting ape whose inherited instinctive behavior is totally inappropriate to his modern environment. This view has great appeal for some people because it rationalizes the individual's alienation from society, and it absolves the individual of personal responsibility for his behavior. Such ideas about inherited characteristics have led to various elitist forms of social policy. In the best of them, the vast mass of mankind is pitied and magnanimously allowed to survive. In the worst, genetics is used to justify racism and genocide. "Social justice" in such systems usually consists of measures to protect the elite.

Giant chromosomes from the salivary glands of a fruitfly.

471

These opinions and their consequences conflict sharply with ideas that developed during the Enlightenment. Philosophers such as Locke, Kant, the French writers and thinkers of the later 18th century, and the American architects of our Declaration of Independence and Bill of Rights viewed men as free and equal. In some cases, "equal" was meant quite literally. Locke views a child's mind as a *tabula rasa*, a blank page upon which experience can write. In other cases, men's differences were recognized but were considered unimportant compared to their fundamental similarities. Presumably, people differed from each other because of inequalities in political and social opportunity. In this case, social justice was viewed as providing educational opportunity and improvement of the environment for all. The intellectual descendants of these men are liberal reformers. Their distaste for policies that depend on simple genetic interpretation of human differences is so strong that they often deny the existence of any heritable differences.

Both groups strive for social progress. The genetic, fatalist, Calvinist group tries to progress by conserving the good and eliminating the unfit. The Enlightenment, reformist, liberal group tries to progress by improving opportunity, particularly for the weak and disadvantaged. But from a biologist's point of view, neither the fatalism of the former group nor the reformism of the latter is fully correct. The biological position usually annoys both groups, since it does not support the ideas of either one exclusively. And unfortunately, proponents of either philosophy often select evidence from biology that supports their personal prejudices and ignore the rest.

Many characteristics of an organism are inherited—that is, they are determined by its genes. The genetic control and limitation of the potentialities of an individual are real. Yet, at the same time, genetic limitations do not wholly determine individual differences, and they certainly do not justify a fatalistic view of those differences. Environment is also important in determining the characteristics of the individual. By themselves, genes don't develop into anything; they can only duplicate, change, and be transmitted to other cells. Organisms develop. And how they develop depends on both their genetic constitution and their environment. Sometimes the environment is so important that organisms with identical genetic constitution become very different. At other times it is less significant.

chromosomal inheritance

Long before DNA was identified as a carrier of genetic information, and even before chromosomes were known, Johann Gregor Mendel described the basic mechanism by which genetic characteristics are distributed from parents to offspring. Mendel published his famous experiments in 1866, but their impact on the scientific world was delayed for a third of a century. One reason is that his paper was published in a rather obscure journal (*Proceedings of the Brünn Natural History Society*). But a more important reason was that Mendel's ideas were different from the dominant opinions of the time.

Mendel's Laws

Mendel was studying plant breeding, as were several of his contemporaries. The current view was that a cross (that is, cross-fertilization) between two plants produced a "blending" or "dilution" of all the characteristics of the parents. Mendel was interested in how the parents' characteristics were passed to the progeny, but he did not attempt to describe changes in the plant as a whole; rather, he selected several isolated characteristics for study. Mendel's shift in focus was crucial. By using specific characteristics that were easily recognized as being either present or absent, he was able to give a quantitative description of the inheritance of single characteristics for the first time. And, while the offspring did indeed share some characteristics with each parent, the inheritance of individual characteristics turned out to be anything but a "blending."

Dominance

Mendel's now-famous paper reported experiments on seven pairs of characteristics in common garden pea plants. Pea plants are easy to grow; they can be self-fertilized or cross-fertilized. One characteristic that Mendel examined was the color of the seed, that is, the pea color.

Mendel began with two separate strains of pea plants. One strain of plants had produced only yellow peas after many generations of self-fertilization; another had all green ones. These strains are called the **parental** generation. Mendel used one strain to fertilize the other and examined all the peas of the resulting generation. That generation and all subsequent ones are termed **filial** (that is, "daughter") generations. In the first filial generation, all of the resulting peas were yellow. Not one pea was green, or had a color that was a blend of green and yellow.

When Mendel grew plants from these first-generation peas and made crosses between them, he obtained both yellow and green peas in the second generation. But, again, none of the peas had an intermediate color. Mendel produced more than 8,000 peas in the second generation; 6,022 of them were yellow, while 2,001 were green. He obtained almost exactly the same result for the other six pairs of characteristics. All of the first generation showed only one trait of the pair. But the second generation showed both characteristics in a ratio of 3 to 1.

To explain these consistent results, Mendel developed the concept of **dominance.** He assumed that each gamete produced by the parent plant contains a factor for the particular characteristic. For example, gametes from the green pea plant contain a green factor (y), and gametes from the yellow pea plant contain a yellow factor (Y). Thus, each plant in the first generation has Yy—a factor for yellow and a factor for green. However, all the peas are yellow. Therefore, the green factor is not being expressed. Mendel called this phenomenon dominance; in this case, the yellow factor is dominant over the green factor.

Each plant in the second generation receives one factor from each parent in the first generation. Since both parent plants have one factor for green and for yellow, the offspring could receive two green factors, two yellow factors, or one of each. In the first case the pea would be green; in the second case it would be yellow. In the last case (Yy) it is also yellow, because the yellow factor is dominant over the green.

This experiment clearly showed that inheritance is not a matter of blending or dilution. Rather, it results from a specific combination of factors derived from each parent. A given factor may or may not be expressed as a visible feature of the new organism. But the factor itself is unchanged during its sojourn in the organism, even if it is not expressed. It can then be passed on to the next generation where it may reappear as a visible characteristic.

Law of Segregation
Mendel developed a second important idea from these experiments. He had studied the mathe-

matics of probability, and recognized that the numerical consistency of his results had a special significance. All the pairs of traits he had studied appeared in the second generation in a ratio of approximately 3 to 1. Mendel was able to account for this ratio by making one simple assumption: that the two factors for pea color in the parent become **segregated** at reproduction. In other words, the factors become separated, and each factor has an *equal chance* of being transferred to the offspring. Mendel then went on to calculate the probability that offspring would contain various combinations of factors.

In the original cross, one parent (P_1) contained two factors for yellow (YY). The other parent (P_2) had two factors for green (yy). Since each parent can contribute only one kind of factor to its offspring, the first generation offspring must all have had the same combination of factors— Yy. But the second generation is more complicated. Now, each parent is Yy, and the offspring can receive either factor from each parent. There are four possible combinations, therefore, depending on which factor is obtained from each parent:

Y from P_1, Y from P_2 = YY in offspring
y from P_1, Y from P_2 = yY in offspring
Y from P_1, y from P_2 = Yy in offspring
y from P_1, y from P_2 = yy in offspring

Offspring have an equal chance of receiving either Y or y from each parent. In other words, the probability of receiving Y is $1/2$, and the probability of receiving y is $1/2$. The probability of receiving a particular *combination* of factors is simply the product of these individual probabilities. The probability of receiving YY, for instance, is $1/2 \times 1/2 = 1/4$.

The same is true of the other three combinations. Each combination has a $1/4$ chance of occurring in a particular offspring. In other words, each combination of factors will appear in one fourth of the offspring. However, three of the four combinations contain at least one Y. Since Y is dominant over y, all three of these combinations will produce yellow peas. Thus, $1/4 + 1/4 + 1/4 = 3/4$, and the overall probability that a pea will be yellow is $3/4$. Phrased the other way, three quarters of the peas will be yellow, and one quarter will be green. This was almost precisely Mendel's experimental result. Of the 8,023 peas, 6,022 were yellow, and 2,001 were green: a ratio of very nearly 3 to 1.

Mendel concluded that, at reproduction, the genetic factors in each pair are segregated from each other and are distributed at random to offspring.

The fact that the factors in the parent separate from each other when they are transferred to offspring has become known as Mendel's **Law of Segregation.**

We can now describe these experiments with modern terminology. Mendel's "factors" are now called genes. The two genes in a pair controlling a single characteristic—for example, the genes for yellow and green peas—are called allelomorphs (Greek for "other form"). This term is usually abbreviated to **alleles.** When both alleles in a pair are identical, the individual is said to be **homozygous** for that characteristic. When the alleles are different, the organism is **heterozygous.** In a heterozygous individual, we use Mendel's term **dominant** to refer to the allele that is physically expressed. The unexpressed allele is described as **recessive.**

A shorthand notation has arisen to indicate the alleles in a gene pair. A pair of alleles is denoted by a single letter, and the dominant allele in the pair is capitalized. In this notation, the alleles for pea color are Y for yellow and y for green, as we have seen—y rather than g to show the dominant–recessive relation between the alleles. The genetic constitution of an individual is termed its **genotype;** the genotype is symbolized by the pair of letters representing that individual's alleles. The appearance of the organism, on the other hand, is termed its **phenotype;** the phenotype is represented by a single letter, that corresponds to the dominant allele, the one that is expressed. All the pea plants in Mendel's first generation were heterozygous, and their genotype was Yy. But all the plants had a yellow phenotype (Y) because all had yellow peas.

Usually the only way to establish an organism's genotype is through breeding experiments: a plant with a Y phenotype and a genotype of Yy, for instance, cannot be distinguished in appearance from a plant with the same phenotype but a genotype of YY. However, the genotype can be established by crossing both plants with a plant that has green peas (yy). Half the offspring of a Yy will have phenotype y. The Yy can contribute either Y or y, while the green plant always contributes y. However, all the offspring of the YY plant will have the Y phenotype (and a Yy genotype).

Mendel's hypothesis about the segregation of genes fits very well with what we now know about DNA, chromosomes, and chromosomal movements during mitosis and meiosis. A diploid organism contains two full sets of chromosomes, that is, two genomes. Each chromosome in one genome can be matched with a similar chromosome in the other genome. These pairs are homologues. The ex-

istence of homologous chromosome pairs agrees with Mendel's idea that individuals contain a pair of factors, or alleles, for each characteristic. One allele is located on each homologous chromosome. The segregation of genes at reproduction and their random distribution to offspring is consistent with the duplication, movement, and distribution and recombination of chromosomes during meiosis and fertilization.

We now know that a gene is a part of a double DNA strand in a chromosome. A gene for yellow peas is a particular section of a DNA molecule; it controls the synthesis of a particular RNA and ultimately of a particular protein. The protein has enzymatic properties, and it may participate in metabolic reactions that produce a yellow color in the pea. Its effects are the same whether one or two dominant alleles are present. But if the dominant yellow gene is absent, that protein will not appear, so the peas will not be yellow.

The results of genetic experiments can always be analyzed as probabilities, as Mendel suggested. However, our modern knowledge about DNA and chromosomes also allows us to analyze breeding experiments by representing the reproductive process more directly. In this type of analysis, each parent organism can be represented by a symbol that indicates both the phenotype and the genotype of the characteristic under investigation. Gametes are also represented by a symbol that shows which alleles they contain. The gametes are arranged along two sides of a square; the possible combinations of alleles in the resulting zygotes are shown inside the square. The genotype and phenotype of each possible offspring is shown—again symbolically—and each is placed below the two gametes from which it was formed (see Figure 15-1).

Independent Assortment

Mendel's second major set of experiments demonstrated the **independent assortment of genes.** Mendel found that the distribution of one gene to an offspring is unaffected by the distribution of another gene. That is, each gene is assorted independently.

Consider two sets of alleles for two different characteristics: pea color (Y = yellow, y = green) and the length of the plant (L = long, l = short). Mendel's original stock of long plants grown from yellow peas had the genotype $LLYY$. He fertilized them with short plants grown from green peas ($llyy$). All the peas of the first generation were yellow, and they grew into long plants ($LlYy$). But the second generation of peas, formed by a cross

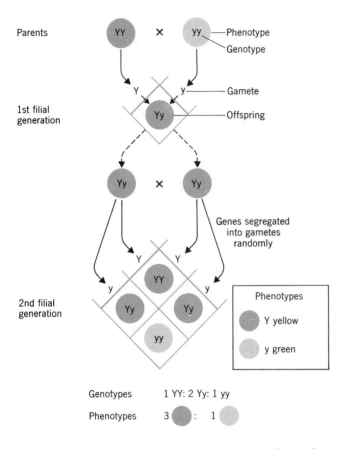

Parents

1st filial
generation

Genes segregated
into gametes
randomly

2nd filial
generation

Phenotypes

Y yellow

y green

Genotypes   1 YY: 2 Yy: 1 yy

Phenotypes   3   :   1

Phenotype
Genotype
Gamete
Offspring

*Figure 15-1 Segregation. Mendel crossed pea plants that
bred true for yellow seeds* (YY) *with ones that bred true
for green seeds* (yy). *All the first filial generation had
genotype Yy and a yellow phenotype. When they were
crossed with each other, the second filial generation
contained three genotypes* (YY, Yy, and yy) *and both
parental phenotypes, yellow and green. Mendel deduced
that his factors* (genes), *Y and y, were randomly segregated into gametes during meiosis, producing the
genotype and phenotype ratios shown.*

between first-filial-generation plants, is more complicated. The overall ratio of phenotypes, of yellow
to green peas, should be 3 to 1, and the overall ratio
of long to short phenotypes should be 3 to 1. But
now there are four phenotypes: long–yellow (LY),
long–green (Ly), short–yellow (lY), and short–green
(ly). If the assortment of alleles for the two characteristics is random and independent, then the ratios
of the four resulting phenotypes can easily be calculated. The probability that a plant will have any
particular combination of characteristics is the

probability that it will have one of them times the
probability that it will have the other. In this case,

Probability of yellow $(Y) = \frac{3}{4}$
Probability of green $(y) = \frac{1}{4}$
Probability of long $(L) = \frac{3}{4}$
Probability of short $(l) = \frac{1}{4}$

Therefore,

Probability of long–yellow $(LY) = \frac{3}{4} \times \frac{3}{4} = \frac{9}{16}$
Probability of long–green $(Ly) = \frac{3}{4} \times \frac{1}{4} = \frac{3}{16}$
Probability of short–yellow $(lY) = \frac{1}{4} \times \frac{3}{4} = \frac{3}{16}$
Probability of short–green $(ly) = \frac{1}{4} \times \frac{1}{4} = \frac{1}{16}$

Notice that of every sixteen plants, twelve will
have long stems and four will have short ones,
while twelve will have yellow peas and four will
have green ones. The overall ratio for each individual characteristic is the expected 3 to 1. But these
characteristics have been combined in four different ways.

Again, this result can be visualized directly by
considering the various combinations of gametes.
There are four kinds of sperm nuclei and four kinds
of eggs, so sixteen kinds of zygotes are possible.
The possibilities are represented by a four-by-four
array of zygote genotypes (Figure 15-2). As you can
see, nine of the sixteen possible zygotes have an LY
phenotype—they have at least one L and at least
one Y in their genotypes, so they will have long
stems and yellow peas. Three of the zygotes are Ly,
and three are lY. They will be long–green and
short–yellow, respectively. Finally, only one zygote
has neither an L nor a Y; it is ly. It will have a short
stem and green peas. We have obtained the same
overall phenotypic ratio of 9:3:3:1. From this diagram we can see that even though $\frac{9}{16}$ of the plants
have a long and yellow phenotype, only $\frac{1}{16}$ of them
would actually breed true for both characteristics.
In other words, only $\frac{1}{16}$ are LLYY in genotype.

This sort of calculation can be done for any
number of independently assorted allele pairs. The
mathematical prediction is easy. Mendel checked
his predictions for three characters, using smooth
and wrinkled seed shape, yellow and green seed
color, and violet and white flowers. He obtained
approximately the predicted ratios. But large
numbers of plants are always necessary to provide a
convincing demonstration. The assortment of
genes, though independent, is a random process,
and the final ratios can only approximate the predicted one. The larger the number of individuals,
the more accurate the result.

The tremendous importance of Mendel's work
was finally recognized in the beginning of the twen-

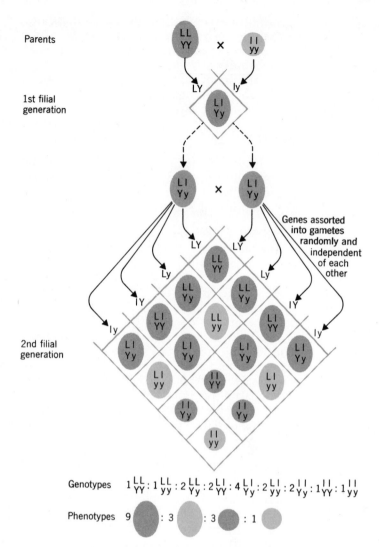

Parents

1st filial generation

2nd filial generation

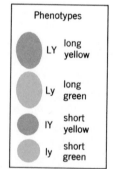

Genes assorted into gametes randomly and independent of each other

Genotypes $1\frac{LL}{YY}:1\frac{LL}{yy}:2\frac{LL}{Yy}:2\frac{LI}{YY}:4\frac{LI}{Yy}:2\frac{LI}{yy}:2\frac{II}{Yy}:1\frac{II}{YY}:1\frac{II}{yy}$

Phenotypes 9 : 3 : 3 : 1

Phenotypes

LY long yellow

Ly long green

IY short yellow

Iy short green

Figure 15-2 Independent assortment of genes. Mendel crossed parental stocks that were homozygous for the dominant and recessive alleles of two different traits. The first filial generation was heterozygous for both traits and phenotypically showed both dominant traits (long and yellow, in this case). The second filial generation had nine genotypes and four phenotypes. Mendel deduced that in meiosis, the segregation of the allele pairs for each of two or more traits occurs both randomly and independently of the segregation of the other allele pair or pairs. Each plant from the first filial generation produces four kinds of gametes with equal probability.

tieth century, and since that time a large volume of additional work has rapidly accumulated. Mendel's findings have led directly to our present understanding of gene action.

Further study of genetics depended on the selection of suitable organisms. Ideally, the subject should have small size, a short life cycle, and large numbers of offspring. A complex morphology with many different conspicuous characteristics also seemed desirable at first (though later this feature proved to be unnecessary). The organism also had to be easy to maintain in the laboratory. Cereal grains were sometimes used because of their great economic importance, even though their life cycle is long. But, the fruit fly, *Drosophila melanogaster*,

was the major organism chosen for genetic analysis (Figure 15-3). *Drosophila* had all the ideal characteristics for genetic studies. Also, as work proceeded and information about it accumulated, it became more and more desirable to use: many different stocks became available on request, and a tremendous backlog of information permitted the investigator to ask very sophisticated questions. Even *Drosophila*, however, has a low reproductive potential and a long life span compared to those of microorganisms. Consequently, bacteria and bacteriophage viruses have now largely replaced *Drosophila* in modern genetic research.

Microorganisms satisfy the requirement for phenotypic variety not with morphological features,

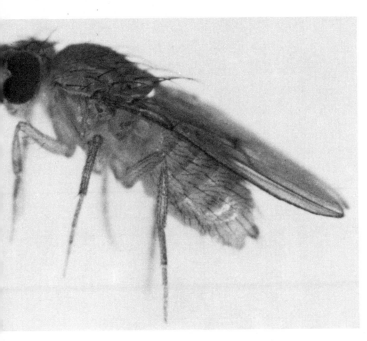

Figure 15-3   The fruit fly, Drosophila melanogaster.

but rather with their various biochemical characteristics. In *Drosophila* geneticists studied characteristics such as eye color, wing shape, and the number of bristles. The bacterial geneticist uses such features as the presence of enzymes, the chemical properties of the cell wall, or resistance to high temperatures as phenotypic characteristics. Geneticists can now take advantage of the tremendous reproductive potential of microorganisms. A researcher working with *Drosophila* might spend his life looking for a particular phenotype that appears once in a million individuals. But this problem is a trivial one for a geneticist working with the bacterium *Escherichia coli*. Billions of *E. coli* can be grown quickly and easily. Microorganisms are the only practical resource for most genetic research.

Complex Chromosomal Inheritance

Drosophila, E. coli, and other organisms have been used in further studies of chromosomal inheritance. Inheritance is usually much more complicated than the independent assortment and segregation of genes that Mendel observed. Not all genes are assorted independently. Genes may be linked together on the same chromosome, or they may be traded between homologous chromosomes. Furthermore, individual traits are often controlled by a number of genes rather than by a single dominant-recessive pair.

Linkages and Crossovers

If alleles for two different characteristics are located on the same chromosome, they are said to be **linked.** Linked genes are assorted as a unit in meiosis. This phenomenon was first found in crosses between sweet peas in 1905. Let us examine the behavior of our two allele pairs, *L* and *l*, and *Y* and *y*, but now assume that the genes are linked: one chromosome carries both *L* and *Y*, and its homologue carries *l* and *y*. If we begin crossing genotypes, *LLYY* and *llyy*, the first generation should have a genotype of *LlYy* and a phenotype of *LY* (Figure 15-4). This result is identical to that predicted for independent assortment. But the second generation is different; it does not show the expected 9:3:3:1 ratio of phenotypes. Rather, only two phenotypes are found, *LY* and *ly*, in a 3:1 ratio. Linkage is equivalent to having both of the characteristics controlled by a single gene. The absence of mixed phenotypes in the second generation—that is, the absence of *Ly* or *lY*—is evidence that the genes are linked.

In fact, genes on the same chromosome rarely behave as though they are completely linked. Their behavior is usually somewhere between complete linkage and independent assortment. The reason is that genes are traded between homologous chromosomes by a process known as **crossing over.** At the tetrad stage in meiosis, crosses, or **chiasmata,** occur when the chromatids break and reassociate (see Figure 15-5).

In the case of our two linked allele pairs, crossover might sometimes separate *L* from *Y*, and *l* from *y*. In the cross between *LLYY* and *llyy*, a crossover during meiosis would have no effect on the gametes, because the alleles on both homologous chromosomes in each parent are identical: the chromosomes would merely exchange one *L* for another, for instance. All of the first generation would have the expected *LlYy* genotype.

Crossover in the gametes that fuse to form the second generation would have an effect, however. The result would be an occasional chromosome that carries *L* and *y* and another that carries *l* and *Y*. The second generation will then contain some *Ly* and *lY* phenotypes (though still not as many as it would if *L* and *Y* were independent). In short, chiasmata break the link between the two genes some of the time (Figure 15-6).

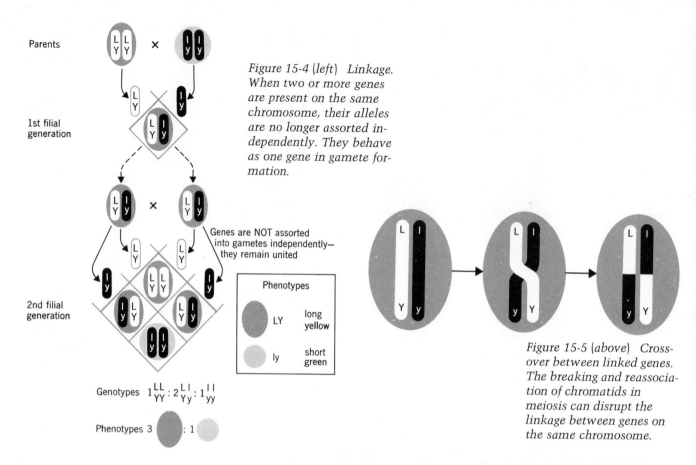

Parents

1st filial generation

2nd filial generation

Figure 15-4 (left) Linkage. When two or more genes are present on the same chromosome, their alleles are no longer assorted independently. They behave as one gene in gamete formation.

Genes are NOT assorted into gametes independently—they remain united

Phenotypes		
	LY	long yellow
	ly	short green

Genotypes $1 \frac{LL}{YY} : 2 \frac{Ll}{Yy} : 1 \frac{ll}{yy}$

Phenotypes 3 : 1

Figure 15-5 (above) Cross-over between linked genes. The breaking and reassociation of chromatids in meiosis can disrupt the linkage between genes on the same chromosome.

Figure 15-6 (right) Whether genes are linked, linked with crossover, or assorted independently, the first filial generation is always the same. But the behavior of the alleles and the expected phenotypic ratios in the second filial generation are quite different. (The L and Y alleles actually are assorted independently in garden peas; the linkage shown is hypothetical.)

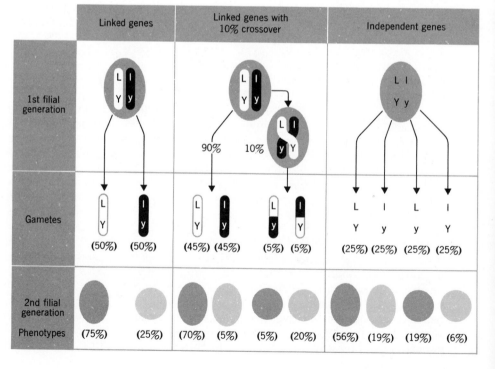

	Linked genes		Linked genes with 10% crossover				Independent genes			
1st filial generation										
Gametes	(50%)	(50%)	(45%)	(45%)	(5%)	(5%)	(25%)	(25%)	(25%)	(25%)
2nd filial generation Phenotypes	(75%)	(25%)	(70%)	(5%)	(5%)	(20%)	(56%)	(19%)	(19%)	(6%)

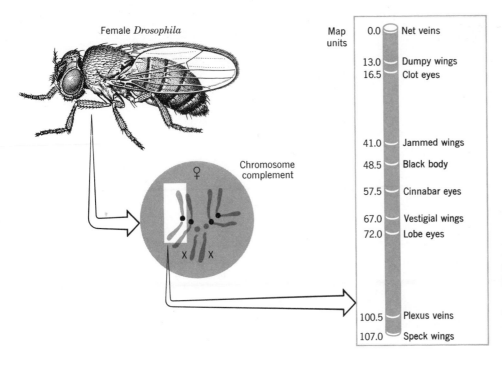

Female *Drosophila*

Chromosome
complement

Map units	
0.0	Net veins
13.0	Dumpy wings
16.5	Clot eyes
41.0	Jammed wings
48.5	Black body
57.5	Cinnabar eyes
67.0	Vestigial wings
72.0	Lobe eyes
100.5	Plexus veins
107.0	Speck wings

Figure 15-7   A map of genes on a chromosome of Drosophila. *Only a few of the known genes are shown. The numbers are the positions of the genes in terms of crossover units.*

Genes that are far apart on the chromosome are more likely to be separated by a crossover than genes that are close together. Thus, the crossover frequency is related to the distance that separates two genes on a chromosome. One can actually map a chromosome—specify the order of the genes—by studying the crossover frequencies. Figure 15-7 is a map of one of the four chromosomes in *Drosophila*. The units along the map are frequencies or probabilities of crossover; 1.0 means a one-percent probability of crossover. Such chromosome maps show the ordering of genes, though not their actual positions on the chromosome. The preparation of such a map takes the cooperative efforts of many investigators over a period of decades. It is not uncommon to examine 100,000 flies in order to establish one detail of the map.

Crossing over and recombination can occur within genes as well as between them. Crossover techniques have been used to analyze fine structure in a single gene. For example, the gene for rapid lysis (*r*II) in bacteriophage viruses has been subjected to extensive analysis. A bacteriophage lyses the bacterial cell in which it grows and multiplies. Phages with the *r*II gene are easily identified and detected because of their ability to lyse bacterial host cells very quickly. When a liquid containing phages is spread on a single plate of bacteria, clear areas, called plaques, develop wherever the bacteria are lysed by phages (see Figure 15-8). Since millions of phage organisms can be spread on a single plate, it is possible to look for crossovers that occur at very low frequencies. Recombinations have been observed within the *r*II gene in one of ten million phage organisms.

This same study shows that some spots in the gene are more likely to break during crossover than others. In some places, the probability of crossover is higher than one would expect from the number of nucleotides in the gene's DNA. This fact means either that breakage and crossing over occur within the nucleotide bases as well as between them (as no one believes), or that breakage and recombination are much more likely at some points in the chain than at others. In other places, crossover probability is very low. The gene can be viewed as a DNA chain with "hot spots," where crossing over is likely, separated by areas where breakage and recombination have a lower probability (see Figure 15-9). The same view applies to the chromosome as a whole.

Sex-linked Inheritance

Our discussion of chromosomal inheritance so far has ignored sex differences between parents and off-

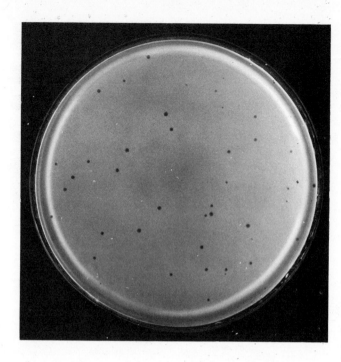

Figure 15-8 As E. coli *bacteria on this culture plate are lysed by a phage virus, clear areas, or plaques, arise.*

Figure 15-9 A map of the rII *gene. The* rII *is a DNA segment with "hot spots"—regions that mutate easily. Each square represents a known spontaneous mutation.*

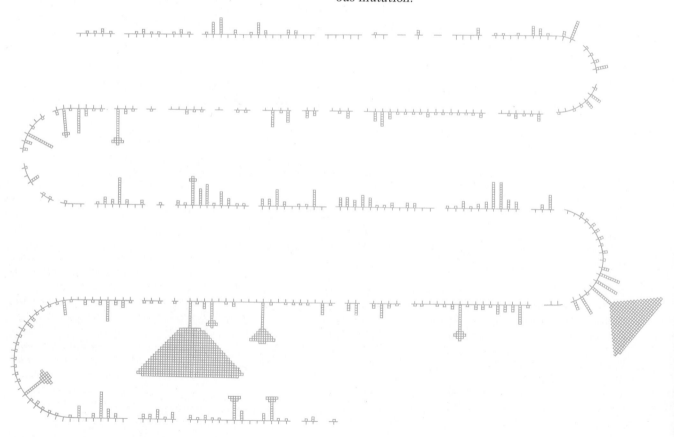

spring. Two chromosomes, the X and Y chromosomes, control sex characteristics. In humans, men have an X and a Y, while women have two X's. But many traits in addition to sexual characteristics are carried on the X or the Y chromosome. Such traits are called **sex-linked characteristics.** A sex-linked characteristic will be inherited along with the sexual characteristics themselves. Sex in humans follows a simple pattern of inheritance: the male parent determines the sex of the offspring. Each parent supplies one sex chromosome, but the female always supplies an X. If the male parent also supplies an X the child is a girl. If he provides a Y the child is a boy (Figure 15-10).

Hemophilia is a good example of a sex-linked characteristic. The blood of a hemophiliac (a person with this condition) does not clot normally because it lacks one of the cofactors needed for the production of thrombin. The gene for the cofactor is recessive and is carried on the X chromosome. The inheritance of hemophilia usually involves a cross between a normal man and a woman who is heterozygous for the allele pair. That is to say, she has one normal allele and one allele for hemophilia. Since the hemophilia gene is recessive, she does not actually have the disease—she merely carries it. Half the woman's eggs will carry the recessive gene for hemophilia. Half her daughters will receive the gene and will also carry but not show the disease; their hemophilia allele will be masked by the normal allele on the X chromosome supplied by the father. But half the sons will be bleeders, because their single X chromosome comes from the mother—no normal allele will be present (Figure 15-11).

Occasionally a daughter is homozygous recessive; that is, she has two hemophilia alleles. In that case, she will have the disease. This condition is extremely rare, however, because both of her parents must carry the hemophilia allele. At one time it was thought that a woman homozygous for the recessive gene could not survive, but a few exceptions have been discovered. Hemophilia has been carefully studied, in large part because it appeared in European royal families. This accident made it possible to construct detailed family trees that revealed the trait's pattern of inheritance. Queen Victoria of England was heterozygous for hemophilia; she passed the gene to at least three of her nine children.

Information from genealogies has enabled geneticists to locate other X-linked genes as well. These genes include the ones that control color blindness,

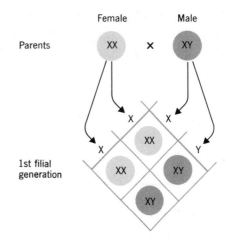

Figure 15-10 Determination of sex in humans. Each egg receives an X chromosome during meiosis; sperm receive either X or Y. Half the zygotes will be female (XX) and half, male (XY).

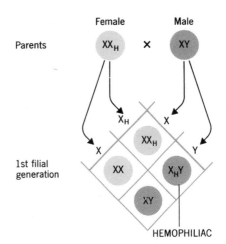

Figure 15-11 Sex-linked inheritance. The gene for hemophilia is recessive, carried on the X chromosome. A mating between a normal male and a heterozygous female is shown. Half the males resulting from this cross will show the hemophilia trait, and half will be normal. Half the female children will be heterozygous carriers with a normal phenotype, and half will be normal both phenotypically and genotypically.

juvenile glaucoma (hardening of the eyeball), double eyelashes, juvenile muscular dystrophy (degeneration of muscles), and mitral stenosis (abnormal heart valve).

Another example of an X-linked trait occurs in calico cats. Cats have the same chromosomal sex determination as humans. The calico color pattern is controlled by two X chromosomes, so it is found only in female cats. One X chromosome bears a gene for black fur and the other, a gene for yellow fur. In the females, patches of black or yellow fur occur randomly, depending on which X chromosome is active in a given group of hair follicle cells. This phenomenon produces a mosaic of color. Since the males carry only one X chromosome, they are either all yellow or all black.

Incomplete Dominance and Multiple Alleles

Such concepts as linkage, dominance, and the segregation and assortment of genes explain inheritance of discrete characteristics of the kind that Mendel studied. Mendel's plants were either tall or short, had seeds that were either round or wrinkled, and had flowers that were either white or violet. But that is one reason that Mendel's work was overlooked. Biologists who might have recognized the importance of Mendel's analysis were themselves looking for a way to explain the more gradual changes that were considered to be a part of evolution. Consequently, ideas about the blending and dilution of parental characteristics, which would give intermediate phenotypes in the offspring, were extremely attractive in Mendel's day. The segregation and independent assortment of discrete characteristics did not seem relevant.

One mechanism that does provide an intermediate phenotype within the framework of Mendelian laws was described in 1912 by Carl Correns. It is called **incomplete dominance.** Mendel had found that individuals heterozygous for a trait (for instance, of Yy genotype) had the same phenotype as individuals homozygous for the dominant allele (YY genotype). The two phenotypes would be distinguished only with breeding experiments. But Correns showed that not all alleles have such a complete dominant–recessive relation. He studied flower color in four-o'clocks. When plants with white flowers (rr) were crossed with a stock with red flowers (RR), all the progeny had pink flowers. Pink plants fertilized by other pink plants have plants with red, pink, and white flowers in a ratio of $1:2:1$ (see Figure 15-12). Apparently the red flower allele (R) is incompletely dominant with

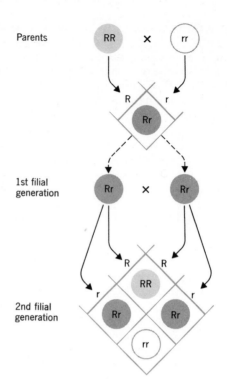

Figure 15-12 Incomplete dominance. In four-o'clocks, the allele for red flowers (R) is incompletely dominant with respect to the allele for white flower color (r); heterozygous individuals are pink.

respect to the white flower allele (r). That is, the Rr genotype has pink flowers. This is a simple example of incomplete dominance. It is one of the ways in which an intermediate phenotype can be produced.

Correns showed that the phenotype of an organism might represent a blending of the phenotypes of the parents. However, it is important to understand that this blending of phenotype was not produced by a blending or dilution of the genotype. Genes are inherited as discrete alleles in these cases, too. The phenotype is blended because of the interaction between alleles; the genes themselves do not change.

When a trait is controlled by several alleles, rather than just a pair, a much wider range of phenotypic expression is possible. In a diploid organism, a particular gene is represented by only two alleles. Therefore, only three genotypes are possible (RR, rr, and Rr, for instance). However, in the whole population there may be more than just two different alleles for the same trait. This variety makes possible a broader spectrum of genotypes.

One example of the control of a characteristic by multiple alleles in the population is the ABO blood grouping in man. Each individual's blood group is controlled by a single gene pair, but that pair can be made up of any two of four possible alleles. The different possible combinations of alleles are responsible for the variety of human blood groups.

A trait can also be controlled by several genes in a single individual. Since each of these genes has alleles, the potential variety of genotype is very great. Skin color in man, for example, is controlled by more than one, gene, but no one is sure how many. Skin color is determined by the number of granules of the pigment **melanin** that are deposited in skin cells. Melanin formation, in turn, is controlled by a system of multiple genes. One hypothesis explains skin pigmentation using only two independent sets of alleles, Aa and Bb. Assume that A and B promote the formation of melanin; they are indistinguishable in the phenotype. A is incompletely dominant over a; and B, similarly, over b.

A cross between a black person who is homozygous for both dominant alleles ($AABB$) and a white person who is homozygous for both recessives ($aabb$) yields offspring with intermediate pigmentation ($AaBb$). This result occurs in the offspring of parents with only black and only white ancestry. But the situation becomes much more complicated in subsequent crosses. For example, children of AaBb parents show nine possible genotypes and five different phenotypes (see Table 15-1). Some offspring may have more skin pigmentation than either parent, because they can receive incompletely dominant alleles from both parents and be $AABB$. The children of a homozygous recessive ($aabb$) and a person with some alleles that favor melanin production can never be more pigmented than the darker parent; but they may be lighter.

This two-gene hypothesis agrees with many observations on inheritance of skin pigmentation. But two allele pairs are really not enough to account for the almost continuous variation that is found in melanin pigmentation. Theories based on four, five, or six genes give satisfactory agreement with observed data. The actual number of genes involved will probably never be discovered by present techniques. One major difficulty is in measuring the actual degree of pigmentation. What is classed as "white" usually includes over ten percent of the whole range of melanin pigmentation. And the

Table 15-1
Control of Skin Color by Multiple Alleles[1]

Genotype	Genotype Frequency	Phenotype	Phenotype Frequency
$AABB$	1	Black	1
$AaBB$	2	Dark	4
$AABb$	2		
$AAbb$	1	Intermediate	6
$AaBb$	4		
$aaBB$	1		
$Aabb$	2	Light	4
$aaBb$	2		
$aabb$	1	White	1

[1] These numbers come from a 4 × 4 array of genotypes written out as a cross between parents with AaBb genotypes. A and B are phenotypically indistinguishable.

individual's level of pigmentation is influenced by his environment (sun and wind, for instance), and by age (it increases to maturity and then declines again in old age).

The exact number of genes that are involved in a single trait can be analyzed by examining the results of controlled cross-breeding. In this way we have learned that two allele pairs control the size of an ear of corn. But in other cases, the number of allele pairs is so large that analysis is impractical. As many as 200 genes can be involved in some traits.

Mendel's basic concepts can be extended to cover more complicated forms of chromosomal inheritance if we consider the effects of multiple alleles, crossovers and linkages, and incomplete dominance. These are simple refinements of his original observations on segregation, assortment, and dominance.

mutation

The mechanisms of chromosomal inheritance occasionally fail. Aneuploidy—abnormal chromosome number—represents such a failure because it produces an unequal distribution of duplicate DNA strands to daughter cells. Another kind of failure is a change in the structure of the DNA itself. These changes are called **mutations** (from the Latin verb *mutare*, "to change"). A mutation is any deletion, addition, or substitution in the base code of the DNA strand.

Mutations can happen in several ways. DNA strands are sometimes disrupted or broken by chemical or physical agents. Some breaks are repaired by the cell, but in the process, the sequence of bases in the DNA code may be altered. Any change in base sequence, of course, means a change in the RNA, and ultimately in the protein product. Mutations were first recognized when organisms showed sudden changes in phenotype that were then passed on to subsequent generations according to normal Mendelian laws.

The disease of **sickle cell anemia** was originally a mutation: one base was substituted for another during the replication or repair of DNA. An individual with sickle cell anemia has a single change in the DNA code for the amino acid sequences in his hemoglobin. A normal hemoglobin chain contains glutamic acid, which is coded for by a base sequence of either GAA or GAG on the original DNA molecule. But the sickle chain contains valine in place of glutamic acid in the polypeptide. Valine is coded for by GUU, GUC, GUA, or GUG. There are two ways a change in a single base of the DNA molecule could have produced the observed change in amino acid sequence: GAA could have become GUA, or GAG could have become GUG. In either case, the substitution of U for A at a single point in the DNA molecule produces a new polypeptide. The new protein is still functional in the hemoglobin molecule, but it has slightly different chemical properties. When the oxygen concentration in blood is low, the hemoglobin molecule can no longer stay dissolved inside the red blood cells. The hemoglobin precipitates out of solution, and the red blood cells become distorted and sickle-shaped (Figure 15-13).

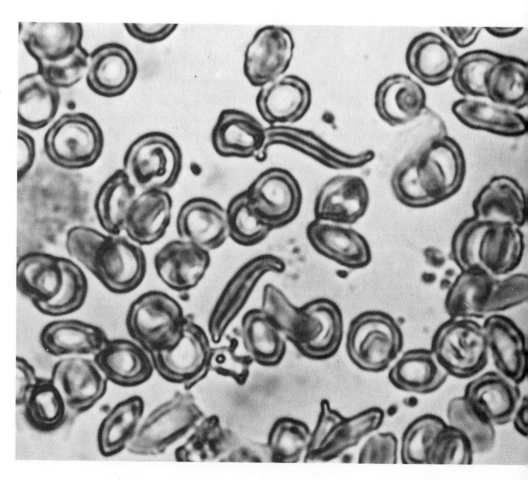

Figure 15-13 Red blood cells from an individual with sickle-cell anemia show the abnormal, elongated "sickle" shape. Magnification 2,000×.

Chromosome Abnormalities

Occasionally, all the cells of an organism contain an abnormal number of chromosomes. This condition is called **aneuploidy** (Greek for "not a true set"). Aneuploidy results from an unequal distribution of chromosomes at meiosis. When chromosomes fail to separate properly, an egg or sperm may end up with either too many chromosomes or not enough. About half the cases of aneuploidy involve sex chromosomes and the other half somatic chromosomes.

In humans, males can be produced with XY (the normal combination), or with XXY, XYY, XXYY, XXXY, or XXXXY. Provided at least one Y chromosome is present, the individual is morphologically male. XXY occurs in one out of 500 newborns. XXY males have Klinefelter's syndrome. (A syndrome is a group of symptoms or defects characteristic of a disease.) Males with Klinefelter's syndrome acquire some female secondary sex characteristics and are usually mentally retarded (see Figure A).

Errors in chromosome assortment at meiosis can also produce female individuals with one to five X chromosomes in each diploid cell. A single X chromosome occurs in one child out of 5,000. These females have what is known as Turner's syndrome. In Turner's syndrome, the ovaries fail to develop; the individual is dwarfed and has low-set ears and a webbed neck (see Figure B).

Extra X chromosomes are particularly easy to recognize because of cellular staining characteristics. In normal women (XX), stained diploid cells from the lining of the mouth (and many other diploid cells) have a dark spot in the nucleus called a **Barr body.** Males (XY), and females with Turner's syndrome (X), have no Barr bodies. But males with Klinefeleter's syndrome (XXY) do. The number of Barr bodies in a cell is simply equal to the number of X chromosomes minus one. This pattern is probably related to the number of func-

Figure A (right) Klinefelter's syndrome is characterized by male genitals, feminine breast development, and mental retardation.

Figure B (far right) Turner's syndrome is characterized by female external genitals, webbed neck, broad chest, small breasts and uterus, and no ovaries.

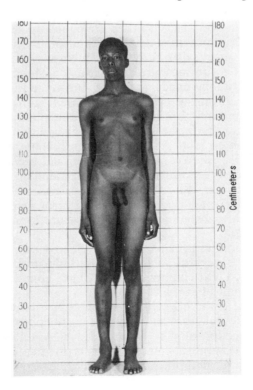
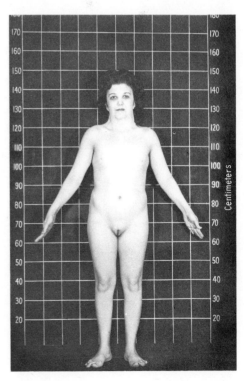

485 *mutation*

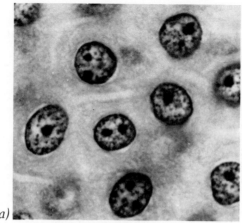

(a)

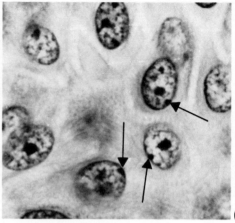

(b)

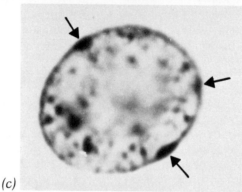

(c)

Figure C   Barr bodies in diploid cells of humans. (a) Cells from a male have no Barr bodies. (b) Cells from a female have one Barr body per cell, indicated by arrows. (c) A cell from an individual with four X chromosomes has three Barr bodies.

tional X chromosomes: any extra X chromosomes may be inactive, staining as dark spots because they are coiled up (see Figure C).

Aneuploidy in somatic chromosomes can be equally serious. For example, Mongolism (also known as Down's syndrome) is associated with an extra somatic chromosome. A victim of Mongolism has a flattened face, small head, stubby hands, and short limbs, and is mentally retarded (see Figure D).

About one of every two hundred humans born shows aneuploidy. Older mothers are more likely to bear aneuploid children. All the cells in the ovary that are destined to become eggs begin their development when the female is still an embryo. At the end of embryonic development all the eggs have undergone the first meiotic division and are at the primary oocyte stage. They now enter a dormant period, remaining inactive until ovulation and fertilization trigger the second meiotic division. The eggs in older women have a longer dormant period, and this factor seems to increase the probability that the chromosomes will fail to separate properly during the second meiotic division.

It is now possible to detect chromosomal abnormalities as early as the twelfth week of pregnancy. The fluid in the amnionic sac that surrounds the embryo will contain cells that have come from the embryo, not the mother. A sample of the fluid is taken with a needle and a chromosome count is made on the embryo's cells. This procedure is called **amniocentesis.** If amniocentesis reveals that an abnormal number of chromosomes is present, therapeutic abortion may be recommended. Amniocentesis also allows the detection of other genetic defects that are present in the embryonic cells.   There is a small but definite risk of inducing abortion (about 1 in 100) associated with

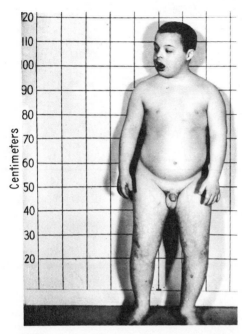

Figure D Mongolism, or Down's syndrome. The individual is short and stocky and is mentally retarded.

the procedure itself; hence it is not justifiable as a routine screening technique. It is valuable, however, where study of the family history reveals a high probability that a detectable abnormality will occur. It may also be indicated for mothers over 40 years of age. At this point, the risk of bearing a Mongoloid child becomes one in 65 or more.

Comparisons between normal and aneuploid individuals indicate that genes in a single genome form a carefully balanced system. The addition of extra alleles (as in Mongolism) is usually extremely harmful. The extra chromosome may interfere with normal development by producing excess quantities of enzymes or other substances. But by contrast, the addition of one, two, or several *complete* genomes in body cells is not nearly as harmful. In fact, it occurs normally in many plants and animals. Genes of all the chromosomes in a genome form a balanced set.

Sickling has serious consequences for homozygous individuals. Their sickled cells may be rapidly destroyed, resulting in anemia; the cells may clump, restricting blood circulation to organs and limbs; and they may collect in the spleen and cause serious damage. The situation is somewhat different in individuals who are heterozygous for the sickling gene. The changes in their hemoglobin are less serious, though still debilitating. But related changes in their cells make them resistant to the parasite that causes malaria, *Plasmodium falciparum*. The parasite is unable to grow within the red blood cells. Since homozygotes are rare in the population, compared to heterozygotes, the disadvantage of having the gene in the population (the occurrence of occasional homozygotes with sickling disease) is more than offset by the benefit (the large number of heterozygotes resistant to malaria). As a result, the gene is maintained at rather high frequencies in populations where malaria is endemic. Sickle cell anemia is very common in Africa. In some African Negro tribes, 40 percent of the population carry the sickling allele. About ten percent of the Negroes in America have the allele. Sickle cell anemia is rare in other racial groups.

Mutation and Variability

Most mutations have a negative impact on the organism. Mutations are analogous to typographical errors in a sentence. The genetic language is made up of three-letter words. Each letter corresponds to a single base in the DNA molecule, and each three-letter (three-base) word corresponds to a particular amino acid. Sentences in the genetic language are made up of many words (a sentence representing a single gene, which produces a single protein). Suppose, then, that we have a sentence that reads:

THE/CAT/AND/DOG/RAN/OUT

If one letter in the sentence is simply replaced by another, a word will be misspelled. Most misspelled words don't mean anything. And even if they do, they are likely to change the meaning of the sentence. In our example, a single substitution either produces a meaningless sentence,

THE/CX T/AND/DOG/RAN/OUT

or changes the meaning:

THE/R AT/AND/DOG/RAN/OUT

The deletion of a single base, or letter, produces nonsense in all the words beyond the deletion. If the "C" is accidentally dropped, the result is:

THE/ATA/NDD/OGR/ANO/UT

Such errors in the DNA chain will change the code and therefore the structure of the protein product. Thus, mutations introduce variability into the genome, but the variations are almost always harmful to the organism; the random deletion or substitution of bases in DNA chains rarely produces a more effective protein. Nevertheless, mutations are the source of the variation that allows organisms to become better adapted to their environments.

Mutation is essential for the evolution of organisms. Very rarely, one or more changes in the DNA base code will produce a protein that functions more effectively, or that serves a new and beneficial function in the organism. Though mutations occur in a disordered way, they should not be compared to a completely random process. Mutation is *not* like an ape pushing typewriter keys until he has written a Shakespeare sonnet, or like a burglar trying 100,000 combinations to a safe. In both these cases, the probabilities of success are very close to zero. If mutation really operated in this way, it could never produce a complicated protein like hemoglobin in a finite length of time. The difference between the way mutations take place and the random attempts by the ape or the safecracker is that mutational success can occur gradually, and in discrete steps. It is not necessary to produce the final hemoglobin molecule in one step. Rather, many small mutations in DNA base code may produce many small functional improvements.

These small improvements can be preserved in the genome as it passes from generation to generation. By contrast, the burglar cannot achieve his goal in small parts. Even if he gets four of five numbers correct, there is no reward. Nor are his four correct numbers preserved and used in the next random try.

This kind of beneficial mutational change is possible because there is a large amount of DNA in the population. If the pool of DNA is large, there is a reasonable probability that favorable mutations will occur occasionally, and a few of them will often be present in the population. If the environment then changes and becomes less suitable for most of the organisms in the population, some individuals might be carrying mutations that increase their survival chances. The interaction between mutation and such environmental changes is the subject matter of evolution.

Large DNA gene pools are achieved in several ways. Microorganisms depend on having enormous numbers of rapidly-reproducing individuals. A vast population will often have a few individuals whose mutations allow them to survive changing conditions. But most multicellular organisms have smaller populations. Multicellular organisms maintain large gene pools by carrying two or more complete genomes in their cells. The extra genome is a source of variability. Often one of the two sets of genes is completely dominant. Thus, the second set is free to provide a store of mutated genes. In diploid, sexually-reproducing organisms, these stored genes provide the whole population with flexibility and variability. Their presence increases the likelihood that the species can survive environmental change. In addition, mutations that require two more modifications of the DNA code are more likely if one of them can persist in one of the two chromosome sets until the second occurs. One change may even be negative when it occurs alone but positive in combination with another.

Frequency of Mutation

The probability of a change in the DNA of any specific gene is about one in a million in each generation of cells. However, this figure is only an average, and some genes are much more likely to mutate than others. Genes with a lower mutation frequency are called stable genes; ones that mutate more frequently are unstable.

Overall mutation frequencies in single genes show a wide variation from organism to organism. The rate is rather low in bacteria (one mutation in

Figure 15-14 Mutator genes in maize kernels are responsible for the dark spots on the light-colored kernels.

ten million cell generations), higher in fruit flies (about one in 100,000), and higher still in man (one in 10,000 to 100,000). Some exceptional genes in plants have a tremendously high mutation frequency. The rates may be so high that a plant is a mosaic of normal and mutated genes; some cells will have the normal trait, while others in the same tissue show the mutated form (see Figure 15-14). Closer investigation indicates that this phenomenon is caused by special **mutator genes,** which render adjacent genes unstable.

These figures for the probability of mutation may seem low when expressed in terms of single genes. But multicellular organisms have a large number of genes, and each one of them is subject to change. Furthermore, each individual represents a considerable number of cell generations. Thus, the probability that any particular fruit fly will carry some sort of mutation is about one in twenty. There are many mutated genes present in natural populations. If wild fruit flies are brought into the laboratory and used as material in breeding experiments, a variety of recessive mutant forms can always be revealed in the wild population. Some of these naturally-occurring mutations are bizarre and obviously deleterious: the mutant "aristopedia," for instance, has a leg in place of an antenna (Figure 15-15).

We don't know what actually causes the normal minimum, or background, mutation rates. But mutation rates can be correlated with other variables. In cold-blooded animals the mutation rate increases as the environmental temperature is increased. Mutation is more likely in stored sperm than in

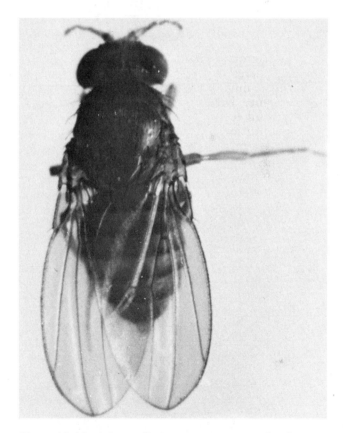

Figure 15-15 Aristopedia is a common mutation in wild Drosophila populations. In a homozygous recessive individual, a leg develops in place of an antenna.

freshly collected sperm (in fruit flies). In humans, as the age of the mother increases, so does the chance of mutation in children she bears.

Mutagens

A number of environmental factors increase the background mutation frequency. Anything that does so is called a **mutagen.** Some chemical substances can act as mutagens. Compounds that resemble normal DNA bases, for example, may be incorporated into the DNA chain by mistake. This alteration in turn causes mistakes in RNA and protein synthesis. There are many other chemical mutagens, but in most cases we have been unable to establish precisely how these compounds produce their effects.

Forms of energy that disrupt the chemical structure of chromosomes—ultraviolet light, X-rays, gamma and beta radiation, and cosmic rays—are also mutagenic. Early studies of the effects of radiation showed a simple linear relation between the total radiation dose and the frequency of mutations; the exposure rate seemed to make no difference. In other words, a given radiation dose had the same effect whether administered over a long time in small amounts or in a single large dose (Figure 15-16). However, it is extremely difficult to determine whether this relationship holds true for very low dose rates, rates barely above the normal background radiation caused by cosmic radiation and ultraviolet light. Extended exposure to low radiation levels may or may not be as mutagenic as short exposure to strong radiation, when the total amount of radiation is the same. The expected increase in mutation frequency is very small and therefore is difficult to measure precisely.

The problem of extended exposure to low background radiation is an important one, because human activities are constantly increasing the background levels of environmental radiation. Considerable public concern has been expressed about the genetic implications of this additional radiation burden. The individual's exposure to high-energy radiation in industrialized countries is about double the natural background radiation. Most of this additional radiation comes from sources such as the X-radiation used in medicine; roughly one percent is from bomb testing and other atomic energy applications. The critical question is whether this level of radiation represents a mutation hazard.

We know that normal background radiation by itself cannot account for the levels of mutation in

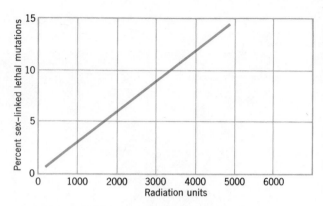

15-16 *The frequency of lethal mutations depends on the amount of radiation administered.*

man and other animals. If we take the data for mutation caused by high dosages, and if we then extrapolate back to the low dosage levels found in the environment, we find that these dosages correspond to a mutation rate lower than the one that actually occurs. This calculation suggests that the observed mutation rate is caused by something more than normal background radiation.

This calculation is the basis of the following argument about atomic energy: (1) since natural background radiation is *not* the most important source of natural mutation, and (2) since man has only doubled that background rate in industrialized countries, then (3) the increase in mutational rates that we are producing must still be—at the very most—less than twice the natural rates. If we add to these considerations the fact that atomic energy sources are only a tiny fraction of the added radiation in industrialized countries, it would seem that public fear about radiation pollution from atomic energy is exaggerated.

There are some flaws in the argument, however. First, additional radiation is not evenly distributed throughout the environment. For example, a radioactive isotope of strontium (Sr^{90}) is actively concentrated in organisms. Sr^{90} is a long-lived by-product of nuclear explosions. The body treats it like calcium and deposits it in developing bone tissue. Radiation from Sr^{90} is thus concentrated and localized near the blood-forming tissues in bone marrow. Obviously we cannot judge the effect of Sr^{90} by assuming that it is evenly distributed in existing background radiation. Many other qualifications must be added to any calculations of changes in the total amount of high-energy radiation.

In conclusion, we really don't know what we are adding to the genetic burden of humanity with high-energy radiation. There is probably some addition, and it may not be "unacceptably" large. But if we increase the mutation frequency at all, the effects will persist for many generations; even a small increase in percentage could affect the lives of thousands of people yet unborn. How, then, can we designate any level of increase as "acceptable"? Obviously, this is a social and political judgment, not a biological one. Decisions about industrialization and all its accompanying effects are based on priorities; and the values of goods, services, and luxuries are considered along with any ethical ideas about increasing the mutation frequency in future generations.

Useful Mutants

The study of mutants is invaluable for understanding the functions of genes. It also has immediate economic importance—it is possible to introduce new genes into a population and thereby create desirable heritable traits. This technique is particularly important in plant and animal breeding. The biggest difficulty is simply devising ways to detect mutants once they have been produced. H. J. Muller devised a number of techniques to detect *Drosophila* mutants created by radiation mutation. This is a complex problem; Muller subsequently received the Nobel prize for his work.

The detection of mutants in microorganisms is a simpler procedure. Bacteria are usually grown on a semisolid medium containing all the nutrients necessary for growth. Agar is added to solidify the medium. A dilute suspension of bacteria is spread over the surface of a flat agar plate. Each viable bacterium then grows and divides until it produces a visible group of bacterial cells, called a colony. If the original suspension was sufficiently dilute, each colony is the product of a single bacterium. Any genetic differences among the original bacteria will show up in every cell of a whole new colony, since they have all arisen by mitosis. A colony produced asexually from a single parent individual is called a **clone.** A certain number of the parent cells, and thus of the resulting clones, will be mutants.

These mutants can be detected in various ways, depending on how the mutation affects a cell. If a mutant can no longer synthesize a certain cell-wall protein, for example, detection may be easy, because the entire colony may look different. But in other cases, mutation affects more subtle metabolic characteristics, and the cells must be subjected to various physical and chemical environments in order to reveal the metabolic change. For example, if samples of different clones are transferred to various growth media (ones lacking certain nutrients or containing certain poisons) or are subjected to different temperatures, some mutant clones might survive where the normal cells cannot.

Some techniques for the genetic manipulation of microorganisms do not depend on random mutations. Specific bacterial mutants can be created using extrachromosomal bodies called **plasmids.** Plasmids are often transferred normally from cell to cell during division, or they can be transferred to microorganisms by a method resembling transduction, the transfer of bacterial DNA by a phage virus. Once in the host cell, the DNA of the plasmid may be duplicated many times, and correspondingly large quantities of a specific protein are obtained from the cells. We can select bacteria with particular genes in the plasmid bodies. In this way, it is possible to create organisms that produce excess quantities of a particular enzyme. Or, we might develop microorganisms that can degrade insecticides or plastics so that they will not persist in the environment. Speculating still farther, it should even be possible to design specific proteins, create DNA strands that code for their production, and then insert these manmade genes into a suitable microorganism. In fact, this has already been done using a simple code (AAAA. . .) for a simple protein containing only lysine amino acids. But our understanding of protein chemistry is still primitive, and we cannot yet design and produce complex, functional proteins.

Mutant cells are currently being created through genetic surgery on individual cells. The cells are grown in culture, and they are harvested when their chromosomes are most visible, in late prophase or metaphase. A fine laser beam is then used to cut off selected parts of chromosomes in certain cells. The DNA is disrupted and deleted at the point of irradiation (Figure 15-17).

Many useful plant mutants have been made with high-energy gamma radiation. Mutant strains of rice and wheat with stronger stems, or more protein, have been produced. Several research centers are involved in plant mutant production. One center in Japan grows plants in a large circular field shielded by a high wall. A source of gamma rays is placed in the center (Figure 15-18).

Some social critics and biologists are afraid that the ability to produce mutants and to control

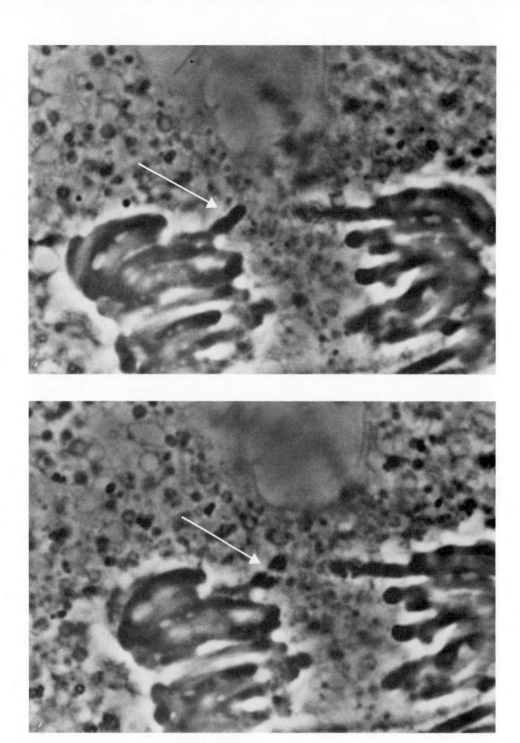

Figure 15-17 Genetic surgery. Portions of a single
chromosome can be removed with a fine laser beam.
The upper photograph shows the intact chromosome;
the lower photograph shows the same chromosome
after its tip has been broken off at the point indicated
by the arrow. Magnification 2,500×.

Figure 15-18   Experimental field of plants at the Institute of Radiation Breeding in Japan. Plants are exposed to a source of radiation at the center of the field; the high wall serves as a radiation shield.

changes in genotype will permit us to intervene in human affairs in unprecedented ways. Our present techniques for "genetic engineering" are still primitive, so in this case, speculations about the creation of breeds of supermen or submen are extrapolations into the distant future. Yet the concern about genetic information and how it interacts with social policy is timely. We will turn to some of these issues later in the chapter.

environment and genotype

The appearance, or phenotype, of the organism is not always a direct expression of its genotype. The environment can have profound effects on physical appearance. There are many examples of how temperature, light, diet, or other environmental stresses can influence the expression of genetically-coded characteristics. The color pattern of Siamese cats is genetically determined, for instance, but its

expression depends on the temperature. If a Siamese cat is kept at a relatively high temperature, all of its fur grows out light in color. However, at cooler temperatures, the characteristic color pattern develops: dark fur on its ears, face, paws, and tail.

In humans, patterns of growth and development are defined by genotype, yet their expression can be strongly affected by an impoverished environment. Diet has a profound effect on human growth. The weight gain of a group of undernourished children from Nigeria was compared to the growth of a genetically comparable group that had been provided with an adequate diet. Both groups were from the same tribe, and both lived in the same town. The weight gain was much lower in the first group (Figure 15-19). Obviously, the individual's genetic potential for growth is not fully expressed in a population with an inadequate diet. And, as we have seen, some of these developmental effects are permanent. Low protein intake early in development produces irreversible mental retardation, and the lost ground is never regained even if protein supplies are sufficient in later life. In a recent essay incisively titled "Food for Thought," Francis Keppel

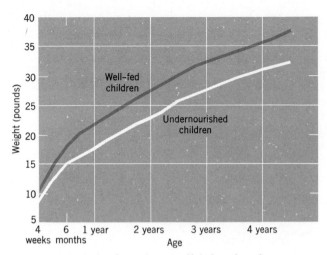

Figure 15-19 *Weight gain in well-fed and under-nourished children. Both groups are from the same tribe in Nigeria.*

estimates that more than two thirds of all human children in the world are "at risk" of such impaired development.[1] Through malnutrition, we are currently producing frail and mentally retarded humans on a massive scale.

Dietary factors are also critical in the expression of genetic diseases such as diabetes. Diabetes appears to be hereditary, though the details of its inheritance are unclear. We do know that many susceptible individuals will not develop a clinical case of diabetes, one that requires insulin treatment, if they follow a special diet. In cases like diabetes, suitable rearrangements or restrictions of the environment permit the individual to develop a normal phenotype.

An individual whose phenotype has been strongly modified by the environment is sometimes called a **phenocopy.** In other words, his phenotype is a copy of the phenotype of an individual with a different genotype. Individuals with dietary control of diabetes are phenocopies of normal individuals. The term is used in the reverse sense, too, to describe individuals with normal genotype whose phenotype mimics that of an affected individual. A

[1] The essay appears in Scrimshaw, N. S., and J. E. Gordon, eds., *Malnutrition, Learning, and Behavior.* Cambridge, Mass.: M.I.T. Press (1968).

diabetes phenotype, for example, can be produced by destroying the beta cells of the pancreas; the resulting individual is a phenocopy of the person whose diabetes develops because of heredity.

Human populations contain many individuals with genetic defects who are phenocopies of normal individuals. It is sometimes argued that mankind is burdened with harmful genes because certain genetically abnormal phenocopies are able to survive and reproduce. Such individuals are kept alive by modern medicine and public health procedures. They are now able to grow to maturity and reproduce, and as a result the human genetic endowment is "deteriorating." It is certainly true that such individuals can now contribute their genes to the genetic pool by producing offspring, whereas formerly they could not. But the idea that they are less "fit" is true only in a hypothetical environment that does not include our present medical and public health facilities.

Individuals with poor eyesight or diabetes can survive very well in an environment that makes eyeglasses and insulin available. Of course, one can argue that if the environment were to change—if a social crisis disrupted medical services, for instance—they would not survive. But other kinds of environmental change would just as surely eliminate or favor other kinds of individuals because of their genotypes. For example, in a cold environment, or under starvation conditions, individuals who tend to be obese due to low metabolic rates could have a survival advantage. But because it is impossible to predict cataclysmic changes in the environment, it is also impossible to say what characteristics will be most important. There is no good reason to focus on some individuals' requirements for medical services, as long as these services are a stable part of the environment. A restriction of breeding by these individuals (which is the usual recommendation), in anticipation of a primitive or impoverished society, seems unjustified.

Interactions between genotype and environment are universal. Genes are not expressed traits; they are genetic instructions. Genes can be expressed only by directing the amino-acid sequences in proteins. And the supporting machinery for this expression is dependent on the environment. The same genotype may produce many different phenotypes, depending on the environment. But at the same time, the range of potential phenotypes is limited by the genotype. These genetic limitations are stringent for some characteristics and broad for others.

Genetics began as an effort to understand individual inheritance patterns. Yet the first genetic information actually came from large populations of organisms, not from individuals. When Mendel wanted to know the frequency of trait Y, he examined a whole population. The answer then provided a model that helped him understand inheritance in the individual. Many questions are related to the genetics of the whole population, rather than to the genotypes of its member individuals.

Stable Populations

A **population** is defined as any group of interbreeding individuals. For this discussion, we will assume that the group is fairly large, and that the nature of the population is unchanging. That is, all the individuals in the population have an equal chance for survival, and breeding is completely random — all possible male-female breeding combinations are equally likely. These restrictions are actually not unnatural; some real populations behave in just this way. A population of marine fishes with coordinated external fertilization would be a good example.

The distribution of genes in a stable population is described by the **Hardy–Weinberg Law.** This law simply states that in a stable population, the overall frequency of genes does not change; that is, the relative numbers of the various genotypes remain the same. The Hardy–Weinberg Law arises from a simple analysis. All genes in the population are part of the population's **gene pool.** If breeding is random, then each adult male and female contributes equally to the pool, and new individuals arise randomly from the pool. The probability that any given gamete will have a particular gene is the same as the frequency of that gene in the gene pool, and therefore in the whole population. No change in gene frequency will occur.

If we know the frequency of particular alleles in the gene pool, then the Hardy–Weinberg Law provides a formula for calculating the frequency of homozygous and heterozygous individuals. Say that p is the frequency of the dominant allele A, and q is the frequency of the recessive allele a. Then the probability that A will occur in a given egg is p, and the probability that a will occur is q. Similarly, the probabilities for the occurrence of A and a in sperm are also p and q. If the gene is not a, then it must be A. Expressed in probabilities, (the chance of A) + (the chance of a) = 1. That is, $p + q = 1$. Or $p = 1 - q$.

If we express these probabilities in the usual 2×2 array, we get the familiar chart:

	p	q
p	pp	pq
q	pq	qq

The frequency of AA zygotes in the population will be $p \times p$, or p^2. The frequency of the Aa's will be equal to $2\,pq$; and the frequency of the aa's will be q^2. These probabilities add up to the mathematical expression $p^2 + 2pq + q^2$, which, we know from algebra, is equal to $(p + q)^2$. Since $p + q = 1$, then $p^2 + 2pq + q^2 = (p + q)^2 = 1$.

Assume that the dominant and recessive alleles for our hypothetical gene have equal frequencies in the population, that is, that p and q both equal $\frac{1}{2}$. Then p^2 and q^2 both equal $\frac{1}{4}$, and $2pq$ equals $\frac{1}{2}$. We can predict that one quarter of the individuals would be genotype AA, half of them would be genotype Aa, and one quarter of them would be genotype aa.

A Hardy–Weinberg calculation is valid for any gene frequency. Suppose we have a recessive phenotype that occurs once in every 100 individuals. This information is sufficient for us to calculate gene frequencies in the whole population:

(1) We have just stated that $q^2 = 0.01$. Therefore $q = 0.1$.
(2) $p = (1 - q)$, or 0.9.
(3) Therefore, $p^2 = 0.81$ and $2pq = 0.18$.

By these calculations, 81 percent of our population is homozygous for the dominant trait, and eighteen percent shows the dominant phenotype but is heterozygous. One percent is homozygous for the recessive trait. Notice that many more of the recessive genes are found in heterozygotes (18%) than in homozygotes (1%). The reason is not that the gene is recessive, but simply that the trait we have chosen has a relatively low frequency. If the situation had been reversed — if only one out of 100 individuals in our hypothetical population had been homozygous for the *dominant* trait — then the numbers would have come out the same except that now 81 percent of the population would have had the recessive phenotype.

This point is important when we consider recessive traits that are harmful. As recessive alleles decrease in frequency in the gene pool, proportionately more of the ones that remain will be found in heterozygotes. As we have seen, if the frequency of the recessive gene $q = 0.1$, then the ratio of heterozygotes to homozygotes is $2pq/q^2$, which we have found to equal 18/1. But when $q = 0.001$, the ratio $2pq/q^2$ is much larger. The value of q^2 becomes 0.000001; and $2pq = 2 \times 0.999 \times 0.001 = 0.001998$.) The ratio therefore is 1,998/1.

For this reason, the parents of homozygous individuals with rare recessive genetic deformities are themselves almost always normal in appearance, because they are almost always heterozygotes.

Traits such as albinism (a gene that blocks the formation of melanin pigment) are found in only one of 20,000 individuals. Yet for each homozygous recessive individual, there are more than a hundred heterozygous individuals in whom the gene is masked by a dominant one. As a result, it is virtually useless to try changing the gene frequency in a population by preventing the breeding of individuals who show a rare homozygous recessive trait.

Inbreeding

The Hardy–Weinberg Law describing gene frequency assumes that all breeding pairs are equally likely. But in most populations breeding is not random. Breeding is more likely to occur between neighboring individuals. This fact means that the gene pool is really broken up into small sub-pools. When individuals in any sub-pool are closely related, inbreeding occurs. **Inbreeding** is the production of offspring by closely related parents; obviously, there are many degrees of inbreeding, depending upon how closely related the parents are.

Inbreeding will tend to decrease the frequency of heterozygous individuals, and therefore will increase the proportion of homozygotes in the population. This frequency change is most obvious in an extreme case, self-fertilization. A corn plant, for example, produces both male and female gametes and is capable of repeated self-fertilization. We start with a plant that is heterozygous for an allele pair (Aa), and we breed it with itself. The plant's male and female gametes will carry A and a in equal numbers, and the first generation will be one-half homozygous (AA and aa) and one-half heterozygous (Aa). This result is just what you would expect in a stable population that obeys the Hardy–Weinberg Law, where $p = \frac{1}{2}$ and $q = \frac{1}{2}$. But when the plants of this first filial (F_1) generation are themselves self-fertilized, there is a different result. The homozygous parents produce homozygous progeny

(since they have only one kind of gamete). But the heterozygous parents segregate gametes in the usual way and produce the expected 1/2/1 ($AA/Aa/aa$) ratio in their progeny. Thus, overall, the progeny in the second filial (F_2) generation show this pattern:

- $\frac{1}{2}$ are homozygous because they are bred from self-fertilized F_1 homozygous individuals
- $\frac{1}{4}$ are homozygous offspring of the heterozygous F_1 individuals
- $\frac{1}{4}$ are heterozygous offspring of heterozygous F_1 individuals.

Homozygous individuals made up half of the F_1 generation; now they are three quarters of the F_2. At each succeeding generation the homozygous individuals produce only homozygous offspring. As long as self-fertilization continues, their offspring cannot become heterozygous. And in the meantime, the heterozygous individuals are adding to the fraction of homozygous individuals in each succeeding generation (Figure 15-20). This effect does *not* change the frequency of genes in the population as a whole. But it does change the ratio of homozygotes and heterozygotes. The Hardy–Weinberg prediction based on a balanced population no longer applies, because the starting gene pool is very small and mating is not random.

Self-fertilization is impossible for most organisms. However, other, more remote patterns of

Figure 15-20 Changes in a population caused by inbreeding. Inbreeding increases the percentage of homozygous individuals in the population; the closer the relation between the parents, the more rapid the increase will be.

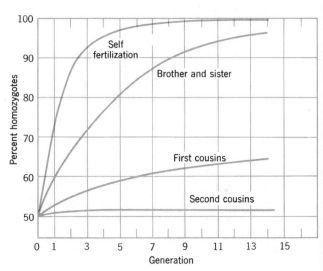

inbreeding (matings of brothers and sisters, first cousins, and so forth) will cause the same steady decrease in the heterozygosity of the population. The effect nearly disappears at the level of second cousins.

This decrease occurs in all the allele pairs of the genome. Consequently, if inbreeding is combined with the selection of progeny having certain characteristics (that is, if one of the two kinds of homozygotes is systematically eliminated), a very uniform population can be obtained. Inbred mouse strains are so uniform that skin transplants and organ transplants between individuals are possible. Inbreeding is also responsible for the stable and predictable characteristics of purebred domestic animals and pets.

Arguments against incest in human populations have a sound biological basis. As we have seen, heterozygous carriers for recessive genes are quite common in the general population, but homozygous recessive individuals are usually rare. Inbreeding, however, changes these numbers. The number of homozygous recessive individuals increases, while the number of heterozygotes decreases. The human population carries a large genetic burden of undesirable recessive alleles, and these recessive alleles would increasingly find expression in homozygous individuals. Taboos against incest tend to minimize the number of homozygotes that appear in the population.

The development and maintenance of inbred populations are important parts of modern agriculture. For example, many inbred strains of corn are maintained to provide seed corn for planting. The inbred strains are often small, low-yield plants, and are not used directly for crops. But they are invaluable for producing a desirable hybrid seed. These inbred plants are genetically uniform, and a large number of their traits are homozygous. As a result, a corn breeder can select a strain with a particularly desirable characteristic, such as a high resistance to a parasite. In addition, when two inbred strains are crossbred, the progeny are often larger and have a higher yield than each of the parental strains. This phenomenon is called **hybrid vigor.** There are a number of hypotheses to explain hybrid vigor, but none is completely satisfactory. If the hybrid itself is inbred, hybrid vigor is lost—the increases in size and yield slowly disappear (see Figure 15-21).

Most corn grown in the United States comes from hybrid seed. It usually starts with two crosses (Strain A × Strain B, and Strain C × Strain D). The offspring of each cross show hybrid vigor and have the specifically desired combinations of yield, disease resistance, growth habits, and soil and water

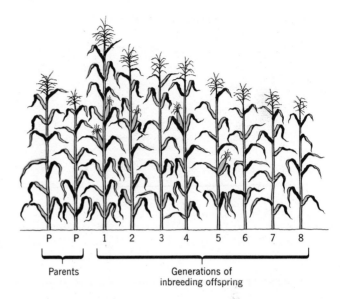

Figure 15-21 *Hybrid vigor results from a cross of two inbred strains. It is lost in subsequent inbred generations.*

requirements appropriate to a particular growing area. Then the two crosses are themselves crossed (AB × CD), and the seed from the resulting plants is used for planting (see Figure 15-22). This sort of hybrid seed has increased corn production significantly.

Breeding is a continuing procedure. It is not possible to achieve a permanent "perfect" seed for an area, because the biological environment in an area gradually changes. New plant parasites may appear. A plant resistant to one strain of mold may be easy prey to another. The plants themselves change in subsequent generations. Hybrid vigor declines and new crosses must be made constantly.

Hybrid strains have new combinations of characteristics that are already present in the parent strains, but the total variety of these characteristics does not change. For this reason, it is important to maintain the largest possible diversity of stocks, so that the breeder can respond to changed requirements with new strains. In 1970, despite the efficiency and sophistication of corn breeding, the United States corn crop still was devastated by a fungus, the Southern corn blight. Fortunately, our collection of inbred strains contains a wide selection of resistant varieties so that new resistant hybrid seed could be developed.

Biologists and agriculture specialists are anxious to preserve as much variety in crop plants as possible. Existing natural populations in various parts

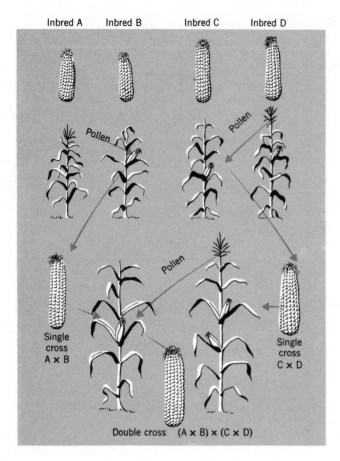

Inbred A   Inbred B   Inbred C   Inbred D

Pollen

Pollen

Single
cross
A × B

Pollen

Single
cross
C × D

Double cross   (A × B) × (C × D)

Figure 15-22   The production of hybrid corn.

way. The sperm are then mixed with eggs from a female. The sperm activate the eggs quite normally, so that the egg nucleus undergoes a final meiotic division resulting in two sets of haploid chromosomes. In the normal course of events one set of chromosomes would be incorporated into a polar body and be lost; the other would combine with chromosomes from the sperm to yield a diploid zygote. However, in this system, the experimenter interrupts the proceedings by plunging the egg cells into freezing-cold water just after the egg nucleus has doubled its chromosome number for division. This shock prevents the completion of cell division in the egg, and no polar body is formed; instead the egg nucleus remains diploid. The sperm nucleus cannot contribute genetic material, because its chromosomes were destroyed. But the egg is diploid and can develop into a normal fish. This fish is entirely homozygous, since it contains two identical sets of chromosomes from its mother (Figure 15-23).

As the homozygous daughter fish mature, they are selected for growth rate, sturdiness, adult weight. (Even though they had the same mother, they will be different, because every egg received a slightly different assortment of haploid chromosomes from the mother.) Particularly desirable homozygous fish can then be selected and the procedure repeated. But this time all the eggs will be identical to their mother and to each other. In this way a purebred line is achieved in just two generations. In essence, the technique activates an egg while it is still diploid, so that it will develop parthenogenetically.

Such work is paving the way for the rapid and economical production of edible fish that can be bred for size, growth rate, palatability, resistance to extreme temperatures, and other qualities. It is quite analogous to the production of inbred corn strains, since only one parent is involved in the breeding program.

human genetics

Mendel's laws of segregation and independent assortment were based on characteristics that were easily recognized. Each trait was controlled by a single allele pair, and one allele of each pair was clearly dominant. The allele pairs were on different chromosomes, so no linkage was observed. But as Mendel's observations were extended and

of the world are still the main reservoir of genetic variability. For example, inbred strains of wheat derived from natural Japanese wheat populations were used to develop a high-yield hybrid wheat. In the 1960's, this hybrid strain was responsible for dramatic increases in food production in Mexico and other countries, including India and Pakistan. The disappearance of such native strains and their total replacement by hybrid seed could be disastrous.

Inbreeding is also important in animal agriculture. A recent example is a technique devised by British scientists, which permits the inbreeding of pure lines of fish in only two generations. The method is ingenious. First they take sperm from adult male fish and expose it to a heavy dose of cobalt radiation. The radiation disrupts the chromosomes but does not harm the sperm in any other

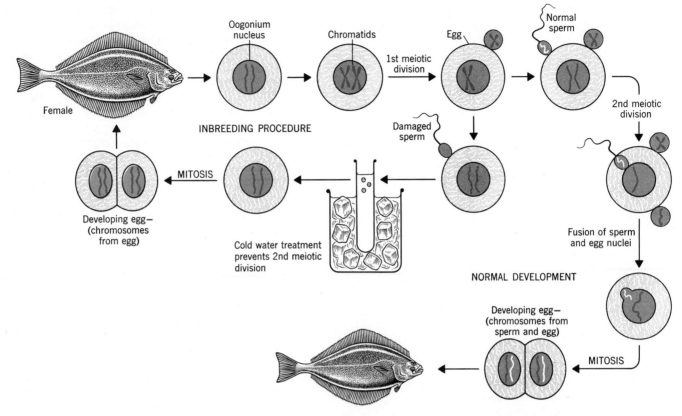

Figure 15-23 *A method for obtaining inbred fish. The egg is activated by damaged sperm. The second meiotic division in the egg nucleus is prevented by exposure to cold. The egg develops as a diploid zygote with both sets of chromosomes derived from the egg.*

thoroughly documented, the observational data became more and more complex. The presence of linkage groups and crossing over made it possible to map chromosomes and even to analyze the fine structure of single alleles. Such research required the processing of millions of individual fruit flies, and the examination of millions of colonies of microorganisms.

Modern ideas of heredity are based on such concepts as incomplete dominance, multiple alleles, and the balance between alleles in the whole genome, but they can still be interpreted in Mendelian terms. The inheritance of some characteristics is so complex, however, that a specific analysis is nearly impossible. Not only are inheritance patterns complicated in themselves, but the phenotypic expression of the characters may be sensitive to environmental influences. At some point it is no

longer practical to seek precise explanations. Suppose, for instance, that there are six allele pairs concerned with skin pigmentation in man. There is, in fact, no practical way in which we can obtain enough data to demonstrate that there are six allele pairs and not five or seven.

The vast number and complexity of organisms, as well as the number of genes in each, impose significant limitations on the study of genetics. There are ten million different kinds of organisms on earth, and the genetic analysis of just one of them, the fruit fly, required thousands of man-years. Obviously, we will never be able to study the genetics of the vast majority of organisms directly. Even organisms very important to us—including ourselves—can not be thoroughly analyzed. A geneticist who tried to conduct human breeding experiments would first have to overcome some rather severe social difficulties and very strong objections to controlled mating. Then he might face even stronger social concerns about incest. And even if his determination and guile overcame these obstacles, he would still have to wait several decades for the outcome of his experiment.

Genealogy and Concordance

Controlled matings, such as brother–sister crosses or backcrosses to the parental stock, are obviously impossible in humans. Such socially disapproved matings do occur, but they are not under the control of any experimenter, and it is very difficult to gather accurate information about them. For this reason, family trees have become very important in the study of human genetics. The study of human genetics is largely historical rather than experimental.

Genealogies—family trees—have enabled us to establish simple inheritance patterns for some characteristics, such as eye color or ABO blood group antigens. And they are particularly important and useful in medical genetics. Through family histories, hundreds of diseases and developmental defects have now been classified as either dominant or recessive alleles, and have been analyzed for sex-linkage. Prospective parents can now consult a genetic counselor to obtain information on the probability that they will transmit a particular trait to their offspring; parents with one defective child can obtain an estimate of the chances that subsequent children will have similar defects.

Genealogies are not very informative, however, about many other characteristics. In some cases, information about a characteristic may be unreliable, or simply unavailable. In other cases, patterns of inheritance are so complicated that we can no longer be sure the characteristic in question is actually inherited. In such circumstances, other techniques may be used to gather information.

The method of concordance, for example, simply compares the rate at which a phenotypic characteristic appears in closely-related individuals with the rate at which it appears in unrelated individuals. **Concordance** itself is defined as the simultaneous presence (or absence) of a trait in paired individuals. For example, if we examine ten pairs of twins, and a certain characteristic appears in both twins in eight of the ten pairs, then that characteristic's concordance is 80 percent. For many characteristics, such as blood pressure or age of puberty,

what constitutes a concordance has to be defined arbitrarily. Usually the characteristic is said to be present if it falls within some narrow range of variation. Figure 15-24 presents the concordance of various traits in identical twins (twins resulting from a single fertilized egg) and fraternal twins (twins resulting from the fertilization of two eggs). Concordance is greater among the identical twins for every trait listed.

By themselves, however, concordance data say very little about patterns of inheritance for the characteristic in question. They do not even provide clear evidence for Mendelian inheritance. Concordance for Mongolism in identical twins, for example, is very high because the twins come from a single fertilized egg. However, we know that Mongolism is actually caused by aneuploidy, not by the inheritance of a specific genetic characteristic. The inheritance of such characteristics as criminality or smoking is even more uncertain. Since identical twins face more nearly identical environments than do fraternal twins, identical twins would be more likely to respond in similar ways. Concordance alone does not allow us to estimate the relative importance of heredity and environment.

Calculations of **heritability** are used to give a mathematical estimate of the amount of genetic, as opposed to environmental, influence on a variable characteristic. **Heritability** is defined as the proportion of phenotypic variability in a population that is due to differences in genotype. For example, the height of individuals in a population shows considerable variability. Some of the variability is controlled by genetic differences among individuals and the remainder is caused by differences in their environment. The actual proportion that is caused by genetic factors is the heritability of height.

Heritability

Estimates of heritability begin by measuring the range and variability of a trait (such as height, weight, or (IQ) in pairs of genetically related individuals. Since we know the degree of kinship between the individuals, we can predict how much variability would result if all of it were caused by genetic differences. If the actual measured variability then turns out to be greater than the prediction, we can assume that the difference between the two values is due to environmental influences.

In heritability studies, the variability of a complex trait in related and unrelated individuals in a

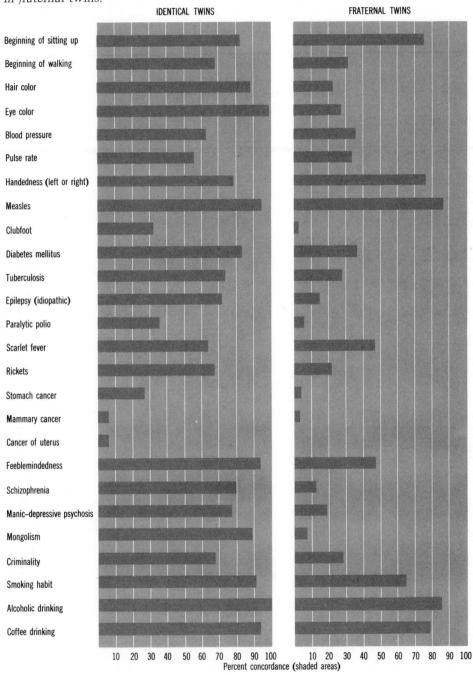

Figure 15-24 The concordance of various traits in twins. Concordance is the percentage of cases in which both twins possessed the trait in question. Concordance in identical twins is always higher than it is in fraternal twins.

IDENTICAL TWINS   FRATERNAL TWINS

Beginning of sitting up
Beginning of walking
Hair color
Eye color
Blood pressure
Pulse rate
Handedness (left or right)
Measles
Clubfoot
Diabetes mellitus
Tuberculosis
Epilepsy (idiopathic)
Paralytic polio
Scarlet fever
Rickets
Stomach cancer
Mammary cancer
Cancer of uterus
Feeblemindedness
Schizophrenia
Manic–depressive psychosis
Mongolism
Criminality
Smoking habit
Alcoholic drinking
Coffee drinking

10 20 30 40 50 60 70 80 90 100   10 20 30 40 50 60 70 80 90 100
Percent concordance (shaded areas)

single population—predicted or measured—is often expressed by a **correlation coefficient.** When the numerical value of one characteristic allows us to predict the exact value of another, then the two characteristics are said to be perfectly correlated. The correlation coefficient is 1. When the value of one characteristic gives us no information at all about the other, there is no correlation, and the correlation coefficient is 0. A specific example will make the concept clear. If we could use a person's height to predict his weight with 100-percent accuracy, the two characteristics would be perfectly correlated and would have a correlation coefficient of 1. If height gave us no idea at all of weight, then there would be no correlation. Actually, of course, neither situation is the case: a person's weight is related to his height, but it cannot be accurately predicted from height alone. These two characteristics therefore have a correlation coefficient somewhere between 0 and 1.0.

Correlation coeffecients are also used to express the variability of the same trait in different individuals. The height of one member of a set of identical twins, for instance, allows a quite reliable prediction of the height of the other member; in this case the correlation is high. By contrast, the height of one person in the population provides no information for predicting the height of an unrelated person.

Table 15-2 gives measured correlation coefficients for mental ability in various pairs of individuals. The closer their relationship, the higher the correlation—that is, the more likely they are to have the same mental ability, as measured by various IQ tests. These numbers are then compared to predicted values calculated on the basis of two different proposed patterns of inheritance. The first value is based on the assumption that mental ability is inherited through incompletely dominant alleles. The second value assumes that mental ability depends on the inheritance of many equally dominant genes. There is reasonable agreement between the measured correlation coefficients and both of the theoretical predictions. This correspondence suggests that genotype is important in defining mental ability. But the agreement is far from perfect. The fact that there are differences indicates that mental ability also depends on environmental influences.

More specifically, the table shows that identical twins reared apart have a measured correlation of 0.75. Since they have identical genes, the predicted correlation would be 1.00. The difference (1.00 − 0.75 = 0.25) is attributed to environmental influ-

ences. At the other extreme are unrelated individuals who have been reared together. In this case the predicted coefficient is 0.0, because the individuals are unrelated. The measured correlation, however, is 0.24. The difference, again, must be the result of the environment. These two estimates of environmental influence—0.25 and 0.24—are remarkably similar. They both say that about 25 percent of the variability in mental ability is due to the environment, and 75 percent to genetic factors. That is, the heritability of IQ is about 75 percent.

Other estimates of the heritability of mental ability are more complicated, because they try to consider various sources of error. First of all, the measures of mental ability—the IQ tests—are themselves somewhat unreliable. Another problem is that the measured correlation coefficients may be too high, because the environment and genotype are not independent variables. For example, a child who has inherited a high mental ability is likely to have parents with similar abilities. The child will be helped and encouraged; it is likely to do better on IQ tests than it would if it lived with people of lower mental abilities. This phenomenon, in which environment and heredity reinforce each other, is called **covariance.** Covariance tends to mask the effects of environment on phenotype, because genotype and environment covary in the same direction.

Mathematical procedures can be used to estimate and separate these sources of error. When the figures in Table 15-2 are adjusted, mental ability turns out to have a heritability of about 0.8. In other words, about 80 percent of the phenotypic variation in mental ability can be assigned to genotypic factors, and about twenty percent to environmental variables. The specific heritability value obtained depends on which test of mental ability is used, and it also depends on the population studied. Heritability may be higher or lower for other populations. Table 15-2 represents about 30,000 paired values summarized from many different studies and several different tests. They come from studies of European and North American white populations, and cannot be generalized to other populations. The table is taken from a famous review of heritability of intelligence by Arthur R. Jensen, published in 1969.

Social Implications

Information on human heritability has a tremendous impact on some kinds of political and

Table 15-2

Correlations for Intellectual Ability[1]

Correlations Between	Number of Studies	Obtained Correlation Coefficient	Theoretical Correlation Coefficient[2]	Theoretical Correlation Coefficient[3]
Unrelated Persons				
Children reared apart	4	−.01	.00	.00
Foster parent and child	3	+.20	.00	.00
Children reared together	5	+.24	.00	.00
Collaterals				
Second cousins	1	+.16	+.14	+.063
First cousins	3	+.26	+.18	+.125
Uncle (or aunt) and nephew (or niece)	1	+.34	+.31	+.25
Siblings, reared apart	33	+.47	+.52	+.50
Siblings, reared together	36	+.55	+.52	+.50
Dizygotic twins, different sex	9	+.49	+.50	+.50
Dizygotic twins, same sex	11	+.56	+.54	+.50
Monozygotic twins, reared apart	4	+.75	+1.00	+1.00
Monozygotic twins, reared together	14	+.87	+1.00	+1.00
Direct Line				
Grandparent and grandchild	3	+.27	+.31	+.25
Parent (as adult) and child	13	+.50	+.49	+.50
Parent (as child) and child	1	+.56	+.49	+.50

[1] After Arthur R. Jensen, "IQ and Scholastic Achievement," in *Harvard Educational Review* Vol. 39, No. 1 (1969).

[2] Assumes non-random mating and partial dominance.

[3] Assumes random mating and only additive genes; in other words, the simplest possible multiple-gene model.

social issues. Often, biological information is misunderstood or misused; it may even be suppressed. One kind of misuse occurs when environmental and genetic effects are confused.

We have already introduced one example. Most people in the world cannot get enough protein for optimum physical and mental development. In such a population the heritability of height, weight, and IQ would be low compared to their heritability in an adequately nourished population. In other words, under the conditions of poverty and deprivation that are typical for most human beings, the availability of adequate food is more significant than heredity.

Comparisons between populations are meaningless unless there is evidence that the heritability of the characteristic in question is similar in both populations. Most biologists are aware of this limitation of the use of heritability. But it is often ignored in nontechnical discussions. Such discussions confuse environmentally-produced differences in one population with heritable differences in another. The result can be a pseudo-scientific justification for racist and elitist social policy.

Another example illustrates how heritability studies can become distorted, and even suppressed, when they bear on controversial social issues. Arthur Jensen's review of intelligence illustrates the difficulties encountered in discussing human heritability. The purpose of his review was to summarize evidence about the success of remedial programs (such as Project Head Start) that try to raise the IQ's of participating students. Jensen considered two major questions: to what extent is IQ modified by environmental factors? What are the characteristics of the IQ measurement itself? Jensen reviewed these topics thoroughly, and took great care to emphasize that many sources of error must be considered when drawing conclusions from calculations of heritability. He also reviewed the evidence for racial differences in IQ, again describing sources of error and withholding final judgment about the meaning of these studies.

Jensen's article produced reactions of explosive violence. Within days, his review was introduced into evidence in the South as an argument against school integration, on the ground that it proved the existence of racial differences in intelligence and that these differences were genetic. Yet Jensen had drawn no such conclusion. Just as quickly, he was personally abused and labelled a racist by liberal and radial political groups. The popular press produced summaries of his review that had little to do with its actual content. Perhaps most interestingly, the National Academy of Sciences decided that it was inappropriate to sponsor more detailed and sophisticated studies of the heritability of intelligence: the issue was, and is, simply too explosive.

We can phrase the basic problem in the following terms. We like to think that our society is egalitarian — each individual should have the same opportunity for the pursuit of happiness. There is a strong correlation between IQ and social and economic status; low IQ is associated with low social-economic status. To the extent that IQ is heritable, we really do not have equal opportunity; persons with a low IQ are destined to be poor. For many individuals, the freedom to pursue happiness is the freedom to fail — a most unpleasant conclusion. One way to reestablish equal opportunity would be to modify society so that discrepancies in social status are reduced. But this process would represent such a sweeping change in our society that most people will not even contemplate the possibility. It is easier to try to equalize opportunity by spending public funds on remedial education while preserving differences in status. If someone then points out that this procedure does very little to solve the problem, he becomes extremely unpopular. This is precisely what has happened to Jensen. His review has enraged almost everyone, whatever his race, by suggesting that attempts to equalize opportunity through efforts to improve education may not equalize social and economic status. A neutral discussion of the heritability of intelligence is impossible, and further research is likely to be discouraged by both political and academic groups for a very long time.

instinct

Other aspects of human behavior, in addition to mental ability, are subject to similar controversies about the relative importance of heredity and environment. Many animal species have elaborate behavior patterns that seem to be unlearned and therefore part of the animal's genetic instructions. When the proper environmental circumstances are encountered, the behavior is triggered. These patterns are fully formed the first time they occur, and they may even appear in animals reared in isolation. Such behavior is called **instinctive behavior.**

There are many fascinating examples of instinctive behavior. A particular kind of hunting wasp, for instance, preys only on a particular kind of grasshopper. The female wasp stings the grasshopper at several points along its main nerve cord in order to paralyze it. The wasp then carries its paralyzed prey to a previously prepared hole and lays an egg on the prey. When a larval wasp hatches from the egg, it has a fresh supply of food. The female wasp has to sting the grasshopper very carefully. It must be completely paralyzed to prevent escape, but if it is killed rather than paralyzed it will rot. The wasp must excavate a proper hole, find the right prey, and sting it in just the right places. She does all of these things correctly the first time she tries. The details of the performance are highly stylized, and different details are characteristic of specific wasp species. Much animal behavior has this stylized and species-specific character. Birds can be recognized by the nests they build or by their songs. Elaborate courtship behavior, such as that of the stickleback (which we discussed in Chapter 12), is yet another example.

Many complicated behavioral sequences are composed of simple component parts. A bird may exhibit many movements during a display of courtship or aggressive behavior. But each component in the sequence is rigidly stylized. The whole sequence, no matter how long and complicated, contains only a limited variety of movements (such as nodding, bowing, extending the neck). These components, too, are species-specific and appear fully formed the first time the behavior is exhibited.

Instinctive behavior may be considerably more flexible than it appears. Behavior that seems rigid and stylized can show some adaptability. Bird nests have a distinctive form, but they can be constructed of many available materials (bits of string and nylon thread as well as strands of grass). Nest construction may also be varied to suit the location. Bird song is also variable. It appears spontaneously in some cases, but in others the young bird chooses its own song when presented with several options. Still other observations suggest that learning is involved in instinct. Female rats instinctively build a nest for a first litter. When the young wander from the nest the mother automatically retrieves them. But if the female rat is reared in a cage with a wire mesh bottom, never encountering shavings or shredded paper until she bears a litter, she will not construct a nest or retrieve her young. Apparently she must learn to recognize the nest-building material before hand. This evidence suggests that learning is important at some point in the behavioral sequence.

The distinction between instinct and learning is like the distinction between heredity and environment — it is artificial and misleading. Every morphological feature of an organism depends on both heredity and environment. Every behavioral feature of an organism depends on both instinct (heredity) and learning (environment). Behavior cannot be categorized precisely as one or the other.

Ethology

Ethology, the study of instinct, is one of several schools of behavior that emphasize the hereditary component of behavior. Ethologists describe the organism as though it contained many kinds of "behavioral energy." Consider the following sequence of behavior: a falcon rises in the midmorning and begins to soar, covering its territory in a random pattern. When it sights potential prey, it begins a sequence of attack behavior. If the prey is a flock of small birds, the falcon may execute a series of sham attacks to isolate one of the birds from the larger flock. The falcon then swoops, captures, kills, plucks, and eats the prey. As it progresses through the hunting sequence its movement becomes more and more predictable. The final swoop and kill is stylized and species-specific.

In classical ethological terms, this incident would be described differently. The falcon has sources of "energy," which activate instinctive behavior patterns associated with various activities, such as feeding, courtship, and aggression. Each kind of energy is associated with a specific kind of final behavioral act (in this case, killing and eating the prey). Thus, the energy is called **reaction-specific energy** (RSE). RSE's increase as time passes. When the RSE for hunting increases, the bird rises and soars. This act is not very stylized. But the increase in RSE explains why the bird begins its hunting pattern rather than simply sitting and sunning. The initial behavior ultimately leads the bird to suitable prey (the flock of small birds). Sighting the prey then activates an **innate releasing mechanism** (IRM). This mechanism permits the reaction-specific energy to generate a more specific behavior pattern.

There are several possible IRM's. Different ones are triggered, depending on whether the falcon sights a crippled gull away from its flock or a mouse running across exposed ground. In each case the falcon's behavior pattern is different. The RSE flows along different channels to activate behavior appropriate to the situation. In this way, the bird's behavior becomes more and more focused. Several more IRM's may be involved. (For example, the isolation of a single bird from the flock would stimulate another IRM that causes precise maneuvering for a strike.) The final swoop and kill is called a **consummatory act.** The occurrence of the consummatory act discharges the RSE that started the sequence. The falcon can now begin another pattern of behavior.

This model accounts for several puzzling aspects of behavior. First, it explains the spontaneity of behavior. The falcon begins to hunt before any external event occurs, and the accumulation of RSE for hunting provides an explanation. Second, the model accounts for the cessation of behavior after the consummatory act: the RSE has been discharged. More generally, much animal behavior is apparently directed toward some goal, and it ceases when some functional event, such as eating or copulation, has been accomplished. Most students of animal behavior find it troublesome to attribute foresight and rational pursuit of a goal to non-human animals. The ethological model bypasses this difficulty; the falcon is not motivated by forethought, it is driven by high levels of RSE. The goal is the performance of the consummatory act rather than eating for its own sake. The occurrence of the consummatory act drains off the RSE and terminates the sequence of behavior.

The ethological interpretation of behavior is supported by the fact that certain behavioral responses show a gradual decrease in strength. For example, when a particular species of male butterfly is presented with a fluttering model of the female, the male responds by rising and following. When the same stimulus is presented a second and third time, the male will fly a shorter distance, or may not respond at all. This reduction in response is interpreted as a decrease in the male's RSE level.

The occurrence of what is called **displacement activity** is also used as evidence for ethological models of behavior. Birds occasionally exhibit behavior that is totally inappropriate to their circumstances. A male, in the midst of aggressive interaction with another male, may suddenly tuck his head under his wing in a stylized sleep posture.

Similarly, aggressive behavior may suddenly intrude into a courtship sequence. According to the ethological model of behavior, these are illustrations of displacement activity. In displacement activity, surplus RSE spills over into an irrelevant channel. It has reached such high levels that it is no longer blocked by the lack of an appropriate stimulus. It must be discharged through a consummatory act, even though the consummatory act is totally irrelevant to the behavior at hand.

The ethological account is appealing because it does not attribute conscious prediction and foresight to the animal. Furthermore, it explains how instinctive behavior can be very plastic at first, gradually becoming more focused through the activation of successive IRM's and finally become rigidly stylized during the consummatory act.

There are several weaknesses in the RSE model, however. Reaction-specific energy is a troublesome concept. The decreases in RSE through repeated presentation of the stimulus can be explained by a gradual accommodation of the nervous system, but the nature and source of the energy remain mysterious. It is not "energy" in the usual sense of that word; it cannot be converted into calories or electrical units and cannot be described in terms of classical nerve impulses. Perhaps changes in blood chemistry (such as hormone balance or level of glucose) will be discovered to provide a biological basis for RSE, but as yet they have not been found.

Ethological interpretations of behavior are also criticized because they view the organism as a "black box"; behavior is reduced to basic unit elements that are then recombined to account for larger, goal-oriented behavior patterns. This kind of analysis draws attention away from the physiological events that must underlie any behavior. The ethologist can certainly say that he is not preventing such physiological analysis; he is merely describing behavior in other terms. He may hope that a detailed investigation of behavior will reveal the physiological basis of his model, but it is not his business to establish such connections.

In fact, ethology originated partly as a reaction to the extreme artificiality of some physiological studies. Animals were studied in the laboratory in circumstances that bore no resemblance to their normal habitats. Responses to abnormal stimuli were interpreted in crude mechanical terms. An animal might be placed in a darkroom for example, and exposed to a point source of light. If it turned toward the light, it turned "because" the light caused a stronger contraction of its muscles on the side toward the light. Such studies dominated the investigation of animal behavior early in the century. Ethologists felt, quite properly, that this sort of explanation had little to do with the richness of natural animal behavior.

Ethologists have made one major attempt to incorporate their black-box analyses into the general body of biological information. As attention was focused on small stereotyped units of behavior, ethologists observed that similar units were present in related animals. This discovery led them to the idea that such units were analogous to genetic traits, such as the flower color and seed shape used by Mendel in classical genetics. The behavioral units were thought to be genetically coded on particular regions of chromosomes, and subject to mutation and selective forces in evolution just as much as any physical characteristics. This is simply another way of saying that behavior, like other animal characteristics, is heritable.

Freud

The ethological account of instinctive behavior and the Freudian account of human behavior have interesting similarities. Classical Freudianism postulates a section of the mind called the **id**, which is a source of energy. Id energy is channeled along instinctive pathways and is drained away in various consummatory acts. Sexual urges are the most important pathways. Each individual faces a fundamental conflict between the need to obtain pleasure through such acts and the social taboos against their performance. The process of acculturation involves rechanneling the energy of the id into socially acceptable behavior. This rechanneling, called **sublimation,** is quite like the displacement activity described by the ethologist.

Freud's interpretation of human behavior, like most theories that emphasize instinct, is rather pessimistic. The discontents of civilization are considered to be inevitable products of sublimation, and are viewed as the price a society must pay to escape a brute existence. Therapy is designed to give the individual a socially acceptable set of sublimations. Happiness is not a realistic goal, according to the Freudian school, since it requires a complete lack of inhibition, and uninhibited behavior is incompatible with social life.

Anthropomorphism and Zoomorphism

Human characteristics have always been ascribed to lower animals. Aesop used imaginary behavior in lower animals as a device for telling moral tales. We still attribute heroism, self-sacrifice, wisdom, and nobility to cats, dogs, and horses; everyone has read such tales. The tendency to treat cats and dogs as mute humans is not confined to elderly maiden ladies. Many popular breeds of cats and dogs have shortened muzzles that mimic human facial appearance and facilitate their humanization. This strong and persistent tendency to attribute human characteristics to other animals is called **anthropomorphism.**

Conversely, man's behavior is sometimes described in terms of lower animals. This tendency is called **zoomorphism.** Zoomorphism is a valuable way to think about our own behavior. After all, the evolution of man from a primate stock is no longer debatable. We must also conclude that human behavior has evolved from the behavior of ancestral primate stocks, and ultimately from mammalian behavior in general. Investigations of the behavior of other animals, particularly nonhuman primates, can therefore offer valuable insights into human behavior. If humans are part of the animal kingdom, as surely they are, then both anthropomorphism and zoomorphism can be useful points of view, but only if we maintain a balanced attitude about the conclusions that can be drawn from comparative studies.

One conclusion that we can draw from studies of nonhuman primates is that their behavior is quite plastic. Differences between their social behavior in zoos and their behavior in natural surroundings are particularly instructive. Apes, monkeys, and baboons in captivity have social relationships that are very unnatural. Patterns of aggression, family organization, and other aspects of behavior are very different in wild populations and in captive groups. Furthermore, even in the wild, different populations of the same species may show local differences in behavior.

Comparative studies of cultural anthropology reveal the same kind of flexibility in human behavior. These observations make fascinating reading. For example, the social customs of South American Indian tribes are determined in large part by the availability of animal protein in the diet. If hunting is easy and meat is readily available, the culture tends to be monogamous. Mild promiscuity and adultery are tolerated, and hunting prowess is not particularly important. But in areas where hunting is difficult, cultural patterns change. Women are treated as a commodity and are deliberately kept scarce to be used as a reward for hunting prowess. Female infanticide is sometimes practiced, and strict rules about promiscuous and adulterous behavior are harshly enforced. Polygamy is sanctioned, but it is reserved for the most successful and vigorous hunters. Human social behavior is, in short, extremely plastic. It is adaptable to a variety of environmental circumstances.

We cannot really separate the portions of human behavior that are genetically controlled from those that are not. All that we can do is observe human societies that are already culturally adapted to their circumstances. It is not possible to isolate a group of individuals from birth through several generations to see what they are able to construct in the way of a language, a religion, kinship laws, and the like. (Some human infants have been isolated a few times to determine whether they could develop such a "natural language," however. King James IV of Scotland tried this experiment in the fifteenth century and optimistically expected the child to speak Hebrew.) In any existing culture, the social structure is reinforced by acculturation of new individuals and will therefore be predictably uniform. Within a culture, social structure will appear to be highly "heritable." But this fact does not necessarily mean that particular kinds of social behavior are instinctive and genetically determined; it may mean simply that social behavior is plastic, and that similarities within a culture are due to the common environment.

It is now very popular to use a limited concept of zoomorphism, based on primate behavior, to explain human behavior. These analyses usually begin with a description of a hypothetical band of hunting apes, supposed to have been our primeval ancestors. (Since these particular apes are no longer available for observation, the account is pieced together from selected studies of nonhuman primates and from anthropology.) It is said that the characteristics that permitted the survival of this hunting band—aggression, the formation of mating pairs, carnivorous eating habits—evolved for several million years as genetically determined behavioral traits. Industrialism, urbanization, and other features of modern Western civilization have arisen in a comparatively brief time. Thus, man is seen as a carnivorous, aggressive ape trying to cope with a modern environment to which he is almost totally unadapted. This theory is the most influential form of zoomorphism.

Pessimistic conclusions about the "beast in man" can be attacked in two ways. First, there are approximately 200 species of primates, and many primate societies, both human and nonhuman, are quite docile. The social organization of some nonhuman primates is nonaggressive and nonterritorial, and the same is true of some primitive human cultures. Bushmen are neither aggressive nor territorial. They can carry all of their possessions on their backs and move freely from one small group to another. One might then argue that human aggression and territoriality are plastic features of behavior. In other words, they arise only after the "hunter-gatherer" stage of culture has been passed.

A second and more effective counter to zoomorphic pessimism is to assert that human behavior is not determined in a simplistic way. This view does not deny our animal heritage or repudiate our primate evolution. But primate behavior is basically flexible and adaptable. The inherited pattern of social relations takes the form of very general genetic instructions. Those general instructions are adapted to the circumstances of a particular culture. This flexibility makes culture itself a valuable adaptive feature for primates. Primates can survive in changing circumstances, not by the slow process of mutation and genetic modification, but by the much more rapid process of altering their cultural structure. There may be some consummatory acts in human behavior, but our behavior is so adaptable and our programming so general that it is difficult to identify any elements of our behavior that might be instinctive.

To restate our earlier conclusion, behavior exists only as a phenotype. It always involves both the genome and the environment of the individual.

Recommended Reading

Jensen, Arthur R., "How Much Can We Boost IQ and Scholastic Achievement?" in *Environment, Heredity, and Intelligence.* Cambridge, Mass.: Harvard Educational Review Reprint Series, No. 2 (1969).

The famous review of the heritability of intelligence, more often praised or maligned than read. It is careful, cautious, and thorough, and deserves a wide audience. Several invited critical papers that discuss Jensen's article are included in the reprint.

Medvedev, Zhores A., *The Rise and Fall of T. D. Lysenko,* tr. by I. Michael Lerner. Garden City, N.Y.: Doubleday (1971).

Lysenko dominated genetics and all of biology in the USSR from 1934 to 1964. Through his misconceptions about the influence of environment and heredity on organisms, he greatly retarded the development of genetics in the Soviet Union. This book discusses the biological and political aspects of his influence and discusses the implications of the episode for the freedom of scientific inquiry.

Saxen, Lauri, and Juhani Rapola, *Congenital Defects.* New York: Holt, Rinehart, and Winston (1969).

Congenital defects in man, based on clinical investigations and experimental studies of laboratory animals.

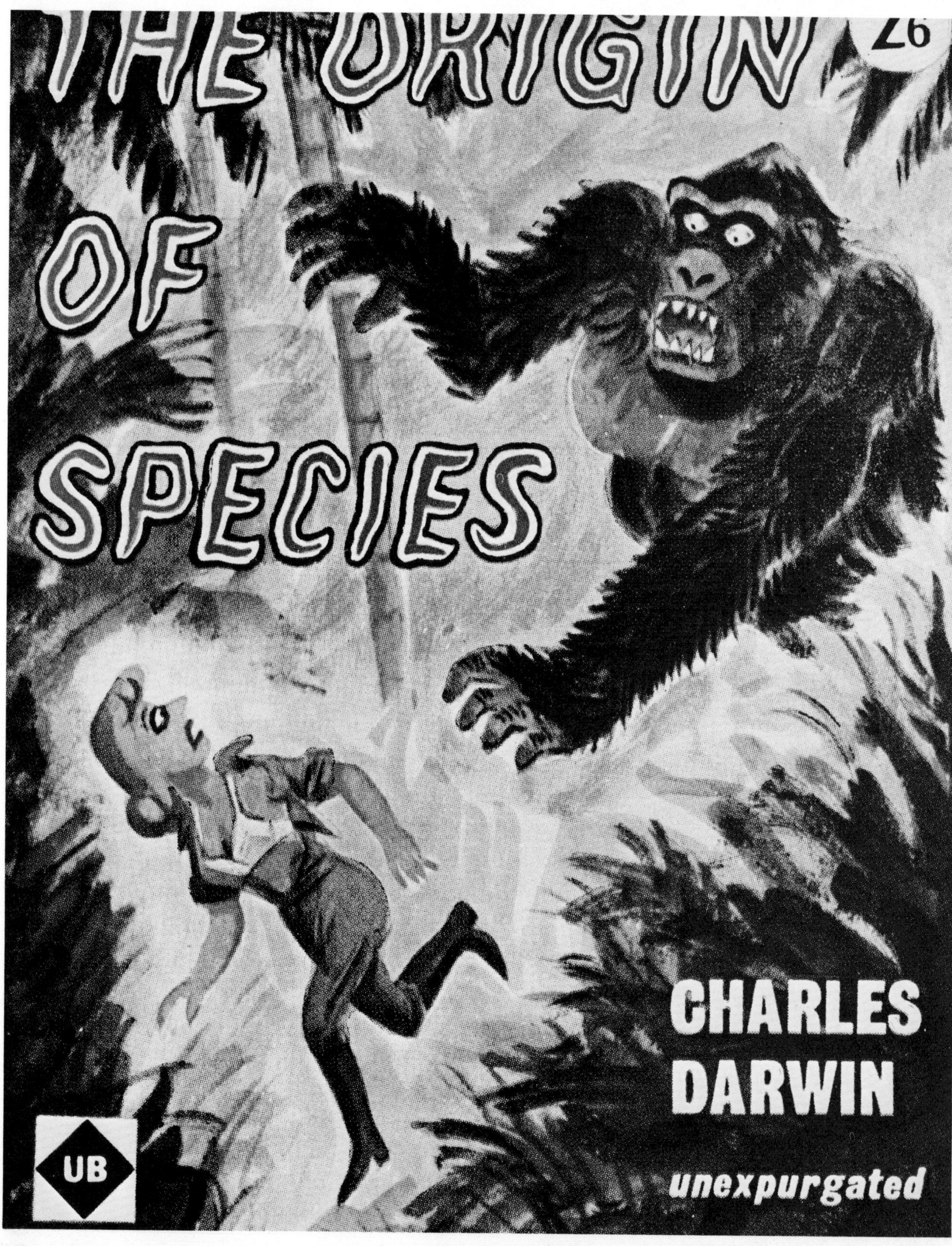
THE ORIGIN
OF
SPECIES
26
CHARLES
DARWIN
UB
unexpurgated

16 evolution

Evolution is the process through which organisms change, the process through which they acquire new physical and physiological characteristics. Another statement of evolution is "descent with modification." As a population of organisms passes through its life history and produces generation after generation of descendants, its characteristics change. Evolution has produced the tremendous variety of living things on earth today.

Evolution is a continuing, ongoing process, not something that has occurred in the past and is now complete. Changes or modifications in populations of organisms usually occur slowly compared to the span of human interest, but this is not always the case: evolution can proceed quite rapidly. In some cases evolution is proceeding so rapidly that it can be directly observed. In addition, there is now a great mass of information about past changes in organisms.

darwinism

The mechanism of evolution was first formulated by Charles Darwin. He called the mechanism **natural selection.** In this chapter we shall explain the concept of natural selection and describe its influence on biology, but we cannot begin to present the overwhelming amount of evidence that exists for Darwin's ideas. Evolution and natural selection are "hypotheses" only in the technical sense that any explanatory concept (gravity, the atomic structure of

Punch's *version of*
"neodarwinism"

511

matter, the revolution and rotation of the earth) is a hypothesis. Among scientists the validity of evolution by means of natural selection is no longer questioned and does not require defense. We are not saying that the concept is universally accepted; certainly most persons have never even heard of evolution and natural selection. However, it would be misleading to pretend that these ideas are still controversial among educated Westerners; debate continues over details, but not over the basic concepts. We shall describe alternate explanations of the variety of life very briefly as a matter of interest.

Natural Selection

The hypothesis of natural selection presented by Darwin is simple. First, he observed that all animals and plants are capable of reproducing more rapidly than is necessary to maintain a constant population. Under favorable circumstances, therefore—new territory, more food, fewer predators, suitable climate—a population rapidly expands.

Second, Darwin observed that there are limits on population growth. No matter how favorable the environment or how well the species tolerates crowding, the world is finite; growth must have a limit. Such limits obviously change, because the limiting factors themselves change. This situation can be illustrated by simple models such as cultures of microorganisms in the laboratory. Organisms are always pressing against their limits by producing more progeny than is necessary to maintain the population size. As a result, some individuals must die without reproducing, or at least will reproduce less than others.

Let us consider a specific population that cannot increase, because of environmental constraints. Suppose each breeding pair in the population has an average of three offspring. If the population size cannot grow further, then only two thirds of the offspring will survive and reproduce. Mortality or reproductive failure will prevent the other third from contributing to the next generation.

Darwin's final observation was that variation does exist among organisms. This fact is obvious. Differences in coloring, weight, and many other features appear in litters of puppies or kittens or pigs, in tulips and wheat. We now can account for such variation with genetic principles, and we can describe how differences in phenotype arise from differences in DNA. This information was not available to Darwin, but his meticulous observa-

tions provided sufficient evidence for the presence of variation in natural populations.

These are the elements of Darwin's natural selection: production of many offspring, natural population limits, and variation. Darwin argued that organisms with the most favorable variations have the best chance to survive, to reproduce, and to transmit the variations to their offspring. Organisms with unfavorable variations are less likely to survive and reproduce; they will not transmit their variations to progeny. In other words, organisms do not die or survive in a random manner. Darwin called this differential survival natural selection. Darwin himself stated the principle of natural selection as follows:

Can it, then, be thought improbable, seeing that variations useful to man have undoubtedly occurred, that other variations useful in some way to each living being in the great and complex battle of life, should sometimes occur in the course of thousands of generations? If such do occur, can we doubt (remembering that many more individuals are born than can possibly survive) that individuals having any advantage, however slight, over others, would have the best chance of surviving and of procreating their kind? On the other hand, we may feel sure that any variation in the least degree injurious would be rigidly destroyed. This preservation of favourable variations and the rejection of injurious variations, I call Natural Selection.

Natural selection easily accounts for many striking features in organisms. The long neck and large size of the giraffe are good examples. The fossil record shows that giraffes developed from smaller ancestors with more normal proportions. At any time, a few of these animals would be slightly larger and would have slightly longer necks than most of the population. These individuals must have had some survival advantage compared to their colleagues. Perhaps they were able to obtain more food because they could reach higher foliage, or maybe they could run faster to escape predators. Whatever the advantage, it increased the probability that these variant animals would reproduce. Hence they contributed more and more genes to the gene pool of the population.

As generations passed, large, long-necked individuals became increasingly common. Still longer necks and even larger individuals continued to have a selective advantage that permitted them to survive and reproduce more successfully. The

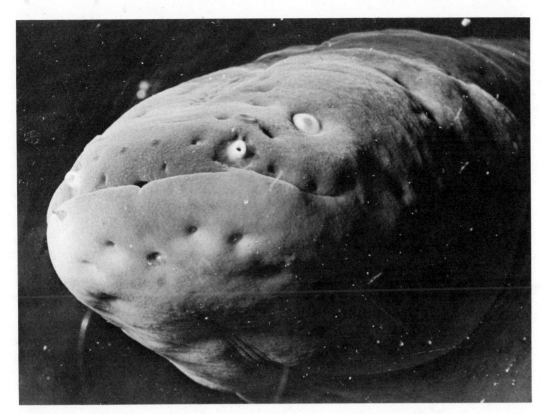

Figure 16-1 *The electric eel*, Electrophorus electricus.

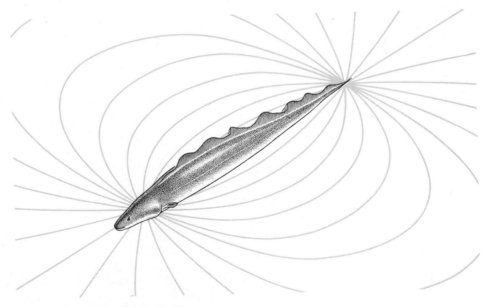

Figure 16-2 *The electric field generated by a "weak" electric fish. When objects enter the field, the fish detects the resulting change in field pattern.*

Charles Darwin

Charles Darwin (1809–1882) is a commanding figure in the intellectual history of Western civilization. He ranks in importance with Copernicus and Freud; the list of his peers is very short. Darwin was an extraordinarily productive scientist. He laid the foundations of a dozen major lines of biological investigation. His interests ranged from responses to light in plants to primate behavior; he studied the origin and development of coral reefs and the classification of barnacles. All of Darwin's insight, curiosity, and prodigious fund of information came to a focus in his major and most influential book, *On the Origin of Species by Means of Natural Selection, or the Preservation of Favored Races in the Struggle for Life,* published in 1859. Darwin did not just state the idea of evolution and natural selection in skeletal form in that work; his presentation drew upon an incredible range of detailed evidence, much of it based upon personal observation. Darwin had circulated other drafts of his ideas among friends and colleagues for more than a decade before the appearance of his major publication, and the intellectual community had been prepared for a major event. The entire edition of 1,250 copies was exhausted the day it appeared.

Evolution immediately became the central issue of discussion and bitter debate throughout Western civilization. In the sciences, Darwin's concepts provided direction and explanations for a variety of subsequent observations. Astronomy, physics, chemistry, geology, anthropology, psychology, economics, and sociology all contributed to and drew from the concept of natural selection. Organized religions and devout individuals were quite concerned about the evolutionary continuity between man and other animals presented in *Origin of Species* and elaborated in *The Descent of Man and Selection in*

Charles Darwin.

Relation to Sex (1871). The idea of progress received tremendous impetus. The Medieval and Renaissance view of a fixed and orderly "great chain of being" (stretching from God through man, the lower animals, and plants, to the inanimate) disappeared. It was replaced by ideas of mutability, progress, evolution, and social change. A history of the reception of the *Origin* is virtually a history of Western civilization; it has been written in many volumes by many authors.

Darwin did not operate in a vacuum. His work had historical roots and depended on the insight of predecessors, a debt that Darwin carefully acknowledged. A. R. Wallace had published a similar hypothesis independently. Geologists such as Charles Lyell had called attention to the implications of the fossil record earlier in the nineteenth century. Perhaps more important, many post-Renaissance writers had already started to talk about progress, change, and man's role in shaping his own culture.

Darwin's concept of natural selection retains its central position in biology more than a century after its formulation. The findings of modern genetics are completely consistent with Darwinian evolution, and they provide a solid explanation for many of Darwin's original observations.

---

tendency toward increased neck length and size continued. The final result after thousands of generations is the modern giraffe. This example is a simple one, because we can easily see how even a small change—a small increase in neck length—might be advantageous to the organism. The final product is just the sum of many small changes, each of which was advantageous.

This progression is more difficult to understand in other organisms, but it still exists. Take the case of the electric eel (Figure 16-1). It is difficult to see how the electric eel evolved gradually. It and several other fish have muscle or nerve or gland tissue that has been modified to discharge several hundred volts as a defense against predators. The electric organs of such fish are massive and occupy a major fraction of the body.

It is impossible to believe that a one-step change in genome could produce such dramatic modifications in structure. Instead, a more gradual sequence of changes must have occurred; the initial genetic change was undoubtedly more modest and, therefore, more probable. But it is hard to see what selective advantage might be found in the early ancestors of electric eels: a discharge of a few volts would hardly have any defensive significance. Yet, in order to be conserved in the gene pool, each step must have contributed to the organisms probability of survival.

A recent analysis of electric organs in several "weak" electric fish has helped to provide the answer. A number of fish have electric organs that discharge only a few volts. These organs are not defensive, but rather are part of an orientation system. The discharge of the electric organ sets up an electric field in the surrounding water (Figure 16-2). The pattern of the field changes if anything enters, because the intruding object has a different electrical conductivity. Receptors along the side of the body register changes in the field pattern, and this information is passed to the brain. In this way the fish can recognize obstacles or predators in dark or murky waters.

These discoveries provide a more believable account of the evolution of electric eels. The first relevant change in the gene pool produced a weak electric organ; this organ was conserved because it was part of a sensory system. On this basis, the subsequent increase in size and voltage becomes more logical. The evolution of the defensive electric organ can now be explained by a process of **relative growth.** There is a changing relationship between the size of particular structures and the size of the organisms as a whole: as an organism gets larger, some of its parts become disproportionately larger. This phenomenon is relative growth. It can be observed in various organisms. Structures such as antlers (in the elk) or claws (in the fiddler crab) become larger as the animal gets larger. But a simple increase in relative size may create a totally new function for the organ. In our present case, larger weak electric fish had larger electric organs. Eventually these organs became so large that their electrical charges began to afford the fish some protection from predators. At this point the sensory organ began to take on an entirely new function, defense. It continued to evolve into a highly successful defensive organ.

Alternative Hypotheses

An alternate formulation of evolution was proposed by a brilliant French biologist, Jean-Baptiste de Lamarck (1744–1829). It is Lamarck's misfortune that he is best remembered for this hypothesis. He argued that changes in organisms resulted from wants, or longings. Giraffes, for example, would stretch their necks through their desire to get at the tender leaves in the higher branches of a forage tree. According to Lamarck, any changes in adults that resulted were preserved and transferred to the offspring. Thus, the giraffe would increase in size, and its neck would gradually elongate, because adults of each generation wanted to have longer necks. Each generation would come closer to achieving that goal and would pass the achievement on to its offspring.

There are some obvious problems with the Lamarckian hypothesis. What is the nature of these longings? A mammal such as a giraffe might well demonstrate convincing desires, but it becomes more and more difficult to think of longings in earthworms, amoebae, palm trees, grass, and bacteria. Either "longing" has its normal meaning, implying consciousness and choice, or Lamarck used it to express a different idea altogether: that living things have unconscious desires that somehow become expressed in their organization. In either case, it is extremely difficult to incorporate the idea into the rest of our assumptions about the nature of life; it cannot be reconciled with the basic position that all properties of living things can be explained by physical and chemical principles.

It is also difficult to understand how acquired characteristics can be inherited. Since it is hard to postulate wants and longings in gametes, Lamarck attributed them to adult organisms. But efforts to demonstrate the inheritance of characteristics acquired by adults have failed to produce clear proof. How might acquired changes in adult characteristics influence the genetic material? We have no idea. Lamarckian theories have a continuing appeal, and there are occasional reports of the inheritance of acquired characteristics. But none of these reports has ever been thoroughly substantiated and verified.

There is, of course, a third hypothesis that attempts to explain the variety of living things: **special creation,** the hypothesis that animals and plants were created at some particular time in the past by an external agency. We are most familiar with the Biblical account of creation in the Book of Genesis, but other religions have comparable stories. Such hypotheses are entirely unlike either Darwinism or Lamarckism, because they are not related to observational evidence in a direct way.

Look, for instance, at the controversy over the significance of the fossil record. Nineteenth-century geologists interpreted fossils to mean that (a) the earth was much older than the Bible indicated, (b) animals now unknown existed in prehistoric times, and (c) gradual changes had taken place in the structures of these prehistoric forms. But this interpretation produced a conflict with advocates of special creation. Genesis allows six unequivocal, sunrise-to-sunset days for the creation of the world. Fortified by prayer, the zoologist Philip Henry Gosse suggested that God created the world with the fossil record intact in its rocks. Such animals had never really lived—God hid the fossils in the rocks to test the faith of geologists. Obviously, such a hypothesis does not interpret physical evidence in the usual way, yet the argument is absolutely irrefutable: empirical evidence becomes entirely irrelevant to anyone with an emotional commitment to such an idea. No matter what evidence is brought forth, it can always be dismissed as merely more false clues from the Creator.

We have three hypotheses, then, to account for the close functional relationship between an organism and its environment. The first is Darwinian evolution, the second is Lamarckian evolution, and the third is special creation. We can set aside special creation here because it is not amenable to scientific discussion. Lamook's hypothesis is troublesome because it postulates non-physical causes (longings and wants) for physical events and is unsupported by observation. By contrast, Darwin's formulation of natural selection has been extremely fruitful from several viewpoints:

1. It organizes, and is supported by, a tremendous body of observation.
2. It is predictive: since evolution is a continuing process, Darwin's hypothesis allows us to forecast the results of laboratory and field observations and then test our predictions.
3. It explains new information: twentieth-century discoveries in genetics and molecular biology are outstanding examples. In turn, new information from genetics and molecular biology helps us understand precisely how evolution can work.
4. Finally, its logic has been fruitfully applied to analogous problems in psychology, sociology, and engineering.

Supporting Evidence

A critical piece of supporting evidence for the theory of natural selection had to come from geology. Is the earth old enough for evolution to have occurred through natural selection? One estimate for the age of the earth came from the Irish clergyman James Ussher in 1654. Ussher used a literal interpretation of the Bible to pinpoint the creation of the world at 4004 B.C. Ussher's figure meant that natural selection could not possibly have any significance. Darwin was encouraged, however, by the discoveries of nineteenth-century geologists, which indicated that the earth had to be much older. Fossils occur in sedimentary rock, and the deposition of sediment on the ocean floor is continuous. Therefore, by comparing the thickness of sedimentary deposits with the measured rates of deposition, geologists were able to estimate how long the process must have lasted to produce the existing deposits. The figure was many millions of years. Other geologists calculated how long it would take to transport enough salt to the ocean to produce its present salinity. Again the answer was many millions of years. Actually, neither of these methods is very reliable, and the second is based on the misconception that the oceans were once fresh water; nevertheless, both gave estimates that were more than enough to allow for the slow process of natural selection.

Geology and Time

Modern geological dating combines various techniques: stratigraphy, dating by radioactive isotope ratios, and observations from astronomy and meteorology. **Stratigraphy** is the study of sedimentary rock layers (see Figure 16-3). Many characteristics give information about the age of such rock layers: the order of their occurrence, their thickness, and the fossil content of each layer. Geologists and paleontologists have prepared a classification system for sediments based on their fossil content. Fossil-bearing rocks are divided sequentially into time periods and epochs. This system provides a linear sequence for the relative ages of various layers, but no absolute time marker. We know that a particular layer was formed earlier than the layers

Figure 16-3 Sedimentary rock layers in the wall of the Grand Canyon.

Figure 16-4   Fossils of organisms from the Cambrian Period (about 600 million years ago).

above it, but we can't tell whether it is 1,000 or 100,000 years old. A great mass of sedimentary rock was deposited below the oldest known fossils. Such rocks cannot easily be classified, because they lack fossils. Also, beds in different localities cannot be compared.

The oldest fossils (of blue-green algae and bacteria) are found in sediments from the Precambrian period; they are probably more than three billion years old. By the time of the Cambrian period or the Ordovician period which immediately followed, the fossil record shows a rich variety of major plant and animal groups (Figure 16-4).

Radioactive isotope dating gives more precise information about the ages of fossils because radioactive decay occurs at a constant rate. The nuclei of radioactive isotopes are unstable. They become stable by decay, that is, by emitting particles and electromagnetic radiation of various kinds. The precise moment of decay for an individual atom cannot be predicted, but all the atoms in a specific isotope have a characteristic probability of decay. Some decay very quickly, others slowly. The decay time of an isotope is usually expressed as its **half-life,** the time required for half of the initial quantity of the isotope to decay. The rate of decay is not influenced by factors that affect the formation of sedimentary rock, such as temperature or atmospheric pressure.

Isotope dating methods are based on the assumption that specific amounts of radioactive elements were incorporated into rock at its formation. A rock's age can thus be estimated by measuring the amounts of the radioactive elements now present in the rock and comparing them with the amounts that were assumed to be present at its formation. In some cases we can also measure the amounts of the decay products that have appeared.

This method provides us with **isotope clocks.** For example, the uranium isotope U^{238} gives off a series of radioactive particles and eventually turns into a form of lead. Uranium clocks are particularly useful. U^{238}, U^{235}, and a thorium isotope, Th^{232}, are often found together in the same minerals. They all decay through a series of steps to different isotopes of lead (Pb^{206}, Pb^{207}, and Pb^{208}, respectively). Their half-lives also differ. The half-lives of U^{238}, U^{235}, and Th^{232} are 4.5 billion years, 713 million years, and 14 billion years, respectively. If we assume that nothing has disturbed the radioactive elements in the rock since its formation, we can calculate the rock's age from the half-lives and decay sequences. If a rock has a Pb^{206}/U^{238} ratio of 0.5, for instance, it is about three billion years old (see Figure 16-5).

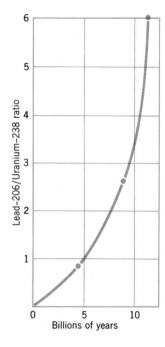

Figure 16-5 *Dating of rocks by isotope ratios. Uranium-238 decays to lead-206 in a series of steps. Since the decay rate is constant, we can tell how old the rock is by measuring the ratio of lead to uranium.*

All three members of the uranium series often occur together. When they do, we have three independent estimates of the rock's age. And when these estimates agree, we can be quite confident of the date.

Radioactive isotopes are useful clocks over time intervals ranging from about one tenth of an isotope's half-life to about ten half-lives. Uranium clocks, with their long half-lives, are consequently very unreliable for short periods and recent events (that is, events from the past few tens of thousands of years). Fortunately, carbon-14 (C^{14}) has a half-life of 5,730 years, which makes it a suitable guide to ages from a few hundred to 40,000 or 50,000 years. In this case, the decay product (N^{14}) is not measured; rather, carbon-14 dating depends on the fraction of the total carbon in the fossil that is C^{14}. (The most common carbon isotope is carbon-12).

C^{14} is continuously produced in the upper atmosphere, and it continuously decays; but the actual C^{14} content of the air is constant because the air is in a steady state. A living organism exchanges carbon with its environment; plants take in carbon dioxide from the air and convert it to organic carbon, which is later used by animals. As a result, the ratios of C^{14} to total carbon in the living organism are the same as in the air. However, when the plant or animal dies, the C^{14} is no longer replaced by more C^{14} from the environment. It begins to decay, and the decrease in the ratio of C^{14} to total carbon is a measure of the time elapsed since the animal died. C^{14} clocks provide important time markers for human prehistory.

The potassium–argon clock is also important. Potassium-40 decays to argon-40 with a half-life of 1.3 billion years. K^{40}/Ar^{40} ratios can be used in rocks that contain potassium but lack uranium ores. When uranium is present, the potassium–argon clock can be used to check on the uranium clock dates.

Isotope dating has been tied to the series of rock strata, so we now have absolute time markers for the rock layers. The best opportunities for dating occur in formations in which lava has flowed between two well-defined strata of sedimentary rock. The lava is clearly positioned in the sedimentary sequence and can be dated with the K^{40} clock. These correspondences can then be used to date the surrounding layers.

Our figure for the overall age of the earth comes from isotope dating of ancient, nonsedimentary rocks. The oldest rocks are 3.6 billion years old. This figure agrees reasonably well with astronomical dates for the origin of our galaxy (six to seven billion years ago), the solar system (five to six billion years) and the earth (four to five billion years). Thus, natural selection had over three billion years overall to evolve the current variety of living things, and at least 600 million years (since the beginning of the Cambrian era) to produce the detailed evolutionary sequence that can be observed in the fossil record.

The Fossil Record

Fossils, when combined with accurate dating techniques, provide a fascinating historical record of natural selection. The preservation of fossil remains is accidental and depends on favorable circumstances. Most fossils are found in the sediments of shallow seas. Swamps, tar pits, and caves also provide important deposits, but they are much more rare. Sometimes whole organisms have been preserved. In other cases only fragments remain, but anatomists can sometimes reconstruct the entire extinct organism using anatomical relations found in present animals. Sometimes the fossilized remains are in the form of casts rather than solid material: impressions of footprints or worm tubes.

We have comparatively little information about soft-bodied animals such as worms. But our fossil record of clams and snails and other shelled dwellers in shallow seas is very good. Fossils are common for some vertebrate groups and exceedingly rare for others. The completeness of the fossil record for different plants and animals depends on their habits and their morphology, and on accident.

Despite its limitations, the fossil record gives us a fairly detailed picture of the changes that have taken place in living things over the last half-billion years. Through fossils we can describe a sequence of dominant organisms over long periods of time, and we can give reasonably precise dates for the appearance of new groups (see Figure 16-6).

A fossil can give us much more than a gross idea of the appearance of the organism. Some fossils have so much detail that their microscopic structure is visible. The pathways of nerves and blood vessels can be identified and compared with those of existing animals, or the details of seed and stem structures of fossil plants can be compared with their contemporary plant descendants. In fact, one really remarkable deduction about the prehistoric environment was based on the fact that the rates of calcium deposition in the skeleton of marine corals vary daily and annually. These variations create ridges and undulations in the coral. The ridge patterns prove that a year had more days in past geological eras than it does now (Figure 16-7). This evidence agrees with other calculations that show that the earth's rotation is slowing down because of frictional drag from lunar and solar tides. The layered structures of marine corals also provide a kind of absolute dating that agrees very well with other dating methods.

Many sequences of fossils illustrate the sort of gradual changes in form that are predicted by natural selection. The best records are for common fossil organisms, such as shallow-water invertebrates and the plants involved in coal formation. Fossil clams and trilobites, cycads and tree ferns are so abundant that variations in their structures can be analyzed statistically. Such techniques give us very detailed lineages. Fossil material for vertebrates is less common, but it has more popular interest, given our prejudices. The rise and fall of the great dinosaurs in the Age of Reptiles is familiar to everyone through children's picture books and lurid fiction. The exploits of paleontological expeditions of the late nineteenth and twentieth centuries are almost as familiar and nearly as lurid as the fictional efforts. Willy Ley has written several fascinating books on the subject.

The ancestors of the modern horse, *Equus*, are particularly well represented in the fossil record. Horses are now the most successful modern perissodactyls (hoofed mammals with an odd number of

Figure 16-6 Major steps in the evolution of life.

Time, years ago	Events	
20–50 thousand	Modern man, modern plants and animals The Atomic Age	
1–60 million	Evolution of higher mammals and plants, even greater diversification and distribution of flowering plants	
50 million	Some mammals enter the oceans, continuing diversification and distribution of flowering plants	
100–150 million	First mammals and first birds, increasing diversification and distribution of flowering plants	
125–200 million	Rise and predominance of reptiles, development of extensive gymnosperm forests, rise of flowering plants	
250–300 million	First amphibians, insects, mosses, ferns; rise of dense swamp forests which later formed major coal beds of the world	
350 million	First succesful invasion of land by animals and plants, appearance of first vertebrates as ancient fish in the sea	
500 million	Ancestors of all major groups of invertebrates present in the oceans, also alga and ancient seaweed	
1 billion	Increasing population of unicellular and simpler multicellular organisms including many invertebrates in the seas, origin of photosynthesis $CO_2 + H_2O$	
2–4 billion	Formation of increasingly complex organic molecules in primeval seas, origin of life	
4–5 billion	Formation of the earth and other members of solar system	
10 billion	Origin of universe?	

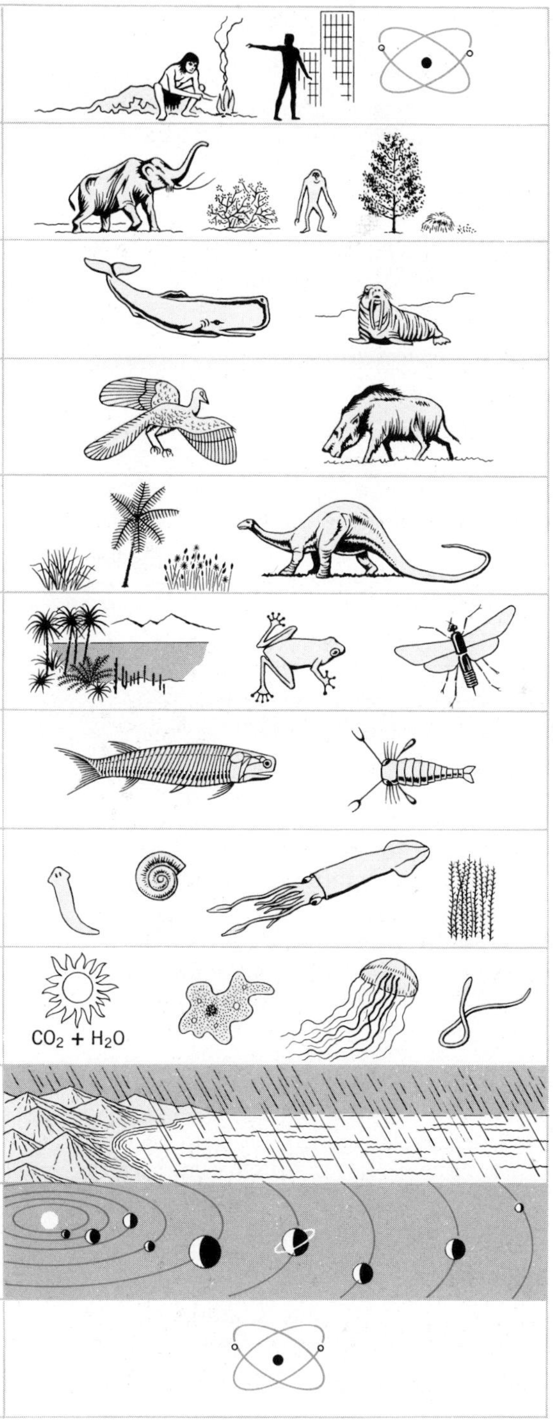

521 *darwinism*

(a)

(b)

Figure 16-7 (a) Fossil corals show both daily and annual changes in growth that appear as ridges in their skeletons. (b) By counting the number of daily ridges in a year's growth, and by knowing the age of the fossil, we can calculate the number of days in a year at various times in the past.

toes). The middle toe is the only functional toe in the modern horse; the lateral toes have been reduced or lost. *Hyracotherium*, the earliest ancestor of the modern horse, first appeared about 60 million years ago in the fossil record. It was the size of a fox terrier and was adapted for running. Similar primitive perissodactyls evolved into the modern tapir, the rhinoceros, and also into various now-extinct groups of browsing animals that had claws rather than hooves. The sequence of fossil forms from *Hyracotherium* to the modern horse shows an increase in overall size, a lengthening of the legs, a loss of lateral toes, a lengthening of the skull, and a modification of the molars and premolars for browsing (Figure 16-8). In modern times, selective breeding has produced many forms of *Equus* (Shetland ponies, draft horses, thoroughbreds, and asses, to name a few). The zebra is also an *Equus*.

The fossil record is entirely consistent with the hypothesis of natural selection. We know that there was enough time for selection to occur. We also can see that animals that were once abundant have become extinct, and this fact supports the idea that they could not compete successfully with more modern animal groups. Sequential changes in struc-

ture agree with the idea that gradual changes in the gene pool result from selection. We must distinguish, however, between the statement that the fossil record is *consistent* with the hypothesis of natural selection, and the statement that it *proves* that hypothesis. The fossil record is equally consistent with Lamarckian ideas (or, for that matter, with the ideas of Gosse, that the fossils were "planted" as tests of faith).

The fossil record is not consistent with all theories, however. One favorite argument of the early anti-evolution literature was that paleontologists had imposed order on a random collection of fossils in order to support their claims that the forms were interrelated by descent. The fossil record contradicts this contention. The time sequence of the record can now be thoroughly documented, and so can some very detailed sequential changes. Changes in organisms have not been random. The fossil record does tell us that evolution has occurred.

Embryology

Diverse groups of organisms often share a common, basic anatomical plan. This fact is also strong evi-

Figure 16-8 *The evolution of the horse.*
(a) The skeleton of **Hyracotherium,** *the*
ancestor of the modern horse. (b) The
skeleton of a modern horse; the two
skeletons are reproduced to scale. (c) A
diagram of the evolutionary sequence;
note the side branches leading to forms
that are now extinct.

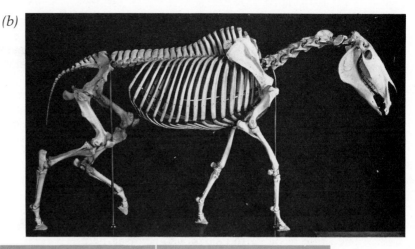

(b)

(a)

(c)

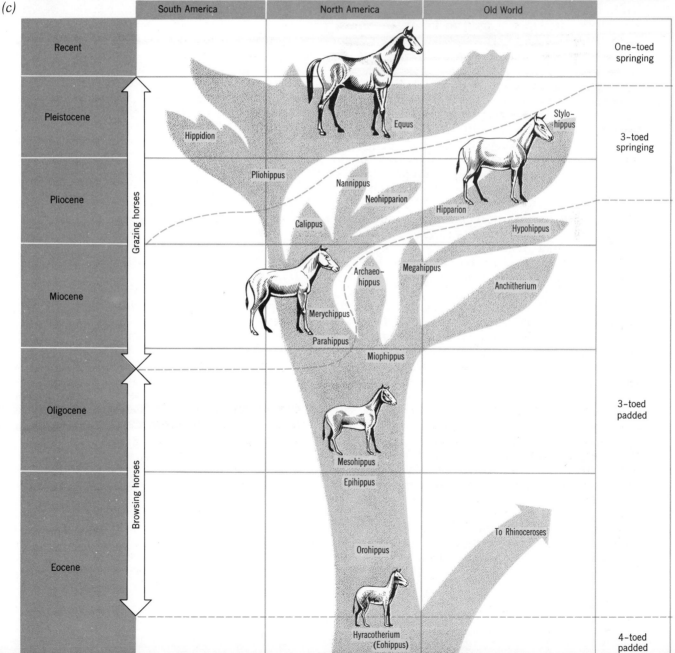

dence that they are related through evolution. The skeletons of mammals provide one clear example. The hand of a human being, the forefoot of a horse, and the wing of a bat all have a basic five-fingered plan. The same plan serves a different function in each animal. In humans it provides an opposable thumb and independent fingers for grasping. In horses it forms the massive middle finger that is known as a hoof. In bats it turns into a light supporting framework for the wing. Despite their enormous functional differences, hands, hooves, and wings all correspond to one another in the overall pattern of their skeletal elements. They all develop from the same basic embryonic pattern and become specialized during the development of the embryo.

The same basic patterns can also be discerned by working back in the fossil record toward the less specialized ancestors of modern forms. Structures with a detailed resemblance and similar embryological development are termed **homologous.** In other cases, structures that serve the same function may represent different basic morphological patterns. A good example is the wing of a moth and the wing of a bat; these structures are termed **analogous.** Much of the variety in living things in based on a small number of basic structural plans. We have already looked at the mammalian skeleton. In fact, mammals are part of a larger group of animals, the vertebrates, all members of which have a roughly similar ground plan.

The mouthparts of insects provide another, similar example. In various insects the same basic plan has been functionally modified for biting or piercing or chewing or sucking. A ground plan may also be modified in different ways in the same organism. For example, foliage leaves and the floral parts in higher plants are homologous. So are the different limbs of crustaceans. For that matter, the forelimbs and hindlimbs of five-fingered vertebrates all have the same ground plan.

Some features of the common mammalian ground plan are disappearing in modern animals. External ear muscles or a tail, for example, may have important functions for some mammals. But in humans these features are not important, and the extent of their development in an individual human being is quite variable. Two hundred such **vestigial** structures are found in human anatomy. Natural selection accounts for this phenomenon quite easily. If these particular structures have no function, then there is no pressure to maintain them in the population; if they do not confer a selective advantage on an individual, then they don't increase the chance of its successful reproduction. They become variable and may slowly drop out of the population. Vestigial structures are features of the ancestral ground plan which are no longer adaptively significant.

Comparative embryology provides a similar line of support for evolution. All vertebrate embryos look alike during their early development; it is almost impossible to distinguish between the early embryos of fish, chickens, and humans. They become recognizable only later in development (see Figure 16-9). The portions of the genome that specialize the vertebrate ground plan do not become active in early development—if they did, sweeping modifications of the ground plan would result. Instead, specialization occurs only later in development, after the basic structural organization has been established. Embryology can be a very useful tool for clarifying questionable homologies between organisms. A foal is born with only one functional digit, for example. But all five mammalian digits are present on the forelimb of the horse early in its embryonic development, dropping out only in later stages. The specialization of the appendages in the horse shows up in its embryology as well as in comparisons with fossil predecessors.

Other common features of embryological development are found throughout the vertebrate group. One such feature is gill slits. Gill slits appear even in the embryos of fully terrestrial organisms, which show no sign of gills as adults. Another feature that shows up in all vertebrate embryos is the circulatory system associated with the gill slits. In human development, six paired arteries arise to supply blood to gills that never fully develop. Our aorta is formed from this primeval arch system; it is the left branch of the fourth pair of gill arteries (see Figure 16-10). Thus, organisms as different as mammals and fish result from the selection and modification of elements of the same primitive ground plan.

The conservative nature of embryological development has been succinctly phrased as Haeckel's Law: *ontogeny recapitulates phylogeny.* **Ontogeny** is the development of the individual; **phylogeny** is the historical development of a whole group of organisms. What Haeckel's Law states is that the embryo, as it develops, passes through the whole sequence of ancestral forms from which its species is evolved. That is, a mammalian embryo in the course of its development would resemble first a fish, then an amphibian, a reptile, and, finally, a mammal. At face value, this statement is certainly untrue. At no time does a mammalian embryo resemble an adult fish or amphibian. But embryo

Figure 16-9   Vertebrate embryos at three equivalent
stages of development.

	1	2	3
Man			
Rabbit			
Chick			
Tortoise			
Salamander			
Fish			

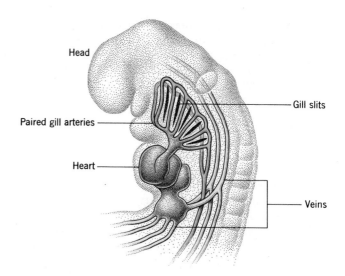

Figure 16-10   Gill slits and gill arteries in a human
embryo. All of the gill slits and three of the sets of arteries disappear in later development. Blood vessels
derived from the remaining three sets of arteries become
the major arteries of the adult.

fishes, amphibians, and mammals do all go through
similar developmental patterns before they achieve
the distinctive structure of the adult. The individual's specializations become clear only as development proceeds. At first, a mammalian embryo
would be hard to distinguish from that of a fish.
Later, it would clearly have developed differently
from a fish but would still resemble the embryo of
an amphibian. Even later still, it would share many
characteristics with a reptile embryo, even though
its differences from fish or amphibians would be
clearer. Only toward the end of its development
would it show the distinct specializations of
mammals and (ultimately) of its own particular
species.

Natural selection predicts such similarities
among the embryos of related organisms. Clearly,
any changes in the adult that improved a species'
reproductive success would be conserved. However,
more basic aspects of the embryological plan would
be less likely to be changed. The degree to which
the embryos of two different species parallel each
other in development is a measure of their common
evolutionary ancestry.

Until now, most of the evidence that we have
looked at involves small, gradual changes in
evolving organisms. The origins of basic groups

that are very different from each other are more speculative. Here the fossil record is not very helpful. Most of the major groups of organisms were already well defined the first time they appeared in the fossil record; few intermediate forms have been found. However, comparisons of embryonic and adult structures can provide some ideas about how basic new patterns in organisms could have evolved.

One mechanism is particularly interesting because it may explain the origin of the vertebrates, the group of which we are members. It is based on the observation that larval forms are often strikingly different from the adult. In many animals, the larva lives a long time and is a major part of the animal's life history. In some cases, what appears to be a larval form even can reproduce sexually—an adult form never appears. In these cases the organizational pattern of the larva has, in effect, become the adult pattern. Natural selection could mold such a new adult pattern to fit a variety of habitats. The result would be a major new group, based on a new ground plan.

The vertebrates may have originated in just this way. Vertebrates are one of three groups of organisms in the phylum Chordata. The other two groups, urochordates and cephalochordates, are marine organisms. They are not very conspicuous and have no economic importance, but they are extremely interesting to biologists. All three subphyla share the defining characteristics of the phylum: gill slits (at some stage in their development), a **notochord** (a primitive skeletal rod of cartilage along the body axis), and a hollow dorsal nerve cord. Urochordates, also called sea squirts, are mostly sessile filter feeders. The adults retain only one feature of the basic chordate pattern, gill slits. However, the larvae show all the chordate characteristics (Figure 16-11). They resemble tadpoles superficially (in fact they are called "tadpole larvae"). By contrast, the cephalochordates are more fishlike in appearance, but they are still quite different from vertebrates (Figure 16-12). They include only a few species, known as "lancelets" or "amphioxus."

Urochordates may be the primitive stock from which vertebrates evolved, while the cephalochordates are simply a minor offshoot. It is argued that the urochordate larva was subjected to selective forces that led to the elaboration of the notochord and nervous system. The tadpole larva makes use of its notochord, its nerve cord, and an associated light receptor in seeking a suitable habitat to establish the new sessile adult. It is easy to visualize how selective forces might have encouraged the development of these organs in the tadpole larva. As they developed, the larvae might have proved to be independently successful. As a result, the sessile adult forms would have become superfluous. They would have dropped from the life history, and the resulting tadpole-like marine organisms would

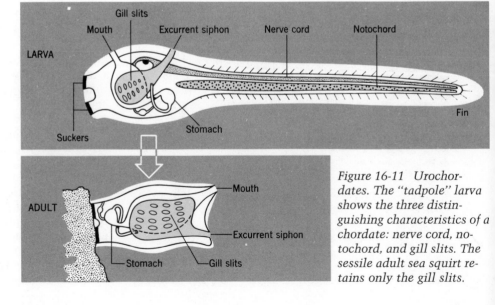

Figure 16-11 Urochordates. The "tadpole" larva shows the three distinguishing characteristics of a chordate: nerve cord, notochord, and gill slits. The sessile adult sea squirt retains only the gill slits.

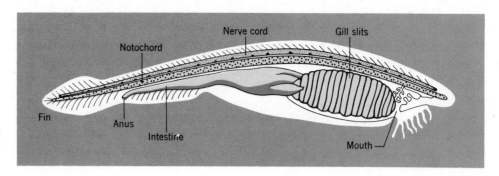

Figure 16-12 A cephalo-chordate adult retains all the chordate characteristics and is fishlike in general appearance.

have been the ones that gave rise to all vertebrates (Figure 16-13).

Mechanisms like this one, in which natural selection acts independently on larval and adult forms, may account for the origin of major new patterns of organization.

Biochemistry

Biochemistry, though a young field, has already contributed much information about evolution. Most biochemical evidence concerns the distribution of various compounds in organisms. Many simple organic molecules—amino acids, nucleic acids, fatty acids, and many metabolic intermediate compounds—are found in almost all living things. But other compounds are found only in restricted groups. This fact has helped in classifying organisms. When a compound is found in two different large groups, it suggests that one may be derived from the other; the photosynthetic pigments in green algae and higher plants are similar, for instance, and this fact is used to relate these groups phylogenetically.

Similarly, the distribution of phosphagens has provided evidence about relations between vertebrate and invertebrate animals. **Phosphagens** are compounds containing high-energy phosphate, which can be transferred to ADP to produce ATP. They are found primarily in muscle tissues. When the first studies were undertaken in the mid-1920's, only two phosphagens were known. Creatine phosphate was apparently found only in vertebrates, and arginine phosphate only in invertebrate animals. This difference seemed to be so basic that it was suggested that the term "vertebrate" be replaced by "creatinate." But additional studies indicated that creatine phosphate was also present in some invertebrates. It was first found in some echinoderms: sea urchins and brittle stars had creatine, but other

echinoderm groups, such as starfish, did not. It was also found in some of the lower chordates. So far, this discovery was exciting because it was evidence that the chordates and echinoderms were related (see Figure 16-13). But further studies revealed creatine in organisms that were only very distantly related to vertebrates: the sea anemones and various marine worms. Furthermore, several other phosphagens were discovered in various invertebrates. Phylogenies based on phosphagen distribution were discredited.

The story has recently been revised, however. If we look for the key enzymes in the synthesis of creatine, and not just for creatine itself, a clear phylogenetic picture does emerge. All of the invertebrate animals that possess creatine lack the enzymes necessary to make it. They must be getting creatine in their diets. Even primitive vertebrates (hagfish and sharks) are unable to synthesize creatine. On the other hand, most vertebrates do not merely use creatine, but they can also synthesize it. Phosphagen distribution supports the same scheme of relationships that we have developed through studies of the organisms' structures.

The occurrence of organic molecules is a very useful tool for constructing phylogenies, provided we examine the molecules in their whole metabolic context. But if we ignore these metabolic patterns, the distribution of compounds may seem random and confusing. In other words, molecules, like larger structures, are not truly homologous unless they have similar patterns of development in relation to the whole organism.

Still, biochemical similarities do not provide absolute proof of relationship. The same small molecule may appear in several kinds of organisms simply because it is uniquely fitted to play a particular chemical role. The visual pigments in animal light receptors are a case in point. The retinal molecule

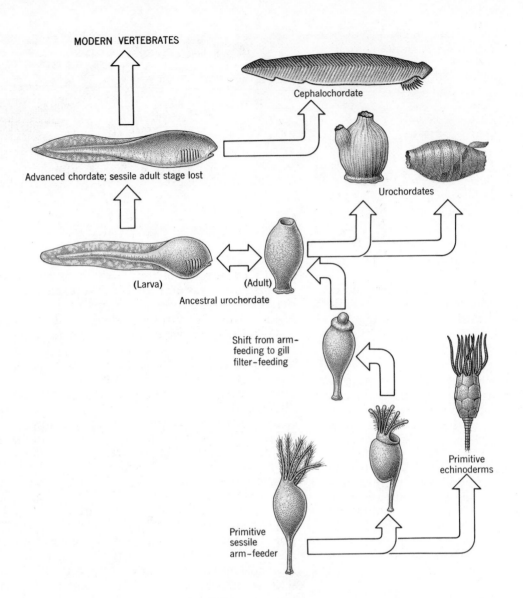

MODERN VERTEBRATES

Cephalochordate

Advanced chordate; sessile adult stage lost

Urochordates

(Larva)   (Adult)

Ancestral urochordate

Shift from arm-
feeding to gill
filter-feeding

Primitive
echinoderms

Primitive
sessile
arm-feeder

*Figure 16-13 The ver-
tebrates are believed to
have evolved from an
echinoderm-like ancestor,
by way of a larval urochordate.*

appears in at least three quite unrelated groups of animals—molluscs, arthropods, and vertebrates— and it functions as the major constituent of the light-sensitive pigment in all three groups. This fact probably reflects nothing more than its unique ability to respond to light; other molecules are not as effective. Selective pressures for vision will be most likely to produce this particular molecule, regardless of the kind of organism.

We now have other more modern and more powerful biochemical techniques for estimating degree of relationship between organisms. For example, as

we discussed in Chapter 4, a test animal can be immunized against a particular protein to stimulate the production of an antibody specific for that antigen. The antibody can then be tested against proteins from other animals. The strength of the interaction between the antibody and the new antigen is a good measure of the similarity of the two antigens. This technique has been applied to primates, using proteins found in breast milk and blood serum as the antigens. The results indicate that human beings are much more closely related to the apes than to monkeys.

A second modern biochemical technique supports this general conclusion. DNA hybridization, described in Chapter 13, measures the similarity between DNA's rather than proteins. DNA fragments from humans, chimpanzees, and rhesus monkeys were tested for their ability to bind with intact human DNA. The binding, or hybridization, is a measure of the similarity in the DNA base sequences. DNA fragments from other humans show the most binding, of course, because they have many complementary base sequences that normally form the double helix. Human DNA was more like that of chimpanzees than that of monkeys.

An even more exciting technique has become available in the last twenty years. We are now able to determine the actual amino-acid sequence of a protein. The technique is routinely used to analyze sequences in homologous proteins from different animals. This work has led to the concept of the "protein clock," which assumes that the number of sequence differences in two protein molecules is a measure of their evolutionary relationship. If the sequences are nearly identical, then the organisms are very closely related; if they have many differences, the proteins must have evolved independently for some period of time. In fact, if we assume that the amino-acid sequences change when the corresponding DNA sequence is altered by random mutation, and that these mutations gradually accumulate as time passes, then we can conclude that the number of differences is directly proportional to the amount of time since the organisms diverged. Results of this sort were summarized in Table 2-3.

The amino acid sequences in hemoglobins from rhesus monkeys, gorillas, chimpanzees, horses, and man are compared in Table 16-1. Hemoglobins from gorillas and chimpanzees are very close to human hemoglobin. Rhesus monkey hemoglobin is quite different from that of larger primates. Horse hemoglobin is very different from all primate hemoglobins, as might be expected.

Surprisingly, these data (and other information as well) indicate that the probability of a change in amino acid sequences is constant in all mammalian lineages: approximately one change every 3.5 million years. Thus, stocks that diverged 3.5 million years ago should differ by two amino acid residues in homologous proteins (one change in each line of organisms). This concept seems to contradict the idea of natural selection. Natural selection predicts that mutations are acted on by environmental pressures. We would expect, therefore, that the number of mutations would be variable; they should not occur at a constant rate. In this case, however, we

Table 16-1

Comparison of Amino Acid Sequences in Various Hemoglobins

Species Compared	Number of Differences
Man–chimpanzee	0
Man–gorilla	2
Monkey–man	12
Monkey–chimpanzee	12
Monkey–gorilla	14
Horse–man	43
Horse–chimpanzee	43
Horse–gorilla	45
Horse–monkey	43

may be dealing simply with changes in relatively noncritical regions of the protein. The changes do not affect the function of the protein, so there is no reason for selection pressures to influence the mutation.

The protein clock concept gives the expected frequency of mutations for which there has been no positive or negative selection pressure. Since we already know, from paleontological evidence, when primate and horse stocks diverged from a common ancestor, we can now date the differences in Table 16-1 and estimate when human and monkey stocks diverged. This figure turns out to be about 30 million years ago. By the same criteria, human lineage branched from that of the chimpanzee and gorilla only about four million years ago.

neodarwinism

Darwin's account predicts that evolution is not a terminal process, something that occurred in the past and is now finished; it is an ongoing feature of the living world. Therefore, we should be able to observe and study evolution. We should also be able to give a detailed account of its mechanism. In particular, we should be able to describe it in terms of molecular biology and genetics. In the *Origin of Species*, Darwin described phenotypic variation in populations, but he had no firm theoretical foundation to explain it. We now have such a foundation, and natural selection can be rephrased using the concepts of genetics. The theory as restated in this way is called **neodarwinism.**

The Evolving Unit

We have already introduced the idea that evolution begins with variation produced by genetic change in individual genes. The basic functional unit of evolution, however, is a population, not a gene. To the geneticist, evolution consists of quantitative and qualitative changes in a gene pool. A change in any particular individual (a mutation, a rearrangement of chromosomes) is significant in evolution only if it changes the population's gene pool in succeeding generations. The change may be fortunate or disastrous for the individual, but if the gene pool is not modified there is no evolutionary change.

A local interbreeding population, which forms the basic evolutionary unit, is called a **deme.** If the boundaries of a deme are poorly defined, there will be frequent breeding encounters between adjacent demes of the same species. Genes will be exchanged, and differences in the two gene pools will tend to average out—though not completely. (If the gene flow is sufficiently rapid between adjacent demes, of course, they are simply part of the same interbreeding population; in other words, there is only one large deme.)

Consider a population spread across a broad geographical area. Sparrows, for example, form a small population that continuously exchanges genes and has an active gene pool. By sampling sparrows from a population, we could describe the frequency of various alleles in that particular gene pool. This actively-interbreeding deme has no sharp geographical limits. Sparrows at the margins of this population breed with sparrows in adjacent regions. In genetic terms, gene flow is rapid between our original population and adjacent populations.

Gene flow would be slower, however, with a second population some distance away. And the two populations might well experience differences in selective pressures: climate, food supply, predators, and available breeding sites, for instance. These factors could result in differences in gene frequencies. Whether or not such differences arise depends on many specific factors: the geographic distribution of the population, its toleration of environmental differences, its mobility, and the rate of gene flow between adjacent populations.

Such differences actually have occurred in populations of the common grass frog in North America. They are all of one species, yet individuals from Canada and from Mississippi cannot produce fertile offspring when brought together. The grass frog is distributed very generally across North America. All adjacent populations can interbreed; at no point are adjacent demes sharply different. But gene flow between adjacent demes is quite slow because frogs have limited mobility. Since frogs cannot move far and must keep close to water, they form partially isolated demes. As a result, demes from northern and southern extremes of this continuous range are so different that they can no longer interbreed. There are several specific reasons. The eggs of the northern frogs, for example, require cooler temperatures for normal development than do the eggs of the southern frogs; cell division is seriously disrupted by improper temperature.

Different demes of the same species are called **varieties, races, breeds** (in domestic animals), or **subspecies.** All of these terms are applied loosely and have no clear biological differences in meaning.

Human demes show many differences in gene frequencies. Barriers of language, religious belief, custom, and social and economic status all help to create local gene pools. Geographical distance is also a strong deterrent to breeding, even with the advent of rapid transportation and easy communication. The probability of marriage is still very high between individuals in an immediate neighborhood, and it falls off very rapidly as the distance between them increases. All these factors tend to divide humanity into numerous demes, whose genetic discontinuity shows up in their gene frequencies.

The ABO blood groups have very different frequencies in different demes. Some South American Indian tribes have only the O allele. In other populations, A ranges from zero to 80 percent, and B from zero to 35 percent. Sharp differences occur in small and culturally isolated groups, such as Eskimo tribes, Indian tribes, or small religious communities. In larger populations, gradual changes can be traced: the B group is common in northern India; it declines steadily as one moves eastward across Europe or northeast through China and into North America. Group A is common in northwest Europe, Australia, and northern North America, and its frequency declines away from these centers (see Figure 16-14). Gene flow is rapid enough to prevent the differences in these larger and more mobile populations from becoming sharp, however.

Sources of Variability

Differences in climate and the availability of food are among the important forces that can produce differences between demes. But before environmental forces can select and favor certain genes over others, genetic variability must exist. Several

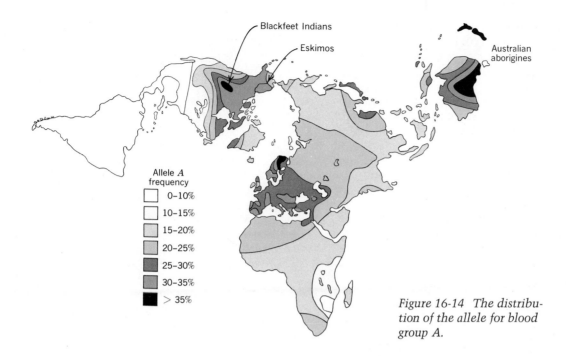

Figure 16-14 *The distribution of the allele for blood group A.*

types of events serve to increase or decrease the genetic diversity in a population.

Mutation

Mutation is the fundamental source of diversity. But mutation operates very slowly. And usually a single mutant allele will eventually be lost from the population.

Assume that a mutant allele (a) occurs in a large population of homozygous alleles (AA). If the one mutant heterozygote (Aa) does not breed at all, of course, the mutation is immediately lost. If it breeds only once, the chance of the allele's loss is ½: the mutant heterozygote can pass on either the A or the a allele to its offspring. If there are two offspring, of course, the chance of losing the mutant gene is reduced to ¼, and so forth. But allowing for a typical distribution of breeding success, the average chance of loss is about 37 percent in each generation. In other words, if we start with 100 mutant genes randomly distributed among members of the population, we can expect only 63 of the mutant alleles to survive to the next generation. Of these, about 40 would remain after two generations, about 25 after three, and so forth until the mutant gene eventually disappears. This calculation assumes, however, that the mutant gene has neither positive nor negative survival value.

Unless the mutation has some adaptive advantage, or occurs very frequently, it will simply be lost from the population.

Now let us assume that the new mutant *is* conserved in the population because it confers some selective advantage on the individuals possessing it; therefore all the mutant genes would be passed to a subsequent generation. In that case, how long would it take the new mutant allele, a, to replace allele A? The time span depends on the frequency of the mutation. Say there is one mutation in every 100,000 generations. In that case, it would require 69,000 generations to obtain a population having equal frequencies of A and a. Mutation is a source of variability in a population, but it is clearly too slow to account for the rapid evolution of genetic differences between adjacent demes.

Genetic Drift

Faster changes in gene frequency result from a mechanism known as **genetic drift.** Genetic drift produces differences in gene pools through statistical sampling errors.

Suppose you have an enormous sack of mixed marbles, 50 percent black and 50 percent white. You draw samples from the sack. If you pick out a thousand marbles, there will be close to 500 of each color—say 483 black and 517 white marbles. If you now make up another sack of marbles with these new frequencies (48.3 percent black and 51.7 percent white) and draw another thousand marbles, you might again get fewer black marbles, or you might reverse that trend. The rules of statistics predict that if you continue this procedure you will get frequencies of black and white marbles that vary on either side of 50 percent, but no long-term trend in either direction.

But this prediction holds true only if the sample numbers are large. If you do the same experiment with only four marbles per sample, there is a reasonable chance ($1/16$) that all four marbles on the first draw will be either black or white. In that case the other color is eliminated right away. If you draw three black and one white on the first draw and then refill the bag accordingly—three-quarters black and one-quarter white—then the next draw has about a one-third chance of producing four black marbles right away. And even if it doesn't, you have a 40 percent chance of drawing three black and one white again so that the same odds will prevail on the following draw. Several draws later you will almost certainly wind up with only black marbles. Of course, once this happens, you have a uniform population of marbles and variability has been eliminated.

When the same thing happens to gene frequencies in small populations of organisms, it is called genetic drift. When a small part of a larger parent population is separated and isolated for several generations, one of two alternative alleles will probably be lost. The smaller the population, the greater the probability of genetic drift. It is truly a random process and has nothing to do with natural selection. There is no way to predict which of two alleles will become fixed in a small population and which will be lost. We can predict only that one or the other will be entirely lost.

Genetic drift reduces genetic variability among individuals, but at the same time it produces diversity between demes that arise from small founder populations. Genetic drift accounts for the wide variation in the frequency of blood groups in small and isolated human populations. It also explains how these differences can arise so rapidly.

Selection Pressures

Selection is another way in which differences in gene frequencies between demes are produced. Perhaps the best example is industrial melanism in moths, which we discussed in Chapter 1. The moths' coloration is protective, serving to hide them from predatory birds (see Figure 1-7, page 16). The color of the carefully-studied *Biston betularia* is controlled by a single dominant allele. This allele for color also controls the moths' selection of what they land on, so that dark moths prefer dark backgrounds. Consequently, as soot has blackened the industrial landscape, dark moths have had a survival advantage.

Experimenters demonstrated the working of these selective forces by releasing dark and light forms together in both polluted and unpolluted environments. About half of the moths with inappropriate coloration were killed by birds within a day after their release. Later, some of the surviving moths were recaptured. The ratios of light to dark had clearly changed. As expected, birds had preyed selectively on the light forms in soot-darkened woodlands and on the dark forms in unpolluted areas. The selective advantage of the dark allele is very great in sooty woods, and its gene frequency in populations near industrial towns is as high as 90 percent. Yet the gene is still quite rare in unpolluted areas. Industrial melanism shows how effectively selective pressures can favor one or the other of two alleles in a particular deme. This sort of pressure is called **directional selection.**

Another kind of selection pressure operates against individuals who vary widely from the population norm. This mechanism is called **stabilizing selection,** because it tends to stabilize phenotype in the species. The classic study of stabilizing selection was carried out on a collection of English sparrows found helpless after a severe snowstorm. The birds were brought into the laboratory, and those that died were compared with those that recovered. Structural features of the dead sparrows deviated more from the species norm than those of the survivors; the latter were quite average. Similarly, mortality is greater in human infants weighing much more or much less than eight pounds at birth than it is in average, eight-pound children. Stabilizing selection, like directional selection, works by reducing variability within a deme; this weeding-out tends to produce a group of individuals of uniformly the "best" genotype.

Natural selection may favor a stable balance of different alleles in the population. In fact, the maintenance of a variety of genotypes in a population is itself an adaptive benefit. A population that contains several different phenotypes may be able to inhabit several kinds of environments, for example. Its versatility would be a selective advantage. Such a variety of phenotypes could be achieved if the gene pool of the species included several alternative sets of genes or groups of alleles, and if these groups were to segregate as a unit. (Such grouped sets of genes are called supergenes.) This process can quickly lead to the existence of two or more demes of the same species in the same geographical area.

Still another source of maintained diversity is the selection for heterozygous individuals. We noted earlier, in Chapter 15, that the gene for sickle-cell anemia protects heterozygous individuals against malaria. A deme living in an area where malaria is endemic will therefore have a high frequency of the sickling gene, even though the gene can be lethal when it becomes homozygous.

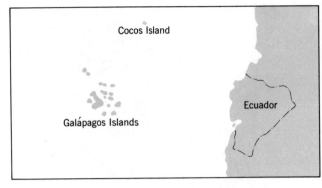

Figure 16-15   *The Galápagos Archipelago.*

Speciation

The genetic constitution of a deme may change for various reasons: mutation, genetic drift, selection pressures, or the migration of individuals into or out of the deme. If individual demes remain in contact and interbreed, gene flow will tend to average out the differences. When demes become isolated, however, the differences between them become more and more extreme. Ultimately, individual demes may diverge so greatly that they become new species. The process of species formation is called **speciation.**

Reproductive Isolation

The isolation of demes may result from drastic events in the physical environment—the formation of a glacier, forest fires, or the division of a land mass by water. The fossil record indicates, for instance, that North America was once continuous with Asia by way of a land bridge across what are now the Bering Straits. When this land bridge was disrupted, many demes became separated. Differences in genome continued to develop over subsequent generations, and demes of the same species gradually diverged, so that interbreeding was no longer possible. The result was the production of two new species, an Asian and a North American one, from each original species. The geographical isolation of demes promotes speciation.

Such catastrophic events are powerful forces in speciation, but they probably were too infrequent to have created the millions of different species that now exist. Discontinuities and differences within stable environments are probably more usual causes of isolation. After all, critical features of the habitat, such as ponds, food plants, and

shelter, are usually unevenly distributed. If migration between demes in separate habitats is rare, then speciation results without the intervention of a major catastrophe.

Darwin's finches on the Galápagos Islands are the most famous example of speciation produced by discontinuous habitats. Darwin spent five years (1831–1836) aboard H.M.S. *Beagle* on a surveying expedition along the coasts of South America and Australia. The Galápagos Islands, about 600 miles off the coast of Ecuador (Figure 16-15), were included in the survey. Many of the things Darwin observed on the Galápagos led him to think about speciation. The most famous occupants of the Galápagos are the finches.

An early observer of the Galápagos finches described them as rather dull, "not remarkable for their novelty or beauty." This judgment about their beauty is correct enough. The birds are short-tailed, with dark plumage, and all are very similar in overall appearance. But their interest and novelty is quite another matter. The birds show a diversity of feeding habits. Some are seed eaters, others feed on insects, others on fruit and buds. This diversity of feeding habits is paralleled by a remarkable diversity of beak form. Some have typical finch like beaks, but others have slender beaks like those of warblers; still others have beaks adapted for probing flowers, for boring wood, or for cracking nuts (Figure 16-16, *a-c*).

The woodpecker finch, *Camarhynchus pallidus,* shows the most remarkable adaptation to its food source. It runs up and down the vertical limbs and branches of trees and bores holes with its stout beak (Figure 16-16*d*). A typical woodpecker bores down to the insect and then extracts it with its

(a)

(c)

Figure 16-16 Darwin's finches. (a) A small ground finch that lives on seeds. (b) A ground finch with a stout beak for cracking larger seeds. (c) A finch specialized for eating cactus fruit. (d–g) A sequence of photographs showing the woodpecker finch, Camarhynchus pallidus, using a cactus spine to dig out a grub.

long tongue. But finches have short tongues. *Camarhynchus*, therefore, holds a cactus spine or a small twig in its beak to probe for its prey. It digs the insect out, drops the cactus spine, and eats its prey. This behavior is a fascinating example of tool use by a bird.

Darwin was much impressed by the Galápagos finches in general. In his *Journal of the Voyage of the Beagle* he says, "Seeing this gradation and diversity of structure in one small, intimately related group of birds, one might really fancy that from an original paucity of birds in this archipelago, one species had been taken and modified for different ends."

The speciation of the Galápagos finches is probably related to both their geographical isolation and the availability of numerous food sources. A primitive population of finches probably arrived at one of the islands from South America. They began as a single interbreeding gene pool. From there they spread to the outer islands, where they formed relatively isolated subgroups. Natural selection then began to produce differences in the genome that enabled the birds to adapt to various food sources on each island. Members of these outer populations would, of course, occasionally return to the central island. But there the selection pressure toward the development of specialized food sources would continue, because the newcomers would have to compete with the local population for food. In this way, the original single species could radiate and form a dozen or more closely related species.

Cocos Island, a solitary island about 600 miles northeast of the Galápagos Archipelago, provides a natural test of this hypothesis. It is far enough away to make exchanges with the Galápagos or the mainland extremely unlikely. Cocos Island has as many food sources as the Galápagos, and no competition from other land birds. Yet only one finch species occurs there. The single small island of Cocos cannot provide the isolated habitats necessary for speciation, and it has not occurred.

(d)
(e)
(f)
(g)

Once new species have been produced by geographical isolation, they will probably remain reproductively separated. In fact, they will most likely stay separated even if interbreeding is possible, and even if they return to the same territory. Hybridization—the production of offspring by members of two different species, races, or varieties—can occur between members of well-defined species. More often, though, hybrids between species are inviable (the offspring fail to develop), or they are sterile. The hybrids of horses and donkeys are a good example. A cross between a male donkey and a female horse yields a mule, a desirable beast of burden. The reciprocal cross, between a stallion and a female donkey, yields a hinny; it is inferior to a mule in size and strength, and is not deliberately bred. Male hinnies and mules are always sterile, and females breed only rarely; the eggs and sperm do not develop normally, because of difficulties in meiosis.

Fruit flies provide another example of hybridiza-

tion. Crosses between different species of *Drosophila* produce vigorous first-generation hybrids. We do not know why, but the offspring of these hybrids always turn out to be weaker than either strain and are rapidly eliminated from a natural population. Because hybrids are almost always inferior to both parental species or races, a species that avoids hybridization has a natural advantage over one that does not. Hybridization is prevented in a number of ways. Flowering plants may remain isolated by producing flowers at different times. Differences in animals' courtship sequences may prevent copulation between members of different species.

Hybridization

Hybridization is often avoided by species that have become reproductively and genetically isolated. But sometimes hybridization may itself lead to speciation. Hybrids are more common among plants than animals because behavioral isolation is more difficult in plants. Plant hybrids are often sterile for the usual reason, the failure of meiosis. The chromosomes from the two distantly related parental stocks cannot come together to form tetrads, because they are not homologous. The chromosomes are thus distributed abnormally to the developing gametes.

If the parent plants have haploid chromosome numbers of three and four respectively, the chromosomes of the hybrid obviously cannot be paired. But if the hybrid plant can persist by asexual reproduction, it may at some point form homologous chromosome pairs by becoming polyploid. DNA duplication without subsequent cytokinesis, early in development, would have this effect. Each chromosome would be simply duplicated, so the total number of chromosomes would be doubled. In this way our hybrid could produce a polyploid flower. It would now have fourteen chromosomes in seven homologous pairs, and meiosis could then occur normally. A new species would have been produced.

This form of speciation is important in plants but extremely rare in animals. Many natural plant species were probably formed by such hybrid–polyploid events. And plant breeders have used this method to create new species. Polyploidy produces major changes in genotype quickly. Similar changes could arise through mutation and selection, but would occur much more slowly.

Species

An important consequence of neodarwinism is a sweeping revision of our concept of a species. Western philosophers have always divided the living world into definite, discrete kinds of organisms. In Greek philosophy, Plato introduced the concept of the ideal, or type: particular cats or chairs or houses were only imperfect examples of a true and changeless type, the ideal cat or chair. The Biblical account of creation describes the origin of specific kinds of organisms, which have passed to the present unchanged. Discoveries that suggested the idea of change and mutability were strongly resisted. Rather than admit that organisms have changed, Biblical scholars used Noah's flood to account for early fossil discoveries. But students of geology soon realized that the fossil record was too complex for this explanation. As many as a hundred catastrophes like the flood had to be introduced, many of them followed by new acts of creation to account for the diversity and apparent change in the fossil record. Yet to many scholars,

such roundabout expedients still seemed preferable to admitting the possibility of change. Obviously, the philosophical and religious idea of fixed and unchangeable organisms is flatly in opposition to theories of evolution.

This same philosophical tradition has contributed to the practice of assigning every organism to a fixed species category. A species was originally considered to be a basic type of organism. Only one individual was needed to define the entire species. A biologist carefully described one **type specimen,** a sample that he considered representative of the species as a whole. He then provided a name for the species and deposited his type specimen in a collection for future reference. Any variations that existed among individual organisms were considered simply to be accidents. They had no significance except as examples of the fallibility of the real world as contrasted with the ideal.

When naturalists began to collect specimens of the same species over a wide area, however, significant problems arose. Animals or plants that would normally have been placed in the same species varied systematically with their location; a good example would be the continuous geographical variation of the North American grass frogs. Clearly a species could no longer be defined on the basis of one typical specimen. The range of variability within a species had to be considered. Type specimens were therefore replaced by **type collections** of many individuals, preferably from several localities.

This extension of the type specimen concept is both impractical and theoretically unsatisfactory, however. The potential range of phenotypic variation is infinite. Even if we collected and preserved all the existing individuals in a species, we would not exhaust all the possibilities. And even the most avid and orderly biologist rejects the idea of preserving the entire world in formalin. In any case, this approach does not solve the basic problem of defining a species. Are demes of grass frogs that can no longer interbreed still members of the same species? Collecting and preserving grass frogs cannot answer this question.

The modern definition of species reflects the influence of evolutionary ideas: a species is now defined as a sequence of populations that are related as ancestors and descendants. One species is separated from other sequences of populations by its evolutionary history, and also because it possesses a separate set of capacities. In other words, a species is a gene pool. A species may become distributed into several population sequences spread over a broad geographical range. However, this relation may be partially or completely interrupted by geographical separation. Whether or not separated populations are still part of the same gene pool must be verified by observation. Thus, the primary criterion of a species concerns the reproductive relation between populations, potential or actual.

The old "type specimen" definition simply cannot cope with the range of variation in a population. But though the modern definition of a species is theoretically satisfying, it is often quite difficult to establish the practical limits of a particular species. The following approaches are used to determine whether two organisms represent separate species:

1. If they coexist in the same area with no evidence of cross-breeding, they probably represent two separate gene pools.
2. If we place members of the two groups together and they do not mate, or they produce weak or sterile hybrids, they probably have separate gene pools.
3. If they are extremely different in appearance, they are probably not members of the same gene pool.

The first criterion is the best, but populations are often separated geographically, so the test does not apply. The second test has several difficulties. Many organisms do not breed in captivity; a failure to reproduce does not necessarily mean that they cannot reproduce. And conversely, viable and healthy hybrids can be obtained in the laboratory even though they are not observed in areas where the populations coexist—the existence of healthy offspring does not necessarily mean that the organisms are members of a common gene pool. The third test is simply a rephrasing of the traditional subjective judgment of a specialist.

Agreement among the three tests is far from perfect. Organisms that look almost identical may be genetically different species, existing in the same area with no observable hybrids. And conversely, populations with several different types (for example, light and dark moths) can be part of the same gene pool.

In practice, most species are identified and named using traditional structural criteria, based on the description of a single type specimen. Unfortunately, this method remains a practical necessity. Probably ten million species of organisms exist at the present time. Only a miniscule number of them have been studied carefully enough to analyze their status as a species according to all three modern criteria. Thus we retain the apparatus of museum collections of type specimens, even though the theoretical basis for this procedure is now shaky. This difficulty is not a crippling one, however. It is simply a necessary consequence of the need to impose order and definition on an external world that is in a state of dynamic change; the imposed order will always be arbitrary and somewhat artificial.

the evolution of man

Humans belong to a group of animals called **primates** ("primary animals"). It includes lemurs, monkeys, apes, and man.

Primate Evolution

Primates first appeared in the fossil record about 80 million years ago. The ancestral primate was a small, shrewlike animal. As later primates evolved, they became arboreal (tree-living) and usually omnivorous. They had limbs and hands and feet (and often, tails) adapted for grasping and agile movement. These general features are reflected in primate bony structures. At the same time, the eyes of primates became large, and they moved forward in the head for binocular vision ("two-eyed" vision, necessary for precise depth perception). Their noses and jaws were shortened in the process. The number of teeth became reduced from the primitive mammalian formula, and the teeth were relatively unspecialized, as would be expected for an omnivorous diet. Their limbs lengthened, their hands became adapted for grasping, and the fingers were protected by nails rather than claws. Their dependence on increased mobility and visual inputs was reflected in a greater brain size.

This general pattern of evolution has produced living primates ranging in size from tiny lemurs (the size of a mouse) to gorillas, the largest living primates (see Figure 16-17). Male gorillas may weigh 800 pounds. Most primate species are agile and unspecialized animals; tarsiers are so agile that they can leap into the air and catch insects in midflight. There are exceptions, however, such as the sluggish loris and the aye-aye, a specialized lemur with huge incisors for husking and opening coconuts.

(a) (b) (c)

(d) (e) (f)

(g)

(h)

(i)

Figure 16-17 Some living primates. (a) Ringtailed lemur. (b) Philippine tarsier. (c) Gray slow Loris. (d) Squirrel monkeys. (e) Spider monkey. (f) Lion-tailed macaque. (g) Baboons. (h) White-headed gibbon. (i) Orangutan. (j) Chimpanzee. (k) Gorilla.

(j)

(k)

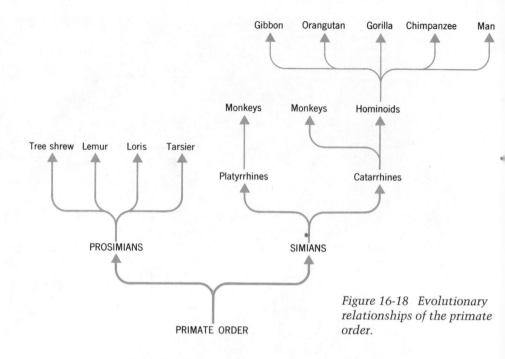

Figure 16-18 Evolutionary
relationships of the primate
order.

Human Evolution

The general picture of human evolution has recently been clarified by isotope dating. The details are still controversial, however, and probably always will be.

Man is most closely related to the chimpanzees and gorillas. The evidence supporting this conclusion includes structural similarities, similarity of blood serum and proteins, and common infectious diseases. Chimpanzees and man also show similarities in behavior. Chimpanzees have a very complicated social life in the wild, which we are just beginning to understand. They use and even manufacture tools. The classical observation of tool use was made by Jane van Lawick-Goodall. She saw a

The Primate order is divided into two suborders, to accommodate the great range of different primates. The **Prosimiae** (prosimians) consist of tree shrews, lorises, tarsiers, and lemurs. The **Simiae** (simians) include monkeys, apes, and man. The simians are further divided into **platyrrhines** (flat noses) and **catarrhines** (narrow noses). The catarrhine group contains the **hominoids** (gibbons, orang-utans, chimpanzees, gorillas, and man). Figure 16-18 is a simplified breakdown of the primate order.

chimp use a twig, thrust it into a termite nest, pull it out, and lick off the termites. The animal went some distance from the termite nest to collect twigs, and broke them into suitable lengths. He even brought back several at once to serve as spares. Many books report the intelligence and complexity of ape behavior. For all these reasons, man is classified zoologically in the same subgroup of primates as the apes and gibbons.

Comparative studies account for some uniquely human characteristics by analyzing trends in primate development. The length of prenatal development is quite similar in many primates. In fact, in both man and monkeys, the limiting factor in the development of the fetus is the relation between its head size and the size of the birth canal. The fit is very close, and birth is hazardous and slow. The newborn have very thin bones that cover a rapidly-growing brain; the junctions between the bones are not completed, and the cranium is not thickened, until some time after birth. In humans, this incomplete cranial development in the newborn is accompanied by a delay in the development of the bony structure of the limbs. (Gorillas and chimpanzees bypass this problem because their newborn are very small compared to their adult size; a newborn gorilla weighs only three or four pounds, and delivery is correspondingly quick and easy.)

These limitations on prenatal development mean that many of the differences between primates may show up only after birth. The higher primates have longer and longer postnatal periods of development. There is a corresponding increase in the length and complexity of the maternal care. The trend reaches its extreme in man. Human brain growth, for example, is incredibly rapid after birth. During the infantile period the brain's weight increases from about 350 grams to more than 800 grams one year later. Ninety percent of brain growth is completed in the first three years after birth. This development represents, in effect, a continuation of fetal development outside the uterus. The extension of the infantile and juvenile periods in humans and other primates has been a key factor in the development of the complicated social behavior that is associated with maternal care and family units.

The Fossil Record

Better information about the immediate ancestors of man comes from fossils. The fossil record for man and other hominoids is rather sparse, and remains are often very incomplete. Important findings may consist of a few teeth or a jaw fragment. However, some dedicated individuals have spent their lives searching out and interpreting fossil material.

Of course, they are not infallible. "Piltdown Man" is famous because it was a deliberate fraud. Parts of a skull, found in a gravel pit at Piltdown, England, in 1911 were estimated to be over a million years old—until British scientists realized that the jaw bone belonged to a modern ape; the fraud wasn't discovered until 1953. Another mistake, this one made in good faith, was the description of a fossil tooth as a new genus of fossil man (*Hesperopithecus*); the tooth was later identified as a worn tooth of a fossil peccary. However, these rare slips are usually quickly corrected.

The fossil record has been dated, analyzed, and organized into a sequence of forms that are progressively less apelike and more manlike. The deciding characteristics are cranial capacity, tooth structure, and various subtle secondary indications. Figure 16-19 is an interpretation of human evolution, beginning with the dryopithecines, a collec-

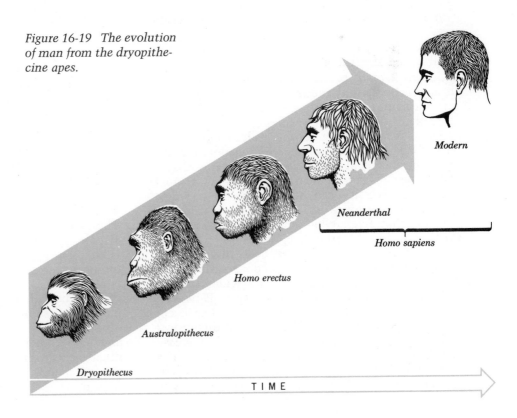

Figure 16-19 The evolution of man from the dryopithecine apes.

Modern

Neanderthal

Homo sapiens

Homo erectus

Australopithecus

Dryopithecus

TIME

tion of apelike animals dating from fifteen to 25 million years ago. They were a very diverse group, and probably gave rise to both the great apes and man. *Dryopithecus* ("tree ape"), the genus for which the group is named, is probably right on the ancestral line of the apes; other dryopithecines (*Ramapithecus* and *Kenyapithecus*) may be the ancestors of man. Still other genera (*Proconsul* and *Oreopithecus*) became extinct and have no direct modern relatives.

The human evolutionary line leads from tree apes to the large and varied genus *Australopithecus*. Detailed interpretation of these forms is based on about 100 fossil finds; most of them are from southern Africa, but occasional remains have been discovered elsewhere. The oldest dates from about two million years ago. The skull of *Australopithecus* is apelike, and its cranial capacity is quite small. However, its pelvic girdle allowed it to stand fully erect. Primitive stone and bone tools have been found with australopithecine fossils at Olduvai Gorge in East Africa.

The genus name *Homo* applies to a variety of fossil forms with increased cranial capacity, a clearly erect carriage, and human dentition. Many were first described under a variety of other names (*Pithecanthropus*, or Java Man; *Sinanthropus*, or Peking Man). These fossils, now all classified as *Homo erectus*, date from about half a million years ago.

Modern man, *Homo sapiens*, first appears in the fossil record about a quarter of a million years ago. All *Homo sapiens* remains can be divided into two groups, Neanderthal and modern. A great number of other names are submerged in this classification: more than 100 different species names have been assigned at one time or another to humanoid fossil remains. This recent simplification results from the modern concept of a species, which states that a wide range of geographical subspecies with great structural diversity can still belong to a single species. Furthermore, our study of living primates indicates that diversity among geographical races is the rule, rather than the exception.

Modern man, after all, has been divided into more than thirty races. A tremendous variation in human gene pools is reflected in a wide range of structural and biochemical phenotypes. However, any decision to designate a particular pool as a separate race is purely subjective. It has no theoretical biological basis; careful studies reveal no decrease in viability or subsequent fertility in matings between different races of man. Hence all modern human beings belong to a single species. (Such variability is characteristic of other hominoids, as well. Chimpanzees have received 21 different generic names and more than seventy species names; they actually are more variable in fundamental skeletal structure than modern man.)

Our modern concept of a species, coupled with the known variability of primates (both living and fossil), suggests that human evolution occurred through a sequence of populations, each descended from an earlier one. The populations in the sequence from tree apes to modern man may have been related to each other in either of two possible ways. The first theory is based on the assumption that at any particular time several geographical populations existed, but they all belonged to a single species. All the coexisting populations must have contained various kinds of individuals, some advanced and some less so. However, all the populations still represented a single gene pool at all stages of the evolutionary sequence.

The second theory, by contrast, assumes that the various geographical races of early man may have become fully isolated, forming separate but coexisting species. Some of these species then evolved into the gene pool of modern man, while others became extinct. On theoretical grounds this second theory seems more attractive, but a good choice between the two will require additional evidence (see Figure 16-20).

Human Behavior

A few characteristics have been singled out as particularly significant in the evolution of human behavior. Three of them are closely related. The first was the evolution of erect posture. The erect stance then freed the hand for further development. And finally, both of these developments in turn were accompanied by an increase in brain size.

Brain Size

The size of the human brain has received much attention in evolutionary theory. Part of the reason is that brain size is one thing that distinguishes humans from the rest of the animal kingdom; but it is also because brain size is one of the few things that can be deduced from fossil skulls. Modern people, in fact, show a considerable range in brain volume, but there is no relation between brain size and any known measure of intelligence. As we might expect, brain function is more dependent on the fine details of neuron interconnections than on total mass.

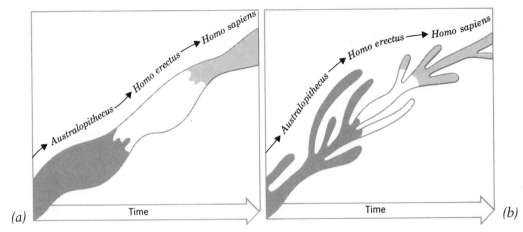

Figure 16-20 Two alternative interpretations of human evolution. (a) Evolution may have proceeded in a direct, linear sequence. A wide range of phenotypes was present at any one time, but all were members of a single gene pool. (b) Evolution may have proceeded in a branching fashion, with populations becoming isolated and having separate gene pools. According to this hypothesis, Australopithecus *and* Homo *could have coexisted.*

Still, despite all these reservations, the increase in brain size in the evolution of man in the last two million years is most impressive. *Australopithecus* had a cranial capacity of about 500 ml; modern man's is about 1400 ml. The increase in brain size was fastest in the transition from *Homo erectus* to *Homo sapiens*; little increase in brain size has occurred during the last 100,000 years, during the transition from Neanderthal to modern man. The overall pace of evolutionary change has been extremely rapid.

Erect Posture and Hands

A trend toward erect posture is characteristic of the primates in general; it is a major feature of human evolution. The primates are adapted for climbing and grasping. Such activities often require the body to be held vertically. Climbing, swinging from a branch, or simply sitting (a popular pastime among most primates) all require an erect posture. Such a posture frees the hands for climbing and also for food handling, body care, and fighting. It also improves the range of vision, because the head can be pivoted to observe the complete horizon. (Most primates can rotate their heads laterally about 90 degrees with respect to their bodies. Apes are more limited by heavy neck muscles, but tarsiers can rotate their heads almost a full 180 degrees.)

The arboreal life style encouraged an erect posture early in primate evolution. The ape can stand unassisted for considerable periods, though it normally moves on all fours. In humans, the pelvis has been modified to support the main weight of the body. The knee and hip joints are straight rather than flexed as in the ape, so in man the shock of locomotion is absorbed by the spinal cord. The spinal cord is flexed, and pads of tissue between the vertebrae further cushion the shock of walking and jumping. The arch in the human foot also absorbs shocks. These adaptations function so well that at the end of an active day, a human being's vertebral column is shortened by perhaps ¼ of an inch, and the foot is elongated.

As early as the australopithecines the hand, freed by erect posture, was highly developed. Careful analysis of an *Australopithecus* forelimb suggests that its hand was as strong as modern man's, and that the thumb could be rotated and brought across the face of the hand as it can in modern man. But the shortness of the thumb and other fingers makes it unlikely that *Australopithecus* could touch the ball of the thumb to the forefinger. This movement is critical for precise use of the hand. It is found only in man. Still, the use of crude tools made of bone and stone had already appeared at this stage of human evolution.

Language

Man's erect posture, dextrous hands, and large brain size are obvious parts of his primate legacy. They represent merely the culmination of trends visible in the entire primate order. But there is one fundamental difference between man and all the other primates: language. The ability to communicate is so much more highly developed in man than in any other primate that the difference is qualitative rather than quantitative. Language permits the rapid and precise transmission of cultural information from one generation to the next. It allows for the storage of the cumulative experience of the group. It underlies the capacity for abstract thought. Such language ability simply does not exist in other primates. Attempts to teach chimpanzees to talk have met with very little success. Deficiencies in brain structure limit the success of such attempts; so does the structure of the tongue and vocal apparatus. Chimpanzees and other living nonhuman primates are physically incapable of producing the range of vowel and consonant sounds that make up human speech.

The upper portion of the human vocal apparatus is a tube formed by the tongue in the floor of the mouth and the mouth cavity itself. The shape and position of the tongue can be precisely controlled. By moving the tongue backward or changing its shape we can produce a considerable range of basic sounds. These sounds are then modified by the muscles of the lip and throat to create the full range of human vowel and consonant sounds (called **phonemes**). The shape of the oral cavity is different in nonhuman primates; the tongue is thinner and cannot alter the diameter of the upper vocal apparatus.

The relationship between vocal structures and the sounds they can produce has been studied by computer. A computer model was used to generate all the possible phonemes for various kinds of vocal structures. When this computer simulation is carried out for macaque monkeys, a rather limited range of vowel and consonant sounds results. These sounds compare well with recordings of the actual vocalizations of macaques. Human infant vocal structure cannot produce the full range of human speech sounds; and again, the range of possible phonemes is correctly predicted by the computer. The vocal apparatus of Neanderthal man has been reconstructed, and it has been subjected to a similar analysis by computer simulation. The range of speech sounds predicted for Neanderthal man is quite limited compared to that of modern man, though greater than the range for monkeys or chimpanzees; it is quite similar to that of newborns of modern man. Neanderthals were able to produce the vowel sounds in the words *bit, bet, but,* and *bat,* but not those in *beet, boot, bought,* and *bite.* Some consonants were physically possible for them (such as *b, d,* and *v*), while others were not (*g* and *k*). If Neanderthal man had a language, it was much more restricted than modern human speech. Figure 16-21 diagrams the vocal apparatus of a newborn human, a Neanderthal, and adult human. The supe-

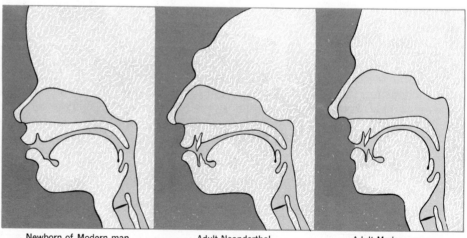

Figure 16-21 The vocal apparatus of a newborn modern human, an adult Neanderthal, and an adult modern human. The infant and the Neanderthal are similar, and both are probably capable of the same range of speech sounds.

Newborn of Modern man   Adult Neanderthal   Adult Modern man

rior vocal apparatus of modern adult man results from a shortened jaw and an increased angle between the skull and the vertebral column. This physical apparatus, of course, had to be accompanied by the evolution of the appropriate neural circuits.

Culture

All these characteristics evolved together, of course; there is no simple cause-and-effect relationship. However, more than any other trait, language seems to be most directly related to evolution of a culture. The physical capacity for modern speech probably arose within the past hundred thousand years. The result was the creation of culture, first as an oral tradition, later in the form of written records.

The long juvenile period of man permits the acculturation of offspring: language, symbols, myths, taboos, mores, and codes of behavior shape the new individual.

The influence of culture on the behavior of the individual is overwhelmingly strong. A human culture has considerable stability, manifested in conservativism and love of tradition. But it is also capable of responding to changing conditions, much more rapidly than a gene pool. We are not saying, however, than culture is somehow nonbiological, nor that it is a substitute for slower changes in the gene pool. Culture and genetic change interact and reinforce each other. When a cultural innovation occurs (for example, the development of agriculture in the river valleys of Mesopotamia) the individuals who can use the new life style are reproductively favored, and the gene pool changes in response. In turn, changes in the gene pool lead to further cultural innovations. Culture is an immensely powerful and successful adaptive device—so successful that man is now the dominant life form on earth.

Recommended Reading

Darwin, Charles, *Origin of Species*. Numerous editions.

> Darwin's original statement of the theory of natural selection. There have been many restatements of the theory, but the first is still enjoyable.

Dobzhansky, Theodosius, *Mankind Evolving*. New Haven and London: Yale University Press (1962).

> A clear, interesting discussion of the evolution of man.

Montagu, M. F. Ashley, ed., *Culture: Man's Adaptive Dimension*. New York: Oxford University Press (1968).

> A selection of articles by prominent anthropologists on the role of culture in the evolution of man.

17 the origin and variety of life

Theories of the Origin of Life

Modern astronomical and geological data indicate that the Earth is about five billion years old, and our galaxy about seven billion years old. Life must have originated on Earth no more than five billion years ago, unless it arrived from an extraterrestrial source.

Some people have, in fact, proposed that life on Earth arose from spores that drifted across space from other planets. The diffusion of life from other sources within our galaxy is not impossible, but it is extremely unlikely. The distances involved are vast, and the time available for the process is relatively short. And no one has seriously suggested that a seeding process could have occurred across the gigantic distances that exist between galaxies. In any case, any such hypothesis simply shifts the location of the origin of life, without changing the time scale very much: a first generation of life must still have occurred somewhere in the galaxy in the past few billion years.

special creation

The most widely accepted and influential theory of the origin of life is special creation. The Christian version is stated in the first two chapters of Genesis. After creating the heavens and the Earth and dividing the land from the water, God created the land plants on the third day of His labors. On the fifth

M. C. Escher, "Mosaic"

547

day, aquatic animals and birds were created. Land animals and man were the work of the sixth day. Variety was a part of creation; each herb and fowl and creeping thing and beast of the earth was created "after his kind." The classification of this variety is discussed in Chapter 2 of Genesis. God installed Adam in the Garden of Eden and brought before Adam each of His creations: ". . . whatsoever Adam called every living creature, that was the name thereof."

Special creation is a simple and attractive hypothesis with a great deal to recommend it. It is irrefutable—all the physical factors that are interpreted in terms of evolution can also be incorporated into special creation. For example, one can state that the fossil record was created *de novo* and does not necessarily prove that the creatures actually lived. The same argument can be applied to isotope ratios, serology and protein structure, and all the observed features of nature: they were simply created that way. Special creation explains the adaptation of living things to their environment, their complexity, and the similarities between organisms: each creature is part of an overall design created by God.

This brief summary does not do justice to this view of the origin of life. In fact, in 1970, the legislature of the State of California admonished the school system to give special creation equal time with evolutionary theories. We shall not attempt that here. Special creation is part of an alternate world view which is separated from biology and the other natural sciences by an entirely different canon of evidence and by different criteria of truth and authenticity. Other sources should be consulted for an adequate account.

spontaneous generation

Until the latter part of the nineteenth century, nearly everyone in the Western world—scientists and laymen alike—adhered to the Christian doctrine of special creation. The work of Louis Pasteur simply seemed to confirm this idea. Pasteur showed that a rich nutrient solution would remain lifeless so long as one killed all the organisms initially present and prevented others from gaining access. He proved that all organisms came from pre-existing organisms. It was concluded that living organisms could not arise spontaneously out of nonliving matter. Divine intervention seemed to be the only alternative explanation for the origin of life.

The idea of spontaneous generation was soon revived, however, by the evolutionary theory of Charles Darwin. Darwin himself, and many of his supporters, felt that spontaneous generation and subsequent evolution could account for the existence of life on Earth. Darwin's theories explained how organisms could change and become more complex through time—how a variety of different creatures could evolve from a single ancestor. There were continuing experimental attempts to generate microorganisms under special conditions. However, many scientists recognized that the chemistry of organic compounds was so poorly understood that it would be impossible to design sensible experiments.

As organic chemistry became more sophisticated, the problem appeared to be even more insoluble; chemists now know that organic compounds could never have formed spontaneously as long as oxygen was present in Earth's atmosphere. All organic material is oxidized by atmospheric oxygen. Reduced carbon in sugars, fatty acids, amino acids, and other organic monomers are oxidized to carbon dioxide and water. This process is usually hastened by micro- or macroorganisms which use some of the chemical potential energy in the organic compounds for their own purposes. However, oxidation will also occur without the intervention of organisms. Any chemical change will proceed toward oxidation, though the rate may be fast or slow depending on the chemical stability of the reduced carbon compound and on the physical environment. The sugar in your cupboard will be there tomorrow morning, but it will eventually oxidize; you can speed the process by raising its temperature so much that it burns. On the other hand, there is simply no way to form sugar spontaneously from a solution of carbon dioxide and water.

Important new information about the Earth's atmosphere began to accumulate early in this century. It was increasingly apparent that Earth has not always had an oxidizing atmosphere. Spectroscopic analyses showed that ammonia, methane, and ice are present in the atmospheres of the outer planets of our solar system; an abundance of hydrogen throughout the universe was also discovered. In addition, astronomers gathered evidence that the planets were formed by the condensation of gases around the new sun. All this information contributed to a reevaluation of Earth's early atmosphere. It was speculated that the original atmosphere formed as the earth condensed from gases, but that it was later driven off into space by a period of high temperatures. Another primitive atmosphere formed secondarily by gases escaping from surface rocks; this atmosphere probably had the composition of present volcanic gases. It contained ammonia and gaseous carbon compounds, but no oxygen.

In 1923, A. I. Oparin in the USSR and J. B. S. Haldane in Britain pulled all this information together and suggested that the atmosphere of the primitive Earth must have been a reducing atmosphere; that is, it was made up of compounds with considerable chemical energy. There was insufficient oxygen (or other oxidizing agents) to oxidize these compounds. Thus, the reactions that form organic compounds were likely to occur, because the starting compounds were already reduced: less energy would be required to synthesize reduced organic molecules. Once synthesized, furthermore, these compounds would not be spontaneously oxidized. But the gases in the reducing atmosphere—ammonia, methane, and others—are quite stable; they undergo chemical reactions only if energy is added to the system. Haldane and Oparin suggested that the external energy could have been supplied by ultraviolet radiation and lightning; radioactive decay and the thermal energy of volcanoes also may have been important. They went on to speculate that any simple organic compounds formed in the atmosphere must have accumulated in the oceans of Earth; eventually the oceans became a thin salty soup of organic compounds that contained all the basic building blocks of life.

Organic Compounds

Oparin and Haldane felt that the formation of organic compounds was not just possible, but was an inevitable consequence of the existing conditions. Surprisingly, it was not until 1953 that these predictions were confirmed by an actual experiment. In that year, S. L. Miller performed the first successful **prebiotic** ("before life") experiment. Miller discharged sparks through a mixture of ammonia, methane, and hydrogen gas in the presence of water (Figure 17-1). An astonishing variety of organic compounds resulted. Miller's first experiments have been repeated and extended, using other energy sources and starting mixtures of gases. All the major amino acids, nucleic-acid bases, and sugars such as ribose, deoxyribose, and glucose have been obtained in these experiments. All biologically-important monomers have now been synthesized from very simple precursors that mimic a

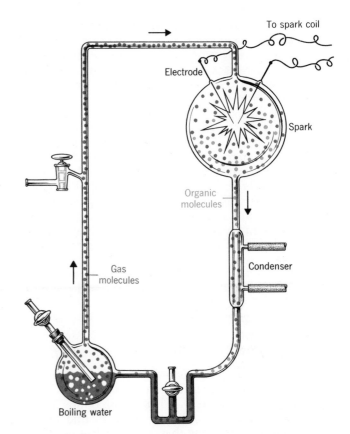

Figure 17-1 Miller's apparatus for experiments on the prebiotic synthesis of organic compounds. Gaseous ammonia, methane, and hydrogen are injected into the apparatus, and water is added to the small flask. The water is boiled, circulating the gases and water vapor through the large flask, where a spark is discharged repeatedly, providing an energy source. Organic compounds accumulate in the water flask.

primitive atmosphere.

The primitive-Earth conditions that produced organic compounds are probably very common during the early history of other planets. Modern theory states that as gases condense during star formation, planets are likely to form. In fact, one of every four stars in our galaxy is surrounded by planets; life, as we know it, could evolve on any planet that meets a few vital conditions. Temperatures cannot be too high, because organic compounds decompose at temperatures of a few hundred degrees. Nor can they be too low, because water must be present in liquid form. The planet probably requires a gravitational pull small enough for it to lose its primary atmosphere, and one large enough for it to retain a secondary atmosphere produced by gases escaping from its surface. According to current theories, such conditions are widely distributed in the universe.

This theory is strongly supported by the recent discovery of extraterrestrial organic material in the cores of meteorites. The best evidence comes from a large meteorite that landed in Australia in 1969. It contained many organic compounds, all of which are likely to be formed under prebiotic conditions. Some simple amino acids were present that are simply not found in the proteins of living organisms on Earth. This fact, and other characteristics of the meteorite, make it quite certain that the organic material came from an extraterrestrial source. Spontaneous generation is now well on its way to becoming a soundly-based theory of the origin of life.

Concentration and Polymerization

The next steps in the origin of life required concentrating the organic compounds that had been formed in the seas. These compounds had to be concentrated in order for them to undergo further chemical reactions that would create polymers and other large molecules.

High concentrations could have occurred in several ways. Many organic compounds are adsorbed on the surfaces of clay particles. This process would have tended to concentrate some of the organic molecules more than others, because only certain molecules bind very tightly to clay. Some compounds would also have been concentrated by evap-

oration in ponds at the edge of the sea. Still other compounds might have become concentrated by the wind and waves, forming surface films much like oil slicks. (In fact, it has recently been hypothesized that primitive oceans were entirely covered by an oil slick several meters thick.) Freezing a water solution is another way to increase the concentration of solutes in the liquid portion. As a solution is cooled, ice crystals begin to form; the dissolved organic molecules are excluded from the ice, and the remaining liquid is more concentrated. In short, there are various natural phenomena that could have produced concentrated solutions (Figure 17-2).

Concentrated organic compounds can undergo many reactions, including polymerization to form larger molecules. The polymerization of amino acids under the conditions of the primitive Earth has been demonstrated in the laboratory. If dry amino acid mixtures are heated moderately, they form spherical **proteinoids** (Figure 17-3); these substances consist of small spheres of amino acids linked together into random polypeptides. Such proteinoids might well have formed along the shorelines of seas warmed by active volcanoes. Proteins can also be obtained by discharging electricity through a solution of amino acids in a methane-and-ammonia atmosphere (the same conditions that produced the amino-acid monomers).

Researchers have also produced hydrocarbons using ultraviolet light as an energy source for the polymerization of methane. (Ultraviolet radiation from the sun easily penetrated Earth's primitive atmosphere.) It is more difficult to form nucleic acids from nucleotide monomers, but there are several promising lines of investigation. Some experiments suggest that mineral particles or small

Figure 17-2 Four natural phenomena capable of concentrating organic material in prebiotic seas.

Water molecules

Organic molecules

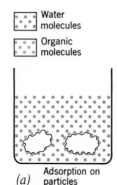
(a) Adsorption on particles

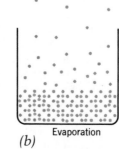
(b) Evaporation

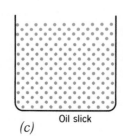
(c) Oil slick

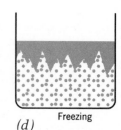
(d) Freezing

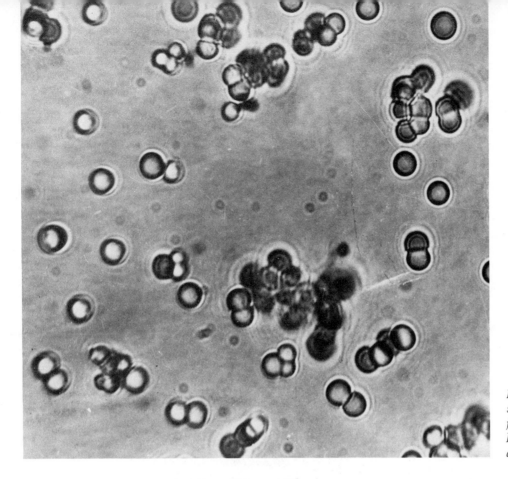

Figure 17-3 Proteinoids are small polypeptide spheres, formed by moderate heating of dry mixtures of amino acids.

organic molecules may catalyze the formation of nucleic acids. Other researchers feel that high-energy phosphate compounds such as ATP may have been present. These molecules might well have been synthesized in warm areas of the Earth: in deserts or near volcanos. These researchers go on to speculate that the high-energy phosphates supplied energy for the production of both the nucleic acids and the proteins by a mechanism like the one used in living cells today.

All in all, it is relatively easy to understand how the conditions of the primitive Earth could have yielded concentrations of various organic compounds, including polymers.

Self-replicating Molecules

The polymers formed out of the prebiotic soup were a far cry from the highly-organized nucleic acids and proteins that occur in modern organisms, yet they had some of the same characteristics. Proteinoids, for example, have some ability to catalyze

chemical reactions. There is also evidence that nucleic acids had the ability to replicate—that is, to direct the synthesis of a new, complementary strand—under primitive-Earth conditions. If our information about bond formation between the nucleotide bases is correct, then replication should occur automatically, as an inevitable consequence of the chemical structure of the bases. Adenine should pair with thymine or uracil, and guanine should pair with cytosine. A simple laboratory experiment confirmed this speculation. A nucleic acid composed only of uracil bases was mixed with single nucleotides of adenine. It organized the adenine units into a double helix. However, it was not able to organize guanine subunits when tested in a similar fashion.

Molecular Natural Selection

At this point it becomes possible to discuss the origin of life in terms of evolution and natural selection. We now have an extremely variable population of organic molecules, some of which had the ability to reproduce themselves. The molecules that reproduced themselves most readily would have become the most numerous in the prebiotic soup. Some of the proteinoids and other substances in the environment had catalytic activity and would have sped the formation of some nucleic acids more than others.

At first the replicating molecule was simply at the mercy of the precursor compounds and catalysts in the environment. But this condition soon changed as the nucleic acids began to control the composition of the molecules around them. They did so by directing the synthesis of proteins, as well as by directing their own replication. It is speculated that the nucleic acid bases formed weak bonds with amino acids in the environment. These amino acids then became linked into proteins by the action of various catalytic processes. This development was the origin of biological protein synthesis. Of course, it was a very imprecise process at first. Unlike the bonds between pairs of nucleic acids, the bonds between nucleic acids and amino acids are not very specific. The primitive nucleic acids could not have distinguished between specific amino acids. Rather, they merely may have distinguished between two or three different *groups* of amino acids, according to general size and shape or, possibly, the way the functional groups on the amino acids are arranged.

In any case, the nucleic acids now had a little control over the composition of some of the proteins around them. These proteins had catalytic activity and could be considered primitive enzymes. No doubt some of them could have catalyzed the replication of the nucleic acids themselves. They might have acted directly, by aiding in the formation of bonds between nucleotide monomers, or indirectly, by catalyzing the formation of nucleotides from yet other organic compounds.

These molecules would have evolved in the usual Darwinian fashion. The nucleic acids whose base sequences synthesized the most effective proteins would have had the best chance of reproducing. Subsequent natural selection would have refined this system. The composition of the proteins would have become more carefully controlled.

Though these are plausible speculations, no particular sequence of events is very clear or detailed. But if we assume that evolution and natural selection occurred at the molecular level, the lack of details becomes understandable. Presumably, many kinds of self-replicating protoorganisms evolved, but

only one kind succeeded. There are no survivors from other ancestral lines. Neither is there any fossil evidence.

This situation can be represented schematically by two cones joined at their tips. Before the origin of life, there was a great variety of organic matter. Then, as competition proceeded at molecular levels, this variety decreased. The seas became dominated by protoorganisms—the molecular systems that had the most efficient mechanisms of protein synthesis, based on a DNA code. At a very early stage, the dominant form of life may have been noncellular. Its functional units may have been adsorbed on clay particles or concentrated in surface films. A protoorganism with cellular membranes may have arisen only after organic molecules became scarce in the environment. In any case, the first protoorganism is represented by the narrow point at which the two cones meet. The upper cone then represents the adaptive radiation of living things from this first successful self-duplicating protoorganism. (This scheme is depicted in Figure 17-7.)

The Genetic Code

This hypothesis assumes that protein synthesis depended on a triplet code very early in the origin of life. There are two lines of argument that support this assumption. The first is the fact that the DNA base code is nearly universal among present-day organisms. The second point is that a triplet code could not have evolved from a simpler code. Suppose that a protoorganism had first worked out a doublet code that used fewer amino acids. A change to a triplet code—reading the bases by threes instead of twos—would reduce all of the preceding doublet information to nonsense. The organization of every protein synthesized by the organism would be destroyed. Selection against such a mutant would therefore be fantastically strong.

The code probably always involved three nucleotide bases. But it may have started with two bases acting as the information carriers, while the third functioned only as a spacer. This mechanism would code for only sixteen amino acids. The third base probably became important later when additional amino acids were added to the code. This theory receives some indirect support: eight of the amino acids are determined by only the first two bases of the triplet. In these cases, the third base is unimportant.

Early Organisms

The first self-duplicating life forms must have used the organic compounds already present in the prebiotic soup as their energy source. Since the atmosphere lacked oxygen they must have been anaerobic. That is, they must have obtained energy from a fermentation process, by oxidizing various organic molecules without using molecular oxygen. But as any particular molecule became rare in the primal soup, a protoorganism capable of producing it from some precursor would have obtained a large selective advantage over its competitors. By synthesizing more and more enzymes, such primitive organisms would have become adapted to utilize all the organic carbon already present. Eventually these organisms would have reduced the availability of organic substrates.

The depletion of organic material set the stage for the appearance of primitive photosynthesizers, who could use light as their energy source. Structures that may be the remains of such photosynthetic microorganisms have been found in Precambrian rocks in South Africa, along with fossil bacteria. The rocks have been dated as 3.4 billion years old. The fossils are minute spheres, resembling single-celled algae (Figure 17-4). Reduced carbon, indicative of photosynthesis, was found in association with these early fossils.

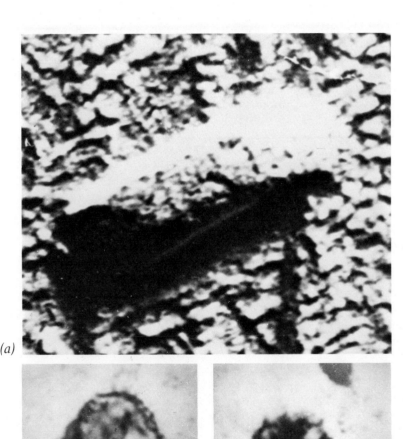

(a)

(b)

Figure 17-4 The oldest known fossils. (a) A fossil bacterium from a rock of the Fig Tree formation in South Africa (about 3.4 billion years old). The bacterium appears as a rectangular shape in a replica of the polished surface of the rock. Magnification 200,000×. (b) Fossil algae from the same formation. Magnification 1,500×.

The first anaerobic photosynthetic organisms probably contained one photosystem, like PS I. This mechanism allowed them to obtain ATP by cycling electrons through their photosynthetic apparatus. Next, they used photosynthesis to transport other electrons from donor molecules "uphill" to other carbon compounds. In this way they acquired reduced carbon compounds and ATP for driving metabolic reactions. But to make this transition they needed a source of electron-donor molecules.

Electrons could have come from organic compounds (sugar or alcohols, for instance) or from various inorganic compounds (H_2S, H_2, or H_2O), but organic compounds were now rare. Water was the most abundant source of electrons, but it had a serious disadvantage: oxygen is produced by the reaction. Molecular oxygen is actually a poison because it is such a strong oxidizing agent. Reduced carbon compounds in organisms are quickly oxidized, and thereby decomposed, by direct exposure to oxygen. But the primitive anaerobic photosynthetic microbes were still able to use water as an electron source, because the molecular oxygen that they produced could not remain long in a re-ducing atmosphere; it would quickly react with and oxidize other compounds in the environment, particularly iron. The geological record suggests that iron was an important sink for oxygen during the early history of life. That is, iron quickly combined with—and removed—molecular oxygen.

Iron is actually found in two forms, depending on how much oxygen is present. If oxygen is available, iron reacts with it to form iron oxides, better known as rust. In this reaction the iron gives up an electron to the oxygen: it becomes oxidized. Oxidized iron is called **ferric** iron. In a reducing atmosphere, iron retains the electron and is in the **ferrous** form. Ferric iron is insoluble in water, but ferrous iron is quite soluble in water.

Presumably, iron in the primitive seas was in the soluble, ferrous state because the atmosphere was a reducing one. Yet insoluble ferric iron is also found in some early Precambrian rocks (3.5 to 1.8 billion years old). It occurs in banded formations (Figure 17-5). These bands are deposits of ferric iron that precipitated and settled out of open water. This fact is interpreted to mean that oxygen from primitive photosynthesizing organisms oxidized ferrous iron to the ferric form. Thus, the appearance of banded

Figure 17-5 Iron occurs in banded formations in Precambrian rocks. The bands of ferric iron oxide may have been precipitated by the action of early organisms that used ferrous iron as an oxygen sink.

1 cm

iron in the geological formations gives us a record of the activity of O_2-evolving photosynthetic organisms. It supports the idea that iron served as an oxygen sink, which kept the photosynthesizers from being poisoned by the oxygen that they themselves produced.

Iron was so abundant in the early oceans that it could aid in other metabolic reactions, as well. It could catalyze electron transport reactions in respiration and photosynthesis. It enabled these critical reactions to take place—however slowly and inefficiently—until the early organisms produced more efficient electron transport systems. These metabolic systems survived only because the atmosphere was oxygen-free.

In the meantime, biotic oxygen production continued and became more efficient. Eventually, all the ferrous iron disappeared from the oceans. This development is shown by a change in the pattern of iron deposition that occurred about 1.7 billion years ago. The later deposits are "red beds": deposits of ferric iron that was transported to sediments as colloidal particles rather than precipitated from ferrous solutions. At this point oxygen began to accumulate in the atmosphere.

The appearance of oxygen in the atmosphere had profound consequences for the further evolution of life. Atmospheric oxygen absorbs ultraviolet radiation. Before oxygen appeared, the ultraviolet radiation from the sun penetrated all the way to the earth's surface. The ultraviolet light supplied energy for the abiotic synthesis of organic compounds, but it also must have disrupted some polymers. DNA is particularly sensitive to ultraviolet light. As a result, life was probably restricted at first to a narrow zone of sediment. It had to be covered by sufficient water to screen out the ultraviolet light, but it also had to remain shallow enough for light of longer wavelengths to penetrate and support photosynthesis. Individuals exposed more directly to the sun would be destroyed by ultraviolet light while those carried to deeper waters would lose their basic source of energy.

This situation changed drastically when oxygen accumulated. Oxygen is converted to ozone in the high atmosphere, and ozone absorbs ultraviolet light very effectively. By the time molecular oxygen reached only one percent of its present level, enough ozone had already formed to intercept the most disruptive short-wavelength ultraviolet radiation. Life could now exist in shallower seas and in surface layers. The total area colonized by living things and the diversity of possible habitats greatly increased.

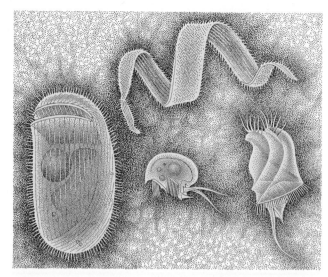

Figure 17-6 *Anaerobic muds and sediments contain a primitive assembly of organisms, such as these protozoa, which cannot survive exposure to oxygen.*

The appearance of free oxygen in the atmosphere had still other consequences. By 1.7 to 1.8 billion years ago, sufficient oxygen was present to oxidize all of the reduced compounds in the ocean. As the oxidation of these compounds took place, primitive photosynthesizers that could obtain electrons from water were favored over those that required reduced substrates. These organisms also had to acquire a tolerance for oxygen, since the protective oxygen sink provided by ferrous iron had been depleted. These developments favored metabolic patterns such as those of blue-green algae, at the expense of those of anaerobic photosynthetic organisms such as sulfur bacteria. These latter organisms were forced to retreat to portions of the environment that remained anaerobic: areas such as sulfur springs, polluted ponds, and muds.

Sediments in marine waters provide an extensive anaerobic environment. A very primitive assembly of anaerobic organisms now dwells in this habitat (Figure 17-6). These organisms either never acquired the ability to cope with molecular oxygen, or at some stage lost their tolerance for it.

Some primitive organisms not only learned to tolerate oxygen but learned to use it in their metabolism. Many modern organisms now use oxygen for the oxidation reactions that provide metabolic energy. But such aerobic respiration is possible only because an elaborate system of electron transport molecules is interposed between an oxidation step and the final transfer of electrons to an oxygen atom. Most anaerobes lack these protective arrangements and are promptly killed by exposure to oxygen. Organisms that were able to use molecular oxygen as an oxidant obtained the advantages of a tremendously accelerated metabolism.

Origin of Eucaryotes

The production of an oxygen-rich atmosphere may have corresponded to the sudden appearance of eucaryotic organisms. Procaryote fossils (particularly fossil bacteria and blue-green algae) are quite common in Precambrian rocks. But no fossil eucaryotes appear until the very late Precambrian or early Cambrian period. At this time they seem to arise very abruptly. All the major kinds (phyla and classes) of modern animals can be found fully-formed in Cambrian or subsequent Ordovician fossils. It is tempting to speculate that the eucaryotes evolved in response to rising levels of atmospheric oxygen at the end of the Precambrian. This sequence of events is shown in Figure 17-7.

Eucaryotes (fungi, protists, and higher plants and animals) must have evolved from procaryotic ancestors. Yet it is difficult to account for this transition, because the procaryotes differ from eucaryotes in many very basic ways. This is the case whether we compare fossil forms or living organisms. Procaryotes do not have membrane-bounded nuclei. Nor do they have other membranous structures such as mitochondria, chloroplasts, or endoplasmic reticulum. Their flagellar structure is simpler than that of typical eucaryotes. Procaryotic cell division is a direct process of binary fission, without mitosis and spindle formation.

The basic structural differences between procaryotes and eucaryotes are mirrored by equally basic chemical differences. The way DNA from various organisms binds with RNA provides an example. RNA from the ribosomes of a simple vertebrate—a toad—binds tightly with DNA from the same cells, because the molecules have homologous regions of complementary base sequences. The toad RNA also has regions that are homologous with DNA from other eucaryotes, including other vertebrates, invertebrate animals, flowering plants, and fungi. But no homologous regions are demonstrable with DNA from procaryotes (blue-green algae or bacteria).

A recent analysis by Lynn Margulis shows how eucaryotes could have arisen from symbiotic com-

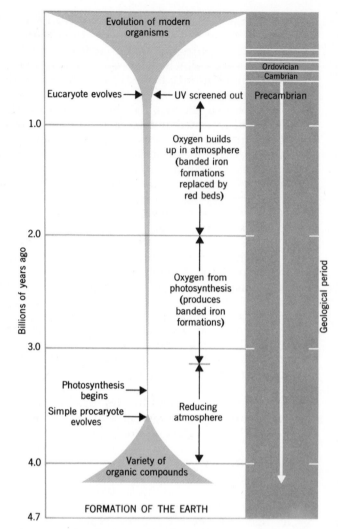

Figure 17-7 *The major events in the origin of life. The width of the colored band represents the diversity of organisms, protoorganisms, or kinds of organic molecules.*

binations of procaryotic organisms. Mitochondria may have arisen through a symbiotic relationship between a procaryotic host cell and an aerobic bacterium. Flagella, centrioles, and the spindle apparatus could have been acquired by a similar symbiotic relationship with spirochaete bacteria. Eucaryotic plant cells could have arisen by acquiring yet another kind of symbiont, a blue-green alga; the algal symbiont eventually became a chloroplast (see Figure 17-8). Once the symbiont was inside the host cell, its growth and reproduction became coordinated with that of the host. In some cases the symbiont genome may have become incorporated into the host's chromosomes. In other cases, however, the symbiont may have retained some or all of its genome and continued to direct the synthesis of some of its own proteins. In fact, it is now known that both chloroplasts and mitochondria contain their own DNA. In addition, chloroplasts have recently been observed to divide independently of host cell division.

This proposal greatly simplifies many of the vexing questions that accompany other proposals about the origin of eucaryotes. The traditional view held that eucaryotic plants are derived from blue-green algae; later, animals and fungi evolved from the eucaryotic plant cells through the loss of chloroplasts. But such a linear sequence must have evolved extraordinarily quickly to account for the diversity of eucaryotes that appears in the Cambrian Period.

The traditional view also has to postulate unknown intermediate forms in order to explain the appearance of various structural elements in subsequent eucaryote groups. Often these structures are of little apparent use to the group that supposedly developed them. For example, primitive plants are set the task of developing and perfecting the eucaryote flagellum. Yet most biologists agree that the flagellum evolved because it allowed a non-photosynthetic unicellular organism to pursue food. Furthermore, many of the proposed ancestral types are totally unrepresented either in surviving organisms or in the fossil record. By contrast, the symbiotic theory allows evolution to proceed more slowly, because it can operate on several groups at once. According to this theory, the basic eucaryote groups developed simultaneously rather than in a long, linear sequence. The parallel formation of the major eucaryote groups fits well with their abrupt appearance in the fossil record.

The symbiotic hypothesis is a recent proposal and does not yet command full agreement among biologists. Nonetheless, it is certainly the best and simplest account of the rise of the diverse eucaryote groups. It organizes a mass of detailed information and does not require us to accept unsupported assumptions.

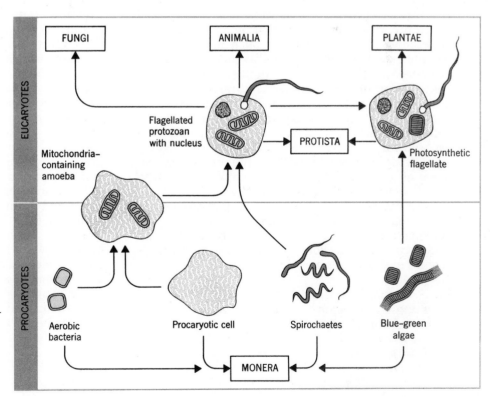

Figure 17-8 The evolution of eucaryotic cell types through symbiosis. The membrane-bounded organelles that characterize eucaryotic cells may have originated in symbiotic relationships among procaryotic cells.

Classification

The variety of living things now extends beyond limits we can readily comprehend. We can certainly never describe all the species that now exist on Earth. There are at least ten million species of organisms; these include some relatively unknown groups such as wasps (half a million species), mites (a million species), and comparable numbers of nematodes, or roundworms. In the last two centuries biologists have described no more than ten percent of this bewildering variety. In any case, more than half of the presently existing species will probably become extinct in the next few decades because of the growth of the human population and the expansion of its environmental manipulation. It is estimated, for instance, that no undisturbed tropical rain forest will exist by the end of this century; this loss alone will eliminate a great, undescribed collection of organisms.

purpose of classification

Biological classification is not really concerned with the unambiguous designation and description of each species of living thing. Such an undertaking would be a practical impossibility and theoretically of only modest interest. Still, the unambiguous designation of a particular species may have great importance, depending on the circumstances. The designation "rat" is adequate for most purposes. However there are more than 500 species and subspecies of rats. The ability to distinguish among them may be extremely important to a student of rodent behavior, or to the physiologist who works with rats.

The purpose and utility of any particular classification system will determine its merits. The following classification of animals, suggested in an essay by Jorge Luis Borges, may well have had utility for a Mandarin tax collector:

". . . animals are divided into
  a. those that belong to the Emperor,
  b. embalmed ones,
  c. those that are trained,
  d. suckling pigs,
  e. mermaids,
  f. fabulous ones,
  g. stray dogs,
  h. those that are included in this classification,
  i. those that tremble as if they were mad,
  j. innumerable ones,
  k. those drawn with a very fine camel's hair brush,
  l. others,
  m. those that have just broken a flower vase,
  n. those that resemble flies from a distance.[1]"

Unfortunately, the biological usefulness of this list is rather limited. The classification of things in the environment may have two different functions; it may help us to identify and to communicate about objects in the world, or it may reflect evolutionary and structural relationships between organisms.

Identification

The identification and naming of objects is absolutely essential for any discourse about the world. After all, we have to be sure what we're talking about. General nouns in a language perform this function. But there is a practical limit to the number of words that can be used in classification. This limit is imposed by the abilities of native speakers to master the language and to use the nouns appropriately. Many nouns describe environmental features of cultural concern, but few are used to describe things of little interest. Eskimos have a large number of nouns to describe kinds of snow, for obvious reasons, but they are unable to distinguish among motor vehicles. An American, however, probably has 50 nouns for describing varieties of cars.

The same principles apply to descriptions of animals and plants. Most human languages contain around 250 to 750 names for plants and animals, regardless of the richness or poverty of the local environment. These nouns also tend to cluster around species or varieties that are culturally or economically important ("thoroughbred," "quarter horse," "appaloosa," for instance). Such clusters of nouns tend to drop out of the language as interests change. Not long ago, for example, any aspiring

[1] Jorge Luis Borges, "The Analytical Language of John Wilkins," tr. by Ruth L. Simms, in *Other Inquisitions.* Austin, Texas: U. of Texas Press (1965).

English gentleman was hopelessly ignorant if he did not know at least twenty or thirty nouns for classifying animal feces (for example, rabbit "scurryings"). He also had an elaborate vocabulary to describe falconry. But with the decay of hunting as a gentle pastime, these words have simply disappeared from the language.

The size of a working vocabulary for classification thus tends to stay constant at a few hundred nouns. The botanist who specializes in orchids will use several hundred nouns for communication with his colleagues. But no single biologist could handle the tens of thousands of names that would be necessary for communicating with all the kinds of technical specialists—in mites, rodents, orchids, yeasts, diatoms, or viruses, for instance. For this reason, the entire biological classification system must reside in a dictionary of terms.

Evolutionary Relationships

A second and quite different function of classification is to express the interrelations of living things. Biological classification is sometimes called **taxonomy** (from Greek words meaning "laws of arrangement") or **systematics,** to emphasize the organizational function of classification. A **taxon** is a grouping, or a particular level of classification. Taxonomists ask how, and on what level, are honeybees related to flies? to horses? to pine trees? Our classification system should provide some answers. This organizational function reflects our views of evolution.

Thus, a specific classification system may serve several functions simultaneously, and not all equally well. Each change in a classification system, for example, hinders its communication function until the users of the system have mastered the change. Yet the system *must* be plastic and changeable, so that new information about relationships can be incorporated into it.

An excellent analogy for the differences between the two functions of biological classification is the card catalogue of a great library. The number of books is similar to the number of organisms, around twenty million. The card catalogue is a guide to users. It should name, identify, and permit the retrieval of any book in the library. It must also provide a unique identification number for each book. But the card catalogue cannot simply list every book in random order; it must somehow be organized. An alphabetical index by author is a convenient beginning, but only if you know the author's name. More general systems, like the Library of Congress code, attempt to organize information by grouping books according to subject matter. Still higher categories of classification are created by librarians and scholars to aid them in retrieving the books they need. This kind of library classification system is essentially a summary and classification of human knowledge. It is analogous to biological classification as a summary of evolutionary relations in organisms.

biological classification

Biological classification arose quite directly from ordinary designation of animals and plants. Carolus Linnaeus (1707–1778), a Swedish botanist, introduced the modern practice of designating organisms by genus and species names. Linnaeus recognized that the whole variety of living plants could not be handled linguistically, so he set himself the task of describing a manageable number of genera. Each genus was to be a distinct group with definite, natural, and universally accepted limits.

Linnaeus listed about 900 genera in his *Genera Plantarum* of *1737,* and he gradually increased this number to over 1,300 in subsequent work. But even this catalogue proved inadequate to characterize the variety of plants. Because the number of genera had become too large to handle, attention shifted to the next higher taxonomic category, the family. Botanists now tend to discuss plant families rather than genera.

Modern biological classification originates with the 1758 edition of Linnaeus' *Systemae Naturae.* International scientific congresses have agreed that any name and description published prior to 1758 has no validity; that is, it is not an appropriate designation for scientific discourse unless it has been reexamined and validated subsequent to 1758. Linnaeus is the universally accepted father of taxonomy.

Binomial Nomenclature

Linnaeus used the genus as his basic taxonomic grouping. But there was still considerable diversity within genera. He therefore subdivided the genera into species. The genus and species names together form a unique designation for each type of organism. This designation is called the **binomial.**

Ladybird beetles, for instance, are of the genus *Coccinella*. The species of ladybird with seven spots on its wing covers was given the name *Coccinella septempunctata*. The name *Coccinella septempunctata* can be applied only to organisms that fit the original published description of that species.

The complete binomial makes each designation unique. Some species names are used in many different genera: names like *vulgaris* ("common") or *viridis* ("green") or *albus* ("white") occur very frequently. But the generic names are all different (except that the same genus name may be used in both the plant and animal kingdoms).

The stability of the classification system is crucial at the binomial level. During the early nineteenth century, biologists indulged in casual name-changing. Latin scholars examined genus and species names and discovered disagreements of case and gender, which they then revised. If a blue-green bird was judged by a subsequent ornithologist to be more nearly green than blue, its species designation was changed from *caerulus* ("blue") to *viridis*. This activity created confusion and led to the adoption of the **priority rule** in 1842; this rule states that once a name and description of a species has been published, that name cannot be changed except under carefully-defined circumstances. Unfortunately, this restriction does not really solve the problem. Many taxonomists then began to search out and argue about the meaning of the vague descriptions in the early literature. Because the original descriptions were difficult to interpret, names would still shift back and forth with the flow of argument. Common and widely-used genus and species names were being changed to antique forms by the application of the very rule that was designed to prevent changes. As one biologist commented a few years ago, taxonomy is one of the few sciences where the work of the incompetent is not ignored; instead it is avidly searched out and preserved.

The most recent Rules of Nomenclature (adopted by the 15th International Congress of Zoology in 1958, and revised in 1964) recognize this difficulty. Common and widely used generic and species names are now exempted from the priority rule. Particular cases are appealed to the International Commission on Zoological Nomenclature, which has final authority.

The Rules of Nomenclature give the impression of legalism, fussiness, and excessive detail. But this rigidity is necessary to maintain stability in classification at the genus and species levels. All the biological disciplines depend on this stability to

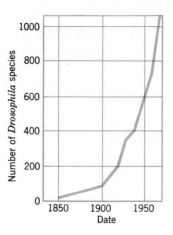

Figure 17-9 Drosophila *species described in the last century. The number of species has increased dramatically because the group has been studied intensively, not because the insect populations have changed.*

enable researchers to identify organisms, express their data, and interpret other people's work.

Yet the more we learn about a particular group of organisms, the more diverse they appear. Figure 17-9 is a graph of the number of described species of the genus *Drosophila*, plotted against the date of their description. Unfortunately, the usual consequence of such proliferation is the division of a single genus into a group of related ones; the revision changes many binomial designations. Yet one can scarcely criticize the systematist for the proper exercise of his organizational function; he simply revises classifications to reflect more modern information.

Higher Categories

Linnaeus used very few higher categories to express the organizational affinities of genera. But as more organisms were described and fitted into the classification system, new categories had to be added to supplement and organize the Linnaean groupings of genus and species. Here are the basic categories into which the living world is divided, along with man's classification in this system:

Category	Man
Kingdom	Animalia
Phylum	Chordata
Class	Mammalia
Order	Primates
Family	Hominidae
Genus	*Homo*
Species	*sapiens*

Sometimes this system is supplemented by other categories designated by prefixes (subfamily, superfamily) and by additional terms (branch, grade, division, and tribe). Originally, structural similarities were the criteria for placing genera in the same family. But later classification came to express the course of evolution; structural similarities were interpreted as reflections of common ancestry.

As we saw in Chapter 16, the criteria defining a species are theoretically clear, though difficult to apply. But the criteria for higher taxonomic categories are subjective. They depend on the judgment of the person who proposes a classification. There is no fast rule for the creation of another genus as there is for another species. The same is true for each higher category in the organizational scheme. Consider a hypothetical group of three related species, which evolved as indicated in Figure 17-10. There are four possible generic arrangements of these three species. Each scheme represents a different opinion concerning their evolutionary relationships.

The classification of human beings is a specific example. Humans, chimpanzees, gorillas, orangutans, and gibbons are related species. This relationship is expressed by their inclusion in the same superfamily (Hominoidea). Gibbons are considered to be relatively distantly related to the others, so they are excluded from the next grouping, the family Hominidae. Humans are placed in a separate subfamily of the Hominidae, to express the view that a considerable evolutionary distance separates them from the other hominids. This classification represents a consensus of the subjective views of primatologists. But its relation to factual information is only indirect. The close similarity of human and ape hemoglobins, for instance, is noted, but most biologists still feel that the differences justify separation into different subfamilies.

Which characteristics are significant for assigning organisms to higher taxonomic categories? They should fulfill certain requirements. First, they should be comparatively stable. Morphology is often a stable characteristic, but not always. With bacteria, for example, it is a very poor one. Single mutations commonly produce changes in the sizes and shapes of individuals or of colonies. Changes in nutritional conditions or in temperature can produce equally dramatic changes in morphology. Instead, therefore, microbiologists traditionally rely on staining characteristics (related to cell-wall chemistry) and metabolic characteristics. The ratio of cytosine and guanine to adenine and thymine in the DNA is a better criterion for assignment to higher taxonomic categories. Another requirement is that the characteristic should occur throughout the group, and should show measureable changes in

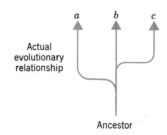

Figure 17-10 Three species (a, b, and c) can be assigned to genera in four possible ways. Each system represents a different judgment concerning their evolutionary relationships.

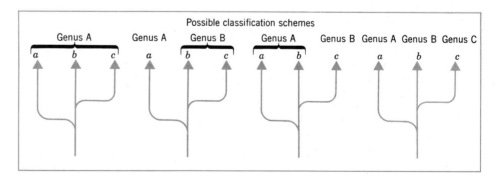

different members of the group. The CG/AT ratio fulfills this requirement as well; it can be determined for all organisms and varies from about ⅓ to ³⁄₁.

The obvious subjectivity of many taxonomic judgments has led to the search for a more objective way to express evolutionary relationships among larger groups of organisms. Biochemical classification (amino acid sequences or nucleic acid ratios, for instance) may be an improvement, but an overall familiarity with the organisms and a subjective evaluation of the data are still required. In assigning organisms to larger taxonomic groupings, there is no substitute for experience with the organisms. The CG/AT ratio for the bacterium *E. coli*, for example, is very close to that of humans, yet no one argues that there is a close relationship between them.

Lately, a more radical break with traditional methods has been proposed. This development is called **numerical taxonomy.** It assumes that all characteristics of organisms are equally meaningful. The technique uses as many objective measurements of the organism as possible: weight, length, various length/width ratios, the number of spines, the levels of enzyme activity—all such characteristics are grist for the mill. These data are then entered into a computer program to evaluate similarities and differences. All the categories are given equal weight. The resulting computer-generated classification agrees very well with assignments of species made on traditional grounds, but only for some groups. For example, the classification of species and genera within an insect family can be accomplished equally well by numerical or traditional approaches. In this case, both the traditional approach and the numerical approach depend heavily on mensurable (countable, weighable, measurable) characteristics. In other groups, numerical taxonomic methods fail.

The classification of the dog family provides a good example of the difficulties encountered. Consider the relationships between the coyote, the wolf, and various domestic breeds of dog. A numerical taxonomist would consider weight, color, length of muzzle, and other mensurable traits. On these grounds he would find that German shepherd dogs are much more closely related to wolves than to miniature poodles or dachshunds. Yet traditional classification methods decree that poodles, dachshunds, and German shepherds belong to one species, *Canis familiaris*, while wolves are a separate species, *Canis lupus*. In this case, the taxonomist has considered the evolutionary history of dogs, as well as their appearance. Ultimately the interpretative and organizational function of classification has to remain a creative activity based on experience and judgment.

In biological classification, the communicative function is seriously at odds with the organizational one. On the one hand, unique and stable designations of species are required for description of the living world. But on the other hand, a classification system attempts to express relations among living things. As our ideas change about these relations, so must our classification. This function requires fluidity; the descriptive function requires stability. Biological classification thus treads a line of uneasy compromise between the two functions. In the long run it will probably be necessary to abandon the use of the same classification system for both purposes. Binomial nomenclature will no longer be used, or it will not be used to reflect the position of an organism in the living world. This suggestion is so radical that it meets with great opposition.

a modern classification of organisms

In the following pages we present a classification of organisms proposed by R. H. Whittaker in 1969. Five kingdoms are used. The most basic division is between procaryotic and eucaryotic organisms. The kingdom Monera contains the procaryotes; the other kingdoms all contain eucaryotes. Within the eucaryotes, multicellular organisms are separated from those which remain unicellular. The kingdom Protista contains single-celled eucaryotic organisms exclusively.

Multicellular organisms are divided into three major groups with fundamentally different organizations. Multicellular algae and higher plants are in the kingdom Plantae; molds and yeasts and mushrooms comprise the kingdom Fungi; invertebrate and vertebrate animals are the kingdom Animalia. This assignment of multicellular organisms to three separate kingdoms reflects the opinion that multicellularity has been achieved more than once in evolution. The chief distinction between them is their main nutritional mode: photosynthesis in plants, absorption of nutrients in fungi, and ingestion of chunks of food in animals.

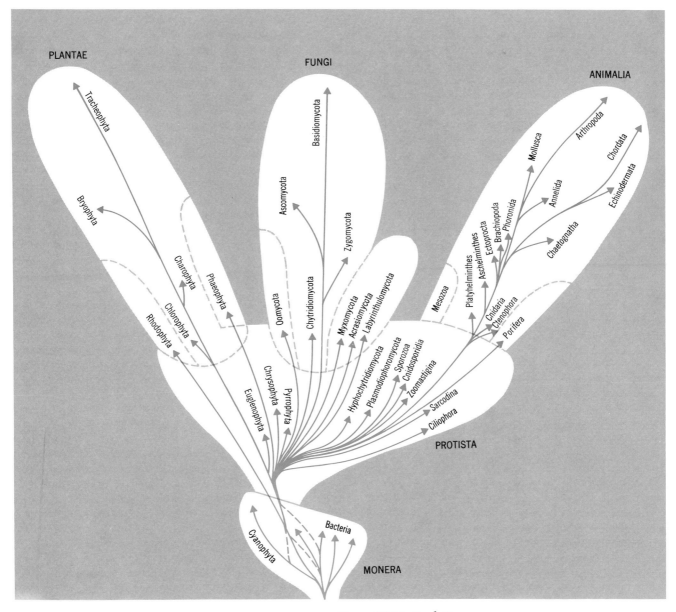

Figure 17-11 Evolutionary relationships in the five-kingdom system of classification.

The postulated evolutionary relationships among the five kingdoms are shown in Figure 17-11. All the phyla in each kingdom are listed, except those that have no living representatives. For the vascular plants and the chordates, we have extended the classification to include classes. An asterisk indicates that the phylum is of interest primarily to specialists.

563 *a modern classification of organisms*

kingdom monera

Procaryotic organisms without nuclear membranes, chloroplasts, or mitochondria. Solitary individuals or colonies, but no tissue differentiation in the colonies. Movement by means of simple flagella or by gliding.

BRANCH MYXOMONERA
Without flagella; motion (if present) by gliding.

Phylum Cyanophyta: blue-green algae (about 1,500 species). Filamentous or spherical; principal mode of nutrition is photosynthesis, but some absorb nutrients. Some species use N_2 as a nitrogen source.
Examples: *Nostoc, Oscillatoria, Micrococcus.*

***Phylum Myxobacteria:** slime bacteria (about 75 species). Resemble slime molds (Phylum Myxomycota), but individuals are bacteria rather than amoebae.
Example: *Myxobacteria.*

BRANCH MASTIGOMONERA
Motion (if present) by simple flagella.

Phylum Eubacteriae: true bacteria (about 3,000 species). Contains most significant rod and coccus bacteria. Includes important symbionts and agents of diseases (such as meningitis, botulism), nitrogen-fixing bacteria of root nodules and soils, and many fermenting forms associated with food spoilage.
Examples: *Escherichia, Clostridium, Nitrosomonas.*

Phylum Actinomycota: mycelial bacteria (about 200 species). Form colonies resembling fungal hyphae. Includes the agents of leprosy and tuberculosis.
Examples: *Mycobacteria, Actinomyces.*

Phylum Spirochaeta: spirochaete bacteria (about 45 species). Found in decaying meat, gingival spaces in oral cavity. Includes the agent of syphilis.
Example: *Treponema.*

kingdom protista

Eucaryotic organisms with a single-cell or simple colonial organization.

Phylum Euglenophyta: *Euglena* and *Euglena*-like organisms (about 500 species). Elongated cells with two flagella (one may be reduced or lost). May feed on particulate matter and also photosynthesize. Most photosynthesize with chlorophylls *a* and *b*; some (*Astasia*) have lost photosynthetic ability. Food reserves stored in the form of paramylum (a starch like carbohydrate).
Examples: *Euglena, Perinema, Astasia.*

Phylum Chrysophyta: golden algae (about 11,000 species). Contain carotenoid pigments which give them a golden or yellowish color. Food stored as lipid rather than carbohydrate. Includes colonial forms of considerable complexity (*Hydrurus*). The most important are the diatoms (planktonic

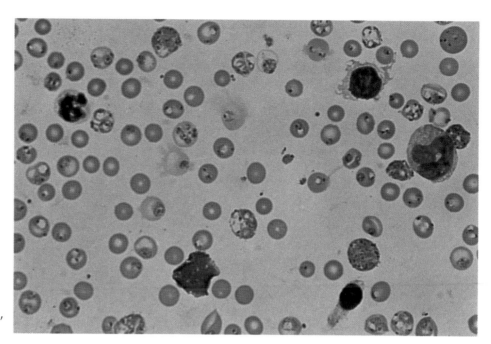

Figure 17-12 Phylum Sporozoa. Plasmodium, *the agent of malaria.*

and benthic organisms in marine and fresh waters).
Examples: *Skeletonema, Nitzschia, Navicula.*

Phylum Pyrrophyta: dinoflagellates (about 1,100 species). Have two flagella: one behind the organism to push it forward, the other in a girdle encircling the cell. May be photosynthetic or may be colorless particle feeders.
Examples: *Gonyaulax, Ceratium, Peridinium.*

***Phylum Hyphochytridiomycota:** (about 15 species) see next entry.

***Phylum Plasmodiophoromycota:** one of two phyla of small, fungus like parasites that dwell inside cells of plants (about 35 species). It and the similar Hyphochytridiomycota are often placed in the kingdom Fungi.

Phylum Sporozoa: immotile, parasitic protozoa (about 4,000 species). Feed by absorption of molecules across the cell membrane. Sexual reproduction by formation of spores. Many protozoologists advocate subdividing them into several phyla. A large and diverse group, including parasites of great medical importance, especially *Plasmodium,* the agent of malaria (Figure 17-12).
Examples: *Plasmodium, Eimeria, Gregarina.*

***Phylum Cnidosporidia:** similar to Sporozoa (about 1,000 species). Fish parasites. Spores contain a polar capsule with a thread that anchors the parasite to the host during the infective process.

Phylum Zoomastigina: animal flagellates (about 6,000 species). A diverse group of protozoans. Many symbiotic. Includes important parasites (*Trypanosoma*) and the cellulose-digesting symbionts of insects.
Examples: *Trypanosoma, Trichomonas, Trichonympha.*

Phylum Sarcodina: *Amoeba* and other unicells that move by pseudopods, protoplasmic extensions (about 12,000 species). Phylum also includes foraminiferans (cells with a hard external shell pierced by minute pores, through

which pseudopods extend for feeding and locomotion), planktonic helio-
zoans and radiolarians, (unicells with needle like pseudopods supported by
silica rods). Some are important parasites; *Entamoeba* causes amoebic dysentery.
Examples: *Amoeba, Elphidium, Actinosphaera.*

Phylum Ciliophora: protozoa that move by means of cilia (about 6,000
species). The cilia may fuse to form complicated feeding and locomotor
structures. Many are multinucleate.
Examples: *Paramecium, Vorticella, Stentor, Euplotes.*

kingdom
plantae

Photosynthetic eucaryotes with cell walls and often vacuoles, primarily multicellular in organization.

Phylum Rhodophyta: red algae (about 4,000 species). Red color due to the
pigments phycoerythrin and phycocyanin. Primarily marine and capable of
living at greater depths than other marine algae. May be bladelike,
branched, or encrusting.
Examples: *Porphyra, Polysiphonia, Porolithon.*

Phylum Phaeophyta: brown algae (about 1,500 species). Olive-green or
brown color due to the carotenoid pigment fucoxanthin. Almost exclusively marine; may be large and complicated.
Examples: *Fucus, Sargassum, Macrocystis.*

Phylum Chlorophyta: green algae (about 8,000 species). Grass-green in
color, with the same photosynthetic pigments as land plants. Synthesize
and store starch. More common in fresh water than in marine habitats, also
occur in moist soil and on damp surfaces.
Examples: *Chlorella, Ulothrix, Ulva.*

Phylum Charophyta: stoneworts (about 250 species). Small aquatic plants
having an elongated stem bearing whorls of leaves. Commonly grown in
freshwater aquaria.
Examples: *Chara, Nitella.*

Phylum Bryophyta: liverworts and mosses (about 2,000 species). Land
plants, lacking conducting tissues and true roots; rhizoids act as absorptive
organs. Lack of vascular tissue restricts them to small size and damp habitats.
Examples: *Marchantia, Polytrichium.*

Phylum Tracheophyta: vascular plants (about 250,000 species). Includes
four subphyla. Three consist of primitive vascular plants, now represented
by only a few species (Lycopsida, the club mosses; Sphenopsida, the horse-
tails; and Psilopsida). The most important tracheophytes are in the subphylum Pteropsida:
  Class Filicinae: ferns.
  Class Gymnospermae: cycads, *Ginkgo,* conifers.
  Class Angiospermae: flowering plants.

kingdom fungi

Mostly multinucleate organisms; eucaryote nuclei are dispersed in a net of hyphae; some are single-celled and uninucleate. Nutrition is absorptive.

Phylum Myxomycota: slime molds (about 500 species). Life history in two distinct phases: a multinucleate amoeboid individual that feeds saprotrophically, and a multinucleate sporangium that produces meiospores. Example: *Physarum*.

Phylum Acrasiomycota: cellular slime molds (about 30 species). Life history in two phases: uninucleate amoeboid individuals, and a sporangium formed from an aggregation of amoeboid individuals. Examples: *Acrasia, Dictyostelium*.

*Phylum Labyrinthulomycota: slime molds which form a multinucleate net like organization (6 species). Parasites of marine plants.

Phylum Oomycota: water molds, downy mildews, white rusts (about 500 species). Produce resistant oospores in sexual reproduction. Includes many important agricultural parasites (such as potato blight). Water molds are confined to fresh water or wet soil; mildews and rusts are disseminated by wind. Examples: *Saprolegnia, Phytophthora*.

Phylum Chytridomycota: chytrid fungi (about 400 species). Have flagellate zoospores that require free water for reproduction. Existence is aquatic, or as plant parasites. Example: *Allomyces*.

Phylum Zygomycota: conjugation fungi (about 1,000 species). Lack motile cells. Sexual reproduction is by conjugation, or fusion, or gametangia, resulting in a zygospore. Widely distributed, some are responsible for fruit rotting and bread mold. Examples: *Pilobus, Rhizopus*.

Phylum Ascomycota: sac fungi (about 15,000 species). Name derived from the sac (ascus) of spores. Includes the yeasts, *Penicillium*, and larger fungi such as morels and truffles. Many are plant and insect parasites (see Figure 17-13). Examples: *Saccharomyces, Penicillium, Tuber, Claviceps*.

Phylum Basiodiomycota: club fungi (about 15,000 species). Named for basidia, characteristic spore-forming structures. Nonparasitic members are terrestrial. Many are parasites (rusts and smuts). Includes mushrooms and puffballs and bracket fungi. Examples: *Puccinia* (wheat rust), *Coniophora* (dry rot), *Agaricus* (mushroom).

Figure 17-13 Phylum Ascomycota. A scarlet cup fungus.

kingdom animalia

Multicellular eucaryotic organisms without cell walls or chloroplasts. Principal mode of nutrition is ingestive.

*Phylum Mesozoa: a small group of minute, ciliated parasites with a simple organization (about 50 species).

Phylum Porifera: sponges (about 5,000 species). Little tissue differentiation, typically colonial in organization. Most are marine (see Figure 17-14).
Examples: *Scypha, Hippiospongia* (bath sponge).

Phylum Cnidaria: anemones, corals, jellyfishes, hydroids (about 6,000 species). Name derived from the stinging cells characteristic of the group.
Examples: *Hydra, Aurelia* (jellyfish), *Metridium* (anemone, Figure 17-15), *Porites* (coral).

*Phylum Ctenophora: comb jellies (about 100 species). A small group similar to cnidarians but lacking stinging cells.

Phylum Platyhelminthes: flatworms (about 13,000 species). Lack a body cavity. Phylum includes stream planarians and other free-living forms. Also includes medically important flukes and tapeworms.
Examples: *Fasciola* (*fluke*), *Taenia* (tapeworm), *Dugesia* (planaria, Figure 17-16).

*Phylum Rhynchocoela: ribbon worms (about 800 species). Similar in general organization to flatworms, but have complete digestive tract with mouth and anus.

Figure 17-14 Phylum Porifera. A marine sponge.

Figure 17-15 Phylum Cnidaria. The sea anemone, Metridium.

Figure 17-16 Phylum Platyhelminthes. The planarian, Dugesia; *magnification 12×.*

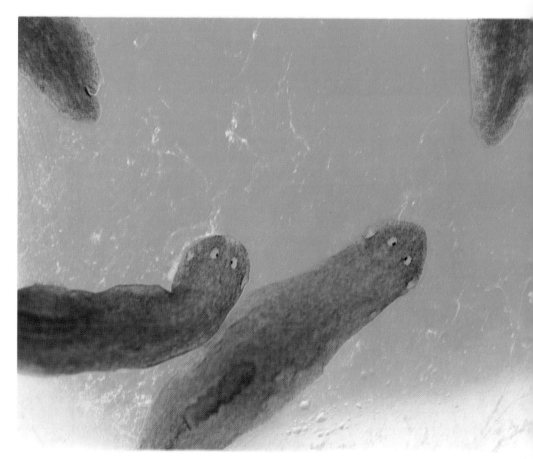

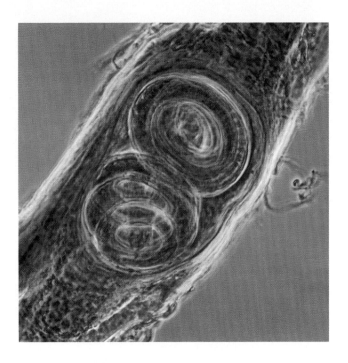

Figure 17-17 Phylum Aschelminthes. A nematode parasite of humans, Trichinella; *magnification 470×.*

Figure 17-18 Phylum Ectoprocta. A bryozoan, Bugula. *(near right, opposite page)*

Figure 17-19 Phylum Brachiopoda. A lampshell, Terebratulina. *(far right, opposite page)*

Phylum Aschelminthes: a large and diverse phylum (about 13,000 species) containing nematodes, rotifers, and other less well known organisms. Nematodes, probably the most important parasitic worms, infest both plants and animals. Also widely distributed as free-living organisms. Examples: *Ascaris, Trichinella* (Figure 17-17), *Necator* (parasites of man).

*Phylum Acanthacephala: a small group of worms (about 500 species). Parasitic in cold-blooded vertebrates.

*Phylum Entoprocta: a small group of sessile, encrusting animals (about 60 species).

Phylum Ectoprocta: moss animals (about 4,000 species). Sessile and colonial; individuals are minute. Feeding is by means of a crown of tentacles called a lophophore (Figure 17-18).

*Phylum Phoronidea: a small group (about 20 species) of wormlike animals allied with ectoprocts and brachiopods because they have lophophores.

Phylum Brachiopoda: lamp shells (about 200 species). So named because they resemble old Roman oil lamps. Clam-like organisms but with shell valves placed dorsally and ventrally around the animal rather than laterally. A small group with many fossil forms. Feeding is by means of a lophophore (Figure 17-19).

Phylum Mollusca: snails, clams, squid, octopus, and related forms (more than 100,000 species). The second-largest animal phylum; mostly marine. Examples: *Littorina* and *Helix* (snails), *Mya* (clam), *Octopus, Loligo* (squid).

Phylum Annelida: segmented worms, leeches (about 9,000 species). Body is a series of segments (the name is derived from the Latin word for "ring").

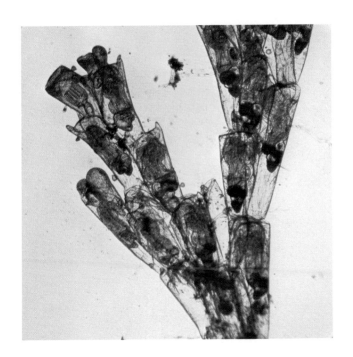

Principally marine. Earthworms are the best known, but marine forms are more colorful and diverse.

Examples: *Nereis* (clamworm), *Terebella* (plume worm), *Lumbricus* (earthworm).

*Phylum Echiuroidea: sausage-shaped marine worms (about 100 species). Lack segmentation.

*Phylum Sipunculoidea: peanut worms (about 250 species). Lack segmentation.

Phylum Arthropoda: joint-legged animals (more than 1,000,000 described species, mostly insects). The largest group of animals. All have exoskeletons and jointed legs. Phylum includes insects, spiders, ticks, mites, crustaceans (crabs, crayfishes, pillbugs), centipedes, millipedes, and other less familiar groups.

*Phylum Pogonophora: a small group (about 100 species) of wormlike marine organisms. Lack a digestive tract.

Phylum Chaetognatha: arrow worms (about 100 species). Transparent, flattened wormlike organisms, important predators in ocean plankton.
Example: *Sagitta, Spadella* (Figure 17-20).

Phylum Echinodermata: sea urchins, starfishes, brittle stars, sea cucumbers (about 6,000 species). A conspicuous and colorful marine group. Thought to be closely related to the chordates (below) because of developmental and biochemical similarities (Figure 17-21).
Examples: *Asterias* (starfish), *Thyone* (sea cucumber), *Arbacia* (sea urchin).

Phylum Hemichordata: acorn worms (about 70 species). Wormlike organisms, formerly placed with the chordates as a subphylum. Closely related to chordates, but their "stomocord" is not a true notochord (see Figure 17–22).

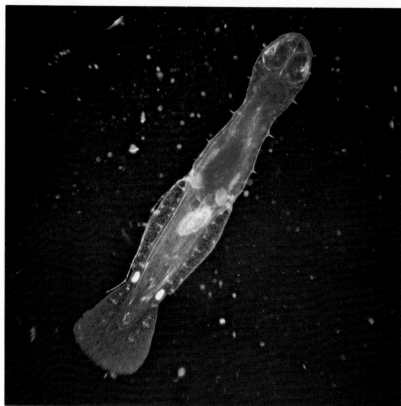

Figure 17-20 Phylum Chaetognatha. Spadella; *magnification 25×.* (right)

Figure 17-21 Phylum Echinodermata. The red sea urchin, Strongylocentrotus. (below)

Figure 17-22 Phylum Hemichordata. The acorn worm Balanoglossus.

Figure 17-23 Phylum Chordata, Class Agnatha. The lamprey, Petromyzon.

Phylum Chordata: the chordates. Distinguishing characteristics include a hollow, dorsal nerve cord, gill slits, and a notochord at some developmental stage. Three subphyla:

SUBPHYLUM UROCHORDATA: sea squirts (about 1,500 species). Marine. Chordate characteristics in the larval stage. Most are attached filter feeders as adults, but some (salps and larvaceans) are planktonic. Example: *Ciona.*

SUBPHYLUM CEPHALOCHORDATA: amphioxus and related forms (about 25 species). Filter-feeding organisms found just below the sand-mud surface in marine waters.
Example: Amphioxus.

SUBPHYLUM VERTEBRATA: chordates with a bony or cartilaginous vertebral column (about 47,000 species).
Class Agnatha: the jawless or roundmouth eels (about 40 species). Rasping and sucking predators on bony fishes, and scavengers that feed on bottom detritus.
Examples: *Myxine* (hagfish), *Petromyzon* (lamprey; Figure 17-23).
Class Chondrichthyes: cartilage fish, including sharks, skates, and rays (about 275 species). As the name implies, the skeleton is cartilage.
Class Osteichthyes: bony fishes (about 25,000 species). The fishes comprise about half of the described species of vertebrates, show great diversity of size and form and habit, and are dominant organisms in both fresh and marine waters.

Class Amphibia: frogs, toads, and salamanders (about 2,500 species). Live both in water and on land. Lose water readily by evaporation and usually require water for reproduction. Mostly restricted to moist, freshwater habitats.

Class Reptilia: snakes, turtles, crocodiles, and lizards (about 6,000 species). Some (various turtles, snakes, and crocodiles) are aquatic, but most are fully terrestrial. Water loss is restricted by skin. Reproduction and development do not necessarily require water.

Class Aves: birds (about 9,000 species). A large, diverse, and successful group. With man and insects, birds dominate the land.

Class Mammalia: mammals (about 4,000 species). Name derived from the characteristic mammary glands, used for suckling the young.

There are several other widely accepted classifications; this is simply one arrangement that reflects a specific interpretation of the variety and evolution of living things. No classification commands total allegiance; new information and new interpretations will generate new classifications.

Recommended Reading

Alexopoulos, C. J., and H. C. Bold, *Algae and Fungi.* New York: Macmillan (1967).

An introduction to the diversity of these groups.

Bold, Harold C., *The Plant Kingdom* (2nd ed.). Englewood Cliffs, N.J.: Prentice-Hall (1964).

A brief discussion of the major groups of plants.

Fingerman, Milton, *Animal Diversity.* New York: Holt, Rinehart, and Winston (1969).

A brief discussion of the functional morphology of vertebrates and invertebrates.

Orgel, L. E., *The Origins of Life.* New York: John Wiley & Sons (1973).

A detailed and lucid presentation of current theories of the origin of life, the supporting data on which they are based, and their present status.

Reprinted courtesy Sawyer Press, Los Angeles, Calif. 90046

18 ecology

The word "ecology" comes from the Greek word *oikos*, meaning "household," and the root *logos*, which means "counting and calculating" or simply, "science." "Household" is used in a broad sense. It includes both organisms and their surroundings, their environment. **Ecology,** therefore, is the study of the household—the study of relations between organisms and their environment.

Ecology is probably the most abused word of the decade. Committees to "improve the ecology," or bumper stickers that read "Ecology Now!" imply that ecology is a particular kind of human environment, or that some ecology is superior to other ecology. Actually, ecology is merely the *study* of organisms' relations with the surroundings—it is not a description of what they ought to be. "Ecology now" makes no more sense than "biology now." The confusion is understandable, however. The influence of man on his environment is certainly part of the subject matter of ecology. When our actions have undesirable effects on the environment, it is natural to ask the ecologist what happened and to seek his advice about how to reverse the effects. But it is important to understand that the advice of the ecologist bears the same relation to conservation and city planning that the advice of the physicist bears to the construction of bombs and reactors; scientists can provide information about consequences of specific policy, and can help execute it, but the framing and initiation of public policy are social and political decisions.

The relationship of organisms to their environment has provided a framework for all the discussions in this book. We have stressed that an organism is an open system, that many· aspects of its chemistry, structure, behavior,

Cartoon by Ron Cobb

577

development, inheritance, and evolutionary history are related to its necessity for continuing interaction with the environment. This point of view is precisely an ecological one, though we have not yet used the word "ecology."

The ecologist simply changes the order of discussion, focusing on the environmental relations first and considering only those aspects of the organism's chemistry, behavior, development, morphology, or physiology that are directly related to the environment. By doing so, ecologists have acquired a body of information and theory about these interactions, which form the subject matter of the science.

An organism without its environment, like a specimen in a museum, is simply a curiosity. The organism becomes much more interesting when we know something about the surroundings in which it functions. Some ecologists concentrate on how an organism gets along in its normal environment. They will observe the organism's behavior, study its physiological responses, and measure features of its environment. Their aim is to understand just what features of the environment are important to the organism—what food is available, or how the external temperature affects it. They also want to know what aspects of the organism's physiology and morphology take advantage of these features. This approach to relations between organism and environment is called **species ecology.**

Studies of plants and animals that live in harsh environments provide good examples of species ecology. Small desert mammals, for example, have been studied extensively (see Figure 18-1). The desert is hot in the daytime and cold at night; water is scarce or lacking altogether, and food is often scarce. Small mammals have very little water to spend on temperature regulation, both because they are so small and because water is scarce in the environment. Kangaroo rats, for example, simply avoid high daytime surface temperatures by being nocturnal. They remain in their burrows, where the temperature never rises above 30° C. Since their body temperature is about 37° C, they do not need water for temperature regulation. These rodents

Figure 18-1 Desert rodents. (a) Kangaroo rat. (b) Antelope ground squirrel. (c) Wood rat on saguaro cactus.

Figure 18-2 Saguaro cactus reaches a height of about 30 feet.

have physiological adaptations that permit them to survive on small quantities of water contained in the seeds that they eat, and on the water produced by their metabolism in oxidation reactions. Their kidneys produce very concentrated urine to minimize water loss. Kangaroo rats can survive even when a complete absence of rain for an entire season has made food extremely scarce. Under these circumstances, the kangaroo rat goes into a torpid, sluggish, state. Its metabolic rate goes down, its body temperature falls, and it requires less food.

Another desert mammal, the white-footed desert mouse, is also nocturnal, but it usually requires a little drinking water. When water is unavailable it enters the same kind of torpor shown by kangaroo rats. In this case, its lower body temperature reduces its requirements for both food and water. Antelope ground squirrels have yet another survival strategy. They remain above ground during the daytime and can thus exploit sources of food that are unavailable to nocturnal rodents. They also escape nocturnal predators such as coyotes. These ground

squirrels survive by being able to tolerate temporary increases in body temperature of several degrees, from 37° C to as much as 43° C, and then to cool off again without losing water. The squirrel dashes about rapidly in the full sun but never leaves the area of its burrow. When its body temperature exceeds the 43° C limit, it retreats into the burrow and flattens itself against the cool floor to lose excess heat. When it is cool again, it returns to the surface and repeats the cycle.

Each of these three animals has a somewhat different set of adaptations to desert conditions. Each has a different species ecology, based on differences in physiology and behavior, though all three are adapted to survival in the desert. Species ecology is often closely related to studies of the animals' physiology. In fact, this kind of ecology often overlaps with comparative physiology.

Knowledge of species ecology can help to solve many kinds of practical problems. For example, the saguaro cactus (Figure 18-2) is disappearing from its namesake, Saguaro National Monument, an area of Southwestern desert that was set aside as a special reserve for this dramatic plant. The disappearance began when cattle were first allowed to graze in the Monument. But the reasons for the destruction are quite complex. The cattle obviously do not consume the saguaro plants directly, though they do trample some of the young ones. The grazing of cattle had several more indirect effects, however. Most of the other young saguaro plants were destroyed because the population of desert wood rats began to increase at the time of the grazing—and wood rats eat young cactus.

The woodrats became abundant for two reasons. First, predators such as coyotes were shot to protect the cattle; this action also protected the rodent populations from predation. And second, the cattle ate many seed plants. As a consequence, rodents such as the kangaroo rats, which survived on seeds, starved to death. The reduction of the kangaroo rat population in turn allowed the growth of the wood rats. The two species had been competing in various ways for living space. With the kangaroo rat population declining, however, the wood rats multiplied. Grazing was discontinued in the Monument in 1958, and the Monument has been extended to include ungrazed areas. However, this action may have come too late to save the saguaro population.

The analysis of the disappearance of the saguaro cactus depended heavily on the judgment and experience of the ecologist, on his familiarity with the saguaro species and with its particular environment. Yet the ecologist is almost inevitably drawn into consideration of other species as well. The disappearance of the saguaro involves cattle, coyotes, wood rats, and other rodents. Organisms simply do not exist independently; groups of species coexist in various environments, and they interact in a variety of ways. Some organisms provide food for others, some are parasites, and others assist in the decay of dead tissues. Organisms must share living space, water, sunlight, and other resources with all the organisms around them. As a consequence, most problems in ecology are concerned with groups of organisms rather than with any one particular species.

Ecologists use several terms to describe assemblies of organisms. A **population** consists of individuals of a particular species that interbreed and live in a particular area. A **community** is composed of populations of different species that live in the same area and interact in various ways. An **ecosystem** includes both the community of organisms and its immediate physical environment.

The limits of an ecosystem are not sharply defined. It can be the whole earth, or it can be a small aquarium or terrarium. What we choose to define as an ecosystem is determined by our particular interests at the time. There are many things that we might want to know about a given ecosystem. We might want a description of it—what organisms are present? what is the physical environment like? Many communities and ecosystems have now been described in great detail. Lists of the species in various communities, and studies of their relative abundance, are numerous. These descriptions all extend and reinforce what we already suspect from simple observation: that certain kinds of organisms often occur together in specific types of environments.

This correspondence is not surprising, of course, since each particular species can exist only in a limited range of environments, and an environment that is suitable for one species should also be suitable for others having similar needs. However, understanding the observed distribution of the various organisms in specific communities is a very complex problem. In fact, one of the most fundamental questions we can ask about an ecosystem is, what controls the distribution and abundance of the organisms in it?

distribution and abundance

The distribution of organisms is obviously limited by their adaptations. Water lilies and duckweed are restricted to water. Seals and shrimp are no threat to the lettuce and peach crop. It should be possible to understand some aspects of the distribution of organisms by searching for the important features of the environment that *prevent* their growth or reproduction. We can examine the limits of a species' tolerance for temperature, light intensity, water availability, specific minerals, and so forth. From this information we can then attempt to explain the observed distribution of the species. In favorable cases this method can produce useful insights. But it is usually a difficult task. Physical factors in the environment interact in rather complicated ways. An organism may be better able to tolerate a water shortage at low temperature than at high temperature, for example. It is hard to weigh more than one or two variables at a time.

Life Zones

One attempt to explain the distribution of organisms was based on the assumption that major types of plants and animals were distributed in **life zones.** Life zones were defined according to average yearly temperature. Temperature decreases both as one goes toward the pole from the equator and as one goes up in altitude. Therefore, the same life zone should occur on a cold mountain top in the southern United States as at a lower altitude closer to the arctic (Figure 18-3). Of course, it was recognized that physical factors such as soil composition and rainfall were also important in controlling the distribution of organisms. But we now know that clearly-marked life zones simply do not exist if one looks closely. A careful study of the distribution of four major kinds of plants at different altitudes shows that altitude is indeed an important factor,

Figure 18-3 Life zones in North America.

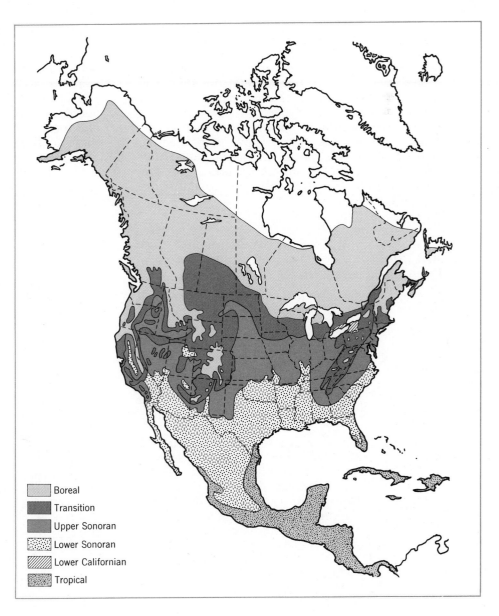

Boreal

Transition

Upper Sonoran

Lower Sonoran

Lower Californian

Tropical

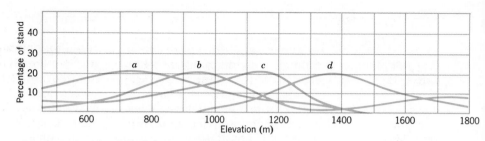

Figure 18-4 The distribution of four plant species on a mountain slope. All of the species overlap considerably; species b occurs at all elevations and shows two peaks of occurrence.

but that there is also a great deal of overlap in distribution between life zones (see Figure 18-4). One species even has peaks of occurrence at two different altitudes.

In fact, it would be rather surprising if most plants and animals were really distributed according to simple physical features of the environment. Physical limitations are important, but they cannot provide a full explanation of actual distribution patterns.

A Model of Population Growth

A second approach to understanding what controls the distribution and abundance of organisms in an ecosystem begins with a simple numerical description of the way a single population grows. It begins with an equation that predicts how rapidly the population will expand under various conditions. This description is then developed into a numerical model that can be used to describe some of the simple interactions among the species in a community.

The model begins with the assumption that the growth (R) of any population of organisms depends on three things: its birth rate (B), its death rate (D), and the number of individuals in the population (N). Birth and death are general terms; we don't care whether the organism is reproducing sexually or asexually, or whether death is due to predation or to senescence. It is usually most convenient to express birth and death rates as probabilities. In other words, the birth rate (B) is the probability that an individual in the population will reproduce in a specific time period. The death rate (D) is the prob-

ability that the individual will die in that same time period. Obviously, not all the individuals in a population really have the same chance of reproducing or of dying. The very young and the very old, for example, are not going to reproduce. But these rates represent an average figure for the whole population.

The growth rate (r) per individual in the population is equal, then, to the birth rate (the chance of individuals' being added) minus the death rate (the chance of their disappearing). In mathematical terms,

$$r = B - D \qquad (18\text{-}1)$$

Obviously these rates are independent of the size of the population. A large population can have the same growth *rate* as a small population. But the total numerical growth of the population (R) does depend on the population's size. If a population is composed of N individuals, and the growth rate per indidivual is r, then

$$R = rN = (B - D)N \qquad (18\text{-}2)$$

Obviously, the increase in the number of individuals over a given time period does depend on how large the population is to start with. Equation 18-2, therefore, actually describes a situation that changes as time passes.

Let us consider what can happen to a population of organisms with N individuals under various conditions. If the birth rate (B) is greater than the death rate (D), then r is positive. This means that R is also positive. The population will grow as time passes, and N will get larger. But as N becomes larger, so does R. In other words, the population will grow faster and faster.

If B is equal to D, then r is zero, and the population neither grows nor shrinks; N stays constant. If

Ecological Models

Any community or ecosystem is extremely complicated—so much so that it is impossible to understand all aspects of it. Ecologists often try to overcome this difficulty by creating simplified models. A model is never intended to be a full description of the situation; rather, it simplifies reality, making it more understandable and easier to study. But at the same time the model must still be complex enough to represent key features of the real world.

Model-building takes several steps. It begins with an assumption about what might control a central feature of the community. For example, we might assume that the size of a population in a community depends on the amount of food available. Then the next step is to explore the consequences of our model, to determine whether its predictions are borne out in the real world. For example, our simple model predicts that if more food becomes available in a region, the population will get larger. We now go back to the real world and see if this is the case. If the predictions are fulfilled, we assume that our model is accurate. But even if the predictions fail to match reality, our model is still useful, because it sends us back to our original assumptions to ask where we oversimplified or went astray.

When we test our food model against reality, we may find that several things control population size in addition to food. Diseases and predators, for example, may be equally important. We can then modify our model to make it fit reality better. Alternatively, we may simply decide that it isn't worth the effort to try to get a better fit with reality. Scientists usually try to develop models that occupy a middle ground; a satisfactory model is one that is realistic enough to represent the most important features of reality, and yet sufficiently simple to be useful in making predictions.

Another example, this time from physics, may help to clarify the relation of models to reality. Classical physics uses simple equations to describe the acceleration of a lead weight, and over short distances it provides a good prediction of where the weight will fall. The science of ballistics introduces some refinements into the model to account for such factors as air resistance and the earth's rotation. These refinements are useful and worth the additional calculation if we are interested in hitting a distant target. But no one would even try to make a similar model to describe the path and acceleration of a dropping feather. Random puffs of wind and other variables make the problem impossibly unwieldy (and no one really cares that much, anyway).

The theoretical model of population growth developed in this chapter can become rather complicated and mathematically sophisticated. However, it conforms to the limitations of all models:

1. A model is always an abstraction of reality and is necessarily simpler than reality. Therefore, it can never be expected to predict and explain all of the details of the natural system.
2. A model is continuously corrected and modified by comparing its predictions with actual observations.
3. Theoretical models are really more useful for explanation than for prediction. We can make general predictions about a natural system, but we are more likely to offer explanations of an existing situation.

Models can be useful in deriving general principles. However, if we hope to understand the forces at work in any particular ecosystem, we must simply get out and look at them. There is no substitute for direct field experience for the would-be ecologist. Theory can inform vision and suggest what to look for, but it cannot replace the looking.

B is smaller than D, then r is negative, and both N and R will get smaller. In short, even though the growth rate (r) is constant, both the size of the population (N) and the growth of the population (R) will change with time. At this point it becomes easier to describe the model with graphs. Our model now takes the form,[1]

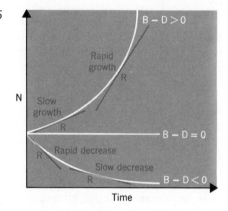

Figure 18-5

The graph shows how the size of the population (N) changes with time under the three conditions we described. The growth of the population (R) at any given time is represented by the slope of the line. The size of R tells us how quickly N is increasing or decreasing. That is, when R is small and positive (when the slope is slightly upward), N is growing slowly. Later, as N gets larger, R is large and positive (sloping steeply upward); the population is now growing rapidly. Similarly, the size of R describes rate of decrease when N is declining. In this case R is negative (the slope is downward).

If we consider the situation a little more carefully, we realize that N is changing in an exponential fashion. Consider a growing population, one in which r is positive. If we begin with ten cells, and if

these cells reproduce by dividing once a day, then after one day the population will have twenty cells. After two days it will have 40; after three days, 80; and so forth. The number of individuals in the population will double every day. The time required for the doubling of a population is called the **generation time.**[2] The same argument applies to any growing population; only the individual numbers are different. The value of r actually determines just how rapidly N increases. When r is large, N increases very quickly, and the generation time is short.

We can now go back into the laboratory or the field and make observations to check the accuracy of the model. The simplest way is to set up a culture flask and introduce some cells of bacteria. We take a small sample of cells each day and count them to keep track of growth. In the early stages our model provides an excellent fit with the observed data. We conclude that bacterial growth is, in fact, exponential. The cells will either grow exponentially, hold their own, or die off, depending on the suitability of the culture medium. We can now use our model to predict how large a population of bacteria will be at any time in the future—provided we can measure N (the present population size), and provided we have established the value of r by previous observations.

But organisms in a flask or anywhere else cannot grow indefinitely. Our model breaks down as the population gets large, because we assumed that B and D (and, therefore, r) were unchanging. This assumption is wrong. We know that both the birth rate and the death rate are likely to vary in real populations. When populations become very large, birth rates tend to decrease, and death rates increase.

We can now go back to our model and modify our assumptions about B and D. Let us define a maximum number of individuals (K) that can exist in a population in a particular environment. (K is sometimes described as the **carrying capacity** of the environment.) As long as the population size is

---

[1] Ecological models like this one are fundamentally mathematical models; interactions between organisms and their environments are presented as numerical relations, and they are expressed in the form of equations. Many of these equations require a knowledge of elementary calculus. But the relations described by the equations can also be represented by simple graphs. We have included the equations here in footnotes, and we intend them only for readers who have sufficient mathematical background to understand them. The mathematical form for the graph in Figure 18-5 is:

$$R = \frac{dN}{dt} = (B - D)N = rN$$
$$N = N_0 e^{(B - D)t} \text{ or } N = N_0 e^{rt}$$

where N_0 = population size at $t = 0$. t = time.

[2] The average growth rate (r) is related to the generation time (GT):

$$r = \frac{\ln 2}{GT} = \frac{0.69}{GT}$$

The reason is that when $t = GT$, $N = 2N_0$;
$$N = N_0 e^{rt}$$
$$2 N_0 = N_0 e^{r(GT)}$$
$$2 = e^{r(GT)}$$
$$\ln 2 = r \times GT$$

small compared to this maximum, r will not change, and the model in Figure 18-5 should hold. However, as the population (N) approaches the maximum (K), the birth rate or death rate (or both) will change as a result of various unfavorable consequences of the larger population. Food becomes scarce, suitable living sites are unavailable, and waste products accumulate, for instance. The growth rate (r) goes down.

In terms of equation 18-2, as N gets larger, r gets smaller, so that population growth (R) slows down and levels off when N = K. Even if N temporarily reaches a level above the normal K, changing birth rates or death rates or both will eventually bring it back down to K. Any influence that slows the growth rate as N gets larger is called a **density-dependent** factor. All populations must ultimately be limited by density-dependent factors in the environment.

We can add this limitation to our model as:[3]

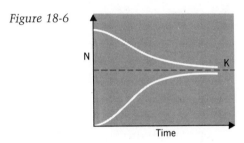

Figure 18-6

Our model now gives a better account of what happens to real populations, including the population in the culture flask, as they get very large. But what about very small populations? Our

model says that the population will adjust to K no matter what size it is to start with. Yet we know that a population is in trouble when it is very small; in fact, a small population in any one particular area is always in danger of becoming extinct. Obviously, in many species, a single individual cannot reproduce alone. In those cases a population of one is automatically doomed. But even in somewhat larger groups the population is in danger. An individual may not be able to find a mate, or there may be too few individuals to hunt successfully, or (in the case of plants) to break up the soil or to retain rainwater and prevent erosion. There is always some minimum population size necessary for the organisms actually to grow and achieve a positive r. If the population drops below this level it is in danger of becoming extinct. Problems like the inability to find a mate are also density-dependent factors. In this case they operate at low population densities. We can define the required minimum population size as M, and include it in the graph of our model:[4]

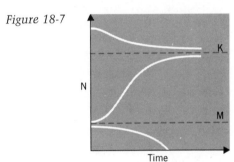

Figure 18-7

At this point, our model is refined enough to test against natural populations that are a little more complex than a culture of microorganisms. For example, we might attempt to understand the growth of a fish population in a small pond. We wish to use the pond for fishing, but we don't want to overfish.

---

[3] $R = \dfrac{dN}{dt} = rN$. But r is no longer constant:

$$r = r_m \left(\frac{K - N}{K} \right), \text{ or } r = \left(r_m - \frac{N}{K} r_m \right).$$

r_m is the maximum growth rate; $\dfrac{N}{K} r_m$ is a density-dependent factor. When N is small, $\dfrac{N}{K} r_m$ is close to zero, and the population will grow at its maximum rate, r_m. When N becomes larger and approaches K, $\dfrac{N}{K} r_m$ gets larger and r gets smaller. When N = K, $\dfrac{N}{K} r_m = r_m$, and r = 0, so growth stops. If N is larger than K (say the food supply has suddenly decreased, lowering the value of K) then $\dfrac{N}{K} r_m$ is larger than r_m, and r becomes negative. N will get smaller until N = K again.

[4] $R = \dfrac{dN}{dt} = rN$, where $r = \left(\dfrac{K - N}{K} \right)\left(\dfrac{N - M}{N} \right) r_m$, or

$$r = r_m - \left(\frac{N - M}{K} \right) r_m - \frac{M}{N} r_m.$$

r_m is the maximum growth rate; $\left(\dfrac{N - M}{K} \right) r_m$ is a density-dependent factor; $\dfrac{M}{N} r_m$ is another density-dependent factor. Here, K and M interact, so that r_m is reduced by two density-dependent factors. The first factor slows growth as N becomes larger and approaches K. The second factor becomes important as N approaches M; when N is less than M, r becomes negative.

In terms of our model, we should not take so many fish that the population (N) is reduced below the minimum (M), or we will have no supply at all. If we keep the population small, the production of new young fish will be small. As the population gets larger, the production of new fish increases as well. But if the fish population gets too large, K begins to become limiting. Production again drops off, as density-dependent factors tend to reduce the birth rate and increase the death rate. Maximum fish production occurs in a population that is somewhere between M and K. Production is equivalent to R. The model predicts that production from our fish pond will be highest when N is halfway between K and M. That is the point where the curve rises most steeply, that is, the point where R is greatest.

Unfortunately, there are still difficulties with our model. For example, we have implicitly assumed that all members of the population are capable of making equal contributions to its growth. However, this assumption is clearly not true. Very young individuals do not reproduce, and old individuals usually decline in reproductive vigor. In fact, we can remove any old, non-reproducing individuals that we catch without changing the production rate of our pond. (Interestingly, it may also be a good idea to take very young individuals in some populations. The reproductive contribution of young individuals in a population may be statistically smaller than that of mature reproductives. This is the case when the young are more likely to die of natural causes. For example, young fish may be eaten or may be unable to survive the winter.) In any case, if we want a better solution to our fish-pond problem, we must further refine our model. More elaborate models are usually written in computer language and are detailed enough to provide real guidance for the management of fisheries and wildlife resources. The construction of an elaborate mathematical model is worthwhile in this case, because it provides rational guidance in solving a real problem.

Our simple model can be modified in various ways to fit specific situations. But if our model has any real validity, it should be applicable to ecosystems more complex than a culture flask or a small fishpond. We should be able to discuss the relationship between a population of organisms and its physical environment in more specific terms, and we should be able to describe the effects that other species in the ecosystem have on the growth of our population.

Species Interactions

Obviously the physical environment operates in various ways to control the total population (N). The growth rate (r) may be influenced by food, by temperature, or by physical hazards that kill members of the population or prevent their birth. Such factors may be density-dependent, or they may operate at all population levels. The physical environment also controls K, by limiting resources such as the number of burrows or caves or the amount of available water. However, as we stated in our discussion of species ecology, physical factors alone rarely control the size of a population. Species interactions are also important. Each particular species has a place, or role, in the ecosystem, described as its ecological **niche.** An organism's niche includes all aspects of the physical environment and all the species interactions that are important to its survival.

Competition

One basic kind of interaction occurs between species whose ecological niches overlap to some extent—that is, who use some of the same sources of food, or require similar habitats for breeding or shelter, or depend on some other feature of the environment for survival. Any such case, in which two species try to share the same niche, is called **competition.** We can express competition between species quite simply in terms of our model. When the requirements of two species are similar, the presence of one reduces the value of K—the carrying capacity of the environment—for the other. This relation operates in both directions. In fact, we can define competition simply as the reduction of K in the two growth equations for the two species.

Two kinds of organisms simply cannot occupy exactly the same niche, however, because the resulting situation is inherently unstable. Let us consider what happens when they try to share the same niche. When both populations are small they will both increase in size for a while. The populations are not identical, however—one will no doubt grow a bit faster or be a bit larger than the other. But when the two populations grow larger and begin to use up the available food or to fill all the nesting sites, the situation changes. Density-dependent factors come into play, and these limitations exert pressure on the growth rates (r) of both populations. This effect will simply increase any existing difference in overall growth (R). Say the value of r is 10 for one species, and 9 for the other. Assume that a density-dependent factor now reduces both growth rates. The reduction will be the same for both species, say 5. Now the values of r are 5 and 4, respectively. In short, what was once a ten-percent difference between growth rates has increased to a twenty-percent difference.

This difference between the two populations will continue to increase as the two populations approach K. The closer they get, the stronger the density-dependent factor becomes. The growth of the smaller population is slowed more and more relative to the large one, and sooner or later its size falls below the M for that species. One species always "wins". It is impossible for both to survive indefinitely in exactly the same niche.

Experiments on competition between species have been performed in the laboratory. The most famous experiment involved placing two kinds of beetles in a jar of flour. Only one species survived, and it was always the same one, no matter how many times the experiment was repeated. The adults of the surviving species always destroyed the young of the other species. However, the outcome of this competition could be changed simply by adding some pieces of broken glass tubing to the jar. Now, the two genera were able to coexist for a long time. The beetle that was heretofore unable to survive now found refuge from its competitor in the tubing. In other words, the niches of the beetles were no longer identical—one lived inside the tubing, the other outside. Both species now could survive because they no longer competed so directly. In fact, this is generally what happens in natural populations. A species that cannot successfully compete for one particular niche in an area does not usually become extinct there. It can probably find alternate resources for survival.

Predation

A second kind of species interaction is **predation.** In predation, one species is killed and used as food by another species. The term includes the cropping of forage grasses and other plants by herbivores, as well as the hunting or trapping of animals for food. Once again, our model of population growth provides an analysis. We begin with populations of two species in the same environment, a predator and its prey. The growth rate of each population is influenced by the size of the other population. That is, the birth rate of the predator population depends on how many prey are available for food; the death rate of the prey population depends on the number of predators in the area.[5]

This interdependency causes the two populations to interact in a complex way. When both populations are growing, the predator will begin to kill more and more prey, and the prey population may begin to decline. However, the population growth of the predator species depends on the availability of prey. So as the prey become rare in the environment, the predator species will also begin to die off. When both the populations are sufficiently small, the prey population may again begin to increase. In that case, however, the predator population will soon begin to grow as well. This simple model predicts that the population size of both prey and predator will fluctuate in a regular, cyclic fashion (Figure 18-8). The cycles of the predator will lag

---

[5] Predator Growth: $R_1 = r_1 N_1 = (B_1 N_2 - D_1) N_1$.
Prey Growth: $R_2 = r_2 N_2 = (B_2 - D_2 N_1) N_2$.
The subscript 1 refers to the predator population; the subscript 2 is the prey population. The predator's birth rate is equal to $B_1 N_2$. The prey's death rate is $D_2 N_1$.

Figure 18-8 The predicted fluctuations in populations of predator and prey, based on a simple model. The oscillations of the predator population lag behind those of the prey.

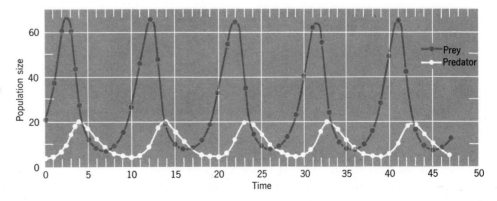

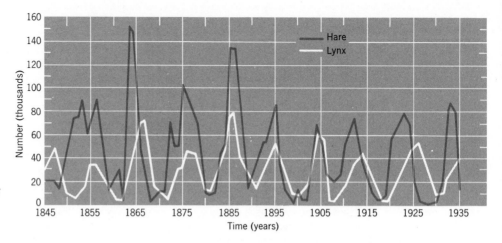

Figure 18-9 The actual fluctuations in populations of a predator (the lynx) and its chief prey (the snowshoe hare) are similar to those predicted by the model. The populations were estimated from the number of pelts received by the Hudson Bay Company.

behind those of the prey, since there is a time delay in the response of predator population growth.

We can verify this prediction by looking for cyclic phenomena in a real environmental situation. We shall never find a situation quite so simple, where single prey and predator species occur together. However, there are some close approximations. One classical example is the relationship between the lynx and the snowshoe hare in northern latitudes. The lynx is the principal predator on snowshoe hares, and the hare is the lynx's principal prey. Cycles in these species' population densities are reflected in records of the number of lynx and hare pelts received by the Hudson Bay Company over a period of about a century (Figure 18-9).

Another prediction we can make from our simple model is that if we remove the same fraction of both populations in a predator–prey system, the population of the prey will recover more rapidly than that of the predator. This result is expected from our model. The prey growth rate should increase (there are fewer predators), while the predator growth rate should go down (there are, at first, fewer prey). Thus, prey growth is initially favored over predator growth, even though the predator/prey ratio is unchanged.

This effect is commonly observed in our attempts to control agricultural pests. Insecticides such as DDT often will kill a large fraction of the pest organisms we are trying to control, but at the same time they also kill the same fraction of the pests' insect predators. The pest population often grows back so rapidly that it becomes even larger than it was before treatment (see Figure B, page 593).

Obviously, this model describes the simplest possible relationship between predator and prey. It assumes that the only thing that influences the growth rate of one population is the population of the other. In particular, it says that while both populations can get very small, they will always recover and continue to oscillate. We know, however, that predator–prey interactions are not always that stable. There is always the danger that the prey or the predator species will suffer a population decline so drastic that it becomes extinct in the area. In fact, this is just what happens in the laboratory if we place predator and prey together in a simple environment. The predator population simply increases until it consumes all the prey. Then the predator itself starves to death. We can change the outcome of the experiment, however, by adding something to the environment that provides a refuge for the prey species. Under these conditions, cyclic oscillations in the predator and prey populations do appear. Still, this system does not validate our simple model at all. It undergoes cycles, but for slightly different reasons. The model predicts that the prey population will recover only because the predators are starving and dying. But in the experiment the prey population recovers not only because the predators are dying, but *also* because the prey have a refuge that keeps them from being wiped out. Since natural environments are nearly always complex enough to provide such refuge, our model must be changed to include this fact.

The revised model can now be applied to relatively complicated predator–prey interactions in nature. The relation between the moth *Cactoblastis* and the prickly pear cactus in Australia is a good example. The cactus was deliberately introduced into Australia at the end of the nineteenth century, both because it is attractive and because it can be planted in rows to form fences. The plants quickly got out of hand, however. They encountered no predators and spread across the land, becoming a great pest. By 1925, millions of acres of Australia were rendered unfit for grazing.

The *Cactoblastis* moth, whose caterpillar stage feeds on the cactus, was introduced into Australia in an attempt to control the prickly pear. The moth spread equally rapidly, and in ten years it had killed off most of the cactus (Figure 18-10). However, neither the cactus nor the moth has become extinct in Australia, because there are refuge areas for the cactus that the moth has not reached. The environment is so large and complicated that a fluctuating balance between predator and prey will probably continue indefinitely.

At this point, our model says that the growth rate, r, of one species is controlled by the population size of the other species *and* by the presence of refuges for the prey. Obviously, more conventional density-dependent factors will also affect r. Factors such as available food and shelter for the prey will cause their growth to cease at some K. In other words, the growth of both populations is also controlled by factors that have nothing to do with the predator–prey interaction. More sophisticated models have been developed to take some of these factors into account.

Diversity and Stability

We can now draw some important conclusions about natural ecosystems. Competitive interactions guarantee that only one species will occupy a particular niche in any area. Therefore, the total number of species that can occupy an area—that is, the ecosytem's **diversity**—is controlled by the number of niches. This fact explains why many organisms that seem to be adapted to a region simply don't appear there. It also explains why attempts to introduce new species into a seemingly suitable area usually fail. In both cases the appropriate niche has already been filled by another species. There have been a few spectacularly successful introductions—starlings into the United States, or prickly pear cactus and rabbits into Australia—but these examples are exceptions.

Our theoretical analysis of competition also leads us to expect that geographically smaller environments would support smaller numbers of species. This relationship is most apparent on offshore islands. Large islands have greater diversity of species than small ones. And in fact, the relationship between size and diversity is surprisingly precise. If the number of species is plotted against the logarithm of the land area, the relationship is a straight line (Figure 18-11). A smaller land area offers less space for each potential niche. On small islands, some niches are simply too small to support a pop-

Figure 18-10 Predator–prey interaction. (a) This farm, like many others in Queensland, Australia, was abandoned after becoming infested with prickly pear cactus. (b) Eighteen months later, the Cactoblastis *moth had destroyed the cactus and made possible the resettlement of the property.*

(a)

(b)

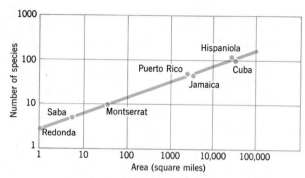

Figure 18-11 *The number of species of amphibians and reptiles on various caribbean islands is proportional to the area of the islands. Larger islands have more niches available.*

ulation above the critical minimum, M. The rate of extinction of populations is also higher on small islands than it is on large islands or on the adjacent mainland. We conclude that larger ecosystems are more diverse, and therefore more stable, than small ones.

A single species in a diverse ecosystem is less likely to suffer a drastic decline. For one thing, more sources of food are available to it. When a predator feeds on several species of prey, the populations of both predator and prey fluctuate less violently. As one type of prey becomes rare, the predator can turn to other species for food. In general, the more channels there are for predator–prey interaction, the more stable each population will be. And, in a diverse community, even when a species does become extinct, the entire system is less likely to be affected since many other species are still present.

We could continue to refine our model to handle even more complex interactions between organism and environment. It is theoretically possible to write a separate equation for the growth of each species in a community, and then to include terms that describe all the interactions of the various species with each other and with the physical environment. However, the model quickly becomes too complex to be useful. Even in its simpler form, however, it gives considerable insight into what controls the distribution and abundance of organisms in an ecosystem.

The numbers of organisms present are of considerable interest in many circumstances, but not in others. One elephant, for example, can have considerable impact on an ecosystem. But we really don't care whether one million or ten million bacteria of a particular species are present in a handful of soil. There are also things we want to know that are not directly related to the numbers of organisms. Often we are more concerned with what organisms are doing in the ecosystem than with how many there are. For instance, how do particular organisms affect soil, temperature, or humidity? A still more fundamental question is how organisms exchange materials and energy with the environment.

Ecology and Agricultural Systems

Any natural ecosystem contains a diverse community, with complex interactions between competitors and among predators and prey. When such land is cleared for agriculture, the diverse and stable community is destroyed and replaced by a large population of a single organism, the crop plant. The yield of the desired crop is raised by clearing and weeding out the competitors. As a result, the K value for the crop plant's population growth will not be depressed by competing species. This result is, of course, the whole point of the agricultural process. The K is increased still more by supplying fertilizer and supplementing natural water supplies. The farmer also tries to preserve the yield by eliminating predators on the crop plants. In practice this last goal is extremely difficult to attain, because agricultural ecosystems are not diverse, and hence are inherently unstable. Many organisms that prey on crops—insects and microorganisms—have very large r values. Even a very small pest population left over from the clearing operation can grow explosively, because the clearing process also decimates the pest's natural predators. Whole agricultural areas are often devastated by a single predator species, such as corn blight or migratory locusts.

Attempts to control agricultural pests often fail because the destruction of one pest merely allows its ecological competitors to expand their populations and to attack the crop themselves. The use of DDT to control pest insects in apple orchards is one of the more spectacular examples. DDT was first used to reduce populations of the apple codling moth. However, DDT also killed a parasite of the woolly apple aphid, as well as competitors and predators of the redbanded leaf roller and red mite. These three types of insect were all resistant to DDT and grew rapidly. Soon they themselves became serious pests of apple orchards. Attempts to control the new pests led to the use of other insecticides such as parathion in combination with DDT. But within two or three years, populations of the codling moth became resistant to DDT, and populations of red mites developed resistance to parathion.

This pattern has become very common. Chemical control of pests is becoming less and less efficient and more expensive. The cost to farmers has been increasing by fifteen percent per year since 1950, while farm production has increased by only two percent per year. In 1950, pesticides cost about 25 cents per acre; they are expected to cost eight to nine dollars per acre in 1975, or about ten percent of the total value of production.

There is another important consequence of such control programs. Changes caused by chemical treatment may have consequences for nonagricultural human activities. For example, mosquitoes that transmit malaria have become resistant to insecticides as a result of the spraying of agricultural crops. This development has created a serious health hazard in areas of the Caribbean. Thus, the future of chemical control of agricultural systems is quite bleak. The pests are adapting to the control agents, and chemical control kills parasites or predators of the pest as well as the pest itself. Futhermore, the chemical may remain in the environment and cause undesirable side effects.

Two new approaches to pest control are being investigated. One, called **cultural control,** attempts to reduce the carrying capacity of the environment for the pest and, it is hoped, to create a permanent reduction in the pest population. Mosquito populations can be reduced by draining swamps and other standing pools of water to remove breeding sites. Similarly, the destruction of crop remnants at the end of the harvest, and the use of crop rotation, may also remove breeding sites and prevent a long-term buildup of pest populations. Common weeds may provide refuges for insect or microbial pests between growing seasons; destruction of the weeds effectively reduces the pest population.

In some cases however, maintaining several kinds of plants in an area is advantageous. This scheme is called **diversification.** There are several ways to diversify an agricultural system. Hedgerows made up of several kinds of plants can be used to divide fields into smaller areas and to provide additional plant species. Similarly, the crop plants themselves can be mixed, or noncommercial species can be planted in alternate rows of the same field. Mixed plants provide a variety of habitats for predators of potential pests (see Figure A).

However, as we have seen, a diversified environment may also provide refuges for pest species. Also, mixed crops are much more difficult to cultivate and harvest than single ones, since each species may have different requirements. Obviously, a program of diversification is going to raise the price of crops significantly. Diversification is perhaps most successful in home gardening, where the individual is willing to supply a considerable amount of hand labor and is not trying to make a profit.

The second method is **biological control.** In this technique, a predator or para-

Figure A Strip cropping. Alternating bands of various crops is one way of diversifying an agricultural system.

site of the pest species is deliberately introduced to reduce the level of the pest population. Biological control has been spectacularly successful in several cases – the control of the prickly pear cactus in Australia, or the use of ladybug beetles to control scale insects in citrus orchards (see Figure B). But these are special cases, in which the pest has been introduced into a new area in which it had no natural enemies. When the enemies were also brought into the area, pest control was spectacular.

Pests and their parasites and predators usually occur together in undisturbed ecosystems. Organisms become agricultural pests only when the simplification of the ecosystem reduces their enemies. These effects are clear when we compare agricultural land with a natural ecosystem. In the USSR, for example, insect communities from undisturbed steppes were compared

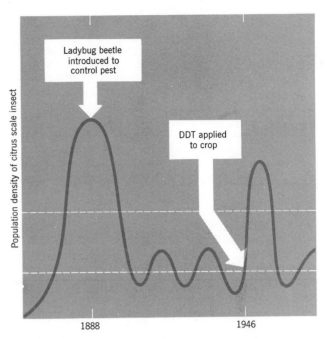

*Figure B   A natural pred-
ator, the ladybug beetle,
was imported around 1888
to reduce the population of
scale insects. The natural
equilibrium between pred-
ator and prey was dis-
turbed by spraying with
DDT in 1946. Populations
of both predator and prey
were reduced, but the prey
(the scale insects) recovered
more rapidly than the
predator.*

with steppes that had been cleared for wheat fields. The number of species in the wheat fields is about 40 percent of the number in the natural ecosystem, even though the number of individuals is doubled or tripled. In other words, many species die out, but those that remain can grow and become pests.

The best long-range strategy for pest reduction in large-scale agriculture requires careful study of the pests of each crop and in each locality, and the use of integrated control systems that make optimum use of cultural, biologi-cal, and chemical control methods. Such integrated systems show great promise, but it will be very difficult to get them implemented on a large scale. They require technical personnel, an effective communication link to a literate farming community, sufficient land to allow plantings of non produc-tive strips of vegetation, and enough government support to convince farmers to use a technique that reduces productivity but achieves long-term stability. Unfortunately, these requirements are lacking in undeveloped countries, where crop failures are most disastrous and often lead directly to famine and death. Only developed countries are likely to implement integrated approaches to agriculture.

energy in ecosystems

The exchange of materials is often studied in order to understand cycling of some material of potential interest: nitrogen, or phosphate, or carbon. But the flow of materials does not have a very coherent or predictable pattern in communities. Minerals enter and leave organisms at very different rates. In mammals alone, calcium exhibits three very dif-ferent flow patterns, depending on whether it is

found in teeth, in bones, or in blood. Furthermore, the same materials are exchanged from organism to environment and back to other organisms.

Energy flow, on the other hand, appears to have a much more fundamental pattern of movement through an ecosystem. Unlike material exchange, the flow of energy through an assembly of orga-nisms proceeds in only one direction, because all of it comes from one source, the sun. Radiant energy is first transformed by photosynthesis into organic compounds. Then, as the energy in these compounds flows from organism to organism, all of it is eventually degraded to heat.

If we examine the organisms in a particular area, we can classify them according to how they obtain energy. Some are photosynthetic, others are herbivores or carnivores, some are scavengers, and still others are decomposers. The photosynthetic organisms are the only ones capable of using radiant energy to produce organic compounds. All other organisms, with negligible exceptions, use the chemical energy from the photosynthesizers for their own needs. Each organism belongs to a particular stage in this transfer of energy, called a **trophic level** (literally, "feeding level"). Plants are called **primary producers;** they represent the first trophic level. Herbivores are animals that feed on plants; they are placed in the second trophic level. Carnivores that prey on herbivores belong to the third level. Carnivores that prey on third-level carnivores are in the fourth trophic level, and so on. The decomposers, organisms that facilitate the decay process, occur at all levels.

The details of energy turnover in different living tissues vary greatly. Many kinds of food are used and metabolized in various ways, and at different rates. But this diversity presents no problems, because we can still determine total energy metabolism by measuring the overall levels of oxygen consumption or carbon dioxide production. This method allows us to examine energy relations among organisms by measuring inputs and outputs of energy at various levels in the ecosystem.

Primary Production

Energy from sunlight flows into the ecosystem through the primary producers. The conversion of sunlight into organic material by photosynthesis is called **gross primary production.** The efficiency of this production is rather low, ranging from only about 0.2 percent to two percent. The highest values are achieved by carefully-managed crop plants and by algal cultures; natural communities have lower production efficiencies.

A large fraction of the organic matter created in photosynthesis is used by the plants themselves to support their own respiration. This fraction is obviously unavailable to other members of the ecosystem. Only the organic material that is not used directly by the plants, that becomes part of their tissues, is available to serve as food for the remainder of the ecosystem. This latter fraction is called the **net primary production.**

Net primary production can be estimated by comparing plants' rates of photosynthesis and respiration, or simply by harvesting and weighing them to see how much they grow. Such measurements show that about half the gross production is utilized by the plants themselves; the other half is net production, available to other organisms in the community. The net efficiency of primary production is therefore half the gross production, or about 0.1 to one percent. About three fourths of the total net production of the entire earth occurs in land communities, and about one fourth in the seas. The first column of Table 18-1 expresses net primary production in various communities as grams of dry plant material produced per square meter per year. As you can see, there is more than a thirtyfold difference between the most productive and the least productive regions.

The reason for low primary production in some areas is quite evident. The arctic tundra receives low radiant energy per square meter and has low temperatures. Both these factors reduce plant growth. In deserts and dry steppes, the lack of water limits production. In yet other communities, net primary production appears to be restricted by lack of nutrients. Nitrogen is particularly scarce in tropical ocean waters. Temperature and incident radiation favor high year-round production in these waters, but the lack of nutrients keeps production at a low level. If tropical waters are supplemented with nitrogen, primary production may increase by a factor of ten. Primary production may also be depressed by local lacks of phosphate or cobalt or vanadium or molybdenum, or of virtually any element that is required for plant growth.

Table 18-1 also shows some interesting relationships between primary production and the total amount of plant tissue present at a particular time. The total amount of plant tissue is called the **standing crop.** Most of the land communities have a standing crop much greater than their annual production. The trees in a forest, for example, are so large and grow so slowly that they will not even double their weight in a year. By contrast, most aquatic communities have small standing crops, because the plants are very small and short-lived. But their production rates are higher than their standing crops. A single algal cell may double its weight and then divide in a single day. The high growth rate tends to make up for the fact that less photosynthesizing tissue is present. Some ocean communities, such as kelp beds (see Figure 18-12), are intermediate between land and plankton communities in terms of standing crop. Their productivity is very high; the adult kelp plant may produce its own weight every few months.

Table 18-1

Production by Plant Communities

Type of Community	Net Primary Production (Grams of dry matter per cubic meter per year)	Standing Crop (Grams of dry matter per square meter)	Net Efficiency of Production[1]
Land			
Arctic tundra	100	500	0.20
Fir forest	700	26,000	0.03
Beech forest	1,300	37,000	0.03
Steppe	420	1,000	0.42
Desert	122	430	0.28
Savanna	1,200	6,660	0.18
Tropical rainforest	3,250	50,000	0.07
Agricultural land	650	1,000	0.65
Ocean			
Kelp bed	3,750	1,500	2.5
Continental shelf	350	10	35.0
Open ocean	125	3	42.0

[1] Primary production/standing crop.

Other Trophic Levels

Early attempts to describe trophic relations between primary producers and other organisms depended upon counting or weighing the individuals at each trophic level in a community: primary producers, herbivores, first-order carnivores, and so forth. This kind of analysis produces numbers that can be arranged as a pyramid. Figure 18-13a shows a pyramid of weight for a very simple ecosystem composed of alfalfa, cattle, and man.

The total weight, or standing crop, of organisms often decreases at each trophic level. As a result, the fourth or fifth trophic level is occupied by animals that are very rare in the ecosystem. Such animals are called **top carnivores.** A tiger is an example. A tiger is safe from other predators simply because it is so rare. An ecosystem can support so few tigers that another carnivore specialized to hunt and kill them simply could not find enough to eat.

Ecosystems do not always show this simple pyramidal organization of weight at successive trophic levels. A pyramid of weight for plankton communities in the ocean would look quite different. The primary producers are minute algae. Many of the

Figure 18-12 A kelp bed in southern California.

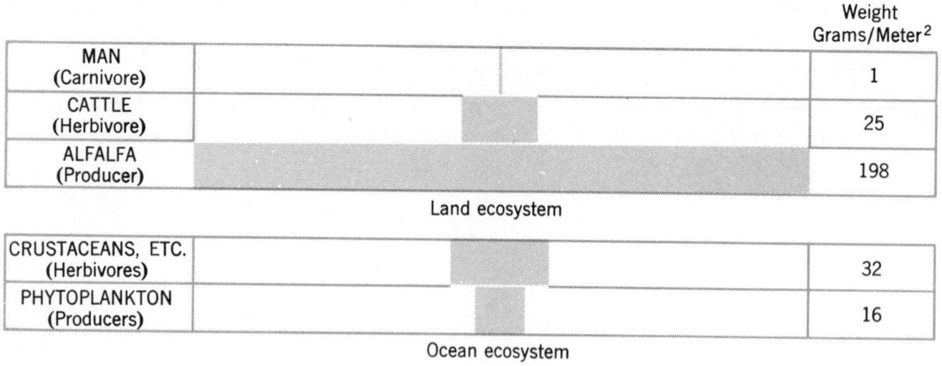

			Weight Grams/Meter^2
MAN (Carnivore)			1
CATTLE (Herbivore)			25
ALFALFA (Producer)			198

Land ecosystem

CRUSTACEANS, ETC. (Herbivores)			32
PHYTOPLANKTON (Producers)			16

Ocean ecosystem

Figure 18-13 Weight of organisms in different trophic levels. (a) A simple ecosystem involving alfalfa, cattle, and man. In most land ecosystems, the total weight of organisms decreases at each higher trophic level. (b) Some aquatic ecosystems, however, may have a greater mass of herbivores than primary producers.

algal cells are consumed by small herbivores, mainly crustaceans. The standing crop of crustaceans in the community often outweighs the standing crop of algae by a factor of two or more (see Figure 18-13b). This relation is possible because the algae have such a high productivity related to their standing crop. Since the algae divide rapidly, they can support a larger population of crustaceans than if they divide more slowly. In planktonic communities, the primary producers are simply dividing so rapidly that they produce sufficient food to sustain a weight of herbivores that is greater than their own standing crop. Obviously, the shape of the pyramid of numbers and weights is subject to considerable variation in different ecosystems.

Numbers and weights show the differences between ecosystems, rather than their similarities. But when we consider energy flow instead of the standing crop, we find that ecosystems are really quite similar. The net primary production of algae or alfalfa or any other green plant is simply a form of energy, some of which can be used by various herbivores. Similarly, the herbivores themselves represent a suitable form of energy for carnivores, and so on, as we ascend through each trophic level.

The actual rates of energy flow through a community are quite difficult to measure, however. Consider a herbivore feeding on a primary producer. Not all of the energy supplied by the producer (say, a tree) will be consumed by the herbivore. Some parts of it, such as wood or bark, will be unsuitable as food. Other parts of the tree will be consumed by

competitors or degraded by decomposers or blown away. Thus only a fraction of the tree's growth will be ingested by the herbivore. Once the food is ingested by the herbivore, several other things can happen. Some portion of the food is simply indigestible or is not digested; this fraction leaves the animal as feces or other excreta. Of the portion that is digested, a fraction is respired to supply energy for the herbivore's activities. The remaining fraction is used for growth, and appears as an increase in body weight. Only this fraction becomes available as food for organisms in the next trophic level in the ecosystem (Figure 18-14).

If we could measure the rate of food consumption of all the herbivores in an ecosystem, and compare it with the rate of food consumption of the first-order carnivores, we would have a direct measure of the total energy flowing through from one trophic level to the next. The ratio of these energy flow rates is described as the **ecological efficiency:**

$$\frac{\text{rate of energy flow into trophic level } T_n}{\text{rate of energy flow into trophic level } T_{n-1}} \times 100$$
$$= \text{ecological efficiency (in percent)}$$

It is usually impossible to make direct measurements in natural ecosystems, but indirect measures can be used. We can measure respiratory rates, the caloric content and quantity of feces, and the growth rates for some organisms in the ecosystem. We can express all of these measurements in terms of kilocalories and use them to calculate rates of food consumption. In other cases, rates of food consumption must simply be estimated. Then each

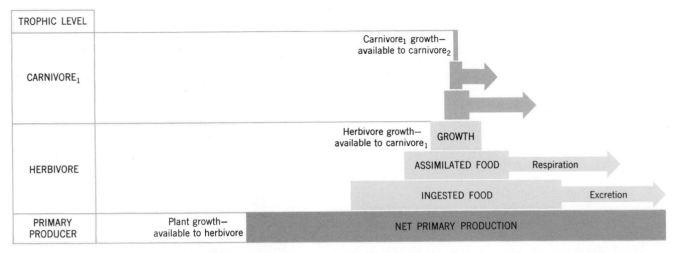

| TROPHIC LEVEL | | | | | |

Carnivore₁ growth—available to carnivore₂

CARNIVORE₁

Herbivore growth—available to carnivore₁ | GROWTH

HERBIVORE

ASSIMILATED FOOD | Respiration

INGESTED FOOD | Excretion

PRIMARY PRODUCER | Plant growth—available to herbivore | NET PRIMARY PRODUCTION

Figure 18-14 The flow of energy through an ecosystem. The net primary production of plants represents the food energy available to the next trophic level, the herbivores. Only a fraction of that total is ingested by herbivoes. A portion of the ingested food is excreted or respired; another portion supports the growth of the herbivores and becomes available as food energy to the next higher trophic level, the carnivores. This process continues through each level; about ten percent of the food available at one level becomes available as food for the next.

organism in the community is assigned to its appropriate trophic level, and the ecological efficiency of energy transfer from one level to the next higher one is calculated.

Many investigators have made estimates of ecological efficiencies in natural ecosystems, and also in simplified laboratory systems where more precise measurements are possible. There is a considerable range in the reported efficiencies, probably because of real differences in the ecosystems and because of difficulties in making measurements. Numbers usually range between five and twenty percent. An ecological efficiency of ten percent for energy transfer from one trophic level to the next higher level is usually accepted as an average figure.

The conception of strict trophic levels is itself a simplification, of course. Tigers prey upon more than one trophic level below them, taking herbivores as well as other, lower-level carnivores. Parasites are very common in natural communities. Decomposers operate at every trophic level. Analysis gets quite complicated, and estimates of the contributions of various trophic levels are often difficult to make. The feeding relations in a natural community are more properly described as a web of interconnections than as a simple progression from level to level (see Figure 18-15).

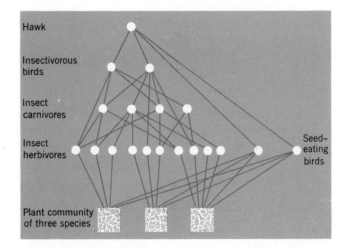

Figure 18-15 The food web of a hypothetical community. Predators in higher trophic levels feed on prey in several levels below them.

Nonetheless, like the numerical model of population growth, the energy-flow analysis helps us to understand various aspects of abundance in ecosystems. It immediately tells us, for instance, why we get relatively little food from the ocean as compared to land. When we harvest fish from the ocean we are using organisms from the third to the fifth trophic levels; this practice is equivalent to harvesting hawks and weasels and tigers on land. If fish come from the fourth trophic level as an average, our ten-percent estimate of ecological efficiency tells us that we can harvest only one kilocalorie of every thousand provided by net primary production. On land we can harvest primary producers directly. Thus, food from the sea is quantitatively much less important than food from the land, even though net primary production is large in both environments.

changes in ecosystems

Up to now, we have treated ecosystems as though they are relatively unchanging except when they are disturbed by human activities. Even our discussion of cyclic oscillations between predator and prey suggests that there are no long-term alterations in the system. But in fact, ecosystems do undergo slow and predictable changes in their organization and structure.

Persistent and stable ecosystems do not arise from the bare earth in their mature form. They have regular patterns of development. The first organisms that are able to invade an area are gradually replaced by other groups of organisms, until a mature and relatively stable community gradually becomes established. A field that has been cleared and is left alone is soon covered by grasses and other fast-growing annual plants. In a few years, the annual plants are replaced with more long-lived shrubs and clumps of young trees. The first trees that appear, perhaps birch or pine, are gradually replaced by other species. Finally, a stable community of hemlock or oak or maple is produced. The sequence itself is called **succession.** The stable community that finally results is called a **climax community** (see Figure 18-16).

The first organisms to colonize a barren area, the **founder species,** are specialized for rapid distribution and growth. In terms of the model for population growth, such species have a very large r. These species produce huge numbers of reproductive

(a)

(b)

(c)

(d)

(e)

(f)

Figure 18-16 Ecological succession in an abandoned cornfield. (a) Crabgrass invades the field the same year it was abandoned. (b) Tall weeds, such as sedge, follow crabgrass in the first and second years after abandonment. (c) Pine trees begin to invade, replacing the sedge. (d) Sedge is killed out under the pines but remains in open areas. (e) Oak trees invade the pines. (f) In time, oak and other trees eliminate the pines. The climax community consists of oak and hickory trees.

bodies, most of which die. They are so numerous, however, that they are dispersed over a huge area. A few will always be around to colonize a barren area as soon as it appears. The founder will then increase its population very rapidly as long as the environment is suitable for growth.

However, the growth of the founder species itself changes the habitat. The metabolic activities of the founder organisms alter the amounts of water, minerals, and sunlight in the environment. In fact, they may change things so much that the environment is no longer favorable for their own growth. But even if they do not directly make the environment unsuitable for their own survival, they often change it enough to favor the survival of other invading species. These secondary species then outcompete the founders. In the terms of our growth model, the K for the founder species is easily depressed by competitors.

The shrubs and clumps of trees that displace grasses and annual weeds in a cleared field are themselves later replaced, because they are intolerant of shading. Birch and cottonwoods and willows cannot survive in competition with hemlock and sugar maples, because the latter can survive and grow in the shade. Very simply, the founder and intermediate species create conditions that selectively favor later species in the succession series. Each successive community contains species that are better able to compete with other organisms. Eventually a community is formed of species that can maintain a stable population size over long periods of time.

The climax organisms do not grow as rapidly as founder species; neither do they produce as many reproductive bodies. But their slow rates of growth and reproduction are more than compensated for by the fact that each individual has a better chance of survival. Each individual of the climax species is a better competitor. And the whole population, though its individuals grow more slowly, will eventually reach a large and persistent size. A growth model for these species has a small r, but rather a large K, because the organisms are well-adapted to the environment. Succession eventually terminates in a climax community, containing the species that are most successfully adapted to the local conditions.

A climax community is both diverse and stable, partly because it tends to have a large standing crop. In land ecosystems, the standing crop forms an elaborate physical structure of stems, roots, branches and leaves. These structures tend to protect the plant from disease or predators, or they may help it to survive environmental changes. This feature makes the plant community relatively stable. The standing crop structures also provide a rich variety of niches for other organisms.

Consider the number of bird species found at various stages of succession in abandoned farmland. Where only two species are found in the grassland and annual weed phases, twenty may be associated with the mature hardwood forest. A forest has specialized niches at various levels of the leafy canopy, in the undergrowth, and also in the roots and soil. In essence, a climax community creates its own environment even more than the communities it has displaced. It is a climax stage because it is relatively insensitive to subsequent environmental fluctuations.

The climax is not really permanent, though, even in very stable communities. Old trees blow down, leaving clearings that allow the succession process to start again. And on a longer time scale, changes in climate allow new groups of species to replace the old communities as the climax types. But replacement of climax communities is most often caused by man's agricultural activities.

Most of the plants in a natural climax system are useless for human food. The herbivores are small and diverse, difficult to harvest, and unpalatable. Clearing a forest and planting wheat or maize essentially returns the land to an earlier stage of succession. Not only does this development increase the production of usable food crops, but it is easier to harvest one or a few species from a young ecosystem: the standing crop is not very large compared to productivity. Of course the land must constantly be tended to prevent other, more stable communities from taking over again. Agricultural systems are unstable both because only one species is present, and because they represent an early step in succession.

There are other kinds of succession that do not produce a stable climax community. Plankton communities are examples. When nutrients are added to coastal water by runoff from the land, a succession of organisms can be observed. First, one or a few algal species will grow and use up most of the nutrients. These species may be displaced by other algae that grow more slowly but can use the remaining nutrients more efficiently. The small animals that feed on algae will also change as the nature of their food supply alters. This succession occurs for much the same reasons as succession in abandoned farmland: the founding species change the conditions of the habitat to favor other species.

However, because the whole system is easily disrupted by wind and currents and temperature, no stable climax results.

Again, kelp communities represent an intermediate situation. They experience a succession of various kinds of attached algae. The founder species are fast-growing but relatively small filamentous algae. These algae are replaced by the aquatic equivalent of shrubs and bushes: leafy or tough red and brown algae. Finally, leafy reds and browns are shaded out by the large brown kelps. The kelps form a climax community that may persist for years, but the kelp community is not as stable as a terrestrial forest.

Other communities do not really experience succession at all. Deserts, for example, do not change very much with time. The species that are capable of tolerating the arid conditions of deserts are simply too sparse to modify those conditions significantly. Thus the founder species do not pave the way for secondary species (with only a few exceptions); instead they continue to survive in an unchanging community.

In summary, we have seen that ecosystems are characterized by short-term and oscillating interactions between different organisms and between organisms and the physical environment. These interactions occur daily and seasonally; they cause ecosystems to be more or less productive, diverse, and stable at any moment. On a slightly longer time scale, a few years to hundreds of years, ecosystems undergo predictable changes that are a direct result of these interactions; this phenomenon is succession. On an even longer time scale, measured in millions of years, ecosystems change so much that all populations of the community eventually become extinct and are replaced by other species.

The extinction of some plants and animals is inevitable, because there is always a possibility that a random fluctuation in the environment will reduce a population so much that it falls below M, the critical minimum number of individuals. (It is estimated that more than 99 percent of the species that have ever existed on earth are now extinct.) The smaller the population in a particular area, the more likely extinction becomes. There are mathematical ways to estimate what the probability of extinction is for any population. These estimates tell us that a very large population may persist essentially indefinitely. Smaller populations have a finite life expectancy. Man, of course, can speed the extinction process considerably.

conservation and preservation

There are many reasons for wanting to preserve the diversity of life and the variety of environments. Many of them may have little to do with ecology in a biological sense. Some things are being preserved simply because they are nostalgic reminders of our national heritage: virgin forest, herds of buffalo, open prairie, and mountain wilderness. In fact, our National Park system was initiated for precisely such reasons. Wild country is being preserved because it is viewed as a solace and balm, an escape from the regimentation and artificiality of modern civilization. The preservation of organisms and environment is usually pursued for its own sake or for aesthetic reasons.

Conservation is somewhat different from preservation as a basic goal. Conservation is an effort to develop environmental resources to provide the maximum long-term benefit for mankind. Conservationists may protect the fur seal, but only because they are attempting to guarantee a continuing fur supply, not out of concern for the seal. Similarly, a conservationist will vote to protect an undisturbed environment in order to preserve genetic diversity, rather than for the environment's own sake.

From an ecological perspective, we don't really have much control over many of the environmental changes that occur. Man has evolved as a very versatile omnivore with a very high population level. Our evolutionary success places pressures on other organisms and drives some of them out of existence. Man as a hunter has eliminated many animal species. Man may even be responsible for the extinction of many large mammals in North America as much as 10,000 years ago; the disappearance of camels, mammoths, sloths, and other mammals from this continent corresponds to the arrival of man in North America (see Figure 18-17).

Man as a farmer has also eliminated many other species. As we clear land for agriculture or urban development we destroy large natural areas. The remaining areas of forest or grassland are reduced to the equivalent of small islands in a sea of developed land. The consequences of this reduction are similar to those described for island communities: the extinction rate increases. Given current high population levels, some of these effects are unavoidable. Food and fiber, for example, are required for immediate survival. No one would choose to starve if

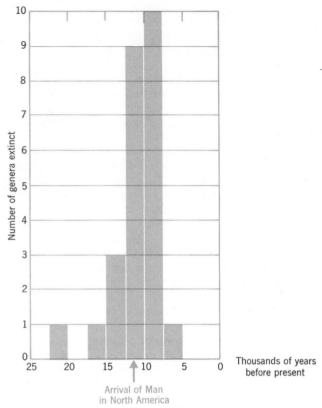

Figure 18-17 The extinction of large mammals in North America.

tions, or to enforce them.

The coming extinction of the great whales is the most famous example of the failure of conservation policies. The blue whale (Figure 18-18) is the largest animal that has ever existed on earth. It is about 25 feet long at birth and may grow to 100 feet and weigh 100 tons as an adult. In the early 1930's, over 25,000 blue whales were caught each year. Blue whales represented 70 percent of the world's total whale catch (Figure 18-19). In 1965, by contrast, one blue whale was harvested. In the past four decades the population has declined from hundreds of thousands of individuals to only a few hundred at most.

The blue whale is peculiarly vulnerable to destruction because of its patterns of feeding and reproduction. The animals must feed in certain areas of the antarctic for eight months of the year. This fact makes them easy to hunt with modern whaling methods. The females do not become sexually mature for about five years, and a mature female can produce one calf every two years at most. She requires a one-year gestation period plus a seven-month nursing period. The low reproductive potential of the blue whale was not calculated until about 1950, when the population was already dangerously low. At that time it became clear that the blue whale could not survive indiscriminate hunting. A nine-year-old female could produce only

faced with a choice between short-term exploitation or destruction of land, and starvation. These choices are the only ones in some areas of the world. More often, however, the choice is between preserving the environment and improving living standards. Historically, the decision has nearly always been in favor of exploitation.

In some of these cases, the extinction of species and destruction of the environment are avoidable. These cases are the province of conservation. But conservation is not really a biological problem; it is a social one. Sometimes it is relatively easy to obtain social cooperation in conservation projects. Most animal species that are hunted for sport are protected by restrictions on the hunting season and by limits on the size and number of animals that can be taken. Government officials can usually enforce such policies well enough to ensure the survival of game species, at least in the near future. But in other cases conservation policies fail because it is virtually impossible to agree on suitable regula-

Figure 18-19 The number of blue whales harvested since 1930. Whaling virtually ceased during World War II, but the whale population did not recover.

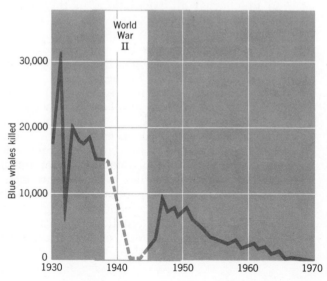

(a)   (b)

Figure 18-18   The blue whale, Balaenoptera musculus *(a) harpooned, and (b) brought alongside a ship for processing.*

two calves—just enough to sustain the population at a constant level—but the average age of the harvested whales was only six years.

This information has had absolutely no impact on the whaling industry. The only existing international organization, the International Whaling Commission, has no power of enforcement or inspection. Furthermore, its members resolutely denied that any restrictions on whaling were required. In fact, the Japanese insisted that they were actually harvesting a species of "pygmy blue whales" in order to justify continued killing of immature blue whales in the 1960's. As the blue whale disappeared, increasing numbers of other whale species were killed. During the 1950's and 1960's, more whales were taken each year, but less oil was produced, since the whales harvested were immature or belonged to smaller species. Each year the efficiency of production in terms of barrels of oil per day per boat has decreased. The whaling industry is clearly doomed and headed for extinction along with the whales.

The large capital investment required for whaling has also contributed to the difficulty. Modern whaling is extremely expensive, and the initial costs must be paid through large catches. Unfortunately, whalers can still make a profit even while the whales are being exterminated. New whaling vessels were still being ordered and built in the 1950's. There has simply been no commercial interest in maintaining a continuously productive whale fishery at lower harvest levels; neither is there any social machinery that can protect these animals from ruthless exploitation for short-term profit.

Political Ecology

"Ecology" movements are usually directed toward one of three basic goals: conserving the environment in order to maximize human benefit, preserving the environment as a good in itself, or improving or at least maintaining the stability and quality of urban existence. Obviously, different organizations or individuals stress one or the other of these goals. For example, city dwellers are more likely to place a high value on the aesthetic aspects of undisturbed environments than individuals who actually live there under marginal economic conditions. It is really only the relatively affluent and the young urbanites who have sufficient leisure to cultivate a value system that concerns itself with preservation of the environment. However, all forms of political "ecology" place a high priority on preservation of selected features of the environment. Leaders and members of "ecology" groups must therefore come into conflict with each other and must effect compromises with other interest groups. Shall we develop the coastline of Maine as a deep-sea port and build oil refineries? To do so would unquestionably impair and partially destroy its beauty. But to do so would also stimulate the economic development of the region. Shall we convert Mineral King Mountain to a ski resort and amusement area, or should we preserve it as a wilderness area? Different groups would assign different priorities to these alternatives.

The distinction between ecology as a branch of biology and "ecology" as a political position is difficult to keep clear in practice. The professional ecologist is often asked to advise society concerning the consequences of land development, pollution control measures, the construction of power plants, and similar matters. But the ecologist to whom we turn for expert advice is likely to be relatively affluent, fond of the out-of-doors, and a city dweller. Almost certainly he will be personally committed to some aspects of political "ecology," in addition to being versed in ecology as a science. Even he cannot always be sure whether he is acting as an ecologist or an "ecologist." However, his assigned task is that of adviser, not advocate.

Recommended Reading

Darnell, Rezneat M., *Ecology and Man.* Dubuque, Iowa: William C. Brown (1973).

A literate presentation of the principles of ecology. The second half of the book is of particular interest for its discussion of the legal and political aspects of preservation and conservation.

Ehrenfeld, David M., *Biological Conservation.* New York: Holt, Rinehart, and Winston (1970).

This book emphasizes the relationship between conservation and ecology, and it analyzes the strengths and hazards of present political, economic, and social policies.

Hamilton, William J., *Life's Color Code.* New York: McGraw-Hill (1973).

Color as it influences the physiology and ecology of animal populations; the book includes a discussion of human skin coloration.

Whittaker, Robert H., *Communities and Ecosystems.* New York: Macmillan (1970).

The study of communities and their interaction with the environment.

19 human ecology

In Chapter 18 we attempted to show how ecologists are exploring some of the relationships between organisms and their environment. In particular, we examined the distribution and abundance of organisms in ecosystems by asking how population growth and population size are regulated. We also examined the exchange of materials and the flow of energy through organisms in an ecosystem.

This chapter is concerned with a more detailed analysis of what controls the growth and size of human populations. Human populations are, of course, subject to the same environmental controls as those of other organisms, but the interaction between human populations and the environment is sometimes obscured by their economic systems. For example, most people in the United States are aware of the relation between the economy and population growth, at least at the level of the family. People are encouraged to have as many children as they can "afford"—and if they cannot "afford" them they should not have them. Or they should work harder and earn more money so they can support a larger family. But these same people fail to see any connection between the economic control of family size and environmental controls. Yet the connection is very direct.

In fact, the description of an economy is partly a description of how a human population interacts with its environment. Does it obtain food by hunting and gathering, or does it have a stable agricultural system? Does it use coal or oil for fuel energy, or must it depend on wood? Does it extract and refine metals from the environment? These interactions impose real limits on the growth and size of populations. Working harder does not change these

607

limits unless it changes the basic nature of the economic system. Human populations are already large, and they are growing so rapidly that man is seriously out of balance with his ecosystem. At this time, human populations are so widely distributed that the human ecosystem now encompasses the entire earth.

human population growth

The modern world is made up of populations with very different patterns of growth. Populations in New York or Tokyo are very different from a village population in India or Zambia; a hunting tribe in South America is profoundly different from both. We can try to understand some of these differences by considering the history of human populations.

History of Human Populations

World population has gone through two distinctive periods of growth from Neolithic times to the present (Figure 19-1). In the Neolithic Age, about 10,000 years ago, the world population must have been no more than a few million individuals. But, at about this time, agriculture arose in river-valley civilizations and began to spread across the world. Local fluctuations in population were great during this period, as civilizations rose and declined to obscurity. But as agriculture developed, the total numbers of mankind gradually increased, reaching several hundred million by about 1000 BC. From 1000 BC to about 1650 AD, local famines and disease epidemics caused repeated population crashes in many areas. Nevertheless, an average population of a few hundred million individuals was sustained throughout this period.

World population began to increase again during the Renaissance in Europe. It doubled between 1650 and 1850, doubled again between 1850 and 1930, doubled a third time between 1930 and 1974, and is currently growing at a pace that will produce an additional doubling every 35 years. The growth of human populations in the last century can only be described as explosive.

Several major historical events have contributed to this dramatic growth. First, European explora-

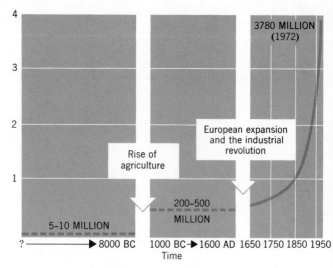

Figure 19-1 *Growth of the human population from Neolithic times to the present. Notice that the time scale for the recent past is much expanded. If the graph were plotted on a uniform time scale, population would rise almost vertically at the extreme right.*

tion and colonization of new land in the seventeenth and eighteenth centuries produced a large population expansion. The great colonial empires were formed at this time. Africa came to be divided among the British, Dutch, Germans, French, Belgians, and Portuguese; South America between the Spanish and Portuguese; North America among the British, French, and Spanish; and so forth.

The second major impetus to population growth was the Industrial Revolution. Colonial empires expanded as industrializing countries sought new sources of raw materials and new markets for manufactured goods. New energy sources were exploited as well. Coal had been used sporadically even in prehistoric times. But the systematic exploitation of all the fossil fuels—coal, oil, natural gas—really began in the 1700's, with the invention of the blast furnace for iron production.

The expansion of energy resources and mechanization increased agricultural productivity tremendously, so populations were able to grow still more. More recently, advances in public health and medicine and continuing agricultural improvements have continued to stimulate population growth.

Western Europe's period of expansion and subsequent industrialization was accompanied by changes in European ideas about the growth and the nature of society. During the Middle Ages and the early Renaissance, stability and balance were prized. Individuals played the social roles that they had acquired at birth. Society placed little value on human labor, and a person could not accumulate wealth or improve his social position by means of individual effort. Instead, land was the source of all wealth, and it was generally passed intact from father to eldest son. Population did not increase, because various traditions discouraged growth. Younger sons, for example, were encouraged to enter the church or the army, where their probability of reproduction was decreased. Stability was still the ideal.

In the eighteenth and nineteenth centuries, these ideas began to change. According to modern economic ideas, the value added to raw materials by human labor was seen as the source of industrial wealth. Capitalist and Marxist thinkers agreed on this point, though they disagreed on how wealth is to be distributed in society. In a very simple and direct way, this change was interpreted to mean that more people could produce more wealth. An ideal of growth gradually displaced the ideal of stability and balance.

Developed and Undeveloped Countries

The agricultural and industrial revolutions did not reach all human populations. A small number of people still live in Neolithic conditions. About two thirds of the world's population is in the agricultural phase of development. Most individuals in these cultures work directly in agricultural pursuits. Their average income is low, and the energy consumption and industrial production per person is small. This group includes most of the former European colonial areas in Africa and South America and Asia, as well as areas that were relatively untouched by western expansion, notably China. These are the "undeveloped" countries of the world. The remaining third of the world's population lives in countries that are more or less industrialized, the so-called "developed" countries. This division of the world into developed and undeveloped countries glosses over very great differences within each category. Nevertheless, it is a useful distinction, provided we remember that it is a simplification.

Birth Rates and Death Rates

Patterns of population growth in developed and undeveloped countries are quite different, but they still follow the same basic model that describes the population growth of other organisms:

$$r = (B - D).$$

The growth rate (r) of human populations is defined as the difference between the birth rate (B) and death rate (D). In human populations, B is called the **crude birth rate** or CBR, and it is defined as the number of births per year per 1,000 total population. The CBR, however, is not always very informative. The total population includes children and individuals past the age of reproduction, neither of whom can contribute to the birth rate. So the CBR does not really tell us whether birth rates are high or low compared to the maximum possible rate.

A second kind of birth rate, the **general fertility rate,** or GFR, is defined as the number of children born per year per 1,000 women of child-bearing age (15–44 years old). The GFR does tell us whether the birth rate is close to its maximum possible rate. The United States has a high CBR, for example, because women of child-bearing age make up a relatively large fraction of the population. Yet birth rates are quite low compared to the maximum, because the fertile women are having relatively few babies. In other words, GFR is low. Both CBR and GFR are measures of fertility. The highest fertility rates found in human populations at the present time are a CBR of about 55 per 1,000 and a GFR of 235 per 1,000.

Neolithic societies probably had CBR's close to the maximum. Since their populations did not grow, however, their crude death rates (D), expressed as deaths per 1,000 total population, must have equalled their CBR's. That is, both numbers must have been about 55 per 1,000. Populations with such high birth and death rates have a characteristic age structure. Life expectancy is short, so most of the individuals are young; about 45 percent of the population is below fifteen years of age. **Infant mortality**—the number of children born alive who die under one year of age—is also high. Individuals only rarely survive to ages of 60 or 70. Since many women in the population die young, the ones who do reach reproductive age must give birth at close to the maximum rate. In other words, the GFR must be very high to maintain a CBR of 55 per 1,000.

Populations in the agricultural stage are more dense, because more people can live off a given area

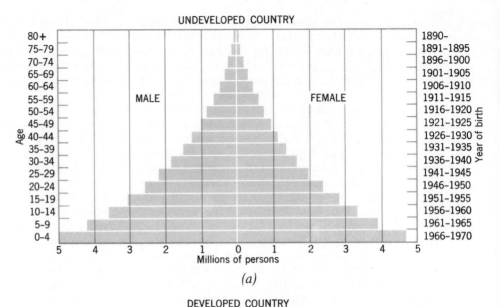

UNDEVELOPED COUNTRY

(a)

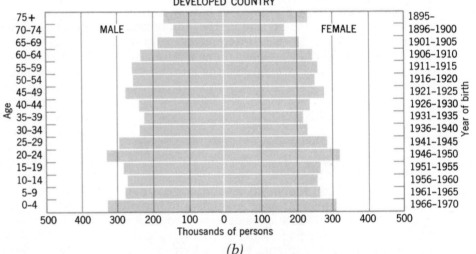

DEVELOPED COUNTRY

(b)

Figure 19-2 The age structures of the populations of India and Sweden (1970). (a) Populations in undeveloped countries such as India contain a large proportion of young people; the number of individuals declines steadily with increasing age. (b) Populations in developed countries such as Sweden have roughly equal numbers of individuals in each age category.

of land, but both their birth rates and death rates remain relatively high. The CBR for most modern cultures in the undeveloped, agricultural stage is between 30 and 50 births per 1,000 of total population. Infant mortality remains high, and the overall age structure of an agricultural population is similar to that of Neolithic populations.

Cultures that are part of the developed world have a very different pattern of population growth. Their CBR's have decreased drastically, ranging from 10 to 25 per 1,000. This change in CBR corresponds to a decrease in the GFR from maximum levels of 235 per 1,000 to a minimum of 60 to 70

per 1,000 women of child-bearing age. The crude death rate also decreases. The age structure of the population changes considerably. The population is no longer young: there are equal numbers of people in all age groups up to about age 60. Figure 19-2 presents the age structures of both a developed country and an undeveloped one.

The Demographic Transition
Developed countries are said to have accomplished the **demographic transition.** That is, high birth rates and high death rates have been replaced by low birth rates and low death rates. Every country

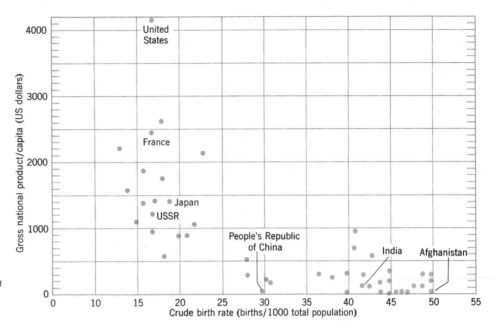

Figure 19-3 The relationship between wealth and crude birth rate. The data are for countries with a population of more than 10 million.

that has made the demographic transition has also acquired the following characteristics:

Increased literacy and education
Increased percentage of urban population
Increased wealth per capita
Increased amounts of food per capita
Increased longevity
Decreased infant mortality

All of these changes are strongly correlated with achieving the demographic transition. For example, almost all countries that have low CBR are relatively wealthy. Conversely, all poor countries have a high CBR. There is virtually no overlap between the two groups (see Figure 19-3).

People of the undeveloped world would like to make the demographic transition, because most of the features that accompany low fertility (wealth, food, longevity, and so forth) are quite desirable. But there is no simple way to achieve it. Patterns of fertility in an undeveloped country are usually based on very ancient traditions, and such traditions are very difficult to change. The population is principally rural and illiterate, and therefore quite isolated from outside influences. Most developed countries have made the transition only very slowly. Birth rates usually decline over a period of many decades. In Sweden, for example, the crude birth rate and crude death rate have been declining

for the past two centuries (Figure 19-4).

The transition can be made more quickly, but only if traditions are forcibly changed. The USSR, for example, accomplished it remarkably quickly, in less than half a century. The strong centralized political system of the USSR was committed to industrialization and was capable of enforcing its wishes. Political leaders introduced sweeping social changes to accelerate the pace of industrialization. Some of the changes—forced collectivization, an emphasis on education—also had the effect of decreasing fertility because they tended to discourage early marriage. (The reduction in fertility was not encouraged directly, because the concept of overpopulation is alien to Marxist ideology. The Marxist believes that problems that are usually attributed to overpopulation are actually caused by unequal distribution of food, goods, and land. Nonetheless, the social changes that encouraged industrialization also reduced fertility.) There was considerable resistance to change, because the forced commitment of labor and resources to the development of industry meant that the individual was worse off than before—at least for a time. The population had to be isolated from the rest of the world during the transition period to keep it from becoming aware of the full extent of the internal privation and forced social change.

The same kinds of changes are now going on in

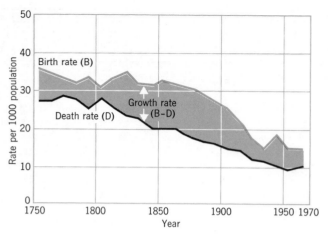

Figure 19-4 *The demographic transition in Sweden.*

severe depressions, and even when employed they were paid wages below the subsistence level. A major factor in the Civil War was the conflict over economic development. Northern states pushed for and achieved rapid industrialization of the entire country, but only by winning the war. The direct cost was half a million human lives and the destruction of vast areas of agricultural land in the South.

Despite the high costs of the demographic transition wherever it has occurred, the suffering in undeveloped cultures is even greater. The cost of the high infant mortality, famine, disease, shortened life expectancy, and the other consequences of an agricultural existence is enormous. It is humane to hope that, in the future, some of the privation and death that has always accompanied a demographic transition can be avoided, but there is no guarantee that it will be possible to do so.

the People's Republic of China. Early marriage is vigorously discouraged because it removes individuals from the labor pool and shortens their education. The postponement of marriage also tends to weaken traditional dependence on the family unit and to strengthen emotional ties to the central government.

The human cost of making the demographic transition has always been high—whether slowly or rapidly achieved—and it has been independent of the political system under which it occurred. In the USSR, the collectivization that drove the peasants from the land produced both famine and revolution. Subsequent purges, terror, and forced labor produced a death toll estimated at ten to twenty million. But the demographic transition in western Europe was no less expensive. Europe's economic development began with colonization, and the colonies relied on forced labor from slaves and local inhabitants. Native populations were unable to maintain their own industry and agriculture, food became scarce, and tens of millions died in subsequent famines. And a conservative estimate of the death toll caused by the black slave trade alone is placed at fifty million.

The United States came into existence after a rebellion against the economic policies of colonial Europe. Yet subsequent economic policies within the United States were just as unpleasant. Southern states depended on black slave labor, and they contributed to the tremendous mortality associated with the slave trade. The labor pool of the more industrialized North also suffered. Much of the wage-earning populus was subjected to repeated

Population Control

We know that fertility is high in undeveloped nations and low in the developed world. The demographic transition means that traditions and eco-

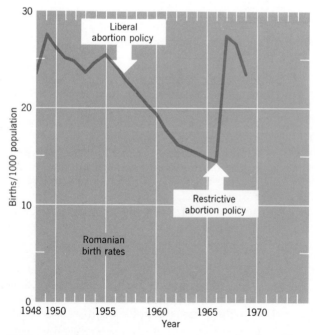

Figure 19-5 *Birth rates in Romania are clearly tied to the availability of abortion.*

nomic factors change, enabling fertility to go down. Let us directly examine the controlling factors. Fertility is under considerable voluntary control among citizens of developed countries. Individual couples weigh their desire for a child against their concern for providing for their present families, their career aspirations, and so forth. But personal decisions about child-rearing are not at all independent judgments. They are strongly influenced by political descisions and government policies. Government policies control fertility by influencing personal decisions.

The availability of contraceptives and birth control information, for instance, has a profound influence on fertility. During 1960 to 1965, about one fifth of the total births in the United States represented children who were unwanted. "Unwanted" simply means that these children would not have been born if the population had had free access to effective contraception. Presumably the fertility of the United States population would have dropped even more precipitately than it did after 1960, had contraceptives been freely available and fully effective. The availability of abortion has the greatest effect on fertility rates. When legal abortion is available, many individuals who might have been unwilling to risk illegal abortion will avail themselves of legal and safe abortion services, and fertility will go down. If abortion policies become more restrictive, fertility will rise again dramatically. These are observations, not predictions. Figure 19-5 presents the changes in birth rate in Romania associated with the availability of abortion.

Policies that modify fertility may take the form of economic sanctions. In the United States, heads of households are given more tax advantages than single individuals. A large fraction of the costs of education are borne by the general population rather than just the parents. All these economic policies encourage marriage and increased fertility. But many other laws tend to decrease fertility and increase the cost of children: child labor is restricted; and parents are legally forced to support their children through various child-welfare, child-support, and inheritance laws.

The impact of central policy may be much more subtle but just as effective. Our society provides very few roles for women as alternatives to marriage and child-bearing. The lack of socially-approved alternatives encourages marriage, and hence increases fertility. On the other hand, in some subcultures, both men and women are encouraged to delay marriage, at least temporarily, in favor of continued education. The balance of all such forces can shift and change fertility very quickly in the fully-literate population of a developed country. But overall fertility rates always remain relatively low.

United States Population

In the United States, the death rate has declined steadily since the beginning of this century. The death rate dropped from 14 per 1,000 to 9.5 per 1,000. But the birth rate has fluctuated up and down dramatically. Before World War II, GFR had reached rather low levels—less than 80 per 1,000. Sociologists predicted that the United States was entering a phase of "incipient population decline." But after the war, birth rates rose dramatically again, reaching a peak in 1960. Between 1960 and 1972, the GFR in the United States steadily declined again, from 120 per 1,000 to less than 75 per 1,000 (see Figure A).

These changes in fertility reflect our recent history in very broad terms. The first decrease in GFR occurred during the Depression. The subsequent increase began at the end of World War II, when the United States emerged as the only major industrialized nation with an intact economy. The second decline, beginning in 1960, coincides with the increasing prominence of foreign and domestic problems and declining optimism about our capacity to solve them easily.

Whatever the exact cause, the increase in fertility between 1945 and 1960 and its subsequent decline have long-term consequences in our society. The surge in GFR between 1945 and 1960 produced a large proportion of babies relative to other age groups in the population. At the same time, infant mortality remained rather low and crude death rates did not increase. Hence most of the additional babies have survived. They remain in the population,

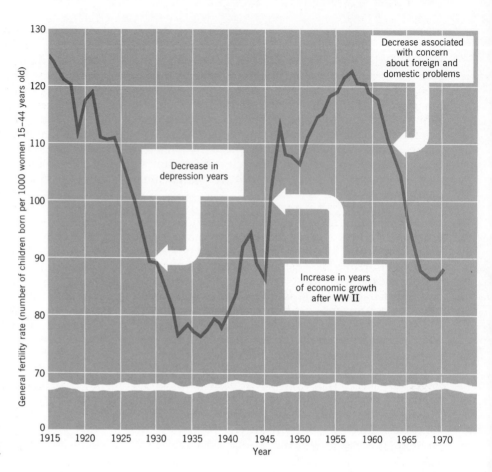

Figure A   Changes in the general fertility rate in the United States. The fluctuations are correlated with important historical events.

Inside figure A:

General fertility rate (number of children born per 1000 women 15–44 years old)

Year

Decrease associated with concern about foreign and domestic problems

Decrease in depression years

Increase in years of economic growth after WW II

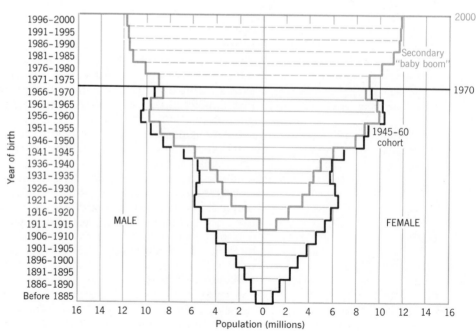

Figure B   The age structure of the United States population in 1970 (black), and the predicted age structure in 2000 (green).

Inside figure B:

Year of birth

Secondary "baby boom"

1945–60 cohort

MALE   FEMALE

Population (millions)

moving as a group through the age structure of the country as a whole. Such a group is called a **cohort.** The impact of this cohort will change as it moves into different positions in the population. In the 1950's and 1960's, elementary and secondary school facilities were crowded. Later, colleges and universities experienced unprecedented enrollments. Now the vanguard of this group has reached an age where they must be assimilated into the labor force of the country. This problem will continue for some years, since the youngest members of the cohort are only in their teens (having been born in 1960). This particular group has influenced the country in yet other ways. It was responsible for a highly visible teenage and subteen culture in the past decade. The subteen and teenage group represented such a large clientele that it quickly became a significant economic and cultural force. (The age structure of the United States population is shown in Figure B.)

The decline in fertility that began in 1960 presages a coming decline in the vigor of this subculture. There will be other consequences as well, such as a temporary reduction in employment in elementary and secondary education —an effect that will extend to colleges and universities in a few years. But as the 1945–1960 cohort begins to reproduce, the crude birth rate will increase again, even if GFR remains at its current low levels, simply because the number of individuals in this cohort is so large. A secondary baby boom will become visible after about 1975 (see Figure B).

Population control in undeveloped countries also begins with personal decisions by individuals. However, the influences that modify their decisions are very different, and nearly all of them encourage the formation of large families. Couples need many children in order to survive. Welfare and old-age assistance must be supplied by their children, not by the government. Children are needed to supply labor in the field. Since infant mortality is high, fertility must be even higher to guarantee the survival of many children. The desired family size in undeveloped countries is significantly larger than in developed countries (Figure 19-6).

We can summarize by saying that the control of GFR in both developed and undeveloped countries always begins with individual choice—but the range of possible choices is almost entirely controlled by the society. A particular society acts on an individual through traditions, economic pressures, distribution (or suppression) of birth control devices, and a number of other channels. But the social pressures are so very different in developed and undeveloped regions that fertility is always low and flexible in the former and high in the latter.

The growth of the population as a whole, however, depends not only on birth rates, but on death rates as well. In developed countries death rates are relatively constant, so the growth of the population is modified primarily by voluntary changes in the birth rate. Growth is slow enough so that new individuals can be added to the population without straining economic resources. That is, the growth does not exhaust the food supply or have any other effect that might increase death rates in the whole population. Most developed countries are growing at a rate of less than one percent per year; the population will double in 70 years or longer. Presumably, voluntary control of fertility could reduce growth to zero or even produce a shrinking population.

Unfortunately, as we have seen, the undeveloped countries have few ways to control and reduce fertility at present. But if their populations are not controlled by limitation of fertility, they will eventually be controlled by an increase in the death rate. (The only alternative would be an unlimited increase in numbers, which is clearly impossible.) The only question open to discussion is that of time. Most undeveloped countries are growing at two to three percent per year. This figure corresponds to population doubling times of 35 and 24 years. Will the populations of the undeveloped world be controlled by high death rates?—or is there still a chance that they can control fertility by making the demographic transition? In other words, is the carrying capacity of the earth for humans large enough to support the growing populations of the undeveloped world while they try to achieve the demographic transition?—or are populations growing so fast and getting so large that

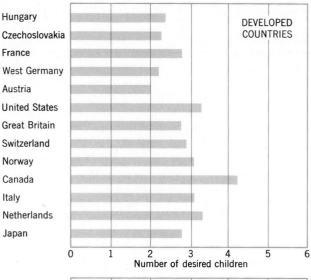

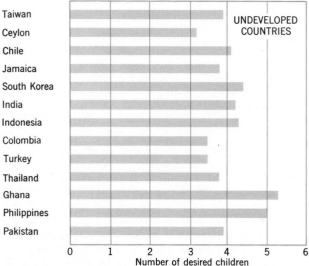

Figure 19-6 Desired family size in developed and undeveloped countries.

pace of fertility reduction. The lowest prediction assumes that fertility will have decreased by 1985 to **replacement** levels — that is, to a level where children born just replace their parents. The other three projections assume that replacement levels will be reached at later dates. No one really expects any of these extrapolations to be accurate, because it is simply impossible for fertility to reach replacement levels by 1985. The assumption that there will be no change in mortality is humane, but also inaccurate.

These extrapolations do show the great inertia of population growth even after replacement fertility is reached. This feature is a consequence of the age structure of the Indian population. Such a large fraction of the population is below age fifteen that the number of individuals of child-bearing age will continue to increase for a long time, regardless of any changes in the fertility rate. In other words, the effects of achieving fertility control will not show up for at least a decade. No matter how rapidly the demographic transition is achieved, population levels of undeveloped countries will still increase greatly if mortality remains constant.

Unfortunately, most students of human population feel that the demographic transition is simply not going on at all in most regions of the undeveloped world. The People's Republic of China is the only significant exception. India, for example, is actually losing ground. India's population was 585 million in 1972 and is growing at a rate of 2.5 percent per year. If this growth rate is sustained, the population will double in 28 years. Food is already in short supply. At present, there is about one acre of arable land per person. Seven out of ten families are engaged in agriculture, partly because production is inefficient, but also because there is no alternative employment to absorb their labor. Indian agriculture will have to double its productivity by the end of the century in order to support its population at present levels. And it must achieve this doubling on the basis of one *half*-acre of arable land per capita, since there is no new land to bring under cultivation. Since the Indian farm population will also double along with the rest, the underemployment that already exists will be exaggerated. Agriculture will therefore have to be even more productive, in order to create the excess of money necessary to create nonagricultural jobs for some of the rural population. One would like to believe that all these developments are possible; however, none of them has yet begun. And as this is being written, India faces a drastic food shortage and famine brought on by drought.

they will be controlled by famine and disease and social disruption long before a transition can occur? Many demographers and other scholars are trying to answer this question by making predictions of population growth based on analysis of present trends.

Consider four forecasts for the population of India (see Figure 19-7). All four assume that death rates in the Indian population will not change. They differ only in their assumptions about the

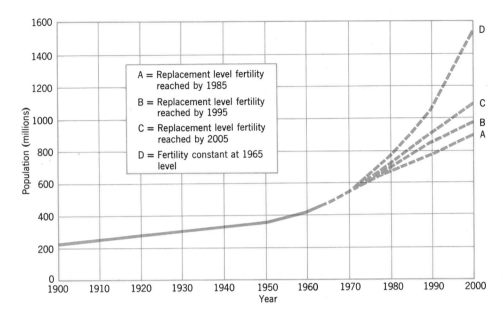

Figure 19-7 Future population growth in India. The four forecasts differ only in the assumptions that are made about fertility; the death rate is assumed to remain constant at the current level.

The pace of fertility control is slow or virtually imperceptible in most of the major undeveloped populations of the world. As we have seen, fertility control is strongly correlated with economic development, and the pace of economic development in undeveloped countries is very slow. The inertia of population growth is consequently very great.

Predictions and Extrapolations

There are many ways to predict future trends: casting dice, reading tarot cards, or examining the entrails of animals. None of these methods is significantly less reliable than demographic projections. This statement does not impugn the competence of demographers; it is simply a fact about the nature of prophecy. Consider the information that is available to the demographer. He has estimates of the present age structure of the population, crude birth rates, crude death rates, general fertility rates; and he has other numbers reflecting immigration and emigration, the size of rural versus urban populations, income and education levels, fertility and mortality among various subgroups in the population, and so forth. If all of this information were reliable (which is unlikely), and if all these rates were to remain constant (which is impossible), the demographer could grind out accurate predictions for any time in the future.

The demographer can compensate for unreliable information to some extent, but unpredictable changes in fertility and mortality are precisely that: unpredictable. In 1950, a demographer could have predicted the GFR in the United States in 1972 only by a flight of imaginative fancy. The Figure presents six projections of future world population, all of them prepared in 1965. We cannot yet decide which is most accurate at the present time, because our estimate of world population is too uncertain.

This disagreement is nothing new. The six predictions diverge so rapidly that they give very different forecasts of world population by the end of this century. And all are based on essentially the same data; all are careful at-

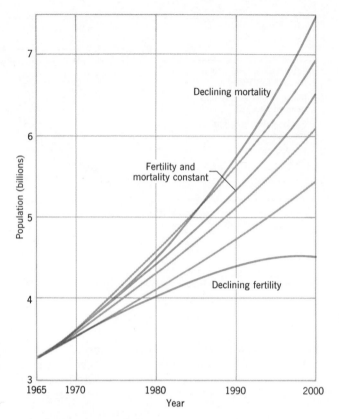

Six different projections of world population growth. The forecasts differ in their assumptions about changes in birth and death rates.

tempts by thoroughly reputable workers to provide a reliable forecast. But they differ in their assumptions about the pace of the demographic transition in undeveloped countries, changes in mortality, and other elements that determine the growth of human populations. The lowest of the six estimates could be either a pessimistic view (if it assumes that there will be a general increase in death rates to match continuing high birth rates), or it could be an optimistic view (if it assumes that the potential for social control of fertility is great). As it happens, it is based on the latter assumption.

Reliable predictions are beyond our capacity, but predictions are still useful. A prediction of future population levels serves two important functions. First, it forces us to examine and to try to understand present trends. Second, it examines the consequences of various assumptions about the future. What would be the result of a gradual decline in fertility? What is the age structure of a stable population? When will population growth level off? How large will the population be when growth ceases? How are these growth rates and population sizes influenced by various methods of fertility control? The accuracy of a particular prediction is not as important as the insights it gives us into the consequences of present trends, and into the influence of any possible changes we might make.

This information can guide present social action, so that we can try to avoid population control based on increased death rate in favor of population control based on decreased birth rate. The consequences of particular *assumptions* about birth and death rates can be predicted quite precisely, even though one cannot predict *which* assumptions will prove to be true. Changes in birth and death rates are at least partly under human control, and we must have clear ideas about the consequences of our choices.

Unfortunately, continued high fertility rates and the resulting world population growth are facts. The question of the ultimate carrying capacity of the earth, therefore, becomes crucial. But the carrying capacity is difficult to estimate. In the first place, it depends on the quality of life we hope to achieve. If we demand a high-protein diet, high consumption of material goods, a long and healthy life, low infant mortality, and the other perquisites of individuals living in the developed countries, the earth's carrying capacity is rather low. Each individual commands a large share of the available resources. In fact, the world probably cannot support its present population at such a standard. If we are satisfied with an existence based on subsistence agriculture, with a low-protein diet and low consumption of material goods, the carrying capacity is considerably greater.

There is no way to set precise limits on the carrying capacity of the earth, but we have good reason to fear that we may be pressing against them. A primary problem is shortage of food. Many populations are already undernourished. We can probably get more food from the environment, but, as we discussed in Chapter 8, it is becoming much more difficult and expensive in terms of energy and resources.

Other basic problems are an inevitable consequence of the sort of industry and technology that supports developed nations. Our technology is really an extension of ourselves. It increases our ability to manipulate the environment. But as a result, this technology has the same open-system characteristics that we have. It works by maintaining exchanges of materials and energy with the environment. Our open-system exchanges are now so vast in scope that they, too, are approaching real limits. For one thing, many of the material resources of the world may be exhausted in the near future. In addition, sources of cheap energy are disappearing. We have been exploiting nonrenewable reserves of energy (fossil fuels) and must now turn to other sources (solar, geothermal, nuclear fission and fusion); but this transition presents many serious problems. Finally, the production and disposal of waste has become a very serious problem. The influx of materials and useful energy into human communities is necessarily accompanied by an outpouring of heat and waste products. More people, combined with increased consumption of raw materials and energy per person, create ever larger amounts of heat and waste. Traditional methods of disposal are no longer feasible in many areas of the world, and some wastes are contaminating the environment on a worldwide scale. Thus, not only food shortages, but also the problems generated by our technology itself, force us to recognize that there are real limits to population, and that we are approaching them.

Materials

Human populations consume two kinds of material resources, renewable and nonrenewable. **Renewable** resources are produced by plants and animals. Food, wood products, and natural plant fibers are examples. Given provident management, such resources are indefinitely renewable. But other raw materials exist in limited quantities and are **nonrenewable.** That is, they can be consumed slowly or rapidly, but when they are exhausted they are gone. Such resources occur in reserves that have been established gradually through long periods in geological history. The consumption of renewable resources is like living off the interest on invested money: it can continue indefinitely. But the use of nonrenewable resources is like spending the capital directly. That practice can continue only until the capital is exhausted.

We can estimate the expected lifetimes of our various mineral and fuel resources under a variety of conditions. If we continue to consume all the known resources at our present rates, gas and oil and many metals will be exhausted within 50 years. But, in fact, our use of our resources is increasing exponentially. The expected lifetimes of known reserves are much less when this rapid increase is taken into account. Of course, the known reserves of some resources may be only a part of all that is available. But even major increases in the size of our reserves will make only small differences in their lifetimes, if our use of them continues to increase exponentially. To be specific, if our actual reserves of nonrenewable resources prove to be five times greater than is known today, their lifetimes will still only be doubled in most cases (see Table 19-1).

These figures clearly show the limited nature of our resources. We are nearing the end of our known stocks of precious metals and of most nonferrous industrial metals. However, the numbers are some-

Table 19-1

Lifetimes of Nonrenewable Resources

Resource	Time Until Exhaustion of Resources (in years)		
	Linear Rate of Consumption (known reserves)	Exponential Rate of Consumption (known reserves)[1]	Exponential Rate of Consumption (5 × known reserves)
Fuels			
Coal	2,300	111	150
Natural gas	38	22	49
Oil	31	20	50
Iron and alloy metals			
Iron	240	93	173
Nickel	150	53	96
Tungsten	40	28	72
Manganese	97	46	94
Nonferrous industrial metals			
Copper	36	21	48
Lead	26	21	64
Zinc	23	18	50
Tin	17	15	61
Aluminum	100	31	55
Precious metals			
Gold	16	9	29
Silver	16	13	42

[1] Note that the exponential rates of consumption do not directly depend on the linear rates. Present gold and silver reserves, for example, will both last sixteen years if we continue to use them at current rates. But our rate of gold use is now *increasing faster* than our rate of silver use. Therefore, if our rates of consumption continue to increase exponentially, gold will be depleted faster than silver.

what misleading, because the extraction of nonrenewable resources will probably not continue in either a linear or an exponential fashion. A more reasonable assumption is that their use will continue to increase exponentially for a while, reach a peak, and then begin to decline as resources become rarer and more difficult to extract from the earth (see Figure 19-8). This assumption extends the expected life of our mineral and fuel reserves somewhat, but not significantly. The inevitable increase in the price of a nearly-depleted resource will also stimulate the use and development of substitute materials. For instance, the decline of metals would force us to use more plastics or glass. But many of these materials, too, are limited.

It is sometimes argued that the impending shortage of mineral reserves is only an apparent shortage, that untapped mineral reserves are present in the earth's crust—we merely have to get at them. This opinion is based on three assumptions. First, it assumes that minerals are present everywhere in the earth's crust in a continuous range of concentrations. That is, a particular mineral is very concentrated in some regions, slightly less concentrated in others, and so forth, down to very low concentrations. Second, it assumes that we will soon have unlimited supplies of cheap energy with which to extract these reserves, presumably nuclear power. And third, it assumes that we can, in fact, mine the portions of the earth where minerals are

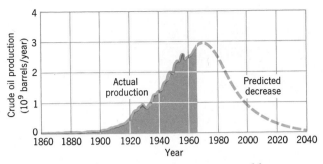

Figure 19-8   *The utilization of a nonrenewable resource is illustrated by the production of crude oil in the United States and nearby coastal waters. Production can be expected to drop as the most easily extracted deposits run out.*

less concentrated, if only we have these unlimited supplies of energy. As we use up the most concentrated ores and minerals, we can simply begin to use less concentrated forms.

Unfortunately, two of the assumptions are simply wrong. First, the concentrations of minerals are not continuous. Some ores, such as iron and aluminum, are present in a wide range of concentrations. But others, such as lead and tin, are not. They occur either very concentrated or very thinly dispersed. There is nothing in between. Second, there is no foreseeable way to extract thinly-dispersed resources from large volumes of rock regardless of how much energy is available. Various methods have been proposed, such as nuclear explosions to fracture rock, or the use of large volumes of chemical solutions to leach minerals out. But none of these proposals is practical and none considers the problem of how to dispose of the wastes that are created—the huge volumes of leaching solutions, the vast quantities of extracted rock.

The sea has also been proposed as an inexhaustible supply of minerals. About two thirds of the known elements are present in seawater. Though most of these elements are present only in extremely low concentration, the seas are huge, so the total amounts of each element are immense. However, the concentrations are so very low that we face the same kind of problem we had on land: it would be necessary to process huge volumes of seawater. It is simply not feasible, even with unlimited energy. Mining the sea floor has become important, particularly as a source of fossil fuels. But the final contribution of the oceans will still be

modest compared to that of land sources. It is certainly not an inexhaustible alternative to present supplies.

Access to new sources of energy, exploitation of marine resources, and substitution of fabricated materials for minerals will increase our available resources to some extent. In addition, as minerals become scarce, they will have to be recovered and recycled more carefully. Their increasing prices will decrease demand, and will therefore decrease their rates of depletion. Nevertheless, none of these developments will change the basic limitations implied by Table 19-1. Industrial societies simply cannot maintain their present rates of resource consumption for more than a few decades.

Energy

There is an excellent correlation between rates of energy consumption and industrialization. Energy is required for the refining of metals, for the fabrication of goods, for transportation, agriculture, food processing and packaging, and other activities. The developed world has achieved its present position of wealth and power by supplementing the labor of humans and domestic animals with other kinds of energy. Most of this supplemental energy is derived from fossil fuels. Hydroelectric power and nuclear power make only a small contribution at present.

We are rapidly moving into an energy crisis, because crude oil and natural gas reserves will not sustain our present rates of energy consumption for more than a few decades. Our coal reserves will last somewhat longer, but they are still limited. Furthermore, most of our remaining known coal reserves are deposited as thin seams that are difficult to mine. The bloom of industrial civilization will be very transient unless we can find other sources of energy.

Let us consider alternative sources of power. Hydroelectric power is one possibility. Hydroelectric energy is derived from water-driven turbines in dams. It might produce considerable energy for a brief time. However, reservoirs behind dams collect silt and fill up so quickly that they can be used only for a few decades (see Figure 19-9). Geothermal power—the energy of volcanoes and hot springs—is too small to be useful. Solar energy is abundant but difficult to convert to useful form, except through photosynthesis. For example, a solar power plant, in order to generate as much electricity as a modern fossil fuel plant, would have to cover about 15

(a)

(b)

*Figure 19-9 Reservoirs
created by dams are temporary features. This reservoir
filled with silt in less than
six years after construction.
Lakes formed by hydroelectric power dams may last
longer, but still they silt in.*

square miles of ground with energy-collecting devices. Solar energy may be used for home heating,
distilling water, and other special purposes, but it is
extremely unlikely that it could ever become a
major new energy source.

Nuclear energy, on the other hand, is a potential
source of energy that dwarfs fossil fuel resources.
Energy is produced by two kinds of nuclear reactions, fission and fusion. **Fission** energy is released
when the atomic nucleus of a heavy element is
split. **Fusion** energy is released when the nuclei of
small elements fuse together to form a heavier element. Both of these reactions release energy be-

cause they are accompanied by a decrease in overall
mass. According to Einstein's famous equation,
mass and energy are interrelated: $E = mc^2$, where E
is energy, m is mass, and c is an enormous constant, the speed of light. This equation says that a
small amount of mass can be converted to a huge
amount of energy.

Fission is a form of radioactive decay which splits
the nucleus of an atom into approximately equal
parts. Energy is released, in the form of electromagnetic radiation and as fast-moving neutrons ejected
from the split nucleus. Radioactive elements such
as uranium and plutonium have very large, heavy

nuclei. These nuclei are quite unstable, so they may undergo fission spontaneously; the elements are said to be highly fissionable. Fissionable elements are thinly distributed in natural rocks, and when an atom undergoes a fission reaction, the released energy is simply absorbed by the surrounding inert material. However, if the unstable element is mined and concentrated, it is then capable of undergoing a sustained fission reaction. In that case, when one atom undergoes spontaneous fission, the ejected neutrons may strike other nuclei of the same element. These nuclei split too, releasing still more neutrons, which in turn strike more nuclei. This process is a **chain reaction.** If the reaction is allowed to proceed uncontrolledly, it quickly becomes a nuclear explosion. However, by returning a certain amount of inert material to the system we can control the reaction.

A nuclear reactor produces a controlled chain reaction in uranium or plutonium. These elements are shaped into rods, called fuel rods. The fuel rods are perfectly safe individually, because they have high surface/volume ratios. Most of the neutrons that are generated by spontaneous fission escape into the surroundings instead of striking other fissionable atoms. But when several rods are close together, ejected neutrons are now more likely to be trapped in a mass of fuel and start a chain reaction. In nuclear reactors, fuel rods are interlaced with movable control rods made of boron or

some other inert substance. When the control rods are raised, the total mass of the fuel rods is sufficient to cause a chain reaction. The inert control rods can be moved up or down to absorb varying numbers of neutrons, thus controlling the rate of the reaction. The chain reaction generates huge amounts of heat. The heat is then used to run steam turbines that generate electricity (Figure 19-10).

At present, nuclear fission reactors supply only about three percent of our electrical energy. However, their use is doubling every two to three years, worldwide. Modern fission reactors use a particular isotope of uranium, U^{235}; nuclear reactors that derive their energy from U^{235} are called **burner reactors.** Unfortunately, this isotope is relatively rare; it represents less than one percent of the total uranium in the world. Most of the uranium occurs as U^{238}, which cannot be used directly as a fuel.

It is possible, however, to construct a reactor that can use U^{238}. The fission of U^{235} liberates neutrons. If these neutrons are trapped by U^{238}, a train of nuclear reactions begins which converts U^{238} to plutonium (Pu^{239}), and Pu^{239} is a suitable reactor fuel. A similar sequence can be used to convert thorium to another fuel, U^{233}. Reactors that convert nonfissionable material to suitable fuel in this manner are called **breeder reactors.** The distinction between burner and breeder reactors has great practical importance. If reactors are built and designed as burn-

Figure 19-10 A nuclear fission reactor and power plant. Heat generated by the fission process is used to heat water and make steam. The steam drives a turbine that generates electricity.

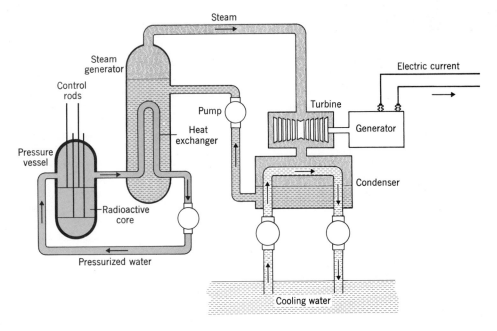

ers, energy can be produced only as long as the U^{235} supply holds out. But U^{235} is in very short supply. If expansion of nuclear generating capacity proceeds as predicted, all our U^{235} will be consumed in a few decades. By contrast, if U^{238} can be used as fuel by conversion to plutonium, and if thorium is similarly converted to U^{233}, our energy reserve is hundreds of times greater than the total energy of all the fossil fuels combined.

Right now almost all our reactors are burners, because the design of breeders is more difficult. To date only a few small experimental breeder reactors have been built. If safe and effective large breeder reactors are built, and the various social problems associated with their siting and operation can be solved, nuclear fission can support our present levels of energy consumption for thousands of years. But if we are forced to rely on burner reactors, the nuclear age will be over almost before it begins.

Less progress has been made with nuclear fusion reactions. The energy liberated by the sun and other stars comes from a fusion reaction, whereby hydrogen atoms are converted to helium. We have mimicked these solar reactions in uncontrolled hydrogen bomb explosions, but we have not yet produced controlled fusion reactions. The technical problems are severe, because the reactions proceed at such high temperatures that they cannot be carried out in material containers. Some progress is being made in developing controlled fusion by containing the reactions in strong magnetic fields.

Many fusion reactions are theoretically possible. One of the best reactions from our point of view is the fusion of lithium and an isotope of hydrogen, deuterium (H^2). The product is helium, and no radioactive wastes are created. Unfortunately, lithium is rather rare, so the total energy derivable from this type of reaction is small compared to that from other possibilities. A second reaction is the fusion of two atoms of deuterium. This reaction is more complicated, and it yields troublesome radioactive by-products. But the supply of deuterium is essentially unlimited. If we can achieve safe and controlled fusion of deuterium, we will have all the energy we need for the foreseeable future.

Wastes

The influx of raw materials and energy into human society must be matched by an equal and opposite outflow of waste materials and heat. Each of our possessions represents the consumption of resources and the production of waste. In the United States, an automobile has an average life span of about five years. While alive, it has a rapid metabolic rate. It consumes refined petroleum and produces heat, waste gases, and, of course, transportation. When it dies, it must be disposed of and replaced with another.

Disposal problems depend on the number of people in the community, on the rate at which they consume raw materials and energy, and on the kind of raw materials and wastes that must be handled. A sparsely-distributed primitive human population with a low standard of living and few possessions has little difficulty in maintaining the flow of materials and energy in and out of society. The inflow is primarily food. Waste, in the form of garbage and human excreta, is not usually abundant enough to constitute a health hazard. All the wastes can readily be decomposed by scavengers and microorganisms.

With the advent of agriculture, populations grow, and the problems of maintaining inflow and outflow become more serious. Agriculture maintains the supply of food inputs. However, the control of waste outflow requires an elaborate social organization. In fact, the waste problem was serious enough to limit the growth of cities for centuries. Cities were able to become large only by establishing elaborate public works, such as road systems, aqueducts, public granaries, and sewers. And most of them still suffered the ravages of disease when waste disposal systems failed. The wastes caused pollution, but only on a local scale, because the waste products were biodegradable. Such wastes were soon decomposed by microbes.

Modern industrial societies have greatly increased the possible size of cities by developing modern transportation systems. The number of people who live in cities of more than one million is increasing at a remarkable rate. In 1950, they housed about four percent of the world's population; the figure is now more than ten percent. Better transportation means that the inflow of raw materials and energy is greatly accelerated; but at the same time, the outflow of wastes increases correspondingly. Most importantly, the nature of the waste materials has changed. Human excreta and garbage must still be eliminated from human communities, of course. But now we must also dispose of detergents, strong acids and bases, heavy metals, radioactive wastes, and perhaps half a million different kinds of chemicals that are produced by an industrial society. Some of these materials can be degraded by decomposer organisms in our surrounding environment, but others cannot; in most cases we simply don't know.

The pollution problems associated with biodegradable wastes are potentially controllable. If too much waste is dumped in too limited an environment, the environmental quality is degraded, at least from a human viewpoint. The oxygen content of waters receiving biodegradable waste decreases, and plants and animals adapted to clean waters are replaced by a community of anaerobic decomposers. We do not yet fully understand how to control biodegradable wastes, but we can reduce their impact considerably with proper waste treatment.

The pollution that results from wastes that are degraded slowly is an entirely different kind of problem. DDT, for example, is very slowly degraded because it is difficult for most organisms to metabolize. It is widely distributed in the environment, because it becomes attached to airborne dust particles. When the dust settles, the DDT is absorbed by plants and passed on to organisms at other trophic levels. The chemical structure of DDT makes it very soluble in fat, so it accumulates in the lipids of plants and animals and becomes increasingly concentrated at each trophic level. Unfortunately, there is a long time lag between the liberation of DDT into the environment and its concentration in animals at the top trophic levels of food chains. If we were to stop using DDT altogether, its concentration in soils would decline immediately. However, soil particles would continue to distribute existing DDT to the oceans and fresh waters. The DDT levels in fish would continue to rise for about ten years, and the DDT levels in fish-eating birds would rise for about two decades.

Other pollutant materials are normally found in the environment in low concentrations but become extremely toxic at high concentrations. Many of these compounds are concentrated for use by industry and then injected back into the environment in a form that can enter food chains very easily. Lead is an example. Levels of lead in the Greenland ice cap have increased steadily over the past two centuries. Since the ice cap itself is remote from centers of industry, the increasing levels of lead there mean that the injection of new lead into ecosystems must be occurring worldwide (see Figure 19-11). We know that lead is toxic and that, like DDT, it is accumulated by organisms, becoming concentrated at higher trophic levels. However, we really can't predict the effects of lead pollution from industry and gasoline fuels.

Lack of information about such pollutants has produced two basic social views about how to deal

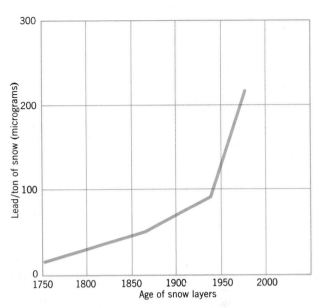

Figure 19-11 *Increase in the lead content of the Greenland icecap. Deeper layers of snow were deposited in the more distant past; the increase can be measured by taking samples at various depths.*

with them. Some people assume that a change from 30 to 300 micrograms of lead in a ton of snow is not very frightening, that the decline of fish-eating birds may or may not be related to DDT levels, and that such a decline may be a pity but is not very important when balanced against the positive aspects of DDT use—increased agricultural productivity and disease control. The alternative view is that the unpredictable effects of long-lived pollutants can be disastrous. Proponents of this view are concerned about the long time lag between the discharge of pollutants into the environment and their accumulation into organisms at high trophic levels. They feel that waste control should begin *before* the negative effects are demonstrated. The ultimate choice or compromise between these views will be determined, again, not by biologists, but by the political process.

In a very real way, our ability to control pollution is quite limited. Collecting potential pollutants after they have been used by industry and disposing of them safely is very expensive. The cost of recovery increases rapidly as we attempt more and more complete recovery. For example, the cost of trapping half the lead coming from a polluting device might be relatively small and could even be balanced by the value of the lead that is recovered. But the cost of removing 90 percent would be much larger. Removal of all of the lead would be virtually impossible and prohibitively expensive.

In many cases, there is simply no way to eliminate pollution and still carry on the industrial process that is generating the pollutant. Most people are unwilling to risk the consequences of stopping the process altogether. Few will even consider outlawing the automobile, for instance, as a solution to lead pollution. The alternative is to establish a set of pollution standards that define "acceptable levels" of pollutants. These standards are intended to minimize the damage to the environment but still to allow us to function as an industrial society. Unfortunately, many of these standards are based on little more than guesses about the potential damage caused by the pollutant.

Our ability to control pollution is also limited by the fact that many undeveloped countries view a concern about pollution as a luxury. They are much more concerned about increasing their industrial and agricultural productivity than about waste generation. Since long-lived pollutants from local sources may be distributed worldwide, a unilateral end to pollution by the developed countries would not stop the injection of long-lived pollutants into the environment.

Radioactive Wastes

Radioactive wastes represent a separate category of long-lived pollutants. The abundance of these wastes will increase rapidly as fission and fusion reactors are developed. Decisions about acceptable levels of radioactivity in the environment and problems of radioactive waste disposal are really part of the overall problem of using nuclear energy as a substitute for fossil fuels.

The elements that are suitable as fuels for nuclear reactors undergo radioactive decay in two ways: by fission, which splits the atom; or by the ejection of small particles from the nucleus. The fission process converts the fuel itself to other isotopes that may be very radioactive. The uranium-235 atom, for example, can split in several ways to produce radioactive isotopes of iodine, krypton, barium, and other elements. These fission products themselves emit various particles; in fact, they are actually more radioactive than the original uranium fuel itself.

The development of nuclear power produces two kinds of waste problems. The first is the leakage of radioactivity during the processing of fuel and during power generation. The second is the disposal of radioactive fission products that remain in the spent fuel rods. Radioactive fuel processing involves the mining and refining of natural ores, the fabrication of fuel elements, power generation, recovery of spent fuel elements, and the isolation of useful isotopes from the spent fuel elements. At each point in the process, some loss of radioactive material to the environment is inevitable.

Two isotopes are particularly difficult to control and may represent a serious hazard. One is **tritium,** a radioactive isotope of hydrogen (H^3). Tritium is short-lived and has a very low emission energy. (Its half-life is 12 years.) However, it is dangerous, because it gets into the water in the cells and tissues of organisms; all the tritium produced by nuclear reactions is eventually liberated during the handling of fission fuels and power generation.

The second problem isotope is krypton-85. Krypton is a gas that does not usually react with other elements, nor is it accumulated in organisms. But because it does not react with other materials, it eventually ends up in the atmosphere. Kr^{85} has a half-life of 11 years. If nuclear generating capacity expands as predicted, radioactive krypton will double the level of radioactivity in the environment in about a century.

The Atomic Energy Commission sets standards for the emission of various isotopes. At present these standards are so stringent that nuclear power plants must often operate well below their full generating capacities in order to avoid exceeding them. These emission rates are as low as we can make them with present technology. But we know that increasing levels of radiation will eventually damage man and other organisms. Therefore the AEC's safety standards are really a statement about how much biological damage we will accept in order to maintain the energy consumption necessary for modern industrialized societies. This is a long-term decision. The krypton-85 that we inject into the atmosphere will remain for many decades. Of course, we may hope that technology will improve, so that less radioactivity will be released into the environment. However, such improvements are in the future, while the demand for more nuclear power is in the present.

The disposal of fuel rods is the second major problem. The fuel for fission reactors is made of uranium compounds packed into a metal casing that is chosen to withstand pressure, heat, and intense radiation. Fission produces radioactive products directly, but the neutrons generated by fission also produce other radioactive isotopes by bombarding materials in the fuel and the metallic casing material. Some of these by-products—such as plutonium—are very useful. Others are simply waste products. When enough of these products accumulate, all the neutrons are absorbed and the chain reaction ceases. At this point the fuel rods must be reprocessed in order to recover and concentrate any unused U^{235}, Pu^{239}, and U^{233}. But roughly half the total radioactivity created by neutron bombardment is waste. The amount of radioactive wastes produced in fuel rods is about a thousand times greater than the leakage of tritium, krypton-85, and other isotopes during processing. If this amount of radioactivity were to escape into the environment, it would be disastrous.

The disposal of such wastes is a formidable problem. The only way to handle it is to find a way to keep them isolated from the environment while they decay; twenty half-lives is usually considered a minimum safe storage period. Strontium-90 is one of the wastes produced. It first attracted public attention as a fallout hazard from bomb testing. This isotope has a half-life of about 27 years. In other words, we must arrange to store it while it decays for 540 years. Other isotopes have much longer half-lives—about 17 *million* years for iodine-129, but fortunately this isotope is only a minor product of fission reactors. The overall composition of the radioactive wastes is such that they must be safely isolated for millenia. Our disposal problem is really a storage problem.

The safe storage of wastes was largely ignored in the United States during the late 1950's and early 1960's, when large quantities of fissionable materials were produced for armaments. It was decided that bombs must be stockpiled quickly for the sake of national security, and that the storage problem could be dealt with later. At present, the United States has almost 100 million gallons of liquid radioactive wastes stored in large underground tanks. The wastes must be constantly stirred and cooled, or the heat released by radioactive emissions will rise the temperature above the boiling point and the containers will burst. Stirring and cooling must continue until the long-lived isotopes decay to safe levels. This is clearly a poor arrangement for storage.

Other disposal schemes are now under discussion. Radioactive wastes could be dispersed in insulating solids such as glass or ceramics and then buried in abandoned mines. Deep salt beds near Lyons, Kansas, were proposed as a storage site, but the AEC was forced to abandon the idea because of opposition by the State of Kansas. West Germany has begun disposal in salt beds. A second idea is to bury the wastes deep in the ice on the Antarctic continent. It has even been suggested that radioactive wastes be put into orbit around the sun, but there is too much waste being generated for this idea to work. The problem is perhaps not unsolvable—but it is unsolved at present.

Still another kind of problem accompanies the development of nuclear fission: the amateur bombmaker. The plutonium produced by breeder reactors, and subsequently used for fuel, is also suitable for bomb construction. Building an efficient and "clean" bomb is difficult; building an inefficient one is not. It simply requires a mass of plutonium large enough to sustain a chain reaction. Unfortunately, the theft of plutonium would be very difficult to prevent. Radioactive materials are transported by commercial carriers in the United States. When the volume of plutonium used for power generation becomes large, the pilfering of small quantities or the hijacking of larger quantities is inevitable. Attempts to steal refined plutonium for private use have already been made. Regardless of the precautions that are taken, the probability of successful theft of plutonium increases as its use and supply increase. The prospect of fissionable materials in private hands is very sobering.

Future Resource Consumption

Our discussion of resources, energy, waste disposal, and food production has not permitted us to assign definite limits to the carrying capacity of the earth. But it is clear from the problems in all of these areas that we must be pressing against that limit right now. The situation is easier to evaluate if we consider the consumption of nonrenewable resources, the production of energy, and the growth of the human population on a longer time scale.

For most of his history, man has consumed only renewable resources. The use of metals for weapons, tools, and ornaments was culturally important, but it represented a trivial depletion of the available resources. The large-scale consumption of nonrenewable resources is a recent development. Major exploitation of mineral resources began only about two hundred years ago, and already many minerals are almost gone. Others will last longer, centuries rather than decades, but the fact remains that these resources are *not* renewable.

We will have no choice, ultimately, but to return

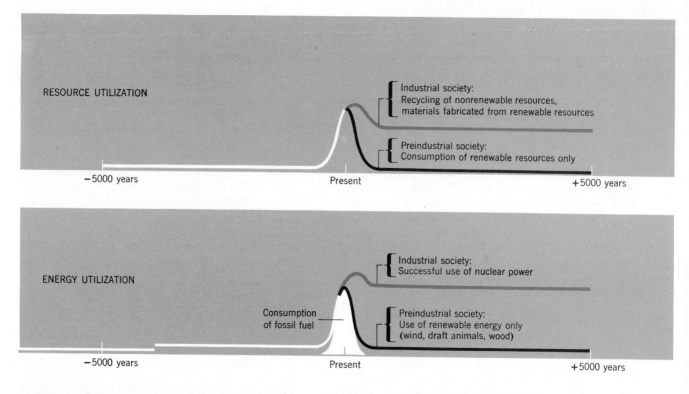

RESOURCE UTILIZATION

Industrial society:
Recycling of nonrenewable resources,
materials fabricated from renewable resources

Preindustrial society:
Consumption of renewable resources only

−5000 years   Present   +5000 years

ENERGY UTILIZATION

Industrial society:
Successful use of nuclear power

Consumption
of fossil fuel

Preindustrial society:
Use of renewable energy only
(wind, draft animals, wood)

−5000 years   Present   +5000 years

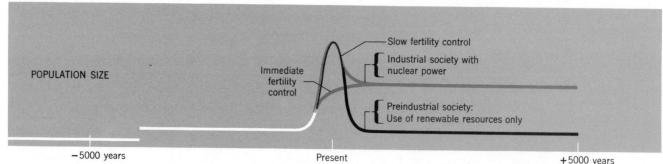

POPULATION SIZE

Slow fertility control

Industrial society with
nuclear power

Immediate
fertility
control

Preindustrial society:
Use of renewable resources only

−5000 years   Present   +5000 years

Figure 19-12   Resource utilization—past, present, and future—with and without the conservation and recycling of critical nonrenewable materials.

Figure 19-13   Energy utilization—past, present, and future—with and without the development of major new energy sources such as nuclear energy.

Figure 19-14   The size of the human population—past, present, and future. Future population size depends on the availability of resources and energy.

to a society that depends primarily on renewable resources. But such a society can operate in two possible ways. The first way is to conserve and recycle the critical nonrenewable materials, such as metals, so that we will be able to set up industries that fabricate materials from renewable resources. For example, paper and lumber and synthetic cloth are all made from plant fiber, a renewable resource. But the sawmill and papermill and

cloth factory need equipment that is made from metal and other nonrenewable resources. The mass-production of cheap food and other consumer goods would also be much more difficult without metals. As a result, the nature and quantity of available food, clothing, shelter, jobs, and transportation would change, but society would retain an industrial base.

The second alternative is to use renewable

resources directly, without benefit of technological modification. We would use logs, not boards, and handmade cotton cloth instead of rayon or nylon. Of course, this society would not be an industrial one. But such a return to a pre-industrial condition may very well be forced upon us if our social organization is disrupted so much that the fabrication of substitute materials is interrupted. These two alternatives are shown in Figure 19-12.

Our use of energy has a similar history. Energy expenditures were low until the exploitation of fossil fuel began. This exploitation, of course, is tied to the rise in our use of other resources. Energy consumption per person in the United States is now about ninety times the basal level of human metabolism, and essentially all the supplementary energy comes from fossil fuels. Fossil fuel reserves are limited. Again, there are two basic possibilities for the future. If nuclear power is developed, and the problems of waste disposal and pollution are solved, we can look forward to an indefinitely large supply of energy. The alternative is that social disruption or technical problems or an unacceptable burden of wastes will force us back to reliance on renewable energy resources: wood for fuel, wind and water and draft animals for power (Figure 19-13). Obviously, predictions about the use of resources and of energy are interdependent. We can establish an industrial society that processes renewable resources only if we manage to develop nuclear power for energy.

The prospects for future changes in the size of human populations are also tied to our use of resources and energy. The most favorable case assumes that fertility control can be achieved immediately, so that the world's population would stabilize at ten to fifteen billion within one or two doubling periods (35 to 70 years). Obviously, this case also assumes that we will command sufficient resources and energy to maintain such huge populations.

A second, more realistic prospect suggests that fertility control will be achieved more slowly. In that case, there will be more people than can be sustained, even with our most optimistic assumptions about resources and energy. A population crash would inevitably bring the numbers down to a level that can be supported. The third prospect is a severe population crash, which would inevitably accompany a return to a pre-industrial society that lacks energy and a sophisticated technology (see Figure 19-14).

Undeveloped countries depend on subsistence agriculture, and they have little industry. Most of them are experiencing rapid population growth. Underemployment resulting from the increased population in rural areas produces a migration to the cities, but urban industry is insufficient to provide work. Population growth has exceeded the capacity to produce adequate food, so most of the population is undernourished and malnourished.

Figure A Rural overpopulation forces people to cities that are increasingly incapable of supporting them. The scene is Bombay, India, 1972.

Figure B Protein malnutrition, Biafra, 1968. The immediate cause of this famine was civil war, but throughout the undeveloped world, population is pressing against the limits of food production.

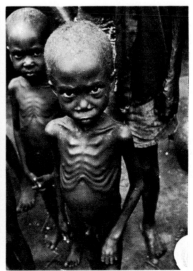

Developed countries have more productive agriculture and more extensive industries. Their problems of unemployment, crowding, and nutrition are less serious. However, increased productivity requires large energy inputs, which are supplied mostly in the form of fossil fuels (coal and oil and their derivatives). The extraction and transportation of these fuels are threats to the quality of the environment.

Many coal reserves are found in thin seams near the surface. The coal is mined by huge bulldozers and giant shovels.

Figure C "Peabody" is a giant shovel used in strip mining coal in Kentucky. The small yellow dot next to the left track of the shovel is a large bulldozer. The shovel handles 120 cubic yards of earth at a time.

Figure D The reclamation of land after strip mining is often difficult. This area of eastern Kentucky is considered an example of good land reclamation.

Accidents associated with the storage and transportation of oil are common occurrences.

Figure E   An oil spill from a leaking tanker in a Japanese harbor.

Figure F   A Pacific cormorant covered with oil. Spills threaten a wide variety of marine organisms.

Figure G   Even with volunteer help, cleaning up after a spill is time-consuming and costly.

Nuclear fuel also presents a serious hazard. Despite great care in handling, radiation escapes to the environment during refining, utilization, and waste disposal.

Figure H   Used nuclear fuel is stored under water.

Industrialized countries produce organic and inorganic wastes in large amounts. In most cases, we do not understand what long-term effects they may have on the environment. However, the overall impact of their discharge is clear enough.

Figure I The Cuyahoga River empties into Lake Erie at Cleveland. By the late 1960's it was so polluted that it was officially declared a fire hazard.

Figure J Dead alewives floating on Lake Michigan. Vast fish kills like this one are the result of artificially accelerated eutrophication through the discharge of organic wastes.

The End of Growth

Notice that we have not even considered the continued growth of population and the increased consumption of energy and resources as possibilities. There can be no argument on this point. The world is finite, and its resources are limited. The question of precisely when exponential growth will cease is open to the usual uncertainty, but growth must end sometime.

The actual end of exponential growth will be determined by many factors: population size, resource limitations, energy, waste production, and economic growth interact in complicated ways. More industrial activity requires more energy and generates more waste and consumes more resources. More people require more food and demand more goods and services. Generation of wastes degrades the agricultural productivity of the environment and affects the health and longevity of the population. It is quite impossible to try to analyze all the possible interactions. However, we can try to predict the end of growth by using some of the most important factors and interactions to construct a model.

The best effort has come from an international team working at the Massachusetts Institute of Technology. They began with several simplifying assumptions. In particular, they treat the world's population as a single unit; they treat all resources as one resource, all pollutants as one pollutant, and so forth. Other assumptions are quite optimistic, because they ignore any possible social and political difficulties: their model assumes that resources are distributed equally within and between populations; it ignores differences between developed and undeveloped countries; and it assumes lasting peace, and rational and benign political leadership. These assumptions make detailed predictions about specific populations impossible; but their predictions of general trends are valid.

Next, the authors made the fundamental assumption that the growth of human populations is controlled by five basic factors: population size, agricultural production, natural resources, industrial resources, and pollution. All of these basic factors are themselves controlled by a long list of variables that interact with each other: birth rates, death rates, longevity, food production per acre, land development rate, industrial output, resource consumption rate, pollution-generation rate, and so forth—about 100 separate variables in all. The authors expressed the relations between the variables in a series of equations in a computer program. In many cases, the quantitative relations between the variables were unknown and had to be estimated. At this point the model was complete.

The next step was to assign a value to each variable and enter it into the computer. The computer then used these values to solve the equations and come up with a series of answers that predict how our basic factors change with time. This model is simply an elaborate extension of our original one that predicted changes in population size based on only two variables, birth rate and death rate.

The MIT team first assigned values to the variables that reflected present trends. That is, they simply assumed that we will continue as we are. Under these conditions, the model predicts a population crash in the middle of the next century. The crash is initiated by the depletion of resources, which limits industrial productivity. But when the authors changed the input assumptions and gave the variables different values, the ultimate prediction did not change very much. Increased resources, unlimited energy supplies, an indefinitely large food supply—none of these assumptions made much difference. The model still predicts population crashes during the next century, though for slightly different reasons each time (see Figure 19-15).[1]

What kinds of assumptions are required to produce a more optimistic prediction? The model predicts that a large population can be supported at a high living standard only if birth rates are set equal to death rates immediately, if resource consumption is reduced, if pollution controls are instituted, and if industrial and economic activities are redirected toward provision of food and services (see Figure 19-16). Obviously some of these changes are impossible to introduce suddenly, and all of them would be subject to political and social opposition. The demand that these things be done immediately and on a worldwide scale is totally unrealistic. One is forced to conclude that a population decline is inevitable. In fact, death due to famine is already common in some undeveloped countries. The exact time of a population crash is, of course, still uncertain, but we can assume that it will take the form of multiple crashes. Stronger nations may survive longer at the expense of weaker ones. Within nations, inequalities of distribution will affect some subcultures more than others. Nonetheless, the basic trends are clearly shown by the model. The longer we postpone changes that are conducive to stability, the more difficult it will be

---

[1] The MIT model is described in Donella H. Meadows and others, *Limits to Growth*. New York: Universe Books (1972).

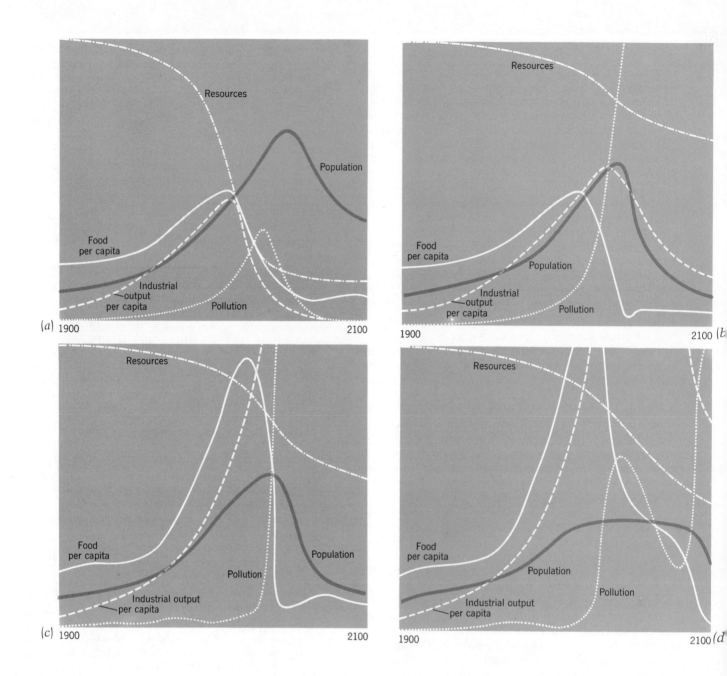

Figure 19-15 Predictions of the M.I.T. model of population growth. (a) Assuming no major changes in present trends. (b) Assuming unlimited nuclear energy, increased availability of nonrenewable resources, and the recycling of materials. Growth is stopped by rising pollution. (c) The same assumptions as in (b), plus a doubling of food yield and partial pollution control. These measures permit more growth, but pollution is again limiting. (d) The same assumptions as in (b) and (c) plus worldwide access to a perfect birth control method permit a temporary achievement of constant population. Still there is a population crash as resources are depleted, industrial growth stops, and pollution accumulates.

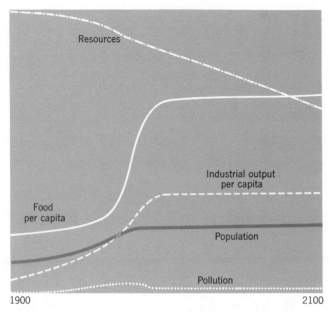

1900   2100

Figure 19-16   The M.I.T. model predicts stable world population if there is a combination of fertility control, reduced utilization of resources, recycling, pollution control, and an emphasis in the provision of food and services rather than industrial output.

to institute them, and the more disappointing and less effective they will be in preventing population crashes.

All too frequently, this kind of careful analysis of the future of human populations is simply rejected as being too pessimistic. The scholars who make the predictions are accused of underestimating human ingenuity, of having no faith in technological advances. Many people seem to be convinced that science and technology will soon overcome problems of food production, pollution, and limited resources. It has even been seriously suggested that a technological solution is imminent in the form of space travel, which will permit colonization of other planets. But we would need to transport about 70 million people a year just to balance our current population increases. Even a brief consideration of the energy and resources required to

send three men to the moon should silence these notions.

Our technology is inventive and imaginative. It can be used to fabricate and synthesize and modify the environment, but only by using energy and resources and by producing waste—it must exchange energy and materials with the environment. These open-system characteristics impose real limits on what our technology can accomplish.

If we accept the consequence of population analyses and the limitations of our technology, we must conclude that modern industrial society is only a transient phase of human history. Our present ideas about progress and expansion and growth should be reexamined to match the limitations we are encountering. This reexamination does not necessarily mean that society must become static, nor that personal development must be blocked. But it does mean that our present ideas of personal development—in the form of consumption of materials and energy—must be replaced by some other pattern of existence, one that places less stress on the environment. Of course, the fact that we can predict the end of growth does not mean that our ideas about progress *must* change. But if they do not, it is difficult to see how we can achieve population stability at a standard of living much above subsistence agriculture.

From a purely biological perspective, however, the qualities of human life associated with a high living standard are at best irrelevant. Evolution is concerned with the fitness and survival of species, not with the fate of individuals, and certainly not with the quality of their lives. The operation of natural selection will not automatically produce a large, stable, well-fed, and ecologically dominant human population for an indefinite future. But at the same time, barring a nuclear catastrophe, extinction is not at hand. Man, as a species, has survived and will continue to survive population crashes. Our tools for survival are the intelligence which provides us with usable knowledge of the world and of ourselves, and the flexibility that has allowed us to create an incredible range of behavior and cultures. Our future is modifiable, within limits, by application of the same intelligence and social flexibility that has brought us to our present position.

Meadows, Donella H., and others, *Limits to Growth.* New York: Universe Books (1972).

The preliminary report of a project undertaken at M.I.T. by the Club of Rome, a group of private individuals concerned about the future. A computer model is used to extrapolate the consequences of present population growth and economic activity and to predict the consequences of possible policy changes.

Murdoch, William W., ed., *Environment: Resources, Pollution, and Society.* Stamford, Conn.: Sinauer (1971).

An outstanding collection of papers on population and resources, environmental degradation, and environment and society. The contributions are authoritative, up-to-date, and fascinating.

U.S. Bureau of the Census, *Pocket Data Book, United States of America.* Published in odd-numbered years by the U.S. Government Printing Office, Washington, D.C.

A brief paperback version of the annual Statistical Abstract of the United States. Tabular information is provided on a wide range of aspects of American society: population and vital statistics, cost of health care, women in the labor force, economic activity, and almost every other imaginable subject.

U.S. Commission on Population Growth and the American Future (Rockefeller Commission), *Population and the American Future.* New York: New American Library (1972).

The Commission was directed to examine the probable extent of population growth in the United States, to assess the consequences of such growth, and to recommend policies to cope with those consequences.

glossary

abortion: the expulsion or removal of the embryo (and placenta) before it becomes independently viable.

abscission: the loss of leaves or flowers through the destruction of tissues at their base.

absorption spectrum: a graph showing the percentage of light absorbed at various wavelengths (colors) of light.

accessory pigments: a variety of plant pigments that assist chlorophyll *a* in trapping light energy.

accommodation: the decline in the output of nerve impulses from receptors in response to a constant input.

acellular organism: an organism without cells, a protozoan.

acetylcholine: a chemical agent that transmits nerve impulses across the synapses between neurons.

acetylsalicylic acid: a component of aspirin; related to methyl salicylate, found in willow bark and oil of wintergreen.

acid: a compound that increases the concentration of hydrogen ions in a solution.

acidity: the concentration of hydrogen ions in a solution.

acrosome: a caplike organelle that covers the head of a sperm cell.

actin: a protein component of muscle filaments.

action potential: the changes in electrical potential across a nerve cell membrane that occur during a nerve impulse.

active transport: the transport of solutes against a concentration gradient.

adrenalin: See epinephrine.

aerobe: an organism that obtains energy by oxidizing sugar (or other organic compounds) in the presence of oxygen.

agglutinin: an antibody that produces the agglutination, or clumping together, of foreign particles.

algae: a large group of plants that contain chlorophyll but lack the roots, stems, and leaves characteristic of more complex plants.

allantois: a membranous sac attached to the embryo, which serves as a repository for waste products that accumulate during development.

alleles: the two (or more) genes controlling a single characteristic.

allergen: an allergy agent.

alpha cells: the cells in the pancreas that produce glucagon.

alpha waves: the pattern of brain waves characteristic of an individual who is awake but resting.

alternation of generations: the cyclic alternation of haploid and diploid organisms in the life cycle of a species.

alveolus: a minute, spherical air sac, which is the site of gas exchange in the lungs.

amino acid: an organic molecule that contains at least one carboxyl group (COOH) and one amino group (NH$_2$).

amnion: the internal sac surrounding the embryo in an egg.

amniotic fluid: the liquid that fills the amnion and encloses and protects the embryo.

anaerobe: an organism that obtains energy by oxidizing organic compounds in the absence of oxygen.

analogous: (in development) a term used to describe structures that serve the same function in different organisms but represent different basic morphological patterns.

anaphase: the third phase of mitosis, during which the chromatids of each chromosome separate and move to opposite ends of the cell.

637

anaphylactic shock: a potentially lethal immune response involving liberation of histamine from cells.

androgen: male sex hormone, secreted by the testis.

androgynous: bearing both male and female sex characteristics; such individuals are neither clearly male nor clearly female.

aneuploidy: a condition in which all the cells of an organism contain an abnormal number of chromosomes.

angiosperms: the flowering plants.

animal pole: the end of an animal egg that has relatively little yolk.

anterior: situated toward the front.

anther: in flowers, the structure at the tip of the stamen in which male meiospores are produced.

anthropomorphism: the tendency to attribute human characteristics to other animals.

antibody: a specific protein molecule produced by an attacked organism in response to the presence of an antigen.

anticodon: the specific nucleotide triplet on tRNA that is complementary to the codon triplet in mRNA.

antigen: a chemical compound that triggers an immune response.

antitoxin: an antibody produced in response to the presence of a toxin.

aorta: a huge artery that supplies blood to all parts of the human body except the lungs.

apical meristem: the region of dividing cells at the tip of a plant shoot.

aquaculture: farming the seas or lakes.

arteriole: a small artery.

arteriosclerosis: the partial blockage of heart circulation by deposits of fatty material or calcium on artery walls.

artery: a blood vessel leading away from the heart.

asexual reproduction: the production of new individuals from the genetic material of one parent cell.

asymmetric: lacking an orderly and balanced arrangement of parts; an asymmetric carbon atom in an organic molecule is covalently bonded to four different groups.

atom: the smallest quantity of an element that can retain the chemical properties of the element; composed of an atomic nucleus (containing protons and neutrons) together with surrounding electrons.

atrium: the compartment of the heart that receives blood from major veins.

autonomic nervous system (ANS): the portion of the peripheral nervous system that regulates the internal environment by controlling various glands or smooth muscles.

auxin: a group of compounds, principally indole-3-acetic acid, that functions as plant hormones.

axenic culture: a culture containing only one kind of organism.

axillary bud: the bud found in the axil, the angle formed by a plant stem and the leaf petiole; also called a lateral bud.

axon: a long fiber extending from a neuron that transmits nerve impulses to other cells.

bacillus: a rod-shaped bacterium.

bacteriophage: a virus that attacks bacteria.

bark: all the layers of a woody plant stem from the outside down to the vascular cambium.

basal metabolism: the metabolism of an organism at rest.

base: a compound that decreases the concentration of hydrogen ions in a solution.

bends: a diving disease resulting from the formation of nitrogen gas bubbles inside the body.

benign tumor: a tumor composed of relatively non-invasive cells.

benthic: living at the bottom of the sea or lakes.

beta cells: the cells in the pancreas that produce insulin.

biochemical oxygen demand (BOD): the amount of oxygen consumed when the organic matter in a body of water is completely oxidized by bacteria and other microorganisms; BOD is a measure of the amount of biodegradable material in sewage.

biodegradable: capable of being oxidized by microorganisms.

binary fission: procaryotic cell division, in which the cell pinches in half; no mitotic spindles are present.

binomial: a two-word designation of a particular kind of plant or animal, consisting of a genus and a species name.

biological control: the introduction of a predator or parasite of a pest organism in order to reduce the level of the pest population.

blastocoel: the cavity in the center of the blastula.

blastopore: the surface indentation on a developing gastrula.

blastula: an early stage of embryonic development, in which the cells form a hollow ball.

botulism: food poisoning caused by the toxin of the bacterium *Clostridium botulinum.*

Bowman's capsule: a curved sac at one end of the kidney nephron surrounding the glomerulus.

brain hormone: an insect hormone, produced by neurosecretory cells, that initiates molting.

brain waves: small changes in electrical potential that result from the coordinated and rhythmic firing of nerve cells in the brain cortex.

bronchi: the branches of the trachea that lead to the right and left lungs.

bronchioles: branches of the bronchi that terminate in minute air sacs.

bud: (in plants) a terminal structure that encloses the apical or lateral meristem of the shoot.

bulbourethral gland: a gland in males that secretes a fluid into the urethra; the fluid neutralizes the acid environment in the vagina.

callipygous: having shapely or beautiful buttocks.

calorie: see kilocalorie.

capillary: the smallest type of blood vessel; the site of the exchange of gases and other dissolved substances between blood and tissues.

carbon cycle: the movement of carbon between organisms and the environment.

carbonic anhydrase: an enzyme in red blood cells that catalyzes the interconversion of carbon dioxide and carbonic acid.

carcinogenic: cancer-producing.

carnivore: an animal that consumes other animals.

cell: the basic unit of life, both in structure and in function.

cellularization: a hypothesis for the origin of multicellularity which proposes that a unicellular organism with many nuclei gave rise to a multicellular organism.

central nervous system: (CNS): the brain and spinal cord.

centriole: a small organelle, lying near the nucleus in the cytoplasm of animal cells, which functions during cell division.

centromere: the point at which two duplicate chromatids are connected to each other.

cephalization: an evolutionary trend which produces a concentration of neurons in the anterior end of an animal.

cerebellum: a portion of the hindbrain that controls and coordinates posture and balance.

cerebral cortex: a major region of the forebrain.

cervix: the base of the uterus; it opens into the muscular vagina.

chain reaction: the physical phenomenon in which particles emitted from atomic nuclei collide with other nuclei and cause the emission of additional particles.

chemical bond: the force that keeps two atoms bound to one another; it results from the interaction of electrons.

chemical reaction: the interaction of electrons in atoms or molecules to form a chemical bond.

chiasmata: crosses formed when chromatids break and reassociate during meiosis.

chitin: a carbohydrate derivative found in insect exoskeletons and other structures.

chlorophyll: the green light-absorbing pigment found in all photosynthetic organisms.

chloroplast: the organelle that carries out photosynthesis in plant cells.

chorion: the external sac surrounding the embryo in an egg.

chorionic gonadotropin: a hormone secreted by the chorionic membrane of the developing embryo; it stimulates the secretion of progestogen.

chromatid: one of the two duplicate strands of a chromosome during mitosis.

chromatin: long, thin strands of unwound chromosomes.

chromatography: an experimental technique for separating a mixture of substances; it depends on the principle that different compounds move through a two-phase system at different rates.

chromosomes: the filamentous or rod-shaped bodies in the cell nucleus that contain the hereditary material, DNA.

chyme: the liquid portion of the stomach contents.

cilia: threadlike organelles that extend out from a cell's surface and undulate to and fro.

climax community: the last and most permanent stage of a natural succession.

clitoris: an external female genitalium, homologous to the penis in the male.

clone: a colony of organisms produced asexually from a single parent individual.

closed system: a system that does not exchange matter (or energy) with its surroundings.

coccus: a spherical bacterium.

codeine: an analgesic, or pain-relieving drug; one of the agents in opium.

codon: a sequence of three nucleotide bases that code for an amino acid during protein synthesis.

cohort: a particular age group in a population.

coitus interruptus: a contraceptive method consisting of withdrawal of the penis from the vagina before ejaculation.

coliform count: the number of *Escherichia coli* bacteria that are present in water samples; it indicates the degree of fecal contamination.

collagen: a major structural protein in animals that maintains the form and elastic properties of organs and tissues.

collecting tubule: a small tube that collects urine produced by the nephron.

colloidal solution: a liquid system in which the particles are large enough to behave as a separate phase but small enough not to settle out.

colon: the large intestine.

colony: (1) a loose and accidental association of organisms; for example, a visible group of microorganisms grown on a solid surface. (2) a group of organisms, the individuals of which have become so specialized that they are no longer self-sufficient.

commensalism: a symbiotic relation from which one organism benefits and the other neither benefits nor is harmed.

community: populations of different species of organisms that live in the same area and interact in various ways.

companion cell: cell next to the sieve tube element in phloem; it carries on metabolic functions for the sieve tube.

competition: (in ecology) a type of species interaction in which two species utilize the same environmental resources.

concentration: usually the quantity of salt, sugar, or other solutes in a known volume of water.

cones: the sensory cells in the retina that are responsible for color vision.

conjugation: a common form of sexual reproduction in bacteria that involves the direct transfer of DNA between two living cells.

contractile vacuole: a vacuole that pumps excess water out of protozoan cells.

copulation: the insertion of the penis into the vagina to effect the transfer of sperm by ejaculation.

cork cambium: meristematic tissue, lying just beneath a plant's epidermis, which produces cork.

coronary artery: an artery that supplies blood to the heart.

corpora quadrigemini: the dorsal region of the midbrain that receives sensory flow from the spinal cord and coordinates it with inputs from the eyes and ears.

corpus callosum: a tract of nerve fibers that connects the two hemispheres of the brain.

corpus luteum: the mass of cells lining the empty egg follicle in the ovary; it secretes hormones.

correlation coefficient: a measure of the predictive value of a set of measurements. For instance, the heights of identical twins have a large correlation coefficient; the height of one twin can be used to predict the height of the other.

cortex: the outer region of a tissue or organ.

co-transport: the movement of amino acids or sugars and sodium ions across a cell membrane by means of a common carrier molecule.

cotyledon: a leaf in the embryo sporophyte of a seed plant.

counter-current exchange: the passive transfer of heat or materials between two fluids moving in opposite directions.

covalent bond: a bond in which an electron pair is equally shared by protons in two adjacent atoms.

cristae: folds in the inner membrane of mitochondria.

cross-fertilization: fertilization between different individuals.

crossing over: the exchange of portions of homologous chromosomes.

crude birth rate (CBR): the number of births per year per 1,000 total population.

cryptobiosis: a state in which an organism retains its structural integrity but is entirely inactive.

cuticle: an external layer secreted by epidermal cells.

cytokinesis: the division of the whole cell.

cytoplasm: a general term for the cell contents, excluding the nucleus.

delta waves: slow, long brain waves that occur during sleep.

deme: a local interbreeding population; the basic evolutionary unit.

demographic transition: the replacement of high birth and death rates in a human population by low birth and death rates.

dendrites: the short, branched fibers on a neuron that receive nerve impulses.

density-dependent factor: any influence that changes the growth rate of a population as it changes in size.

dermis: the subsurface layer of skin.

detritus: particles of nonliving organic material.

diastole: a resting phase during which the heart fills with blood.

diastolic pressure: the pressure in the large arteries during the resting phase of the heartbeat.

diatom: a unicellular alga enclosed in a cell wall constructed of silica.

dictyosome: a cell organelle composed of stacks of membranous sacs and vesicles; it functions in the secretion of cell products.

diencephalon: the central deep-lying region of the forebrain.

differentiation: the formation of specialized cells and tissues.

diffusion: the tendency of a dissolved material to move from a region of high concentration into a region of low concentration.

digestion: the mechanical and chemical breakdown of food particles.

digitalis: a specific heart stimulant; found in foxglove.

dinoflagellate: a planktonic unicellular alga with an armorlike cell wall and two flagella.

diploid: having two sets of chromosomes per cell.

directional selection: (in evolution) selection pressure that favors one or the other of two alleles in a deme.

distal convoluted tubule: the region of the kidney nephron that empties into the collecting tubule; it regulates solutes in the filtrate.

diversity: (in ecology) the total number of species that occupy an area.

dominant: (in genetics) the allele in a pair which is expressed in the phenotype of an organism without modification by its allele.

dorsal root ganglion: a cluster of receptor cell bodies that lies outside each segment of the spinal cord.

dorsal surface: (of an organism) the surface that is oriented upward.

double fertilization: the occurance of two separate nuclear fusions in angiosperm fertilization.

duodenum: the first region of the small intestine, lying between the stomach and the jejunum.

ecdysis: the replacement of an old insect exoskeleton with a new, larger one; molting.

ecdysone: the hormone that controls molting in insects; it is produced by the prothoracic gland.

ecological efficiency: the efficiency with which food energy moves through a trophic level in an ecosystem; in other words, the ratio of the total energy flowing into two successive trophic levels.

ecology: the study of relationships between organisms and their environment.

ecosystem: an interacting community of organisms and its immediate physical environment.

ectoderm: the embryonic cell layer, or germ layer, that gives rise to the outermost tissues of the body: the skin and its derivatives and the nervous system.

effector: a general name for cells whose activity produces an effect on the internal or external environments; in other words, glands and muscles.

electronegative: having a strong attraction for electrons.

emulsifier: a compound that has polar and nonpolar regions on the same molecule; it stabilizes colloidal solutions by binding with both water and fat molecules.

endemic: native to a particular area, constantly present.

endoderm: the embryonic cell layer or germ layer that gives rise to the innermost tissue of the body, including the digestive tract and related organs.

endodermis: a tube of cells surrounding the stele in the root of a plant.

endometrium: the lining of the uterus.

endoplasmic reticulum: membranes within the cytoplasm of a cell that provide a large surface area for metabolic reactions.

endosperm: nutritive tissue in the seed; it nourishes the developing embryo sporophyte.

engram: a general term for a physical change that may occur in the nervous system during learning.

enzyme: a protein substance that catalyzes chemical reactions between molecules.

epicuticle: the outer layer of an exoskeleton.

epidermis: a surface layer of flattened cells.

epididymis: a coiled duct attached to the surface of the testis where sperm are stored while they mature and await ejaculation.

epinephrine: a hormone product of the adrenal medulla.

erythrocyte: a red blood cell.

esophagus: the tube leading to the stomach.

estrogen: a sex hormone secreted by the ovary; it directs the repair and regrowth of the uterine lining after menstruation.

ethology: the study of behavior, with particular emphasis on instinctive behavior.

eucaryote: an organism that contains a distinct nucleus and possesses numerous membrane-bounded organelles and complex internal membrane systems.

eutrophication: the accumulation of plant nutrients in natural waters.

evolution: the composite of changes in populations after many generations of natural selection.

exoskeleton: an external skeleton; the hardened, outside covering of arthropods.

fat: a compound composed of three fatty acids joined to glycerol.

fatty acid: a molecule composed of a long chain of carbon atoms (most of which are linked to hydrogen atoms) linked to a carboxyl group.

fermentation: the oxidation of sugar without the use of oxygen.

fertilization: the fusion of haploid gametes.

fibrin: an insoluble protein that polymerizes with other fibrin molecules forming a netlike blood clot; it is produced from fibrinogen.

fibrinogen: a soluble plasma protein that is transformed into an insoluble protein in the clotting process.

filter-feeder: an animal that feeds on fine particulate matter in lakes and oceans.

filtrate: fluid that has passed through a filter.

flagella: threadlike organelles that extend out from a cell's surface and undulate to and fro, moving the cell or its environment.

fluorescence: the rapid reemission of a photon (light energy) by a molecule.

follicle: a small pocket that is lined with cells.

follicle stimulating hormone (FSH): a hormone produced by the anterior pituitary, which stimulates the growth and maturation of oocytes and sperm.

food vacuole: a type of vacuole found in unicellular organisms, containing engulfed food that is being digested.

founder species: the first organisms to colonize a barren area.

fruit: layers of tissue, produced by the ovary of an angiosperm, which cover the seed.

frustule: the elaborate cell wall of a diatom.

functional group: a group of atoms in an organic molecule that readily participates in chemical reactions.

gametangium: the specialized cell or structure in a gametophyte that produces gametes.

gametes: haploid cells that fuse to produce a diploid cell in sexual reproduction.

gametogenesis: the formation of gametes.

gametophyte: a haploid plant that produces gametes.

gastrin: a hormone that stimulates the production of gastric juice.

gastrula: the two- or three-layered embryo formed from the blastula.

gastrulation: a process in animal development, in which the blastula is transformed into a gastrula.

gel: a colloidal system in which the solid phase is continuous.

gene: a DNA segment, bounded by start and stop signals, that codes for one protein or a small group of related proteins; it is responsible for the development of some particular trait.

gene pool: the total of all the genes in a population.

general fertility rate (GFR): the number of children born per year per 1,000 women of child-bearing age (15–44 years old).

generation time: the time required for the doubling of the number of individuals in a population.

generator potential: a slow change in electrical potential at a localized region of the receptor neuron's membrane.

genetic drift: the random change in gene frequencies in a small population.

genitalia: the organs and ducts of the reproductive system.

genome: a single, complete set of DNA instructions.

genotype: the genetic constitution of an individual.

germinal cells: special cells in the gonad that give rise to gametes.

germ layers: in multicellular animals, embryonic cell layers from which adult tissues and organs are formed.

gestation period: the duration of pregnancy.

gizzard: a region in the gut of a bird specialized for grinding.

globulins: a class of proteins in blood plasma, some of which function as antibodies.

glomerulus: a ball of capillaries that delivers blood fluid to the nephron from the circulatory system.

glucagon: a sugar-regulatory hormone produced by the pancreas.

gonad: an organ that produces gametes.

gonadotropins: hormones produced by the anterior pituitary that act on the gonads; they stimulate oogenesis and ovulation in the female and spermatogenesis in the male.

Gram staining: a staining procedure used to distinguish between two fundamental types of bacteria, those with and those without lipids in their cell walls.

grana: stacks of membrane sacs found within chloroplasts.

green revolution: a general term for the introduction of new varieties of wheat and rice into undeveloped regions.

grey matter: the central part of the spinal cord; consists of nerve cell bodies.

gross primary production: the total amount of organic material produced by photosynthesis; the term includes both the organic material used by the plants themselves as an energy source and the plant material that becomes available to other organisms.

growth rings: alternating layers of spring and summer wood.

guard cells: specialized epidermal cells that border each leaf stoma and control its opening and closing.

gymnopserms: the conifers and their relatives.

half-life (of a radioactive isotope): the time required for half of the isotope initially present to decay.

haploid: having a single set of chromosomes in a cell.

Hardy–Weinberg Law: a law in genetics that states that in a stable population, the overall frequency of genes does not change.

hemoglobin: a red blood pigment that combines with and transports respiratory gases.

hemophilia: a sex-linked condition, in which blood does not clot normally because it lacks one of the cofactors needed for the production of thrombin.

herbivore: an animal that eats plants.

heritability: the proportion of phenotypic variability in a population that is due to differences in genotype.

hermaphrodite: an organism that possesses both male and female sex organs.

heroin: a narcotic formed by the addition of two acetyl groups ($=COCH_3$) to the morphine molecule.

heterozygous: having different alleles in a pair of homologous chromosomes.

histone: a protein with basic properties that coats DNA molecules.

homeostasis: the maintenance of a constant internal environment in animals.

homologous chromosomes: a pair of chromosomes that both contain alleles for the same sequence of genes.

homologous structures: structures with a detailed resemblance and similar embryological development.

homozygous: (in genetics) having two identical alleles for a particular gene located on homologous chromosomes.

hormone: a chemical agent that controls the functioning of one or more tissues.

host: usually the larger of the two associates in a symbiotic relationship.

hybridization: (1) in biochemistry, the formation of hydrogen bonds between complementary strands of DNA, or between DNA and RNA. (2) In genetics, the production of offspring by a cross between two different species or strains.

hybrid vigor: the production of progeny that are often larger and more vigorous than each of the inbred parental strains.

hydrocarbon: an organic compound containing only carbon and hydrogen.

hydrogen bond: a weak attractive force that occurs between an electronegative atom and a hydrogen atom that is covalently bonded to another electronegative atom.

hymen: a membrane that surrounds the external opening of the vagina.

hypersensitivity: extreme sensitivity to certain antigens.

hyphae: the branched filaments of a fungus.

hypothalamus: a region of the diencephalon in the brain that is closely associated with the control of internal functions.

ileum: the region of the small intestine that lies between the jejunum and the large intestine.

immotile: motionless; lacking a means of propulsion or movement.

immune response: the production of specific antibodies in response to an antigen.

immunoglobulin: blood antibody.

incomplete dominance: a condition in which the phenotypic expression of one allele is modified by the presence of a second allele.

independent assortment: a law, formulated by Mendel, that states that the distribution of one gene to an offspring is unaffected by the distribution of another gene.

industrial melanism: the spread of the dark form of various moths in industrial areas of Britain and North America.

infant mortality: the number of children who die in their first year of life per 1,000 live births.

inherited behavior: see instinctive behavior.

innate releasing mechanism (IRM): (in ethology) a genetic mechanism that produces a specific behavior pattern when triggered by a particular event in the environment.

inorganic compound: a compound that does not contain carbon–hydrogen bonds.

instinctive behavior: inherited behavior patterns that are fully formed the first time they occur, and that may even appear in animals reared in isolation.

insulin: a sugar-regulatory hormone produced by the pancreas.

intercalary meristem: masses of meristematic cells that occur at intervals along the plant stem.

interphase: the stage in the cell division cycle between mitoses.

intrauterine device (IUD): a contraceptive device consisting of a small plastic or metal object to be inserted into the uterus; it prevents the implantation of the embryo.

ion: an atom or molecule that has either acquired one or more extra electrons (giving it a net negative charge), or lost one or more electrons (giving it a net positive charge).

ionic bond: a chemical link between atoms in which the electron pair is attracted strongly to one atom in the pair.

ionization: the production of ions.

isometric contraction: the contraction of a muscle with both ends held in a fixed position; its tension increases but its length does not change.

isotonic contraction: the shortening of a muscle with no change in muscle tension.

isotopes: atoms having the same number of protons and electrons but different numbers of neutrons.

jejunum: the region of the small intestine that lies between the duodenum and the ileum.

juvenile hormone: an insect hormone that causes epidermal cells to secrete a larval cuticle; it inhibits development to a more advanced stage.

keratin: a tough protein secreted by epidermal cells.

kidney: an organ involved in maintaining the composition of solutes in the blood, eliminating dissolved waste products, and regulating water content.

kilocalorie: 1,000 calories; a metric unit of heat representing the amount of heat necessary to raise the temperature of 1,000 grams of water one degree.

kinins: a group of plant growth regulators.

labia majora: lips or folds of skin that surround and enclose the labia minora, the clitoris, and the urinary and vaginal openings.

labia minora: lips or folds of skin that surround and enclose the clitoris and the urinary and vaginal openings.

larva: a developmental stage that does not resemble the adult form of the organism.

latency: the time between a stimulus and the response.

leaf primordia: tiny embryonic leaves that cover the apical meristem.

leaf scale: a structure that protects dormant buds.

learning: changes in behavior that result from encounters with the environment.

leukocyte: a white blood cell concerned with the destruction of foreign particles.

lichen: fungi and algae living entwined together in a single structure.

linked genes: alleles for two different characteristics located on the same chromosome.

lipase: a fat-digesting enzyme.

lipid: a compound that is soluble in non-polar solvents but relatively insoluble in water.

loop of Henle: the long U-shaped region of the nephron.

lumen: the central cavity of the digestive tract.

luteinizing hormone (LH): a hormone produced by the anterior pituitary that causes development of the corpus luteum in the female and stimulates androgen production in the male.

lymphatic system: a system of vessels that returns protein and other macromolecules in the tissue fluid to the blood.

lysin: an antibody that promotes the lysis, or breaking open, of a cell carrying an antigen.

lysosome: a membrane-bounded droplet of enzymes that lyse the major components of a cell.

malignant tumor: a tumor composed of cells that spread and invade many tissues, setting up secondary tumors.

malnourishment: inability to obtain sufficient amounts of one or more essential nutrients.

Malpighian tubules: the excretory organs of insects.

mammals: a class of animals that nourish their young from the secretion of mammary glands and are covered with hair.

mastication: chewing.

matrix: a general term for material that is secreted external to a cell and in which the cell is embedded.

medulla: the inner region of a tissue or an organ.

medulla oblongata: a segment of the hindbrain; the junction of the spinal cord and the brain.

megaspore: the female meiospore in seed plants.

meiosis: a type of nuclear division that produces daughter nuclei with half the number of chromosomes of the parent cell.

meiospore: a haploid spore that arises by meiosis.

melanin: a pigment deposited in skin cells.

memory: the part of learning that involves information storage and retrieval.

menstrual cycle: the sequence of events in the human female that produces a mature egg and prepares the uterus for receiving a fertilized egg; it is controlled by hormones.

menstruation: the periodic discharge of uterine tissue and blood through the vagina.

meristematic tissue: tissue in a plant that retains its capacity for cell division.

mesoderm: the middle embryonic cell layer, or germ layer. It produces the skeleton, muscles, circulatory system, urogenital system, and most of the cartilage and connective tissue.

mesophyll: the middle layer of leaf cells between the epidermal layers.

messenger RNA (mRNA): RNA that is synthesized off the DNA sense strand; it serves as a template for protein synthesis.

metabolic rate: a measure of the overall rate of energy utilization by an organism.

metabolism: the totality of chemical reactions that occur in an organism.

metamorphosis: the transformation of a larva into an adult.

metaphase: the second phase in mitosis, during which the chromosomes become oriented at the center of the cell.

metaphase plate: the plane in which all the chromosomes lie at the end of metaphase.

method of concordance: a comparison of the rate at which a phenotypic characteristic appears in closely-related individuals and the rate of its appearance in unrelated individuals.

micelle: a tiny droplet of partly digested fats.

microfilament: a protein fibril that is probably contractile and may serve as an internal cell "muscle."

microorganism: any organism too small to be seen with the unaided eye.

micropyle: an opening at the end of the ovule in seed plants, through which the pollen tube enters at fertilization.

microspore: the male meiospore of seed plants.

microtubule: a hollow tube formed from a single layer of stacked protein molecules that may function as an internal cell "skeleton."

microvilli: fine tubes of protoplasm, produced by folding of a plasma membrane, that increase a cell's surface area for absorption.

mitochondrion: a spherical or elongated cell organelle that is the site of respiration.

mitosis: the duplication of a cell nucleus.

molecule: a discrete group of atoms joined by covalent bonds.

molting: the shedding of an old exoskeleton.

monomer: a molecule that exists by itself, or in association with other, similar, molecules to form a polymer.

monosaccharide: a simple sugar, a monomer of a carbohydrate.

morphine: an analgesic, or pain-relieving drug; one of the agents in opium.

mosaic eye: a type of eye found in invertebrates, which is composed of many hexagonal receptors; it produces an image formed of hexagonal light and dark spots.

motile: having a means of movement or of self-propulsion.

motor cortex: the region of the cerebral cortex concerned with producing motor responses.

motor nerve: a nerve that sends impulses to a muscle.

mucosa: the inner layer of epithelium that lines the gut.

mutagen: anything (chemicals, radiation, etc.) that increases mutation frequency.

mutation: any change in a gene that results in a permanent change in some characteristic of the individual; a deletion, addition, or substitution in the base code of the DNA strand.

mutator gene: a gene that renders adjacent genes unstable and likely to mutate.

mutualism: a symbiotic relationship in which both participants benefit.

myosin: a protein component of muscle filaments.

natural selection: the differential survival and reproduction of individuals in a population; the process by which organisms with variations less suited to an environment are eliminated.

neodarwinism: a modern restatement of Darwin's theory of natural selection, using the concepts of genetics.

nephron: the small functional unit of the kidney.

nerve fiber: an extension of a nerve cell that transmits a nerve impulse; an axon.

net primary production: the organic material from photosynthesis that is *not* used directly by plants, but instead becomes part of their tissues and is available as food for other organisms.

neural tube: nervous tissue in an embryo that develops into brain and spinal cord.

neuromuscular junction: the junction between a neuron and a muscle cell.

neuron: a nerve cell.

neurosecretory cell: a nerve cell that secretes hormones in addition to conducting impulses.

niche: all aspects of the physical and biological environment that are important to the survival of an organism.

nitrification: the oxidation of ammonia (NH_3) to nitrate (NO_3^-) by microorganisms.

nitrogen cycle: the movement of nitrogen between organisms and the environment.

nitrogen narcosis: a diving disease caused by breathing nitrogen at high pressures.

non-homologous chromosomes: chromosomes that contain alleles for different genes.

nonrenewable resource: a raw material that exists in limited quantity and cannot be replaced.

noradrenalin: see norepinephrine.

norepinephrine: a hormone product of the adrenal medulla.

notochord: a primitive skeletal rod of cartilage.

nucleolus: a spherical body found within the cell nucleus; it is rich in ribonucleic acid and is believed to be the site of rRNA synthesis.

nucleotide: the repeating monomer in a nucleic acid, it contains a five-carbon sugar molecule, a phosphoric acid, and a ring-shaped base (a molecule that contains nitrogen and has basic properties).

nucleus: (1) in a cell: an organelle, bounded by a double membrane, that contains most of the DNA of the cell. (2) In the brain: a cluster of nerve cell bodies.

nymph an immature form of a terrestrial insect.

oligotrophy: nutrient deficiency in natural waters.

oogenesis: the development of a mature egg from germinal cells in the female gonad.

oogonium: a germinal cell in the ovary that gives rise to an egg.

open system: a system that exchanges matter and energy with its surroundings.

operator site: a region on the DNA molecule that can bind with a specific repressor protein and prevent transcription.

opsin: the protein component of the rhodopsin molecule.

optic chiasma: the point in the brain at which fibers from each eye cross over and go to the opposite hemisphere of the cortex.

organ: a group of one or more tissues that together perform a specific function.

organelle: a membrane-bounded structure within a cell specialized to perform a particular activity.

organic compound: a chemical compound that is normally synthesized by living organisms; it contains carbon and hydrogen, and very often other elements, especially oxygen, nitrogen, sulfur, and phosphorus.

orgasm: the final stage of sexual excitement, characterized in females by rhythmic vaginal contractions and characterized in males by ejaculation of semen.

osmosis: the tendency of water to move from one solution to another across a membrane.

osmotic concentration: the concentration of solute particles in a solution.

ovarian cycle: the process of egg maturation; part of the menstrual cycle.

ovary: the female gonad, which produces eggs.

ovulation: the release of the mature egg from the ovarian follicle into the oviduct.

ovule: the specialized structure in seed plants that contains the egg cell.

ovum: the egg.

oxidation: the removal of electrons from an atom or molecule.

oxygen poisoning: a diving disease caused by breathing oxygen at high pressures.

oxytocin: a hormone from the posterior pituitary that stimulates the onset of muscle contractions in childbirth.

palisade layer: an orderly array of elongated cells in the upper region of the leaf mesophyll.

pancreas: a gland, located near the stomach, that produces digestive enzymes and the hormones insulin and glucagon.

parasitism: a symbiotic relation in which one partner (the parasite) benefits to the detriment of the other (the host).

parasympathetic division: the part of the autonomic nervous system that is concerned with regulating the internal environment in order to maintain homeostasis.

parthenogenesis: the development of the egg into a complete organism without fertilization.

partial pressure: the portion of the total air pressure contributed by a particular gas.

pasteurization: moderate heating and cooling of a liquid in order to kill all active bacteria that are dangerous to man.

penis: the male sex organ, consisting of a mass of erectile tissue surrounding the urethra.

pericycle: a layer of cells around the xylem and phloem of a root.

peripheral nervous system: nerves outside the spinal cord and brain that transmit impulses to and from the central nervous system.

peristalsis: waves of contraction that move along the gut.

petal: a flower part consisting of a modified leaf; usually brightly colored.

petiole: the specialized region at the base of a leaf by which the leaf is attached to the stem.

phage: see bacteriophage.

phagocytosis: the engulfing of microorganisms, other cells, and foreign particles by a cell such as a white blood cell in a multicellular organism.

phase: any homogeneous and physically distinct part of a system which is separated from other parts of a system by definite surfaces.

phase-contrast microscope: a light microscope with an optical arrangement that emphasizes small differences in the speed with which light passes through different structures.

phenocopy: an individual whose phenotype is a copy of the phenotype of an individual with a different genotype.

phenotype: the outward physical manifestation of the genotype of an individual; all the observable characteristics of an organism.

phloem: the conducting tissue for the movement of organic materials throughout a vascular plant.

phosphagen: a compound containing a high-energy phosphate that can be transferred to ADP to produce ATP.

phosphate bond: a covalent bond between a phosphate group and another molecule.

phosphorescence: the slow reemission of photons (light energy) by an excited molecule.

photochemistry: the study of chemical changes caused by light.

photon: the elemental unit or packet of light energy; also called a quantum.

photosynthetic unit: the smallest structure that contains all the molecules necessary for the first chemical reactions of photosynthesis; it includes the light-absorbing molecules of the photosystems and the reaction center.

photosystem: a group of pigment molecules that absorbs the light energy used in photosynthesis.

phytoalexin: a toxic compound produced by plants in response to an invader.

phytochrome: a plant pigment specialized for absorbing light and controlling flowering and other responses.

phytoplankton: the plant members of the plankton community.

pigment: a chemical compound that absorbs certain wavelengths of light.

pineal gland: an outgrowth of nerve tissue from the roof of the midbrain; it is involved in the environmental control of reproduction through a hormonal system triggered by light.

pistil: the flower part in which eggs are produced.

pith: the center of the stem of a plant.

placenta: tissue formed by extra-embryonic membranes and the uterine wall, which serves as a region for the exchange of nutrients, gases, and wastes between the mother and the embryo.

plankton: the aquatic community of small, free-floating plant and animal life.

plaque: a clear area that develops on a plate of bacteria wherever the bacteria are being lysed by a clone of phages.

plasma: the fluid portion of the blood.

plasma cell: a special lymphoid cell that produces antibodies.

plasma membrane: the membrane that forms the outer limit of a cell.

plasmid: an extrachromosomal body containing DNA in bacteria.

plasmodesmata: minute cytoplasmic tubes that pass through pores in plant cell walls and connect the cytoplasm of adjacent cells.

platelet: a tiny cell in the blood that contains one of the substances needed for the formation of blood clots.

polar body: a nonviable cell formed during oogenesis.

polarity: an unequal distribution of some property, such as the electrical charge on a molecule.

polymer: a large molecule built up by repetition of simple units.

polyploid: having three or more sets of chromosomes in a cell.

polysaccharide: any carbohydrate polymer.

polytene chromosome: a giant chromosome in certain insect tissues, consisting of hundreds of parallel strands of DNA.

population: individuals of a particular species who interbreed and live in a particular area.

portal vein: a vein that carries blood from one capillary system to another.

postural reflex: a reflex that helps maintain posture by preventing changes in muscle length.

prebiotic: before life.

precipitin: an antibody that forms a precipitate with a soluble antigen.

predation: a type of species interaction in which one species is killed and used as food by another species.

primary growth: growth in the length of a plant by means of apical or intercalary meristems.

primary producer: an organism that produces organic matter by photosynthesis.

procambium: cells that give rise to the vascular cambium.

procaryote: an organism that lacks a membrane-bounded nucleus, mitochondria, chloroplasts, and complicated internal membrane systems; the bacteria and blue-green algae.

procuticle: the inner layer of an exoskeleton.

progestogen: a sex hormone secreted by the ovary that prepares the endometrium for the implantation and maintenance of the developing embryo.

prolactin: a hormone secreted by the anterior pituitary that stimulates milk secretion during late pregnancy and during lactation.

promoter site: a control site on the DNA molecule that provides a point of attachment for RNA polymerase and allows the transcription of structural genes to begin.

pronucleus: the haploid nucleus in egg or sperm, just before fertilization.

prophase: the first phase of mitosis, during which the chromosomes coil into thick strands.

prostaglandins: a family of chemically-related compounds that modify metabolic control systems; some may induce abortion by causing uterine contractions.

prostate: a gland that secretes into the urethra a fluid thought to activate sperm.

protein: a polymer of amino acids; proteins function as enzymes or as structural material.

prothallium: the flat, green, heart-shaped gametophyte in the fern life cycle.

prothoracic gland: an insect gland that secretes ecdysone, the insect molting hormone.

protoplasm: the colloidal aggregation of complex molecules and structures that makes up the substance of living cells.

protozoa: unicellular (or acellular) animals found in water and in soil, and as parasites of multicellular organisms.

proximal convoluted tubule: the region of the kidney nephron that is closest to Bowman's capsule; it regulates solutes in the filtrate.

puberty: a period of human development characterized by a surge of growth and by visible changes in sexual differentiation.

pulmonary: having to do with the lung.

pupa: a stage in insect development between the larva and the adult.

quantasome: a type of particle found on the grana of chloroplasts; thought to contain a photosynthetic unit.

quantum: the elemental unit or packet of energy; also called a photon.

radioactive decay: the emission of particles or electromagnetic radiation from the nuclei of atoms.

reaction center: in photosynthesis, the center that receives the energy trapped by the photosystems; it contains the necessary enzymes for the subsequent steps of photosynthesis.

reaction-specific energy (RSE): in ethology, a kind of "instinctive drive" or "energy" that causes an organism to perform a particular behavioral act.

reafferenz: a term that describes perception as an indirect and secondary inspection of an abstraction of the original sensory input.

receptor: a sensory structure that detects a particular form of energy in the environment: light, sound, mechanical pressure, heat, etc.

recessive: the allele (of a pair) which is not expressed in the phenotype of an organism.

red drop: the discrepency between photosynthetic absorption and action spectra in red light.

reduction: the addition of electrons to an atom or molecule.

reduction-division: see meiosis.

reflex: the initiation and conduction of a nerve impulse to produce a muscle contraction.

reflex arc: a simple nervous pathway consisting of a sensory receptor cell, a postjunction nerve cell, and a muscle cell.

refractory period: a recovery period after the passage of a nerve impulse, during which time the membrane cannot be stimulated to produce another impulse.

regulator site: a control site on the DNA molecule that synthesizes the repressor protein.

relative growth: an increase or decrease in the size of particular structures relative to the size of the organism as a whole.

releasing factor: a substance produced by neurosecretory cells that stimulates hormone release from the pituitary.

renewable resource: a resource produced by plants and animals that, with provident management, will last indefinitely.

replacement level: the fertility level at which births just equal deaths.

repressor: a protein molecule produced by the regulator site on the DNA that can bind to the operator site on DNA and prevent the transcription of a particular portion of the genome.

resolution: the ability to distinguish between separate objects of small size.

respiration: the burning or oxidation of organic compounds to produce energy.

respiratory chain: an integrated series of compounds through which electrons are transferred; ATP is produced in the process.

reticular formation: a mass of nerve cell bodies in the hindbrain and midbrain that activates brain responses.

retina: the layer of sensory cells in the eye.

retinal: part of the rhodopsin molecule; synthesized from vitamin A.

Rh factor: an antigen found on the red blood cells of certain individuals; it was originally discovered in the blood of rhesus monkeys.

rhodopsin: a visual pigment found in the rod cells of the retina; also called visual purple.

rhythm method: a method of contraception that attempts to prevent the union of egg and sperm by restricting copulation to "safe" times, when no egg is present in the female genital tract.

ribonucleic acid (RNA): a polymer of ribonucleotides; named for the sugar component (ribose) of the nucleotides.

ribosomal RNA (rRNA): the nucleic acid component of ribosomes.

ribosome: a small, spherical particle made up of ribosomal RNA and protein; it provides a surface for protein synthesis.

rods: the sensory cells in the retina that are responsible for black and white vision, vision at low light intensities.

root cap: the cells that cover and protect the root tip.

root crops: plants that are cultivated for their roots or underground stems.

root hair: an extension of an epidermal cell near the root tip that increases its surface available for absorption.

root pressure: pressure resulting from the osmotic movement of water into the root.

ruminant: a herbivore with an elaborate stomach that contains and makes use of microorganisms.

scavenger: an animal that feeds on waste material or dead organisms.

sarcomere: the short functional unit of a muscle fibril.

sarcoplasmic reticulum: a network of fine tubes that penetrates muscle cells; it delivers calcium ions to muscle fibrils, thereby stimulating muscle contraction.

saturated fatty acid: a fatty acid in which each carbon atom of the chain is linked to two hydrogen atoms; it has no double or triple bonds between carbon atoms.

scrotum: the sac of skin and smooth muscle that contains the testes.

seaweeds: multicellular marine algae.

secondary growth: an increase in the girth of a stem produced by the activity of secondary meristems.

secondary sex characteristics: patterns of hair growth and changes in voice, muscular strength, and fat deposition that are caused by sex hormones but are not directly involved in reproduction.

segregation: the seperate transfer of genetic factors from each parent to its offspring.

semen: a mixture of sperm and various secretions from the male reproductive tract.

seminiferous tubule: a long, coiled tubule in the testis that produces sperm.

sensory cortex: the region of the cerebral cortex that is concerned with sensory signals.

sepal: a flower part consisting of a modified leaf; it forms the lowest whorl of the flower.

serosa: a layer of flattened epithelial cells covering the gut.

serum: the clear portion of a biological fluid separated from its particulate elements; e. g. the liquid portion of blood after coagulation.

sessile: sedentary; attached to a surface.

sex chromosomes: the pair of chromosomes containing the DNA instructions that determine the individual's sex.

sex-linked characteristic: a trait that is carried on a sex chromosome and that is inherited along with the sexual characteristics themselves.

sexual reproduction: the production of offspring from the genetic material of two parent cells.

sickle cell anemia: a genetic disease that causes the synthesis of abnormal hemoglobin.

sieve tube: the basic functional element of phloem tissue.

skeletal muscle: muscle, attached to the skeleton, which is responsible for body movements.

smooth muscle: a type of muscle that contracts slowly; formed by sheets of spindle-shaped muscle cells; it is found in various internal organs such as the digestive tract and the bladder.

sol: a colloidal system in which the continuous phase is liquid and the dispersed phase is solid particles.

somatic chromosomes: all the chromosomes except the sex chromosomes.

somatic nervous system: the part of the peripheral nervous system that links the central nervous system with sensory receptors, skeletal muscles, and glands.

somatotype: the size, shape, and proportion of the body.

speciation: the process of species formation.

spectrophotometer: a machine that measures a substance's absorption of light of different wavelengths.

spermatogonium: the germinal cell in the testes that produces sperm.

spindle apparatus: an arrangement of microtubules radiating from two poles at opposite ends of the dividing cell; it aids in the separation of chromosomes at mitosis.

sporangium: a special structure in multicellular organisms in which haploid spores are formed by meiosis.

spinal nerves: the nerves that branch off from each segment of the spinal cord.

spiracle: a surface pore that controls gas exchange in insects.

spiral cleavage: a complicated asymmetrical pattern of cell division that occurs in the developing egg of some invertebrates.

spirillum: a helical bacterium.

spleen: an ovoid organ that collects and breaks down aged erythrocytes; located behind the stomach.

spontaneous generation: the creation of biological entities from non-biological materials.

sporophyte: a diploid plant produced from a zygote.

stabilizing selection: selection pressure that tends to stabilize the phenotype in the species by eliminating individuals with characteristics that deviate from the norm.

stamen: a flower part, the site of pollen formation.

standing crop: the total amount of plant tissue present in an area at a particular time.

steady state: a continuous dynamic interchange of a system with its surroundings.

stele: the center of the root, containing the conducting tissues.

stigma: the tip of the pistil in a flower; it is specialized for the reception of pollen.

stomata (singular, *stoma*): the small pores that pierce the surface of the leaf and permit gas exchange.

stone cell: a type of thick-walled cell that occurs as a strengthening element throughout a plant.

stratigraphy: the study of sedimentary rock layers.

stretch receptor: a sensory receptor, located in muscle, that responds to stretching.

stroma: a protein matrix substance that fills the clear regions between the lamellae in chloroplasts.

style: the slender column of tissue that arises from the top of the ovary in a flower, and through which the pollen tube grows.

submucosa: a tissue layer in the gut, lying beneath the mucosa.

substrate: a compound acted upon by an enzyme.

substratum: the surface on which an organism lives.

succession: temporal changes in the community of species that inhabits an area.

suspension: a liquid system in which the particles are so large that they will eventually settle out.

symbiont: usually the smaller of the two associates in a symbiotic relationship.

symbiosis: any intimate association between organisms that involves a physiological exchange.

sympathetic division: the part of the autonomic nervous system that supports vigorous physical activity and prepares the animal for an emergency.

synapse: the functional connection between two neurons.

synaptic cleft: the narrow gap at the synapse between two neurons.

systematics: biological classification.

systole: the contraction of the heart.

systolic pressure: the pressure in the large arteries during heart contraction.

taxon: a grouping, or a particular level of classification.

taxonomy: biological classification.

tegmentum: the ventral region of the midbrain, involved with the coordination of movement.

telophase: the last phase of mitosis, during which nuclear membranes form around the two separate sets of chromosomes.

testis: the male gonad.

testosterone: a hormone secreted by the testis that promotes the development of the male ducts and external genitals.

tetanic contraction: a smooth increase in muscle tension.

tetrad: a set of four DNA strands resulting from the duplication and pairing of homologous chromosomes during meiosis.

thrombin: an enzyme that initiates blood clotting.

thrombosis: an internal blood clot.

thymus gland: the gland that activates and maintains the immune system.

tissue: any group of similarly specialized cells that together perform particular functions.

toxin: a poisonous substance.

tracheal system: the respiratory system of insects; composed of a complex of branching tubes.

tracheid: an elongated element of xylem tissue.

tract: (in the nervous system) a bundle of parallel nerve fibers.

transduction: (1) in energy relations, the conversion of one form of energy into another form. (2) In genetics, the transfer of genetic material between bacteria, using a virus as the transmitting agent.

transfer RNA (tRNA): any of at least twenty structurally similar species of RNA; one portion of the tRNA attaches to a specific amino acid, and another portion attaches to the corresponding mRNA codon.

transformation: the incorporation of a DNA fragment from dead bacteria into the chromosome of a living bacterium.

transpiration: the loss of water vapor from a leaf's surface through evaporation.

trichomes: protective hairs derived from the epidermal cells of higher plants.

tritium: a radioactive isotope of hydrogen (H^3).

trophic level: a particular stage in the transfer of energy from organism to organism in an ecosystem.

tropism: the directional response of a plant to environmental stimuli.

tubal ligation: a sterilization procedure, consisting of the surgical interruption of the woman's oviduct.

tumor: a cell mass formed by rapidly-dividing abnormal cells.

turgor pressure: the pressure exerted by a plant cell against the cell wall.

twitch: the contraction of a single muscle cell, or the simultaneous contraction of many muscle cells, produced by a single nerve impulse.

type specimen: a preserved specimen designated as typical for a given species and used as an aid in classification.

undernourished: unable to secure an adequate caloric intake.

unsaturated fatty acid: a fatty acid in which two or more carbons are joined together by double bonds, rather than linked to hydrogen atoms.

ureter: the duct that transports urine from the kidney to the urinary bladder.

urethra: the duct that drains urine from the bladder.

uterine cycle: changes in the uterus that occur during the menstrual cycle.

uterus: an internal female reproductive organ; it houses the developing embryo.

vaccine: a weakened preparation of disease-causing microorganisms that confers immunity on an individual by stimulating antibody production.

vacuole: a membrane-bounded region within the cytoplasm of a cell that contains various substances.

vascular cambium: a meristem that produces phloem and xylem.

vas deferens: a major duct that transports sperm from the epididymis to the urethra during ejaculation.

vasectomy: a sterilization procedure; the surgical interruption of the man's vas deferens.

vegetal pole: the end of an animal egg that is quite yolky.

vegetative reproduction: the production of a new organism from a fragment of the old organism.

vein: a blood vessel that moves blood toward the heart.

vena cava: the large vein that delivers blood to the heart.

venereal disease: a disease that is often transmitted through sexual contact.

ventral root: the part of the spinal nerve that contains motor nerve fibers that activate muscles.

ventral surface: (of an organism) the surface that is oriented downward.

venule: a small vein.

ventricle: the compartment of the heart that pumps blood into the arteries.

vertebral column: the set of hollow bone segments that encloses the spinal cord.

vertebrate: an animal having a vertebral column.

vesicle: small membrane-bounded sphere.

vestigial: a term used to describe structures that are in a degenerate stage, are no longer functional, but have not yet been lost from a species.

villi: small finger-like extensions of the mucosa.

virogene: viral DNA that has been inserted into the chromosomal DNA of a host cell.

virus: a small parasitic particle composed of nucleic acid and protein molecules.

visual cortex: the region of the cerebral cortex that is concerned with visual input.

visual purple: a visual pigment found in rod cells of the retina; also called rhodopsin.

vitamin: a specific molecule, required in small amounts, that participates in metabolic reactions.

water of oxidation: water produced by the oxidation of carbon compounds with oxygen.

white matter: the outer layer of the spinal cord; composed of nerve cell axons that run to and from the brain.

xylem: the conducting tissue responsible for water movement throughout the plant.

xylem vessels: cylindrical cells joined end to end to form xylem tubes.

Z scheme: a series of photochemical events, driven by energy from the reaction centers associated with the photosystems, that results in the net movement of electrons from water to $NADPH_2$.

zonation: the distribution of organisms into zones, or regions, where first one and then another kind of organism is most conspicuous.

zoomorphism: the tendency to describe man's behavior in terms of the behavior of lower animals.

zoospore: an asexual spore propelled by flagella.

zygote: a diploid cell formed by the fusion of two gametes.

photo credits

Cleveland. From R. D. Goldman, *J. Cell Biol.*, *51*, 745 (1971). **Figure C,** p. 87: Pamela E. Williams, University of Hull, Yorkshire. From P. E. Williams *J. Cell Sci.*, *9*, 767 (1971). **Figure 3-23:** Paul Chien, California Institute of Technology, Pasadena. **Figure 3-24:** Charles J. Brokaw, Kerckhoff Marine Laboratory, California Institute of Technology, Corona del Mar. From C. J. Brokaw, *J. Cell Biol.*, *43* (1), 68 (1965). **Figure 3-25 (c):** G. Benjamin Bouck, University of Illinois, Chicago. From G. B. Bouck, *J. Cell Biol.*, *50*, 373 (1971). **Figure 3-25 (d):** Peter Satir, University of California, Berkeley. **Figure 3-28:** Photo by the late Professor P. A. Roelofsen, Technische Hogeschool, Delft. **Figure 3-31:** Carl Struewe/Monkmeyer. **Figure 3-32:** Walter Dawn/National Audubon Society. **Figure 3-34:** J. T. Bonner, Princeton University. From J. T. Bonner, *Cellular Slime Molds* (2nd ed.), Princeton University Press (1967). **(a)(b)(c):** K. B. Raper. **(d):** W. Francis. **(e)(f):** J. T. Bonner. **Figure 3-34 (g):** Marica Miller, Research Fellow, and Professor Jean-Paul Revel, Division of Biology, California Institute of Technology, Pasadena.

CHAPTER 4. Figure 4-1: Eric V. Gravé. **Figure 4-2:** James G. Hirsch, Rockefeller University, New York. From J. G. Hirsch, *J. Exp. Medicine, 116,* 827 (1962). **Figure 4-7:** June Almeida, Bernard Cinader and Allan Howatson, Institute of Immunology, University of Toronto, Toronto. **Figure 4-9 (a)(b):** Julius Weber. **Figure 4-11:** Jacques F. A. P. Miller, Institute of Cancer Research, Chester Beatty Research Institute, London. **Figure 4-12:** Professor Charles C. Capen. From the Department of Veterinary Pathobiology, The Ohio State University, Columbus.

CHAPTER 5 OPENER: Gregory Antipa, Argonne National Laboratory, Argonne, Illinois. From Small, Antipa and Marszalek, *Acta Protozoologica, 9,* 275 (1972). **Figure 5-1:** Walter Dawn. **Figure 5-2:** Lee D. Simon, Institute for Cancer Research, Philadelphia. **Figure 5-3 (c):** T. F. Anderson and L. D. Simon, The Institute for Cancer Research, Philadelphia. From T. F. Anderson and L. D. Simon, *Virology, 32,* 294 (1967). **Figure 5-4 (b)(c)(d):** E. Kellenberger, Biozentrum der Universität, Basel. **Figure 5-5 (a)(b)(c):** From Stanier et al, *The Microbial World* (3rd ed.), Prentice-Hall (1970). **Figure 5-6 (a):** G. Cohen-Bazire, Institute Pasteur, Paris. **Figure 5-7:** Mary Mennes Allen, Wellesley College, Wellesley. **Figure 5-8 (a):** Hugh Spencer. **Figure 5-8 (b):** M. C. Noailles. **Figure 5-8 (c):** Hugh Spencer. **Figure 5-8 (d):** Walter Dawn/National Audubon Society. **Figure 5-9 (a-h):** Eric V. Gravé. **Figure 5-10 (a):** F. T. R. Taylor, Institute of Oceanography, The University of British Columbia, Vancouver. From Taylor, *J. Phycology, 7,* 253 (1971). **Figure 5-10 (b):** John D. Dodge, Birkbeck College, University of London. From J. D. Dodge, *J. Phycology, 6,* 138 (1970). **Figure 5-10 (c):** F. E. Round, University of Bristol. From F. E. Round, *Phycologia, 11* (2), 113 (1972). **Figure 5-10 (d):** T. L. Hufford and G. B. Collins, "The Stalk of the diatom *Cymbella cistula:* SEM observations", *J. Phycology, 8,* 208(1972). **Figure 5-11 (a)(b):** David Pramer, Rutgers University, New Brunswick. From D. Pramer, *Science, 144,* 384 (1964), copyright © 1964 by the American Association for the Advancement of Science. **Figure 5-13:** W. D. Russell-Hunter. © W. D. Russell-Hunter, *Aquatic Productivity,* Macmillan Company (1970). **Figure 5-15:** Theodor Rosebury and The American Society for Microbiology. **Figure B,** p. 149: David Linton. From *Scientific American,* June 1959. **Figure 5-18:** G. Bond, University of Glasgow, Scotland. From W. D. Stewart, *Nitrogen Fixation,* Humanities Press (1966).

CHAPTER 6 OPENER: Bruce Roberts/Rapho Guillumette. **Figure 6-8 (a)(b)(c):** G. Cohen-Bazire, Institute Pasteur, Paris. **Figure 6-8 (d):** W. Menke, Max-Planck-Institut, Vogelsand. From *Zeitschrift für Naturforschung, 166,* 543 (1961). **Figure 6-8 (e):** Norma Lang, University of California, Davis. **Figure 6-8 (f):** James Cronshaw, University of California, Santa Barbara. **Figure 6-8 (g):** Elisabeth Gantt, Smithsonian Institution. From *J. Phycology, 4,* 66 (1969). **Figure 6-9:** Courtesy William T. Hall, Electro-Nucleonics Laboratories, Inc., Bethesda, Maryland, 20014.

CHAPTER 7 OPENER: Graphische Sammlung Albertina, Vienna. **Figure**

7-1 (a): W. J. North, California Institute of Technology, Pasadena. **Figure 7-2 (a):** Carl Struewe/Monkmeyer. **Figure 7-2 (b)(c)(d):** Walter Dawn. **Figure 7-2 (e):** Eric V. Gravé. **Figure 7-5:** R. H. Noailles. **Figure 7-6:** William Randolph Taylor University of Michigan, Ann Arbor. From W. R. Taylor, *Marine Algae of the Eastern Tropical & Subtropical Coasts of the Americas*, University of Michigan Press (1960). **Figure A,** p. 183: Robert Perron. **Figure 7-8:** William Randolph Taylor, University of Michigan, Ann Arbor. From W. R. Taylor, *Plants of Bikini*, University of Michigan Press (1950). **Figure 7-9:** Orion Press, Tokyo. **Figure 7-10:** W. J. North, California Institute of Technology, Pasadena. **Figure 7-11:** C. Lavett Smith, American Museum of Natural History, New York. **Figure 7-12:** Helen Finlayson. **Figure 7-13 (a):** Hugh Spencer/National Audubon Society. **Figure 7-15:** T. Elliot Weier, University of California, Davis. From Weier, Stocking Barbour, *Botany* (4th ed.), Wiley (1970). **Figure 7-18 (a–e):** T. Elliot Weier, University of California, Davis. From Weier, Stocking, Barbour, *Botany* (4th ed.), Wiley (1970). **Figure 7-20 (a):** Courtesy Professor J. Heslop-Harrison and Dr. Yolande Heslop-Harrison, Royal Botanic Gardens, Kew. From "Scanning Electron Microscopy of Leaf Surfaces," *Proc. 2nd Engis Sterioscan Colloquium*, Chicago (1969). **Figure 7-22 (b):** Michael Shaw, The University of British Columbia, Vancouver. From *The New Phytologist*, 53, (1954). **Figure 7-29 (b):** H.-D. Behnke, University of Heidelberg, Heidelberg. From *J. Ultrastructure Res.*, 36, 493 (1971). **Figure 7-30 (a)(b)(c):** Martin H. Zimmerman, Harvard Forest, Petersham. **Figure 7-32:** Robert Boardman/National Audubon Society. **Figure 7-34:** T. P. O'Brien, Monash University, Australia. From O'Brien and McCully, *Plant Structure & Development*, Macmillan (1969). **Figure 7-34 (b):** T. Elliot Weier, University of California, Davis. From Weier, Stocking, Barbour, *Botany* (4th ed.), Wiley (1970). Photos by Lewis Feldman and Elizabeth Cutter. **Figure 7-34 (c):** T. Elliot Weier, University of California, Davis. From Weier, Stocking, Barbour, *Botany* (4th ed.), Wiley (1970). **Figure 7-36:** T. Elliot Weier, University of California, Davis. From Weier, Stocking, Barbour, Botany (4th ed.) Wiley (1970). **Figure 7-40 (a)(b):** Professor Brian, Academic Press. **Figure 7-41 (a)(b):** Runk/Schoenberger for Grant Heilman. **Figure 7-42:** T. P. O'Brien, Monash University, Australia. From O'Brien and McCully, *Plant Structure & Development*, Macmillan (1969). **Figure 7-43 (a)(b):** Professor J. Heslop-Harrison and Dr. Yolande Heslop-Harrison, Royal Botanic Gardens, Kew. **Figure 7-45 (a–d):** Photographs by L. P. Brower, Amherst College.

CHAPTER 8 OPENER: New York Public Library Picture Collection. **Figure 8-2 (a):** Grant Heilman. **Figure 8-2 (b):** Ross E. Hutchins. **Figure 8-3 (a)(b):** Rada and Neville Dyson-Hudson, Johns Hopkins University, Baltimore. **Figure 8-7:** Charles Perry Weimer. **Figure 8-8:** F. Caracciolo-F. Banoun/M. Grimoldi, Rome. **Figure 8-9:** Robert W. Krauss, Oregon State University, Corvallis. From R. W. Krauss, *American J. Botany*, 49, 425, (1962). **Figure 8-10:** USDA. **Figure 8-11 (a):** Pan American Coffee Bureau. **Figure 8-11 (b)(c):** USDA. **Figure 8-14:** USDA photo by Murray Lemmon. **Figure 8-15:** M. K. McCarty, Research Agronomist, Agricultural Research Service, USDA. **Figure 8-16:** C. H. Muller, University of California, Santa Barbara. From *Bulletin of the Torrey Botanical Club*, 93, 334 (1966).

CHAPTER 9 OPENER: Graphische Sammlung Albertina, Vienna. **Figure 9-6 (c):** Jerome B. Jacobs, University of Massachusetts Medical School, Worcester. **Figure 9-7 (a)(b):** R. L. Pardy, University of California, Irvine. **Figure 9-8:** Walter Deas, Seasports, Sydney. From Deas *Seashells of Australia*, Rigby (1972). **Figure 9-9 (b):** Alvin H. Rothman, California State University, Fullerton. **Figure 9-10 (a)(b)(c):** P. B. Armstrong, D. W. Deamer, J. J. Mais, University of California, Davis. **Figure 9-15 (a)(b):** Courtesy Dr. Jeanne M. Riddle, Department of Pathology, Wayne State University School of Medicine, Detroit. **Figure 9-19:** Vance A. Tucker, Duke University, Durham. From *J. Exp. Biol.*, 48, 67 (1968). **CHAPTER 10 OPENER:** The Trustees of the British Museum. **Figure 10-3 (b):**

M. C. Noailles. **Figure 10.4:** R. H. Noailles. **Figure 10-6:** David N. Menton, Washington University, St. Louis. From *J. Ultrastructure Res., 35,* 258 (1971). **Figure 10-7:** B. W. Zweifach, University of California, San Diego. **Figure 10-10:** Francois Morel, California Institute of Technology, Pasadena. From *J. Cell Biol., 48,* 91 (1971). **Figure 10-12:** Keith R. Porter, University of Colorado, Boulder. **Figure 10-16 (a):** Robert C. Hermes/National Audubon Society. **Figure 10-16 (b):** William M. Stephens. **Figure 10-20:** Johannes A. Kylstra, Duke University Medical Center, Durham.

CHAPTER 11 OPENER: "Encounter," 1944, by M. C. Escher. Courtesy Escher Foundation—Haags Gemeentemuseum—The Hague. **Figure 11-7:** The New York Academy of Medicine Library. **Figure 11-10 (c):** Lesnick E. Westrum, University of Washington School of Medicine, Seattle. **Figure 11-12 (c):** R. I. Birks, McGill University, Montreal. Birks, Huxley & Katz, "The fine structure of the neuromuscular junction of the frog," *J. Physiology, 150,* 134 (1960). **Figure 11-30 (b):** William H. Ittelson, City University of New York. **Figure 11-30 (c):** Reprinted with permission of *Science Digest,* copyright © The Hearst Corporation, 1969, all rights reserved. **Figure 11-31:** Nina Leen/Life Magazine, copyright © Time Inc. **Figure 11-35:** Courtesy I. M. Diab, Ph.D., Department of Pharmacology, University of Chicago. From *Science, 173, 1023* (1971), copyright © 1971 by the American Association for the Advancement of Science. **Figure A, B,** p. 352: S. K. Song and E. Rubin, Mount Sinai School of Medicine. From *Science. 173,* 327 (1972), copyright © by the American Association for the Advancement of Science.

CHAPTER 12 OPENER: Mia Tegner, University of California. San Diego. From M. J. Tegner and David Epel, *Science, 179,* 687 (February 16, 1973), copyright © 1973 by the American Association for the Advancement of Science. **Figure 12-4 (a):** Peter K. Hepler, Stanford University, Stanford. From P. K. Hepler, W. T. Jackson, *J. Cell Biol., 38,* 437 (1968). **Figure 12-4 (b):** Myron C. Ledbetter, Brookhaven National Laboratory. **Figure 12-6:** C. Robinow, The University of Western Ontario, London, Ontario. **Figure 12-7:** Courtesy The Fleischmann Laboratories, Standard Brands, Inc. **Figure 12-8 (a):** Runk/Schoenberger for Grant Heilman. **Figure 12-8 (b):** Ross E. Hutchins. **Figure 12-9:** Runk/Schoenberger for Grant Heilman. **Figure 12-12:** C. Robinow, The University of Western Ontario, London, Ontario. **Figure 12-16 (b):** Grace M. Donnelly and Arnold H. Sparrow, Brookhaven National Laboratory. **Figure 12-19:** Douglas P. Wilson, Marine Biological Laboratory, Plymouth. **Figure 12-24 (a)(b):** Everett Anderson, Harvard Medical School, Boston. From *J. Cell Biol., 37,* 514 (1968). **Figure 12-25:** Mia Tegner, University of California, San Diego. From M. J. Tegner, David Epel, *Science, 179,* 687 (February 16, 1973), copyright © 1973 by the American Association for the Advancement of Science. **Figure 12-35:** T. Elliot Weier, University of California, Davis. From Weier, Stocking, Barbour, *Botany* (4th ed.), Wiley (1970). **Figure 12-37:** Runk/Schoenberger for Grant Heilman. **Figure 12-41 (a)(b):** M. W. F. Tweedie/National Audubon Society. **Figure B,D,** p. 394: C. H. Dodson, University of Miami, Coral Gables. **Figure E,F,H,** p. 395: C. H. Dodson, University of Miami, Coral Gables. **Figure I,J,** p. 396: C. H. Dodson, University of Miami, Coral Gables, **Figure L,** p. 396: Joseph Arditti, University of California, Irvine. **Figure M,N,O,P,** p. 397: Bertil Kullenberg, University of Uppsala.

CHAPTER 13 OPENER: Jen & Des Bartlett/Bruce Coleman. **Figure 13-10:** A. A. Tuffery, Ewell County Technical College, Ewell, England. From *J. Cell Science, 10,* 131 (1972). **Figure 13-11:** Malcolm S. Steinberg, Princeton University. From *Science, 141,* 406 (1963), copyright © 1963 by the American Association for the Advancement of Science. **Figure 13-15 (a)(b):** T. Elliot Weier, University of California, Davis. From Weier, Stocking, Barbour, *Botany* (4th ed.), Wiley (1970). **Figure 13-18 (a–q):** John A. Moore. **Figure 13-26 (b):** Jerome Gross, Massachusetts General Hospital, Boston. **Figure 13-27 (a)(b):** Alex Comfort, University College, London. Photos by V. N. Nikitin, University of Kharkov, USSR. **Figure 13-29 (a)(b):** Merle Mizell, Tulane University, New Orleans. From *Science, 161,* 284 (July 19, 1968),

copyright © 1968 by the American Association for the Advancement of Science. **Figure 13-30 (a)(b):** American Cancer Society.
CHAPTER 14 OPENER: The Metropolitan Museum of Art, Fletcher Fund, 1919. **Figure 14-2:** Lester V. Bergman & Associates, Inc. **Figure 14-4:** Lester V. Bergman & Associates, Inc. **Figure 14-12 (a):** Courtesy of the American Museum of Natural History. **Figure 14-12 (b)(c):** Department of Embryology, Carnegie Institution of Washington. **Figure 14-13 (a)(b):** Department of Embryology, Carnegie Institution of Washington. **Figure 14-14 (a)(b)(c):** William M. Moss, University of California Medical School, Irvine. **Figure 14-16 (left):** Robert Perron. **Figure 14-16 (center, right):** Planned Parenthood of New York City, Inc. **Figure 14-17 (a)(b)(c):** Department of Embryology, Carnegie Institution of Washington. **Figure A (a)(b)(c),** p. 468: Courtesy William H. Sheldon. From W. Sheldon, C. W. Dupertuis, E. McDermott, *Atlas of Men,* Hafner, (1970).
CHAPTER 15 OPENER: R. A. Boolootian, Ph.D. **Figure 15-3:** William P. Nye, USDA Bee Laboratory, Utah State University, Logan. **Figure 15-8:** Lee D. Simon, Institute for Cancer Research, Philadelphia. **Figure 15-13:** Walter Dawn/National Audubon Society. **Figure A,B** p. 485: Victor A. McKusick, Johns Hopkins University School of Medicine, Baltimore. **Figure C (a,b),** p. 486: M. L. Barr, The University of Western Ontario, London, Ontario. **Figure C (c),** p. 486: Courtesy *American Scientist.* From James German "Studying Human Chromosomes Today" *American Scientist, 58* (1970). **Figure D,** p. 487: Victor A. McKusick, Johns Hopkins University School of Medicine, Baltimore. **Figure 15-14:** Courtesy Professor Myron C. Neuffer, University of Missouri, Columbia, and Crop Science Society of America. **Figure 15-15:** William P. Nye, USDA Bee Laboratory, Utah State University, Logan. **Figure 15-17:** M. W. Berns, University of California, Irvine. **Figure 15-18:** Orion Press, Tokyo.
CHAPTER 16 OPENER: Copyright *Punch,* London. **Figure 16-1:** Alan Band/Keystone Press. **Figure A,** p. 514: Courtesy of the American Museum of Natural History. **Figure 16-3:** Josef Muench. **Figure 16-4:** Courtesy of the American Museum of Natural History. **Figure 16-7 (a):** J. W. Wells, Cornell University, Ithaca. **Figure 16-8 (a)(b):** Courtesy of the American Museum of Natural History. **Figure 16-16 (a–g):** I. Eibl-Eibesfeldt, Arbeitsgruppe für Hamanethologie. AM Max-Planck-Institut Fur Verhaltensphysiologie. **Figure 16-17 (a):** Ylla/Rapho Guillumette. **Figure 16-17 (b)(c)(d):** Arthur W. Ambler/National Audubon Society. **Figure 16-17 (e):** New York Zoological Society. Figure 16-17 (f): Ylla/Rapho Guillumette. Figure 16-17 (g): J. Allan Cash/Rapho Guillumette. **Figure 16-17 (h):** Arthur W. Ambler/National Audubon Society. **Figure 16-17 (i):** Jeanne White/National Audubon Society. **Figure 16-17 (j):** Ylla/Rapho Guillumette. **Figure 16-17 (k):** San Diego Zoo Photo.
CHAPTER 17 OPENER: "Mosaic I," 1961 by M. C. Escher. Courtesy the Escher Foundation—Haags Gemeentemuseum—The Hague. **Figure 17-3:** S. W. Fox, Institute of Molecular & Cellular Evolution, University of Miami, Coral Gables. **Figure 17-4 (a):** E. S. Barghoorn, The Biological Laboratories, Harvard University. From *Science, 152* (3723), 758 (1966), copyright © 1966 by the American Association for the Advancement of Science. **Figure 17-4 (b):** E. S. Barghoorn, The Biological Laboratories, Harvard University. From Schopf and Barghoorn, *Science, 156,* 508 (1967), copyright © 1967 by the American Association for the Advancement of Science. **Figure 17-5:** A. Wallner, University of Uppsala, Institute of Geology. **Figure 17-12:** Manfred Kage/Peter Arnold. **Figure 17-13:** L. West. **Figure 17-14:** Douglas Faulkner. **Figure 17-15:** Runk/Schoenberger for Grant Heilman. **Figure 17-16:** Runk/Schoenberger for Grant Heilman. **Figure 17-17:** Walter Dawn. **Figure 17-18:** Walter Dawn. **Figure 17-19:** Walter Dawn. **Figure 17-20:** Giuseppe Mazza. **Figure 17-21:** Douglas Faulkner. **Figure 17-22:** Walter Dawn. **Figure 17-23:** Runk/Schoenberger for Grant Heilman.
CHAPTER 18. Figure 18-1 (a): Woodrow Goodpaster/National Audubon Society.

catarrhines, 540
Catesetum, 395, **395**
Cattleya, 393, **394**
Caulerpa, **181**
 reproduction in, 385
Cecropia, 244
cells, 63–68
 differentiation of, 406, 408–410
 division, 357–364, **359, 361,** 369, **371,** 404–406, **405**
 growth, 363, **363,** 404–406, **405**
 membrane, *See* plasma membrane
 membranes in, 67–68, **68**
 movement of, 66, 410–411, **412**
 specialization of, 102
 structure of, 67–89, **90, 91**
 types of, 86, 90–91, **90, 91**
 water in, 92–94, **93, 95**
cell plate, 361–362, **362**
cellular slime mold, 95–96, **97,** 567
cellularization, 96–98, **98**
cellulose, 30–31, **30, 31**
central nervous system, 310
 structure of, 329–333
centriole, 86, 361, 556
cephalization, **328**
cephalochordates, 526–528, **527, 528,** 573
Ceratium, **133**
cereal grains, 221–222, **222**
cerebellum, 329, **330–331**
cerebral cortex, 329–334, **330–331, 332**
Chaetoceros, **178**
chain reaction, 623
chemical bonds, 24
chiasmata, 477, **478**
childhood, 465
chimpanzees, **539,** 540, 544, 561
 behavior of, 540, 544
chinch bug, metamorphosis in, 430
chitin, 281
Chlamydomonas, **170**
 reproduction in, 385, **385**
Chlorella, 167
 as food, 234, **234**
 as symbiont, 255–256, **256**
chloride ion, 311–312, 318–319
Chlorohydra, 255–256, **256**
chlorophyll, in photosynthetic organisms, 162
 in photosystems, 163–165
 structure of, 172
Chlorophyta, 180–181, **180, 181,** 566
chloroplast, 79–83, **83,** 90, 169–172, **171,** 556, **557**
cholera, 138, 141, 147
chordates, classification of, 573–574
chorion, 421–424, **424–426**
chorionic gonadotropin, 460–461, **462**

chromatin, 76, **77**
chromatography, **167,** 168, **168**
chromosomal inheritance, 472–483
chromosomes, 76, **77**
 abnormalities of, 485–487
 crossing over, 367, 369, **370**
 DNA replication and, 358–359, **359**
 homologous, 369, **370**
 human, 369, **370**
 in meiosis, 369, **370–371**
 in mitosis, 360–361, **361**
 mapping of, 479, **479–480**
 polytene, 406, **408, 470**
 sex, 372
 somatic, 372
cilia, 96
 of filter feeders, 250–251, **251**
 structure of, **87,** 88
circulation, in insects, 281–282, **282**
 in mammals, 286–294, **289**
clam, 570
 feeding, 250–252, **251**
 reproduction, 372–373
 Tridacna, 256, **257**
classical conditioning, 341–342
classification, five-kingdom system, 122, 562–574, **563**
 of man, 560–561
 nomenclature in, 559–562
 origin of, 559
 purpose of, 558–559
 role of subjective judgment in, 561–562, **561**
Clematis, **414**
clitoris, 444, **444,** 445
clone, 491
closed system, 5, **9**
Clostridium botulinum, 145, 147
clotting, blood, 291–292, **293**
coal, 620, **630**
coccus, 124, **127**
codeine, action of, 352
 in opium, 240
 structure, **241**
codon, 44–45, **46,** 49–50
coffee, 238, **239**
coitus interruptus, 455
collagen, 432, **433**
colloids, 25–26, **26**
colon, 265
colonization, 17, **18**
colony, 94–96, **95, 96, 97,** 135
commensalism, 255
communities, 580
 climax, 598, 600
 succession of, 17, **18,** 598, **598–599,** 600–601
 trophic levels in, 594–598, **597**
competition, 586–587

implantation, of human zygote, 452–453, **454**

inbreeding, 496–498, **496, 498**

incomplete dominance, 482, **482**

independent assortment, 474–475, **476**

India, population growth in, 616, **617**

individuality, development of, 116

induction, in embryonic development, 410, **410**

  of enzyme synthesis, 58–59, **59, 60**

  in plant development, 415, **415**

industrial melanism, 16, **16,** 532

Industrial Revolution, 608–609, **608**

infant mortality, in developed countries, 462, **463**

  in undeveloped countries, 609

influenza, 140

inguinal hernia, 464

inhibition, end-product, 58, **58**

  lateral, 339–340, **339, 340**

  in nervous system, 318–319

innate releasing mechanism, 505

inorganic compounds, 27

insects, circulation in, 281–282, **282**

  classification of, 571

  development of, 424, 426–430, **428, 429**

  evolution of, 521

  exoskeleton, 280–281, **281**

  larvae, 105, 150, 424, 426–430

  mouthparts of, 259, **260**

  respiration in, 282–283, **283**

  as scavengers, 256

  water regulation in, 283–284

instinct, 309, 333, 504–508

instructional theory, 111, **111**

insulin, absorption of, 268

  and blood sugar, 294–295, **295**

intercalary meristem, 204

intertidal environment, 186–187, **187**

intestine, **81,** 262, **263,** 264

intrauterine device (IUD), 456–457, **457**

invertebrates, 521, 527. *See also individual organisms*

ions, 25

  in blood, 27

ionization, 25

iron, in primitive photosynthesis, 554–555, **554, 556**

  in respiratory chain, 52, **54**

isotope clocks, 519–520, **519**

Java Man, 542

jejunum, 262, **263**

jellyfish, 568

Jenner, Edward, 138

Jensen, Arthur R., 502–504

juvenile hormone, effects of, 244–245, **244**

  in metamorphosis, 429, **431**

  as a pesticide, 244–245

kangaroo rat, 578

kelp, 184, **185,** 186

  productivity of, 594–595, **595**

  reproduction in, 385, **385**

Keppel, Francis, 493

keratin, 284, 285

kidney, **338**

  development of, 417–418, **418**

  solute regulation by, 295–296, **298**

  structure of, 295–296, **297**

  transplants, 115

  water regulation by, 296–298, **298**

kilocalories, definition of, 10, 226

  and exercise, 274–275

  requirement for, 226–229, 270–276, **270**

kingdom, 122, 560–563

kinins, 416

Kinsey Report, 447

Klinefelter's syndrome, 485, **485**

Koch, Robert, 137–138

labor, 461

lactose, 58–59, **59**

ladybug beetles, 592, **593**

Lamarck, Jean-Baptiste de, 516

lamellae, photosynthetic, 169, **170**

lamprey, **573**

lampshell, **571**

Landsteiner, Karl, 113

Langmuir trough, 72, **72**

language, evolution of, 544–545, **544**

  and nomenclature, 558–559

large intestine, 262, **263**

larvae, 421, **423**

  in evolution, 526–527, **526–528**

Lashley, K. S., 346

lateral inhibition, 339–340, **339, 340**

leaf, primordia, 208

  stomata, 191–194, **193, 194**

  structure of, 191–192, **192**

  types of, 191, **191, 416**

  water conservation by, 191–194, **193**

learning, 341

  in autonomic nervous system, 346–348, **348**

  measurement of, 340–341

  and protein synthesis, 344

  range of, 346–348

  in split-brain animal, 346, **347**

legumes, 223–224, **224**

lemur, 537, **538,** 539

leprosy, 139

leukocyte, 290

Ley, Willi, 520

LH (luteinizing hormone), 449–452, **451, 465–466, 466**

libido, 448

Librium (chlordiazepoxide), 353–354

lichens, 254–255, **255**

geographical distribution of, 580–582
interrelation with environment, 5–18, 577–578
organ transplants, 115–116
orgasm, 444–446, **445–447**
origin of life, major events in, **556**
  special creation, 547–548
  spontaneous generation, 548–557
Origin of Species, 514, 529
Ornithocercus, **133**
Ornithoscatoides dicipiens, 254, **254**
Oryza sativa, 222
Oscillatoria, **131**
oscilloscope, 313–314
osculum, 99
osmosis, 92–93, **93**
osmotic concentration, 93
ovarian cycle, 449, **449**
ovary, 398, 443, **443**
ovulation, 376
oxidation, 52, 54, **54**, 56
oxygen, diffusion of, 63, **64**
  in gas exchange, 301–302, **303**
  in respiratory chain, 52, **54**
oxygen debt, 305
oxytocin, 461

pancreas, 263, 266
Papaver somniferum. See opium
Paramecium, 566
  contractile vacuole of, 93–94, **94**
parasitism, in human populations, 258
  nutritional basis of, 255, 257–259, **259**
parasympathetic division, 334–336, **335, 337**, 338
parthenogenesis, 378, 498, **499**
pasteurization, 147
Pasteur, Louis, 137–139, **137**
pasturage, 223–224
Pavlov, Ivan, studies of conditioning, 342–343
  studies of digestion, 265–266
pea plant, 472–475, 477
Peneroplis, **132**
Penicillium, 135, **135**, 567
penis, 442, **443**
peptide bond, 36, **37**
peptidyl transferase, 47
perception 339–342, **340**
pericycle, 205, **206, 207**
Perinema, **132**
peripheral nervous system, 310, 334
peristalsis, 264–265, 336
perspiration, 306–307, 353
pest control, 244–245, 590–593, **592, 593**
pesticides, 244–245
petiole, 191
Petromyzon, **573**
Phaeophyta, 184–186, **185, 186**

phage, description of, 123–124, **123**
  genes of, 479, **480**
  replication of, 124, **125, 126**
phagocytosis, 106–107, **107, 108, 109**
phase, 25–26
phase-contrast microscope, 65, **67**
phenocopies, 494, **494**
phenotype, 474
phenylketonuria (PKU), 15
phloem, aphid studies of, 199–200, **201**
  structure of, 195, **196**, 198–199, **200**
  production of, **205**
  transport of molecules in, 199–202, **201, 202**
phonemes, 544
phosphagens, 527
phosphate bond, high-energy, 52, 54, **55**, 56–57, **57**, 83
  in nucleic acids, 41, **41**
phosphate pollution, 182
phosphorescence, 163
photochemistry, 162–163
photon, 160
photosynthesis, action spectrum of, 162
  carbon dioxide fixation in, 166–169, **167, 169**
  general formula for, 160
  light absorption in, 161–165, **165, 172**
  in nature, 172–173
  by primitive organisms, 553–554
  production of ATP and NADPH$_2$ in, 165–166, **166, 172**
photosynthetic organisms, 160, **170, 171**, 172–173, **173**
  primitive, 553–554
photosynthetic unit, 80, 164, 172
phycobilisomes, **170**, 172
phycocyanin, 129
Physarum, **365**
phytoaxelins, 106
phytochrome, 400
Phytophthora infestans, 145
phytoplankton, 176–180, 252
  in food chain, 232–233
pigments, in photosynthesis, 161, 164, **164**, 168, **168**
  respiratory, 290, 301–303, **302**
Piltdown Man, 541
pineal gland, 398–399, **399**
pine tree, reproduction in, 387–388, **389**
Pinguicula, 215
pinta, 142
pituitary gland, **331**, 381, **381**, 449–450, **451**
placenta, 424, **426**, 453
plague, 141
planarian, 344, **569**
  learning in, 344
plankton, 130, 136, **136**, 252, 432

and origin of multicellularity, 96–98, **98**
psilocybin, 352
psychosomatic illness, 348
psychotherapy, 337, 506
puberty, 465–466, **466**
Puccinia graminis, **366**
pyrethrins, 244

quantasome, 80, **83,** 172
quantum, 160

rabies, 139, 140
radiant energy, 7
  and photosynthesis, 160–161, **161**
  and mutation, 490–491, **490**
radioactive decay, 519
radioactive wastes, 626–627, **631**
radiolaria, **62, 132**
Rauwolfia, 240
reaction center, 164–166
reaction-specific energy, 505
reafferenz, 341
receptor, field, 340–341, **341**
  sensory, 314–315
  stretch, 326, **327**
  visual, 339–342, **339, 340, 341**
recessive allele, 474
rectum, 262, **263**
red beds, 555, **556**
red blood cells, 290, **290**
red drop, 163–166, **165**
red tide, 176, **177**
reduction 52
reflex, 326–327, **326–327**
reflex arc, 326, **326–327,** 334
refractory period, 319
regeneration, 435–436, **435**
regulation, of internal environment, 279–280
  of temperature, 306–307
regulator site, 58, **60**
relative growth, 515
releasing factors, 381, **381,** 450, **451**
REM sleep, 349–351
renewable resources, 619–621, 627–629, **628**
replication, of DNA, 42, **44**
  of primitive nucleic acids, 551
repression, of enzyme synthesis, 58–59, **59, 60**
repressor, 58–59, **60,** 406
reproduction, in animals, 372–382
  asexual, 364–366
  and behavior, 381–382, **383**
  environmental control of, 15–16, 397–400, **398, 399**
  hormonal control of, 380–381, **381,** 448, 449–450, **451,** 452

human, 449–462
  in plants, 385, 397
  sexual, 366–400
reproductive isolation, 533–535, **533**
reproductive tract, in hermaphroditic animals, 372, **372**
  in humans, **380,** 442–444, **442, 443,** 463–464, **464**
reptiles, 118, 380, 521, 574
reserpine, 240
respiration (cellular), 52–56, 63
respiration (gas exchange), in insects, 282–283
  in mammals, 299–305
respiratory chain, 52, 54, **54**
reticular formation, 334
retina, 316
retinal, 316, 317, 527–528
retrograde amnesia, 345
rheumatoid arthritis, 118
Rh factor, 113–114
Rhizobium, 155, **155**
  in legumes, 223
Rhodophyta, 181–184, **184, 185**
rhodopsin, 316–317, **316, 317**
rhythm, and birth control, 454, **455**
ribonucleic acid. *See* RNA
ribosomes, 47–49, **48, 49,** 76–78, **79**
Riccia, **188**
rice, 222
RNA (ribonucleic acid), 41
  and memory, 344
  messenger (mRNA), 44, 47, **47, 49**
  polymerase, 45, **47,** 58–59, **60**
  ribosomal (rRNA), 47–49, **48**
  role in induction, 410
  transfer (tRNA), 51, **49, 50**
  in viruses, 123–124
rods, 316, 340
root, 412
  growth of, 205–206, **206, 207**
  pressure, 195, **198**
  structure of, 195, 205–206, **205, 206**
root crops, 222–223
rotenone, 244
rumen, 252–253, **253**
ruminants, 252–253, **253**

saguaro cactus, 579–580, **579**
salamander, 299, 380, **422–423,** 435, 525
salivation, 264, 265–266
Salmonella, 146–147
sarcomere, 323, **323**
sarcoplasmic reticulum, 324, **324**
Sargassum, 186, **186,** 566
salt, 24, **24**
satiety center, 271–272
saturation (active transport), 73, **75**
scavengers, 254